Y0-BTD-406

CHEMICAL ANALYSIS

CHEMICAL ANALYSIS

A SERIES OF MONOGRAPHS ON ANALYTICAL CHEMISTRY AND ITS APPLICATIONS

VOLUME 31

WILEY-INTERSCIENCE

a Division of John Wiley & Sons

New York · London · Sydney · Toronto

To my loved ones
For you must know I care for thee

PREFACE

Specialization in the various fields of organic analysis has resulted in the accumulation of a vast quantity of photometric methodology but has also led to the paradoxical result of the separation of that knowledge into the more manageable but artificial analytical compartments of air pollution, biochemistry, clinical chemistry, pesticides, molecular biology, toxicology, and so on.

Actually organic analysts in each of these fields do have common interests—their analytical techniques. In addition, each field has valuable knowledge that could benefit the other fields. With the burgeoning literature, the increasing complexity and cost of instrumentation, increasing specialization, and the development of new words and terms in each discipline, cross-fertilization has become more difficult and more fruitful. The aim of this book is to simplify this process and to furnish the beginner and the specialist, the analyst and the researcher, the organic chemical and practical spectrophotometric background basic to all these disciplines.

Few of us would admit that we just do not comprehend the theoretical and mathematical treatments of the increasingly sophisticated and highly tribalized jargons of our various highly specialized fields of chemistry. Changes are taking place at such a rapid rate that the language our graduate students speak is different from the language we spoke as students. All too soon students become foreigners to the teachers who taught them in this frenzied man-made evolution, this mutation of language, skills, instruments, chemicals, and knowledge. The old languages, the old theories, and the most glamorous research fields are fading into the background of our civilization's palimpsest while new, more glamorous disciplines can be seen emerging and seemingly dominating the outer edges and surfaces of our scientific world. The energetic disciples of these new fields are now solving questions we never knew existed, finding answers to our seemingly insoluble problems, sometimes puzzling over questions that have been solved, and occasionally, out of touch with the past of a different jargon, rediscovering phenomena in terms of a more sophisticated language and a more imposing, enshrouding mass of theoretical wordage and equations.

To understand and control our environment we shall have to use mind and hands, equipment and know-how and theories and equations to characterize and determine the chemicals in our internal and external environments and their short- and long-term effects. Too often theory is a smog of complicated words and equations that hide our lack of knowledge of the problem at hand. Empirical investigations cannot be replaced by pretentious interpretation or by massive bureaucratic organization.

Although our facts are usually indisputable, they are always incomplete, and the interpretation of incomplete facts leads to theories that are, at most, only an approach to the truth.

In this book involved equations and theoretical discussions have been omitted so as to simplify and facilitate an understanding of the utility of the basic empirical background of electronic absorption spectra to the investigations of organic trace analysts, biochemists, organic chemists, and others.

I have presented some of our problems in the solution of which absorptiometry could play an important role. In the remainder of the book the empirical aspects of ultraviolet–visible absorption spectrophotometry are discussed. These two aspects – the practical and the basic – are placed in juxtaposition to show that before these practical problems are solved we shall need much more basic research in the spectral fields (one phase of which is presented in this book) and in the many other fields of scientific instrumentation and technique. We must never forget that in this age of change when we were most practical we strove to improve the horse and buggy while the basic researchers in their ivory towers puttered around with steam, electricity, light energy, nuclear fission, space travel, and other types of impractical nonsense.

The principles discussed in this volume will be extended and applied in the next volume to a better understanding of the absorptiometric, fluorimetric, and phosphorimetric techniques of organic trace analysis useful (or potentially useful) in studying the chemical composition of living tissue, its by-products, and the human environment.

Cincinnati, Ohio EUGENE SAWICKI
January 1970

CONTENTS

CHAPTER 1

INTRODUCTION

I. Origins of Spectra

The ultraviolet–visible absorption spectral tool described in this book is not a scientist's only tool and, frequently, not even his best tool for many studies. However, for some types of investigation, especially in trace analysis, it is a superb problem solver. Photometric techniques have proven to be indispensable extensions of our eyes, making it possible for us to examine the darkness around us. Because of these tools we have acquired insight into our environment, into life- and mind-threatening processes, and into life itself. The value of many other investigations has been increased through the complementary use of spectral procedures with the various other types of techniques.

Since trace amounts of chemicals can perform (damage or interfere with) important biochemical functions in the various processes of life, researchers are constantly developing electronic spectral methods of analysis for determining picogram to milligram amounts of organic compounds in the complex mixtures around and in us.

Important steps in the solution of many of the problems that life faces in its attempt to thwart disease, old age, and death are the determination of the chemical composition of living cells and of all mixtures in the human environment and the acquisition of knowledge concerning the interplay of the various chemicals in and outside the cell and the results of that interplay on the multitudinous facets of life. Before the quantitative chemical makeup of cells and heterogeneous mixtures can be determined, the qualitative composition of these mixtures must be ascertained. For this kind of analysis quick, efficient types of separation, spot testing, and photometric examination are of great value.

The simplest situation in an analysis is the direct spectral characterization and determination of a trace chemical without preliminary reaction with a reagent and without separation. In many instances the compound or a breakdown product is allowed to react with a reagent to give a new derivative that absorbs, fluoresces, or phosphoresces in a wavelength region relatively free of interfering background.

However, when trace amounts of individual members of a family of compounds are to be determined, separation methods have to be used. For the more complicated chemicals separation and spectral methods have to be used in conjunction and there are many instances when the techniques of cleanup and organic synthesis are necessary before photometry can be brought to bear on the problem.

In this book we consider ultraviolet–visible absorption spectral studies, understanding that this is only one tool of the many inter-related electronic spectral tools available to organic chemists of various types. The author believes that a knowledge of the absorption spectral empirical laws will be invaluable in developing fluorimetry and phosphorimetry into even more powerful trace analytical tools.

The photometric tool with which this book is concerned can be derived from the simplified energy-level diagram of electronic states shown in Figure 1.1. Each molecule has only one ground state, and this is the state it is normally present in, as depicted by the singlet ground state S_0 in Figure 1.1 The process of absorption, as depicted by A, consists of the absorption of a photon by a molecule in the ground state and the change of the molecule to the lowest excited state, S_1^*. There can be other excited singlet states, none of which is shown in the diagram.

Usually the A involving the lowest energy singlet π, π^* state is the one of interest in analytical work. Methods of analysis involving ultraviolet, visible, and near-infrared electronic absorption spectra are based on this particular process.

The next step, vibrational relaxation, is denoted by R, the relaxation of the molecule to the lowest vibrational level of S_1^* with its loss of vibrational energy. Photon emission can then occur to a vibrationally

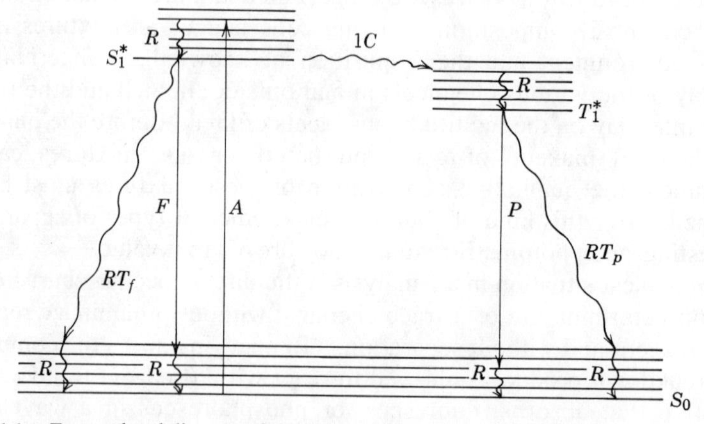

Fig. 1.1. Energy-level diagram of electronic states. Symbols: A = absorption; F = fluorescence emission; P = phosphorescence emission; RT_f = radiationless transfer from excited state to ground state; R = relaxation, or loss of vibrational energy, to lowest vibrational level; S_0 = singlet ground state; S_1^* = singlet first excited state; T_1^* = triplet first excited state; IC = intersystem crossing.

excited level of the ground state. This process is known as fluorescence, depicted by F in the diagram. Since the change in energy depicted by F is almost always smaller than the change in energy depicted by A, due to the vibrational relaxation R after both absorption and emission, fluorimetric-emission-wavelength maxima are at longer wavelengths than the absorption-wavelength maxima. The room-temperature fluorimetric technique has been found extremely useful in solving many types of problems.

The molecule in the lowest vibrational level of the excited singlet state can also lose its energy through radiationless transfer, RT_f in Figure 1.1. In such a case the molecule is nonfluorescent. Even fluorescent molecules can be made to lose their absorbed energy in this fashion and thus become nonfluorescent. For some compounds an increase in the dielectric constant of the solvent, for others a change to a non-hydrogen-bonding solvent, can cause a loss of fluorescence. Any fluorescent molecule can have its fluorescence quenched through the selective effect of a particular type of solvent or reagent on the particular functional group of the molecule. This type of functional group analysis has been called quenchofluorimetric analysis (1).

In all these cases radiationless transfer RT_f has supplanted fluorescence. With decreased concentrations of the quencher the fluorescence intensity of the molecule can be decreased and both processes, F and RT_f, can take place.

For some compounds the radiationless transition process can be supplanted gradually by the fluorescence process by lowering the temperature of the solution of the test substance. This change can become a major factor when the temperature is lowered enough to cause the solution to solidify. In this case some nonfluorescent, nonphosphorescent compounds become highly fluorescent. Low-temperature fluorimetric-analysis techniques can then be used for the analysis of these compounds. In addition a change in solvent can cause a drastic change in the proportion of the two processes of RT_f and F. This drastic change in fluorescence intensity caused by a solvent effect has never been thoroughly studied, even though its analytical usefulness is obvious.

For many molecules a favorable condition can prevail whereby vibrational coupling can take place between the excited singlet S_1^* and excited triplet T_1^* states. Thus, after absorption and vibrational relaxation, intersystem crossing IC to the first excited triplet state can take place. After vibrational relaxation, photon emission can then occur to a vibrationally excited level of the ground state. This process is known as phosphorescence, shown by P in Figure 1.1. Because the absorbed energy can be used up in some of the preliminary processes, fewer molecules phosphoresce than fluoresce. In addition, since the first excited triplet state is of lower energy than the first excited singlet state, the phosphorescence-emission bands are always at longer wavelength than the fluorescence-emission bands. Since fewer compounds phosphoresce than fluoresce or absorb, phosphorimetric analysis is much more highly selective than either fluorimetric or absorptiometric analytical methods. With proper procedures to ensure that radiationless transfer does not take place to any large extent and that intersystem crossing is encouraged through organic synthetic methodology, the usefulness of phosphorimetric analysis will increase at a rapid rate. At the present moment phosphorimetry has been used without the help of organic synthesis.

The phosphorescence of a molecule can be decreased or even eliminated through the proper encouragement of internal conversion between one of the excited states and the ground state or through interference with intersystem crossing. This can be done through the use of solvent effects, or quenchophosphorimetry (2). With this latter method highly selective analysis of functional groups is possible. Indications are that this technique is the most highly selective of all the spectral methods available to the trace analyst.

In addition, these various tools—absorptiometry, fluorimetry, low-temperature fluorimetry, quenchofluorimetry, phosphorimetry, and quenchophosphorimetry—can be used for the direct analysis of paper and thin-layer chromatograms and pherograms, cellular material, and living tissue.

II. Problems Amenable to the Photometric Approach

This treatise is based on the belief that the use, amplification, and perfection of the described spectral methodological tools will make it possible to ask meaningful questions of nature and to understand and eventually solve many of the problems that life faces in this universe. With the help of these tools we have reached the stage of development where we now know of the existence of many once-unacknowledged problems. In the darkness around us we can dimly distinguish the shape of many other problems. We have reached the point where we can imitate nature in the causation of some of the problems. Thus we can cause cancer, diabetes, allergic attacks, mental depression, etc. The question is whether we can alleviate and cure what nature and we can cause.

First we shall briefly discuss some of the major human problems whose solution many scientists believe can be solved with the help of spectral tools.

A. Health

The most important problem for many individuals is their physical and mental health, for without either of these life can become difficult or even meaningless and unbearable.

1. Cancer

As the variety and quantity of chemicals in our environment increase, the problem of the relationship between chemical structure and chemical disease becomes all the more pressing. Hueper and Conway(3) have stated the belief that, because of the unrestrained and reckless increase in contamination of the human environment with chemical and physical carcinogens and with chemicals supporting and potentiating their action, the stage is being set for the future occurrence of an acute, long-lasting, catastrophic epidemic of human cancer. These authors emphasized that such epidemics have occurred among baby chicks and trout brought up on commercial feeds containing carcinogens.

A large number of chemicals are known to cause cancer in animals (3–9). These include polynuclear aromatic hydrocarbons, polynuclear aza heterocyclic compounds, polynuclear imino heterocyclic compounds, polynuclear aromatic amines(10–12), polynuclear nitro arenes, aminoazo dyes(13), aliphatic carbamates, nitro olefins, epoxides(14, 15), lactones (14, 15), peroxy compounds(14, 15), alkylating agents(16, 17), pesticides (18), nitrosoamines(9, 19), lactones(20), some solvents, and mycotoxins(9).

Some of these chemicals or mixtures containing them have been used industrially and found to be carcinogenic in man(3, 21–29).

With the help of column chromatography and ultraviolet-absorption-spectra analysis the carcinogen benzo[*a*]pyrene has been found in the air of all United States communities(30), as well as in the atmospheres of cities in various areas on the globe(31). Benzo[*a*]pyrene and other hydrocarbon carcinogens have been found in tobacco leaves and smoke, snuff, asbestos, commercial solvents, peat, carbon black, coal gas, wood soot, charred biscuits, petroleum wax, petrolatum, shale oil, creosote oil, stack gases from pulp mills, incinerator effluents, automobile and diesel-engine exhaust fumes, bituminous clays, nonurban soil, river water and mud, snow, pyrolyzed cellulose, lignin, pectin, etc., plants, fossils, marine flora, oysters, barnacles, and other marine fauna, lung and other human tissues, urine from city dwellers, human hair wax, smoked foods, roasted coffee beans, and house dust(29).

The widespread presence of this type of carcinogen in the human environment indicates the necessity of quick, reliable methods of assay for these compounds. Such spectral methods are available.

Other types of carcinogens have also been shown to be present in urban atmospheres (31), for example, polynuclear aza heterocyclic and ring carbonyl compounds. Combustion processes add many types of carcinogens to the environment. Some carcinogens, usually in a mixture, are added directly to the environment for some useful purpose—for example, asphalt on roads, coal-tar pitch on roofs and pipes; carcinogenic pesticides in homes, gardens, waterways, forests, or farms; and mold- or plant-produced carcinogens. Many more examples of carcinogens in the human environment have been listed in the excellent book by Hueper and Conway (3). Methods of analysis for many of these carcinogens are available. However, the composition of the mixtures is mainly unexplored.

The mechanism of chemical carcinogenesis is being studied with ever-increasing vigor at the molecular level. Miller and Miller (32) have stated the belief that the formation of alkylating or arylating structures from the carcinogens is a central, but not universal, step in chemical carcinogenesis. Many workers in this field believe that an important step in the solution of the carcinogenesis process would be the elucidation of the structures of the ultimate carcinogens and their cancer-initiating product with the appropriate macromolecules. If this is so, the chemical nature of the reactive metabolites and the reactive sites on the macromolecules (proteins, nucleic acids, or?) need to be determined. Whatever the mechanism of carcinogenesis, sophisticated, sensitive methods of separation and analysis will be necessary to elucidate this life-destroying mechanism.

Reliable screening tests are needed for the early detection and study of the various human cancers. Spectral methods have been used to a relatively slight extent. Some examples are the use of routine screening for urinary lactic dehydrogenase and alkaline phosphatase activities for the early detection of many occult, potentially fatal diseases of the genitourinary system [patients with renal-cell carcinoma and carcinoma of the prostate show increased activity of both enzymes in 80 and 60–80% of the cases, respectively (33)]. Since urinary lactic dehydrogenase is elevated in almost all patients with bladder cancer (33), assay for this enzyme is of some value in studying this disease. The measurement of 4-hydroxy-3-methoxymandelic acid has to some extent supplanted catecholamine assay in the diagnosis of catecholamine-secreting tumors (34). Assay of homovanillic acid is being performed with increasing frequency because of its value in the detection of dopamine-secreting tumors (34). The use of spectral techniques in the early diagnosis and study of malignant diseases is still in its infancy, but the potential of such methods is indicated by the usefulness of the presently available methods.

Can the environment exert its effect during the prenatal period or even before the conception of an individual? Such a suggestion has been advanced by Miller(35) to explain the early onslaught of leukemia in children. Miller has also suggested that an environmental agent may induce Wilms' tumor, a neoplasm of childhood, before the end of the first trimester of pregnancy. The excessive rate of leukemia among mongoloids(36, 37) and the discovery of an extra chromosome in mongolism has contributed to the belief that certain forms of leukemia are related to chromosomal aberrations. The question is whether leukemogens—such as benzene, chloramphenicol, phenylbutazone, or some of the viruses—could also produce cytogenetic abnormalities(35).

It has been suggested that a leukemogen(s) was introduced into the environment or became effective in the 1920s(38). Here again methods of assay for the leukemogens in the human environment and a thorough study of the mechanism of leukemogenesis is necessary from the viewpoint of the ultimate leukemogens, DNA changes, etc.

A large amount of controversy is present in research investigations in the front lines of chemical disease studies. This disagreement is especially apparent in the attempts to identify the ultimate physiological reactant in the processes of allergy, cancer, mental disturbances, and so on.

In quite a few instances the change in concentration or the presence of a key metabolite is a vital part of a theory explaining the chemical disease. The controversy is usually reinforced by the widely varying results obtained by different workers, especially when the method of assay is difficult or close to the questionable borderline of selectivity or sensitivity. Of help to the solution of such a controversy would be the development of simpler, more sensitive, and more selective methods of assay for the various metabolites of possible importance in the disease process. The author believes that this method development will be helped by knowledge of the empirical laws of ultraviolet–visible absorption spectroscopy.

Most of the carcinogenesis theories involve the nucleic acid molecule. Even the viral theory of cancer is no exception, since the main component of a virus is a nucleic acid. However, there is still no evidence that any human malignant tumor is caused by a virus(39). If there is a change in the nucleic acid of a malignant cell, separation of the pure acid and assay are of prime importance. Since naturally occurring cancers of lower animals are usually caused by an RNA virus, separation and assay of pure RNA molecules of this type may be of vital importance in cancer studies. This is now possible since successful extractions of the Rauscher murine leukemia(40), Rous sarcoma(41), and Newcastle disease viruses(42) for apparently intact RNA have been reported. In this respect study of DNA is also necessary since cells experimentally transformed to malignancy by DNA viruses retain noninfectious viral genes that are continuously transmitted in subsequent generations in a manner stated to be reminiscent of bacterial lysogeny(39).

Similarly the structure of the DNA in all tumorigenic cells would be of interest in studying the mechanism of carcinogenesis. In this respect it has been postulated that independent fragments of viruslike genetic material, perhaps assisted at certain stages of their existence by viruses proper, are the key to many naturally occurring tumors(43).

2. *Cardiovascular Diseases*

Fifty four percent of all deaths during 1963 in the United States were caused by cardiovascular diseases(44). That year 1 million Americans died of this disease and 23 million were afflicted with some form of it. Consequently methods of assay have been suggested that would help to indicate the individuals who would experience an early onset of this disease. Methods of assay for plasma β-lipoproteins(45, 46), triglycerides(44), and cholesterol(44), have been suggested. The results are controversial (44, 47) so that it would seem that improved analytical aids to the prognosis of cardiovascular disease are needed.

3. *Genetic Diseases*

A moderate number of genetic diseases are known (48, 49). With increasing knowledge many more will be discovered and, as has been shown in the past(50), many thought to be one entity will be found to be several clinically similar but fundamentally distinct disorders. Examples of such diseases include phenylketonuria(51), maple syrup disease(52), alcaptonuria(53), tyrosinosis(54), glycogen storage disease(55), the mucopolysaccharidoses(50), hereditary intestinal polyposis(56), albinism (57), chondrodystrophy(58), spinocerebellar ataxia(59), brachydactyly (60), congenital deafness(57), and retinitis pigmentosa(61). If more information could be obtained on the metabolic errors in individuals with genetic diseases, more and better early diagnostic assays could be developed. With the help of the additional knowledge and early diagnosis, the physical and, in some cases, mental deterioration experienced by many of these individuals could be retarded or even halted through proper treatment. This objective has been obtained with some phenylketonuriacs whose treatment with a low phenylalanine diet prevents the subsequent development of a severe mental defect(51).

4. *Allergy and Asthma*

A large portion of the population suffers from allergy, a fairly large number from asthma. Actually more people suffer from ragweed pollen than from smog. Allergic reactions can be obtained in a large variety of ways. Thus foods, bacteria, fungi, house dust, insect bites, drugs, medicines, pollen, contact with plant and animal products and chemicals, etc., can cause allergic reactions. Asthmatic attacks in New Orleans are believed to be triggered by some air pollutant(62, 63). Tokyo–Yokohama asthma experienced by Americans in that heavily industrialized air is believed to be an air pollution disease(64).

Allergens in the human environment are under intensive investigation. The structures of the allergens in house dust are being investigated(65). The phototoxic and photoallergic reactions of drugs are also under study (66). Since pollen counting is so unsatisfactory as an indicator of hay-fever outbreaks, the development of chemical tests for aeroallergens and coallergens should be pursued.

In addition, simpler tests for allergy in individuals should be developed. These could be based on assay of the appropriate antibody or the chemical released during the allergic reaction. The chemical released could be histamine; a polypeptide breakdown product, such as bradykinin; or a mucopolysaccharide. Thus, for example, a test for penicillin allergy is based on the assay of released histamine(67). The question has been asked whether contact with air pollutants in the grain industry may not cause asthma in some people(68). Obviously much more work needs to be done on the characterization of the environmental allergens. Methods of assay for these various chemicals are certainly badly needed.

5. *Pregnancy*

Although much needs to be done in this field, only a few of the spectacular problems can be mentioned. In the pregnant woman pregnanediol excretion reflects chiefly the development of the placenta(69). Its estimation in urine is of value in the control of induction of ovulation and for assessing some types of fetal distress(70) and for the early detection of pregnancy(71, 72). However, a simpler, more sensitive, and more reliable chemical test for the early detection of pregnancy is still badly needed. Another chemical test that is needed is one that could reliably determine the sex of an unborn child.

6. *Medicines and Drugs*

One of the puzzling aspects of the human environment is its saturation by medicines. Pills are taken for headaches, colds, allergies, insomnia, motion sickness, depression, birth control, pep, and for every conceivable kind of sickness and situation. In addition salves, syrups, and injections are also used in the attempt to control the human situation. A large variety of tranquilizing, antihistaminic, cardiovascular(73), soporific, motion sickness (74), and other drugs are available.

Methods of assay for controlling the concentration of these chemicals in the medicines and in the body are necessary and are available to some extent. Methods of characterization and assay are also needed for the metabolites of these chemicals. The side effects, the interactions between medicines, and the long-term effects (especially) of many of the medicines are not known; thus constant control and study are necessary. The discovery of occasional adverse effects from some aspect of the chemical environment emphasizes these points. Thus thalidomide, a mild and supposedly safe sedative taken by pregnant women, has deformed the limbs of several thousand infants (75).

7. *Toxicology*

With the increasing availability of medicines and other types of chemicals, death by accident or design is much closer to the individual. Even in the garden death can be a few nibbles away. Bittersweet bush berries, Jimson weed, castor-oil plant beans, foxglove foliage, monkshood, the christmas rose, larkspur foliage and seeds, English and American holly, lily-of-the-valley, deadly nightshade, mountain laurel, rhododendron, oleander, dumbcane, and rhubarb leaves are a few of the dangers. Methods of analysis for some of the toxic agents have been described (76, 77).

8. *Pesticides*

The danger of pesticides to animal life has been amply covered by Carson in her book (78). In 1961 there were 261 human deaths due to pesticides (79). Pesticides have contaminated water, soil, air, food, and other places at some time or other (80). The big problem is their long-term effect since many of them have a carcinogenic potential (18). Gas chromatography methods have been used mainly in the analysis of these chemicals, although absorptiometric methods are still used and in many cases necessary to characterize the unknown compound (especially in controversial pollution cases).

9. *Food, Water, and Air Pollution*

The biological consequences of the environmental shift caused by the increasing adulteration, contamination, and decomposition of food is one of the problems faced by civilization. To follow these upward shifts in concentrations of increasing numbers of biologically active agents (81) a large variety of assay methods have been developed.

Not a single river system in the entire United States is free of pollution. American lakes are rapidly becoming polluted. We are ignorant of the physiological effects caused by absorption of minute amounts of substances in drinking water over long periods of time. Water pollution by organic pesticides occasionally becomes a major problem (82). Massive fish kills and foul-tasting water have been blamed on pesticides. Assay for these contaminants is a necessity so that the problem can be put in a proper perspective for adequate controls to be established and enforced.

In the industrialized countries energy is obtained mainly through the combustion of coal, gas, and oil. The United States is powered by about 20% coal, 40% gas, and 40% oil. Wherever combustion occurs, pollutants are released into the atmosphere. The pollutants derived from automobile exhaust (83) and those found in the air (84, 85) have been thoroughly discussed from the chemical, engineering, medical, and meteorological viewpoints. In addition, it has been estimated that approximately 2×10^8 tons of organic matter is released into the atmosphere from plant life (86). Under the influence of sunlight and oxidizing and reducing agents, innumerable syntheses involving these organic materials take place in the air. Smog and lachrymatory agents are examples of atmospheric syntheses. The situation is worsening with each succeeding year, as is indicated by the fact that the visibility on a cloudless day of the ground from an airplane becomes poorer.

Indoor air pollution has never been investigated thoroughly. In many cases pollution indoors could be worse than that outdoors. Air conditioning and the lack of ventilation could be factors in spreading pollutants or stagnating them in one area. The latter factor is important in pollution inside cars in heavy traffic where pollutant concentrations have been reported to be 10 times greater than those measured off the street (87).

Cigarette smoking is a special type of air pollution. Statistical studies indicate that smokers are more prone to lung and bladder cancer, emphysema, and heart disease (82). In addition, smokers are more susceptible to respiratory disease in the polluted air of the Tokyo–Yokohama area; removal from the area alleviates the symptoms (89). Since the composition of cigarette smoke is not completely known, it is difficult to ascertain what components of this smoke are physiologically dangerous.

Other air pollution diseases are the pneumoconioses caused by exposure to moldy vegetable dusts. These include mill fever (organic dusts), byssinosis (cotton), bagassosis (moldy sugar cane), grain asthma (grain), tarimand asthma (tarimand seed), weaver's cough (moldy cotton yarn), and farmer's lung (moldy hay, grain, straw, and silage)(90). The problem here is to analyze the polluted air and to determine whether allergens are involved and, if so, what their structures are. The carcinogens and allergens in urban atmospheres have been already discussed.

Air pollution can be lethal to some human beings. A large number of deaths have resulted from exposure to high pollution levels in the Meuse Valley in France and Belgium in 1930; Donora, Pennsylvania, in October 1948; Poza Rica, Mexico, in November 1950; and London in December 1952, 1956, and 1962. Badly needed is a more thorough investigation of the composition of polluted air.

B. Brain

1. Memory, Creativity, and Emotions

Of all human physical and mental activities the one that gives the most intense pleasure is the act of creativity. Probably the most godlike activity is the laborious mental attempt to create much more and at ever more difficult levels of attainment. However, before creativity can occur, extensive memories have to be accumulated. What is memory—and what chemical and physical processes are involved in the collection, storage, and retrieval of information? An increasing number of studies have implicated ribonucleic acid in the storage of information in code as molecules of itself or as a protein whose synthesis is controlled by ribonucleic acid(91, 92). One suggested theory is that this code is similar to the genetic code and depends on the arrangement of bases in a molecule of RNA(93). The theory is attractive because it could provide a common mechanism for the storage of instincts and learning.

Evidence has been presented for the hypothesis that memory formation involves the formation of covalent chemical bonds(94, 95). In line with this are the observations demonstrating that rats given enriched experience develop, in comparison with restricted littermates, greater weight and thickness of cortical tissue and an increase in the total acetylcholinesterase activity of the cortex and the rest of the brain(96). The latter change suggests the formation of new synaptic connections among neurons.

If memory is stored as a code in some type of chemical structure, methods of assay for memories should be possible. The following questions are then relevant: What are short- and long-term memories? Could a memory be mutated by time or powerful chemicals so that parts of the memory would consist of something that never happened? Could parts of memory recordings be erased? Is there a change in structure involved in the fading of memory? How does repetitiveness strengthen memory? Could a firm belief in a religion, a superstition, or a prejudice be isolated from a dead human brain and then unequivocally characterized and assayed physically and chemically? Obviously a high order of selectivity and sensitivity would be necessary.

Much controversy exists in the field of memory. Thus various chemicals injected into animals have been reported to help learning and memory, whereas others have been stated to cause a loss of memory (97) or to interfere with the learning of a new task (98). Ribonucleic acid extracted from the brains of trained animals and injected into naive animals has been reported to produce positive (99) and negative (100) transfer of learning. The controversial aspects of these studies have been discussed (95).

Obviously much more work needs to be done in this field. In addition, highly selective and sensitive microchemical techniques for the direct analysis of the appropriate cellular tissue or for the analysis of single cells will have to be developed to further these investigations.

In respect to creativity the main problem is how some stored memories are caused to interact so as to form a memory that had never before existed. In this sense creativity is a retrieval with interest.

In respect to emotions the question is, what chemicals and chemical reactions are involved in onset of the various emotions? It should be possible to induce the various emotional reactions through varieties of chemicals. For example, patients with anxiety neurosis show an excessive rise in lactate, and infusions of lactate can induce anxiety symptoms and attacks (101). A copulation-reward site has been found in the posterior hypothalamus (102). The injection of minute amounts of sex hormones near or in the hypothalamus has elicited sexual behavior in rats, and the injection of salt solutions has regulated thirst (103). Once this field is at an advanced stage, analytical methods for therapy, control, and study will become necessary.

2. *Psychoses*

Controversy is to be expected in all front lines of research on the various phases of life and its problems. This is due to a large extent to the use of inadequate separation and spectral assay methods. In the latter case the selectivity and sensitivity of many of the assay methods are inadequate. In the mental health field especially these techniques are being used at the limit of their capabilities. Thus smaller amounts of increasing numbers of chemicals need to be determined in ever smaller amounts of material and, in some cases, as close to the natural living state of the material as possible.

The most prevalent of the mental diseases is schizophrenia, which afflicts about 1% of the population and has hospitalized about 250,000 Americans. Although many theories have been advanced to explain the symptoms, there is no agreement among investigators as to the underlying nature of schizophrenia and its cause(104). Because many chemicals sometimes cause symptoms mimicking various aspects of the disease, many investigators believe that schizophrenics are victims of a subtle brain poison that produces the bizarre psychological changes in perception, thought, mood, personality, and behavior so common to the disease.

The hallucinogenic symptoms in schizophrenia have been hypothesized as being caused by the accumulation in the brain of one or more abnormal methylated compounds(105), of psychotomimetic adrenochrome and adrenolutin(106), a decrease or increase in brain serotonin(107), a depletion of aldehydic substances vital to the psyche and so on.

The strict periodicity of periodic catatonia is stated to indicate "a chain reaction with an accumulation of a toxic substance (amine?) which on reaching a certain level in the brain, puts the patient into his psychotic state and his sympathetic nervous system into action as a part of the stress due to the intoxication"(109).

In 1884 Thudichum advanced the hypothesis that insanity could be the result of toxic substances produced within the brain by faulty metabolism (110). Quite a few mental diseases are now known to result from some biochemical effect on the brain. Some of these diseases and their characteristics are listed in Table 1.1. In addition, biochemical research in epilepsy(107) and in Parkinson's disease indicates that psychotoxic chemicals released in the brain may be causing some of the symptoms of these diseases.

Now that a large number of psychotomimetic drugs are known, it would seem reasonable to believe that mental illness may be caused by the psychotoxic effect of some chemical in the brain. On this basis a simple biochemical test to detect schizophrenia early in life should be possible. Methods of early diagnosis and prognosis for various mental diseases will be described. Simpler spectral methods need to be developed.

One wonders whether the intense feelings of depression and elation in sane people can be due to less drastic changes in the concentration of some chemical(s) in the brain. Could biochemical tests be developed to detect and study these and other emotional problems?

TABLE 1.1
Some Diseases that Affect the Brain: Their Causes and Chemical Characteristics

Disease	Cause	Chemical characteristics	Ref.
Phenylketonuria	Deficiency of liver enzyme, phenylalanine hydroxylase	Accumulation of phenylalanine and metabolites	111, 112
Maple syrup urine disease		High level in blood and urine of leucine, isoleucine, and valine and their keto acids	113
Histidinemia	Deficiency of the enzyme, histidine α-deaminase	Very high level of histidine in plasma and urine, and excessive excretion of the imidazolepyruvic, imidazolelactic and imidazoleacetic acids	114
Argininosuccinic aciduria		High level of argininosuccinic acid in blood and urine	115
Hartnup disease		Increased excretion of alanine, asparagine, glutamine, histidine, isoleucine, leucine, phenylalanine, serine, threonine, tryptophan, tyrosine, and valine	116, 117
Lowe syndrome		Renal amino aciduria	118
Galactosemia	Lack of galactose-1-phosphate uridyl transferase	Accumulation of galactose-1-phosphate in tissues and galactose in urine	107, 119
Tay–Sachs disease		Accumulation of gangliosides in the central nervous system and in the ganglia of peripheral nerves	120, 121
Niemann–Pick disease		Accumulation of sphingomyelin in the reticuloendothelial and in the central and peripheral nervous systems	120, 121
Gaucher's disease		Accumulation of cerebrosides in the nervous and reticuloendothelial systems	120, 121
Gargoylism		Accumulation of sphingolipids in the brain	120, 121
Cretinism	Lack of iodine in diet or lack of an enzyme concerned with thyroxine synthesis	Deficiency of thyroxine	107
Myxedema		Deficiency of thyroxine	107
Pellagra	Inadequate nicotinic acid		107

3. Mental Retardation

A substantial number of genetically determined biochemical disorders in infants and young children produce mental deficiency in early life (123, 124). A few of these diseases are listed in Table 1.1. Approximately thirty-five syndromes of mental retardation based on disturbances in the metabolism of amino acids have been discussed(124). These disorders are derived from the faulty metabolism of tyrosine, histidine, proline, tryptophan, sulfur-containing amino acids, amino acids of the urea synthesis cycle, etc. Other biochemical disorders that cause mental deficiency include galactosemia, cretinism, Wilson's disease, hereditary fructose intolerance, Hurler's syndrome, and the cerebro-oculo-renal syndrome of Lowe(123), as well as diseases of sphingolipid metabolism—such as Gaucher's disease, Niemann–Pick disease, Fabry's disease, metachromatic leukodystrophy, Tay–Sachs disease, and generalized gangliosidosis(125).

With prompt detection of these diseases, effective therapy can be undertaken to prevent the crippling effects of mental defect, or, where this disease cannot be overcome, the parents can be told so that they can take the action they deem desirable. Screening and follow-through chemical tests are useful in this respect. Such tests of urine and blood are available(49, 123). However, they can be improved in terms of simplicity, sensitivity, and reliability. But many more such tests are needed. It would be even more desirable to develop screening tests of expectant mothers for these defects. These various tests should be of considerable diagnostic and prognostic usefulness, as well as being helpful adjuncts for the selection of the proper therapeutic approach.

C. Biochemistry of Various Life Forms

Many problems in the fields of animal, insect, plant, bacterial, and viral biochemistry will be more readily studied with the development of better methods of assay. With the increased improvement of some of the highly sensitive spectral techniques, direct methods of assay for important components of living tissue will become available.

In any study of a living organism questions about enzymes always arise. One set of problems is concerned with enzymic sites and miscellaneous chemicals involved in the regulation of the synthesis and activity of enzymes. What is the precise structure of the active site of enzymes and the reason for the catalytic function of the active site? Can smaller chemicals mimicking the specific catalytic reactions of enzymes be used in analytical research?

Probably as important as the reaction between enzyme and substrate is the series of unknown chemical reactions that produce the physiological effects of hormones. One wonders what type of products are formed in the reaction between hormones and their receptors.

Another biological problem is photosynthesis. Some of the questions that are being investigated are the nature of the mechanism of energy transfer, the route to oxygen, and the identity of the various light-absorbing pigments that participate in photosynthesis. The spectral aspects of the primary processes in photosynthesis have been discussed by Kamen(126), as have the controversies in this field.

Research in insect biochemistry is in a revolutionary phase(127). Research has been accelerated in the fields of repellants and attractants (128, 129) and insecticides. The problem has been crop damage by some insects and the carrying of virulent human diseases by others. The big problem is the selective control of insects so that their damage is reduced while their benefits are not decreased. The control has to entail little or no risk to mankind and other forms of life beneficial to man. Use of the 900 pesticides registered in the United States(130), as well as the many other chemicals being developed, means that assay methods will have to be continually improved so as to ensure a safe, selective use of these chemicals.

For a better understanding of insects, ourselves, and all other forms of life a more thorough study of the biological clocks controlling life's changes and rhythms is necessary. We need to know the chemical reactions and what initiates these reactions in such phenomena as the flowering and leafing of plants, the autumn defoliation of many types of plants, the migration of various types of birds and insects, insect metamorphosis, the maturation of an individual, the onset of old age, the fading of memory, the rise of the sex drive with adolescence followed by its gradual fading in later years, and all the other changes to which living creatures are prone. One wonders what the secret of the long life of the sequoia tree is.

Research is only beginning on understanding the cellular mechanism for measuring time used by various plants and animals(131, 132). The occurrence of oscillatory phenomena in a wide variety of organisms has raised the question of the possible relationship of oscillating enzymic systems to biological clocks(133).

Although some of these processes are seemingly irreversible, one wonders whether they can be reversed or even delayed. Is old age, with its concomitant loss of sexual and mental creative powers, an irreversible process? Is senescence a more efficient organization of the cell's polymeric components similar to the aging of colloid systems(134), essentially a decreased molecular state of chaos(135)? In man's drive toward increased creativity at ever higher levels of sophistication and achievement these questions must be asked—and answered with the help of every available tool. One of the important tools is absorptiometry.

D. Origin and Synthesis of Life

One way of studying the beginning and the evolution of life is through the application of trace-analysis tools involving separation and photometric techniques to organic geochemical investigations(136). A second way is through similar examination of meteorites for organic matter(137). Because of the difficult problem of contamination, a method of collecting meteorites before they reach the earth's atmosphere is necessary. A third way is through the detection and study of extraterrestrial life(138). Sensitive methods of detection and assay, combined with automated equipment capable of converting the response to a signal and sending it through space and receiving it, will be needed.

Another aspect in studies on the beginning of life concerns laboratory investigations of prebiological syntheses under "natural" conditions. The attempt is to find the critical chemical syntheses and combinations that were preliminary to the formation of the first life forms. The method involves the syntheses of complex biopolymers and mixtures, and the trace analysis of the components of these mixtures.

The question arises whether such prelife structures and crude life forms are being synthesized in some of the environments around us. If such structures exist, they are probably competing with living things. This means that many of them could be harmful and/or beneficial to some forms of life and even to man himself. One wonders whether viruses are breakdown products of life forms—for example, breakoff pieces of the long, relatively indestructable DNA molecule and/or products of spontaneous generation through molecular evolution—and whether viruses are being synthesized today in the environment from simpler chemicals or as breakdown products of cells.

Before a living cell could be synthesized, the chemical composition of the cell would have to be known, and extremely sensitive and reliable microphotometric methods of analysis of the components of single cells would have to be developed. When one considers that a bacterial cell weighs several picograms, contains about 10 femtograms of DNA; 40 femtograms of RNA; 20 femtograms each of lipids, phospholipids, and polysaccharides; and a fraction of a picogram of proteins, it is apparent that a combination of microscope and photometer is necessary for analysis(139).

A vital part of the solution of this problem is a complete understanding of the genetic code. Investigations into the molecular biology of the gene(140) need to be accelerated with the help of every useful tool. Badly needed are better methods for the separation of complex nucleotide mixtures into their components and much more highly sensitive and selective methods for the assay of nucleic acids, nucleosides, and nucleotides. At the heart of the whole problem is the development of a simple reliable method for the determination of the base sequence in the various types of DNA, messenger RNA, ribosomal RNA, and transfer RNA. Once this knowledge of the cell and its components is available, the proper chemicals could be put together in the proper environment, and natural forces would take over to form a pre-cell whose evolution into a living cell could be hastened through laboratory methods.

With the help of this knowledge some day man will synthesize a vast variety of life forms much more varied and useful than plastics are today. There will be dreadful by-products and new problems, but this we now expect with each advance. The by-products will have to be channeled in useful directions, and the new problems will have to be investigated by generations to come.

It must be emphasized that only a few of the problems facing man have been briefly mentioned in this chapter. The photometric trace-analysis techniques described and discussed in this treatise are only one type of powerful tool used to solve, understand, and/or control some of the aspects of these problems. Like the other tools, they need further improvement, development, and application to penetrate the infinite darkness around us about which we have such an abysmal ignorance.

The various photometric processes have been briefly described to emphasize that they are all interconnected. Consequently any better understanding of the absorptiometric process will benefit the other fields of photometric analysis.

References

1. E. Sawicki, T. W. Stanley, and H. Johnson, *Mikrochim. Acta*, 178 (1965).
2. E. Sawicki and J. Pfaff, *Mikrochim. Acta*, 322 (1966).
3. W. C. Hueper and W. D. Conway, *Chemical Carcinogenesis and Cancers*, Thomas, Springfield, Ill., 1964.
4. G. M. Badger, *Brit. J. Cancer*, **2**, 309 (1948).
5. N. P. Buu-Hoi, *Arzneimittel-Forsch.*, **11**, 812; 896 (1961).
6. J. L. Hartwell, *Survey of Compounds Which Have Been Tested for Carcinogenic Activity*, 2nd ed., U.S. Public Health Service Publ. No. 149, U.S. Government Printing Office, Washington, D.C., 1951.
7. D. B. Clayson, *Chemical Carcinogenesis*, Churchill, London, 1962.
8. P. Shubik and J. L. Hartwell, *Survey of Compounds Which Have Been Tested for Carcinogenic Activity*, Suppl. I, U.S. Government Printing Office, Washington, D.C., 1957.

9. J. H. Weisburger and E. K. Weisburger, *Chem. Eng. News*, **44**, 124 (1966).
10. T. S. Scott, *Carcinogenic and Chronic Toxic Hazards of Aromatic Amines*, Elsevier, Amsterdam, 1962.
11. E. K. Weisburger and J. H. Weisburger, in J. P. Greenstein and A. Haddon, Eds., *Advances in Cancer Research*, Vol. 5, Academic Press, New York, 1958, pp. 331–431.
12. E. Sawicki, H. Johnson, and K. Kosinski, *Microchem. J.*, **10**, 72 (1966).
13. J. A. Miller and E. C. Miller, in J. P. Greenstein and A. Haddon, Eds., *Advances* in Cancer Research, Vol. 1, Academic Press, New York, 1953, pp. 339–396.
14. B. L. Van Duuren, N. Nelson, L. Orris, E. D. Palmes, and F. L. Schmitt, *J. Natl. Cancer Inst.*, **31**, 41 (1963).
15. B. L. Van Duuren, L. Orris, and N. Nelson, *J. Natl. Cancer Inst.*, **35**, 707 (1965).
16. A. L. Walpole, *Ann. N.Y. Acad. Sci.*, **68**, 750 (1958).
17. W. C. J. Ross, *Biological Alkylating Agents: Fundamental Chemistry and Design of Compounds for Selective Toxicity*, Butterworths, London, 1962.
18. H. L. Falk, S. J. Thompson, and P. Kotin, *Arch. Envir. Health*, **10**, 847 (1965).
19. P. N. Magee and R. Schoental, *Brit. Med. Bull.*, **20**, 102 (1964).
20. F. Dickens, *Brit. Med. Bull.*, **20**, 96 (1964).
21. M. Barsotti and E. C. Vigliani, *Arch. Ind. Hyg. Occup. Med.*, **5**, 234 (1952).
22. E. Boyland, *The Biochemistry of Bladder Cancer*, Thomas, Springfield, Ill., 1963.
23. R. E. Eckardt, *Industrial Carcinogens*, Grune and Stratton, New York, 1959.
24. M. W. Goldblatt, *Brit. Med. Bull.*, **4**, 405 (1947).
25. T. S. Scott and M. H. C. Williams, *Brit. J. Ind. Med.*, **14**, 150 (1957).
26. I. S. Temkin, *Industrial Bladder Carcinogenesis*, Pergamon, New York, 1963.
27. A. L. Walpole and M. H. C. Williams, *Brit. Med. Bull.*, **14**, 141 (1958).
28. S. A. Henry, *Cancer of the Scrotum in Relation to Occupation*, Oxford University Press, London, 1946.
29. E. Sawicki, *Chemist-Analyst*, **53**, 24, 28, 56, 88 (1964).
30. E. Sawicki, W. C. Elbert, T. R. Hauser, F. T. Fox, and T. W. Stanley, *J. Am. Ind. Hyg. Assoc.*, **21**, 443 (1960).
31. E. Sawicki, paper presented at the American Medical Association Air Pollution Medical Research Conference, March 2–4, 1966, Los Angeles.
32. E. C. Miller and J. A. Miller, *Pharmacol. Rev.*, **18**, 805 (1966).
33. W. E. C. Wacker, *Clin. Chem.*, **11**, 825 (1965).
34. M. Sandler and C. R. J. Ruthven, *Pharmacol. Rev.*, **18**, 343, (1966).
35. R. W. Miller, *Yale J. Biol. Med.*, **37**, 487 (1965).
36. W. Krivit and R. A. Good, *J. Dis. Child.*, **94**, 289 (1957).
37. N. Wald, W. H. Borges, C. C. Li, J. H. Turner, and M. C. Harnois, *Lancet*, 1228 (1961).
38. W. M. C. Brown and R. Doll, *Brit. Med. J.*, 981 (1961).
39. A. B. Sabin, *Am. J. Dis. Child*, **111**, 1 (1966).
40. P. T. Mora, V. W. McFarland, and S. W. Luborsky, *Proc. Natl. Acad. Sci. U.S.*, **55**, 438 (1966).
41. W. S. Robinson, A. Pitkanen, and H. Rubin, *Proc. Natl. Acad. Sci. U.S.*, **54**, 137 (1965).
42. P. H. Duesberg and W. S. Robinson, *Proc. Natl. Acad. Sci. U.S.*, **54**, 794 (1965).
43. M. Stokers, *Endeavor*, **25**, 119 (1966).
44. H. J. Sanders, *Chem. Eng. News*, **43**, 130 (Mar. 8, 1965).
45. J. W. Gofman, B. Strisower, O. De Lalla, A. R. Tamplin, H. B. Jones, and F. T. Lindgren, *Mod. Med. (Minneapolis)*, **7**, 119 (1953).

46. J. W. Gofman, M. Hanig, H. B. Jones, M. A. Lauffer, E. Y. Lawry, L. A. Lewis, G. V. Mann, F. E. Moore, F. Olmstead, and J. F. Yeager, *Circulation*, **14**, 691 (1956).
47. G. L. Mills, C. E. Taylaur, and P. A. Wilkinson, *Clin. Chim. Acta*, **13**, 527 (1966).
48. J. E. Seegmiller, *Clin. Chem.*, **14**, 554 (1967).
49. D. Y. Hsia, in M. Stefanini, Ed., *Clinical Pathology*, Vol. I, Grune and Stratton, New York, 1966, pp. 493–511.
50. V. A. McKusick, D. Kaplan, D. Wise, W. B. Hanley, S. B. Suddarth, M. E. Sevick, and A. E. Maumanee, *Medicine*, **44**, 445 (1965).
51. S. W. Wright, *J. Am. Med. Assoc.*, **165**, 2079 (1957).
52. J. Menkes, P. Hurst, and J. Craig, *Pediatrics*, **14**, 462 (1954).
53. C. Bödeker, *Z. Rat. Med.*, **7**, 130 (1859).
54. G. Medes, *Biochem. J.*, **26**, 917 (1932).
55. R. A. Field, in J. B. Stanbury, J. B. Wyngaarden, and D. S. Frederickson, Eds., *The Metabolic Basis of Inherited Disease*, McGraw-Hill, New York, 1960, p. 156.
56. V. A. McKusick, *J. Am. Med. Assoc.*, **182**, 271 (1962).
57. W. Tietz, *Am. J. Human Genet.*, **15**, 259 (1963).
58. P. Rubin, *Radiol. Clin. N. Am.*, **1**, 621 (1963).
59. J. G. Greenfield, *The Spino-Cerebrellar Degenerations*, Thomas, Springfield, Ill., 1954.
60. D. V. Haws and V. A. McKusick, *Bull. Johns Hopkins Hosp.*, **113**, 20 (1963).
61. A. Sorsby, *Acta Genet. Med.* (*Rome*), **13**, 20 (1964).
62. H. Weill, M. M. Ziskind, R. C. Dickerson, and V. J. Derbes, *J. Air Poll. Control Assoc.*, **15**, 467 (1965).
63. R. Lewis, M. M. Gilkeson, and R. O. McCaldin, *Publ. Health Rept.*, **77**, 947 (1962).
64. R. B. W. Smith, E. J. Kolb, H. W. Phelps, H. A. Weiss, and A. B. Hollinden, *Arch. Env. Health*, **8**, 805 (1964).
65. J. H. Morris, L. Berrens, and E. Young, *Clin. Chim. Acta*, **12**, 407 (1965).
66. J. Rothenstein, K. Schwarz, M. Schwarz-Speck, and H. Storck, *Int. Arch. Allergy*, **29**, 1 (1966).
67. W. B. Shelley and J. S. Comaish, *J Am. Med. Assoc.*, **192**, 122 (1965).
68. D. W. Cowan, H. J. Thompson, H. J. Paulus, and P. W. Mielke, Jr., *J. Air Poll. Control Assoc.*, **13**, 546 (1963).
69. W. B. Ober and G. A. Kaiser, *Am. J. Clin. Pathol.*, **35**, 297 (1961).
70. M. J. Levell and T. K. Cottam, *Clin. Chim. Acta*, **23**, 231 (1969).
71. D. Waldi, *Klin. Wochschr.*, **40**, 827 (1962).
72. H. O. Bang, *J. Chromatog.*, **14**, 520 (1964).
73. H. J. Sanders, *Chem. Eng. News*, **43**, 74 (Mar. 22, 1965).
74. J. J. Brand and W. L. M. Perry, *Pharmacol. Rev.*, **18**, 895 (1966).
75. H. B. Taussig, *Scientific American*, **207**, 29 (1962).
76. A. S. Curry, *J. Pharm. Pharmacol.*, **7**, 969 (1955).
77. C. P. Stewart and A. Stolman, *Mechanisms and Analytical Methods*, Vol. 1, 1960, Vol. 2, 1961, Academic Press, New York and London.
78. R. Carson, *Silent Spring*, Houghton Mifflin, Boston, 1962.
79. W. J. Hayes and C. J. Pirkle, *Arch. Env. Health*, **12** 43 (1966).
80. I. West, *Arch. Env. Health*, **9**, 626 (1964).
81. F. Bicknell, *Chemicals in Your Food*, Emerson, New York, 1961.
82. S. D. Faust and O. M. Aly, *J. Am. Water Works Assoc.*, **56**, 267 (1964).
83. *Motor Vehicles, Air Pollution, and Health*, A Report of the Surgeon General to the U.S. Congress in Compliance with Public Law 86–493, the Schenck Act, House Document No. 489, U.S. Government Printing Office, Washington, D.C., 1962.

84. *Proceedings of the National Conference on Air Pollution, Washington, D.C.*, U.S. Department of Health, Education, and Welfare, U.S. Public Health Service, 1958.
85. A. C. Stern, Ed., *Air Pollution*, Vols. I and II, Academic Press, New York, 1968.
86. F. W. Went, *Proc. Natl. Acad. Sci. U.S.*, **46**, 212 (1960).
87. R. I. Larsen, *Air Pollution From Motor Vehicles*, paper presented at a meeting of the New York Academy of Sciences, April 6, 1966.
88. *Smoking and Health*, Report of the Advisory Committee to the Surgeon General of the United States, U.S. Department of Health, Education, and Welfare, U.S. Public Health Service, Washington, D.C., 1964.
89. H. W. Phelps, *Arch. Env. Health*, **10**, 147 (1965).
90. D. E. Van Wormer, *Arch. Env. Health* **10**, 71 (1965).
91. E. R. John, *Bull. Atomic Scientists* **21**, 12 (1965).
92. L. B. Flexner, J. B. Flexner, and R. B. Roberts, *Proc. Natl. Acad. Sci. U.S.*, **56**, 730 (1966).
93. J. Gaddum, *Perspectives in Biology and Medicine*, **8**, 436 (1965).
94. B. W. Agranoff, *Perspectives in Biology and Medicine*, **9**, 13 (1965).
95. S. Rose, *New Scientist*, **16**, 781 (1965).
96. E. L. Bennett, M. C. Diamond, D. Krech, and M. R. Rosenzweig, *Science*, **146**, 610 (1964).
97. L. B. Flexner, J. B. Flexner, G. De La Haba, and R. B. Roberts, *J. Neurochem.*, **12**, 535 (1965).
98. W. Dingman and M. B. Sporn, *J. Psychiat. Res.*, **1**, 1 (1961).
99. F. R. Babich, A. L. Jacobson, S. Bubash, and A. Jacobson, *Science*, **149**, 656 (1965).
100. M. Luttges, T. Johnson, C. Buck, J. Holland, and J. McGaugh, *Science*, **151**, 834 (1966).
101. F. N. Pitts, Jr., *Scientific American*, **220**, 69 (1969).
102. A. R. Caggiula and B. G. Hoebel, *Science*, **153**, 1284 (1966).
103. S. P. Grossman, *Discovery*, **27**, 19 (1966).
104. D. D. Jackson, *Scientific American*, **207**, 65 (1962).
105. H. Osmond and J. Smythies, *J. Mental Sci.*, **98**, 309 (1952).
106. A. Hoffer, H. Osmond, and J. Smythies, *J. Mental Sci.*, **100**, 29 (1954).
107. D. W. Woolley, *The Biochemical Bases of Psychoses*, Wiley, New York and London, 1962.
108. J. Behrman, *Clin. Chem.*, **12**, 211 (1966).
109. L. R. Gjessing, *Scand. J. Clin. Lab. Invest.*, **17**, 549 (1965).
110. J. L. W. Thudichum, *A Treatise on the Chemical Constitution of the Brain*, Bailliere, Tindall, and Cox, London, 1884, p.13; through S. S. Kety, *Nature*, **207**, 1252 (1965).
111. W. E. Knox, in J. B. Stanbury, J. B. Wyngaarden, and D. S. Fredrickson, Eds., *The Metabolic Basis of Inherited Disease*, McGraw-Hill, New York, 1960, pp. 321–382.
112. L. E. Woolf, in *Advances in Clinical Chemistry*, **6**, 97–230 (1963).
113. N. G. Woody and C. D. Hancock, *Am. J. Dis. Child.*, **106**, 578 (1963).
114. H. Ghadhimi, M. W. Partington, and A. Hunter, *Pediatrics*, **29**, 714 (1962).
115. B. Levin, H. M. M. Mackay, and V. G. Oberholzer, *Arch. Dis. Child*, **36**, 622, (1961).
116. K. Halvorsen and S. Halvorsen, *Pediatrics*, **31**, 29 (1963).
117. J. B. Jepson and M. J. Spiro, in J. B. Stanbury, J. B. Wyngaarden, and D. S. Fredrickson, Eds., *The Metabolic Basis of Inherited Disease*, McGraw-Hill, New York, 1960, pp. 1338–1364.
118. D. P. Greaves, *Proc. Roy. Soc. Med.*, **52**, 26 (1963).
119. K. J. Issellbacher, in J. B. Stanbury, J. B. Wyngaarden, and D. S. Fredrickson,

Eds., *The Metabolic Basis of Inherited Disease*, McGraw-Hill, New York, 1960.
120. G. Rouser, G. Kritchevsky, and C. Galli, *J. Am. Oil Chemists Soc.*, **42**, 412 (1965).
121. D. S. Fredrickson, in J. B. Stanbury, J. B. Wyngaarden, and D. S. Fredrickson, Eds., *The Metabolic Basis of Inherited Disease*, McGraw-Hill, New York, 1960, pp. 553–636.
122. A. Barbeau, *J. Neurosurg.*, **23**, Suppl., 162 (1966).
123. T. L. Perry, S. Hansen, and L. MacDougall, *Can. Med. Assoc. J.*, **95**, 89 (1966).
124. D. O'Brien, F. Ibbot, and D. Rodgerson, in M. Stefanini, Ed., *Progress in Clinical Pathology*, Grune and Stratton, New York, 1966, pp. 512–599.
125. R. O. Brady, *Clin. Chem.*, **14**, 565 (1967).
126. M. D. Kamen, *Primary Processes in Photosynthesis*, Academic Press, New York and London, 1963.
127. E. E. Smissman, *J. Pharm. Sci.*, **54**, 1395 (1965).
128. M. Jacobson, *J. Am. Oil Chemists Soc.*, **42**, 681 (1965).
129. M. Jacobson, *Ann. Rev. Entomol.*, **11**, 403 (1966).
130. L. A. Richardson and M. J. Foter, *J. Milk Food Technol.*, **29**, 148 (1966).
131. B. M. Sweeney, in *Plant Biology Today*, Wadsworth, California, 1963.
132. E. B. Hillman, *The Physiological Clock*, Holt, Rinehart and Winston, New York, 1962.
133. R. Frenkel, *Arch. Biochem. Biophys.*, **115**, 112 (1966).
134. V. Ruzicka, *Arch. Mikroscop. Anat. Entwicklungsmech.*, **101**, 159 (1924).
135. I. G. Fels, *Gerontologia*, **12**, 109 (1966).
136. I. A. Breger, Ed., *Organic Geochemistry*, Macmillan, New York, 1963.
137. B. Mason, *Scientific American*, **208**, 43 (1963).
138. G. Mamikunian and M. H. Briggs, *Current Aspects of Exobiology*, Pergamon, New York, 1966.
139. D. Glick, *Quantitative Chemical Techniques of Histo- and Cytochemistry*, Vols. I and II, Interscience, New York, 1961, 1963.
140. J. D. Watson, *Molecular Biology of the Gene*, Benjamin, New York, 1965.

CHAPTER 2

NOMENCLATURE

The nomenclature used in this treatise is summarized in this chapter. Terms that have several definitions, some of which have become obsolete due to the accretion of data inconsistent with these definitions, have been redefined to make the definition consistent with the facts. In some cases new terms consistent with present nomenclature in allied fields have been introduced.

I. Photometric Nomenclature

Absorbance (A). Logarithm to the base 10 of the reciprocal of the transmittance: $A = \log_{10}(1/T)$.

Absorptivity (a). Absorbance divided by the product of the concentration c of the test substance (in grams per liter) and the sample path length b (in centimeters): $a = A/bc$.

***A* Value** ($A^{1\%}_{1\,\text{cm}}$). Absorbance of a 1% solution in cell length 1 cm, used where the molecular weight of the compound is unknown. Usually signified as $E^{1\%}_{1\,\text{cm}}$.

Beer's law. The absorbance of a dissolved substance contained in a cell of constant path length is directly proportional to its concentration: $A = kc$. In other words, the absorptivity of a substance is a constant with respect to changes in concentration.

Inflection (i). A slight bump on the slope of a wavelength maximum.

Millimolar absorptivity (mϵ). See molar absorptivity. The millimolar absorptivity ($\epsilon \times 10^{-3}$) is used throughout this treatise.

Molar absorptivity (ϵ). Product of the absorptivity and the molecular weight (MW) of the pure compound. It has units of square centimeters per mole:

$$\epsilon = a \times \text{MW} = \frac{A \times \text{MW}}{c(\text{g/l}) \times b} = \frac{A}{c(\text{moles/l}) \times b}$$

where b is the path length in centimeters of the light through the cuvette.

Near infrared (NIR). Region of electromagnetic spectrum from 800 to 2000 mμ.

Sample path length (b). Path length in centimeters of light through the solution being examined spectrally in the cuvette.

Shoulder (s). A strong outward bulge on the slope of a wavelength maximum.

Solvent polarity (E_t). Essentially the transition-energy value for a solvent obtained from the long-wavelength maximum (usually) of an extremely solvent-sensitive dye dissolved in the solvent. The value is calculated by the equation

$$E_t = \frac{hc}{\lambda} = \frac{28{,}600}{\lambda_{\max}(\mathrm{m}\mu)}$$

where h is Planck's constant and c is the velocity of light. The standard dye picked to determine solvent polarity should ideally be soluble in all solvents; that is, polar solvents, such as water and phenol, as well as nonpolar solvents, such as heptane and carbon tetrachloride. The maximum ΔE_t value—that is, the difference in the E_t values of a highly polar solvent and a nonpolar solvent—should be as large as possible, so as to make the E_t values of various solvents meaningful.

Three standards should be available, each of which should contain either a Z_n, Z_z, or Z_b resonance structure. Charge-transfer complexes could also be used as standards. An example is 1-ethyl-4-carbomethoxypyridinium iodide (1), which has a measurable ΔE_t of 23.2 kcal/mole between chloroform ($\lambda_{\max}$ 452.5 mμ) and 70% ethanol ($\lambda_{\max}$ 331 mμ). Some of the standard dyes that have been recommended include the betaine of 2,4,6-triphenyl-*N*-(3,5-diphenyl-4-hydroxyphenyl)pyridinium perchlorate, a Z_b-structure, which has a ΔE_t of 27.8 kcal/mole between diphenyl ether ($\lambda_{\max}$ 810 mμ) and water ($\lambda_{\max}$ 453 mμ) (2). Another is 4-[5-(5-ethyl-3-methyl-1,2,3,5-tetrahydropyrido[1,2-*a*]benzimidazolyl)-2,4-pentadienylidene]-2,2-dimethyl-1,3-cyclobutanedione, a Z_z-structure, which has a ΔE_t of 27.2 kcal/mole between water ($\lambda_{\max}$ 458 mμ) and 2,6-lutidine ($\lambda_{\max}$ 723 mμ) (3). The third standard is 1,3-diethyl-5-[5-(2,3,6,7-tetrahydro-1*H*,5*H*-benzo[*i,j*]quinolizin-9-yl)-1,3-neopentylene-2,4-pentadienylidene]-2-thiobarbituric acid, a Z_n-structure, which has a ΔE_t of 12.6 kcal/mole between isooctane ($\lambda_{\max}$ 562 mμ) and 20 vol % lutidine in water ($\lambda_{\max}$ 747 mμ) (3).

Empirical parameters of the polarity of solvents have been discussed (4).

Transmittance (T). The ratio of the radiant power P transmitted by a sample to the radiant power P_0 incident on the sample, both being measured at the same spectral position and with the same slit width:

$$T = \frac{P}{P_0} = 10^{-kbc}$$

Ultraviolet (UV). Region of electromagnetic spectrum from 200 to 400 mμ.

Visible (VIS). Region of electromagnetic spectrum from 400 to 800 mμ. Visible to the human eye.

Wavelength (λ). The distance, measured along the line of propagation, between two points that are in phase on adjacent waves. The unit is the millimicron. However, attempts are being made to displace this by nanometer (nm). In view of the resolution normally attainable, the wavelength unit for absorption in the ultraviolet and visible regions is the millimicron.

Wavelength maximum (λ_{max}). The wavelength value in millimicrons at an absorption maximum. All wavelength values in this book are reported in terms of millimicrons. Wave-number or frequency values are not used.

Wavelength minimum (λ_{min}). The wavelength value in millimicrons at an absorption minimum.

II. Absorption-Band Nomenclature

Anionic resonance band. A spectral band derived from an anionic resonance structure. These structures can be formed in two main ways: by the addition of an anion to a molecule, for example,

$$O_2N\text{-}C_6H_3(NO_2)_2 + OH^- \longrightarrow O_2N\text{-}C_6H_3(H)(OH)(NO_2)(NO_2^-)$$

or by the substraction of a proton, for example,

$$O_2N\text{-}C_6H_4\text{-}NH_2 + OH^- \longrightarrow O_2N\text{-}C_6H_4\text{-}NH^- + H_2O$$

Aromatic resonance bands. Spectral bands associated with neutral aromatic hydrocarbons. The structures involved here are essentially a special case of zwitterionic resonance. The most successful and thorough study of the properties of these bands has been published by Clar (5). Platt and his collaborators have also done noteworthy work in this field (6). The various nomenclatures and the theoretical aspects have been discussed by Jaffe and Orchin (7). These bands and their properties will not be discussed except where their analytical usefulness is definitely potential or self-evident.

The monocyclic and polycyclic aromatic hydrocarbons can also form anionic and cationic resonance structures, as well as free radicals and π-complexes, from each of which distinct spectra can be obtained.

Cationic resonance band. A spectral band to which a cationic resonance structure is a contributor. These structures can be formed in two ways: by the subtraction of an anion, for example,

$$C_6H_5-CH_2OH \xrightarrow{H^+} C_6H_5-CH_2^+$$

or by the addition of a proton (or a cation) to the negative resonance terminal, for example,

$$C_6H_5-CHO + H^+ \longrightarrow C_6H_5-CH=\overset{+}{O}H$$

Free-radical resonance band. A spectral band derived from a molecule that contains an unpaired electron. Depending on the molecule, the resonance can be either anionic, cationic, or zwitterionic. These various types of free radical can be obtained by oxidation (subtraction of an electron or a hydrogen atom from a molecule), as in

$$\text{phenoxazine (N–H)} + HONO \xrightarrow{-H^\cdot} \text{phenoxazinyl radical } (\dot{N}) + H_2O + NO$$

or by reduction (addition of an electron or a hydrogen atom to a molecule), as in

$$\text{xanthone (C=O)} \xrightarrow{+H^\cdot} \text{xanthenol radical } (\dot{C}-OH)$$

In the first example, if an electron were subtracted, the final product would have a cationic resonance structure; in the second example addition of an electron would result in an anionic resonance structure. The appropriate band systems would result.

π-Complex band. Intermolecular electron-transfer band. This band is derived from a π-complex formed between a Lewis acid (electron acceptor) and a Lewis base (electron donor). Polynitroarenes and aromatic hydrocarbons form such complexes. Since these complexes are weak and are easily broken in solution and have low millimolar absorptivities ($m\epsilon' = 0.1$ to ~ 5.0), they are rarely used in wet analytical methods. Thus in the determination of pyrene with tetracyanoethylene in chloroform–methylene chloride solution an mϵ' of 0.14 is obtained at 720 mμ (8). On the other hand, a solution of 1,3,5-trinitrobenzene in mesitylene gives a relatively high mϵ' value of 3.68 (9), still fairly insensitive for trace analysis purposes.

Transition. An electronic change in state as represented by the change from a ground state to an excited state on absorption of light energy. The electronic transitions that are involved in the ultraviolet–visible regions include $\sigma \rightarrow \sigma^*$, $n \rightarrow \sigma^*$, $n \rightarrow \pi^*$, and $\pi \rightarrow \pi^*$. Compounds involved in the $\sigma \rightarrow \sigma^*$ transition absorb below 200 mμ and so are not useful in organic trace analysis. Compounds that contain nonbonding electrons on singly bonded nitrogen, oxygen, sulfur, or halogen atoms are capable of showing low intensity violet shift bands derived from $n \rightarrow \sigma^*$ transitions, whereas compounds that contain nonbonding electrons on doubly bonded nitrogen, oxygen, or sulfur atoms are capable of showing low intensity violet shift bands derived from $n \rightarrow \pi^*$ transitions. Bands derived from these last two transitions are not useful in organic trace analysis because of their low intensity. Consequently we shall be interested in bands derived from the much more intense $\pi \rightarrow \pi^*$ transitions.

Zwitterionic. In many molecules the zwitterionic resonance structure is the main contributor either in the ground state (Z_z), the excited state (Z_n), or in both states (Z_b). An example of a Z_z-structure is the chromogen obtained in the determination of phenol with 4-aminoantipyrine (10):

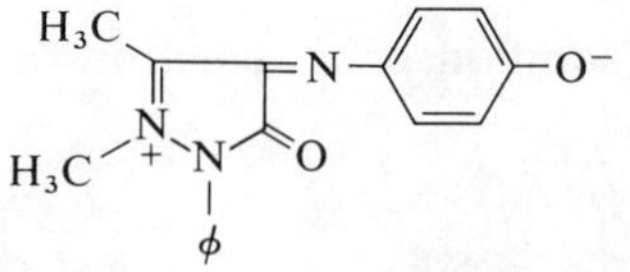

An example of a Z_n-structure is the chromogen obtained in the determination of aniline with 4-dimethylaminobenzaldehyde (11):

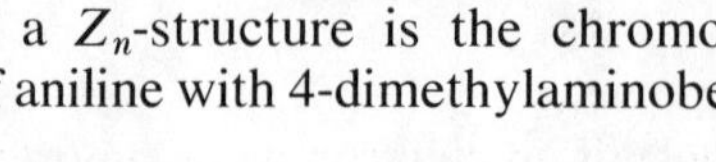

An example of a Z_b-structure is found in pyridinium cyclopentadienylide (12):

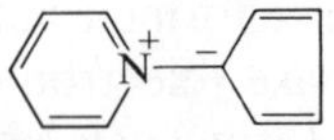

A compound with a Z_n-structure has its dipolar form contributing mainly to the excited state. This is shown by the increased dipole moment in the excited state. Thus 4-dimethylamino-4′-nitrostilbene and 4-amino-4′-nitrostilbene have dipole moments in the ground and excited states of 7.6 and 32, and 6.0 and 22.2, respectively(13,14). On the other hand, a compound with a Z_z-structure, such as

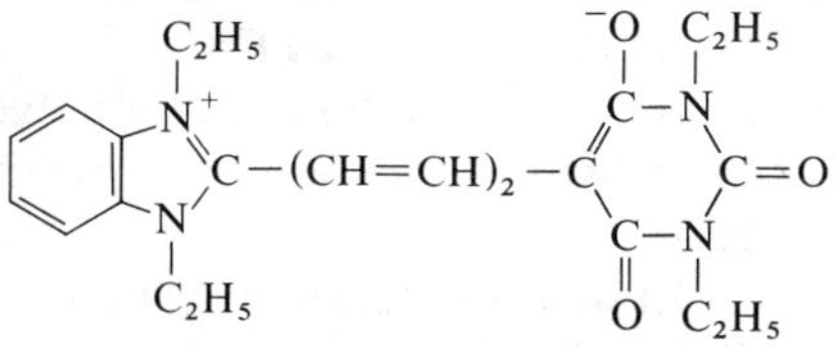

has a ground-state dipole moment of 17.7(11) and could be expected to have a lower dipole moment in the excited state.

III. Photometric Structural Terms

Acidity, internal, or intramolecular. The charged minus neutral stabilization of a negative resonance terminal in an anionic resonance structure is defined as the relative attraction of the uncharged form of an anionic resonance terminal for a negative charge.

Annulation. The fusion of one or more benzene rings to a heterocyclic or aromatic ring. An example would be 1-naphthylamine as compared to aniline.

Basicity, internal, or intramolecular. Essentially the "onium minus ide" stabilization of a positive resonance terminal in a cationic resonance structure. It is defined as the relative attraction of the uncharged form of a cationic resonance terminal for a positive charge.

Bathochromic shift. Refer to "red shift".

Chromogen. The final compound prepared from the test substance through organic microsynthetic methods and subsequently measured colorimetrically.

Chromophore. The smallest component of a molecule (test substance or chromogen) containing the conjugation system and resonance terminals from which a particular electronic transition and its resultant absorption band are derived.

Cross-conjugation. A condition in which part of a conjugation system is common to two different conjugation systems. Separate bands can be obtained for the various conjugation systems. For example, consider 3-amino-4-nitrodibenzothiophene

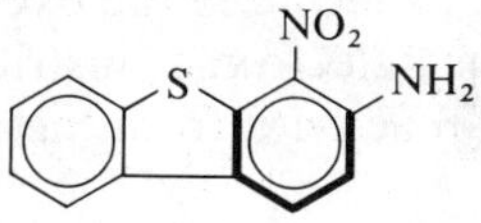

The common conjugation system of the two *o*-nitrobenzene chromophores are outlined in heavy lines. The *o*-nitroaniline chromophore in this molecule dissolved in ethanol has a long-wavelength maximum at 445 mμ, whereas the *o*-nitrophenylsulfide chromophore has its long-wavelength maximum at 362 mμ (15).

Coupling. The connection of polymethine and/or polyene chromophores by a π-bond across which interaction can take place, leading to a strong red-shift effect.

Deviation concept (16). In any zwitterionic, cationic, or anionic resonance compound that contains a linear conjugation chain connecting two resonance terminals where the limiting resonance structures are of unequal energy there is a deviation of the wavelength maximum of the dye from the average wavelength maximum of a symmetrical dye containing the same chain connecting one resonance terminal of the original dye at each end and the other symmetrical dye containing the same chain connecting the other resonance terminal of the original dye at each end. Where the limiting resonance structures are of equal energy, there is no deviation. For example, the dye **1** absorbs at 524 mμ in methanol, whereas the average wavelength maximum of the model dyes **2**, λ_{max} 557.5 mμ, and **3**, λ_{max} 532 mμ, is 545 mμ. The deviation is then 21 mμ.

Many examples of this concept have been given (16).

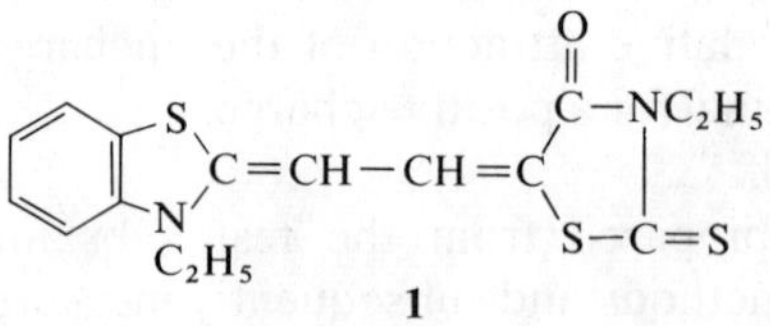

1

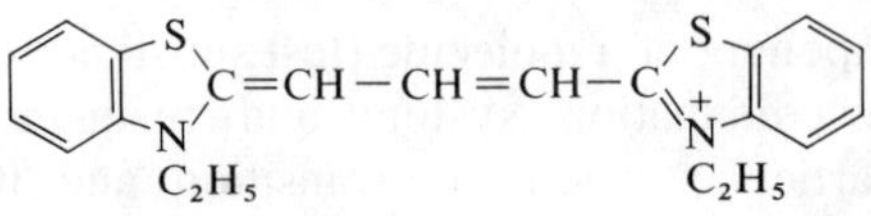

2

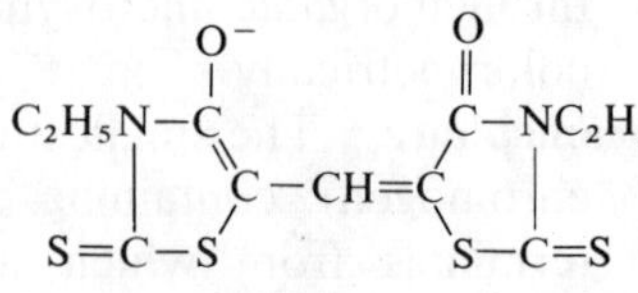

3

Excited state. The state of a molecule after absorption of light. There are many possible excited states, the type of which depends on the kinds of orbitals involved. This treatise is concerned with singlet excited states. A molecule in the excited state will have a different geometry, different electronic distribution, and consequently a different chemical reactivity from the same molecule in the ground state.

Excited-state proton shift. A change in the basicity or acidity of a molecule caused by the absorption of a photon and excitation of the molecule to the excited singlet state. The pH of the solution is such that a proton is added to or subtracted from the molecule after the absorption of a photon. Because of this phenomenon the fluorescence excitation spectrum will be that of the neutral compound (or its salt), whereas the fluorescence emission spectrum will be that of the salt (or its neutral compound).

Extraconjugation. The additional conjugation involving either or both of the resonance terminals, a conjugated system outside of the conjugated group of atoms involved in the main resonance structure. Extraconjugation results in a red shift.

Fine structure (fs). Component bands of an electronic transition band. They are derived from vibrationally excited states and are especially prevalent in the absorption spectra of aromatic hydrocarbons. These bands are useful in characterization.

Ground state. The electronic and vibrational state of a chemical in a bottle, on a chromatogram, or in solution before the absorption of light. There is only one ground state for a molecule.

Homoconjugation. Interaction between two seemingly isolated conjugation systems wherein two "insulated" molecular π-orbitals have a crosswise orbital overlap is shown by $\pi \rightarrow \pi^*$ absorption at longer wavelengths than is found for either conjugation system and by the fact that a zwitterion involving the two conjugation systems can be drawn as an important contributor to the excited state.

Hyperchromic shift. An increase in molar absorptivity of some compound in comparison with a reference compound or of a given compound under varying conditions.

Hypochromic shift. A decrease in molar absorptivity, etc., as for hyperchromic shift.

Insulation. An atom or a group of atoms which separate two conjugation systems and through which resonance cannot take place or is considerably decreased.

Isoenergetic point. The solvent composition in which the wavelength maximum of a compound dissolved in mixtures of a polar and nonpolar solvent reaches a point of longest wavelength and/or highest intensity. A greater increase or decrease of the solvent polarity shifts the wavelength maximum to shorter wavelength and/or lower intensity. This is the point at which, it is believed, the two limiting resonance structures are of approximately equal energy(16). Examples of this phenomenon have been given(17).

Limiting resonance structures. Structures found in chromogens in which there are two main resonance terminals. Thus the two limiting resonance structures of *p*-nitroaniline are

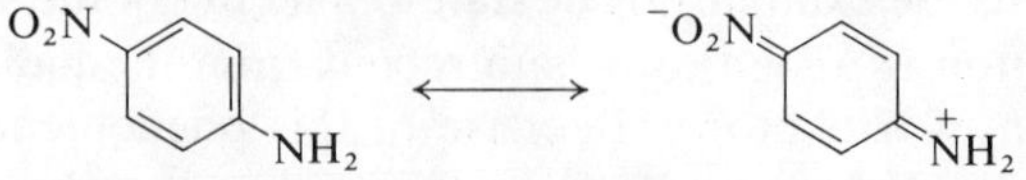

O_2N–C_6H_4–NH_2 ⟷ $^{-}O_2N$=C_6H_4=$\overset{+}{N}H_2$

Negative resonance terminal. The terminal atom (which accepts the negative charge) in a conjugation chain of an anionic or zwitterionic resonance structure. This atom holds the negative charge in one of the limiting resonance structures of the $\pi \rightarrow \pi^*$ transition with which the long-wavelength band is usually associated.

Positive resonance terminal. The terminal atom (which accepts the positive charge) in a conjugation chain of a cationic or zwitterionic resonance structure. This atom contains the positive charge in one of the limiting resonance structures of the $\pi \rightarrow \pi^*$ transition with which the long-wavelength band is usually associated; for example, the nitrogens in the structures below are positive resonance terminals.

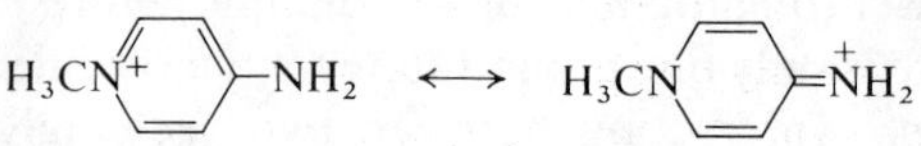

$H_3C\overset{+}{N}$–C_5H_4–NH_2 ⟷ H_3CN=C_5H_4=$\overset{+}{N}H_2$

Red shift. A shift in the wavelength maximum toward the infrared end of the spectrum. Also called bathochromic shift.

Resonance nodes. These are the atoms that make up the resonance chain between the resonance terminals. The terminals determine the positivity or negativity of the nodes. Thus, as can be seen from the various intermediate resonance structures of 4-nitroaniline, carbon atoms 1, 3, and 5 are positive resonance nodes, whereas 2, 4, and 6 are negative resonance nodes. Two of these structures are shown.

$^{-}O_2N$=(4 3 2 1 5 6, + at C-3)–NH_2 ⟷ O_2N–(4 3 2 1 5 6, − at C-2)=$\overset{+}{N}H_2$

Resonance terminals. In many chromogens, fluorogens, and phosphorogens there are electron-donor and electron-acceptor groups usually situated at each end of a chain of conjugation. Thus in a simple example, such as *p*-nitroaniline, the nitro oxygen is the negative, or electron-acceptor, terminal and the amino nitrogen is the positive, or electron-donor, terminal.

Sensitivity rule. This rule, as applied to the deviation concept, states that a small change in the energetic asymmetry of the limiting resonance structures produces a small effect on the deviation of an energetically symmetrical or nearly symmetrical dye, but a much greater effect on the deviation of a highly asymmetrical dye(16).

In changing from a solvent of moderate polarity (e.g., methanol) to one of lower polarity the result will be an increase in energy of the dipolar structure of a merocyanine. For weakly polar merocyanines the asymmetry of the extreme structures and the deviation increase, whereas for a strongly polar merocyanine the asymmetry of the extreme structures and the deviation decrease.

Spiroconjugation. A special type of homoconjugation involving two perpendicular conjugation systems joined by a common tetrahedral atom.

Transannular conjugation. Interaction between two seemingly isolated conjugated systems wherein two "insulated" molecular π-orbitals have either parallel end-to-end or parallel face-to-face overlap. This overlap is shown by $\pi \rightarrow \pi^*$ absorption at longer wavelength than found for either conjugation system and by the fact that a zwitterion involving the two conjugation systems can be drawn as an important contributor to the excited state.

Violet shift. A shift of the wavelength maximum toward the vacuum-ultraviolet region or toward shorter wavelengths. Also called blue shift and hypsochromic shift.

IV. Empirical Spectral Rules

Acidity (or basicity) of resonance terminals and the spectrum of an anionic (or cationic) resonance structure. The greater the difference of internal acidity (or basicity) between the two resonance terminals of an asymmetrical anionic (or cationic) resonance structure, and therefore the greater the difference in energy between the two limiting resonance structures, the greater will be the deviation for the compound under examination and the greater the degree of convergence with increasing chain length of the series.

Brunings–Corwin effect. Substitution into a symmetrical or nearly symmetrical cationic resonance dye so as to cause a crowding of the two halves of the molecule out of planarity results in a red shift of the long-wavelength band.

Conjugation chain length and the absorption spectrum. An increase in the length of conjugation results in red and hyperchromic shifts.

Coupling. The connection of two conjugation systems by one or more π-bonds forms a structure that absorbs at much longer wavelengths than either of the conjugation systems.

Forster's rule. The absorption wavelength maximum of a compound will shift toward longer wavelength as the contributions by any of the intermediate ionic excited structures decrease. Two ways of doing this are to negativize the positive node or to positivize the negative node.

Negativizing the negative node. Substitution of an electronegative group on a negative node causes a violet shift.

Negativizing the positive node. Substitution of an electronegative group on a positive node causes a red shift.

Ortho–para effect. This is a special example of Forster's rule where the resonance terminals are attached to a benzene ring or to some other aromatic hydrocarbon. Substitution of an electron-donor group *ortho* or *para* to the positive resonance terminal causes a red shift and an intensity decrease.

Positivizing the negative node. Substitution of an electropositive group on a negative node causes a red shift.

Positivizing the positive node. Substitution of an electropositive group on a positive node causes a violet shift.

Proton addition at negative resonance terminal of a zwitterionic resonance structure causes a red shift for Z_n-structures and a violet shift for Z_z- and Z_b-structures.

Proton addition at positive resonance terminal of a zwitterionic resonance structure causes a violet shift.

Proton addition at the resonance terminal of an anionic or cationic resonance structure causes a violet shift.

Solvent-basicity effect on anionic resonance bands. An increase in the basicity of a solvent results in a red shift of the anionic resonance band combined with a hyperchromic effect.

Solvent effect on fine structure. The absorption spectra of conjugated hydrocarbons and many of their neutral derivatives show increased fine structure with decreasing polarity of the solvent. Thus the fine structure of a polynuclear aromatic hydrocarbon is accentuated in pentane and considerably decreased in ethanol.

Solvent effect on a Z_n resonance band. A Z_n-structure absorbs at longer wavelength in a polar solvent, such as ethanol, as compared with a nonpolar solvent, such as hexane.

Solvent effect on Z_b and Z_z resonance bands. A Z_b or Z_z-structure absorbs at shorter wavelength in a polar solvent than in a nonpolar solvent.

Steric effect on absorption spectra. Some steric hindrance usually causes a decrease in intensity. With increased steric hindrance in the excited state as compared with the ground state a violet hypochromic shift takes place. With increased steric hindrance in the ground state as compared with the excited state a red hypochromic shift takes place.

Straight-line distance between resonance terminals and the molar absorptivity. The intensity of an absorption band is proportional to the square of the transition moment; that is, the change of dipole moment during the transition. Consequently the integrated intensity (measured by the area under band envelope) should be proportional to the square of the straight-line distance between resonance terminals.

Temperature effect on fine structure. A decrease in temperature causes an increase in fine structure in the absorption spectra.

References

1. E. M. Kosower, *J. Am. Chem. Soc.*, **80**, 3253 (1958).
2. K. Dimroth, C. Reichardt, T. Siepmann, and F. Bohlmann, *Ann.*, **661**, 1 (1963).
3. L. G. S. Brooker, A. C. Craig, D. W. Heseltine, P. W. Jenkins, and L. L. Lincoln, *J. Am. Chem. Soc.*, **87**, 2443 (1965).
4. C. Reichardt, *Angew. Chem., Intern. Ed.*, **4**, 29 (1965).
5. E. Clar, *Polycyclic Hydrocarbons*, Vols. I and II, Academic Press, London and New York, 1964.
6. J. R. Platt, *J. Chem. Phys.*, **17**, 484 (1949).
7. H. H. Jaffé and M. Orchin, *Theory and Applications of Ultraviolet Spectroscopy*, Wiley, New York and London, 1962.
8. G. H. Schenk, M. Santiago, and P. Wines, *Anal. Chem.*, **35**, 167 (1963).
9. L. N. Ferguson and A. Y. Garner, *J. Am. Chem. Soc.*, **76**, 1167 (1954).
10. F. W. Ochynski, *Analyst*, **85**, 278 (1960).
11. E. V. Gernet and A. A. Russkikh, *Zavodskaya Lab.*, **26**, 58 (1960).
12. E. M. Kosower and B. G. Ramsey, *J. Am. Chem. Soc.*, **81**, 856 (1959).
13. E. Lippert, *Z. Elektrochem.*, **61**, 962 (1957).
14. J. Czekalla, W. Liptay, and K. O. Meyer, *Ber. Bunsenges. Physik. Chem.*, **67**, 465 (1963). See also J. Czekalla and G. Wick, *Z. Elektrochem.*, **65**, 727 (1961).
15. E. Sawicki, *J. Org. Chem.*, **18**, 1492 (1953).
16. L. G. S. Brooker et al., *J. Am. Chem. Soc.*, **73**, 5332 (1951).
17. L. G. S. Brooker, G. H. Keyes, and D. W. Heseltine, *J. Am. Chem. Soc.*, **73**, 5350 (1951).

CHAPTER 3

ABSORPTION SPECTRA OF ZWITTERIONIC RESONANCE STRUCTURES

I. BAND CHARACTERIZATION EFFECTS OF TEMPERATURE AND STRUCTURE

With the elegant methods of separation now available to the trace analyst and other investigators relatively pure compounds can be separated from complex mixtures. Consequently the electronic spectrum of a molecule, as affected by its geometry and environment, is of utmost importance to the analyst since this property can be used in identifying and determining, after separation, the important trace chemicals in living things and the environment around them.

The compounds treated in this chapter are mainly those that give spectacular spectral changes with a change in structure or environment. This interest in the spectacular is a necessity to the research analyst and the people who use his techniques since most colorimetric methods of analysis are based on spectacular spectral changes.

The classification of many bands as Z_n or Z_z is an oversimplification, because there are many bands that could be considered as being on the borderline or as belonging to some other classification.

Electronic transitions that are involved in the ultraviolet-visible region are of the following kinds: $\sigma \rightarrow \sigma^*$, $n \rightarrow \sigma^*$, $n \rightarrow \pi^*$, and $\pi \rightarrow \pi^*$. The $\sigma \rightarrow \sigma^*$ transition, as found in saturated hydrocarbons, does not show absorption in the ultraviolet region. The $n \rightarrow \sigma^*$ transition is derived from compounds that contain nitrogen, oxygen, sulfur, or halogen atoms with their nonbonding electrons. Examples are trimethylamine (λ_{max} 227 mμ, mϵ 0.9, as the vapor) and methyl iodide (λ_{max} 258 mμ, mϵ 0.38, in hexane). The $n \rightarrow \pi^*$ transition is derived from compounds that contain unsaturated groups, one of whose atoms is a hetero atom containing nonbonding electrons (e.g., C=O, C=S, C=N, N=O, and N=N). Although Burawoy called the band derived from this transition an *R*-band, he was the pioneer who most thoroughly investigated the properties of this transition from 1939 through the early 1960s.

However, since the bands derived from all these transitions, except $\pi \rightarrow \pi^*$, absorb in the vacuum ultraviolet or with very low intensity, they have not proven useful in colorimetry.

In this book we are interested in the $\pi \rightarrow \pi^*$ transition, which can be understood in terms of an interplay of three limiting states. One of these is the aromatic state, whose ultraviolet–visible absorption spectra have been thoroughly described by E. Clar in a large number of papers. This state is characterized by maximal bond equalization, equal π-electron densities at all atoms, and high delocalization energies. The polyene state is characterized by alternating bond orders and equal π-electron densities, whereas the polymethine state is characterized by equal bond orders, alternating π-electron densities, and high delocalization energies (1). An essential criterion of the long-wavelength absorption of polymethines is the alternation of the π-electron density in the ground state. Consequently in this book we concern ourselves with the $\pi \rightarrow \pi^*$ transition and, in particular, with the large or spectacular change.

I. Band Types and Their Characterization by Solvent and Acidity Effects

Zwitterionic resonance, or *Z*, bands are long-wavelength bands in the π-electronic spectra of molecules to whose hybrid structure dipolar resonance forms contribute. Many of the chromogens measured in colorimetry have this type of structure, which usually contains resonance terminals at both ends of the conjugation chain. The positive resonance terminal is an electron-donor group, and the negative resonance terminal is an electron-acceptor group. Examples of electron-donor and electron-acceptor groups are listed in Table 3.1. Included in the table are amphoteric groups, which can act as electron donors or electron acceptors. The bands derived from aromatic hydrocarbons can be considered as a special class of zwitterionic resonance bands.

Brooker et al.(2) have stated that no single structural formula can adequately account for the absorption of dyes. Their absorption spectra are intelligible only in terms of the energy relationship of the contributing structures, and this relationship can sometimes be varied between wide limits by changing the environment. However, many of the chromogens obtained in colorimetry that have a zwitterionic resonance structure have either the uncharged or charged structure as the dominant one in the ground state. For a few dyes a solvent mixture can be prepared in which the dipolar and uncharged limiting resonance structures have energetic equivalence. In such a mixture a dye is at its isoenergetic point(1), and either an increase or decrease in solvent polarity will cause the same type of wavelength shift.

For many compounds analyzed by ultraviolet absorptiometry the positive and negative resonance terminals are weak electron-donor and/or electron-acceptor groups, respectively. For these compounds the spectral effect of a change in solvent polarity would be slight.

TABLE 3.1
Electron-Donor and Electron-Acceptor Groups

Electron donors	Electron acceptors
$—NR_2 \leftrightarrow =N^+R_2$	$—N^+(O^-)=O \leftrightarrow =N^+(O^-)—O^-$
$—SR \leftrightarrow =S^+R$	$—C(R)=O \leftrightarrow =C(R)—O^-$
$—OR \leftrightarrow =O^+R$	$—S(=O)(R)=O \leftrightarrow =S(=O)(R)—O^-$
$—Cl \leftrightarrow =Cl^+$	$—S(R)=O \leftrightarrow =S(R)—O^-$
$—CH_3 \leftrightarrow =CH_2H^+$	

Amphoteric groups		
$—C_6H_5^{+}$ (ring, +) $\longleftrightarrow$	$—C_6H_5$ (ring) $\longleftrightarrow$	$=C_6H_5^{-}$ (ring, −)
$—C^+R_2 \leftrightarrow$	$=CR_2 \leftrightarrow$	$—C^-R_2$
$—N^+R \leftrightarrow$	$=NR \leftrightarrow$	$—N^-R$
$=N^+=O \leftrightarrow$	$—N=O \leftrightarrow$	$=N—O^-$
$—N^+(R)=O \leftrightarrow$	$=N^+(R)—O^- \leftrightarrow$	$—N(R)—O^-$

Three extreme types of Z-bands are differentiated. The nonionic type, or the Z_n-band, has a lower dipole moment in the ground state as compared with the excited state; the zwitterion type, or the Z_z-band, has a higher dipole moment in the ground state; the betaine type, or the Z_b-band, has permanent positive and negative charges because of the geometry of the molecule. On the other hand, in amphizwitterionic resonance structures resonance terminals can be either positively or negatively charged(3), as shown below.

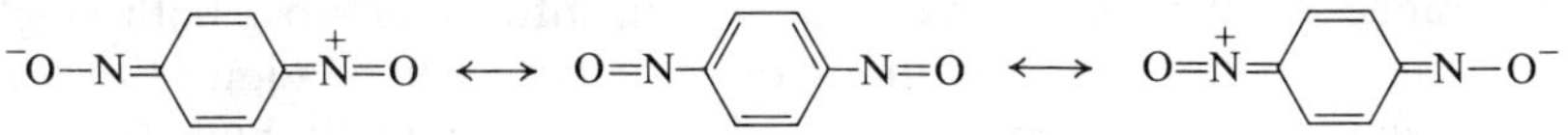

$$^{-}O{-}N{=}\langle\,\rangle{=}\overset{+}{N}{=}O \longleftrightarrow O{=}N{-}\langle\,\rangle{-}N{=}O \longleftrightarrow O{=}\overset{+}{N}{=}\langle\,\rangle{=}N{-}O^{-}$$

λ_{max} 405 mμ, mϵ 3.58, in ethanol(4) (3.1)

This type of band system with amphoteric resonance terminals can be either Z_n, Z_b, or Z_z.

A. Z_n-Band

1. *Aromatic Resonance Bands*

A special type of Z_n-structure is found in benzene and in the polynuclear aromatic hydrocarbons. The presence of definite resonance terminals in aromatic hydrocarbons is difficult to depict because of the seemingly circuitous nature of resonance, the presence of alternative pathways of conjugation, and the lack of a hetero atom.

Aromatic hydrocarbons as diverse in complexity as benzene, picene, and coeranthrene have at least three bands in common(5). The position and intensity of these bands are dependent mainly on the structure of the molecule and the band type. The bands stem from the presence of six π-electrons in the benzene ring. The difficulties and the controversy in the classification and assignment of these bands have been discussed (6–9). Three to six principal band systems have been assigned to benzene and the polynuclear hydrocarbons. The nomenclature problem can be appreciated by a comparison of the band assignments for benzene, as shown below.

Band assignments for Benzene

	Reference				
λ_{max}(mμ)	10, 11	12	13, 14	15	16
184	I		Second primary	^{1}Ba, b	β
206	II	*K*	First primary	^{1}La	*para*
254	III	*B*	Secondary	^{1}Lb	α

Other systems of nomenclature have been used by Coulson(17) and Moffitt(18).

Doub and Vandenbelt have studied a very large number of mono-, di-, and tri-substituted benzene derivatives(13). The absorption of benzene and its derivatives has been represented as an approximately regular progression of at least three bands—the secondary, first primary, and second primary bands. The spectral position of these bands is apparently determined by the displacing value of substituent groups, both singly and in combination. By far the most extensive compilation of experimental data on the comparative spectra of the aromatic hydrocarbons has been reported by Clar(16). According to Clar the absorption spectra of the hydrocarbons consists of at least three types of band.

1. α-Bands are the least intense ($\log \epsilon = 2\text{–}3$) and are the long-wavelength bands in the absorption spectra of benzene, naphthalene, phenanthrene, and the higher phenes. These bands shift to the red on linear and angular annulation in the acene and phene series and are hidden, or partly hidden, by the more intense *para* bands in the higher acenes, beginning with anthracene. The α-bands are longitudinally polarized in the polyacenes but are polarized parallel to the symmetry axis in phenanthrene. These bands have a sharp vibrational structure, with 5 or 6 individual peaks about 300 cm^{-1} in half-width, the strongest one commonly being the first (longest wavelength) or the third. The average frequency separation (ΔV_{av}) between the individual vibrational bands of many polynuclear hydrocarbons in paraffin solution has been reported to be 700 cm^{-1}(19). These ΔV_{av} values have been tentatively assigned to the out-of-plane C—H bending vibration. In a study of the infrared absorption spectra of over 100 aromatic compounds Cannon and Sutherland have assigned the out-of-plane C—H bending vibrations to the 700 cm^{-1} area(20). In uniplanar aromatic hydrocarbons the ratio of the frequencies of the α- and β-bands is constant, $V_\alpha : V_\beta = 1 : 1.350$. In the case of hydrocarbons with a high degree of overlapping of hydrogen atoms, the loss of resonance energy in the strained molecule becomes apparent by a decrease in this ratio; for example, chrysene, 1 : 1.348; and benzo[*c*]-phenanthrene, 1 : 1.324(16).

2. β-Bands are the most intense of the three main aromatic bands ($\log \epsilon \simeq 5.0$) and are usually further in the ultraviolet than the other two bands. The β-bands show the same annulation effect as the α-bands. This has been shown by the fact that in uniplanar aromatic hydrocarbons the ratio of the frequencies of the α- and β-bands is constant, $V_\alpha : V_\beta = 1 : 1.350$. In naphthalene and anthracene these bands are longitudinally polarized. For some 16 hydrocarbons the log of the frequency of this band system has been shown to have a linear relation to the log of the greatest length of the conjugated system(21). Since the vibrational structure is very diffuse, often no clear minima are seen. The average frequency separation between the individual vibration bands is stated to be approximately 1400 cm^{-1}(19). These ΔV_{av} values have been tentatively assigned to in-plane C—H bending vibrations(22).

3. Para bands have a moderate intensity with a log ϵ of 3.5–4.5. The vibrational structure usually consists of 5 or 6 bands about 500 cm^{-1} in half-width and more diffuse than the α-band. The first (longest wavelength), second, or third peak is usually the strongest. The average frequency separation between the individual vibrational bands in paraffin solution is on the order of 1400 cm^{-1}(16, 19, 23). This has been tentatively assigned to in-plane C—H bending vibrations.

The long-wavelength bands in anthracene and the higher acenes are *para* bands. This band system moves to the red the fastest with increasing length (linear annulation), and its interval from the β-band increases at the same time. In polyacenes the band is transversely polarized, whereas in phenanthrene it is polarized perpendicular to the symmetry axis. With increasing linear annulation the *para* band shifts to the red, and the chemical reactivity of the hydrocarbon increases; that is, addition of halogen, hydrogen, alkali metal, maleic anhydride, and oxygen is easier(16).

A quantitative relation between the relative affinity of the methyl radical for aromatic hydrocarbons with the relative position of their *para* bands has been reported(24). A relationship has been found between the phosphorescence of aromatic hydrocarbons at 77°K and the *para* band(15, 25). The half-wave potentials of 78 polycyclic hydrocarbons have been measured in β-methoxyethanol(26). This constant measures the energy needed to place one (or more) electrons in the lowest unoccupied molecular orbital(27) and has been shown to be directly related to the wave numbers of the *para* bands.

2. *Polymethine Bands*

Some zwitterionic resonance structures have been examined and found to have greater dipole moments in the excited state than in the ground state(28–30). Examples of such compounds with their ground- and excited-state dipole moments (μ) include such typical Z_n-structures as 4-dimethylamino-4′-nitrostilbene, with $\mu = 7.6$ and 32; 2-amino-7-nitrofluorene, with $\mu = 7$ and 17.8; and 4-amino-4′-nitrobiphenyl, with $\mu = 6.6$ and 23.1. In such a structure the energy of the excited state can be decreased relative to the ground state when the dipolar form is stabilized by hydrogen bonding or electrostatic interaction with a polar solvent, or when a proton is added to the negative resonance terminal.

The Z_n-band absorbs at longer wavelength in a polar solvent, such as ethanol, than in a nonpolar solvent, such as heptane (Table 3.2). Such factors as the dielectric constant of the solvent and the intermolecular hydrogen bond between solvent and solute probably account for the

TABLE 3.2
Solvent Effect on Z_n-Bands

Compound	Nonpolar solvent	λ_{max}(mμ)	Polar solvent	λ_{max}(mμ)	Ref.
Acetophenone	*n*-Heptane	238	Water	245	31
p-Hydroxyacetophenone	*n*-Heptane	258	Water	278	31
N-p-Nitrophenyltrifluoroacetamide	*n*-Heptane	285	95% EtOH	301	
p-Nitrophenol	*n*-Hexane	290	95%EtOH	320	32
N-Methyl-*N*-*p*-nitrophenylcyanamide	Cyclohexane	305	95% EtOH	313	33
p-Dimethylaminobenzaldehyde	*n*-Heptane	319, 326	95% EtOH	340	
p-Nitroaniline	*n*-Heptane	325	95% EtOH	377	
2,6-Diethyl-4-nitroaniline	*n*-Heptane	333	95% EtOH	381	
Acetaldehyde 2,4-dinitrophenylhydrazone	Carbon tetrachloride	343	95% EtOH	357	34
N,*N*-Dimethyl-*p*-nitroaniline	Isooctane	351	Methanol	389	35
p-Dimethylaminocinnamaldehyde	*n*-Heptane	360	95% EtOH	390	
o-Nitroaniline	*n*-Hexane	381	Water	408	32
5-Amino-2,1,3-benzoselenadiazole	*n*-Heptane	392	95% EtOH	426	
4-Dimethylaminoazobenzene	*n*-Heptane	398	95% EtOH	408	
4-Dimethylamino-β-β-dicyanostyrene	*n*-Heptane	410	95% EtOH	432	
4-Dimethylamino-4′-nitrostilbene	Cyclohexane	417	Pyridine	455	36
4-Dimethylaminofuchsone	Cyclohexane	445	Methanol	552	37
4-Dimethylamino-4′-nitroazobenzene	Cyclohexane	447	*m*-Cresol	523	38
2-4′-Dimethylaminobenzal-1,3-indandione	*n*-Heptane	457	95% EtOH	486	
4-Tricyanovinyl-*N*,*N*-dimethylaniline	Isooctane	478	*m*-Cresol	545	39
Dehydrolycopene	*n*-Hexane	504	CS_2	557	40
(1-Methyl-1,4-dihydropyridine)-4-azino-4-(1,4-dihydro-1-oxobenzene)	Benzene	512	Benzyl alcohol	558	41
Phenol blue	Cyclohexane	552	Water	668	42

red shift. The different factors that play a role in various types of solvent spectral shifts have been discussed by Bayliss and McRae(43). Usually the effect of hydrogen bonding swamps that of other less specific solvent interactions(44). In the case of a Z_n-structure in alcoholic solution the intermolecular hydrogen bond probably decreases the energy of the excited state, thus causing a decrease in the transition energy and a red shift.

The spectra of quite a few compounds have been investigated in a large number of solvents. Correlations between particular solvent properties and the wavelength maximum have not been very satisfactory. Past work indicates that the solvent effect involves several properties of a solvent and different properties in different types of solvent. However, a knowledge of the red-shift effect of various solvents would be of value in spectral analysis. The chromogen 4-tricyanovinyl-*N*,*N*-dimethylaniline (**1**), formed in the determination of tetracyanoethylene, has the following wavelength maxima (in millimicrons) in various solvents: isooctane, 478; ether, 498; benzene, 507; acetic acid, 509; ethanol, 514; acetone, 514; acetonitrile, 517; ethyl benzoate, 519; *N*,*N*-dimethylaniline, 519; methyl iodide, 519; dimethylformamide, 525; pyridine, 527; and *m*-cresol, 545(39).

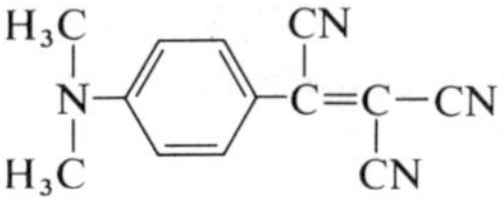

H_3C, N, H_3C, CN, C=C—CN, CN

1

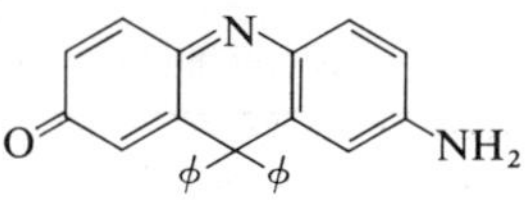

N, O, NH_2, ϕ, ϕ

2

In compound **2** the solvent effect is also striking; the wavelength maxima (in millimicrons) are as follows: benzene, 500; chloroform, 508; carbon disulfide, 508; ether, 512; acetone, 548; pyridine, 564; methanol, 591; and ethanol, 595(45).

A more extensive study has been made on the effect of solvents on the long-wavelength bands of phenol blue (**3**) and the mercocyanine **4**:

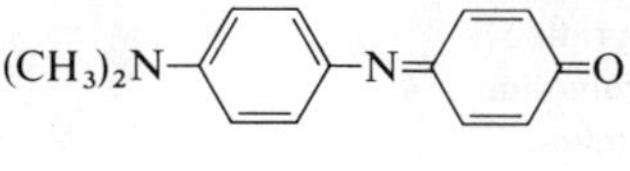

$(CH_3)_2N$—, —N=, =O

3

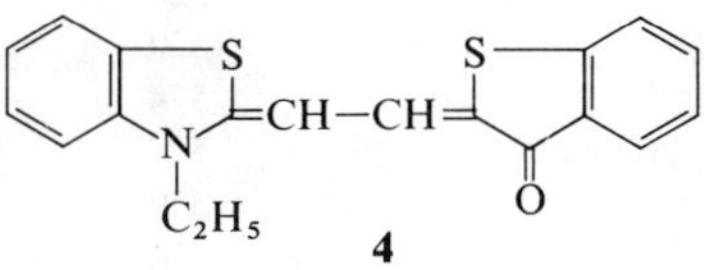

S, N, C_2H_5, =CH—CH=, S, O

4

In compound **4** an increase in the polar moment of the basic solvent series causes a red shift(46). Thus triethylamine, piperidine, and *o*-chloroaniline have the following polar moment and corresponding wavelength maxima: 0.8 and 517 mμ, 1.17 and 532 mμ, and 1.77 and 565 mμ, respectively. There were some exceptions to this rule. The spectral data are presented in Table 3.3. A similar study of phenol blue is presented in Table 3.4. The position of the long-wavelength band shifts toward the red with an increase in the refractive index of hydrocarbon and aryl halide solvents. Practically all other solvents cause a shift of the long-wavelength band of phenol blue to a longer wavelength than a hydrocarbon or aryl halide of the same refractive index. Solvation by hydrogen bonding between solute and solvent causes the largest red shifts.

An extremely sensitive solvent effect on the Z_n-structure shown in Table 3.5 has been investigated with a wide variety of solvents by Brooker et al.(48). The authors have defined a polarity effect in terms of the transition energy, X_R, which is equivalent to $28{,}600/\lambda_{max}(m\mu)$ in kilocalories per mole, where the wavelength maximum is that of the compound in the solvent for which X_R is reported. Since solvents are of such vital importance in photometric analytical research, the data are fully reported.

As can be seen by comparing the data in Tables 3.3, 3.4, and 3.5, there is an overall correlation between each solvent and its relative wavelength position. The correlation is imperfect, but, in spite of this, knowledge of the relative polarity of solvents should be of value in improving photometric analytical methods.

TABLE 3.3
Solvent Effect on the Long-Wavelength Band of Compound **4**(46)

Solvent	$\lambda_{max}(m\mu)$	Solvent	$\lambda_{max}(m\mu)$
n-Pentane	510	*n*-Propylamine	532
n-Hexane	510	Ethylamine	535
n-Dodecane	510	Acetone	535
n-Octane	514	Diethyl ketone	535
n-Decane	515	Di-*n*-propyl ketone	535
Cyclohexane	515	Methyl *n*-hexyl ketone	535
n-Tetradecane	517	α-Picoline	542
Triethylamine	517	Carbon disulfide	545
Diethyl ether	520	Ammonia	548
Dibutyl ether	520	Pyridine	550
Carbon tetrachloride	524	Quinoline	553
Diethylamine	527	Ethanol	555
Methyl acetate	530	*n*-Propanol	555
Ethyl acetate	530	*n*-Butanol	555
n-Propyl acetate	530	Isoamyl alcohol	555
n-Butyl acetate	530	*n*-Heptanol	555
n-Amyl acetate	530	Methanol	558
Piperidine	532	Aniline	560
n-Amylamine	532	*o*-Chloroaniline	565
n-Butylamine	532		

TABLE 3.4
Solvent Effect on the Long-Wavelength Band of Phenol Blue (42, 47)

Solvent	$\lambda_{max}(m\mu)$	Solvent	$\lambda_{max}(m\mu)$
Heptane	549	Anisole	586
Octane	550	*p*-Chlorotoluene	586
Cyclohexane	552	Acetal	588
Dodecane	557	Phenyl ether	589
Ethyl ether	564	Bromobenzene	590
Butyl ether	566	*o*-Iodotoluene	593
Isopropyl ether	567	Pyridine	598
Amyl ether	569	Chloroform	598
Dioxane	571	Butanol	606
Benzene	577	1-Iodonaphthalene	607
Acetone	582	Acetic acid	610
Carbon disulfide	584	Methanol	612
Chlorobenzene	584	Water	668
Bromocyclohexane	585		

TABLE 3.5
Solvent Effect on the Long-Wavelength Band (48) of

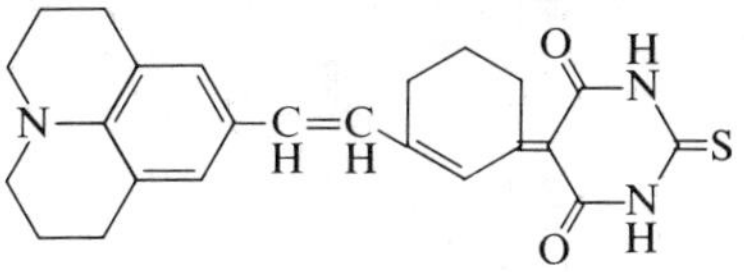

Solvent	X_R[a]	$\lambda_{max}(m\mu)$	Solvent[b]	X_R[a]	$\lambda_{max}(m\mu)$
Isooctane	50.9	562	Dioxane-isooctane:		
			1:9	50.2	570
n-Hexane	50.9	562	2:8	50.0	572
n-Heptane	50.9	562	3:7	49.5	578
Methylcyclohexane	50.1	571	4:6	49.5	578
Cyclohexane	50.0	575	5:5	49.1	582
Triethylamine	49.3	581	6:4	49.1	582
Carbon tetrachloride	48.7	589	7:3	48.8	586
Isopropyl ether	48.6	590	8:2	48.7	587
n-Butyl ether	48.6	590	9:1	48.5	590
p-Dioxane	48.4	592	2,6-Lutidine-methyl cyclohexane:		
Ethyl ether	48.3	593	1:9	49.3	580
p-Cymene	48.0	596	2:8	48.5	590
m-Xylene	47.8	600	3:7	47.8	600
p-Xylene	47.7	600	4:6	47.3	605
Cumene	47.6	601	5:5	46.7	614
Mesitylene	47.5	602	6:4	46.3	619

TABLE 3.5 (*cont'd*)

Solvent	X_R[a]	λ_{max} (mμ)	Solvent[b]	X_R[a]	λ_{max} (mμ)
n-Butyl acetate	47.5	602	7:3	45.8	625
o-Xylene	47.3	605	8:2	45.5	630
Toluene	47.2	606	9:1	45.0	636
Ethyl acetate	47.2	606	Dioxane-water:		
Benzene	46.9	610	9:1	46.4	617
Tetrahydrofuran	46.6	615	8:2	45.0	636
Ethyl *N*,*N*-dibutyl-carbamate	46.0	622	7:3	44.0	650
2-Octanone	45.8	625	6:4	42.3	676
Acetonitrile	45.7	626	2,6-Lutidine-methanol[c]		
Acetone	45.7	626	9:1	44.3	645
p-Chlorotoluene	45.6	628	8:2	44.0	650
1-Octanol	45.4	631	7:3	43.9	652
Butyronitrile	45.4	631	6:4	43.7	655
2-Butanone	45.4	631	5:5	43.8	654
Chlorobenzene	45.2	635	4:6	43.7	655
Dichloromethane	44.9	639	3:7	43.6	656
2-Butoxyethanol	44.8	639	2:8	43.5	657
2,6-Lutidine	44.7	640	1:9	43.3	660
Bromobenzene	44.6	642	Pyridine-water:		
Isopropanol	44.5	644	9:1	42.1	680
n-Butanol	44.5	644	8:2	40.8	701
Diethyl phthalate	44.3	646	7:3	39.9	716
Cyclohexanone	44.3	646	6:4	39.3	729
Chloroform	44.2	648	5:5	39.2	730
n-Propanol	44.1	650	2,6-Lutidine-water[c]:		
2-Ethoxyethanol	44.1	650	9:1	42.2	678
Nitromethane	44.0	651	8:2	40.8	701
Pyridine	43.9	653	7:3	39.8	720
Ethanol	43.9	653	6:4	39.2	730
Dimethylformamide[c]	43.7	656	5:5	38.9	735
2-Methoxyethanol	43.5	659	4:6	38.8	737
Benzonitrile	43.3	661	3:7	38.6	741
Methanol	43.1	664	2:8	38.3	747
Dimethylacetamide[c]	43.0	665			
Nitrobenzene	42.6	672			
γ-Butyrolactone	42.6	672			
Dimethyl sulfoxide[c]	42.0	681			
Aniline	41.1	696			
Ethylene glycol[d]	40.4	710			
Propargyl alcohol	38.9	736			
o-Cresol	(35.5)[e]	760			
m-Cresol	(33.6)[e]	770			

[a] $X_R = 28{,}600/\lambda_{max}(\text{m}\mu)$ = transition energy in kilocalories per mole.

[b] All solvent mixtures are volume in volume.

[c] Acetic acid (3–5 drops) was added to the dye solution to stabilize the dye.

[d] The solution contained 2% of 2.6-lutidine.

[e] Extrapolated values.

For some compounds the Z_n or Z_z resonance structure predominates in the ground state, depending on the polarity of the solvent system (Table 3.6)(1). Thus in solutions of the various merocyanines the presence of a greater percentage of water than is present at the energetic equivalence point (i.e., an increase in solvent polarity) made the Z_z-structure predominate in the ground state. The Z_n-structure is much more important in pyridine solutions containing less water than is present at the equivalence point; that is, a decrease in solvent polarity increases the contribution of the Z_n-structure to the ground state.

Another method of characterizing chromogens to which a Z_n-structure(s) is the main contributor is through the red spectral shift in acid solution when protonation takes place at the negative resonance terminal (Table 3.7). A violet shift takes place when the proton adds to the positive resonance terminal (Table 3.8). But here again the Z_n-structure of the nonionized compound can be readily deduced from the close resemblance of the absorption spectrum of the salt to that of a readily obtained *iso*-π-electronic compound; for example, the spectrum of *p*-nitroaniline sulfate closely resembles that of nitrobenzene or *p*-nitrotoluene.

The chromogens obtained in the determination of alkylating agents with 4-nitrobenzylpyridine (65a), enzymically demethylated *p*-chloro-*N*-

TABLE 3.6
Change in Energetic Equivalence Point with Decreasing Basicity of a Positive Resonance Terminal (1)

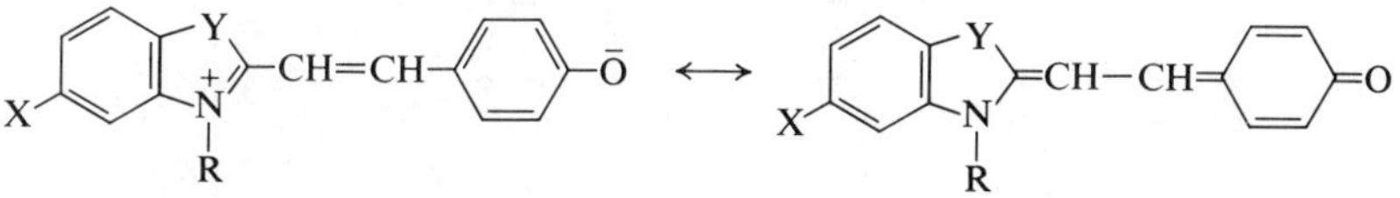

Substituent			Energetic equivalence[a] (% pyridine)
X	Y	R	
H	S	CH_3	92.5
Cl	S	C_2H_5	86
Cl	S	$C_6H_5CH_2$	75
H	$(CH_3)_2C=$	CH_3	50
H	$(CH_3)_2C=$	C_6H_5	20

[a]With a greater percentage of water than is present at the energetic equivalence point the Z_z-structure predominates; with a lower percentage the Z_n-structure predominates.

TABLE 3.7
Effect of Protonation at the Negative Resonance Terminal on the Z_N Band

Compound	Z_n − Band				Monocation Band			
	Solvent	λ_{max}(mμ)	mϵ	Ref.	Solvent	λ_{max}(mμ)	mϵ	Ref.
p-Anisaldehyde	95% EtOH	278	15.0		H_2SO_4 (conc.)	349	16.0	
		283	15.1					
4-Nitrodiphenylsulfide	95% EtOH	337	13.5	49	H_2SO_4 (conc.)	480	20.0	50
4-Ethylaminoazobenzene	95% EtOH	410	27.8		1 *N* HCl in H_2O–EtOH (1 : 1)	514	36.3	
2-(2-Anilinovinyl)-benzothiazole	Methanol	394	38.0	51	Methanol (acidic)	414	55.0	51
2-(6-Anilino-1,3,5-hexatrienyl)benzothiazole	Methanol	485	68.0	51	Methanol (acidic)	613	76.0	51
9-Formylacridine *p*-methoxyphenylhydrazone	95% EtOH	470	—	52	AcOH	610	—	52
5-Dimethylamino-2,1,3-benzoselenadiazole	95% EtOH	449	6.8	53	1 *N* HCl in H_2O–EtOH (1 : 1)	503	5.90	
β-4-Chloroanilinoacrolein *p*-chloroanil	95% EtOH	368	57.0	54	Aqueous acid	392	—	
2,3-Dimethyl-6-methylthioquinoxaline	95% EtOH	360	7.3	55	6 *N* HCl in H_2O–EtOH (1 : 1)	419	7.4	55
2-Hydroxyanthraquinone	Methanol	368	3.9	56	H_2SO_4 (conc.)	590	—	

TABLE 3.8

Effect on the Long-Wavelength Band of Proton Addition to the Positive or Negative Resonance Terminals of Z_n-Structures

				Monocationic salt added to					
				Positive resonance terminal			Negative resonance terminal		
Compound	Base solvent	λ_{max}(mμ)	Ref.	Solvent	λ_{max}(mμ)	Ref.	Solvent	λ_{max}(mμ)	Ref.
4-Methylthionitrobenzene	95% EtOH	338	57				H_2SO_4 (conc.)	522	
p-Nitroaniline	95% EtOH	375	58	H_2SO_4 (conc.)	263				
4-Methoxyazobenzene	H_2O–EtOH (1 : 1)	350	59				H_2SO_4 (conc.)	500	
4-Dimethylaminoazobenzene	95% EtOH	410		EtOH (acidic)	320		1 *N* HCl in H_2O–EtOH (1 : 1)	516	
Chalcone	95% EtOH	310	60				95% EtOH–H_2SO_4 (4 : 1)	390	61
4′-Dimethylaminochalcone	Methanol	419	62	Methanol (acidic)	294	63			
$C_6H_5(CH{=}CH)_6C(C_6H_5){=}O$	95% EtOH	429	60				95% EtOH–H_2SO_4 (4 : 1)	735	61
4-Amino-2,1,3-benzoselenadiazole	95% EtOH	462	63	EtOH (acidic)	333	64			
5-Amino-2,1,3-benzoselenadiazole	95% EtOH	426	63				1 *N* HCl in H_2O–EtOH (1 : 1)	459	63
2,3-Diphenyl-5-amino-7-chloroquinoxaline	95% EtOH	402	55	EtOH (acidic)	355	65	1 *N* HCl in H_2O–EtOH (1 : 1)	520	55
4-Methylacetophenone	95% EtOH	260	64				H_2SO_4 (conc.)	310	64

methylaniline reacted with *p*-dimethylaminobenzaldehyde(66), and 4-aminoantipyrine reacted with *p*-dimethylaminocinnamaldehyde(67) are examples of Z_n-structures obtained in trace analysis. The long-wavelength band of the first chromogen has been shown to move to the red with increasing polarity of the solvent(65a).

B. Z_z-Band

Where the dipole moment is higher in the ground state of a zwitterionic resonance structure than in the excited state, the long-wavelength band of this structure will be a Z_z-band. This type of molecule will have a dipole moment in the ground state of at least approximately 15 Debye units, as compared with a maximum of 6.5 units for a highly polar Z_n-structure, such as *p*-nitroaniline.

Whereas Z_n-bands have been shown to undergo a red shift in a hydroxylated or polar solvent as compared with a nonpolar solvent, Z_z-bands under similar conditions show a violet shift (Table 3.9). This violet shift is due to the strongly zwitterionic ground state being stabilized much more by the more polar solvent or the strong hydrogen bond with the hydroxylated solvent than is the less polar excited state. The result is an increase in the transition energy and a shift to the violet.

An example of this phenomenon is the violet shift of the long-wavelength band of compound **5** with increasing polarity of alcoholic solvents (2); λ_{max} for this compound is 486 mμ in methanol, 513 mμ in ethanol, 526 mμ in *n*-propanol, and 546 mμ in isopropanol.

$H_3C\overset{+}{N}$–(pyridine ring)–CH=CH–(benzene ring)–O^-

5

Consider the following three dyes (70):

H_3C, CH_3 / C / C=CHCH=CHCH=C / N / C_2H_5 ; O / C–N–C_2H_5 / S–C=S

6

H_3C, CH_3 / C / C=CHCH=CHCH=C / N / C_2H_5 ; O, C_2H_5 / C–N / C=S / C–N / O, C_2H_5

7

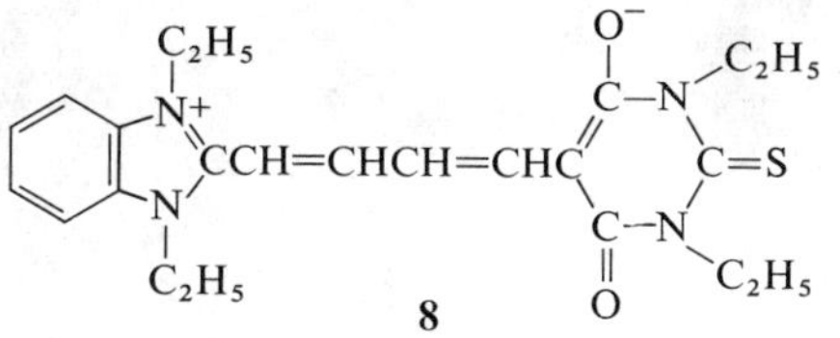

8

No.	Dipole moment μ ($\times 10^{-18}$)	Solvent	$\lambda_{max}(m\mu)$
6	6.91	Benzene–cyclohexane	548
		Aqueous ethanol	588
7	9.7	Benzene–cyclohexane	585
		Aqueous ethanol	590
8	17.7	Benzene–cyclohexane	555
		Aqueous ethanol	427

Compound **6** has one contributing dipolar structure and a dipole moment of 6.91; from the solvent effect its long-wavelength band can be considered as a Z_n-band. Compound **8** has four contributing dipolar structures and a dipole moment of 17.7; from the solvent effect its long-wavelength band can be considered as a Z_z-band. These two bands are the extremes, whereas compound **7** is an intermediate case.

Another compound with a Z_z-structure that shows a strong violet shift and intensity decrease with increasing amounts of water in the pyridine–water solvent is shown in Figure 3.1. Here the ground state is stabilized by increasing amounts of water, and the transition energy is thus increased.

Of the compounds showing a Z_z-band in solution 4-[5-(5-ethyl-3-methyl-1,2,3,5-tetrahydropyrido[1,2-*a*]benzimidazolyl)pentadienylidene]-2,2-dimethyl-1,3-cyclobutanedione (**9**) is the most sensitive to solvent polarity (48). Brooker et al. (48) suggest that the transition energy (in kilocalories per mole) of this dye in a given solvent, or solvent mixture, be designated X_β, and used as a solvent-property criterion of that solvent or solvent mixture (48). The data are presented in Table 3.10.

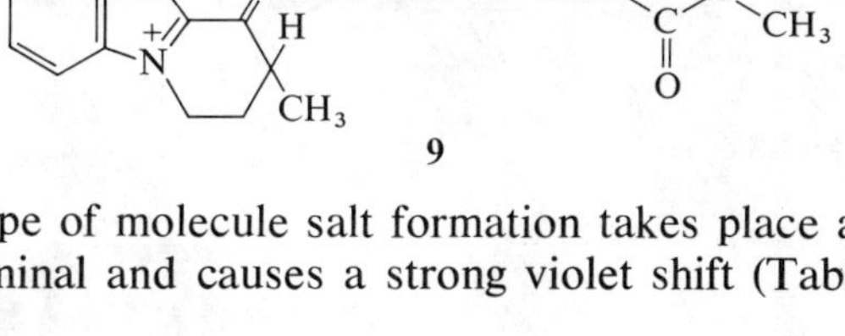

9

In the Z_z-type of molecule salt formation takes place at the negative resonance terminal and causes a strong violet shift (Table 3.11). The

TABLE 3.9
Solvent Effect on Z_z-Bands

Compound	Solvent	$\lambda_{max}(m\mu)$	Solvent	$\lambda_{max}(m\mu)$	Ref.
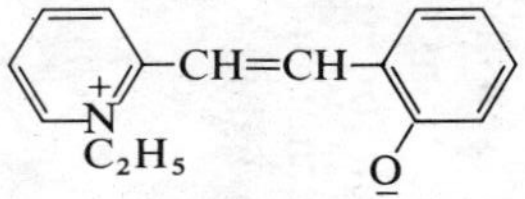 $\overset{+}{N}$ C_2H_5 –CH=CH– O^-	Pyridine	568	Aqueous methanol	430	68[a]
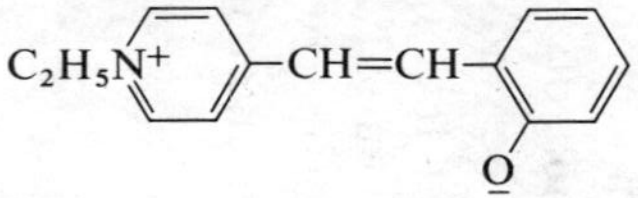 $C_2H_5N^+$ –CH=CH– O^-	Pyridine	620	Aqueous methanol	470	68
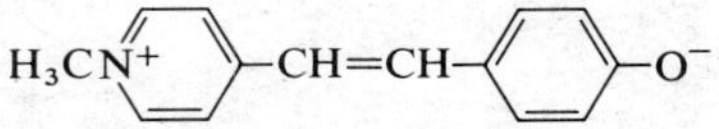 H_3CN^+ –CH=CH– O^-	Pyridine	604	Water	443	1[a], 69[a]
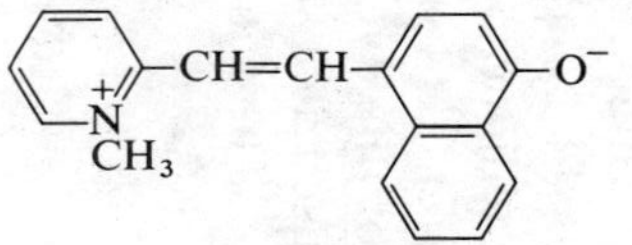 $\overset{+}{N}$ CH_3 –CH=CH– O^-	Benzene	584	95% Ethanol	430	69
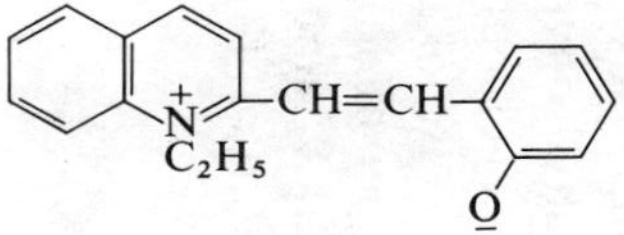 $\overset{+}{N}$ C_2H_5 –CH=CH– O^-	Pyridine	637	Water	513	70[a]
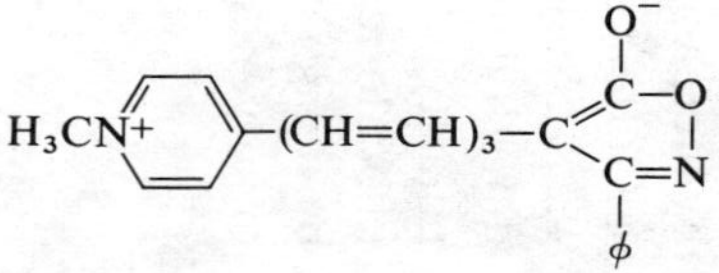 H_3CN^+ –$(CH=CH)_3$–C(C(O⁻)–O, C(ϕ)=N)	Pyridine	710	Water	448	70

Compound	Solvent	λ_{max}	Solvent	λ_{max}	Ref.
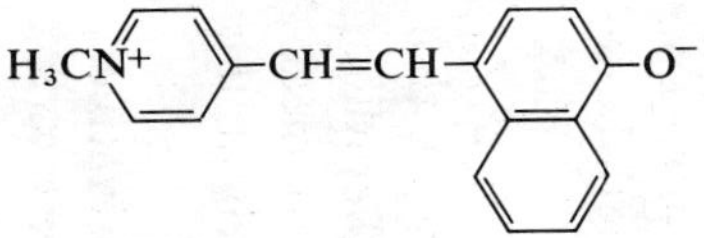 $H_3C\overset{+}{N}$(pyridinium)–CH=CH–(naphthyl)–O^-	Chloroform	643	Methanol	570	71
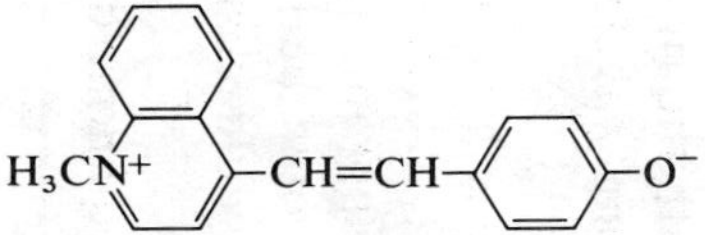 $H_3C\overset{+}{N}$(quinolinium)–CH=CH–(phenyl)–O^-	Benzene	615	Methanol	565	71
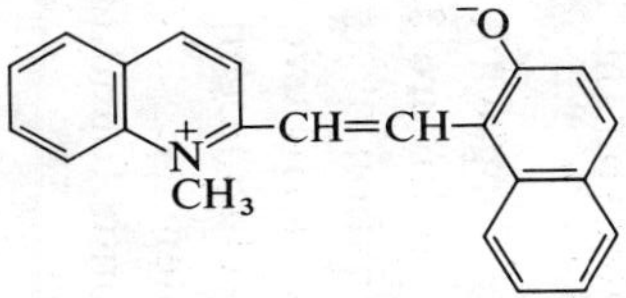 (N-CH_3 quinolinium)–CH=CH–(naphthyl)–O^-	*o*-Dichloro benzene	603, 635	Trifluoroethanol[b]	555	

[a]And other compounds of this type.
[b]Containing 1% of 10% aqueous tetraethylammonium hydroxide.

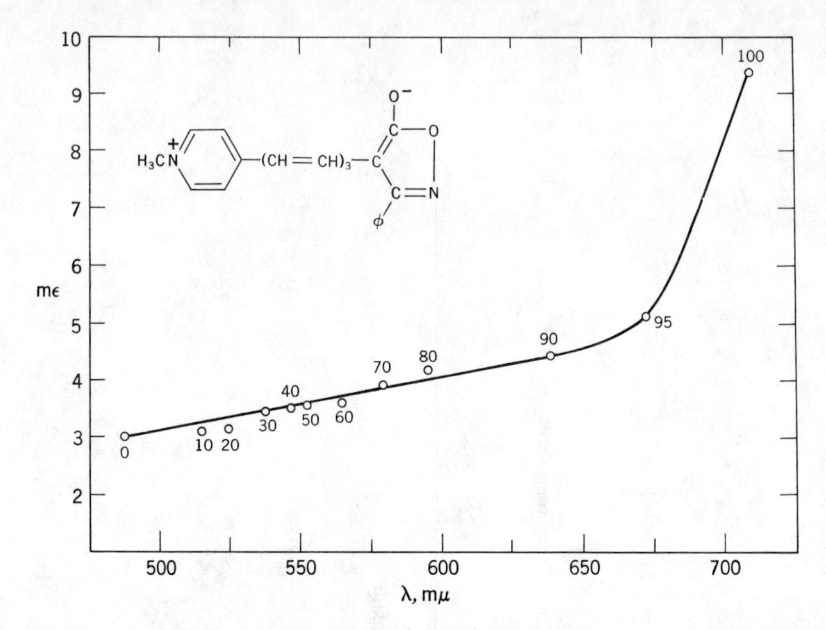

Fig. 3.1. Change in position of wavelength maximum and intensity with change in solvent composition. Figures on the curve indicate parts by volume of pyridine in 100 volumes of solvent (70).

2′- and 4′-hydroxy derivatives of 4-stilbazole methiodide undergo a strong red shift on deprotonation, and the 3′-hydroxy isomer (λ_{max} 343 mμ) on deprotonation gives a zwitterion whose λ_{max} is 345 mμ. In the 3′-isomer the oxy group is not in conjugation with the ring nitrogen. Consequently deprotonation has no effect on the long-wavelength band of this compound.

An example of the violet shift of a long-wavelength band of a Z_z-structure on protonation is presented in Figure 3.2, which shows that the chromogen formed in the determination of an appropriate lepidine, quinaldine, or hydroxynaphthaldehyde derivative would be best measured in alkaline solution, where the chromogen would absorb at the longest wavelength and with greatest intensity.

Another example of this phenomenon is found in quinone derivatives, as shown below (79):

$^{-}$O—C$_6$H$_4$—C(S)$_2^{+}$ $\xrightarrow{H^+}$ HO—C$_6$H$_4$—C(S)$_2^{+}$

10	**11**
λ_{max} 444 mμ, mϵ 20.0	λ_{max} 403 mμ, mϵ 46.8

TABLE 3.10
Solvent Effect on the Long-Wavelength Band (48) of

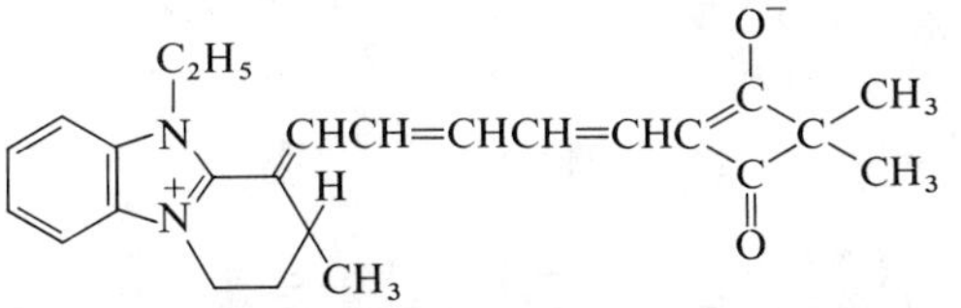

C_2H_5 N CHCH=CHCH=CHC C C CH_3 CH_3 O^- O + N H CH_3

Solvent	X_β[b]	$\lambda_{max}(m\mu)$	Solvent[a]	X_β[b]	$\lambda_{max}(m\mu)$
Toluene	41.7	685	Dioxane–water:		
2,6-Lutidine	43.2	662	9:1	55.0	520
Dichloromethane	47.5	602	8:2	58.6	488
Pyridine	50.0	575	7:3	59.8	478
Acetone	50.1	571	6:4	61.2	468
Dimethylformamide	51.5	555	5:5	61.0	469
Acetonitrile	53.7	532	3:7	63.4	451
Isopropanol	56.1	510	1:9	66.0	434
n-Butanol	56.8	504	Pyridine–water:		
Ethanol	60.4	474	9:1	55.4	516
Methanol	63.0	455	8:2	58.1	492
Water	68.9	415	7:3	59.8	478
2,6-Lutidine–methanol:			6:4	61.0	469
9:1	48.4	592	5:5	61.5	465
8:2	51.9	550	4:6	62.3	459
7:3	54.9	520	3:7	63.1	453
6:4	57.4	498	2:8	64.0	446
5:5	59.2	479	1:9	65.3	438
4:6	60.3	474			
3:7	61.1	467	Methanol–water		
2:8	61.6	464	9:1	63.1	453
1:9	62.2	460	8:2	63.4	451
2, 6-Lutidine–water:			7:3	64.1	446
9:1	52.1	549	6:4	64.7	442
8:2	56.8	504	5:5	65.3	438
7:3	58.8	486	4:6	66.0	434
6:4	60.1	475	3:7	66.5	430
5:5	61.1	468	2:8	67.1	426
4:6	61.8	463	1:9	67.8	422
3:7	62.4	459			
2:8	63.1	453			
1:9	64.4	444			

[a]All solvent mixtures are volume in volume.
[b]X_β = Transition energy in kilocalories per mole = $28{,}600/\lambda_{max}(m\mu)$.

TABLE 3.11

Effect of Protonation at the Negative Resonance Terminal on the Z_z-Band

Compound	Z_z-Band		Monocation band		Ref.
	Solvent	$\lambda_{max}(m\mu)$	Solvent	$\lambda_{max}(m\mu)$	
$H_3C\overset{+}{N}C_5H_4$–CH=CH–C_6H_4–O^-	Aqueous alkali	445	Aqueous acid	375	72
3-CH_3-benzothiazolium-2-C–N=N–C_6H_4–O^-	Aqueous alkali	525	Aqueous acid	505	73
$H_3C\overset{+}{N}C_5H_4$–CH=CH–C_6H_4–O^- (ortho)	Aqueous alkali	451	Aqueous acid	366	72
1-C_2H_5-quinolinium-2-CH=CH–naphthyl–O^-	95% Ethanol	597	6 *N* HCl in H_2O–EtOH	460	74
HC–S–C(–CH=CH–C_6H_4–O^-)=$\overset{+}{N}(C_2H_5)$–CCH_3 (thiazolium)	95% Ethanol	510	6 *N* HCl in H_2O–EtOH (1 : 1)	400	74
3-C_2H_5-benzothiazolium-2-C–CH=CH–$C_6H_3(O^-)(NO_2)$	95% Ethanol	525	6 *N* HCl in H_2O–EtOH (1 : 1)	395	74

Compound	Solvent	λ_{max}	Solvent	λ_{max}	Ref.
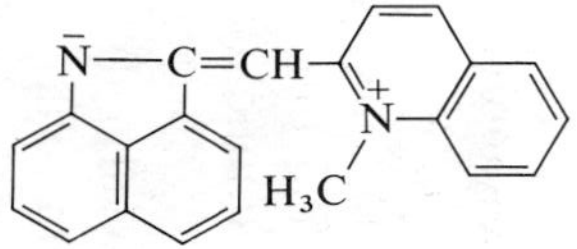 $\bar{N}$—C=CH— (naphthalene-fused) / quinolinium $\overset{+}{N}$–H_3C	95% Ethanol	620	Aqueous acid	577	75
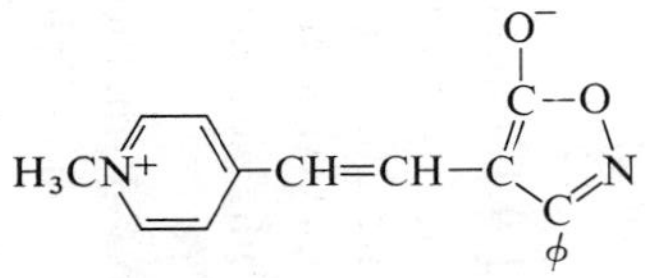 $H_3C\overset{+}{N}$(pyridine)—CH=CH—C(C(O⁻)–O–N=C(φ))	Water	436	Aqueous acid	382	76
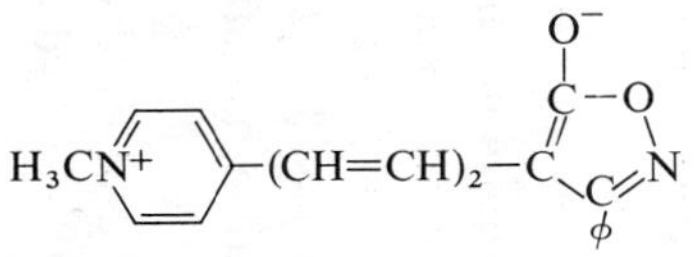 $H_3C\overset{+}{N}$(pyridine)—$(CH{=}CH)_2$—C(C(O⁻)–O–N=C(φ))	Water	469	Aqueous acid	403	76
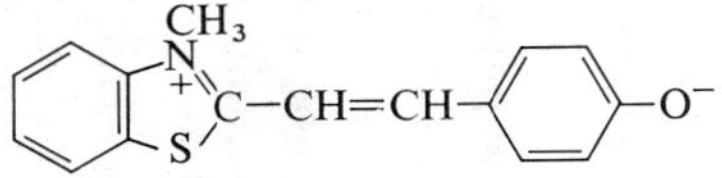 benzothiazolium (N^+–CH_3, S)—C—CH=CH—C_6H_4—O⁻	Aqueous alkali	493	Water (pH6)	406	76
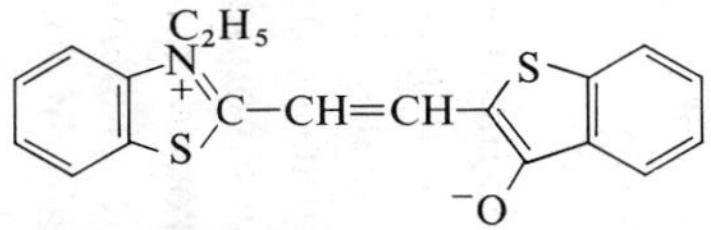 benzothiazolium (N^+–C_2H_5, S)—C—CH=CH—benzothiophene (S, ⁻O)	95% Ethanol	555	Aqueous acid	460	46
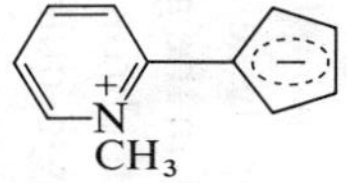 pyridinium ($\overset{+}{N}$–CH_3)—cyclopentadienide (−)	Isooctane	410– 425_s[a]	6 *N* HCl in H_2O–EtOH (1 : 1)	342	77
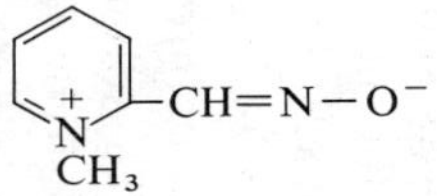 pyridinium ($\overset{+}{N}$–CH_3)—CH=N—O⁻	Aqueous alkali	336	Aqueous acid	293	78

[a]s = Shoulder.

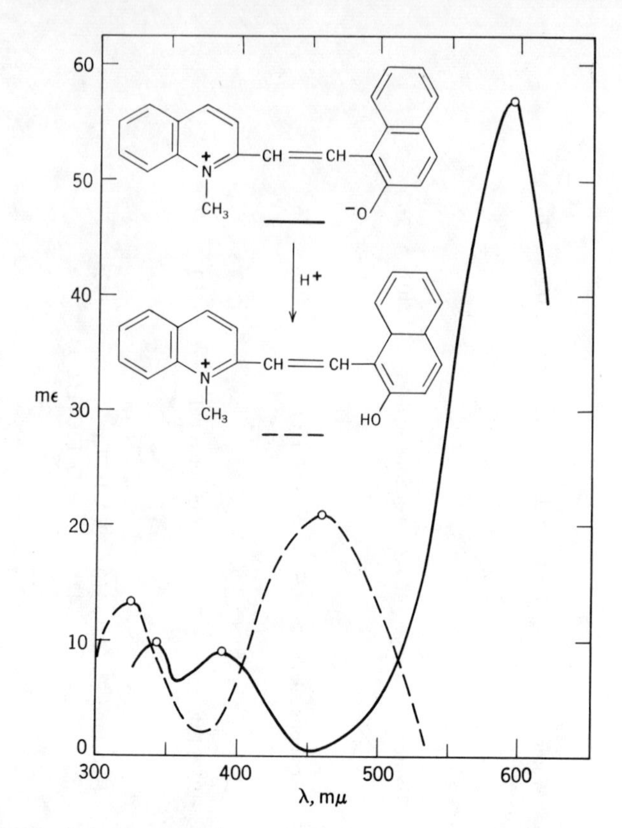

Fig. 3.2. Absorption spectra of the product of condensation of quinaldinium iodide and 2-hydroxynaphthaldehyde in alkaline solution (——) and acid solution (---).

On the other hand, the 1,4-naphthaquinone and 9,10-anthraquinone analogs show a very strong red shift on protonation; for example, λ_{max} for the naphthalene analog shifts from 415 mμ, mε 26.9, to 459 mμ, mε 33.9; for the anthracene analog from 413 mμ, mε 10.7, to 585 mμ, mε 8.9. For these compounds annulation has stabilized the quinonoid Z_n-structures in the ground state.

The solvatochromic chelating agent 2-(4′-hydroxy-3′-methoxystyryl) -5-(8-hydroxy-5-quinaldylmethyl)-1-methyl-8-hydroxyquinolinium chloride is an example of a compound with a Z_z-structure that absorbs at 414 mμ, mε 28.8, in 0.1 *N* hydrochloric acid, at 478 mμ, mε 31.6, in 0.01 *N* potassium hydroxide, and at 610 mμ in chloroform or pyridine (80).

One other group of molecules that shows a strong violet shift on protonation are the azulenes and pseudoazulenes. The meager data available for these compounds seem to indicate that these molecules are much less sensitive to solvent polarity changes than are the molecules discussed in this chapter. Thus the azulene bands show little shift in ethanol as compared with cyclohexane (81). However, in acid solution all the bands in the 500 to 700 mμ region are lost and a new intense long-wavelength band at 352 mμ is formed (82). The heterocyclic (83) oxa- (84) and thia- (85) pseudoazulenes show a similar phenomenon. For example, *N*-methyl-1,2,5,6-dibenzazalene has a broad long-wavelength band at 530 mμ, mε 2.73, in dioxane and at 372 mμ, mε~10, in ethanol–75% perchloric acid (1 : 1) (83).

C. Z_b-Band

The long-wavelength bands of compounds that contain the Z_b-type of structure show a violet shift as the polarity of the solvent is increased (Table 3.12). This solvatochromic effect has been demonstrated in the spectra of the betaine **12** in hydroxylated solvents (91). The position of the more intense shorter wavelength band is found at 302–311 mμ, mε 29.5–38.7, in the various hydroxylic solvents.

In an important study the wavelength maxima, molar absorptivities, and transition energies of a betaine (see Table 3.13) have been reported in a large number of solvents (91). Similar results have been reported for the betaine **13** in a large number of solvents, with wavelength maxima ranging from 560 mμ in ethylene glycol to 867 mμ in toluene (91).

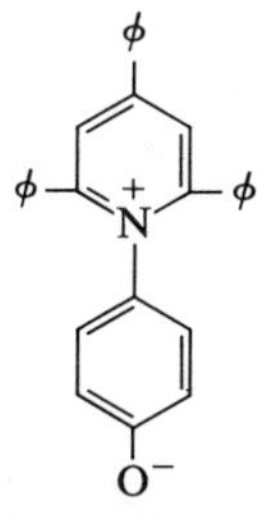

ϕ, ϕ, ϕ, N+, O−

12

ϕ, ϕ, ϕ, N+, H_3C, H_3C-C, H_3C, CH_3, $C-CH_3$, CH_3, O−

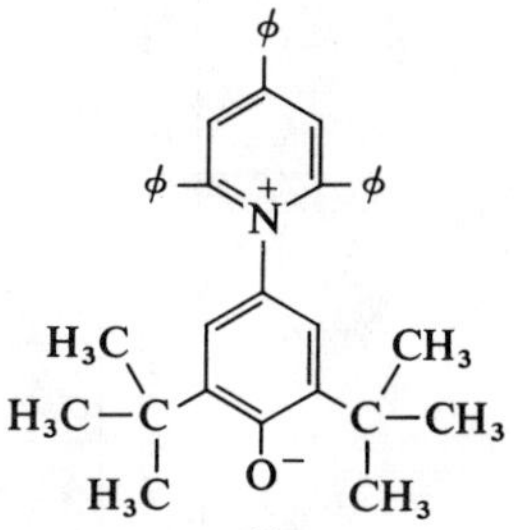

13

Solvent	λ_{max}(mμ)	mε	Solvent	λ_{max}(mμ)	mε
Water	412	2.48	*n*-Butanol	484	3.16
Glycol	437	2.55	*i*-Propanol	501	3.32
Methanol	442	2.75	Isoamyl alcohol	516	3.30
Ethanol	467	2.33	*tert*-Butanol	560	3.50
n-Propanol	477	2.60			

Table 3.12
Solvent Effect on Z_b-Bands

Compound	Solvent	$\lambda_{max}(m\mu)$	mϵ	Solvent	$\lambda_{max}(m\mu)$	mϵ	Ref.
$^{+}$N–O^{-}	Carbon tetrachloride	283	14.4	Water	254	12.0	86
H_3C– $^{+}$N–O^{-}	Carbon tetrachloride	288	31.0	Water	256	14.3	87
$^{+}$N– (–)	*n*-Heptane	524 546		Aqueous alkali	453	20.0	88
$\overset{+}{N}$–CH_3, O^{-}	Benzene	595		Water	428		89, 90
ϕ, ϕ, ϕ, $^{+}$N–, –O^{-}	Chloroform	615	3.22	Water	412.4	2.29	91
ϕ, ϕ, ϕ, $^{+}$N–, ϕ, ϕ, –O^{-}	Phenyl ether	810	12.2	Water	453		91

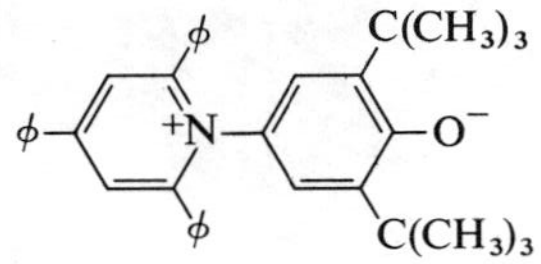	Chloroform	792	12.5	Methanol	563	3.4	91
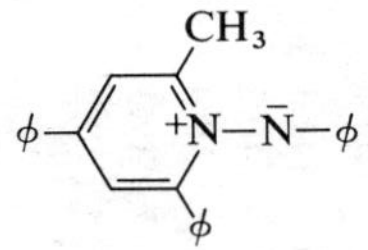	Benzene	585	9.76	Methanol	545	2.74	92
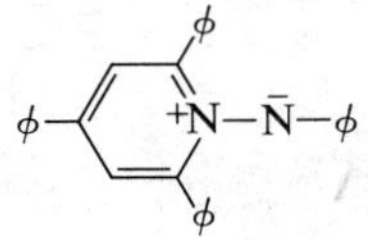	Benzene	620	5.65	Methanol	589	3.13	92
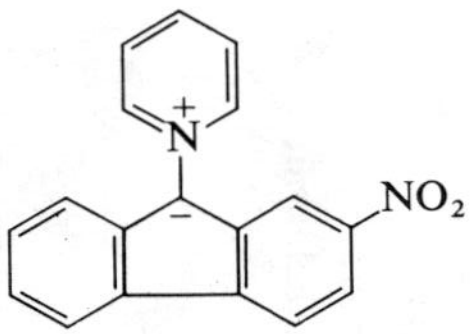	*o*-Dichlorobenzene	630		Methanol	611		
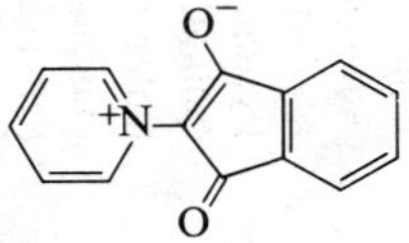	95% Ethanol	440		Water	425		93

TABLE 3.13
Absorption Spectra and Transition Energies in Various Solvents (91) of the Betaine

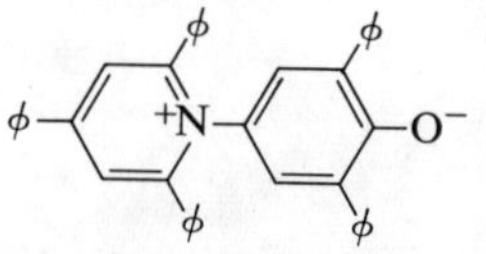

Solvent	$\lambda_{max}(m\mu)$	$m\epsilon$	$E_{t_{30}}$
Water	453	–	63.1
2,2,3,3-tetrafluoropropanol	481	–	59.4
Formamide	505	2.27	56.6
Ethylene glycol	508	2.04	56.3
Methanol	515	3.51	55.5
Methyl formamide	528	3.72	54.1
Monoglyme	547	3.61	52.3
Ethanol	550	4.43	51.9
Benzyl alcohol	563	4.12	50.8
n-Propanol	564	5.16	50.7
n-Butanol	570	4.14	50.2
Isopropanol	587	5.45	48.6
2,6-Dimethylphenol	600[a]	–	47.6
Isopentanol	608	4.93	47.0
Nitromethane	618	4.74	46.3
Acetonitrile	622	5.41	46.0
Dimethyl sulfoxide	635	5.24	45.0
Aniline	645	5.54	44.3
tert-Butanol	652[b]	5.83	43.9
Dimethylformamide	653	5.86	43.8
Dimethylacetamide	654	5.34	43.7
Acetone	677	6.74	42.2
Nitrobenzene	680	6.14	42.0
Benzonitrile	680	6.70	42.0
Acetophenone	692	6.47	41.3
Methylene dichloride	695	8.46	41.1
Pyridine	710	8.87	40.2
Quinoline	726	9.20	39.4
Chloroform	730	8.25	39.1
Diglyme	748	10.60	38.2
Ethyl acetate	750	8.48	38.1
Fluorobenzene	751	9.03	38.1
Iodobenzene	754	7.63	37.9
Bromobenzene	762	8.37	37.5
Chlorobenzene	762	8.11	37.5
Tetrahydrofuran	765	9.2	37.4
Anisole	769	11.5	37.2

TABLE 3.13 (*cont'd*)

Solvent	λ_{max}(mμ)	mϵ	E_{t30}
2,6-Lutidine	772	11.3	37.0
p-Dioxane	795	12.3	36.0
Diphenyl ether	810[b]	12.2	35.3
Benzene	—	—	~34.1
Toluene	—	—	~33.7

[a]Measured at 55°C.
[b]Measured at 30°C.

A somewhat different type of betaine structure is represented by the sulfonium phenacylides(94). The violet-shift effect is shown in the absorption spectrum of a representative of this series, $C_6H_5(CH_3)$ $=S^+—CH=C(C_6H_5)—0^-$, which shows the following wavelength maxima and millimolar absorptivities:

Solvent	λ_{max} (mμ)	mϵ
Absolute ethanol	310	15.1
Chloroform	320	8.9
Acetonitrile	320	10.2
Tetrahydrofuran	326	11.5

On the other hand, the violet shift is slight in the betaine obtained 9-(1-pyridinium)fluorene dissolved in alkaline *o*-dichlorobenzene (λ_{max} 606mμ) and alkaline ethanol, (λ_{max} 598mμ).

In cases where an hydroxylic solvent causes a violet shift, so will protonation at the negative (or positive) resonance terminal (Table 3.14). For the Z_b type of molecule this shift can be very large.

Some of the various types of chromogens obtained in a variety of colorimetric methods are described in Table 3.15. The effect of solvents and pH on these chromogens has been little studied.

II. Temperature Effects

The use of controlled high and low temperatures as a routine analytical tool in colorimetric measurement is still a relatively untouched field. Spectrometry at fairly high temperatures and at very low temperatures would use new solvents and new reagents. For example, biochemical analytical methods could be developed for use at temperatures so low that almost all intermolecular chemical reactions would be arrested within the

TABLE 3.14
Effect of Protonation on a Z_b-Band

Compound	Solvent	λ_{max} (mμ)	Solvent	λ_{max} (mμ)	Ref.
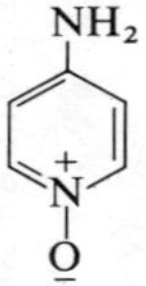	Water	277	Aqueous acid	270	95
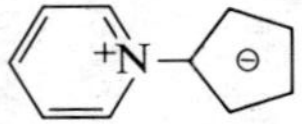	95% Ethanol	496	6 *N* HCl in H_2O–EtOH (1 : 1)	377	88
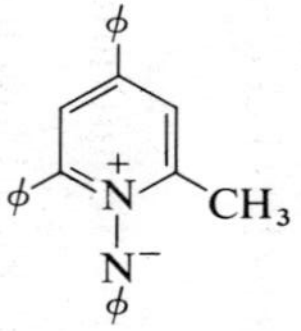	Methanol	545	Aqueous acid	Yellow	92
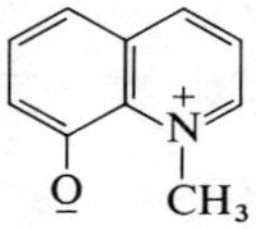	Aqueous alkali	432	Aqueous acid	357	90
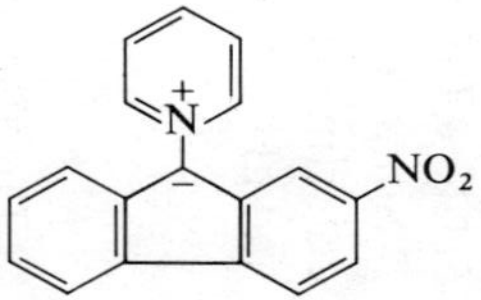	95% Ethanol	611	6 *N* HCl in H_2O–EtOH (1 : 1)	320	

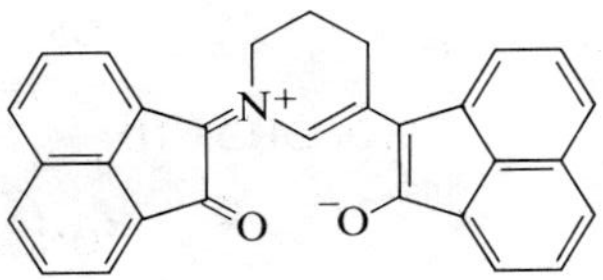	Dimethylformamide	650	11 *N* HCl	477	96[a]
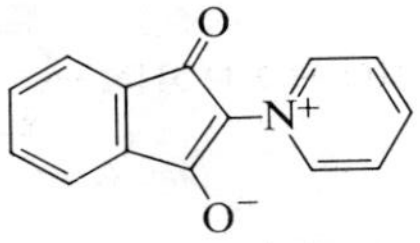	Water	425	Aqueous acid	240, 290	93
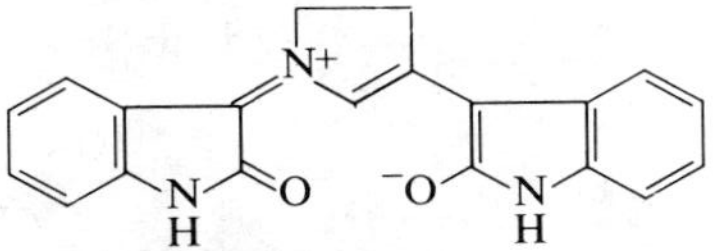	Dimethylformamide	596	4 *N* HCl	484	96
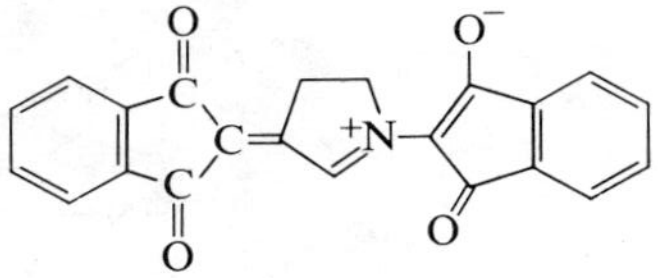	Water	550	Aqueous acid	515	97

[a]Many other compounds are covered in this reference.

TABLE 3.15
Examples of Chromogens with Z_n-, Z_z-, and Z_b-Structures

Chromogen	λ_{max} (mμ)	Reagent	Test substance	Ref.
$O_2N-C_6H_4-NHCH_3$	405	4-Nitrofluorobenzene	Methylamine	98
$O_2N-C_6H_3(NO_2)-NH-CH(R)-COOH$	360	2,4-Dinitrofluorobenzene	Amino acids	99
$O_2N-C_6H_3(NO_2)-NH-R$	353	2,4-Dinitrofluorobenzene	Primary amines	100
$O_2N-C_6H_3(NO_2)-NHCH_2CH_2OH$	340	2,4-Dinitrofluorobenzene	*N*-Acyl ethanol-amine	101
$O_2N-C_6H_2(NO_2)_2-NH-CH(R)COOH$	340	2,4,6-Trinitrobenzene-sulfonic acid	Amino acids	102

 O, O, $-NHCH_2CH_2COOH(CH_2OH)$	465	1,2-Naphthoquinone-4-sulfonic acid	Pantothenic acid, pantothenic alcohol	103
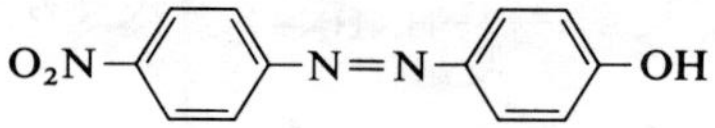 $O_2N-C_6H_4-N{=}N-C_6H_4-OH$	358	4-Nitrobenzenediazonium salt	Phenols	104
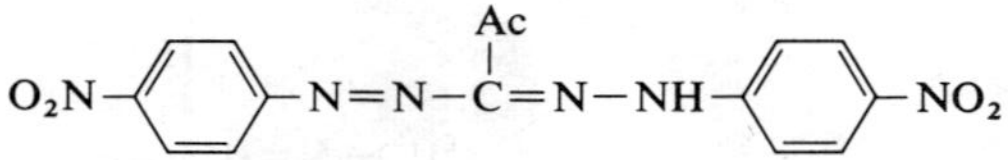 $O_2N-C_6H_4-N{=}N-C(Ac){=}N-NH-C_6H_4-NO_2$	440	4-Nitrobenzenediazonium salt	Acetoacetate	105
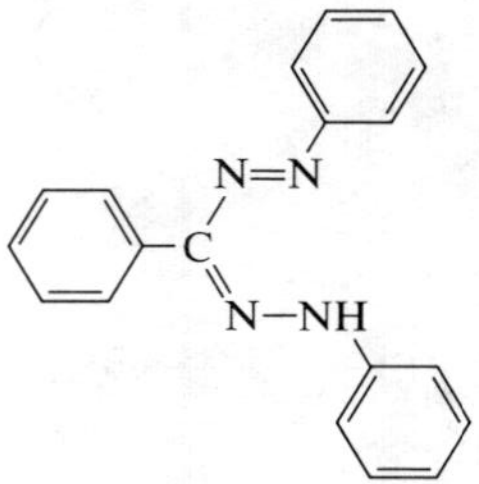 N=N, C, N–NH		2,3,5-Triphenyltetrazolium salt	Reducing substances	106
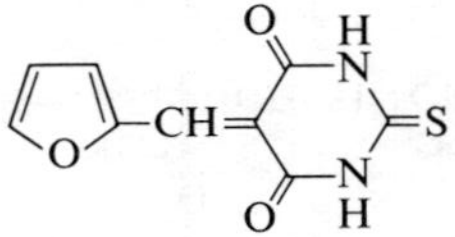 O, CH=, O, H, N, S, N, H, O	424	Thiobarbituric acid	Sugars	107

TABLE 3.15 (*cont'd*)

Chromogen	$\lambda_{max}(m\mu)$	Reagent	Test substance	Ref.
Thiobarbituric acid–derived =CHCH=CH– dimer (S=, NH, O, HO substituted rings)	532	Thiobarbituric acid	2-Deoxyribose	108
$(CH_3)_2N-C_6H_4-CH=N-C(=O)-N=CH-C_6H_4-N(CH_3)_2$	435	4-Dimethylaminobenzaldehyde	Urea	109
3,5-Br$_2$-2-OH-$C_6H_2$$-CH=N-CH_3$	400	3,5-Dibromosalicyladehyde	Methylamine, etc.	110
$O_2N-C_6H_3(NO_2)-NHNH-C(=O)-R$	330	2,4-Dinitrophenylhydrazine	Aliphatic acids	111
Indigo (bis-indole, C=O, NH)	660	Disodium indoxyphosphate	Alkaline phos-phatase	112

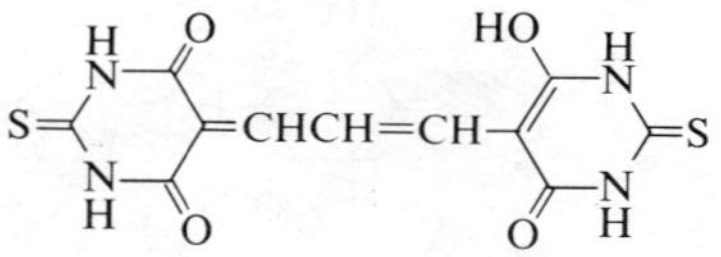

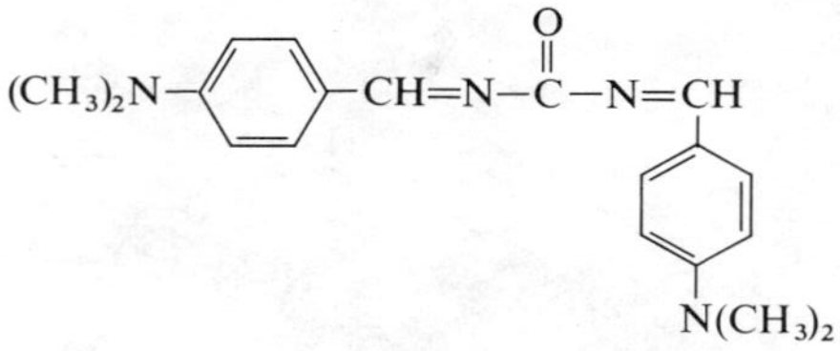

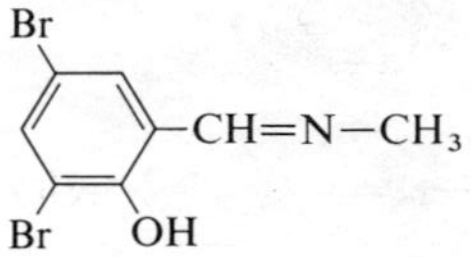

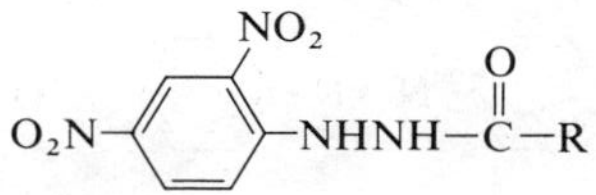

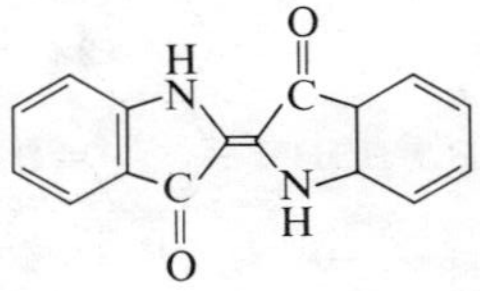

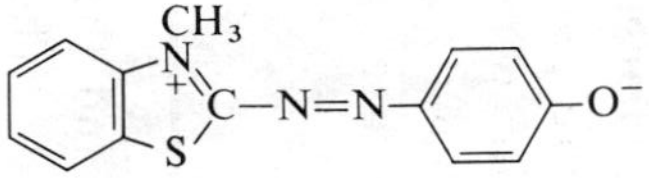	525	3-Methyl-2-benzothiazolinone hydrazone	Phenols	73
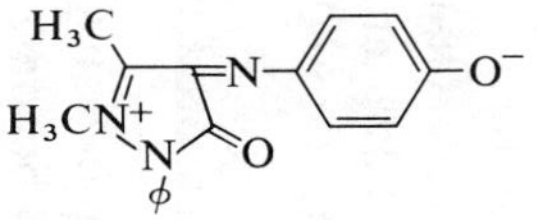	510	4-Aminoantipyrine	Phenols	113
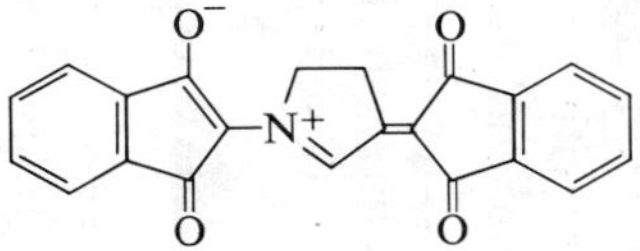	635	Ninhydrin	Proline	114
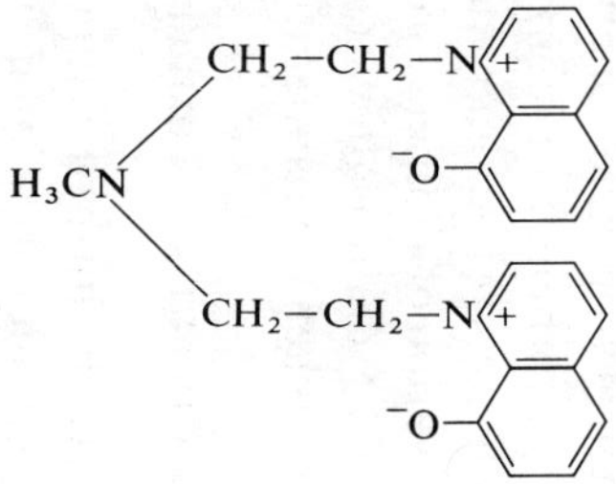	468	8-Hydroxyquinoline	Bis(β-chloroethyl)-methylamine	115

living or just-killed tissue. Appropriate types of light energy and reagents would be needed for this type of low-temperature assay.

Data are available on the changes in spectra of some compounds with changes in temperature. Some of these data have been used in colorimetric analysis, and many have potential use. In this section we discuss the data.

A. Temperature Dependence of Absorbance

Since so many spectral measurements are taken at room temperature and room temperatures do vary with the laboratory, the time of day, the season of the year, and so on, the effect of changing temperature on absorbance needs to be known. Another factor that can affect the absorbance is the increasing temperature in the uncooled cell compartment as the instrument "warms up" during a working day. The temperature dependence of absorbance in the ultraviolet and visible regions for some molecules and ions influences the reproducibility of analytical spectrophotometric applications (116–121). The temperature coefficient of absorbance is strongly dependent on solvent and solute.

The variations of absorbance with temperature of the sample in the ultraviolet spectra of 19 organic compounds have been studied at increments between 5 and 33°C, employing methanol and isooctane as solvents (*121*). The decrease in absorbance with increasing temperature ranges from 0.0% per degree Celsius for acetone and 2,5-hexanedione to 0.74% per degree Celsius for toluene. For benzene, toluene, and the xylenes the temperature coefficient of absorbance was approximately the same in methanol and isooctane, but for 2,4-pentanedione and ethyl acetoacetate the decrease in absorbance per degree Celsius was much greater in methanol than in isooctane. Phenol, nitrobenzene, and aniline also show changes in their temperature coefficients of absorbance, with a change from acidic methanol to alkaline methanol solution. The change in absorbance of benzene with temperature is shown in Figure 3.3. Solutions of sodium nitrophenolates show a deepening of color with elevation of temperature (122). The drastic change possible is shown by the 3,5-dinitrosalicylate anion, which shows an absorbance increase of 50% when the temperature is raised from 10 to 30°C (123).

B. Effect of Temperature Changes on Fine Structure

At diminished temperatures rotational transitions decrease, and the electronic-vibrational bands tend to become sharply defined. The absorption spectra of benzene, naphthalene, anthracene, diphenyl, perylene, pyrene, phenanthrene, and triphenylene have been investigated

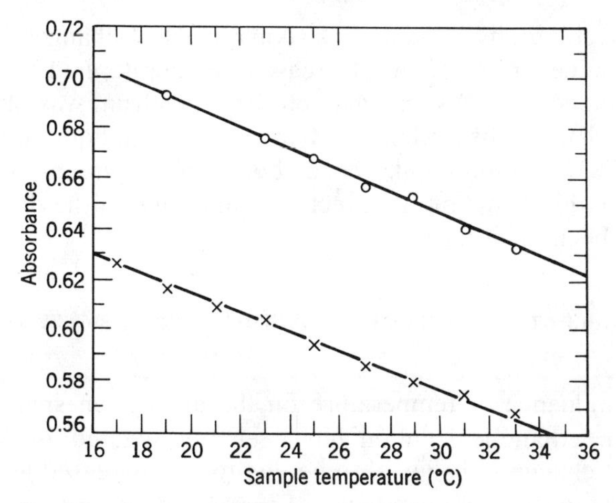

Fig. 3.3. Effect of temperature on the absorbance of benzene (measured in 100-mm quartz cells) in isooctane (2.8248×10^{-5} g/ml) (○——○) and in methanol (2.8804×10^{-5} g/ml) (×——×).

in solid solution at about −170°C(124). The absorption bands are much narrower than at room temperature, and many more are found; in addition, the difference in intensities between maxima and minima is more than doubled. At liquid-nitrogen or liquid-air temperatures, as compared to room temperature, hemoproteins(125, 126) and cytochrome *c* and benzene(127) show a sharpening, splitting, and intensification of absorption bands. Many other compounds should show this phenomenon. An increase in fine structure should improve the possibilities of characterization and assay.

C. Temperature and Dielectric Constant

Since an increase in temperature generally lowers the dielectric constant of a solvent(128) and so reorients the solvent dipoles around a solute, the temperature of a solution could affect the absorption. In the Z_n type of molecule an increase in solvent temperature would tend to cause a violet shift, whereas in the Z_z and Z_b types a red shift would result. This effect is shown in the ethanolic solution of compound **14**, a Z_z molecule(129), which is crimson at 0°C and violet at the boiling point.

$H_3C\overset{+}{N}C_5H_4$—CH=CHCH=CH—(isoxazole ring: O^-, O, N, ϕ)

14

Even chlorophyll *b* shows striking spectral changes in ether or isopropylamine solutions at decreased temperatures(128). Thus in isopropylamine at 230°K the characteristic spectrum was obtained, a green color with peak absorption at 470 mμ, whereas at liquid-nitrogen temperature (77°K), chlorophyll, the embodiment of greenness, became red. Whether this temperature effect is mainly due to the dielectric constant has not been ascertained.

D. Temperature Effect on Acidity or Basicity of Solvent or Solute

The influence of temperature on the absorption spectra of merocyanines in ethanolic solution has been investigated(74, 130, 131). The spectral change is believed to be due to the increased acidity of ethanol with decreased temperature. Thus 1,1-dicyano-6-(*p*-dimethylaminophenyl)-1,3,5-hexatriene shows the spectrum of the cation (proton addition to the positive resonance terminal) in ethanolic acid solution or in ethanol at low temperatures(130). Similarly bis(1,3,3-trimethyl-2-methylenylindoline)ketone shows the presence of the cation (proton addition to the negative resonance terminal) in the spectra of this compound in acidic methanol and in ethanol at low temperatures(130). In the former compound the new band is at shorter wavelength, whereas in the latter compound it is at longer wavelength, as expected.

Triarylcarbinols can ionize at higher temperatures. Data have been reported for the following:

Compound	Solvent	Color at room temperature	Color at elevated temperatures	Ref.
4-(4′-Aminoxenyl)-diphenylcarbinol	Fused benzoic acid	None	Deep green	132
Bis-4-(4′-dimethyl-aminoxenyl)phenyl-carbinol	Acetic acid	Pale yellow	Green	133
Bis(4-aminophenyl)-phenylcarbinol	Dilute mineral acid	None	Red	133
Tris-2-anisyl-carbinol	Acetic acid	None	Violet	134

When a colorless solution of tris-2-anisylchloromethane in benzene is cooled to −70°C, a solid purple "ice" of the cation is formed which becomes colorless when warmed to room temperature(135).

E. Thermochromic Spiropyrans

Thermochromic spyropyrans have a spiro carbon atom as the insulation between two systems of conjugation. With heat the bond between the spiro carbon atom and the ring oxygen atom is broken to form a Z_z-structure (usually) with a positive charge on a hetero atom from one system of conjugation and a negative charge on the phenolic oxygen atom from the other system, both systems in conjugation (**15** and **16**). Examples of thermochromic spiropyrans are listed in Table 3.16.

The chemical and thermochromic properties of many spiropyrans have been reviewed(136, 137). An important factor in the tendency of spiropyrans to exist in a colored form is the ease with which the colored form can lie in a plane. Dinaphthospiropyran (**15**) is a typical thermochromic compound. Whereas **15** is nonplanar, **16** is coplanar. The lack of thermochromism in 3,3′-dialkylated dinaphthospiropyrans(146) can be explained by the large amount of steric strain that would be present in the ring-opened compound. Dibenzospiropyran (**17**) is not thermochromic since the benzopyran ring is not broken under these conditions to form a phenolic structure (e.g., **18**).

TABLE 3.16
Effect of Heat on Spiropyrans[a]

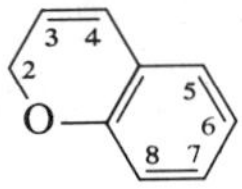

2-Spiro	Other substituents	Color[b] in boiling solvent or (melt)	Ref.
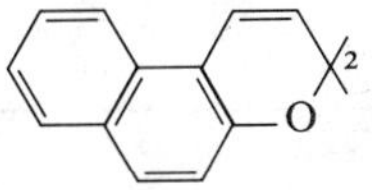	(c)	Violet (purple)	137
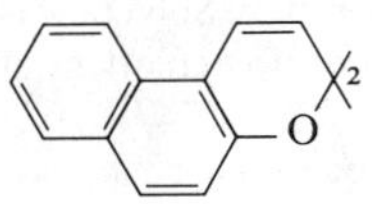	5,6-Benzo[c]	Violet (blue)	137, 138
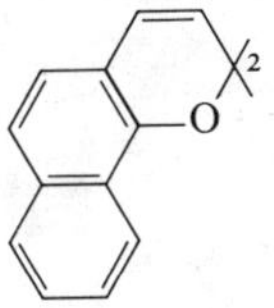	5,6-Benzo[c]	(Purple)	137, 139

TABLE 3.16 (*cont'd*)

2-Spiro	Other substituents	Color[b] in boiling solvent or (melt)	Ref.
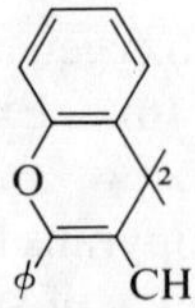 [structure: 2H-chromene, spiro at C-2, 3-CH₃, 4-ϕ]	5,6-Benzo[c]	Purple (purple)	137
[structure: xanthene/thioxanthene, spiro at 9] X = O	5,6-Benzo	Red (red)	140–142
X = S	5,6-Benzo	Red (pink)	143
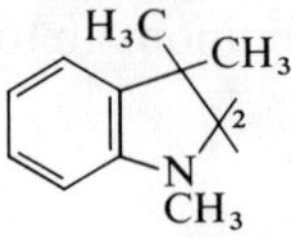 [structure: 1,3,3-trimethylindoline, spiro at 2]	[(c)]	Violet	144, 145
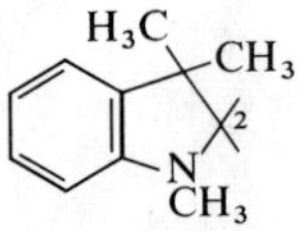 [structure: 1,3,3-trimethylindoline, spiro at 2]	5,6-Benzo	Blue (red)	138
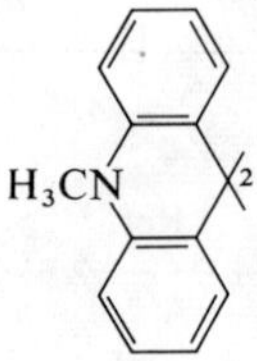 [structure: N-methylacridan, spiro at 9]	8-Methoxy	Violet	144

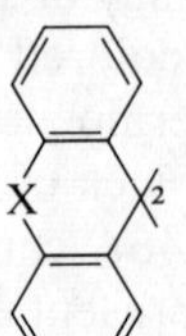

[a]See references 136 and 137 for more detail.

[b]Colorless to much lighter color in solvent at room temperature. Solvent was benzene, xylene, anisole, or diphenyl ether. Solid material was colorless at room temperature.

[c]Derivatives of the parent ring system also showed thermochromic changes.

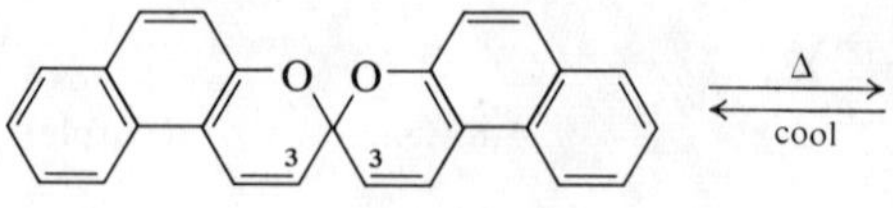

$\xrightarrow[\text{cool}]{\Delta}$

15

Colorless in xylene

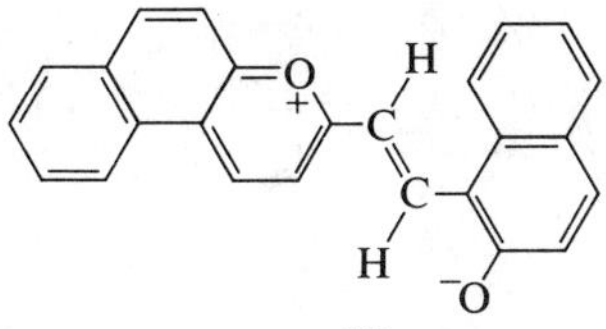

(3.3)

16

Purple

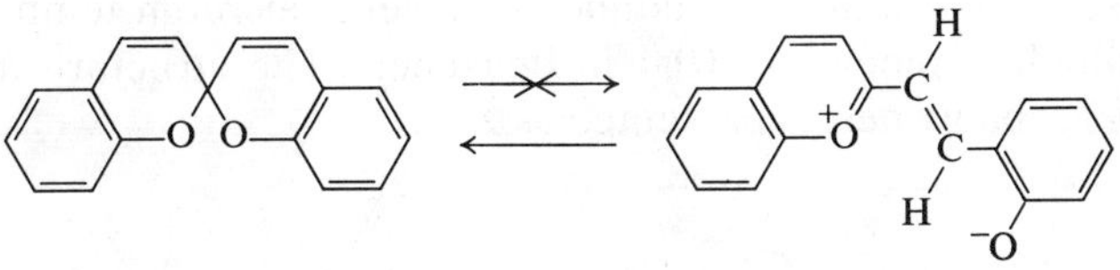

(3.4)

17 **18**

These considerations apply to 3- and 3′-methylbenzo-β-naphthospiropyrans (**19** and **21**, respectively)(147). The former is thermochromic, because its ring-opened form (**20**) is flat without strain, but the latter is not thermochromic, because steric requirements of the methyl group prevent structure **22** from forming.

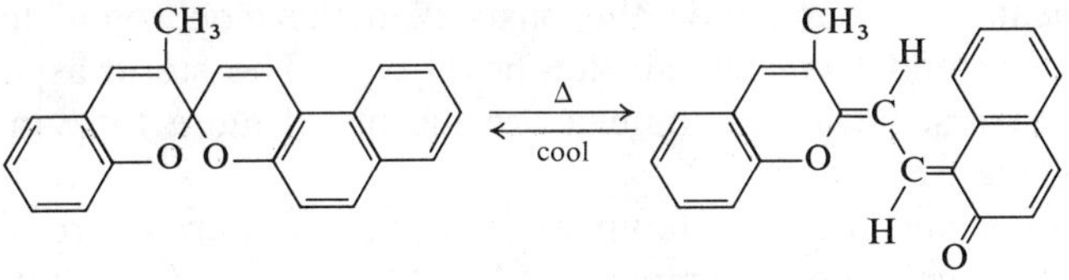

(3.5)

19 **20**

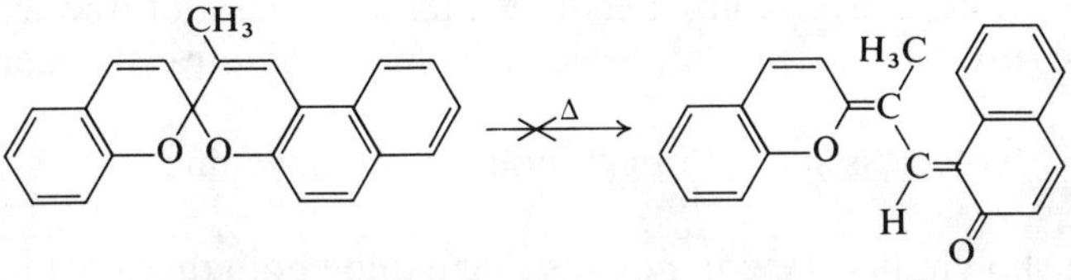

(3.6)

21 **22**

Let us consider the effect of change of structure on three polycyclic spirans(148). Dibenzospiropyran (**17**) is a colorless solid, forms a colorless melt, is colorless in solution, and has a dipole moment of 1.20 in benzene. 1,3,3-Trimethylindolino-β-naphthopyrylospiran is a colorless solid, forms a blue melt, and gives pink solutions at room temperature and intense red solutions in nonpolar solvents on heating. This compound has a dipole moment of 1.38 in benzene, so its structure is best represented as **23** at room temperature and **24** at an elevated temperature.

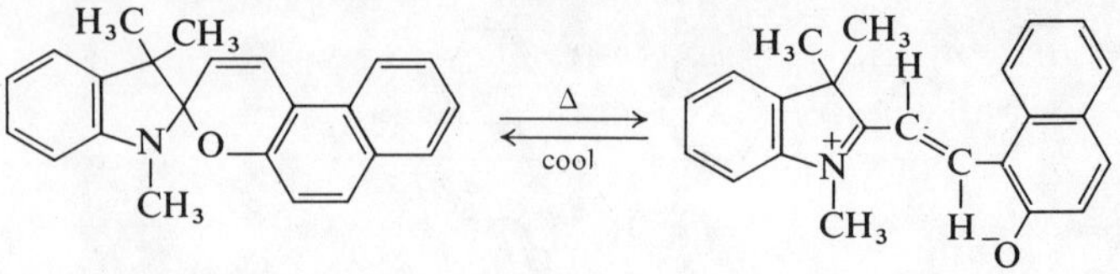

N-Methylquinolino-β-naphthopyrylospiran is a blue-green solid that forms intense blue solutions in nonpolar solvents at room temperature and has a dipole moment of 10.4 in benzene. The structure for this compound is probably best represented as **25**.

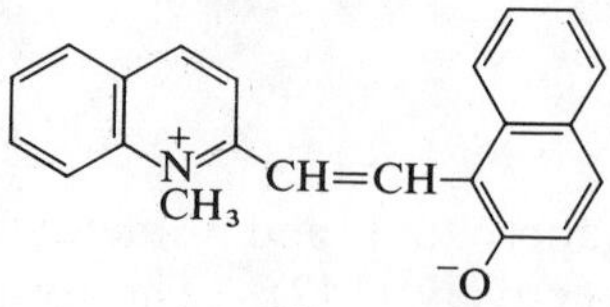

The formation of a dark green cationic resonance structure (λ_{max} 650 mμ) from which a colorless spiropyran can be formed through moderate heat has been made the basis of a thermochromic test for methyl ketones and some cyclic ketones (149). The steric aspects of these and other spectral phenomena are discussed more fully in other parts of this treatise.

The photochromism at low temperatures and the thermochromism of the spiropyrans have been compared (142, 150–153). The results indicate the occurrence of consecutive and concurrent phototransformations and thermal interconversions between stereoisomers of the colored zwitterionic resonance structure, besides the main phototransformation of

$$\text{colorless spiropyran} \rightleftharpoons \text{colored merocyanine.}$$

It has been shown that the compounds that exhibit both thermochromism and photochromism form spectroscopically identical colors (154). Whether solutions of spiropyrans are irradiated by ultraviolet light at low temperatures or heated to the boiling point, they undergo an intramolecular ionization process. In thermochromism, because of thermal equilibrium, only a small amount of the colorless form can undergo ring opening, whereas ultraviolet irradiation at low temperatures brings about a considerable transformation from the colorless to the colored dipolar form (153).

Reverse photochromism is exhibited by 1,3,3-trimethyl-spiro-[indoline-2,2′-benzopyran]carboxylic acid(155). In this case the colored zwitterion is stable in organic solvents, is converted to the colorless spiro derivative by irradiation with visible or ultraviolet light, and reverts to the colored state spontaneously. Temperature affects this equilibrium, as has been shown for the hydroxylated compound with the same parent ring system—for example, 1,3,3-trimethyl-6′-hydroxy-spiro-[indoline-2,2′-benzopyran] (**26**) (156).

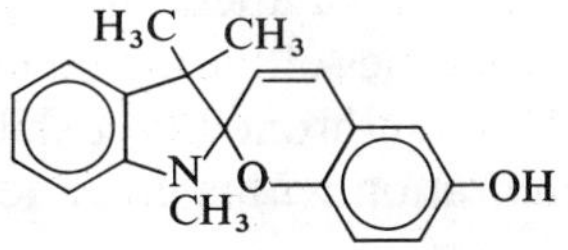

26

In compound **26** reverse photochromism takes place at low temperatures in the presence of a hydrogen donor. The thermal equilibrium of the concentrations of the colored ring-opened Z_z-structure and the colorless spiro structure **26** (λ_{max} 280 mμ) varies with solvent polarity. The concentration of the Z_z-structure (λ_{max} 550 mμ) increases with solvent polarity (e.g., $m\epsilon = 29$ in ethanol, 17 in acetone, 8 in benzene, and 5.5 in dioxane). This colored state is not destructible by irradiation with light. However, decreasing the temperature of a solution of this colored compound in a hydrogen-donor solvent, such as alcohol, results in a reversible increase in the concentration of the cationic resonance structure (λ_{max} 450 mμ) formed by protonation of the charged oxygen atom of the Z_z-structure. The new colored species can be destroyed by irradiation with visible light and caused to reappear by irradiation with ultraviolet light.

F. Thermochromic Ethylenes

The summary of the various theories postulated over the years to explain the color changes or lack of color change of the thermochromic ethylenes and their derivatives is fascinating reading(136). In most cases the molecules at room temperature consist of two systems of conjugation twisted out of a coplanar position by steric hindrance at the ethylene bond. Thus the conjugation systems are partially insulated from each other. It has been hypothesized that at an increased temperature (or on irradiation at low temperatures) the molecule acquires a greater degree of planarity through a reversible change in structure and consequently absorbs at longer wavelength(154). In line with this hypothesis evidence has been obtained to show that a loss of thermochromic properties results when either (*a*) the molecule is substituted in a ring *ortho* to the central ethylene group so that planarity is hindered at high temperatures or (*b*) overcrowding is decreased so that a molecule can approach planarity at room temperature (157).

The compound $\Delta^{9,9'}$-bixanthene (**27**) is a strongly thermochromic substance; the crystals are colorless in liquid air, but the melt is deep blue green, and the solution in boiling diphenyl ether is light blue green (158). In boiling dimethyl phthalate this compound has a weak band at 600 mμ (158). Substituents in a $\Delta^{9,9'}$-bixanthene that hinder planarity interfere with the thermochromic properties of the molecule (Table 3.17) (159, 160).

In solution yellow $\Delta^{10,10'}$-bianthrone (**28**) exhibits a reversible thermochromism involving equilibration with a green form whose concentration

27 **28**

TABLE 3.17
Thermochromism of Some Bixanthene Derivatives[a] (136, 159, 160)

Compound	Thermochromic effect	Compound	Thermochromic effect
$\Delta^{9,9'}$-Bixanthene	Strong	2,3,2′,3′-Tetramethyl	Strong
2,2′-Dichloro	Strong	2,4,2,4′-Tetramethyl	Strong
4,4′-Dichloro	Strong	1,1′-Dichloro	Very weak
Di-3,4:3′,4′-dibenzo	Strong	1,4,1′,4′-Tetramethyl	Negative
2,2′-Dibromo	Strong	1,3,1′,3′-Tetramethyl	Negative
4,4′-Dibromo	Strong	1,1′-Dichloro-4,4′-dimethyl	Negative
2,2′-Dimethyl	Strong	Di-1,2:1′,2′-dibenzo	Weak

[a]Light yellow in diphenyl ether solution, becoming blue green on boiling the solution. Light yellow solid, giving blue-green melt.

increases with temperature(136, 161). The same green species is obtained by irradiation of the solution at −77°C and by the application of pressure to the solid(161, 162).

Examination of the thermochromic $\Delta^{10,10'}$-bianthrones has led to similar conclusions(163). For example, $\Delta^{10,10'}$-bianthrone (**28**) has its long-wavelength band at about 680 mμ in hot dimethyl phthalate(158). All the investigated dimethyl, dibromo, and dimethoxy derivatives of $\Delta^{10,10'}$-bianthrone, $\Delta^{9,9'}$-bixanthene, and 9-xanthylidene-10-anthrone, with the exception of those substituted in the sterically hindered positions (e.g., 4,4′- in $\Delta^{10,10'}$-bianthrone), show pronounced thermochromism (164, 165). When the substituents were in the 4,4′-positions of $\Delta^{10,10'}$-bianthrone, the thermochromic behavior depended on the size of the substituent. Thus the 4,4′-difluoro derivative is thermochromic at a much higher temperature than the parent compound(166), and the 4,4′-dibromo, dimethyl, dimethoxy, and 3,4:3′,4′-dibenzo derivatives are not thermochromic. Many of the thermochromic derivatives were also found to exhibit the phenomena of low-temperature photochromism—that is, reversible color changes produced by irradiation with ultraviolet light of a cold solution(158, 167).

When solutions of $\Delta^{9,9'}$-bixanthene, $\Delta^{10,10'}$-bianthrone, 10-(9-xanthylidene)anthrone, and their derivatives are irradiated at temperatures between −70 and −150°C with light whose wavelength does not exceed 450 mμ, reversible formation of colored modifications takes place. The visible absorption spectra of these colored modifications are similar to the spectra of the colored forms obtained reversibly on heating dimethyl phthalate solutions of the same compounds. However, the heated solutions have only about 1% the intensity of the cold irradiated solutions. It is postulated that the colored compounds produced by both procedures are identical(158, 168). For 10-(9-xanthylidene)anthrone the colored compound in both cases has its wavelength maximum at 700 mμ. Other ethylenic compounds that have shown thermochromic effects are $\Delta^{4,4'}$-biflavylene(169), 1,2-bis(10,10′-anthronylidene)ethane(170), 2-(9-xanthylidene)indan-1,3-dione(154), anthraquinone mono-*p*-dimethylaminoanil(154), 10-(diphenylmethylene)anthrone(154), 10-(9-fluorenylidene)anthrone(171), $\Delta^{5,5'}$-binaphthacene-12,12′-dione(172), $\Delta^{7,7'}$-dibenzo[*a*]anthracene-12,12′-dione(172), and so on.

Many ethylenic compounds show piezochromic properties. The color changes in this case are probably also caused by thermal effects resulting from pressure or friction. Crystals of $\Delta^{4,4'}$-biflavylene change from yellow to dark red under pressure(169). Brown-yellow crystals of 10-(9-fluorenylidene)anthrone become dark violet when pressed strongly with a glass pestle(171). The color changes back to brown yellow in the presence of ether vapor. Many other examples have been reported in the literature.

G. Temperature and Depolymerization

Another type of thermochromic effect has been reported for the cyclopentadienones, which can be present as monomer, dimer, or as a mixture (Table 3.18)(173).

2 [cyclopentadienone] ⇄ [dimer, C=O] (3.8)

29 Red **30** Colorless

The nondissociating dimers, such as 2,3,4,5-tetrachlorocyclopentadienone, are colorless in solution regardless of temperature, have the expected molecular weight, and do not add to dienophiles in benzene The dissociating dimers, such as 2,3,5-triphenylcyclopentadienone, are colorless solids that give colored solutions, the intensity increasing to a maximum as the temperature is increased. The molecular weights are intermediate between monomer and dimer. The compounds add to dienophiles in benzene solution. The monomers, such as tetraphenylcyclopentadienone, are colored compounds that give bright red solutions regardless of the temperature. They add to dienophiles in benzene solution.

H. Miscellaneous Effects

Photochromism can also take place through ring closing, as in the phenylfulgides (i.e., α-phenyl-dimethylenesuccinic anhydrides)(174). Thus irradiation of α,ϵ-diphenylfulgide at -196°C in a rigid glass of 2-methyltetrahydrofuran with light corresponding to its absorption band at 355 mμ changed the long-wavelength maximum to 485 mμ (1,1*a*-dihydro-1-phenylnaphthalene-2,3,-dicarboxylic anhydride).

Changes in temperature can affect *cis–trans* isomerization and even cause tautomerism and thus affect the absorption spectrum.

Quantum yields of the *cis–trans* photoisomerization of azobenzene and related compounds have been measured spectrally(175). The yields for the *trans* → *cis* transformation were found to decrease sharply on cooling, whereas those for the *cis* → *trans* reaction change but little in this temperature range. These results indicate the existence of energy

TABLE 3.18
Effect of Temperature on the Color of Cyclopentadienones (173)

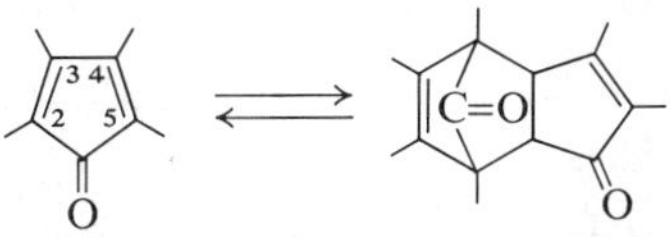

Substituent in position				Substituent in position			
2	3	4	5	2	3	4	5
Nondissociating Dimers[a]				*Dissociating Dimers*[c]			
H	H	H	H[b]	CH_3	ϕ	ϕ	CH_3
Cl	Cl	Cl	CH_3	CH_3	ϕ	ϕ	$n\text{-}C_3H_7$
Cl	Cl	CH_3	Cl	ϕ	ϕ	H	ϕ
CH_3	Cl	CH_3	Cl				
Cl	Cl	Cl	Cl	*Monomers Only*[d]			
H	ϕ	ϕ	H				
H	ϕ	ϕ	CH_3	CH_3	ϕ	ϕ	ϕ
H	ϕ	ϕ	$n\text{-}C_5H_{11}$	C_2H_5	ϕ	ϕ	C_2H_5
H	ϕ	ϕ	$n\text{-}C_{10}H_{21}$	C_6H_{13}	ϕ	ϕ	C_6H_{13}
H	ϕ	ϕ	CH_2COOH	ϕ	ϕ	ϕ	ϕ
H	ϕ	ϕ	ϕ				

[a]Colorless in solution regardless of temperature. They show expected molecular weights. They do not add to dienophiles in benzene solution.

[b]Oxime of.

[c]Colorless solids that give colored solutions, the intensity increasing to a maximum as the temperature is increased. Molecular weights are intermediate between monomer and dimer. They add to dienophiles in benzene solution.

[d]Colored monomers that give bright red solutions regardless of the temperature. They add to dienophiles in benzene solution.

barriers somewhere between the electronically excited singlet state of one isomer (formed by light absorption) and the ground state of the second isomer.

Irradiation of 2-(2′,4′-dinitrobenzyl)pyridine (**31**) turns the solid (176) or the solution (177) deep blue. At room temperature the color is observed as long as the irradiation is in progress. Irradiation of the solution at liquid-nitrogen temperature gives a characteristic blue color that lasts indefinitely at this temperature. A thin film of the irradiated crystalline solid gives a maximum at 610 mμ, with shoulders at 545 and 575 mμ (176). An ethanolic solution irradiated at −60°C has a band at 568 mμ. The long-wavelength maxima (in ethanol) of the tautomers are compared below to that of 1-methyl-2-(2′,4′-dinitrobenzal)-1,2-dihydropyridine (**33**)(177).

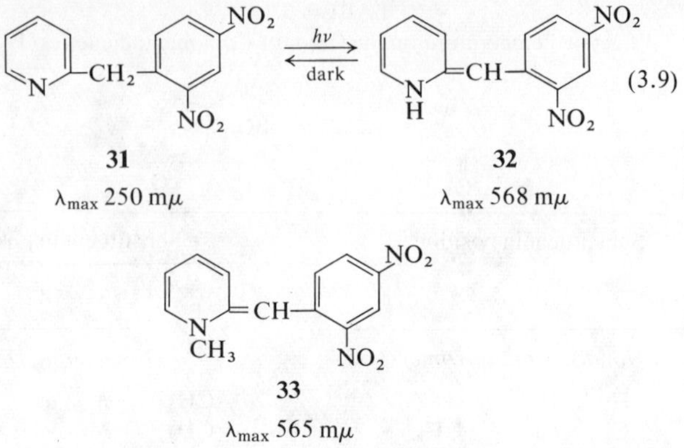

$$\mathbf{31}\ (\lambda_{max}\ 250\ m\mu) \underset{\text{dark}}{\overset{h\nu}{\rightleftharpoons}} \mathbf{32}\ (\lambda_{max}\ 568\ m\mu) \quad (3.9)$$

33

λ_{max} 565 mμ

The millimolar absorptivity of **33** is 45.0 in acetone and 12.0 in ethanol.

4-(2′,4′-Dinitrobenzyl)pyridine also shows photochromotropism in solution(178). At $-180°C$ irradiation of an ethanolic solution gives an intense purple color ($\lambda_{max} \simeq 575\ m\mu$). The crystalline solid is not photochromic.

A reversible thermochromic change attributed to lactam–lactim tautomerism has been described(179). The violet photoaddition products obtained on irradiation of benzo[*a*]phenazine-3,4-quinone (**34**) with benzaldehyde or 4-methoxybenzaldehyde show reversible thermochromic changes in ethyl benzoate, attributed to the following reactions(179):

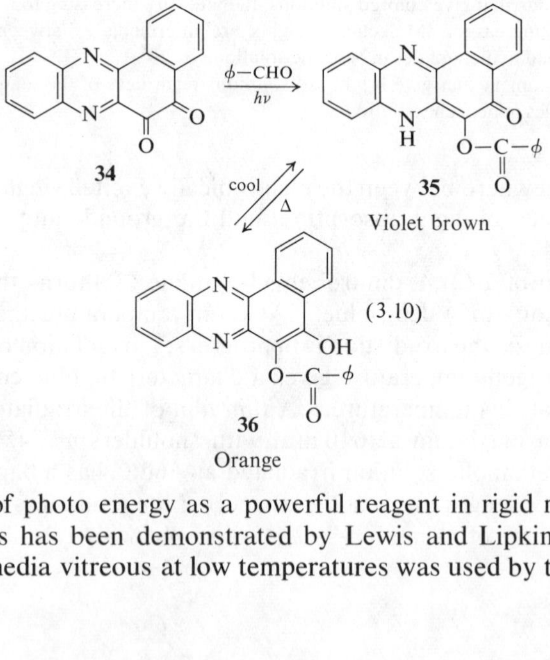

$$\mathbf{34} \xrightarrow[h\nu]{\phi\text{—CHO}} \mathbf{35}\ \text{(Violet brown)} \underset{\text{cool}}{\overset{\Delta}{\rightleftharpoons}} \mathbf{36}\ \text{(Orange)} \quad (3.10)$$

The use of photo energy as a powerful reagent in rigid media at low temperatures has been demonstrated by Lewis and Lipkin(180). The rigidity of media vitreous at low temperatures was used by these authors to limit diffusion and to eliminate bimolecular reactions. Monomolecular reactions of complex organic molecules, involving photodissociation, photoionization, and photo-oxidation, were examined spectrally at about 100°K. Lewis and Lipkin demonstrated that a molecule may be dissociated by light into two radicals, into positive and negative ions, and into a positive ion and an electron. The capabilities of this powerful spectral technique have been, as yet, inadequately studied in trace analysis research.

III. Steric Effects

The wavelength position and especially the intensity of ultraviolet and visible absorption bands are sensitive to the steric requirements of the ground and electronically excited states.

Maximum resonance interaction occurs in a compound in which the entire conjugation system is coplanar. Where substituents in a conjugated molecule can force part of the conjugation system out of the plane of the molecule, the resonance interaction and the molar absorptivity can be decreased in comparison with those in a reference compound that is not sterically hindered. Some of the mechanisms by which steric interference may be reduced are penetration of van der Waals radii, loss of planarity, increase or decrease in bond length or angle, twisting of a bond, or assumption of an alternative configuration. More specifically, steric strain may be effectively relieved through the stretching, bending, or twisting of a conjugating bond connecting two conjugated systems or a conjugated system and a resonance terminal. At the same time these two parts of the molecule lose their coplanarity and the resonance interaction is reduced, as shown by the electronic spectra. Where the strain is at a maximum so that the planes of the two essential parts of the molecule are at right angles to each other, the resonance interaction is negligible and the spectrum is a composite of the two "unsaturated" parts.

In molecules with steric strain either the ground or excited state can be the more sterically hindered or both states can be more hindered than the comparable electronic states of the reference compound. Because the change in energy (ΔE) between the ground and the excited state of the sterically hindered molecule can be smaller, larger, or the same as the ΔE of the reference, the wavelength maximum of the former compound can show no shift or be shifted to the red or the violet.

However, considering the wide variety of substitutions that can be made in conjugated molecules, it should be possible to have steric effects capable of causing all combinations of violet or red shifts with hyperchromic or hypochromic shifts. Thus a sterically caused decrease in conjugation or a sterically caused decrease or increase in the electron-donor or electron-acceptor strengths of the resonance terminals should be capable of causing these various spectral results.

Excellent reviews are available on many aspects of stereochemistry of interest to the analytical chemist(181).

A. Ortho Substituents

An example of increasing steric hindrance and the resultant hypochromicity is shown in Table 3.19. In the benzaldehyde chromophore, as in its simple derivatives (e.g., acetophenone and 4-methylacetophenone), the interplanar angle between the benzene ring and the carbonyl group is zero; that is, the molecules are uniplanar. Although alkyl substitution in the ring can cause a slight hypochromic effect, an increase in the interplanar angle causes a definite hypochromic shift.

The buttressing effect can increase steric hindrance since it is due to the crowding of a sterically hindering group by an adjacent group. This effect has been discussed by Forbes and Mueller(183). Examples are 2,3,4,5,6-pentamethylacetophenone, which has an mϵ of 12.0 at the wavelength inflection of 216 mμ as compared with 2,4,6-trimethylacetophenone (λ_{max} 242 mμ, mϵ 3.6) and 2,3,5,6-tetramethylacetophenone,

TABLE 3.19
Steric Effects in some Aromatic Ketones(182)

Compound	λ_{max}(mμ)	mϵ	θ (degrees)[a]
C_6H_5CHO	242	14.0	0
$C_6H_5COCH_3$	242	13.0	0
4-$CH_3C_6H_4COCH_3$	252	15.0	0
$C_6H_5COC_2H_5$	242	13.5	0
1-Indanone	243	12.3	17
$C_6H_5COC_5H_{11}$	238	12.0	19
1-Tetralone	248	11.6	22
$C_6H_5COC(CH_3)_3$	242	9.1	34
$C_6H_5COC(CH_3)_2C_2H_5$	239	8.3	38
Benzocycloheptan-1-one	246	8.1	39
2-$CH_3C_6H_4COCH_3$	242	8.5	40
$C_6H_5COCCH_3(C_2H_5)_2$	238	7.1	43
Benzocyclooctan-1-one	247	6.5	46
2,6-$(CH_3)_2C_6H_3COCH_3$	251	5.0	57

[a]Interplanar angle of twist between benzene ring and carbonyl group calculated from $\cos^2\theta = \epsilon/\epsilon'$, where ϵ' is molar absorptivity of hindered compound and ϵ is molar absorptivity of reference compound.

which has an mϵ of 11.5 at the wavelength inflection of 212 mμ, as compared with 2,6-dimethylacetophenone (λ_{max} 251 mμ, mϵ 5.6). The absorption bands of the sterically hindered trimethyl and dimethyl acetophenones are derived from the phenylcarbonyl chromophore, whereas the ultraviolet absorption bands of the tetramethyl and pentamethyl acetophenones are derived from a composite of the polymethylbenzene and methyl carbonyl conjugation systems. The former system dominates the spectra, as shown by the close resemblance of the spectrum of 2,3,5,6-tetramethylbenzene (λ_{infl} 214 mμ, mϵ 8.5) to that of the strongly hindered acetophenones. In the same way the spectrum of the strongly hindered bimesityl closely resembles that of mesitylene(184).

Increasing the angle of twist of the two sterically affected parts of the conjugation system of a test substance can decrease either the intensity of the long-wavelength band derived from the entire conjugation system or the yield of chromogen obtained in colorimetric analysis of the test substance. For example, consider aniline derivatives with substituents on the nitrogen and in the *ortho* position.

Klevens and Platt(185) investigated the spectra of *ortho*-substituted dimethylanilines from 170 to 350 mμ. From the spectra they determined the "effective spectroscopic angle of twist." Wepster(186) corrected their values on the basis that benzoquinuclidine (**38**) more closely resembled tetralin or benzene than did 2,6-dimethyl-*N*,*N*-dimethylaniline (the standard 90° angle of twist of Klevens and Platt). On this basis, with benzoquinuclidine representing a 90° angle of twist, *N*,*N*,2,6-dimethylaniline has a 65° angle of twist, *N*,*N*,2-trimethylaniline a 49° angle, 2-bromo-*N*,*N*-dimethylaniline a 49° angle, 2-chloro-*N*,*N*-dimethylaniline a 45° angle, and 2-fluoro-*N*,*N*-dimethylaniline a 30° angle of twist.

A comparison of the spectra in isooctane of *N*,*N*-dimethylaniline (**37**) and benzoquinuclidine (**38**) indicates that the free electrons on the nitrogen atom of benzoquinuclidine react very little, if at all, with the π-electrons of the adjacent benzene ring.

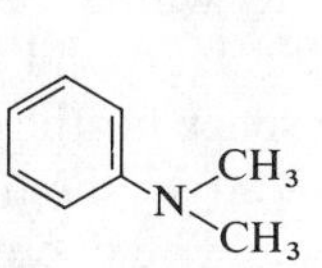

37 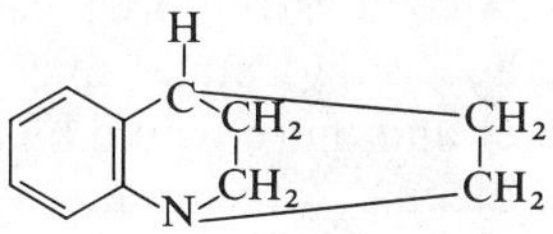 **38**

This isolation of the benzene ring from the amino group is also indicated by the close spectral resemblance of benzoquinuclidine, its salt, and tetralin as shown in Figure 3.4(186).

The effect of steric hindrance on the analysis of *N*,*N*-dialkylanilines can be shown by the reaction of 4-nitrobenzenediazonium fluoborate in

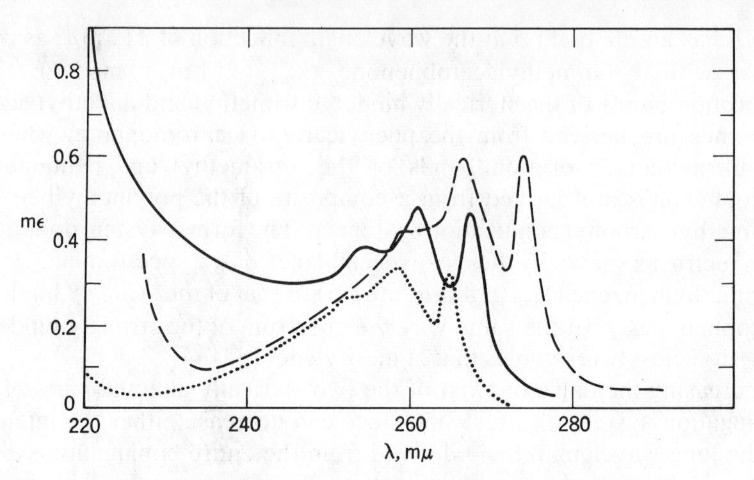

Fig. 3.4. Ultraviolet absorption spectra of benzoquinuclidine in isooctane (——), tetralin in isooctane (– – –), and benzoquinuclidine in ethanol–0·1 *N* HCl (· · · ·)(186).

sodium acetate solution with some aromatic amines: *N,N*-dimethylaniline, vivid color; *N,N*,2,6-tetramethylaniline, weak color; *N,N*-dimethyl-2-*tert*-butylaniline, weak color; and benzoquinuclidine, no color. In these compounds steric hindrance decreases both the yield and the molar absorptivity of the chromogen. The spectral limit of determination for sterically hindered test compounds would be poorer than that for analogous nonhindered reference compounds.

A different type of steric effect is shown by indigo, which absorbs at 595 mμ in xylene, whereas its *N*-methyl and *N,N*-dimethyl derivatives absorb at 639 and 644 mμ, respectively(187).

The loss of intramolecular hydrogen bonding would tend to cause a violet shift, and alkylation of the positive resonance terminal would cause a red shift. However, the substitution of the first methyl group would cause steric hindrance at the C_2—C_2' bridge bond. The excited state to which dipolar extreme structures contribute would be less sterically hindered than the ground state since the bridge bond would be more single in character and there would be more space for the methyl group(s) in the excited molecule. In the methyl derivatives the transition energy between the ground and the excited states would be decreased relative to indigo, and the molecules would be expected to show a red shift.

In isoenergetic dyes (e.g., symmetrical cyanines) any loss of coplanarity of the resonance terminals caused by steric hindrance will result in a shift to longer wavelength, accompanied by a decrease in intensity, the magnitude depending on the degree of twisting(188–190). This effect is discussed more fully in Chapter 9. However, such a phenomenon has been reported in the pyrazolone azomethine dyes(191). An example are the compounds **39**.

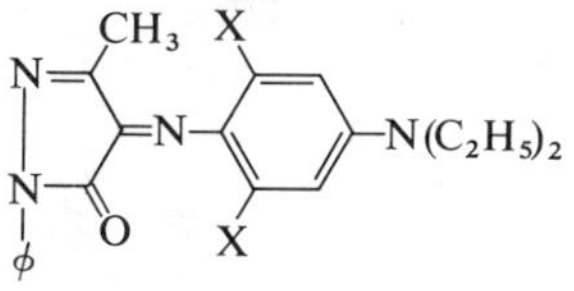

39

X	λ_{max} (mμ)	mϵ
H	520	~36
CH_3	640	12.0

An increase in steric hindrance by substituting $X = CH_3$ for H has caused a red shift and an intensity decrease. It would seem that there was less steric strain in the excited state than in the ground state.

Many examples of the hypochromic effect of steric hindrance on the absorption spectra of conjugated compounds are available in the literature (Table 3.20)(182, 192–224). Severe crowding can result in complete steric inhibition of resonance so that two chromophoric systems can be formed with their planes at approximately 90° to each other at the twisted single bond. This type of insulation can give rise to an additive spectrum of the two isolated chromophores; for example, the ultraviolet spectra of bimesityl and mesitylene or isodurene are similar but different from that of biphenyl(184, 225).

B. Effects of Ring Size

Where two conjugation systems or a conjugation system and a resonance terminal are connected by a single bond and also by a saturated grouping, for example,

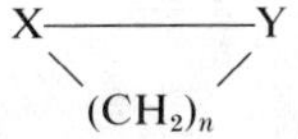

the geometry of the molecule, as well as the spectrum, will be strongly affected by the size of the partially saturated ring. This dramatic spectral change can be seen in Figure 3.5. Other examples are listed in Table 3.21. Comparison with reference compounds shows that an increase in n in the *ortho*-substituted compounds or a decrease in n in the *para*-substituted compounds causes the planes of the two conjugated parts of each molecule to approach perpendicular positions, with twisting and bending of the conjugating single bond.

TABLE 3.20
Effect of Steric Hindrance on Absorption Spectra

Reference compound	Solvent	λ_{max}	mϵ	Hindered compound	λ_{max}[a]	mϵ	Ref.
Butadiene[b]	Cyclohexane	217	21.0	1,1,3-Trimethylbutadiene	232	8.5	182
Cyclohexadiene[b]	—	256	8.0	1,3-Cyclononadiene	220	2.5	182
Styrene[b]	95% EtOH	248	14.6	α-Isopropylstyrene	234	7.8	192
Styrene[b]	95% EtOH	244	17.0	α,2-Dimethylstyrene	nm[c]	—	193
2-Methylstyrene[b]	—	245	12.6	2,4,6-Trimethylstyrene	245	7.0	182
1-Phenylcyclopentene	—	254	12.6	1-(2-Tolyl)cyclopentene	248	9.8	194
1-Phenylcyclohexene	—	248	11.1	1-(2-Tolyl)cyclohexene	nm[c]	—	194
1-Phenyl-2-methylcyclohexene	—	~241	5.6	1-(2-Tolyl)-2-methylcyclohexene	nm[c]	—	195
1-Phenyl-1,3-butadiene	95% EtOH	280	28.3	1-(2-Tolyl)-1,3-butadiene	281	25.0	196
Biphenyl[b]	—	249	14.5	2-*n*-Butylbiphenyl	233	9.0	197
Biphenyl[b]	95% EtOH	249	17.3	2,2′-Dichlorobiphenyl	230*i*	6.6	198
4,4′-Dimethylbiphenyl[b]	Petroleum ether	255	21.0	2,2′-Dimethylbiphenyl	227*i*	6.1	199
9,10-Dihydrophenanthrene	—	264	17.0	4,5-Dimethyl-9,10-dihydrophenanthrene	260	12.6	182
N,*N*-Dimethyl-4-aminostilbene[b]	95% EtOH	346	30.8	*N*,*N*-2,2′-Tetramethyl-4-aminostilbene	338	21.3	200
2,4,6-Trimethylstilbene	95% EtOH	289	18.9	2,4,6,2′,4′,6′-Hexamethylstilbene	264	8.3	200
4,4′-Dimethoxystilbene[b]	95% EtOH	307	29.8	α,α',2-Trimethyl-4,4′-dimethoxystilbene	236	17.5	201
1-(1-Cyclohexenyl)naphthalene	Cyclohexane	282	12.0	1-(6-Methyl-1-cyclohexenyl)naphthalene	283	6.8	202
2-(1-Cyclohexenyl)naphthalene	Cyclohexane	286	12.0	2-(6-Methyl-1-cyclohexenyl)naphthalene	284*s*	6.2	202
1-Acetylcyclohexene[b]	95% EtOH	232	12.4	1-Acetyl-2-*n*-butylcyclohexane	245	4.0	203
1-Acetyl-2-methylcyclohexene[b]		245	6.5	1-*tert*-Butyroyl-2-methylcyclohexene	239	1.3	204

1-Acetylcyclohexene semicarbazone	95% EtOH	260	24.0	1-Acetyl-2-methylcyclohexene semicarbazone	240	10.4	203
4-(Cyclohexen-1-yl)-3-buten-2-one[b]		281	20.8	3-Methyl-4-(2,6,6-trimethyl-2-cyclohexen-1-yl)-3-buten-2-one	278[d]	4.5	205
4-Methylacetophenone[b]	95% EtOH	253	14.1	2,4,6-Trimethylacetophenone	247	2.5	206[e]–208
α-Tetralone[b]	–	243	11.5	2,3-Benzocyclooctanone	243	6.5	209
Propiophenone[b]	–	242	13.5	α,α-Dimethylpropiophenone	239	8.3	182
Benzylideneacetone	–	286	22.0	2,6-Dimethylacetophenone	291	10.5	210
Benzophenone[b]	95% EtOH	253	18.0	2,2′,6,6′-Tetramethylbenzophenone	262	9.4	211
2-Isopropylbenzophenone[b]	95% EtOH	251	15.2	2,2′,4,4′,6,6′-Hexa-isopropylbenzophenone	264	11.7	211
Nitrobenzene[b]	Isooctane	250	8.9	2,3,5,6-Tetramethylnitrobenzene	250	0.99	212
N-Ethylaniline[b]	Isooctane	245	13.2	2,4,6-Tri-*tert*-butyl-*N*-ethylaniline	246	5.2	213
N,*N*-Dimethylaniline[b]	Isooctane	251	15.5	2-Isopropyl-*N*,*N*-dimethylaniline	248	4.3	213
Benzidine[b]	95% EtOH	350	25.0	2,2′-Dichlorobenzidine	350*s*	6.0	214
4-Nitroaniline[b]	95% EtOH	376	15.3	3,5-Di-*tert*-butyl-4-nitroaniline	401	0.54	213
3-Methyl-4-nitroaniline[b]	95% EtOH	~380	13.2	2,3,5,6-Tetramethyl-4-nitroaniline	395	1.56	212
N,*N*-Dimethyl-4-ethoxycarbonylaniline[b]	95% EtOH	~380	28.0	*N*,*N*-2,6-Tetramethyl-4-ethoxycarbonylaniline	~380	5.5	212
Acetanilide[b]	95% EtOH	242	14.4	2-Methylacetanilide	230	6.28	215
4-Nitrophenylacetonitrile	95% EtOH	262	66.0	3,5-Dimethyl-4-nitrophenylacetonitrile	250	14.8	216
4-Nitrophenol	Aqueous acid	317	10.0	4-Nitro-3,5-xylenol	359	1.7	217
4-Nitrophenol	95% EtOH	314	13.0	4-Nitro-2,6-xylenol	323	11.0	218
N,*N*-Dimethyl-4-aminobenzaldoxime		302	23.7	*N*,*N*-2,3-Tetramethyl-4-aminobenzaldoxime	289	11.4	219

TABLE 3.20 (*cont'd*)

Reference compound	Solvent	λ_{max}	mϵ	Hindered compound	λ_{max}	mϵ	Ref.
N,*N*-Dimethyl-4-cyanoaniline		294	27.2	*N*,*N*-2,3-Tetramethyl-4-cyanoaniline	293	12.3	219
2-Phenyl-5-sulfo-2*H*-naphtho-[1,2]triazole[b]		340	19.3	2-*o*-Chlorophenyl-5-sulfo-2*H*-naphtho[1,2]triazole	330	9.4	220
5-Nitro-2-furfural semicarbazone[b]	Water	375	15.8	5-Nitro-2-acetylfuran-2-methyl semicarbazone[f]	360[g]	8.1	221
N,*N*-Dimethylindoaniline	Cyclohexane	552	15.8	*N*,*N*-3,5-Tetramethylindoaniline	562	8.3	222
N,*N*-2-Trimethylindoaniline[b]	Acetonitrile	630	19.8	*N*,*N*-2′,6′-Tetramethylindoaniline	610	2.3	223
N,*N*-Dimethyl-4-phenylazoaniline[b]	95% EtOH	407	27.5	*N*,*N*-2,3-Tetramethyl-4-phenylazoaniline	377	17.4	219
N,*N*-Dimethyl-4-(4-nitrophenylazo)aniline[b]	95% EtOH	478	32.1	*N*,*N*-2,3-Tetramethyl-4-(4-nitrophenylazo)aniline	425	17.9	219
N,*N*-3-Trimethyl-4-phenylazoaniline	95% EtOH	375	18.2	*N*,*N*-Dimethyl-3-*tert*-butyl-4-phenylazoaniline[h]	428	0.79	224
N,*N*-3-Trimethyl-4-(4-nitrophenylazo)aniline[b]	95% EtOH	423	19.5	*N*,*N*-Dimethyl-3-*tert*-butyl-4-(4-nitrophenylazo)aniline[i]	450	1.5	224

[a] *i* = inflection; *s* = shoulder.
[b] Spectral data also given for other spectrally hindered derivatives of this compound.
[c] nm = No maximum in the 220 to 400 mμ region.
[d] Dienone band. Enone band at 228 mμ, mϵ 11.6, present also.
[e] Data from reference 206. References 207 and 208 also present data on steric hindrance in acetophenone derivatives.
[f] Solvent is acetonitrile–phosphate buffer (1 : 4), pH 6.3.
[g] Also bands derived from 2-nitrofuran chromophore.
[h] Absorption spectrum of this compound in ethanol closely resembles that of azobenzene.
[i] Absorption spectrum of this compound in ethanol closely resembles that of 4-nitroazobenzene.

TABLE 3.21
Effect of Ring Size on Absorption Spectra

Compound	Solvent	n	λ_{max} (mμ)[a]	mϵ	λ_{max} (mμ)[a]	mϵ	Ref.
Acetophenone			238	13.2	278	0.89	209
					286	0.71	
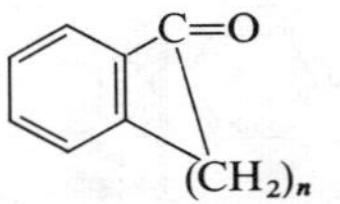 C=O, $(CH_2)_n$	*n*-Heptane	2	239	12.7	284	2.65	209, 227, 228
			244	10.6	292	2.55	
		3	243	11.5	286	1.55	
					296	1.43	
		4	240	9.0	281	1.2	
		5	243	6.5	286	1.1	
Acetophenone oxime			246	10.0			
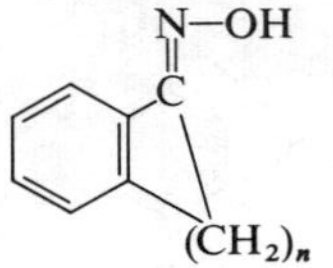 N—OH, C, $(CH_2)_n$	95% EtOH	2	253	12.3	288*fs*	5.0	227
		3	255	12.0	290*s*	2.5	
		4	237	9.6			
		5	~230	~3.0			
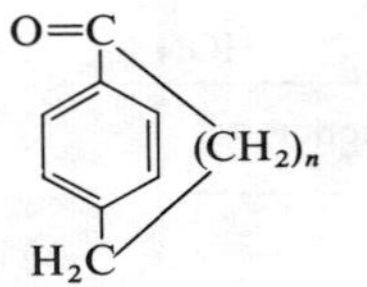 O=C, $(CH_2)_n$, H_2C	95% EtOH	7	250	6.6			209
		8	255	11.8			
		9	255	12.4			
		10	255	13.2			
		11	256	13.9			
		12	256	14.5			
4′-Methylacetophenone			253	14.7			

TABLE 3.21 (*cont'd*)

Compound	Solvent	n	λ_{max} (mμ)[a]	mϵ	λ_{max} (mμ)[a]	mϵ	Ref.
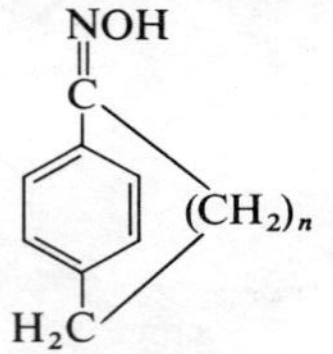 NOH=C–(p-phenylene)–H_2C, bridged by $(CH_2)_n$	95% EtOH	7	230	8.6			209
		8	241	11.3			
		9	243	12.2			
		10	245	12.4			
		11	246	12.3			
		12	246	13.0			
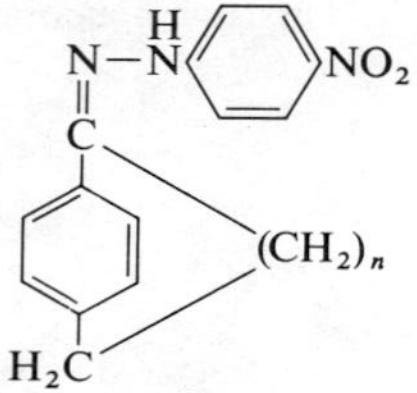 O_2N–C_6H_4–NH–N=C–(p-phenylene)–H_2C, bridged by $(CH_2)_n$	95% EtOH	7	393	27.1			209
		8	398	27.8			
		9	401	28.7			
		10	404	27.7			
		11	407	28.8			
		12	407	30.1			
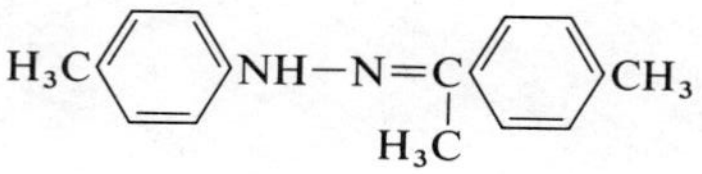 H_3C–C_6H_4–NH–N=C(CH_3)–C_6H_4–CH_3			407	29.3			
2′-Acetylacetanilide			258	31.6	333	7.08	229
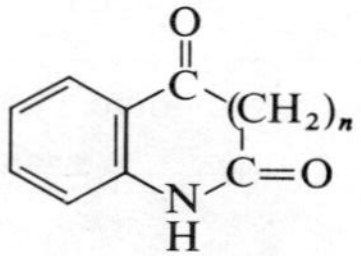 C_6H_4–C(=O)–$(CH_2)_n$–C(=O)–NH– (ring)	Ether	3	229	27.5	310	2.75	230
			250	8.2			
		4	~220	~13.9	280*s*	0.79	
			245*i*	~5			

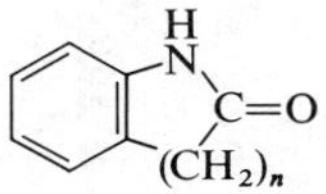	95% EtOH	1	~250	~9	280	~1.3	231
		2	~255	~12	280*i*	~2	
		3	~240	~11.5	267*s*	~1.3	
		4	~230	~9	268	~1	
		5	230*i*	~4	~265	~0.8	
Acetanilide			242	14	280	0.50	215
o-Toluidine			235	7.9	285	1.6	232
o-Xylene			210*s*	8.3	263	0.3	233
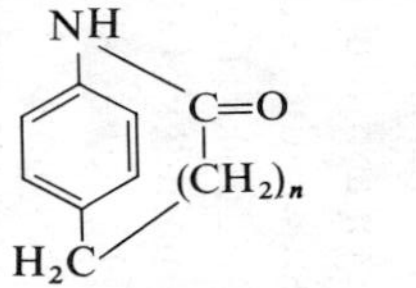	95% EtOH	8	228	9			234
		9	227	8.8			
		10	236	8.3			
		11	250	12.6			
		12	250	16.4			
		13	249	17.4			
4-Methylacetanilide			245	15.7			
p-Toluidine			236	7.9	291	1.26	232
p-Xylene			216	7.6	274*fs*	0.62	235
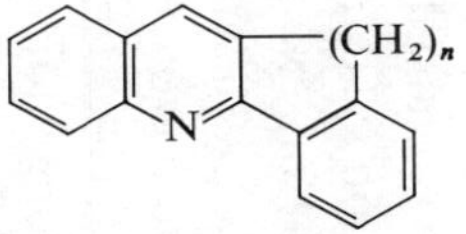	95% EtOH	1	~260	~43	~343*fs*	~18	226
		2	~265	~37	~345*fs*	~12	
		3	~240	~30	~328*fs*	~8	
			~255	~29			
		4	~238	~42	325*fs*	~6	
1,2,3,4-Tetrahydroacridine			~232	~33	318*fs*	~5	226
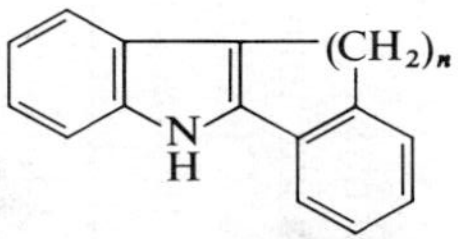	95% EtOH	1	~245	~23	~320	~25	226
		2	~250	~22.5	~330	~22.5	
		3	~230	~19.5	~317	~22.6	
			~245	20			
		4	~234	~25.5	~302	~18.0	
2,3-Dimethylindole			~230	~11	~285	~7.3	226

[a] *fs* = Fine structure; *s* = shoulder; *i* = inflection.

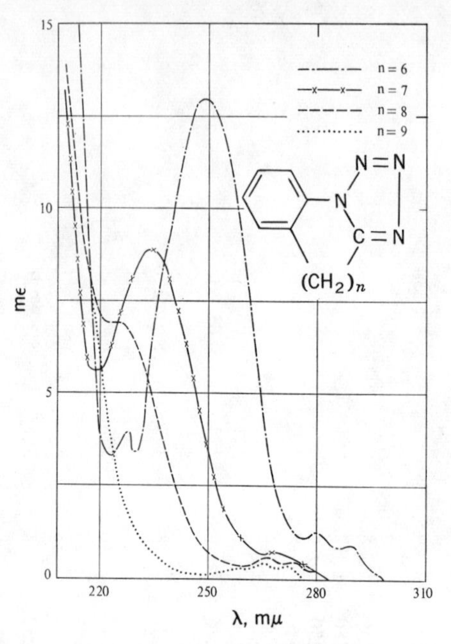

Fig. 3.5. Ultraviolet absorption spectra of 5,2′-polymethylene-1-phenyltetrazole in ethanol (226).

The absorption spectrum approaches that of a sum of the spectra of the two shorter conjugation systems. Most bands show the expected violet and hypochromic shifts.

IV. Insulation

A. Insulation by Saturated Carbon Atom(s)

Where one part of a chromophore is insulated completely or partially from another part the absorption spectrum will be affected. One common type of insulation is caused by a saturated carbon atom(s) interposed between two conjugation systems or between a conjugation system and a resonance terminal. For example, consider the compound $(C_2H_5)_2$-$NCH{=}CHCOC_2H_5$ (λ_{max} 307 mμ, mε 24.6) (236). Inserting a methylene group between the amino group and the remainder of the conjugation system, as in $(C_2H_5)_2NCH_2CH{=}CHCOCH_3$, gives a λ_{max} of 218 mμ, mε < 13.0 (237). Strong violet and hypochromic shifts result, and the new compound spectrally resembles $CH_3CH{=}CHCOCH_3$, λ_{max} 224 mμ, mε 9.78 (238). The long-wavelength bands of allylbenzene, 4-phenyl-1-butene, and ethylbenzene are identical in position and intensity, and differ in these respects from styrene (239). Other examples of insulation are

compounds **40** and **41**(240); the former is green blue in acid, the latter is colorless.
The carbonyl and amino groups are in conjugation in the colored compound **40** but not in the colorless one (**41**).

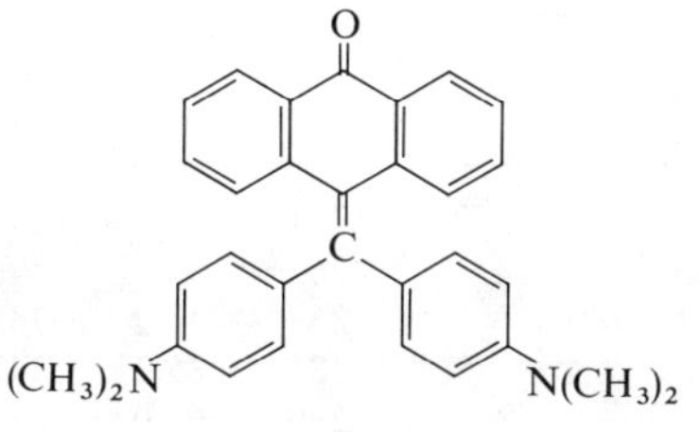

40 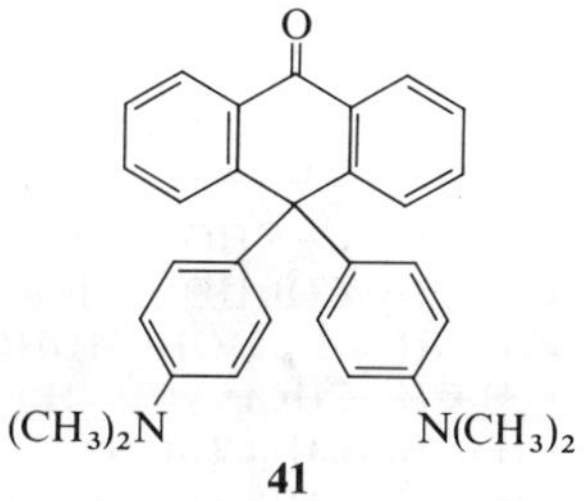 **41**

Comparing compound 2 in Table 3.22 with compounds 1, 3, and 4, 6 with 5 and 7, and 8 with 9 indicates that there is some interaction across one saturated carbon atom and much less across two. Where the molecule consists of two identical chromophores separated by an insulating group, the molar absorptivity is approximately twice that found in the molecule with only one of the chromophores.

However, it is possible for two chromophoric groups in a compound to be insulated and neutral in the ground state and oppositely charged in the excited state. Two types have been described(244). In transannular conjugation there can be orbital overlap in the usual π fashion (parallel orbitals) when the electron-acceptor and electron-donor groups are nonconjugated in the classical sense but are suitably oriented, as, for example, in equation 3.11.

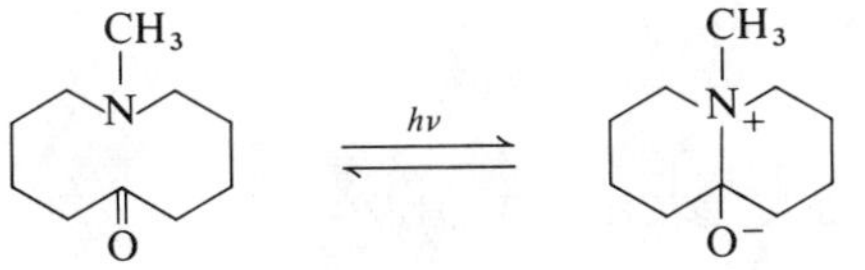

$$\mathbf{42} \underset{}{\overset{h\nu}{\rightleftharpoons}} \mathbf{43} \qquad (3.11)$$

λ_{max} 221 mμ, mϵ 5.61

Homoconjugation is stated to arise when the orbital overlap of an unsaturated carbon of the chromophore containing the electron-acceptor grouping and an unsaturated carbon of the chromophore containing the electron-donor grouping is crosswise, that is, partially sigma in character. Enhanced $\pi \rightarrow \pi^*$ bands result. An example is the following:

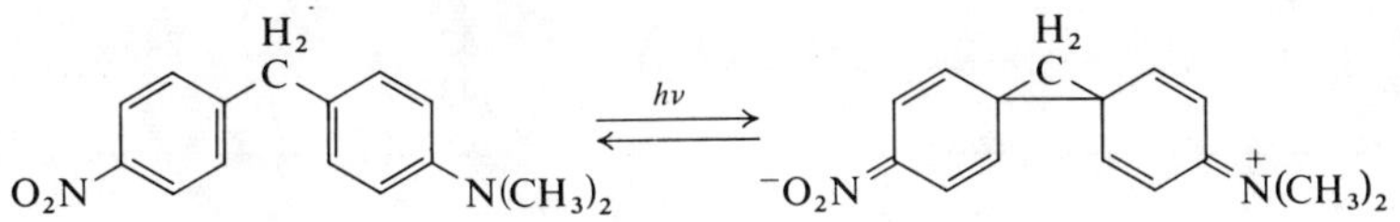

$$\mathbf{44} \underset{}{\overset{h\nu}{\rightleftharpoons}} \mathbf{45} \qquad (3.12)$$

TABLE 3.22
Reaction across an Insulating Carbon Group

	Compound	λ_{max}(mμ)	mϵ	Ref.
1.	$C_6H_5CH{=}CHCHOH$	251	16.4	241
2.	$C_6H_5CH{=}CHCHOHCH{=}CHC_6H_5$	263	33.2	241
3.	$C_6H_5CH{=}CHCHOHCHOHCH{=}CHC_6H_5$	257	37.2	241
4.	$C_6H_5CH{=}CHCH{=}CHCH_2OH$	286	35.2	241
5.	$CH_3(CH{=}CH)_4CH_2OH$	299	64.0	241
6.	$CH_3(CH{=}CH)_4CHOH(CH{=}CH)_4CH_3$	321	–	241
7.	$CH_3(CH{=}CH)_4CHOHCHOH(CH{=}CH)_4CH_3$	304	–	241
8.	$C_6H_5CH{=}NCH_3$	247	17.2	242
9.	$C_6H_5CH{=}NCH_2CH_2N{=}CHC_6H_5$	247	29.8	242
10.	$C_6H_5CH{=}NC_6H_5$	263	16.8	242
11.	$C_6H_5CH{=}N{-}C_6H_4{-}N{=}CHC_6H_5$	267	33.8	242
12.	$C_6H_5CH{=}CHC_6H_5$	295	26.3	242
13.	$C_6H_5CH{=}CHC_6H_4CH{=}CHC_6H_5$	298	52.7	242
14.	$C_6H_5N{=}NC_6H_5$	318 442[a]	22.0 0.54	243
15.	$C_6H_5N{=}NC_6H_4N{=}NC_6H_5$	320 435[a]	34.0 1.45	243
16.	$C_6H_5N{=}NC_6H_4N{=}NC_6H_4N{=}NC_6H_5$	320 433[a]	46.0 2.75	243

[a] $n \rightarrow \pi^*$ Band.

From an analyst's point of view one important factor in these phenomena is the large increase in intensity in many of these molecules. This hyperchromic shift makes these compounds more readily analyzable. Other examples are listed in Table 3.23, in which several compounds are compared with reference substances so that the red and hyperchromic shifts can be more readily comprehended.

B. Insulation by Hetero Atoms

Hetero atoms can insulate two chromophoric systems in some compounds and in others can act as a partly sterically hindered —CH=CH— group. Thus in 4-nitrophenylsulfide and in 4-nitro-4′-aminophenylsulfide the long-wavelength band is at 337 mμ, mε 13.5, and 344 mμ, mε 12.6 (250). Since 4-nitro-4′-aminostilbene has its long-wavelength band at 405 mμ, mε 25(251), the conjugative properties of the —S— compared with the —CH=CH— are negligible in this case.

On the other hand, in such compounds as divinyl sulfide the sulfide group appears to have conjugate properties that are similar to those of the —C=C— group (Table 3.24)(252). (This conjugative property of hetero groups is discussed in Section VII of Chapter 6.)

C. Insulation through Meta Substitution

Some examples of insulation through *meta* substitution are listed in Table 3.22. A classical example of insulation can be found in the *m*-polyphenyls (Table 3.25). However, some type of interaction takes place when a strong electron-donor and a strong electron-acceptor group are present as *meta* substituents (Table 3.26). By no means is the spectrum of *m*-nitroaniline the summation of the spectra of nitrobenzene and aniline nor is the spectrum of *m*-phenylazoaniline the summation of the spectra of nitrobenzene and aniline.

D. Miscellaneous Insulation

Consider the case of an allene group separating two conjugating systems. In the allene group the three carbons lie in a straight line since the middle carbon uses *sp* hybrid orbitals. The remaining two *p*π-orbitals on the middle carbon atom are at right angles to each other and overlap the analogous orbitals of the carbons 1 and 3 on each side. In order to form the two π-bonds at right angles, carbon groups 1 and 3 are forced into right angles to each other. The central carbon atom of the allene serves as an electronic insulator between the two resulting conjugated systems (259, 260). The effect is to break the conjugation; one double bond in

TABLE 3.23
Spectral Data for Compounds Showing Electronic Interaction between Nonconjugated Groups

Compound[a]	$\lambda_{max}(m\mu)$[b]	$m\epsilon$	Ref.
CH_3COCH_3	195	0.9	244
$H_2C{=}CH_2$	174	5.0	245
1. $H_2C{=}$(cyclobutane)${=}O$	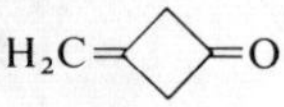 214	1.5	244
2. (cyclodecenone, O)	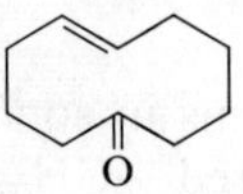 215	2.3	244
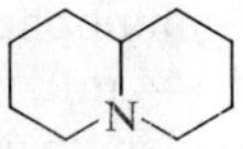 (N, bicyclic amine)	215	3.1	246
3[c]. (CH_3–N, O)	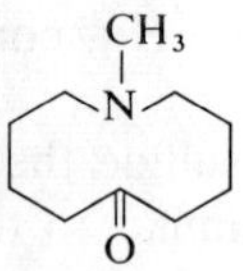 221	5.6	244
$(CH_3)_2S$	210 229*s*	1.02 0.14	247
4. (O=C, S ring)	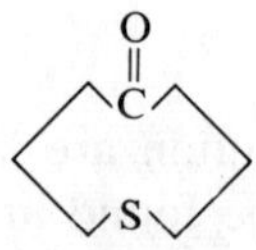 238	2.54	244
5. (bicyclic, O)	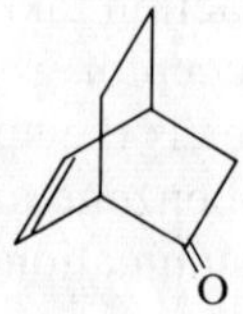 202	3.0	244
6. (bicyclic, O, O)	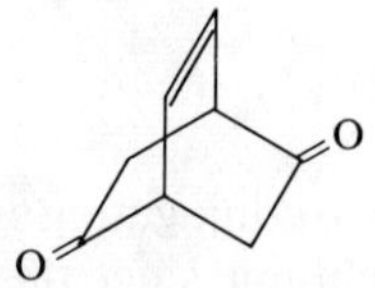 202	3.0	244
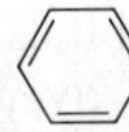 (benzene)	198	7.9	248
7. (O, benzobicyclic)	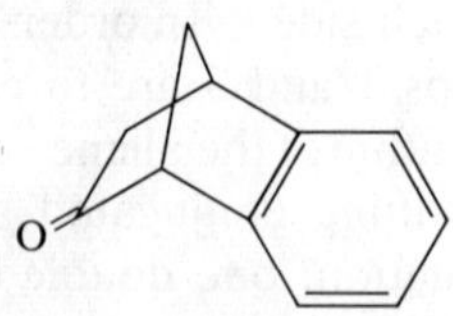 210	10.7	244

TABLE 3.23 (*cont'd*)

	Compound[a]	$\lambda_{max}(m\mu)$[b]	$m\epsilon$	Ref.
	$O_2N-C_6H_4-CH_3$	274	9.5	244
	$H_3C-C_6H_4-OCH_3$	277 286	2.19 1.79	244
8.	$O_2N-C_6H_4-CH_2-C_6H_4-OCH_3$	280 287	24.4 16.9	244
	$H_3C-C_6H_4-OCH_3$ (meta)	278	1.65	244
	$H_3C-C_6H_4-NO_2$ (meta)	264	7.75	244
9.	$O_2N-C_6H_4-CH_2-C_6H_4-OCH_3$ (meta, meta')	266 282	14.9 7.2	244
	Dibenzobicyclo compound with CH_2-CH_2 bridge	[d]		249
	Dibenzobicyclo compound with $C{=}O-CH_2$ bridge	[d]		249
10.	Dibenzobicyclo compound with $C{=}O-C{=}O$ bridge	245 253	28.0 46.0	249

[a]Compounds that are not numbered are reference compounds.
[b]*s* = Shoulder.
[c]Compound **42** in text.
[d]No bands present in the ultraviolet–visible region with $m\epsilon > 3.50$.

TABLE 3.24
Conjugation in Divinyl Sulfide(252)

Compound	$\lambda_{max}(m\mu)$	mϵ	Compound	$\lambda_{max}(m\mu)$	mϵ
$CH_2{=}CH{-}S{-}CH{=}CH_2$	220	2.0	$CH_2{=}CH{-}S{-}CH_3$	225	16.0
	255	5.6		240*s*	10.0
	260	4.0	$CH_2{=}CHCH{=}CH_2$	217	20.0
	275	5.0			
$CH_2{=}CHCH{=}CHCH{=}CH_2$	248	56.0			
	258	79.0			
	268	68.0			

the allenic group is conjugated with an unsaturated system in one plane, whereas the other allenic group is conjugated with another planar system rotated by 90° from the former plane. Thus tetraphenylallene which absorbs at λ_{max} 267 mμ, mϵ 12.0, is closely similar spectrally to 1,1-diphenylethylene (λ_{max} 250 mμ, mϵ 11.0) and not to *trans,trans-diphenyl-*1,3-butadiene (λ_{max} 328 mμ, mϵ 56.0)(260). A glance at one of the dipolar resonance structures of tetraphenylallene (**46**) clearly shows the insulation of the central carbon atom and its role as a resonance terminal.

TABLE 3.25
Insulation in the *m*-Polyphenyls
(253, 254)

n^a	$\lambda_{max}(m\mu)^b$	mϵ	m$\epsilon/n+1$
0	252	18.3	18.3
1	252	44.0	22.0
2	249[c]	59.0	19.7
3	249[c]	79.5	19.9
4	249[c]	102.0	20.4
6	249[c]	146.0	20.9
7	253	184.0	23.0
8	253	213.0	23.7
9	253	215.0	21.5
10	253	233.0	21.2
11	253	252.0	21.0
12	253	283.0	21.8
13	254	309.0	22.1
14	255	~320.0	21.3

[a]n = number of m-phenylene groups.
[b]Measured in chloroform unless otherwise noted.
[c]In cyclohexane.

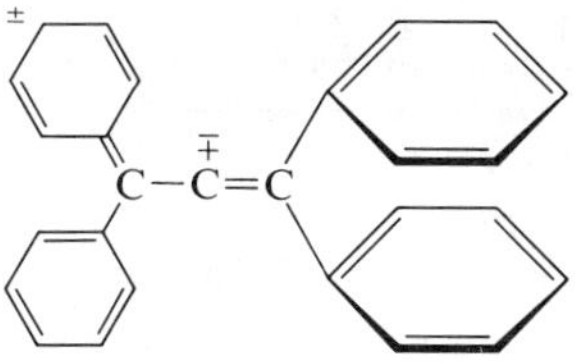

$\overset{\pm}{}$ C$-\overset{\mp}{C}$=C

46

Another example of an allene group as an effective insulator can be shown for mycomycin when it is compared with an ene-di-yne, a triene, and a diene-tri-yne (Table 3.27). The absorption spectrum of mycomycin more closely resembles the spectra of the ene-di-yne and the triene than it does that of the somewhat similarly unsaturated isomycomycin. A classical zwitterionic resonance structure cannot be written for the entire chain of unsaturation in mycomycin.

So far evidence would indicate that the substitution of a silicon atom for a carbon atom in a chain confers insulation at that point; for example, $[(CH_3)_2NC_6H_4]_3SiCl$, unlike the carbon analog, is colorless in all examined solutions(263). In the same way attempts to form a colored triaryl silicon cation from $[4\text{-}(CH_3)_2NC_6H_4[(C_6H_5)_2SiOH$ have not been successful(264).

Partial insulation by interposing a *para*-substituted benzene or other ring between two conjugated systems or by replacing a double bond with a triple bond is discussed in the sections on quinonoid and conjugation chains.

TABLE 3.26
Interaction between *Meta* Substituents

Compound[a]	$\lambda_{max}(m\mu)$		(mε)	Ref.
Nitrobenzene	260 (7.9),	340*s* (0.2)		255
Aniline	234 (11.5),	285 (1.74)		256
o-Nitroaniline	231 (16.6),	276 (4.9),	404 (5.3)	257
m-Nitroaniline	235 (15.9),		373 (1.5)	257
p-Nitroaniline	229 (6.3),		372 (15.2)	257
Azobenzene	228 (14.0),	318 (21.9),	442 (0.54)[b]	243
o-Phenylazoaniline	~275 (6.6),	313 (16.0),	414 (6.6)	258
m-Phenylazoaniline	~262 (8.9),	~313 (12.6),	414 (1.33)[c]	258
p-Phenylazoaniline	248 (8.9)		384 (23.4)	243

[a]Except for *o*- and *m*-phenylazoaniline, all in 95% ethanol.
[b]$n \rightarrow \pi^*$ Band.
[c]Partial contribution from hidden $n \rightarrow \pi^*$ band.

TABLE 3.27
Insulation by an Allene Group

Compound	λ_{max}(mμ)	mϵ	Ref.
H(C≡C)$_2$CH=C=CH(CH=CH)$_2$CH$_2$COOH[a]	256	35.0	260
	267	61.0	
	281	67.0	
H(C≡C)$_2$CH=CHCH$_3$	235	5.0	261
(in ether)	248	9.8	
	263	14.8	
	279	10.2	
CH$_3$(CH=CH)$_3$H	248	33.9	262
(in isooctane)	258	46.8	
	268	38.0	
CH$_3$(C≡C)$_3$(CH=CH)$_2$CH$_2$COOH[b]	246	23.0	260
(in ether)	258	58.0	
	267	11.0	
	288	13.0	
	306	77.0	
	324	42.0	
	347	34.0	

[a]Mycomycin.
[b]Isomycomycin.

V. Geometry

A. Monoene and Polyene Stereoisomers

The spectra of interconvertible geometrical isomers show some striking differences (Table 3.28). The simplest of this type are the *trans* and *cis* isomers. During a *trans* → *cis* rearrangement the long-wavelength band, derived from the entire π-electron system, usually undergoes a hypochromic shift, sometimes with a loss in fine structure. An example is given by *trans*-stilbene (**47**) and its *cis* isomer (**48**).

47 **48**

Since *trans*-stilbene is coplanar there is complete conjugation across the molecule. The *cis* isomer is not coplanar because the *ortho*-hydrogen atoms would overlap. This forces the phenyl rings to be rotated out

TABLE 3.28
Absorption Spectra of Geometrical Isomers

Compound[a]	λ_{max} (mμ)[b]	mϵ	Ref.
cis,cis-2,4-Hexadien-1-ol	231	17.0	265
cis,trans-2,4-Hexadien-1-ol	230	23.0	265
trans,cis-2,4-Hexadien-1-ol	230	21.4	265
trans,trans-2,4-Hexadien-1-ol	228	25.1	265
cis-3,5-Hexadien-2-ol	224	20.0	266
trans-3,5-Hexadien-2-ol	223	28.0	266
cis-5,7-Octadien-2-one	228	15.1	267
trans-5,7-Octadien-2-one	228	22.4	267
5-Chloro-*cis*-3-hexen-1-yne[c]	226	12.3	268
5-Chloro-*trans*-3-hexen-1-yne[c]	227	15.1	268
cis,cis-2,4-Decadienol	233	20.0	269
cis,trans-2,4-Decadienol	232	24.0	269
trans,cis-2,4-Decadienol	232	22.0	269
trans,trans-2,4-Decadienol	230	31.0	269
cis-3-Decen-2-one	223	9.1	270
trans-3-Decen-2-one	226	12.3	270
cis-2-Octadecenoic acid[d]	213	16.0	271
trans-2-Octadecenoic acid[d]	213	20.0	271
N-Isobutyl-*cis*-2-decenamide	226	10.0	272
N-Isobutyl-*trans*-2-decenamide	226	10.2	272
α-Chloro-*cis*-crotonic acid	228	7.08	273
α-Chloro-*trans*-crotonic acid	222	10.5	273
β-Chloro-*cis*-crotonic acid[d]	226	43.7	274
β-Chloro-*trans*-crotonic acid[d]	221	20.0	274
4-Methoxy-*cis*-3-penten-2-one[c]	245	12.3	275
	262	14.1	275
4-Methoxy-*trans*-3-penten-2-one[e]	247	12.0	275
	254	13.5	275
Methyl β-acetylthio-*cis*-acrylate	269	18.6	276
	277	17.0	
Methyl β-acetylthio-*trans*-acrylate	267	18.2	276
cis-α-Ionone	234	6.6	277
trans-α-Ionone	228	14.1	277
cis,cis-2,4-Decadienoic acid	260	12.9	269
cis,trans-2,4-Decadienoic acid	258	17.4	269
trans,cis-2,4-Decadienoic acid	259	16.2	269
trans,trans-2,4-Decadienoic acid	257	27.5	269
N-Isobutyl-*cis,cis*-2,4-decadienamide	258	17.4	269
N-Isobutyl-*cis,trans*-2,4-decadienamide	259	23.4	269
N-Isobutyl-*trans,cis*-2,4-decadienamide	258	26.3	269
N-Isobutyl-*trans,trans*-2,4-decadienamide	258	29.5	269
cis-4-Hexen-2-ynal[d]	256	8.3	278
	264	10.7	
trans-4-Hexen-2-ynal[d]	254	10.5	278
	262	13.5	
	274	11.2	

TABLE 3.28 (*cont'd*)

Compound[a]	λ_{max} (mμ)[b]	mϵ	Ref.
cis-3-Hexen-5-yn-2-one	256	10.5	279
trans-3-Hexen-5-yn-2-one	258	16.6	279
	330	0.056	
cis,cis-Sorbic acid	259	20.0	280
cis,trans-Sorbic acid	257	20.9	280
trans,cis-Sorbic acid	260	22.4	280
trans,trans-Sorbic acid	254	26.9	280
cis,cis-Muconamic acid	264*fs*	22.9	281
cis,trans-Muconamic acid	264*fs*	28.8	281
trans,cis-Muconamic acid	264*fs*	21.4	281
trans,trans-Muconamic acid	264*fs*	30.9	281
anti-Furfural oxime	270	15.9	282
syn-Furfural oxime	265	17.8	282
anti-Ethyl 2-furyl ketoxime	262	14.0	282
syn-Ethyl 2-furyl ketoxime	266	24.0	282
cis-β-Methylstyrene[c]	242	13.0	283
	280	0.32	
trans-β-Methylstyrene[c]	249	17.0	283
	283	1.0	
cis-Cinnamic acid	257	12.9	284
trans-Cinnamic acid	268*fs*	20.4	284
α-Methyl-*cis*-cinnamic acid	257	15.9	284
α-Methyl-*trans*-cinnamic acid	261	16.6	284
β-Methyl-*cis*-cinnamic acid	248	7.6	284
β-Methyl-*trans*-cinnamic acid	257	14.5	284
Methyl *cis*-β-benzoylacrylate	251	12.0	285
Methyl *trans*-β-benzoylacrylate	235	12.0	285
	268	8.9	
cis-Chalcone[f]	247	14.0	286
	289	8.9	
trans-Chalcone[f]	225	12.0	286
	298	23.4	
cis-α,β-Diphenylacrylic acid	220	16.0	287
	274	14.0	
trans-α,β-Diphenylacrylic acid	228	13.0	287
	294	24.0	
cis-1,2-Dibenzoylethylene	260	18.2	286
trans-1,2-Dibenzoylethylene	269	18.2	286
cis-1-Phenylbutadiene	267	20.0	288
trans-1-Phenylbutadiene	280*fs*	32.0	288
cis-1-Naphthylacrylic acid[g]	307	6.9	289
trans-1-Naphthylacrylic acid[g]	317	11.5	289
cis-2-Naphthylacrylic acid[g]	300	13.0	289

TABLE 3.28 (*cont'd*)

Compound[a]	λ_{max} (mμ)[b]	mϵ	Ref.
trans-2-Naphthylacrylic acid[g]	308*fs*	25.0	289
cis-Stilbene	280	10.5	290
trans-Stilbene	295*fs*	28.9	290
cis-α-Fluorostilbene[d]	285	12.3	291
trans-α-Fluorostilbene[d]	291*fs*	28.2	291
cis-α-Nitrostilbene	316	12.0	292
trans-α-Nitrostilbene	281	21.4	292
cis-2-Nitrostilbene	260	13.0	290
trans-2-Nitrostilbene	274	20.0	290
cis-4-Nitrostilbene	329	7.9	290
trans-4-Nitrostilbene	342	18.6	290
cis-4-Methoxy-4′-nitrostilbene	260	15.5	251
	353	9.0	
trans-4-Methoxy-4′-nitrostilbene	248	12.0	251
	372	27.5	
cis-4-Amino-4′-nitrostilbene	282	18.0	251
	395	7.0	
trans-4-Amino-4′-nitrostilbene	290	11.0	251
	410	25.0	
cis-Benzalphthalide	324	12.8	293
trans-Benzalphthalide	340*fs*	20.8	293
cis-1-Phenyl-1-buten-2-one[g]	279	9.5	294
trans-1-Phenyl-1-buten-2-one[g]	279	21.0	294
cis,cis-1,4-Diphenyl-1,3-butadiene[d]	299	29.5	295
cis,trans-1,4-Diphenyl-1,3-butadiene[d]	240	16.0	295
	312	32.0	
trans,trans-1,4-Diphenyl-1,3-butadiene[d]			295
	327*fs*	63.0	
cis-(1)-1,6-Diphenyl-1,3,5-hexatriene[d]	337	49.0	296
cis-(3)-1,6-Diphenyl-1,3,5-hexatriene[d]	350	51.3	296
cis,cis-(1,3)-1,6-Diphenyl-1,3,5-hexatriene[d]	335	46.8	296
all-*trans*-1,6-Diphenyl-1,3,5-hexatriene[d]	351	72.4	296
cis-*p*-Distyrylbenzene[d]	328	40.0	297
trans-*p*-Distyrylbenzene[d]	350*fs*	60.0	297
cis-*p*-Styryl-1,4-diphenyl-1,3-butadiene[d]	330*fs*	32.0	297
trans-*p*-Styryl-1,4-diphenyl-1,3-butadiene[d]	367*fs*	72.0	297
all-*trans*-1,8-Diphenyloctatetraene[d]	372*fs*	108.0	298
middle-*cis*-1,8-Diphenyloctatetraene[d]	274	28.0	298
	283	28.0	
	370*fs*	76.5	
1-*cis*-1,8-Diphenyloctatetraene[d]	272	13.0	
	280	13.0	
	364	70.0	298
1,3(or 1,5-)-di-*cis*-1,8-Diphenyloctatetraene[d]	360	67.0	298
cis-9-Cinnamylidenefluorene[d]	358	34.7	299
trans-9-Cinnamylidenefluorene[d]	373*fs*	46.8	299
11,11′-di-*cis*-β-Carotene[h]	400	39.8	300

TABLE 3.28 (*cont'd*)

Compound[a]	λ_{max} ($m\mu$)[b]	mϵ	Ref.
15,15′-*cis*-β-Carotene[h]	450*fs*	100.0	300
all-*trans*-β-Carotene[h]	450*fs*	159.0	300
cis-Thioindigo[i]	490	20.0	301
trans-Thioindigo[i]	546	16.0	301
cis-*N*,*N*′-Diacetylindigo[i]	431	4.2	302
trans-*N*,*N*′-Diacetylindigo[j]	520	5.3	302
cis-*N*,*N*′-Bis(trifluoroacetyl)indigo[j]	405	4.3	302
trans-*N*,*N*′-Bis(trifluoroacetyl)indigo[j]	535	6.45	302
cis-Azobenzene[k,l]	281	5.26	303
	433[m]	1.5	
trans-Azobenzene[k,l]	320	21.3	303
	444[m]	0.51	
cis-4-Phenylazoaniline[k,l]	332	9.2	304
	450[m]	2.7	
trans-4-Phenylazoaniline[k,l]	377	25.8	304
cis-*N*,*N*-Dimethyl-4-phenylazoaniline[l]	362	12.0	304
	460[m]	4.3	
trans-*N*,*N*-Dimethyl-4-phenylazoaniline[l]	410	28.3	304
cis-4-*p*-Tolylazophenol[l]	304	7.7	304
	446[m]	1.7	
trans-4-*p*-Tolylazophenol[l]	347	25.6	304
	440	0.9	
syn-α-Methoxypropiophenone 2,4-dinitrophenylhydrazone[i]	383	28.1	305, 306
anti-α-Methoxypropiophenone 2,4-dinitrophenylhydrazone[i]	258	13.6	305,
	362	24.7	306

[a] In 95% ethanol, unless otherwise specified.
[b] *fs* = Fine structure.
[c] Solvent not specified.
[d] In *n*-hexane.
[e] In *n*-heptane and methanol, respectively.
[f] In isooctane.
[g] In methanol.
[h] In petroleum ether.
[i] In chloroform.
[j] In carbon tetrachloride.
[k] And many other *cis* and *trans*-azobenzene derivatives.
[l] In benzene.
[m] $n \rightarrow \pi^*$ Band.

of each other's plane and produces violet and hypochromic shifts. The decrease in the straight-line distance between the resonance terminals contributes also to the hypochromic shift. This effect is discussed in Chapter 7.

Examination of the data in Table 3.28 indicates that the most important differences between the *cis* and *trans* and the *syn* and *anti* isomers is the difference in the intensities of the long-wavelength bands. The *trans* isomer usually has the higher molar absorptivity. This is probably because the straight-line distance between resonance terminals is longer in the *trans* isomer (see Chapter 7). In nonhindered compounds, in which this distance is shorter in the *trans* isomer, the *cis* isomer has a higher molar absorptivity, as in the *trans* (**49**) and *cis* (**50**) isomers of thioindigo, where the S is the positive resonance terminal and the C=O is the negative resonance terminal.

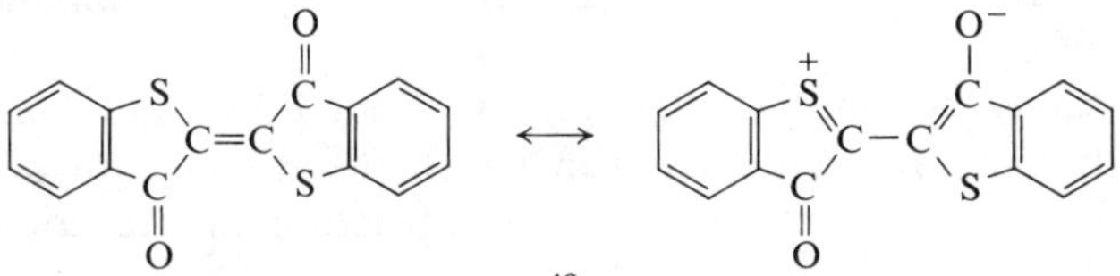

49

λ_{max} 546 mμ, mϵ 16.0

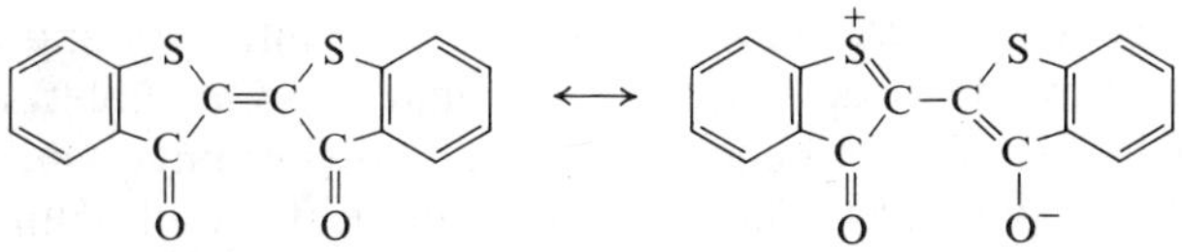

50

λ_{max} 490 mμ, mϵ 20.0

For each isomer only one of the contributing dipolar structures is shown.

In addition, when steric hindrance is present in the *cis* and not in the *trans* isomer, the *cis* isomer has a lower molar absorptivity. If the steric hindrance is large enough, the *cis* isomer shows a violet shift also, as shown in the *cis* isomers of cinnamic acid, chalcone, 1-phenylbutadiene, naphthylacrylic acid, stilbene, benzalphthalide, 1,4-diphenyl-1,3-butadiene, 1,6-diphenyl-1,3,5-hexatriene, *p*-distyrylbenzene, and azobenzene as compared to their *trans* isomers. They usually have less fine structure in their long-wavelength bands than do nonhindered *trans* isomers.

Since polyene molecules, such as the carotenes, have a long chain of conjugation, a single *trans* → *cis* shift can drastically alter their overall shape. The rearrangement of all-*trans* carotenoids yields mainly mono-

cis and di-*cis* forms(307) and these would be the unhindered types. The hindered types (with a methyl group in a hindering position), such as **51** would take much more energy to form. In the case of β-carotene only 20 unhindered isomers can be expected and 12 have been observed(308).

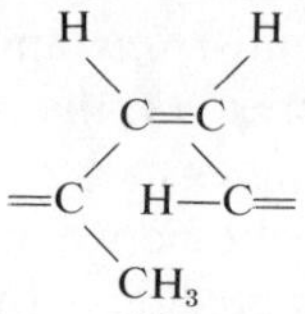

```
      H     H
       \   /
        C=C
       /   \
   =C   H—C=
     \
      CH3
```

51

Many of the carotenoid stereoisomers are sensitive to heat to a varying extent. In the presence of light all known sterically hindered *cis*-carotenoids are extremely sensitive to iodine. They reach stereochemical equilibrium at a faster rate than do the all-*trans*, or unhindered, *cis* isomers(309).

Iodine-catalyzed photoisomerizations about ethylenic double bonds are a fairly general phenomenon in which the wavelength of the light, solvent system, temperature, pH, and structure of the compound all play a vital role.

An invaluable help in characterizing a polyene is to record its spectra before and after catalysis by iodine. In any all-*trans* → *cis* rearrangement the visible spectral curve is altered. The intensity and the degree of fine structure decrease, and the maxima show a violet shift. The validity of this statement concerning the long-wavelength shift has been tested with lycopene of which 40 spectroscopically well-defined isomers are known(310).

The all-*trans* isomer of lycopene absorbs at wavelengths from 4 to 72 mμ further into the visible than do the various *cis* isomers. In addition the molar absorptivity of the all-*trans* compound is higher than that of any of its *cis* isomers. This rule has no general validity, as can be seen from the data in Table 3.28 and the report that the coplanar all-*cis*-deca-2,4,6,8-tetraene absorbs at longer wavelengths than does the all-*trans* form(311).

When the solution of an all-*trans* carotenoid is treated with catalytic amounts of iodine and then exposed to light, a "*cis* peak" appears in the 320- to 380-mμ region(309). The best media for the observation of these peaks are the nonpolar hydrocarbon solvents. The *cis* peak of C_{40}-carotenoids containing 10 or 11 conjugated double bonds is shifted toward the violet 142 ± 2 mμ (hexane solution) as compared with the long-wavelength maximum of the all-*trans* isomer(309).

The intensities of the long-wavelength and *cis* bands for all mono-*cis* isomers lie between the extremes of the all-*trans* and central mono-*cis* forms, the *cis* band being most intense in the central mono-*cis* isomer. As more *cis* double bonds are introduced, the *cis* peak will decrease in intensity and disappear as the molecule straightens out. For example, all-*trans*-lycopene has its long-wavelength band at 472.5 mμ, mϵ 186.0, and no *cis* band; central mono-*cis*-lycopene has its long-wavelength band at 500 mμ, mϵ ~71, and a *cis* band at 360 mμ, mϵ 65; and most of the poly-*cis*-lycopenes have no *cis* band(309). These results indicate that the electronic transition from which the *cis* band is derived has its dipole moment perpendicular to the larger dipole moment of the main transition derived from the entire conjugation system.

B. Azobenzene Stereoisomers

In the azobenzene as in the polyene series stereoisomers can be separated by chromatography. Consequently spectral data for the stereoisomers can be obtained. However, the half-lives of *cis* isomers vary widely, depending on the compound, solvent, solution temperature, and irradiation effects. Half-lives of many *cis* species are known; they vary from 3.3 min for *p*-phenylazophenol to more than 100 hr for azobenzene (312). In benzene or acetone solutions the half-life of the *cis* isomer of *p*-phenylazoanisole is 600 times longer than that of *p*-phenylazophenol. The addition of traces of methanol to benzene solutions of *p*-phenylazophenol or *p*-phenylazoaniline greatly reduces the half-lives of the *cis* isomers(312).

The importance of hydrogen bonding in the stabilization of the *trans* isomers can be seen from the following reports. The phototropism of *p*-phenylazophenols and *p*-phenylazoanilines is suppressed by hydroxylic solvents as a result of the increased rate of thermal isomerization to the *trans* isomer in these solvents(313). The *cis* isomer of sodium *m*-(4-hydroxyphenylazo)benzenesulfonate has a half-life of 21 sec in acetone; addition of small amounts of water decreases the half-life [e.g., 0.25 and 1.25% water gave half-lives of 14 and 5 sec, respectively(313)]. On the other hand, the *cis*-methoxy analog has a half-life of about 1 week in water at room temperature, whereas the half-life in acetone at room temperature is about 36 hr. Similarly *cis*-azobenzene is more stable in water than in acetone(312).

The rapid photochemical isomerization of azo dyes in benzene solution has been described(304). The percentage of *cis* isomer increases with increasing difference in spectra between *cis* and *trans* forms and decreasing rate of thermal *cis*-to-*trans* reaction. The effective filter combinations for optimum conversion are those transmitting light of a wavelength at which there is the greatest percentage difference between the absorbance of the two forms(313). In the absence of thermal isomerization the concentration of the two isomers of various azo compounds at photochemical equilibrium is largely determined by the relative absorbancies of the two forms at the wavelength of illumination used(313–316).

Hartley's results indicate the importance of solution temperature in the *cis* → *trans* conversion of the azo compounds(317). The half-life of *cis*-azobenzene is strongly dependent on temperature; for example, in benzene solution at 56.5°C $t_{1/2} = 150$ min and at 76.6°C $t_{1/2} = 21$ min. Similar results have been reported for *cis*-1,1′-, *cis*-2,2′-, and *cis*-1,2′-azonaphthalenes. The half-life of the 1,1′-*cis* isomer was 173 and 29.5 min at 0 and 12.5°C, respectively; that of the 2,2′-*cis* isomer, 150 and 19 min at 29 and 47°C, respectively; and that of the 1,2′-*cis* isomer, 104 and 5 min at 18.5 and 45°C, respectively(318).

The solvent effect on the photochemical *trans* → *cis* isomerization of 10 azo dyes has been studied spectrophotometrically(319). Ethanol suppresses isomerization, whereas hydrocarbon solvents favor it. When solutions of *cis* and *trans* isomers of azobenzene and its derivatives are heated, the equilibrium shifts toward the more stable *trans* modification (312, 319). Of the various azo dyes studied only *trans*-1-(2′-hydroxyphenylazo)-2-naphthol (**52**), which is stabilized by two hydrogen bonds, resisted isomerization.

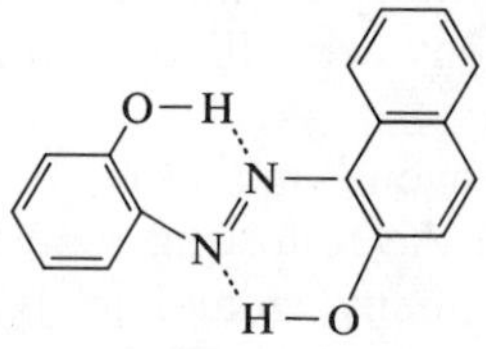

O−H
N
N
H−O

52

It is obvious that analysis of azo dye chromogens (analogous in structure to some of the compounds discussed in this section) in weakly polar, and nonpolar solvents could cause difficulty unless the analyst was cognizant of the spectral changes that could take place.

C. Hydrazones

Many carbonyl compounds are analyzed through their hydrazone derivatives. One of the most popular of these reagents is 2,4-dinitrophenylhydrazine. The geometry of these derivatives is of some importance in analysis, especially when isomers are separated by chromatography. The presence of *syn* and *anti* isomers of 2,4-dinitrophenylhydrazones in different spots or fractions of a chromatogram, as has been reported (320), presents some difficulty in analysis.

With the exception of the symmetrical ketone derivatives, hydrazones should be capable of existing in *syn* and *anti* forms. Purified 2,4-dinitrophenylhydrazones of aldehydes were stated to exist in the coplanar *syn* forms (320). When these derivatives are dissolved in solvents containing a trace of hydrochloric acid, interconversion takes place to give an equilibrium mixture of about 75% *syn* form and 25% *anti* form. The stereoisomers could be separated by thin-layer chromatography, using phenoxyethanol on kieselguhr or dimethylformamide on alumina and developing twice with heptane. The *syn* forms were isolated as orange crystals, the *anti* forms as light yellow liquids or liquid crystals. The *anti* form (**53**) of the 2,4-dinitrophenylhydrazone of pentanal absorbed at 352 mμ, whereas the *syn* isomer (**54**) absorbed at 358 mμ (320).

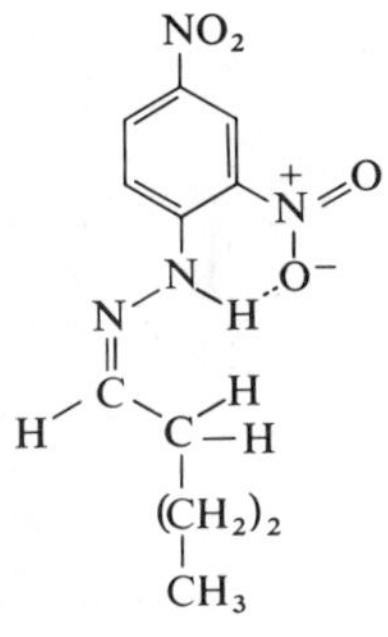

53 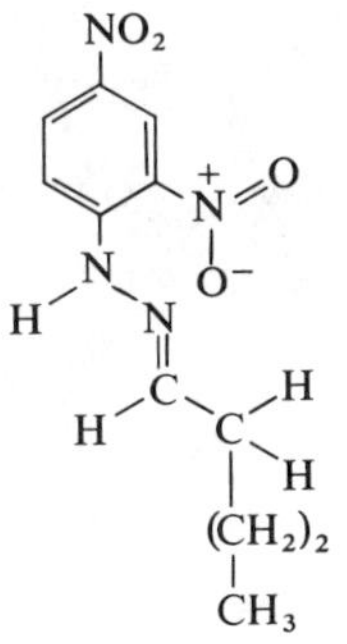 **54**

As can be seen by comparing **53** and **54**, the *anti* compounds of the *n*-alkanal 2,4-dinitrophenylhydrazone are more crowded around the NH group but show no serious steric hindrance. Substitution of an alkyl group for the hydrogen of the NH group or a bulkier aldehyde for the *n*-pentanal variety would cause a steric effect that would be reflected in the ultraviolet absorption spectrum.

Ramirez and Kirby (305) used bulkier aldehydes and were able to readily characterize the *syn* and *anti* forms of the 2,4-dinitrophenylhydrazones by their absorption spectra and other physical properties. For example, the stereoisomers of α-methoxypropiophenone 2,4-dinitrophenylhydrazone have been assigned configuration on the following basis (306). The *syn* isomer (**55**) has the following properties. Its infrared spectrum has a broad strong band at 3.10 μ derived from the N—H stretching vibration and apparently involved in hydrogen bonding. The long-wavelength band shows a violet shift of +5 mμ as compared with *syn*-propiophenone 2,4-DNPH. The absorption maximum at 258–260 mμ (mϵ = 10.0–13.0) present in the sterically hindered *anti*-phenylcarbonyl 2,4-DNPH is apparently derived from the conjugation of the benzene ring with the hydrazone group. The long-wavelength band shows the extraconjugative effect of the benzene ring.

The *anti* isomer (**56**) of α-methoxypropiophenone 2,4-DNPH has a sharp band in its infrared spectrum at 3.00 μ derived from the non-hydrogen-bonded N—H stretching vibration. The *syn*-propiophenone 2,4-DNPH shows the same band. The $\Delta\lambda$ of the long-wavelength band of **56** is about 16 mμ as compared with *syn*-propiophenone 2,4-DNPH. In addition, **56** has the band at 258–260 mμ expected of a sterically hindered *anti*-phenylcarbonyl 2,4-DNPH. The position of the long-wavelength band is that expected of an aliphatic carbonyl 2,4-DNPH, which one would expect in the sterically hindered **56**.

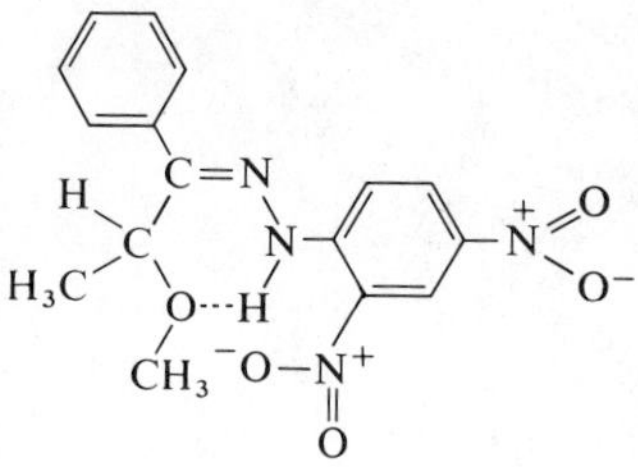

55

λ_{max} 383 mμ, mϵ 28.1

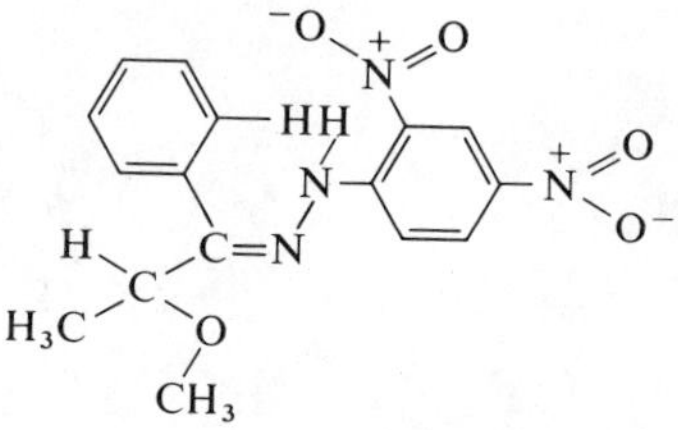

56

λ_{max} 258, 362 mμ, mϵ 13.6, 24.7

D. Formazans

A large number of analytical methods are available that depend on the formation of a formazan to analyze for some trace constituent. For example, a formazan is formed in the analysis of acetoacetic acid with diazotized *p*-nitroaniline (105).

Four stereoisomers are possible for each formazan. The hydrogen-bonded *trans-syn* form is apparently the most stable isomer. The assignment of structure for the four stereoisomers of triphenylformazan is shown in Figure 3.6 (321, 322). The large amount of steric hindrance in the *cis-syn* isomer is apparent.

E. Indigoid Isomers

Obtaining *cis*-indigo has proven difficult because of the high degree of stabilization of the *trans* isomer (**57**) by two hydrogen bonds (323).

On the other hand, *N,N'*-dimethylindigo undergoes rapid *trans* → *cis*

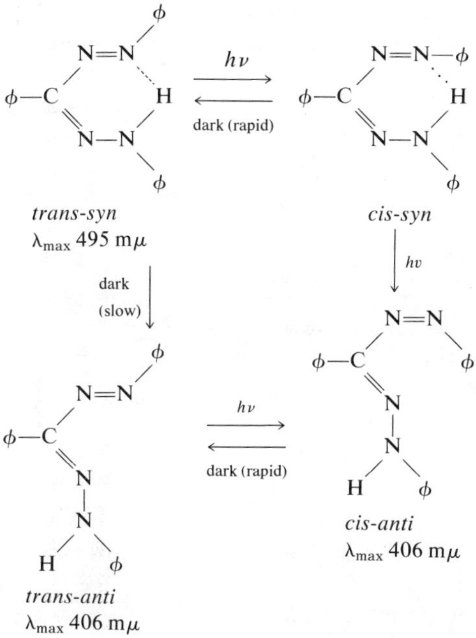

Fig. 3.6. The stereoisomers of triphenylformazan(321, 322).

isomerization when it is irradiated with yellow or red light; in the dark reversal takes place almost instantly(324). Other *N,N'*-disubstituted indigos and thioindigo can also form *cis* and *trans* isomers under proper

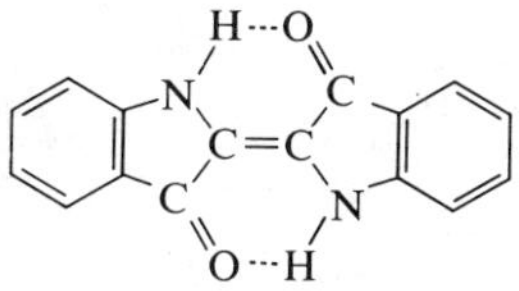

57

thermal and irradiative conditions (Table 3.28). When a chloroform solution of selenoindigo (λ_{max} 570 mμ, mϵ 11.0) is treated with light of wavelengths greater than 550 mμ, a large amount of the *cis* isomer (λ_{max} 500 mμ, mϵ ~7.0) is obtained(325).

F. Photochromic and Thermochromic Stereoisomers

Photochromic and thermochromic stereoisomers show drastic changes in spectra. Many factors affect the geometrical change—for example, structure of the molecule, temperature, type and intensity of radiation, basicity of the resonance terminals, and steric-hindrance effects. This extremely complicated topic has been discussed in Section II.

58

Yellow

59

60

However, some spiro compounds can be readily changed to their highly colored stereoisomers. For example, consider the remarkable color changes in solutions of 1,3,3-trimethyl-6′-nitroindoline-2-spiro-2′-benzopyran (**58**) with increasing solvent polarity (68). Some of the structures that contribute to the ground and excited states of the nonspiro compound in varying degree are dependent on the solvent polarity.

The spiro compound **58** is present in heptane solution; in ethanol ring opening takes place and a red solution is formed; in pyridine the ring-opened compound is blue. The colored compound can be classified as a Z_z-structure on the basis of solvent properties.

References

1. S. Dahne and D. Leupold, *Ber. Bunsenges. Physik. Chem.*, **70**, 618 (1966).
2. L. G. S. Brooker, G. H. Keyes, and D. W. Heseltine, *J. Am. Chem. Soc.*, **73**, 5350 (1951).
3. R. Robinson, *Tetrahedron*, **1**, 170 (1957).
4. J. Boyer, U. Toggweiler, and G. Stoner, *J. Am. Chem. Soc.*, **79**, 1748 (1957).
5. E. Clar and D. G. Stewart, *J. Am. Chem. Soc.*, **74**, 6235 (1952).
6. J. R. Platt, *J. Optical Soc. Am.*, **43**, 252 (1953).
7. H. B. Klevens and J. R. Platt, *J. Chem. Phys.*, **17**, 741 (1949).
8. E. Clar, *J. Chem. Phys.*, **17**, 741 (1949).
9. W. F. Forbes and R. Shilton, *Symposium on Spectroscopy*, ASTM Special Technical Publ. No. 269, 1959, p. 176.
10. G. M. Badger, R. S. Pearce, and R. Pettit, *J. Chem. Soc.*, 3199 (1951).
11. E. A. Braude, *Ann. Rpts.*, **42**, 123 (1945).
12. A. Burawoy, *J. Chem. Soc.*, 1068 (1939).

13. L. Doub and J. M. Vandenbelt, *J. Am. Chem. Soc.*, **69**, 2714 (1947); ibid., **71**, 2419 (1949); ibid., **77**, 4535 (1955).
14. B. G. Gowenlock and K. J. Morgan, *Spectrochim. Acta*, **17**, 310 (1961).
15. J. R. Platt, *J. Chem. Phys.*, **17**, 484 (1949).
16. E. Clar, *Polycyclic Hydrocarbons*, Vol. I, Academic Press, New York, 1964.
17. C. A. Coulson, *Proc. Phys. Soc.*, **60**, 257 (1948).
18. W. Moffitt, *J. Chem. Phys.*, **22**, 320 (1954).
19. N. Coggeshall and A. Pozefsky, *J. Chem. Phys.*, **19**, 980 (1951).
20. C. Cannon and G. Sutherland, *Spectrochim. Acta*, **4**, 373 (1951).
21. H. B. Klevens and J. R. Platt, *J. Chem. Phys.*, **17**, 470 (1949).
22. C. Sandorfy and R. N. Jones, *Can. J. Chem.*, **34**, 888 (1956).
23. R. N. Jones, *Chem. Rev.*, **41**, 353 (1947).
24. M. Levy and M. Szwarc, *J. Chem. Phys.*, **22**, 1621 (1954).
25. E. Clar and M. Zander, *Ber.*, **89**, 749 (1956).
26. I. Bergmann, *Trans. Faraday Soc.*, **50**, 829 (1954).
27. A. Maccoll, *Nature*, **163**, 178 (1948).
28. E. Lippert, *Z. Elektrochem.*, **61**, 962 (1957).
29. J. Czekalla, W. Liptay, and K. O. Meyer, *Ber. Bunsenges. Physik. Chem.*, **67**, 465 (1963).
30. D. E. Freeman and W. Klemperer, *J. Phys. Chem.*, **40**, 604 (1964).
31. E. Lippert in *Hydrogen Bonding*, Pergamon Press, London, 1957.
32. L. Dede and A. Rosenberg, *Ber.*, **67**, 147 (1934).
33. R. Huisgen and H. Koch, *Ann.*, **591**, 200 (1955).
34. R. Ellis, A. M. Gaddis, and G. T. Currie, *Anal. Chem.*, **30**, 475 (1958).
35. J. H. P. Utley, *J. Chem. Soc.*, 3252 (1963).
36. E. Lippert, *J. Phys. Radium*, **15**, 627 (1954); through *Chem. Abstr.*, **52**, 17953 (1957).
37. S. Hunig, S. Schweiberg, and H. Schwarz, *Ann.*, **587**, 132 (1954).
38. S. Hunig and K. Requardt, *Ann.*, **592**, 180 (1955).
39. B. C. McKusick, R. E. Heckert, T. L. Cairns, D. D. Coffman, and H. F. Mower, *J. Am. Chem. Soc.*, **80**, 2806 (1958).
40. L. Fieser, *J. Org. Chem.*, **15**, 130 (1950).
41. S. Hunig and H. Hermann, *Ann.*, **636**, 32 (1960).
42. L. G. S. Brooker and R. H. Sprague, *J. Am. Chem. Soc.*, **63**, 3214 (1941).
43. N. Bayliss and E. McRae, *J. Phys. Chem.*, **58**, 1002 (1954).
44. G. C. Pimental and A. L. McClellan, *The Hydrogen Bond*, Freeman, London, 1960, p. 159.
45. F. Kehrmann, H. Goldstein, and F. Brunner, *Helv. Chim. Acta*, **9**, 222 (1926).
46. S. Sheppard, P. Newsome, and H. Brigham, *J. Am. Chem. Soc.*, **64**, 2923 (1942).
47. A. L. LeRosen and C. E. Reid, *J. Chem. Phys.*, **20**, 233 (1952).
48. L. G. S. Brooker, A. C. Craig, D. W. Heseltine, P. W. Jenkins, and L. L. Lincoln, *J. Am. Chem. Soc.*, **87**, 2443 (1965).
49. A. Mangini and R. Passerini, *J. Chem. Soc.*, 1168 (1952).
50. L. Bartolotti and R. Passerini, *Ric. Sci.*, **25**, 3 (1955).
51. L. G. S. Brooker, F. L. White, G. H. Keyes, C. P. Smyth, and P. F. Oesper, *J. Am. Chem. Soc.*, **63**, 3192 (1941).
52. A. Kharkharov, *Izvest. Akad. Nauk SSSR, Otdel. Khim. Nauk*, 326 (1955); through *Chem. Abstr.*, **49**, 12126 (1955).
53. E. Sawicki and A. Carr, *J. Org. Chem.*, **22**, 507 (1957).
54. J. Nys and A. Van Dormael, *Bull. Soc. Chim. Belges*, **65**, 809 (1956).
55. E. Sawicki, B. Chastain, H. Bryant, and A. Carr, *J. Org. Chem.*, **22**, 625 (1957).

56. R. H. Peters and H. H. Sumner, *J. Chem. Soc.*, 2101 (1953).
57. E. A. Fehnel and M. Carmack, *J. Am. Chem. Soc.*, **71**, 2889 (1949).
58. W. Kumler, *J. Am. Chem. Soc.*, **68**, 1184 (1946).
59. G. M. Badger, R. G. Buttery, and G. Lewis, *J. Chem. Soc.*, 1888 (1954).
60. R. Wizinger and H. Sontag, *Helv. Chim. Acta*, **38**, 363 (1955).
61. J. Thomas and G. Branch, *J. Am. Chem. Soc.*, **75**, 4793 (1953).
62. E. Katzenellenbogen and G. Branch, *J. Am. Chem. Soc.*, **69**, 1615 (1947).
63. E. Sawicki and A. Carr, *J. Org. Chem.*, **23**, 610 (1958).
64. L. Flexser and L. P. Hammett, *J. Am. Chem. Soc.*, **60**, 885 (1938).
65. E. Sawicki, D. F. Bender, T. R. Hauser, R. M. Wilson, Jr., and J. E. Meeker, *Anal. Chem.*, **35**, 1479 (1963).
66. D. Kupfer and L. L. Bruggeman, *Anal. Biochem.*, **17**, 502 (1966).
67. M. Strell and S. Reindl, *Arzneimittel-Forsch.*, **11**, 552 (1961).
68. E. Knott, *J. Chem. Soc.*, 3038 (1951).
69. A. Kiprianov and E. Timoshenko, *Ukrain. Khim. Zh.*, **18**, 347 (1952); through *Chem. Abstr.*, **49**, 984 (1955).
70. L. G. S. Brooker, G. H. Keyes, R. H. Sprague, R. Van Dyke, E. Van Lare, G. Van Zandt, F. L. White, H. Cressman, and S. G. Dent, Jr., *J. Am. Chem. Soc.*, **73**, 5332 (1951).
71. S. Hunig and O. Rosenthal, *Ann.*, **592**, 161 (1955).
72. A. Phillips, *J. Org. Chem.*, **14**, 302 (1949).
73. M. Umeda, *Yakugaku Zasshi*, **83**, 951 (1963).
74. Y. Hirshberg, E. B. Knott, and E. Fischer, *J. Chem. Soc.*, 3313 (1955).
75. G. E. Ficken and J. D. Kendall, *J. Chem. Soc.*, 1537 (1960).
76. E. G. McRae, *Spectrochim. Acta*, **12**, 192 (1958).
77. J. A. Berson, E. M. Evleth, and Z. Hamlet, *J. Am. Chem. Soc.*, **82**, 3793 (1960).
78. J. R. May, P. Zvirblis, and A. A. Kondritzer, *J. Pharm. Sci.*, **54**, 1508 (1965).
79. R. Gompper, *Tetrahedron Letters*, 165 (1968).
80. J. W. Faller and J. P. Phillips, *Anal. Chim. Acta*, **32**, 586 (1965).
81. E. Kloster-Jensen, E. Kovats, E. Eschenmoser, and E. Heilbronner, *Helv. Chim. Acta*, **39**, 1051 (1956).
82. W. Simon, G. Naville, H. Sulser, and E. Heilbronner, *Helv. Chim. Acta*, **39**, 1107 (1956).
83. W. Treibs, W. Schroth, H. Lichtmann, and G. Fischer, *Ann.*, **642**, 97 (1961).
84. M. Los and W. H. Stafford, *J. Chem. Soc.*, 1680 (1959).
85. R. Mayer and H. Russ, *Ber.*, **95**, 1311 (1962).
86. T. Kubota, *Ann. Rpts. Shionogi Res. Lab.*, **6**, 31 (1956).
87. R. Wiley and S. Slaymaker, *J. Am. Chem. Soc.*, **79**, 2233 (1957).
88. E. M. Kosower and B. G. Ramsey, *J. Am. Chem. Soc.*, **81**, 856 (1959).
89. J. P. Phillips and R. Keown, *J. Am. Chem. Soc.*, **73**, 5483 (1951).
90. J. W. Faller and J. P. Phillips, *Talanta*, **11**, 641 (1964).
91. K. Dimroth, C. Reichardt, T. Siepmann, and F. Bohlmann, *Ann.*, **661**, 1 (1963).
92. K. Dimroth, G. Arnoldy, S. von Eicken, and G. Schiffler, *Ann.*, **604**, 221 (1957).
93. W. Stafford, *J. Chem. Soc.*, 580 (1952).
94. H. Nozaki, M. Takaku, and K. Kondo, *Tetrahedron*, **22**, 2145 (1966).
95. H. Hirayama and T. Kubota, *J. Pharm. Soc. Japan*, **73**, 140 (1953).
96. A. W. Johnson and D. J. McCaldin, *J. Chem. Soc.*, 3470 (1957).
97. G. K. Summer and J. A. Hawes, *Proc. Soc. Exptl. Biol. Med.*, **112**, 402 (1963).
98. H. Suhr, *Ann.*, **687**, 175 (1965).
99. K. Satake, A. Matsuo, and T. Take, *J. Biochem. (Tokyo)*, **58**, 90 (1965).

100. M. Richardson, *Nature*, **197**, 290 (1963).
101. J. J. Wren and D. S. Merryfield, *J. Chromatog.*, **17**, 257 (1965).
102. K. Satake, *J. Biochem. (Tokyo)*, **47**, 454 (1960); ibid., **50**, 293 (1961).
103. M. Schmall and E. G. Wollish, *Anal. Chem.*, **29**, 1509 (1957).
104. G. A. L. Smith and D. A. King, *Analyst*, **90**, 55 (1965).
105. P. G. Walker, *Biochem. J.*, **58**, 699 (1954).
106. H. Austerhoff, *Deut. Apotheker-Ztg.*, **102**, 765 (1962).
107. R. Zimmerman, H. Ruttloff, and K. Taufel, *Nahrung*, 680 (1961).
108. V. S. Waravdekar and L. D. Sashaw, *Biochim. Biophys. Acta*, **24**, 439 (1957).
109. G. W. Watt and J. D. Chrisp, *Anal. Chem.*, **26**, 452 (1954).
110. T. Tsukamoto and K. Yuhi, *Yakugaku Zasshi*, **79**, 1294 (1959).
111. A. C. Thompson and P. A. Hedin, *J. Chromatog.*, **21**, 13 (1966).
112. K. C. Tsou and H. C. F. Su, *Anal. Biochem.*, **11**, 54 (1965).
113. F. W. Ochynski, *Analyst*, **85**, 278 (1960).
114. W. Troll and J. Lindsley, *J. Biol. Chem.*, **215**, 655 (1955).
115. E. G. Trams, *Anal. Chem.*, **30**, 256 (1958).
116. G. H. Ayres and B. L. Tuffly, *Anal. Chem.*, **23**, 788 (1951).
117. R. Bastian, *Anal. Chem.*, **25**, 259 (1953).
118. G. W. Haupt, *J. Res. Natl. Bur. Std.*, **48**, 414 (1952).
119. G. Kortüm and A. Halban, *Z. Phys. Chem.*, **170**, 212 (1934).
120. B. K. Mukerji, A. K. Bhattacharji, and W. R. Dhar, *J. Phys. Chem.*, **32**, 1834 (1938).
121. V. A. Yarborough, J. F. Haskin, and W. J. Lambdin, *Anal. Chem.*, **26**, 1576 (1954).
122. T. Davis and J. Richmond, *J. Am. Chem. Soc.*, **62**, 756 (1940).
123. R. Bottle and G. Gilbert, *Chem. Ind. (London)*, 575 (1956).
124. E. Clar, *Spectrochim. Acta*, **4**, 116 (1950).
125. D. Keilin and E. F. Hartree, *Nature*, **164**, 254 (1949).
126. D. Keilin and E. F. Hartree, *Nature*, **165**, 504 (1950).
127. G. F. Doebbler and W. B. Elliot, *Biochim. Biophys. Acta*, **94**, 317 (1965).
128. S. Freed, *Science*, **150**, 576 (1965).
129. L. G. S. Brooker, *J. Am. Chem. Soc.*, **62**, 1116 (1940).
130. D. Lauerer, M. Coenen, M. Pestemer, and G. Scheibe, *Z. Phys. Chem.*, **10**, 236 (1957).
131. E. Lippert, in *Hydrogen Bonding*, Pergamon Press, London, 1958.
132. W. Theilacker and H. Muller, *Ber.*, **89**, 984 (1956).
133. W. Theilacker, W. Berger, and P. Popper, *Ber.*, **89**, 970 (1956).
134. A. Baeyer and V. Villiger, *Ber.*, **35**, 3013 (1902).
135. C. Marvel, J. Whitson, and H. Johnston, *J. Am. Chem. Soc.*, **66**, 415, (1944).
136. J. H. Day, *Chem. Rev.*, **63**, 65 (1963).
137. A. Mustafa, *Chem. Rev.*, **43**, 509 (1948).
138. O. Chaude and P. Rumpf, *Compt. Rend.*, **233**, 405 (1951).
139. I. M. Heilbron, R. N. Heslop, and F. Irving, *J. Chem. Soc.*, 430 (1933).
140. R. Heiligman-Rim, Y. Hirshberg, and E. Fischer, *J. Chem. Soc.*, 156 (1961).
141. Y. Hirshberg, *J. Chem. Phys.*, **27**, 758 (1957).
142. Y. Hirshberg and E. Fischer, *J. Chem. Soc.*, 3129 (1954).
143. A. Mustafa, *J. Chem. Soc.*, 2295 (1949).
144. R. Wizinger and H. Wennig, *Helv. Chim. Acta*, **23**, 247 (1940).
145. E. Berman, R. E. Fox, and R. D. Thomson, *J. Am. Chem. Soc.*, **81**, 5605 (1959).
146. R. Dickinson and I. M. Heilbron, *J. Chem. Soc.*, 1704 (1927).
147. C. Koelsch, *J. Org. Chem.*, **16**, 1362 (1951).
148. E. Bergmann, A. Weizmann, and E. Fischer, *J. Am. Chem. Soc.*, **72**, 5009 (1950).

149. E. Sawicki, *Chemist-Analyst*, **48**, 4 (1959).
150. Y. Hirshberg and E. Fischer, *J. Chem. Soc.*, 297 (1954).
151. Y. Hirshberg, E. H. Frei, and E. Fischer, *J. Chem. Soc.*, 2184 (1953).
152. R. Heiligman-Rim, Y. Hirshberg, and E. Fischer, *J. Phys. Chem.*, **66**, 2470 (1962).
153. Y. Hirshberg, *J. Am. Chem. Soc.*, **78**, 2304 (1956).
154. A. Schonberg, A. Mustafa, and W. Asker, *J. Am. Chem. Soc.*, **76**, 4134 (1954).
155. I. Shimizu, H. Kokado, and E. Inoue, *Kagaku Zasshi*, **88**, 1127 (1967).
156. I. Shimizu, H. Kokado, and E. Inoue, *Kagaku Zasshi*, **89**, 755 (1968).
157. J. F. D. Mills and S. C. Nyburg, *J. Chem. Soc.*, 927 (1963).
158. Y. Hirshberg and E. Fischer, *J. Chem. Soc.*, 629 (1953).
159. A. Schonberg, A. Mustafa, and E. Sobhy, *J. Am. Chem. Soc.*, **75**, 3377 (1957).
160. A. Mustafa and M. Sobhy, *J. Am. Chem. Soc.*, **77**, 5124 (1955).
161. G. Kortüm, *Angew. Chem.*, **70**, 14 (1958).
162. E. Wasserman and R. E. Davis, *J. Chem. Phys.* **30**, 1367, (1959).
163. Y. Hirshberg, E. Loewenthal, E. Bergmann, and B. Pullman, *Bull. Soc. Chim. France*, **18**, 88 (1951).
164. W. T. Grubb and G. B. Kistiakowsky, *J. Am. Chem. Soc.*, **72**, 419 (1950).
165. W. Theilacker, G. Kortüm, and G. Friedheim, *Ber.*, **83**, 508 (1950).
166. E. Harnik, *J. Chem. Phys.*, **24**, 297 (1956).
167. G. Kortüm, W. Theilacker, and V. Braun, *Z. Physik. Chem.*, **2**, 179 (1954).
168. Y. Hirshberg and E. Fischer, *J. Chem. Phys.*, **23**, 1723 (1955).
169. A. Mustafa and A. E. Hassan, *J. Am. Chem. Soc.*, **79**, 3846 (1957).
170. A. Schonberg, A. Mustafa, and S. M. Zayed, *Science*, **119**, 193 (1954).
171. A. F. A. Ismail and Z. M. El-Shafel, *J. Chem. Soc.*, 3393 (1957).
172. W. Theilacker, G. Kortüm, and H. Elliehauser, *Ber.*, **89**, 1578 (1956).
173. C. Allen and J. Van Allan, *J. Am. Chem. Soc.*, **72**, 5165 (1950).
174. A. Santiago and R. S. Becker, *J. Am. Chem. Soc.*, **90**, 3654 (1968).
175. S. Malkin and E. Fischer, *J. Phys. Chem.*, **66**, 2482 (1962).
176. A. E. Chichababin, B. Kuindshi, and S. V. Benevalenskü, *Ber.*, **58**, 1580 (1925).
177. R. Hardwick, H. S. Mosher, and P. Passailaigue, *Trans. Faraday Soc.*, **56**, 44 (1960).
178. H. S. Mosher, C. Souers, and R. Hardwick, *J. Chem. Phys.*, **32**, 1888 (1960).
179. A. Schonberg, A. Mustafa, and S. M. A. D. Zayed, *J. Am. Chem. Soc.*, **75**, 4302 (1953).
180. G. N. Lewis and D. Lipkin, *J. Am. Chem. Soc.*, **64**, 2801 (1942).
181. W. Klyne and P. B. D. de la Mare, Eds., *Progress in Stereochemistry*, Vols. 1 and 2, Academic Press, 1954 and 1958, New York.
182. A. E. Braude, *Experientia*, **11**, 457 (1955).
183. W. F. Forbes and W. A. Mueller, *J. Am. Chem. Soc.*, **79**, 6495 (1957).
184. L. W. Pickett, G. F. Walter, and H. France, *J. Am. Chem. Soc.*, **58**, 2296 (1936).
185. H. B. Klevens and J. R. Platt, *J. Am. Chem. Soc.*, **71**, 1714 (1949).
186. B. M. Wepster, *Rec. Trav. Chim.*, **71**, 1159 (1952).
187. E. B. Knott, *J. Soc. Dyers Colourists*, **67**, 302 (1951).
188. K. J. Brunings and A. H. Corwin, *J. Am. Chem. Soc.*, **73**, 5332 (1951).
189. L. G. S. Brooker, F. L. White, R. H. Sprague, S. G. Dent, Jr., and G. Van Zandt, *Chem. Rev.*, **41**, 325 (1947).
190. R. A. Jeffreys and E. B. Knott, *J. Chem. Soc.*, 1028 (1951).
191. E. B. Knott and P. J. S. Pauwells, *J. Org. Chem.*, **33**, 2120 (1968).
192. C. Overberger and D. Tanner, *J. Am. Chem. Soc.*, **77**, 369 (1955).
193. Y. Hirschberg, *J. Am. Chem. Soc.*, **71**, 3241 (1949).
194. G. Baddeley, J. Chadwick, and H. Taylor, *J. Chem. Soc.*, 451 (1956).

195. R. B. Carlin and H. P. Landerl, *J. Am. Chem. Soc.*, **75**, 3969 (1953).
196. E. A. Braude, E. R. H. Jones, and E. S. Stern, *J. Chem. Soc.*, 1571 (1953).
197. M. O'Shaugnessy and W. Rodebush, *J. Am. Chem. Soc.*, **62**, 2906 (1940).
198. D. M. Hall and F. Minhaj, *J. Chem. Soc.*, 4584 (1957).
199. E. A. Johnson, *J. Chem. Soc.*, 4155 (1957).
200. A. E. Haddow, R. J. C. Harris, G. A. R. Kon, and E. M. F. Roe, *Phil. Trans. Roy. Soc. (London)*, **A241**, 147 (1948).
201. W. H. Laarhoven, R. J. F. Nivard, and E. Havinga, *Rec. Trav. Chim.*, **79**, 1153 (1960).
202. L. H. Klemm, H. Ziffer, J. W. Sprague, and W. Hodes, *J. Org. Chem.*, **20**, 190 (1955).
203. R. Turner and D. Voitle, *J. Am. Chem. Soc.*, **73**, 1403 (1951).
204. E. A. Braude and C. J. Timmons, *J. Chem. Soc.*, 3766 (1955).
205. E. A. Braude and E. R. H. Jones, *J. Am. Chem. Soc.*, **72**, 1041 (1950).
206. W. Schubert, W. Sweeney, and H. Latourette, *J. Am. Chem. Soc.*, **76**, 5462 (1954).
207. W. Forbes and W. Mueller, *Can. J. Chem.*, **33**, 1145 (1955).
208. E. A. Braude, F. Sondheimer, and W. F. Forbes, *Nature*, **173**, 117 (1954).
209. G. Hedden and W. Brown, *J. Am. Chem. Soc.*, **75**, 3744 (1953).
210. E. A. Braude and F. Sondheimer, *J. Chem. Soc.*, 3733 (1955).
211. R. F. Rekker and W. T. Nauta, *Spectrochim. Acta*, **8**, 348 (1957).
212. B. M. Wepster, *Rec. Trav. Chim.*, **76**, 335 (1957).
213. J. Burgers, M. A. Hoefnagel, P. E. Verkade, H. Visser, and B. M. Wepster, *Rec. Trav. Chim.*, **77**, 491 (1958).
214. A. J. Bilbo and G. M. Wyman, *J. Am. Chem. Soc.*, **75**, 5312 (1953).
215. H. Ungnade, *J. Am. Chem. Soc.*, **76**, 5133 (1954).
216. A. Bruylants, E. Braye, and A. Schonne, *Helv. Chim. Acta*, **35**, 1127 (1952).
217. H. Kloosterziel and H. Backer, *Rec. Trav. Chim.*, **72**, 185 (1953).
218. A. Burawoy and J. Chamberlain, *J. Chem. Soc.*, 2310 (1952).
219. R. T. Arnold, V. Webers, and R. Dodson, *J. Am. Chem. Soc.*, **74**, 368 (1952).
220. H. Foster, *J. Am. Chem. Soc.*, **82**, 3780 (1960).
221. R. Raffauf, *J. Am. Chem. Soc.*, **72**, 753 (1950).
222. P. Vittum and G. Brown, *J. Am. Chem. Soc.*, **69**, 152 (1947).
223. S. Hunig and P. Richters, *Ann.*, **612**, 282 (1958).
224. A. Van Loom, *Relation between Steric Hindrance and Mesomerism in the Absorption Spectra of Aminoazobenzene Compounds in Neutral and Acid Solution*, Ph.D. Thesis, 1959, Delft University, The Netherlands.
225. E. Marcus, W. M. Lauer, and R. T. Arnold, *J. Am. Chem. Soc.*, **80**, 3742 (1958).
226. R. Huisgen and I. Ugi, *Ann.*, **610**, 57 (1957).
227. R. Huisgen, W. Rapp, I. Ugi, H. Walz, and E. Mergenthaler, *Ann.*, **586**, 1 (1954).
228. P. Ramart-Lucas and J. Hoch, *Bull. Soc. Chim. France*, **2**, 327 (1935).
229. A. E. Lutskii and V. V. Dorofeev, *Zh. Obshch. Khim.*, **27**, 1059 (1957).
230. B. Witkop, J. B. Patrick, and M. Rosenblum, *J. Am. Chem. Soc.*, **73**, 2641 (1951).
231. R. Huisgen, I. Ugi, H. Brade, and E. Rauenbusch, *Ann.*, **586**, 30 (1954).
232. P. Grammaticakis, *Bull. Soc. Chim. France*, 220 (1951).
233. L. Doub and J. M. Vandenbelt, *J. Am. Chem. Soc.*, **71**, 2414 (1949).
234. R. Huisgen and I. Ugi, *Ber.*, **93**, 2693 (1960).
235. E. A. Braude, E. R. H. Jones, and E. S. Stern, *J. Chem. Soc.*, 1087 (1947).
236. F. Grunstone and A. Tulloch, *J. Appl. Chem. (London)*, **4**, 291 (1954).
237. C. B. Clarke and A. R. Pinder, *J. Chem. Soc.*, 1967 (1958).
238. K. Bowden, I. M. Heilbron, E. R. H. Jones, and B. C. L. Weedon, *J. Chem. Soc.*, 39 (1946).
239. N. R. Jones, *Chem. Rev.*, **32**, 36 (1943).

240. F. Mason, *J. Chem. Soc.*, **123**, 1546 (1923).
241. F. Bohlmann, *Ber.*, **85**, 1144 (1952).
242. L. N. Ferguson, *Electron Structures of Organic Molecules*, Prentice-Hall, New York, 1952, 280.
243. H. Dahn and H. Castelmur, *Helv. Chim. Acta*, **36**, 638 (1953).
244. L. N. Ferguson and J. C. Nnadi, *J. Chem. Educ.*, **42**, 529 (1965).
245. J. R. Platt, H. B. Klevens, and W. C. Price, *J. Chem. Phys.*, **17**, 466 (1949).
246. N. Leonard and D. Locke, *J. Am. Chem. Soc.*, **77**, 437 (1955).
247. E. A. Fehnel and M. Carmack, *J. Am. Chem. Soc.*, **71**, 84 (1949).
248. K. Bowden, E. A. Braude, and E. R. H. Jones, *J. Chem. Soc.*, 948 (1946).
249. R. C. Cookson and N. Lewin, *Chem. Ind. (London)*, 984 (1956).
250. A. Mangini and R. Passerini, *Gazz. Chim. Ital.*, **84**, 606 (1954).
251. M. Calvin and H. W. Alter, *J. Chem. Phys.*, **19**, 765 (1951).
252. C. Price and H. Morita, *J. Am. Chem. Soc.*, **75**, 4747 (1953).
253. A. E. Gillam and D. H. Hey, *J. Chem. Soc.*, 1170 (1939).
254. R. Alexander, Jr., *J. Org. Chem.*, **21**, 1464 (1957).
255. R. Morton and A. Stubbs, *J. Chem. Soc.*, 1347 (1940).
256. H. Ungnade and I. Ortega, *J. Org. Chem.*, **17**, 1475 (1952).
257. W. Schroeder, P. Wilcox, K. Trueblood, and A. Dekker, *Anal. Chem.*, **23**, 1740 (1951).
258. M. Martynoff, *Compt. Rend.*, **235**, 54 (1952).
259. W. Oroshnik, A. D. Mebane, and G. J. Karmas, *J. Am. Chem. Soc.*, **75**, 1050 (1953).
260. W. O. Celmer and I. A. Solomons, *J. Am. Chem. Soc.*, **74**, 1870, 2245, 3838 (1952); ibid., **75**, 1372 (1953).
261. F. Bohlmann and J. J. Mannhardt, *Ber.*, **89**, 2268 (1956).
262. H. Fleischacker and G. F. Woods, *J. Am. Chem. Soc.*, **78**, 3436 (1956).
263. U. Wannagat and F. Brandmair, *Z. Anorg. Allgem. Chem.*, **280**, 223 (1955).
264. H. Gilman and G. E. Dunn, *J. Am. Chem. Soc.*, **72**, 2178 (1950).
265. A. Butenandt, E. Hecker, and H. G. Zachau, *Ber.*, **88**, 1185 (1955).
266. E. A. Braude and J. A. Coles, *J. Chem. Soc.*, 2085 (1951).
267. L. Crombie, S. H. Harper, and F. C. Newman, *J. Chem. Soc.*, 3963 (1956).
268. I. Bell, E. R. H. Jones, and M. C. Whiting, *J. Chem. Soc.*, 2597 (1957).
269. L. Crombie, *J. Chem. Soc.*, 1007 (1955).
270. G. Martin, *Ann. Chim. (Paris)*, **4**, 541 (1959).
271. G. S. Myers, *J. Am. Chem. Soc.*, **73**, 2100 (1951).
272. L. Crombie, *J. Chem. Soc.*, 2997 (1952).
273. L. N. Owen and M. U. S. Sultanbawa, *J. Chem. Soc.*, 3105 (1949).
274. O. Dann, *Ber.*, **80**, 427 (1947).
275. B. Eistert and W. Reiss, *Ber.*, **87**, 108 (1954).
276. L. N. Owen and M. U. S. Sultanbawa, *J. Chem. Soc.*, 3109 (1949).
277. Y. R. Naves, *Compt. Rend.*, **241**, 1209 (1955).
278. E. R. H. Jones, L. Skattebl, and M. C. Whiting, *J. Chem. Soc.*, 1054 (1958).
279. I. Bell, E. R. H. Jones, and M. C. Whiting, *J. Chem. Soc.*, 1313 (1958).
280. J. L. H. Allen, E. R. H. Jones, and M. C. Whiting, *J. Chem. Soc.*, 1862 (1958).
281. J. A. Elvidge, R. P. Linstead, and P. Sims, *J. Chem. Soc.*, 1793 (1953).
282. L. Lang, Ed., *Absorption Spectra in the Ultraviolet and Visible Region*, Publishing House of the Hungarian Academy of Science, Budapest, 1959.
283. C. G. Overberger, D. Tanner, and E. M. Pearce, *J. Am. Chem. Soc.*, **80**, 4566 (1958).
284. A. Mangini and F. Montanari, *Gazz. Chim. Ital.*, **88**, 1081 (1958). See also M. Hirota and Y. Urushibara, *Bull. Chem. Soc. Japan*, **32**, 703 (1959).

285. A. W. Nineham and R. A. Raphael, *J. Chem. Soc.*, 118 (1949).
286. L. P. Kuhn, R. E. Lutz, and C. R. Bauer, *J. Am. Chem. Soc.*, **72**, 5058 (1950).
287. J. F. Codington and E. Mosettig, *J. Org. Chem.*, **17**, 1027 (1952).
288. O. Grummitt and F. J. Christoph, *J. Am. Chem. Soc.*, **73**, 347 (1951).
289. K. A. Jensen, A. Kjaer, and S. C. Linholt, *Acta Chem. Scand.*, **6**, 180 (1952).
290. G. Berti and F. Bottari, *Gazz. Chim. Ital.*, **89**, 2371 (1959).
291. J. Bornstein and M. R. Borden, *Chem. Ind. (London)*, 441 (1958).
292. J. P. Freeman and T. E. Stevens, *J. Org. Chem.*, **23**, 136 (1958).
293. G. Berti, *Gazz. Chim. Ital.*, **86**, 655 (1956).
294. E. Merkel, *Ber.*, **86**, 895 (1953).
295. J. H. Pinckard, B. Wille, and L. Zechmeister, *J. Am. Chem. Soc.*, **70**, 1938 (1948).
296. K. Lunde and L. Zechmeister, *J. Am. Chem. Soc.*, **76**, 2308 (1954).
297. J. Dale, *Acta Chem. Scand.*, **11**, 971 (1957).
298. L. Zechmeister and J. H. Pinckard, *J. Am. Chem. Soc.*, **76**, 4144 (1954).
299. E. F. Magoon and L. Zechmeister, *J. Am. Chem. Soc.*, **77**, 5642 (1955).
300. O. Isler and M. Montavon, *Chimia*, **12**, 1 (1958).
301. G. M. Wyman and W. R. Brode, *J. Am. Chem. Soc.*, **73**, 1487, (1951).
302. G. M. Wyman and A. F. Zenhausern, *J. Org. Chem.*, **30**, 2348 (1965).
303. P. P. Birnbaum, J. H. Linford, and D. W. G. Style, *Trans. Faraday Soc.*, **49**, 735 (1953).
304. W. Brode, J. Gould, and G. Wyman, *J. Am. Chem. Soc.*, **74**, 4641 (1952).
305. F. Ramirez and A. F. Kirby, *J. Am. Chem. Soc.*, **75**, 6026 (1953).
306. F. Ramirez and A. F. Kirby, *J. Am. Chem. Soc.*, **76**, 1037 (1954).
307. L. Zechmeister and F. J. Petracek, *J. Am. Chem. Soc.*, **74**, 282 (1952).
308. A. Polgár and L. Zechmeister, *J. Am. Chem. Soc.*, **64**, 1856 (1942).
309. L. Zechmeister, Cis-Trans *Isomeric Carotenoids, Vitamins A, and Arylpolyenes*, Academic Press, New York, 1962.
310. E. F. Magoon and P. Zechmeister, *Arch. Biochem. Biophys.*, **69**, 535 (1957).
311. D. Holme, E. R. H. Jones, and M. C. Whiting, *Chem. Ind. (London)*, 928 (1956).
312. G. A. Hartley, *J. Chem. Soc.*, 633 (1938).
313. M. N. Inscoe, J. H. Gould, and W. R. Brode, *J. Am. Chem. Soc.*, **81**, 5634 (1959).
314. E. Fischer, M. Frankel, and R. Wolovsky, *J. Chem. Phys.*, **23**, 1367 (1955).
315. E. S. Lewis and H. Suhr, *J. Am. Chem. Soc.*, **80**, 1367 (1958).
316. E. Fischer and Y. Frie, *J. Chem. Phys.*, **27**, 328 (1957).
317. G. A. Hartley, *J. Chem. Soc.*, 638 (1938).
318. M. Frankel, R. Wolovsky, and E. Fischer, *J. Chem. Soc.*, 3441 (1955).
319. W. Brode, J. Gould, and G. Wyman, *J. Am. Chem. Soc.*, **75**, 1856 (1953).
320. H. M. Edwards, Jr., *J. Chromatog.*, **22**, 29 (1966).
321. R. Kuhn and H. M. Weitz, *Ber.*, **86**, 1199 (1953).
322. I. Hausser, D. Jerchel, and R. Kuhn, *Ber.*, **82**, 515 (1949).
323. W. R. Brode, E. G. Pearson, and G. M. Wyman, *J. Am. Chem. Soc.*, **76**, 1034 (1954).
324. J. Weinstein and G. M. Wyman, *J. Am. Chem. Soc.*, **78**, 4007 (1956).
325. R. Pummerer and G. Marondel, *Ber.*, **93**, 2834 (1960).

CHAPTER 4

ABSORPTION SPECTRA OF ZWITTERIONIC RESONANCE STRUCTURES II. TAUTOMERISM OF KETONES AND PHENOLS

By tautomerism is meant a dynamic reversible isomerism in which an atom or a group is shifted from one part of a molecule to another because of a change of solvent or temperature. The most common type of tautomerism involves a proton shift.

I. Keto ⇄ Enol Tautomerism

Examples of the keto ⇄ enol type of tautomer are listed in Table 4.1.

Simple aliphatic carbonyl compounds enolize to an insignificant extent (45). The enol content in the undiluted state has been reported to range from 0.00015% for acetone to 0.92% for 2-octanone. Acetophenone contains only 0.035% enol. Aqueous solutions of the cyclic ketones also contain negligible amounts of the enol tautomer; for example, solutions of cyclopentanone and cyclohexanone contain 0.0048 and 0.02%, respectively, of the enol isomer (46).

The long-wavelength bands of the aliphatic carbonyl compounds are derived from the singlet–singlet $n \rightarrow \pi^*$ transition involving excitation of a nonbonding oxygen electron to an antibonding π-orbital between the carbon oxygen of the carbonyl group (47). The intensities of the $n \rightarrow \pi^*$ bands are low and are especially weak in the aliphatic carbonyl compounds (e.g., acetone, λ_{max} in heptane 280 mμ, mϵ 0.013)(48). Consequently analysis for these types of compounds in complex mixtures has usually involved the formation of a conjugated chromogen with an intense long-wavelength $\pi \rightarrow \pi^*$ band (e.g., acetone 2,4-dinitrophenylhydrazone in chloroform, λ_{max} 364 mμ, mϵ 22.4)(49).

When a carbonyl compound enolizes, a long-wavelength Z_n-band is formed; it is derived from a singlet–singlet $\pi \rightarrow \pi^*$ transition. This type of band is much more intense than a $n \rightarrow \pi^*$ band and in addition shows the expected solvent effects of a molecule that has a higher dipole moment in the excited state than in the ground state (Table 4.1).

TABLE 4.1
Absorption Spectra of Tautomers and Reference Compounds

			Keto ⇄ Enol			
	Compound[a]	Solvent	$\lambda_{max}(m\mu)$[b]	$m\epsilon$	Remarks	Ref.
1.	Malonaldehyde	Aqueous acid	245	13.0		1
	1-Methoxy-1-pentene-3-one	95% EtOH	249	11.2		2
2.	2-Ethoxymalonaldehyde	Water	275	16.6	In 95% EtOH λ_{max} 256 $m\mu$	3
	Reductone	Water	286	13.5	After 10 hr, 5°C + H_2 atmosphere, λ_{max} 320 $m\mu$, $m\epsilon$ 0.68	4
3.	2,4-Pentanedione[c]	Water	274	1.5	15% Enol in water[d]	6
		Methanol	274	7.9	84% Enol in 95% EtOH[d]	7
		n-Hexane	271	10.0	96% Enol in *n*-hexane[d]	6
	cis-4-Methoxy-3-pentene-2-one	Methanol	262	14.1		8
		n-Hexane	245	12.3		
	trans-4-Methoxy-3-pentene-2-one	Methanol	254	13.5		8
		n-Hexane	247	12.0		
	3,3-Dimethyl-2,4-pentanedione	95% EtOH	313[e,f]	0.17		9
4.	3-Methyl-2,4-pentanedione[c]	*n*-Hexane	287	7.1	Intramolecularly H-bonded *cis* form	9
		67% Methanol	292	1.1	Intramolecularly H-bonded *cis* form	
5.	3-Ethyl-2,4-pentanedione[c]	*n*-Hexane	286	4.3	Intramolecularly H-bonded *cis* form	9
		67% Methanol	293	1.0	Intramolecularly H-bonded *cis* form	
6.	3-*n*-Propyl-2,4-pentanedione[c]	*n*-Hexane	292	4.0	*cis* Form	9
			253*s*	1.6	*trans* Form	
		67% Methanol	295	1.1	*cis* Form	
			264	1.2	*trans* Form	

TABLE 4.1 (*cont.*)

Compound[a]	Solvent	Keto ⇄ Enol		Remarks	Ref.
		$\lambda_{max}(m\mu)$[b]			
7. 3-*n*-Butyl-2,4-pentanedione[c]	*n*-Hexane	288	5.0	*cis* Form	9
		252*s*	1.8	*trans* Form	
	67% Methanol	295*s*	1.1	*cis* Form	
		265	1.5	*trans* Form	
8. 3-*i*-Butyl-2,4-pentanedione[c]	*n*-Hexane	292	4.3	*cis* Form	9
		255*s*	1.7	*trans* Form	
	67% Methanol	295	1.3	*cis* Form	
		262	1.5	*trans* Form	
9. 3-*i*-Propyl-2,4-pentanedione[c]	*n*-Hexane	251	3.7	*trans* Form	9
	67% Methanol	263	4.5	*trans* Form	
10. 3-*sec*-Butyl-2,4-pentanedione[c]	*n*-Hexane	253	2.4	*trans* Form	9
	67% Methanol	264	3.0	*trans* Form	
11. 2,6-Dimethyl-3,5-heptanedione	Cyclohexane	274	11.1	100% Enol	6
	Water	274	1.6	~14% Enol	
12. 2,2,6,6-Tetramethyl-3,5-heptanedione	Isooctane	274	12.7	100% Enol	6
	Water	278	1.8	~14% Enol	
13. 1,2-Cyclopentanedione	*n*-Hexane	246	10.0		10
		300[e]	0.08		
	95% EtOH	252	7.9		11
2-Methoxy-2-cyclopentene-1-one	80% Dioxane	246	9.1		12
		308[e]	0.05		
2,3-Butanedione	*n*-Hexane	276[e,f]	0.016		13
		447*fs*[e,f]	0.02		
	95% EtOH	286[e,f]	0.025		14
		417[e,f]	0.01		
14. 2,4-Dimethyl-1,3-cyclobutanedione	Aqueous acid	245	13.8		15

	1-Methoxy-2,4-dimethyl-1-cyclobutene-3-one	Methanol	242.5	13.0		8
15.	1,3-Cyclopentanedione	Aqueous acid	242	20.9		16
	Cyclopentene-3,5-dione	H_2O-EtOH (1:1)	220[f]	15.8		17
			310[e,f]	0.02		
			360[e,f]	0.02		
16.	1,2-Cyclohexanedione	*n*-Hexane	263	8.3		10
			305*s*[e]	0.13		
		95% EtOH	266	2.6	Mainly diketo form	18
	3,3,6,6-Tetramethyl-1,2-cyclohexanedione	95% EtOH	298[e,f]	0.029		19
			380[e,f]	0.01		
17.	1,3-Cyclohexanedione	95% EtOH	253	22.4		20
		o-Dichlorobenzene	294[e,f]	~0.2		
	3-Ethoxy-2-cyclohexen-1-one	95% EtOH	249	18.6		21
18.	5,5-Dimethyl-1,3-cyclohexanedione	Aqueous acid	259	15.1		15
	2,2,5,5-Tetramethyl-1,3-cyclohexanedione	Methanol[g]	288[e,f]	0.06		22
19.	5,5-Spiro-cyclohexyl-1,3-cyclohexanedione[h]	Methanol	252	16.7		22
		n-Heptane	237	4.7	Mainly diketo form	
20.	1,2-Cycloheptanedione	*n*-Hexane	280[e,f]	0.02		10
			405*fs*[e,f]	0.5		
21.	Ethyl acetoacetate	Vapor	263[e,f]	0.3	Mainly keto form	23
			238	16.0		
		n-Hexane	242	6.3	52% Enol[i]	24
		95% EtOH	248	1.6	7.1% Enol in methanol[j]	
		Water	255[e,f]	0.1	Mainly keto form	26, 27
	β-*tert*-Butoxyacrylic acid	95% EtOH	237	15.5		28
22.	Diethyl malonate	Methanol	235[e,f]	0.05		27
23.	Oxalacetic acid	Petroleum ether	260	8.8	In water 8% enol[k]	29

TABLE 4.1 (*cont.*)

	Compound[a]	Solvent	Keto ⇄ Enol		Remarks	Ref.
			$\lambda_{max}(m\mu)$[b]	$m\epsilon$		
24.	1-Phenyl-1,3-butanedione	Ether	248	6.0		31
			310	15.0		
		Water	250	10.0		
			313	5.5		
25.	1-Phenyl-2-methyl-1,3-butanedione	95% EtOH	247[f]	11.6		32
			284	1.5		
			310[e,f]	0.2		
			228*i*	4.5	Reverts to keto form in time	
			309	12.2		
		n-Hexane	230	4.5		31
			310	12.0		
	1-Phenyl-2,2-dimethyl-1,3-butanedione		237[f]	14.5		32
			278*s*	1.05		
			310[e,f]	0.1		
	4-Methoxy-4-phenyl-3-buten-2-one	*n*-Heptane	272	16.2		33
	3-Methoxycrotonophenone	*n*-Heptane	248	9.8		33
			278	13.5		
26.	1,3-Diphenyl-1,3-propanedione	95% EtOH	251	8.4		32
			345	23.0		
		n-Heptane	250	7.2		34
			338	22.4		
		Water	254	12.5		32
			358	11.8		
	cis-1,3-Diphenyl-1-methoxy-1-propen-3-one	*n*-Heptane	249	10.3		34
			284	10.3		

No.	Compound	Solvent	λ (nm)	ε × 10⁻³	Remarks	Ref.
	trans-1,3-Diphenyl-1-methoxy-1-propen-3-one	*n*-Heptane	242	10.7		34
			306	16.5		
27.	1,3-Di-2-furyl-1,3-propanedione	95% EtOH	284	8.3		35
			365	30.5		
	1,3-Di-2-furyl-2-propen-1-one	95% EtOH	256	4.3		36
			354	60.3		
28.	1,1,2-Tribenzoylethane[h]	EtOH (abs.)	248[f]	40.2		37
			282*s*[f]	3.5		
		Isooctane	244	21.8	Intramolecularly	
			312	10.8	H-bonded *cis* form	37
29.	3-Benzoylcamphor[h]	Methanol	247[f]	14.9	One day in dark → 42%	22
			290[f]	1.7	enol	
			320*s*[f]	1.3		
			230	7.1	*n*-Heptane solution 3 days	
			308	17.9	in dark → 92% enol	22
	Acetophenone	95% EtOH	242[f]	16.3		32
			279[f]	1.2		
			320[e,f]	0.05		
30.	2-*p*-Fluorobenzoylcyclohexanone[h]	95% EtOH	248[f]	14.0	Enol content at equili-	38
			315[f]	1.7	brium ~ 25%	
			248	5.2	Enol content at equili-	38
			315	12.4	brium ~ 25%	
31.	2-*m*-Chlorobenzoylcyclohexanone[h]	95% EtOH	244[f]	14.8	Completely enolic at	38
			292[f]	2.3	equilibrium	
			244	4.8		
			315	11.2		
	1-Methoxy-2-*o*-chlorobenzoylcyclohexene	95% EtOH	290	13.7		38
	α-Methoxy-2-*o*-Chlorobenzalcyclohexanone	95% EtOH	280	9.2		38

TABLE 4.1 (*cont.*)

	Compound[a]	Solvent	Keto ⇄ Enol λ_{max} (mμ)[b]	mϵ	Remarks	Ref.
32.	2-Benzoylcyclohexanone	95% EtOH	245[f]	9.7	In methanol 3% enol at equilibrium	38
			245	3.8		
			314	10.1		
	2-Benzoylindanone[h]	95% EtOH	248	10.5	In methanol 78% at equilibrium	38
33.	α-Propionyl-*p*-chlorophenylacetonitrile[c]	Cyclohexane	226[f]	11.7	5% Enol	39
			274[f]	1.0		
		95% EtOH[g]	274	18.7	98% Enol	39
		Aqueous alkali	250	9.0	Iso-π-electronic to compound 34	39
			310	17.0		
	1-Cyano-1-*p*-chlorophenyl-2-ethoxy-1-butene	95% EtOH	274	19.0		39
34.	α-(*N*,*N*-Dimethylaminoacetyl)-*p*-chlorophenylacetonitrile	95% EtOH	250	8.5	Dipolar structure; H^+ on nitrogen	39
			310	16.4		
		95% EtOH[g]	273	15.5	This salt iso-π-electronic to the enol of compound 33 in 95% EtOH	
35.	ω-Cyanoacetophenone	95% EtOH	245[f]	13.0	14% Enol in 95% EtOH, 6% in cyclohexane	39
	trans-β-Methoxy-β-phenylacrylonitrile		275–280	2.0		
		95% EtOH	272	15.0		39
36.	1-Cyano-2-indanone	95% EtOH[g]	274	8.5	Completely enolized in all solvents	39
	2-Methoxy-3-cyanoindene	95% EtOH	272	9.5		39
	1-Cyano-1-methyl-2-indanone	95% EtOH			No bands above 230 mμ	39
37.	2-Acetyl-5,5-dimethylcyclohexane-1,3-dione[h]	95% EtOH[c]	232	10.0		40
			276	9.7		

Compound	Solvent	λ	ε	Remarks	Ref.
	Cyclohexane	230	10.2		
		276	9.8		40
38. 2,4,6-Heptanetrione	Methanol	274	8.5		41
		314	4.8		
39. 3-Acetyl-2,4-pentanedione	Methanol	278	8.1		42
40. α-Ethyltetronic acid	Aqueous acid	233	12.0		22
	Ethylene chloride	225	10.0		22
41. α-Methyltetronic acid[h]	95% EtOH	232	12.3		43
42. γ-Methyltetronic acid	Aqueous acid	227	15.8		22
	Ethylene chloride	219	0.79	Mainly keto form	22
43. 2,2-Dimethyl-1,3-dioxane-4,6-dione	Methanol	259	20.5		44
44. 2,2,5-Trimethyl-1,3-dioxane-4,6-dione	Methanol	273	22.0		44
2,2,5,5-Tetramethyl-1,3-dioxane-4,6-dione	Methanol			No band above 230 mμ	44
Malonic acid	Methanol			No band above 230 mμ	44
Diethyl 3-methylmalonate	Methanol			No band above 230 mμ	44

[a]Unnumbered compounds are reference substances.
[b]s = Shoulder; fs = fine structure; i = inflection.
[c]Keto–enol equilibrium in other solvents studied.
[d]Data from reference [5].
[e]$n \rightarrow \pi^*$ band.
[f]Bands derived mainly from ketone forms; all others from enols.
[g]Contains some hydrochloric acid.
[h]And many other keto–enol tautomers studied.
[i]From reference 25, which contains data on the keto–enol tautomers of many other alkyl acetoacetates studied in six solvents.
[j]Data from reference 25.
[k]Data from reference 30.

A. Electronic Effects of Substituents

When an electron-acceptor group, a carbonyl group, and a hydrogen atom are attached to a carbon atom, enolization can take place readily through ionization of the hydrogen atom followed by an electron-pair shift and the attachment of the proton to the carbonyl oxygen atom. The electronegativity strengths of the electron-acceptor and carbonyl groups play an important role in the extent of enolization of the molecule, as can be seen from the data in Table 4.2.

In agreement with these data is the report that no evidence for enolization has been found for such alkylsulfonyl derivatives as $RSO_2CH_2SO_2R$

TABLE 4.2
Percentage of Enolization of a Variety of Ketones in the Undiluted State

$$X—CH_2—\overset{\overset{\large O}{\|}}{C}—Y$$

Substituent			
X	Y	Enol (%)[a]	Ref.
H	OC_2H_5	0	
H	CH_3	1.5×10^{-4}	45
C_2H_5OOC	OC_2H_5	7.7×10^{-3}	45
NC	OC_2H_5	2.5×10^{-1}	45
CH_3	CH_3	1.2×10^{-1}	45
H	$COCH_3$	1.1	45
C_6H_5	CH_3	2.9	45
CH_3CO	OCH_3	4.7	50
CH_3CO	OC_2H_5	8.0	45
		6.0	51
CH_2ClCO	OCH_3	7.1	50
CH_2FCO	OC_2H_5	7.2 ± 0.2	51
CH_2ClCO	OC_2H_5	10.9	50
C_6H_5CO	OC_2H_5	17.7	45
CCl_3CO	OC_2H_5	40–50	50
CHF_2CO	OC_2H_5	53 ± 4	51
CH_3CO	CH_3	76.4	45
CF_3CO	OC_2H_5	89	51
C_6H_5CO	CH_3	89.2	45

[a]Some of the values (45) extrapolated to undiluted state. For the fluoro derivatives and ethyl acetoacetate (% enol = 6.0) nuclear magnetic resonance was used with pure liquids.

and RSO_2CH_2COR(52), which apparently results from the weak electronegativity of the alkylsulfonyl group. Somewhat similar results have been obtained for the esters of β-keto phosphonic acids(27). Such compounds as $(C_2H_5O)_2P(O)CH_2COCH_3$ and $(C_2H_5O)_2P(O)CH_2COC_2H_5$ are not enolized in methanol, hexane, or water.

The effect of an electron-donor group on the enolization of 1,3-diphenyl-1,3-propanedione is in agreement with the electronegativity concept. Thus, whereas 1,3-diphenyl-1,3-propanedione (**1**) and its *p*-methyl and *p*-methoxy derivatives are completely enolized in alcoholic solution, the analogous 1,3-diphenyl-2-hydroxy-1,3-propanedione (**2**) and its *p*-methyl and *p*-methoxy derivatives are present mainly as the diketo tautomers(53).

However, the electronegativity principle is not so simple, as shown by the report that 1,3-dimethoxyacetone is enolized 6.2% in water, whereas many other alkoxy and hydroxy derivatives of acetone are enolized to a much larger extent than acetone(54).

No.	Solvent	$\lambda_{max}(m\mu)$	$m\epsilon$
1.	Ethanol	254	10.0
		345	31.6
2.	Methanol	234	15.9
		360	4.0

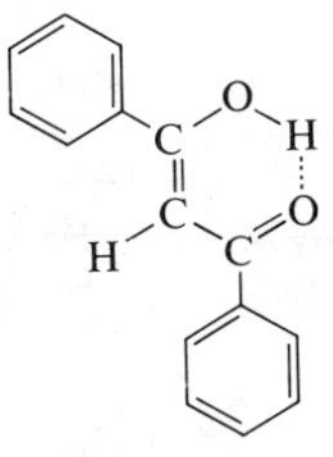

1 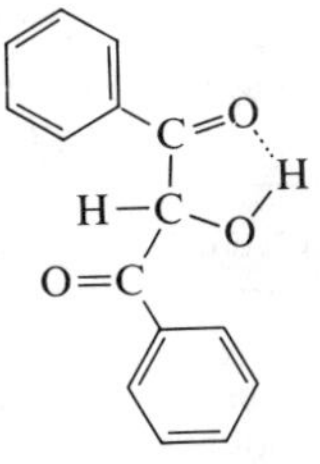 **2**

The extent of enolization of a carbonyl compound depends on other factors besides the relative acidity of the electronegatively substituted C—H groupings. Thus ω-nitroacetophenone is present in the enol form to the extent of 10.3% in toluene and 2.7% in 67% aqueous methanol(55). The enolization is much less than one would expect when the compound is compared with 1-phenyl-1,3-butanedione (see Table 4.2). Thus, although the nitro group is more electronegative than the benzoyl group, 1,3-diphenyl-1,3-butanedione, or essentially ω-benzoylacetophenone, enolizes more readily than does ω-nitroacetophenone.

B. Solvent Effects

When an enol tautomer can form an intramolecular hydrogen bond as in **4**, the percentage of the *cis*-enol form is increased in nonpolar solvents and decreased in polar solvents.

$$H_3C{-}CO{-}CH_2{-}CO{-}CH_3\ (\mathbf{3}) \underset{\text{polar solvent}}{\overset{\text{nonpolar solvent}}{\rightleftharpoons}} H_3C{-}C(OH){=}CH{-}C(=O){-}CH_3\ (\mathbf{4}) \tag{4.1}$$

This is probably because of the lower polarity of the hydrogen-bonded *cis*-enol as compared with the diketo form. Consequently polar solvents should stabilize the diketone form. The enol form (e.g., **4**) would be stabilized in nonpolar solvents in which the intramolecular hydrogen bond would not have to compete with the solute–solvent intermolecular hydrogen bonds. This type of solvent effect is demonstrated in Table 4.1 for a large variety of compounds. The same type of solvent effect is shown for ethyl acetoacetate (Table 4.3) and 3-methyl-2,4-pentanedione (Table 4.4). Other compounds that show this phenomenon are 1,2-cyclopentanedione, 1,3-cyclohexanedione, 3-benzoylcamphor (Fig. 4.1),

TABLE 4.3
Percentage of Enolization of Ethyl Acetoacetate at 20–24°C in Various Solvents

Solvent[a]	Solvent polarity, X_R	Enol (%)	
Water	<35		0.4
Acetic acid			5.7
Methanol	43.1	7.1	6.9
Acetone	45.7		7.3
Pure liquid		7.5	7.7
Chloroform	44.2		8.2
Nitrobenzene	42.6		10.1
Ethanol	43.9		10.5
Benzene	46.9	20.0	16.2
Ether	48.3	32.0	
Carbon disulfide		42.0	
Cyclohexane	50.0	52.0	
n-Hexane	50.9	52.0	46.4

[a]Somewhat similar results for 15 other alkyl acetoacetates as shown in the third (25) and fourth columns (26).

TABLE 4.4
Tautomeric Equilibrium Constants[a] in 1% Solutions of 3-Alkyl-2,4-pentanedione[b] (Bromometric Evidence)(7)

Solvent	Methyl	*sec*-Butyl
67% Methanol	0.19	0.26
Methanol	0.41	0.26
Benzene	0.61	0.26
Carbon tetrachloride	0.64	0.27
Ethanol	0.66	0.27
Ether	0.69	0.26
Hexane	0.65	0.29

[a]For the methyl derivative $K_T(cis) =$ *cis*-enol/ketone $= EL$, where E is the *cis*-enolizability of the keto–enol, equal to the relationship between the constant of the keto–*cis*-enol equilibrium and that of the constant of keto–enol equilibrium of ethyl acetoacetate in the same solvent, and L is the *cis*-enolizing capacity of the solvent equal to the keto–enol equilibrium constant of ethyl acetoacetate (as the standard of comparison) in the particular solvent. For the *sec*-butyl derivative $K_T(trans) =$ *trans*-enol/ketone = constant.

[b]About eight other alkyl derivatives studied in eight solvents.

and 1-phenyl-1,3-butanedione. In all cases it would appear that the intramolecular hydrogen bond, the increased resonance of the enol as compared to the keto tautomer, and the decreased polarity of the Z_n resonance structure in a nonpolar, as compared with a polar, solvent all tend to increase the stability of the *cis*-enol form in nonpolar solvents.

One cannot expect a tautomeric equilibrium involving two or more species to respond in a simple way to solvent change. This complexity can even be seen in the keto–enol tautomerism of ethyl acetoacetate whose equilibrium is largely determined by the interaction of the keto form with the solvent. The equilibrium does show a relationship to the solvent polarity (Z value)(56) or the X_R value (Table 4.3), although some deviations in the correlation are apparent in Table 4.3 and in Kosower's data(56).

Some compounds are enolized in all solvents; for example, the 2-acyl-1,3-cyclohexanediones studied by Chan and Hassal have approximately identical molar absorptivities in ethanol–hydrochloric acid and in

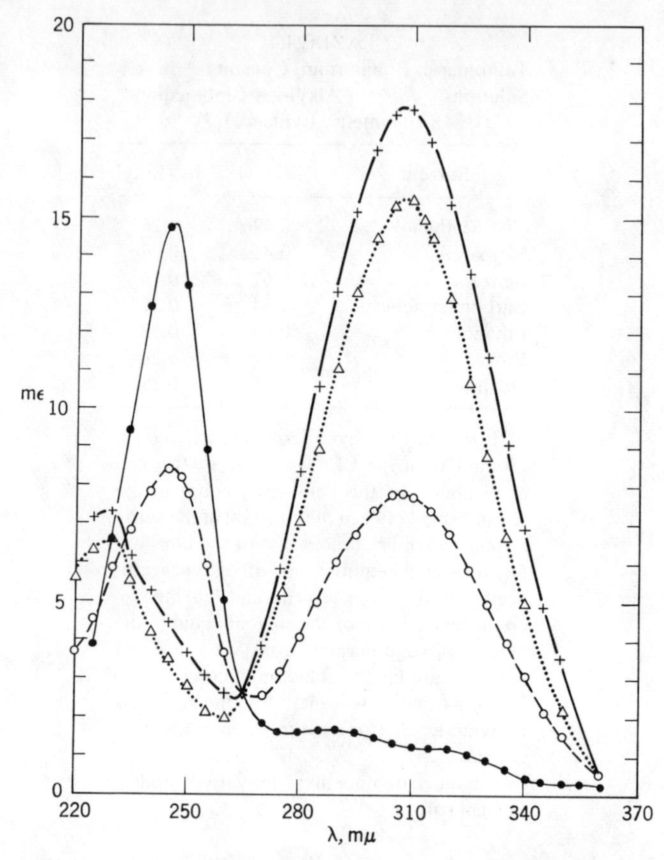

Fig. 4.1. Ultraviolet absorption spectra of benzoylcamphor (5×10^{-5} or 10^{-4} M): diketo form in methanol (· · ·), measured immediately; enol form in methanol (—+—), measured immediately; equilibrium in methanol (—○—); equilibrium in *n*-heptane (—△—)(22).

cyclohexane(40). All these compounds have unusually strong intramolecular hydrogen bonds, as indicated by their infrared and visible absorption spectra(40). A few 2-acylcyclohexane-1,3-diones have been isolated from natural sources (e.g., calythrone, humulone, lupulone, usnic acid, leptospermone, angustione, dehydroangustione, flavaspidic acid, and albaspidin)(40). These compounds absorb in alcohol at approximately 235 and 280 mμ with spectra characteristic of an intramolecularly hydrogen bonded enol tautomer.

Whereas *cis*-enolization of 3-alkylacetylacetones is solvent dependent, *trans*-enolization is independent of the nature of the solvent (Table 4.4)(7).

A fourth type of solvent effect is found in the tautomeric carbonyl compounds whose enol concentration is greater in polar than in nonpolar solvents. In these compounds the chelation of the enol is prevented or cannot take place. Examples of this class of compound are 1,3-cyclohexanediones and α-acylphenylacetonitriles. Tetronic acids, which can form an intramolecularly hydrogen bonded five-membered ring, may also show this phenomenon since a hydrogen bond in this type of ring is under greater strain that is such a bond in a six-membered ring. In line with this is the report that γ-methyltetronic acid is almost completely enolized in weakly acidic aqueous solution and is present in the keto form in ethylene chloride (22).

Examples of this phenomenon are shown in Table 4.5. In these compounds there are two important factors; namely, the decrease in enol content with increasing solvating power of the solvent and with a change to a less polar solvent (39, 57). The first trend is regarded as resulting from the solvation of the keto form, forcing the tautomeric equilibrium in this direction. The second is derived from the low solubility of the nonchelated enol in nonpolar solvents. In summary, Russell (57) states that, whereas solvation of the keto tautomer is the important factor determining the extent of enolization in hydroxylic solvents, in nonhydroxylic solvents the ability to form a chelate enol is decisive.

TABLE 4.5
Solvent Effect on the Enolization of β-Ketonitriles[a] (39, 57)

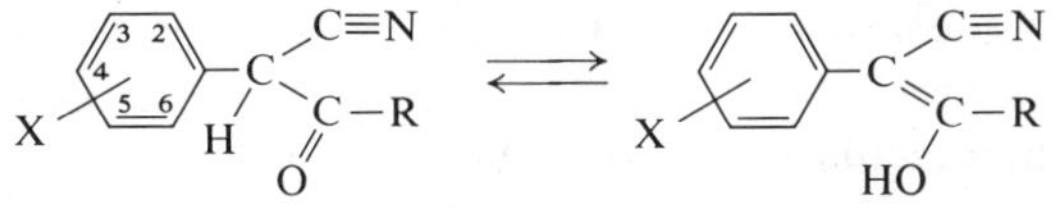

Solvent	Enol (%)[b]			
	X = H, R = C_6H_5	X = H, R = 2-furyl	X = 3,4-Dichloro, R = 2-furyl	X = 4-Chloro, R = C_2H_5
Water	47	34		33[c]
Methanol	60	60	87	84[c]
Ethanol	67	61	91	98[c]
n-Pentanol	76	68		60
Ether	55	43	83	
Ether–hexane (1 : 1)	39		81	
Hexane–ether (9 : 1)	20		50	
Chloroform				7.5
Cyclohexane				5.0
Hexane				

[a]Concentration 10 μg/l.
[b]From the enolic structure it can be seen that *cis* and *trans* forms are possible.
[c]Solution, approximately 0.01 *N* hydrochloric acid.

For many compounds containing two or three different acyl groups two (and sometimes three or more) enolic tautomers are possible. Examples of these types of compounds can be found in Table 4.1. The correlation of structure with spectral bands and the attempt to determine the relative proportions of three or more tautomers in various solvents is a difficult matter. An example of such a compound is α-formyl-α-benzoyltoluene (**6**), which has even been isolated in two enolic forms (58). With the help of infrared and ultraviolet spectroscopy and a study of the reactions of the two forms with diazomethane the structures of the two isomers, as well as their spectra in solution, have been determined (59). The authors state that in nonpolar solvents the aldehyde-enol (**5**) predominates, in polar solvents the keto-enol (**7**).

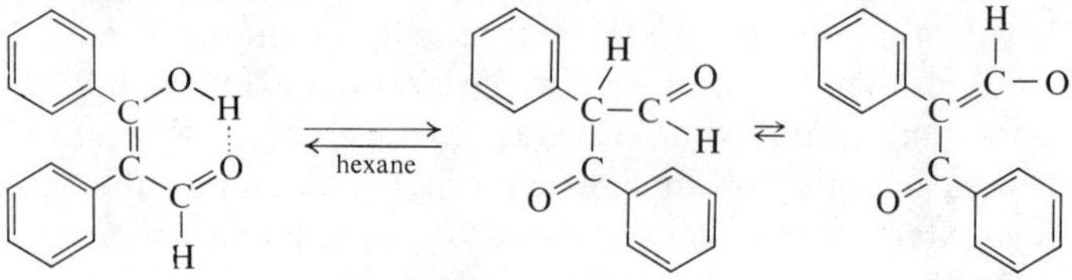

hexane (4.2)

5 **6** **7**

λ_{max} 238, 323 mμ
mϵ 14.0, 9.5

λ_{max} 245 mμ, mϵ 13.5

In addition to the postulated structures shown for **5** and **7**, geometric isomers should also be possible. Structure **7** could also be capable of having an intramolecularly hydrogen bonded form.

The high solubility of **5** in nonpolar solvents and of **7** in polar solvents is stated to be in accord with the Van't Hoff–Dimroth hypothesis (60), which states that the ratio of the concentrations of the forms of a tautomeric compound in any solvent is proportional to the ratio of the solubilities of the forms in that solvent.

C. Steric Effects

Another factor of importance in tautomerism is steric hindrance. Either the keto or the enol form may be favored, depending on which form is less sterically hindered. Consider the equilibrium in equation 4.3:

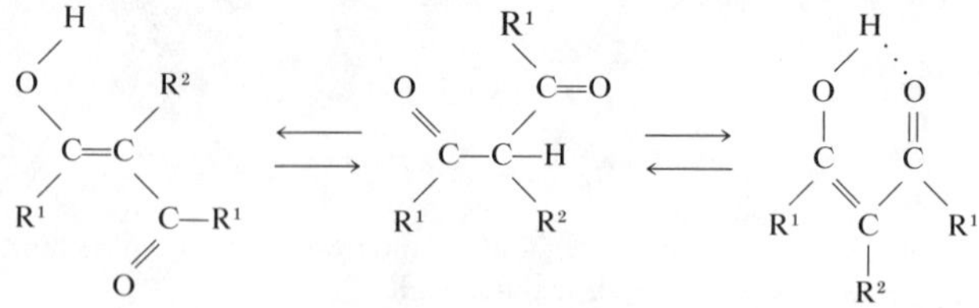

(4.3)

When $R^2 = H$, increasing the size and branching of R^1 would shift the equilibrium toward the hydrogen bonded *cis*-enol. For example, when $R^1 =$ methyl, isopropyl, or *tert*-butyl, the enolization in acetonitrile is 58, 64, 86%, respectively(6). However, the polarity of the solvent can still affect the equilibrium, as can be seen by the fact that all three compounds are almost completely enolized in cyclohexane or isooctane and only 14% enolized in water.

In the structures in equation 4.3, when $R^1 = CH_3$, increasing the size and branching of R^2 creates steric hindrance, opposing *cis*-enolization and favoring *trans*-enolization(9, 61, 62) (see Table 4.4). Examination of the ultraviolet and infrared absorption spectra of nine 3-alkyl-2,4-pentanediones in eight solvents has shown that, when the 3-alkyl is a primary group, the solution usually contains a mixture of three tautomers: the keto, *cis*-enol, and *trans*-enol(9); when the 3-alkyl is a secondary group, the solution contains the keto and *trans*-enol forms. Contrary to these derivatives, the 3-cyclopentyl analog does not enolize in any solvent.

D. Effect of Ring Size

In the alkyl-2-cyclanone carboxylates (Table 4.6) enolization in ethanol is strongly dependent on ring size(63). Fresh ethanolic solutions of 1,2-cyclopentanedione, 1,2-cyclohexanedione, and 1,2-cycloheptanedione contain 100, 100, and 2% enol, respectively; the same solutions after 36 hr at 25°C contain 96, 45, and close to 0% enol, respectively(10).

TABLE 4.6
Effect of Ring Size on the Tautomerism of Methyl[a] 2-Cyclanone Carboxylates in Ethanol(63)

```
      (CH2)n
     /      \
 H2C          C=O
  |           |
 H2C---------CHCOOC2H5
```

n	Enol (%)	n	Enol (%)
1	5	5	15
2	57	6	50
3	12	7	9
4	40	8	5

[a] $n = 1$, 2, and 3 are ethyl esters.

TABLE 4.7
Enolization of 2-Benzoylcyclanones (64)

2-Benzoyl derivative of	Enol (%)
Cyclopentanone	39
Cyclohexanone	3
Cycloheptanone	9
Indanone	78
Tetralone	25
Benzosuberone	48

Examination of the enolization of the 2-benzoyl derivatives of cyclopentanone, cyclohexanone, cycloheptanone, indanone, tetralone, and benzosuberone in methanol disclosed that the five-membered-ring derivatives exhibited the greatest tendency to enolize, followed by the seven- and six-membered-ring derivatives (Table 4.7)(64). The dibenzoylmethane-type derivatives show a tenfold greater tendency to enolize than those of the benzoylacetone type. Steric effects play a large role in this phenomenon.

E. Effect of Concentration

Since the photometric analysis of many types of compounds involves the separation and concentration of a compound either as a solid on a chromatogram or in the form of a very concentrated or very dilute solution, an understanding of the effect of concentration on spectra could prove to be of vital importance to the solution of a problem.

Meyer and Kapelmeier (65) have shown that the enolization of ethyl acetoacetate in various solvents depends on concentration to some extent. Thus in the following solvents enolization is affected by increasing the given percentage of ethyl acetoacetate: ethanol, 5%; benzene, 5%; carbon disulfide, 2%; hexane, 1%.

The method of Dieckmann (66) was used in the determination of enolization. Table 4.8 demonstrates the change in the equilibrium with concentration.

Many examples have been given showing the extent of enolization of a carbonyl compound in the solid state as compared with the soluble state. The alkyl acetoacetates enolize approximately 10% (range of about 7–16% for 15 compounds) in the pure form or in methanolic solution (25). In cyclohexane they are approximately 50% enolized (range 40–58%).

Because of the major role of oxalacetic acid in intermediate metabolism, the tautomerism of this acid and its derivatives has been studied (30). Oxalacetic acid appears to be present as the enol in water, methanol, and

TABLE 4.8
Effect of Concentration of 3-*n*-Butyl-2,4-pentanedione on Its Enolization[a] in Solution (9)

	Enolization (%)		
	Concentration (%)		
Solvent	1	0.1	0.02
67% Methanol	19	27–33	41–50
Ethanol	40	38–45	56–70
Hexane	61	100	–

[a]Determined by Dieckmann's procedure (66).

the pure compound to the extent of approximately 8, 21, and >21%, respectively. Diethyl oxalacetate is 50% enolized in methanol and 79% enolized in the pure form, whereas diethyl fluorooxalacetate is not enolized as the pure compound.

II. Ketol ⇄ Enediol Tautomers

In benzoin we have an example of a compound that is present in solution in the ketol form, as shown by its spectral similarity to acetophenone:

Compound	Solvent	λ_{max}(mμ)[a]	mϵ	Ref.
Benzoin	Methanol	248	13.8	
		318	0.28	
	n-Hexane	243	10	67
		283 *fs*	1.3	
Acetophenone	Methanol	242	10	68
		320 *s*	0.08	
	n-Hexane	238	15.9	69
		248	10	
		279 *fs*	1.3	

[a]*fs* = Fine structure; *s* = shoulder.

On the other hand, 9,9′-phenanthroin is stated to be present in solution as the enediol, as shown by the close resemblance of its ultraviolet absorption, visible absorption, and fluorescence spectra to *sym*-di-(9-phenanthryl)ethylene (70, 71). The enediol is postulated to be a *trans* derivative since irradiation with ultraviolet light causes a spectral change postulated to be a *trans* → *cis* isomerization(70). The two tautomers have been isolated(72).

In addition, benzoins are stabilized in their enediol form, provided that at least one phenyl group is highly substituted, as in the mesityl or isoduryl group(73).

When intramolecularly hydrogen bonded enediols can be formed, the enediol form is the stable one. Some examples of these compounds are shown in Table 4.9(74–81). This type of molecule absorbs at wavelengths much further into the visible than the analogous ketol form or the corresponding benzil-type derivative. The importance of intramolecular hydrogen bonding in stabilizing the tautomeric form of the heterocyclic benzoins can be seen in α-pyridoin. The structures and the spectra in cyclohexane of the ketol (**8a**), *trans*-enol (**9**), and the derived acetate (**8b**), methiodide salt (**10**), and α-pyridil (**11**) are compared(75). The importance of the intramolecular hydrogen bonds in shifting the long wavelength band toward the red is demonstrated by comparing the absorption spectra of the enediol **9** and its diacetate (λ_{max} 300 mμ; mε 20.0). Two forms of the latter compound were isolated, possibly *cis* and *trans* isomers.

In the molecules discussed in this section steric hindrance and intramolecular hydrogen bonding tended to stabilize the enediol form.

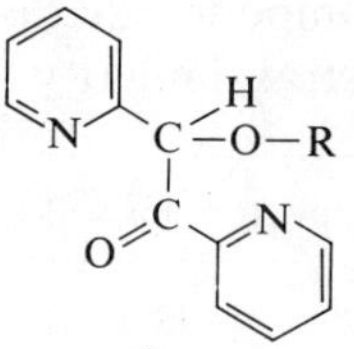

8 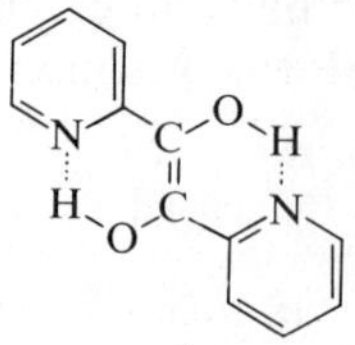 **9**

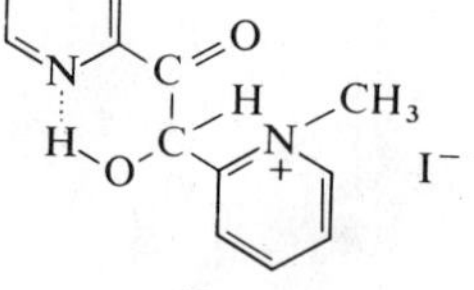

10 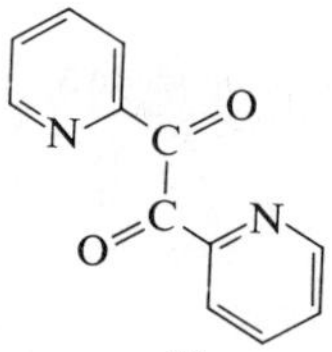 **11**

No.	R	λ_{max} (mμ)	mε
8a	H	unstable	
8b	Ac	263	6.5
9		387	20.0
10		225, 273	17.0, 12.6
11		235, 270	17.0, 8.5

TABLE 4.9
Some Enediols Stabilized by Intramolecular Hydrogen Bonds

Compound	Solvent	Color or $\lambda_{max}(m\mu)$	$m\epsilon$	Ref.
1-Benzoyl-2-α-pyridyl-1,2-ethanediol	*n*-Hexane	405	15.9	74
2,2′-Pyridoin	Cyclohexane	387	20.0	75
2,2′-Quinaldoin	Chloroform	440	~62	76,77
2,2′-Thiazoloin		Orange		78
3,3′-(5,6-Benzoquinoxoloin)	Pyridine	Red		79
2,2′-Benzothiazoin	Chloroform	406 *fs*[a]	~40	80
2,2′-Benzoselenazoin	Chloroform	417 *fs*[a]	~33	81

[a]Fine structure.

III. Thioketo ⇄ Thioenol Tautomerism

The thiocarbonyl group is a much more powerful electron acceptor (and thus more electronegative) than the carbonyl group, as shown by the absorption spectra in ethanol of *N*,*N*,*N*′,*N*′-tetramethyl-4,4′-diaminobenzophenone (λ_{max} 359.5 mμ, mϵ 33.5) and its thiobenzophenone analog (λ_{max} 433.5 mμ, mϵ 43.8)(82).

Examination of several thiocarbonyl compounds has indicated that these compounds are more enolized than the analogous carbonyl derivatives(83, 84). Thus ethyl thioacetylacetate is 41% enolized in ethanol, whereas ethyl acetoacetate is 11% enolized. Ethyl thiobenzoylacetate shows the same phenomenon; however, it is enolized 87% in ethanol and about 95% in isooctane. An increase in the electronegativity of the thiobenzoyl group can be expected with the addition of a *p*-nitro group—and a decrease with a *p*-ethoxy group. As shown in Table 4.10, the enolization of compounds 1, 2, and 3 is then as to be expected. The spectra of these compounds are compared with that of compound 4 in Table 4.10, which is present only as a thiocarbonyl compound and consequently shows a full-strength thiocarbonyl $n \rightarrow \pi^*$ transition. On the other hand, compound 5 is only present as the thioenol and so has $\pi \rightarrow \pi^*$ transitions but no band near 560 mμ derived from a thiocarbonyl $n \rightarrow \pi^*$ transition.

Examples of various thioketone compounds and the extent of their enolization are presented in Table 4.11. The extent of *cis–trans* isomerization of the thioenol forms or the presence of enolic forms has not been investigated thoroughly as yet. Model compounds were used to prove the structures of the compounds in Table 4.11. The proof was somewhat similar to that used in Table 4.10. Iodometric titration was used to determine the percentage of enolization of many of the compounds (83–85).

TABLE 4.10
Enolization of Some Ethyl Thiobenzoylacetates

No.	Substituent X	Substituent R	Solvent	Thioenol (%)	λ_{max} (mμ)	mϵ	Ref.
1	NO_2	H	Isooctane	97	258	17.2	83
					310	9.15	
2	H	H	Isooctane	87	233	37.1	84
					250	42.5	
					297	39.7	
					575[a]	0.0016	
3	OC_2H_5	H	95% EtOH	82	550[b]	0.017	83
4	H	CH_3	95% EtOH	0	565[b]	0.10	84
5	$C_6H_5C(SC_2H_5)=CH-C(=O)OC_2H_5$		Isooctane	"100"	238	7.5	84
					295	7.8	

[a] $n \rightarrow \pi^*$ Band. Compound 2 in ethanol, λ_{max} 564 mμ, mϵ 0.002.
[b] $n \rightarrow \pi^*$ Band. Ultraviolet bands not reported.

The compounds 2-pentanone-4-thione, 1-phenyl-1-thione-3-butanone, and 1,3-diphenyl-1,3-propanedithione have been found to exist as enols, by means of a study of their nuclear-magnetic-resonance spectra(87).

IV. Phenol Tautomers

In this section we consider the tautomerism of carbocyclic and oxygen and sulfur heterocyclic phenols. As in most tautomers, the shift of a hydrogen atom in a phenol can cause a drastic change in absorption spectra. The molecular structure and the solvent play a role in the equilibrium.

The tautomeric transformations of the various types of phenol have been discussed by Ershov and Nikiforov(88).

TABLE 4.11
Percentage of Enolization of Some "Thioketone" Compounds[a]

Substituent				
R	R′	R″	Enol (%)	Ref.
CH_3	C_2H_5	H	41	85
C_6H_5	C_2H_5	H	87	84
C_6H_5	C_2H_5	H	95[b]	84
$4\text{-}C_2H_5OC_6H_4$	C_2H_5	H	82	83
$4\text{-}O_2NC_6H_4$	C_2H_5	H	97	83
C_6H_5[c]	H	H	~100	86
$C_6H_5CH{=}CH$	H	H	~100	86
CH_3	C_2H_5	CH_3	63	85
CH_3	C_2H_5	$CH_2CH(CH_3)_2$	64[d]	85
CH_3	C_2H_5	$CO_2C_2H_5$	61	85

[a]In 95% ethanol, unless otherwise indicated.
[b]In isooctane.
[c]Similarly the 3-thienyl, 2-furyl, 1-naphthyl, 2-naphthyl, and 3,4-dimethoxyphenyl analogs are also ~ 100% enolized in alcohol solution.
[d]At 60°C.

A. Phenol and its Derivatives

The evidence for the possible presence of the keto form of the simple phenols has been discussed in detail by Thomson(89). No ketonic tautomers of the mono-, di-, and trihydroxy phenols have been isolated nor do the ultraviolet or infrared absorption spectra of these compounds indicate any appreciable amounts of the keto isomers. Although phloroglucinol has an estimated total energy (ΔE) of the enol form minus that of the keto form of +3 kcal/mole (the ΔE of phenol and of 9-anthrol is −18 and +1 kcal/mole, respectively)(90), only in some of its chemical reactions does phloroglucinol appear to have a keto form(89).

The separation and isolation of the keto (**12**) and enol (**13**) tautomers of 1,2,3,4-tetrahydroxybenzene have been reported(91). The spectral bands reported for these compounds in dioxane solution are shown in equation 4.4. The spectrum of the enol form closely resembles that of the model compound, 1,2,3,4-tetramethoxybenzene (**14**).

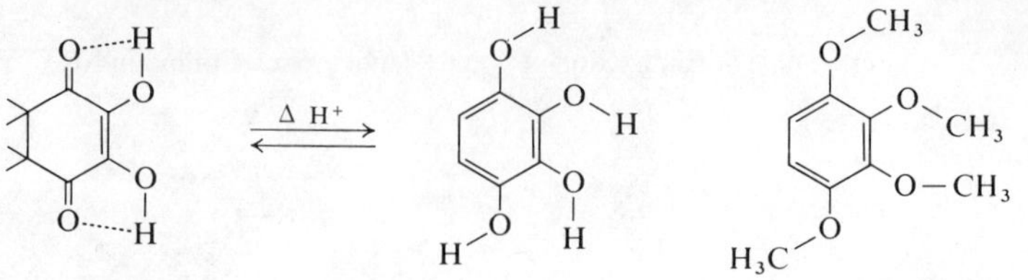

$$\mathbf{12} \underset{}{\overset{\Delta\ H^+}{\rightleftarrows}} \mathbf{13} \qquad \mathbf{14} \qquad (4.4)$$

12: λ_{max} 226, 308 mμ; mε 5.75, 17.4

13: λ_{max} 220, 282 mμ; mε 8.9 1.66

14: λ_{max} 220, 284 mμ; mε 8.9, 2.5

The keto tautomer (**12**) is stabilized by two intramolecular hydrogen bonds and resonance.

B. Naphthols

The monohydroxy naphthols are present in solution as the phenolic form, although in some of their chemical reactions they appear to react as ketones (89). When two hydroxy groups are present in the same ring, the diketo form can be isolated, provided that both carbonyl groups are conjugated with the second benzene ring. Both the dienol (**15**) and the diketone (**16**) can be crystallized unchanged (92).

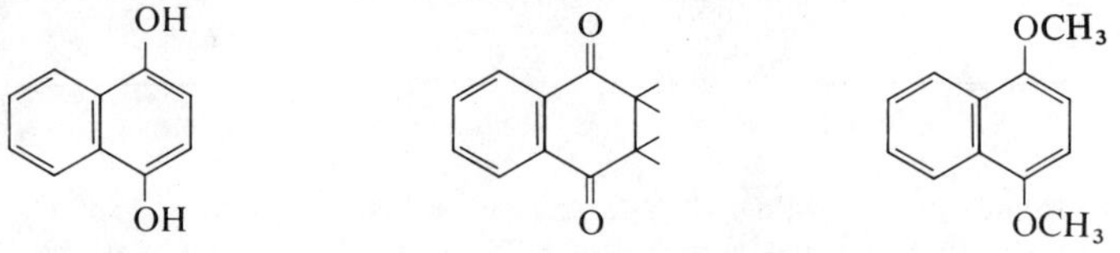

15   **16**   **17**

No.	Solvent	λ_{max} (mμ)	mε	Ref.
15	95% EtOH, 5% HCl	244	15.1	93
		275*s*	1.2	
		327	5.3	
		334	5.3	
16	95% EtOH	228	31.6	92
		253	10.0	
		294	1.6	
17	95% EtOH	240	20.0	94
		320	5.0	
		335	5.0	

Comparison of the two tautomers with **17** demonstrates which isomer is the dienol.

Introduction of hydroxy groups at the peri positions (5 and/or 8) of 1,4-naphthalenediol stabilizes the diketo system through increased resonance and strong intramolecular hydrogen bonding. Examples of such compounds that have been isolated as 1,4-diketo tautomers are listed in Table 4.12 (92, 95–97). As can be seen, the absorption spectra of the two tautomers differ from each other and also from the analogous quinone, as demonstrated by the spectral properties of 5,8-dihydroxy-1,4-naphthoquinone (**18**) and 5,8-dihydroxy-1,2,3,4-tetrahydro-1,4-naphthoquinone (**19**).

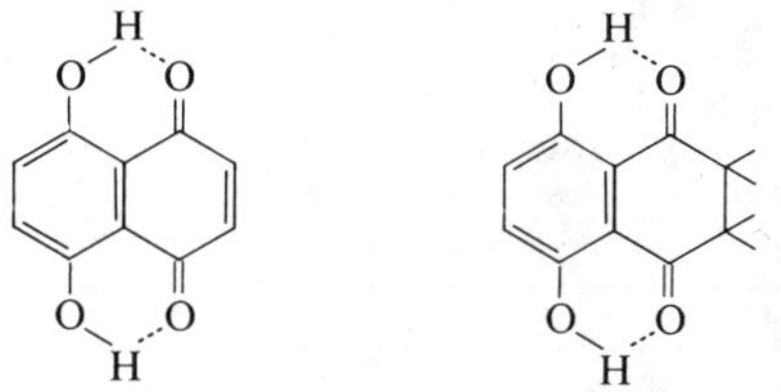

18 **19**

No.	Solvent	λ_{max} (mμ)	mϵ	Ref.
18	Methanol	270	15.9	98
		330	1.6	
		490	12.6	
		515	12.6	
19	Ethanol	228	14.1	96
		255	11.0	
		395	7.2	

One monohydroxynaphthalene derivative has been reported to form keto and enol tautomers (**20** and **21**, spectra in 95% ethanol) (99).

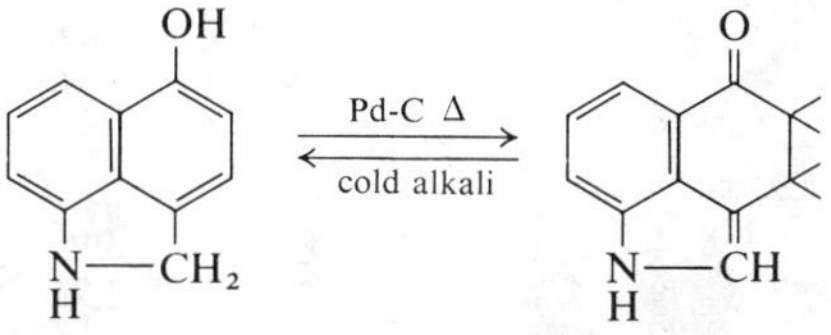

Pd-C Δ / cold alkali (4.5)

20: λ_{max} 335, 405 mμ; mϵ 9.6, 0.24

21: λ_{max} 224, 245, 327, 360 mμ; mϵ 15.5, 15.9, 4.3, 4.4

Resonance involving the NH and C=O resonance terminals could account for the stability of the carbonyl tautomer (**21**). On the other hand, the 405-mμ band in the enol tautomer (**20**) is difficult to account for since 5-amino-1-naphthol absorbs at λ_{max} 228 and 308 mμ (mϵ 47.9 and 7.08, respectively) in ethanol (99).

C. Polynuclear Phenols

The tautomerism of the larger ring phenols has not been studied to any large extent, neither has the effect of solvents on the spectral properties of

TABLE 4.12
1,4-Naphthalenediol Tautomers

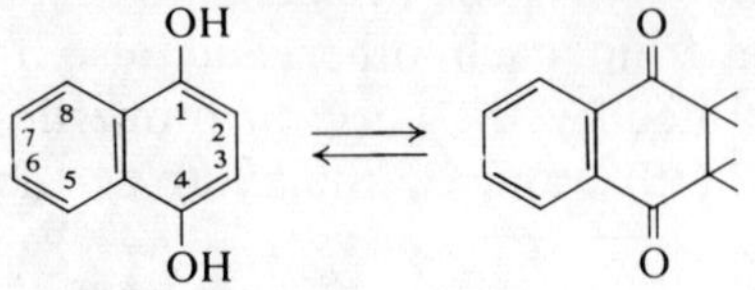

	Diol[a]			Dione[b]		
Substituent	λ_{max} (mμ)[c]	mϵ	Ref.	λ_{max} (mμ)[c]	mϵ	Ref.
None	244	15.1	93	228	31.6	92
	275*s*	1.2		253	10.0	
	327	5.3		294	1.6	
	334	5.3				
2-CH_3	245	30.9	93	226	31.6	92
	266	2.5		247	10.0	
	325	5.1		295	1.6	
	333	5.0				
5-HO	222	50.0	93	230	21.4	92,95
	316	6.5		245*s*	12.6	
	348	8.3		347	5.1	
5-HO-2-CH_3				230	19.1	95
				243*s*	11.0	
				260*s*	4.3	
				348	5.1	
5-HO-7-CH_3				236	22.9	95
				247*s*	11.2	
				270	5.9	
				348	5.4	
5,8-diHO				228[d]	14.1	96
				255	11.0	
				395	7.2	
5,8-diHO-6-CH_3				234	16.6	96
				263	9.8	
				395	7.6	
5,8-diHO-6-Anilino				265	19.5	96
				308	8.7	
				339	7.2	
				417	14.5	
2,5,8-triHO				250	12.3	96
				264	12.3	
				303	8.5	
				356	8.1	
				399	9.6	

[a]In 95% ethanol + 5% HCl.
[b]In 95% ethanol.
[c]*s* = Shoulder.
[d]In water similar spectrum (97).

the few known compounds of this type. Thus it would appear from data in the literature that the following monohydroxy compounds assume a carbonyl form with an increasing number of rings:

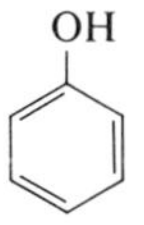

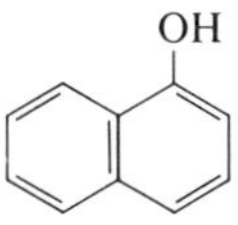

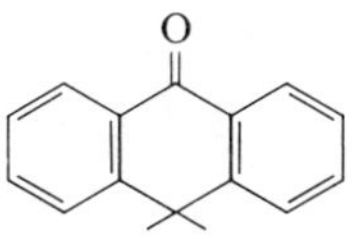

OH OH O O

22 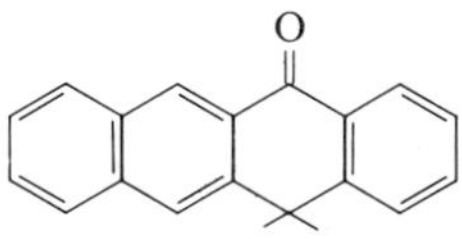 23

O

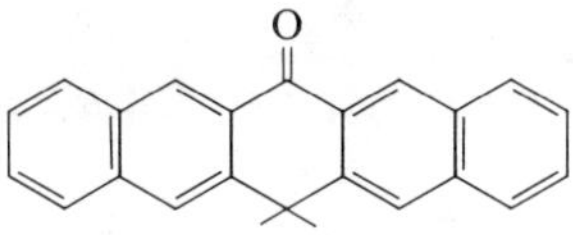

24

9-Hydroxyanthracene can be obtained from anthrone (**22**) with alkali but rearranges readily to anthrone. The kinetics of isomerization of anthranol to anthrone has shown that the reaction is first order in all solvents and that the hydroxy compound isomerizes easily in methanol, ethanol, benzene, and toluene(100).

In its ultraviolet absorption spectrum in methanol anthrone is closely similar to 10-hydroxy-10-methylanthrone in methanol(101). Spectrally anthrone is also closely similar to benzophenone, both in ethanol(102). These various spectra are entirely different from that of 9-ethoxyanthracene in hexane(103), which has the anthracene type of spectrum moved to slightly longer wavelengths.

Through absorption and fluorescence spectrophotometry it was found that the addition of increasing amounts of triethylamine leads to a gradual shift of the equilibrium toward the enol form; this indicates the formation of a hydrogen bond between anthranol and the amine in isooctane or benzene(104). It is concluded that the base-catalyzed isomerization takes place through an intramolecular proton transfer in which the base acts as a proton carrier(105).

5-Hydroxynaphthacene can be obtained in solution by treatment of the keto form **23** with alcoholic alkali, followed by neutralization, but it is so unstable that it could not be isolated in a pure state(106). The scant data reported in the literature do not show that the pentacenone (**24**) can be enolized at all(107). The stability of the hydroxy compound decreases with an increase in the number of linearly fused rings and high-energy *o*-quinonoid bonds. The carbonyl tautomer (e.g., **22**, **23**, and **24**), is much more stable since the ring containing the carbonyl group acts as a saturated insulator and the molecule contains fewer *o*-quinonoid bonds.

In this respect the ultraviolet absorption spectra in ethanol of the naphthacenone compounds **25a** and **b** closely resemble the spectrum of the carbonyl compound **25c**(108).

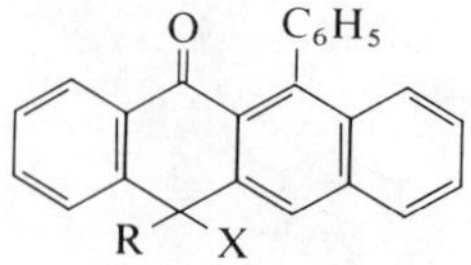

25

No.	R	X	λ_{max} (mμ)	mϵ
25a	H	H	Similar to 25b	
25b	H	C_6H_5	228	49.0
			316	13.5
			365	3.6
25c	OH	C_6H_5	228	49.0
			268	29.5
			316	11.8
			367	2.9

Relatively little is known about the dihydroxyanthracenes. 10-Hydroxy-9-anthrone can be rearranged to the 9,10-dihydroxyanthracene with boiling pyridine(106). The tautomeric equilibrium of the 9,10-dihydroxyanthracenes depends on the substituent(109). Thus in a 50% alcoholic solution buffered at pH 9.4 and at an ionic strength of 0.1 the following equilibrium takes place:

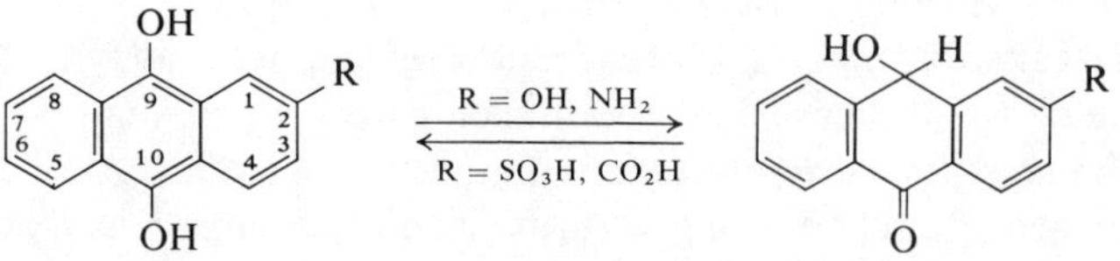

R = OH, NH_2 →
← R = SO_3H, CO_2H (4.6)

With an electron-acceptor group in the 2-position the formation of the ketone is hindered; with an electron-donor group in the 2-position the equilibrium lies on the side of the keto form.

An electron-donor group in the 2-position would increase the electron-density of the 9-position through a resonance effect, as shown in **26**, and thus cause the equilibrium to shift to the keto form. Once the ketone is formed, it would tend to be stabilized by resonance, such as shown in **27**. An electron-acceptor group in the 2-position would stabilize the dihydroxy form through the help of resonance structure **28**.

For the equilibrium shown in equation 4.7 there is a linear relation between the logarithm of the equilibrium constant and the half-wave potential $E_{1/2}$ or the Hammett sigma constant, σ_p (Table 4.13)(109).

TABLE 4.13
Equilibrium Constants in Equation 4.7(109)

R	K	σ_p	$E^0_{1/2}$ (volt)
H	0.13	0.00	−0.71
C_2H_5	0.44	−0.15	−0.74
HS	1.31	−0.48	−0.78
H_2N	2.5	−0.66	−0.79
HO	13.9	−1.0	−0.88

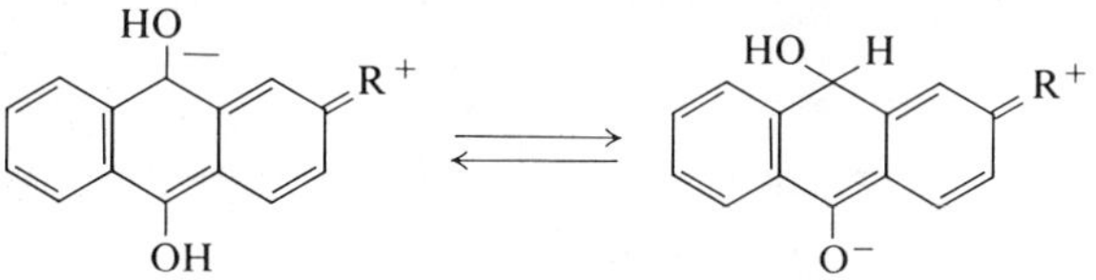

HO, R⁺, OH **26** ⇌ HO H, R⁺, O⁻ **27** (4.7)

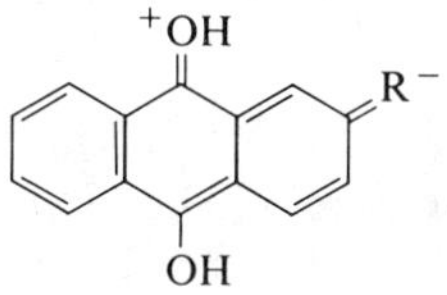

⁺OH, R⁻, OH **28**

1,4-Dihydroxyanthracene is converted into the diketo form **29a** by fusion(96), whereas the nonfluorescent diphenyl derivative **29b** is formed by simply treating a solution of the fluorescent 1,4-dihydroxy-9,10-diphenylanthracene with 10 *N* hydrochloric acid(110).

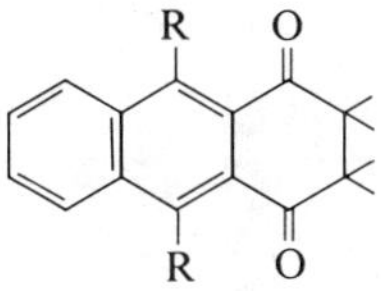

R O R O **29**

a: R = H; **b**: R = C_6H_5; **c**: R = OH

As in the naphthalene compounds, stability of the keto form is increased if peri hydroxy groups are present and an intramolecular hydrogen bond can be formed—as in **29c**, which is easily obtained by reduction of quinizarin in hot acid solution(111). By treatment of the **29c** with alkali and then acid, the tetrahydroxyanthracene is obtained; it rearranges to the more stable diketo compound during crystallization.

The various 1,4-dihydroxyanthracene tautomers should have widely varying spectra. Confirmation of these structures should be obtained through intensive spectral examination in a wide variety of solvents.

Examination of the spectra of 9-formylanthrone in several solvents has shown the presence of three isomers and the anion in various proportions (Table 4.14)(112). The model compounds used in the study were anthrone, 9-methoxy-10-anthraldehyde, and 10-methoxymethylene-9-anthrone. Different spectra were shown by each isomer and the anion.

TABLE 4.14
Tautomeric Equilibrium of 9-Formylanthrone in Various Solvents(112)

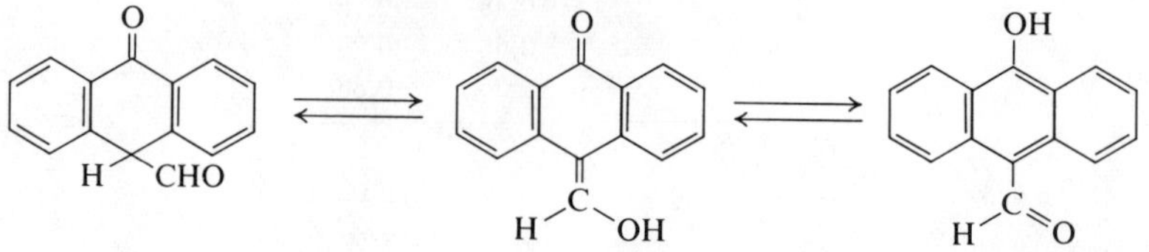

I ⇌ II ⇌ III

Solvent	Tautomeric equilibrium (%) I	II	III
Ethanol[a]	–	80	–
Ether	15–20	70–75	10
Chloroform	90	10	–
Benzene	94	6	–

[a]Five percent ionization also.

Study of the spectrum of 9-anilinomethyleneanthrone in ethanol discloses that the compound is present as the tautomer **30**, and not as **31** or **32a**(112).

a: R = OH; **b**: R = OCH_3

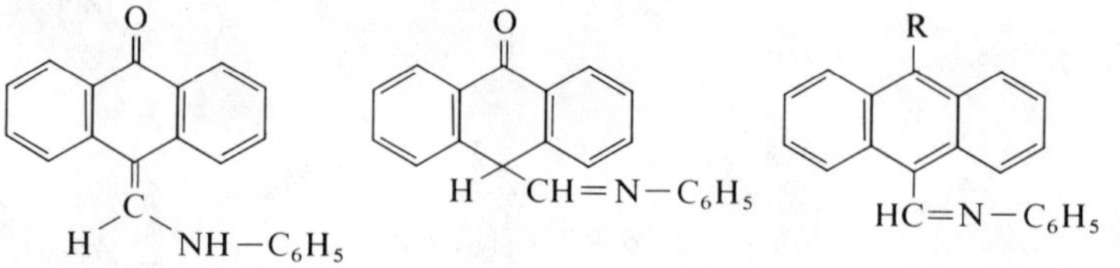

30 **31** **32**

No.	λ_{max} (mμ)	mε	No.	λ_{max} (mμ)	mε
30	241	38.0	**32b**	260	80.0
	275	17.0		~320	3.1
	346	12.4		405	11.2
	459	23.4			

Thus, the spectrum of **30** is not that of an anthrone, such as **31**, or an enol, such as **32a**. The tautomer **32a** would have a spectrum closely similar to that of **32b**.

Two tautomers of the oxime of 10-formyl-9-anthrol have been isolated and shown to have the structures **33** and **34**(112). The absorption spectra were measured in ether.

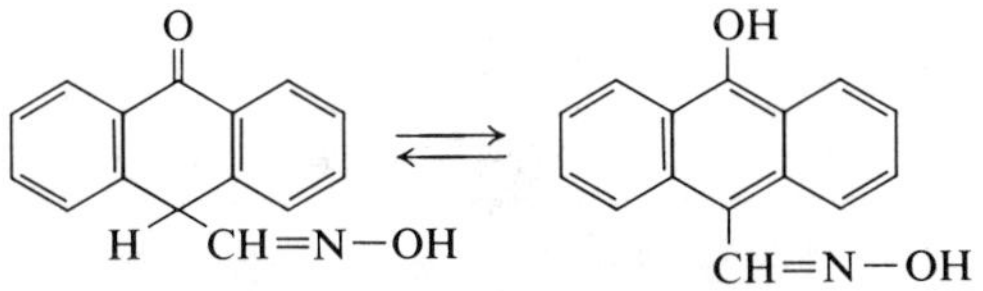

(4.8)

33

λ_{max} 260, ~292, ~304 mμ
mε 19·0, 3·9, 3·6

34

λ_{max} 262, 354, 375, 398 mμ
mε 99.0, 2.75, 5.3, 6.9

A solution of either tautomer in ethanol after 20 hr gave a spectrum closely similar to that of **34**, with indications of the presence of **33** (shoulder at 305 mμ): λ_{max} 260, 305*s*, 370, and 398 mμ (mε 47.0, 3.0, 2.35, and 2.95, respectively).

Since linear enol structures contain more quinonoid rings than angular enol structures with the same number of rings, the linear enol structures are less stable. Thus the keto form of 9-phenanthrol is unknown, whereas both tautomeric forms of 7-hydroxybenz[*a*]anthracene can be isolated and crystallized unchanged from benzene(113). However, in acetone the enol isomerizes fairly rapidly.

D. Quinonoid Tautomers

A simple example of the quinonoid type of tautomeric equilibrium is the following(114):

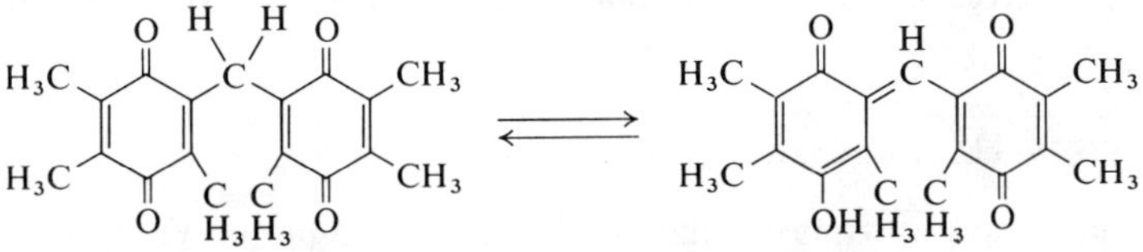

35
Yellow

36
Red

(4.9)

Both compounds were isolated.

4-Benzyl-1,2-naphthoquinone can be enolized by strong acid or base, but the reverse transformation is more difficult(115).

An equilibrium for which some spectral data are available is the enolization(116)

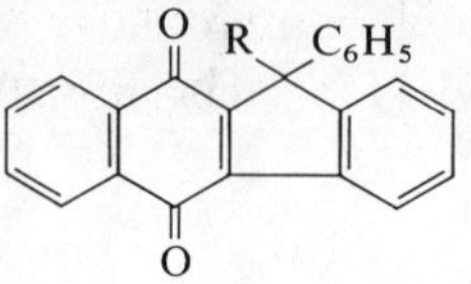

37

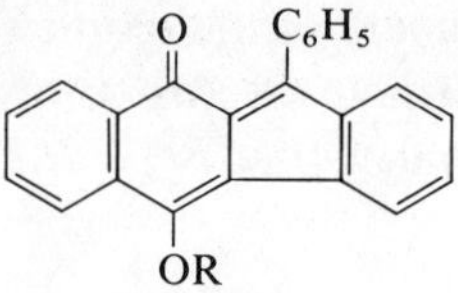

38

No.	R	Solvent	λ_{max}(mμ)	mϵ
37a	H	Chloroform	260	28.8
			308	8.3
			413	3.2
37b	OH	EtOH	258	21.9
			308	7.2
			417	2.6
38a	H	EtOH containing 0.1% AcOH	283	38.9
			330*i*[a]	5.9
			475	10.2
38b	CH_3	EtOH	280	48.9
			325*i*[a]	5.4
			450	13.5

[a]Inflection.

The assignment of structures was accomplished with the help of the model compounds **37b** and **38b**. Compound **37a** remains unchanged for hours in pure chloroform; addition of methanol transforms it to **38a**. The quinone **37a** is stable in dioxane solution but enolizes readily in acetone

Bis[1-(9,10-anthraquinonyl)]amine is present in chlorobenzene solution as the intramolecularly hydrogen bonded enol, as suggested by a comparison of the spectra of the tautomer (λ_{max} 363, 516 mμ, mϵ 13·4, 12·5, respectively) and 1-aminoanthraquinone (λ_{max} 463 mμ, mϵ 7·2) (117).

E. Oxygen Heterocyclic Phenols

Except when intramolecular hydrogen bonding is possible, as in **39**(89), hydroxyfurans exist only in the keto form. Thus 2-hydroxyfurans exist as but-β-enolides or crotonolactones, and 2,5-dihydroxyfurans exist as succinic anhydrides (89).

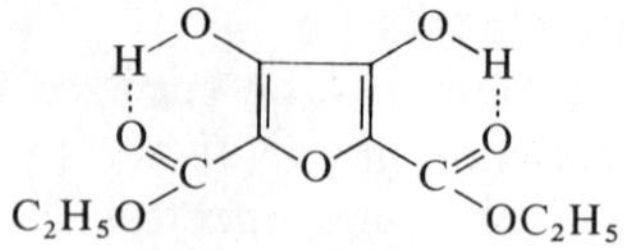

39

Examination of the spectra of 4-hydroxy coumarins in dioxane solution and in the solid state has disclosed that the compounds have the coumarin structure **40**, and not the chromone structure **41**(118). The longer chain of conjugation in the coumarin probably helps to stabilize this tautomer.

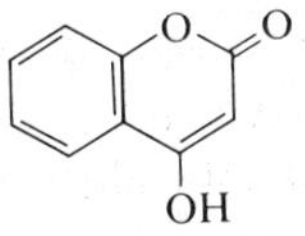

40 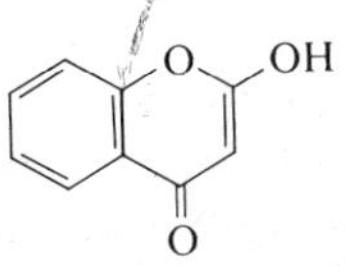 **41**

F. Hydroxythiophenes

In 2-hydroxythiophenes the following equilibrium could take place:

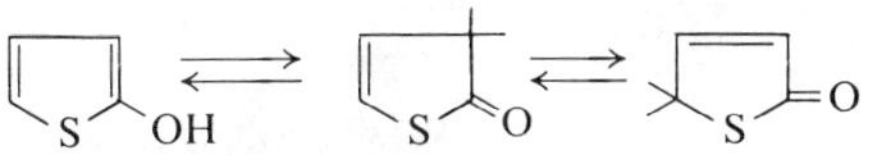

(4.10)

Nuclear magnetic resonance spectroscopy has shown that most 2-hydroxythiophenes exist as α,β- and β,γ-unsaturated γ-thiolactones (119). Infrared spectroscopy studies indicate the presence of carbonyl and enolic tautomers in 2-hydroxythiophene(120). The ultraviolet absorption spectrum in water shows the presence of a conjugated thiolactone and a shoulder at 320 mμ (mϵ 0.1) apparently derived from the carbonyl $n \rightarrow \pi^*$ transition(120).

The proportion of enol and keto forms of 2-thienol has been calculated with the help of paper chromatography data(121). In isopropyl ether the proportion of enol to keto forms was 2:3; in the stationary phase, triethylene glycol, the proportion of enol to keto was 7:3.

3-Hydroxythiophene is present as a mixture of the enol and keto isomers(122). The 2-methyl-3-thienol derivative is present in carbon disulfide solution mainly in the enolic form, the 2-*tert*-butyl derivative contains approximately equal amounts of the tautomers, and the 2,5-dimethyl derivative contains only 30% of the enol(123).

On the other hand, β-(3,4-dihydroxy-2-thienyl)propionic acid is reported to be entirely dienolic(124). Chelated formyl(125), acetyl, and carbethoxy derivatives of 2-hydroxy(126) and 3-hydroxythiophene(127) exist in the hydroxy form in the pure liquid state and in solution.

The thiophenethiols exist predominantly as the thiol tautomer(128, 129). Even 3,4-thiophenedithiol (λ_{max} 269 mμ, mϵ 3.98) is believed to exist as the dithiol tautomer in hexane solution(130) since its ultraviolet absorption spectrum is closely similar to that of methyl 3-thienyl sulfide (λ_{max} 273 mμ, mϵ 3.80) in ethanol(129). Infrared and nuclear magnetic resonance spectra are in line with the postulated structure.

Proton-magnetic-resonance spectral studies have also shown that 2- and 3-thiophenethiols substituted with methylmercapto or bromo groups have all been shown to be –SH compounds (131).

The tautomerism of 5-phenyl-2- and 3-thiophene-ols has been studied spectrally (132). Chemical and physical evidence indicates that these compounds exist in both keto and enol forms.

In both equilibria the keto form is favored by nonpolar solvents, whereas the enol form is favored by polar solvents. This is anticipated on the basis of the empirical equation of Dimroth (133), as verified by Meyer (134):

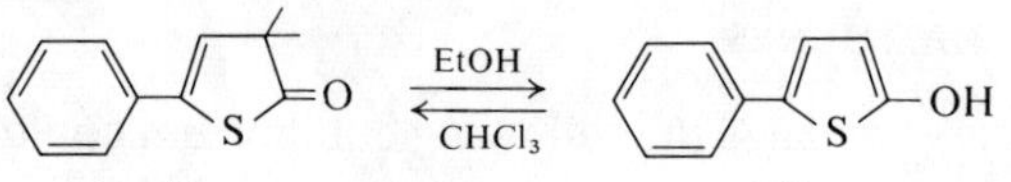

42 $\underset{CHCl_3}{\overset{EtOH}{\rightleftharpoons}}$ **43** (4.11)

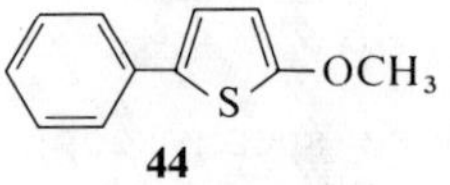

44

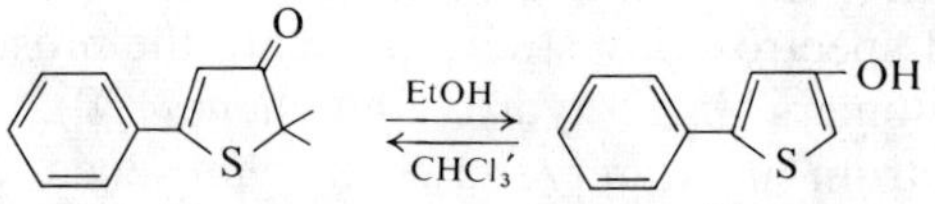

45 $\underset{CHCl_3}{\overset{EtOH}{\rightleftharpoons}}$ **46** (4.12)

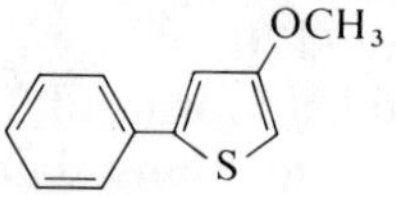

47

No.	Solvent	$\lambda_{max}(m\mu)^{a,b}$	$m\epsilon^{a}$
42	Chloroform	244	13.2
		258*s*	10.5
		280*s*	3.2
43	Ethanol	307	6.8
		258*i*	6.3
44	Methanol	305	16.0
45	Chloroform	270	15.0
		340	7.1
46	Ethanol	263	11.2
		300	10.3
47	Ethanol	263	11.0
		297	9.6

[a] All data from curves.
[b] *s* = Shoulder; *i* = inflection.

$$\frac{C_{\text{enol}}}{C_{\text{keto}}} = K \frac{S_{\text{enol}}}{S_{\text{keto}}} \tag{4.13}$$

where K is a constant characteristic for the system but independent of the solvent, C is concentration, and S is solubility. The enol structure would be expected to have greater solubility in polar solvents, the keto structure greater solubility in nonpolar solvents.

On the other hand, 3-hydroxybenzo[*b*]thiophene is said to exist only as the keto tautomer(135, 136). 2-Hydroxybenzo[*b*]thiophene also exists primarily as the keto tautomer, as shown by its close spectral resemblance to 2-methylthio-5-methylacetophenone and no spectral resemblance to 3-methoxybenzo[*b*]thiophene(137). 3-Hydroxybenzo-[*b*]selenophene is also present predominantly as the keto tautomer in neutral solution(138). Caution must be exercised in any analytical investigations of the hydroxythiophenes since these compounds are much less stable than the simple phenols.

The pyrrole analogs of the hydroxythiophenes have also been studied. The simple 2-hydroxypyrroles are best represented as the keto tautomers(139). 3-Hydroxypyrroles substituted in the 2-position with carbonyl groups usually exist as the intramolecularly hydrogen bonded hydroxy tautomer, whereas the 3-hydroxy-4-carbonylpyrroles are usually found as the keto isomer.

G. Schiff-Base Phenols

The combined effects of resonance involving strong electron-donor and electron-acceptor terminals and intramolecular hydrogen bonding in ketamines make them more stable than the isomeric enol-imine or ketimine forms(140). The compounds derived from a 2:1 condensation of a β-diketone and a diamine are present in solution to an extent $\geqslant 95\%$ in the ketamine form, with only a small dependence of the equilibrium on solvent.

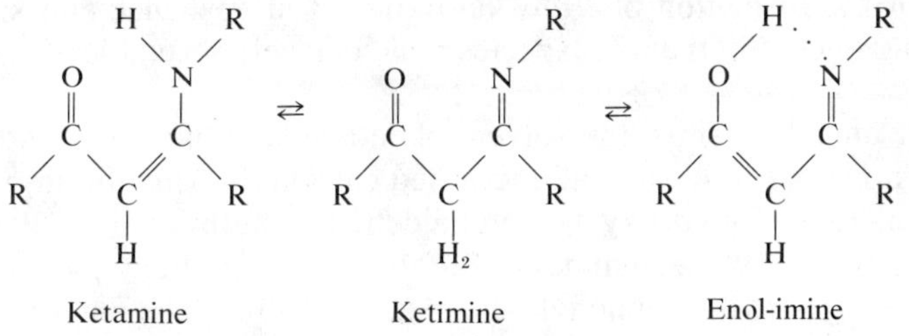

Ketamine ⇄ Ketimine ⇄ Enol-imine (4.14)

The magnitude of the resonance and intramolecular hydrogen bonding effects can be seen in the report that *N*-(*o*- and *p*-hydroxybenzylidene)-anils exist in alcoholic solvents partly in the ketamine form, a tautomer (141) in which a high-energy benzoquinonoid structure (e.g. **49A**) is involved(141). The equilibrium in 2,2,2-trifluoroethanol is as follows:

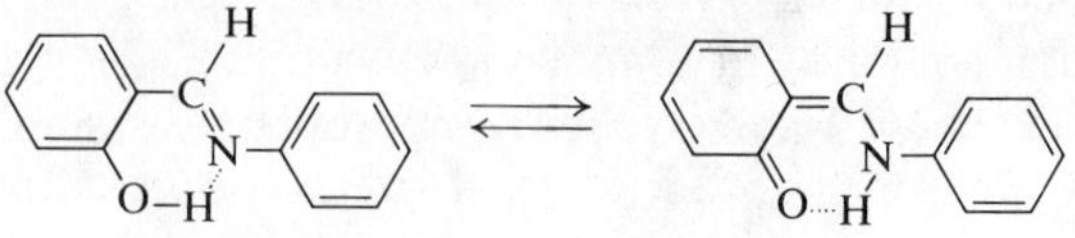

48 ⇌ **49A**

λ_{max} 342 mμ, mε 10.8 — λ_{max} 437 mμ, mε 7.8

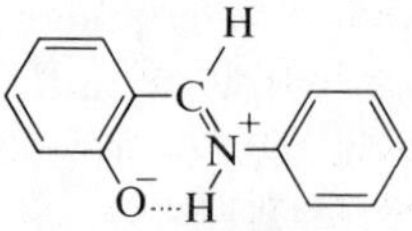

49B (4.15)

Whether structure **49A** or **B** contributes most to the ground state from which the band near 430 mμ is derived is difficult to determine. The polarity of the solvent has little effect on the wavelength position of the 430 mμ band. However, the band near 430 mμ shifts to shorter wavelength on protonation of the negative resonance terminal and thus resembles the Z_Z-bands. Examples of compounds of this type apparently present in tautomeric forms are listed in Table 4.15.

Evidence that indicates that the visible band is derived from the quinonoid tautomer of the anil has been summarized(141). Isosbestic points in cyclohexane–ethanol mixtures demonstrate two absorbing species. Spectra of various molecular structures show that the *N*-(*o*- or *p*-hydroxybenzylidene)amine structure is required (Table 4.15). Increasing the polarity of the solvent increases the concentration of the quinonoid tautomer (Table 4.16).

The addition of a small amount of an organic acid to a salicylaldehyde anil dissolved in a nonpolar solvent results in the conversion of some enol to *cis*-keto anil as the temperature of the solution is lowered to 77°K(148). The irradiation of most enol forms caused photocoloration, which is the result of the formation of *trans*-keto anil. Fluorescence emission from the enol is actually that of a species which closely resembles the *cis*-keto form.

Other anils that show the solvent phenomena of increased enol formation in nonpolar solvents and increased carbonyl formation in polar solvents include 2-hydroxy-1-naphthaldehyde methylimine(149), di-*N*-salicylidene ethylenediamine(150), 2-(*o*-hydroxyphenyl)benzothiazole (150), and 2-(*o*-hydroxyphenyl)benzoxazole(150). The majority of anils

TABLE 4.15
Long-Wavelength Bands of Schiff-Base Phenol Tautomers

Structure	Solvent	Enol		Keto		Ref.
		$\lambda_{max}(m\mu)$[a]	$m\epsilon$	$\lambda_{max}(m\mu)$	$m\epsilon$	
–CH=N–, OH	Methanol	338	13.4	432	0.28	141
	H_2O–EtOH (9 : 1), pH 5	?		445	0 16	142
–CH=N–, OH, OH	Methanol	349	9.8	444	1.6	141
	H_2O–EtOH (9 : 1), pH 5	335	5.0	450	1.26	142
–CH=N–, OH, OH	Methanol	333	16.7	420	0.28	141
–CH=N––OH, OH	Methanol	349	17.0	435	0.40	141
HO––CH=N––OH	Methanol	313	16.5	415	0.15	141
–CH=N––OH	Methanol	336	14.0	—	—	141
–CH=N–, OH	Methanol	345	9.5	—	—	141

TABLE 4.15 (*cont.*)

Structure	Solvent	Enol		Ketol		Ref.
		λ_{max} (mμ)[a]	mϵ	λ_{max} (mμ)	mϵ	
m-HO-C_6H_4-CH=N-C_6H_5	Methanol	316	10.3	—	—	141
o-HO-C_6H_4-C(CH_3)=N—CH_3	Methanol	~322		396[b]		143[c]
(o-HO-C_6H_4-CH=N-CH_2-$)_2$	95% EtOH	320	6.3	410	1.20	145
	Methanol	317	1.8	405	3.5	146
(p-HO-C_6H_4-CH=N-CH_2-$)_2$	95% EtOH	290	26.3	—[d]	—	145
(3-Cl-2-HO-C_6H_3-CH=N-CH_2-$)_2$	95% EtOH	323	5.0	410	4.6	145
(3-Br-2-HO-C_6H_3-CH=N-CH_2-$)_2$	95% EtOH	343	5.0	410	2.0	145

$(3\text{-I-2-OH-}C_6H_3\text{-}CH{=}N{-}CH_2{-})_2$	95% EtOH	330	6.3	410	4.0	145
$(3\text{-}O_2N\text{-2-OH-}C_6H_3\text{-}CH{=}N{-}CH_2{-})_2$	95% EtOH	343*s*	3.2	440	11.0	145
$(2\text{-}OCH_3\text{-}C_6H_4\text{-}CH{=}N{-}CH_2{-})_2$	95% EtOH	320	8.7	—	—	145
2-OH-1-naphthyl-$CH{=}N{-}R$[e]	Methanol	?		400		147

[a]Enol and keto bands found in same spectrum. Dash signifies keto band not present; ? signifies band not reported in the available literature.

[b]Estimated mϵ 5.9 for pure compound (144).

[c]Spectra of many other alkylimine derivatives of salicylaldehyde and acetophenone reported in various solvents.

[d]Compound reported to have a light yellow color, which might indicate the presence of a small amount of keto tautomer.

[e]Spectra of many alkyl derivatives reported in various solvents.

TABLE 4.16
Solvent Effects (141) on the Long-Wavelength Enol and Keto Bands of

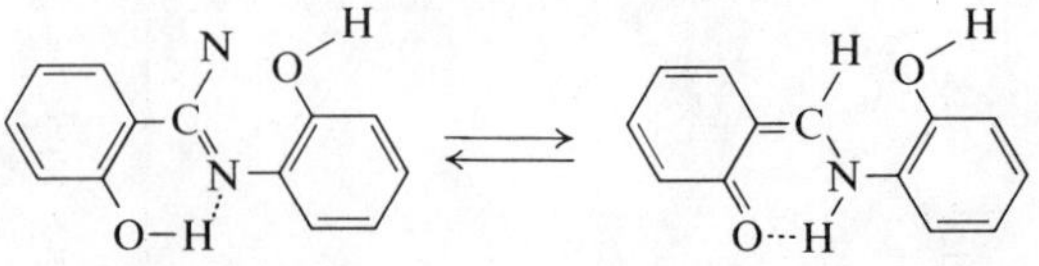

	Enol		Keto[a]	
Solvent	$\lambda_{max}(m\mu)$	$m\epsilon$	$\lambda_{max}(m\mu)$	$(m\mu)$
Cyclohexane	358	10.8	—	—
Carbon tetrachloride	358	13.4	—	—
1,4-Dioxane	352	8.9	—	—
Ethyl acetate	349	13.6	—	—
Acetonitrile	348	11.4	—	—
Chloroform	354	10.8	440*s*	0.1
tert-Butanol	350	9.6	432	0.5
Isopropanol	348	12.0	446	1.0
Ethanol	350	12.0	448	1.4
Methanol	349	9.8	444	1.6
Ethanol–water (1 : 4)	333	5.8	438	3.2
2,2,2-Trifluoroethanol	342	10.8	437	7.8

[a]Although the intensity of the quinonoid band is much weaker, the same relative change in intensity is obtained for the *p*-hydroxybenzylidene isomer.

not containing the chelated ring system are nonfluorescent. For those that are chelated and fluorescent the emission is the mirror image of the long-wavelength carbonyl tautomer band, regardless of the wavelength of the exciting light (150).

The same effect of solvent polarity on the intensity of the quinonoid band of the alkylimines of salicylaldehyde and *o*-hydroxyacetophenone has been described by Kazitsyna et al. (143).

Intermolecular hydrogen bonding appears to provide a route for the mechanism through hydrogen transfer and the necessary stability for the quinonoid tautomer. Solid *N*-(*o*-hydroxybenzylidene)anils show a phototropism that has been explained as due to a reversible tautomerism with the quinonoid structure of the anil (151).

In the attempt to understand transamination reactions, the structures of Schiff bases of pyridoxal and its analogs have been studied. In the 3-hydroxypyridine-2-aldehyde amino acid Schiff bases the keto enamine species predominate (73%), whereas in the benzenoid Schiff bases and those derived from 3-hydroxypyridine-4-aldehyde about equal proportions of the keto and enol tautomers are present in dioxane solution(152).

For the Schiff bases that have been studied the keto tautomer (analogous to **49**) has a pair of absorption bands in neutral dioxane between 404 and 425 mμ, and between 265 and 280 mμ; the enol tautomer (analogous to **48**) has a pair of absorption bands between 316 and 330 mμ, and between 238 and 256 mμ. Examples of such compounds and the solvents used in the spectral studies are *N*-(3-hydroxy-2-(or 4)-pyridylmethylene)valine in dioxane, methanol, and solid KBr(152), *N*-salicylidenevaline in dioxane(152), *N*-(3-hydroxy-2-(or 4)-pyridylmethylene) glycine in methanol and solid KBr(152), *N*-(3-hydroxy-2(or 4)-pyridylmethylene)alanine in methanol and solid KBr(152), *N*-(3-hydroxy-2-(or 4)-pyridylmethylene)phenylalanine in methanol and solid KBr (152), *N*-(3-hydroxy-2(or 4)-pyridylmethylene)glutamic acid in methanol and solid KBr(152), pyridoxylidenevaline in methanol(153), pyridoxal phosphate valine Schiff base in methanol(153), and pyridoxal glycine butyl ester Schiff base in ethanol(154). The position of the long-wavelength enol band was relatively unaffected in a change from dioxane to methanol, but the long-wavelength keto band was shifted to shorter wavelengths.

H. Nitrosophenol ⇆ Quinone Oxime Equilibrium

For some of this type of compound the following equilibrium can take place, especially where the molecular structure and solvent situation are favorable:

$$\mathrm{HO{-}Ar{-}N(O^-){-}\overset{+}{N}(=O){-}Ar{-}OH}$$
$$\upharpoonleft\downharpoonright$$
$$\mathrm{O{=}Ar{=}NOH \rightleftarrows HO{-}Ar{-}NO \xrightarrow[H^+]{ROH} RO{-}Ar{-}NO} \qquad (4.16)$$
$$\mathrm{ROH} \downharpoonleft\upharpoonright$$
$$\mathrm{^-O{-}Ar{-}NO}$$

Complications due to dimerization, etherification, and ionization have been discussed by Schors et al.(155) in their study of this type of tautomerism.

Where weak interaction of an aromatic ring with a nitroso group exists, dimerization is favored by concentration or lowering the temperature of the solution, for example, as with nitrosomesitylene. In methanolic or acetone solutions the probable presence of small amounts of alkaline impurities could cause some ionization. Addition of a small amount of concentrated acid to alcoholic solutions to repress ionization catalyzes ether formation.

It is obvious that in analyzing phenols through nitrosation solvent and pH effects must be controlled and understood to obtain optimal results in the analysis. Nitrosophenol chromogens have been formed in the trace analysis of 4-methoxyphenol(156) and total phenols in waste water(157).

1. p-*Nitrosophenols*

Extensive spectral investigations of these compounds have been reported(155, 158–160). The more recent studies indicate that these compounds are present mainly as *p*-benzoquinone oximes(161). Infrared absorption spectral studies indicate that they are also present in the solid state as the oxime(161, 162).

It has been postulated that an anion of a *p*-nitrosophenol is an intermediate in the equilibrium of the two tautomers in solvents that allow partial dissociation into ions(161, 163). For example, *p*-nitrosophenol in methanol has bands at 301 mμ (mϵ 13.6) derived from *p*-nitrosophenol; 1,4-benzoquinone oxime has bands at 405 mμ (mϵ 5.8) derived from the anion and 715 mμ (mϵ 0.0074) derived from nitrosophenol(161).

For the few alkylated derivatives that have been investigated ionization and percentage of nitroso tautomer decrease with alkylation. The amount of nitroso isomer was calculated from the intensity of the nitroso $n \rightarrow \pi^*$ band of *p*-nitrosoanisole (Table 4.17)(161).

Somewhat similar results have been obtained for *p*-nitrosophenol and

TABLE 4.17
Percentage of Nitroso Tautomer in Methanolic Solutions of *p*-Nitrosophenol(161)

Substituent in O=⟨1 2 3 4 5 6⟩=NOH	λ_{max}	mϵ	R–NO (%)
None	715	0.0074	22
3,5-Dimethyl	760	0.0018	5
3,5-Diethyl	755	0.0021	6
2,5-Dimethyl	770	~ 0.0002	~ 0.5
2,6-Dimethyl	—	< 0.0001	< 0.3
2,6-Diethyl	—	< 0.0001	< 0.3
2,6-Di-*tert*-butyl	—	< 0.0001	< 0.3
2,3,5,6-Tetramethyl	—	< 0.0001	< 0.3
p-Nitrosoanisole	730	0.0338	100.0
p-Benzoquinone oxime, methyl ether	—	0.0	0.0

its 2-chloro- and 2,6-dimethyl derivatives in dioxane solution(164). The percentage of the nitroso derivative dropped from 25% for the unsubstituted derivative to 3% for the 2-chloro derivative and $< 0.2\%$ for the 2,6-dimethyl derivative.

The *syn–anti* isomerism in the *p*-nitrosophenol-*p*-benzoquinone monoxime system has also been investigated(165).

2. *o*-Nitrosophenols

Unlike the *p*-nitrosophenols, the ultraviolet–visible absorption spectra of 2-nitrosophenol in such solvents as hexane, carbon tetrachloride, ether, benzene, chloroform, ethanol, and water show that this compound exists in solution as the true nitrosophenol, although the spectra do not exclude the presence of small but undetectable amounts of the oxime tautomer(166, 167). For example, in carbon tetrachloride *o*-nitrosophenol has the following λ_{max} and mϵ values: 400 mμ and 2.6 ($\pi \rightarrow \pi^*$), 697.5 mμ and 0.065 ($n \rightarrow \pi^*$). In the same solvent the λ_{max} of *o*-nitrosoanisole is 369 ($\pi \rightarrow \pi^*$) and 777 mμ ($n \rightarrow \pi^*$). In alcoholic solution *o*-nitrosophenol has bands at 389 and 715 (mϵ 3.4 and 0.053, respectively), whereas *o*-nitrosoanisole has comparable bands at 377 and 761 mμ (mϵ 3.4 and 0.05, respectively). The differences in wavelength position between the phenol and the anisole are the result of intramolecular hydrogen bonding. Pure *o*-benzoquinone oxime would have no $n \rightarrow \pi^*$ bond.

However, when the spectra of 5-methoxy-2-nitrosophenol and 5-dimethylamino-2-nitrosophenol were compared with the spectra of the corresponding 2-nitrosoanisoles and *o*-benzoquinone oxime methyl ethers in the same solvents, it was found that the former compounds contain an intramolecular hydrogen bond and exist in solvent-dependent tautomeric equilibria.

5-Methoxy-2-nitrosophenol crystallizes from benzene in green prisms (mainly the nitroso tautomer) and from ethanol as yellow-brown prisms (predominantly the quinone oxime tautomer)(166). The tautomeric equilibrium is solvent dependent; the relative amounts of nitroso tautomer decrease in the following order as shown by the decreasing intensities of the violet-shifting $n \rightarrow \pi^*$ bands(166):

Solvent	λ_{max}(mμ)	mϵ
Ether	646	0.032
Benzene	643	0.035
Chloroform	622	0.029
Ethanol	614	0.015
Aqueous ethanol (1:1)	610	0.010

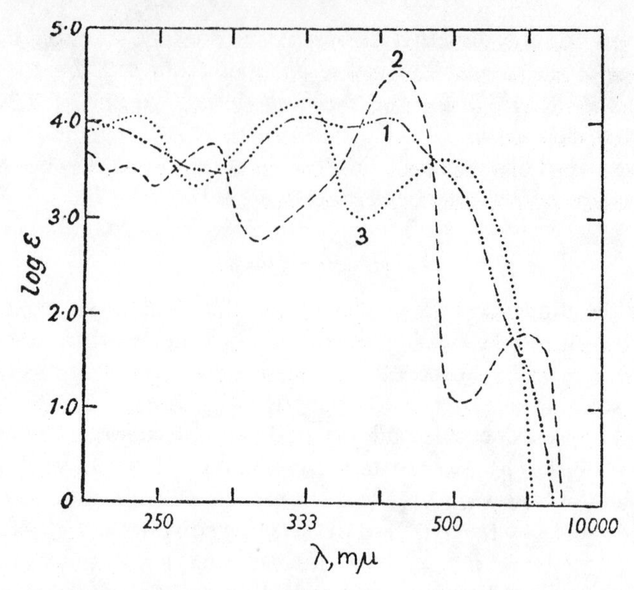

Fig. 4.2. Absorption spectra in ethanol of (1) 5-dimethylamino-2-nitrosophenol (— · · · —), (2) 5-dimethylamino-2-nitrosoanisole (– – –), and (3) 3-dimethylamino-6-methoxyiminocyclohexa-2,4-dienone (· · ·)(166).

A comparison of the absorption spectra of 5-dimethylamino-2-nitrosophenol, 5-dimethylamino-2-nitrosoanisole, and 5-dimethylamino-1, 2-benzoquinone-2-oxime methyl ether in ethanol solution clearly show that the nitrosophenol is in tautomeric equilibrium with the quinone oxime (Fig. 4.2)(166).

3. *Nitrosonaphthol ⇄ Naphthoquinone Oxime*

1-Nitroso-2-naphthol exists in all solvents as 1,2-naphthoquinone-1-oxime, as shown by the absence of the low intensity $n \rightarrow \pi^*$ band derived from the nitroso group(166, 168–172). The reported spectra of 2-nitroso-1-naphthol appears to indicate that the oxime tautomer predominates(170, 172). However, the spectrum of a carbon tetrachloride solution of the compound does show the presence of a band at 725 mμ derived from the nitroso $n \rightarrow \pi^*$ transition(162). The slight solubility of the compound and the ready solubility of 1,2-naphthoquinone-1-oxime in carbon tetrachloride would reinforce the evidence that the former compound is present mainly as 2-nitroso-1-naphthol and the latter as the oxime in carbon tetrachloride.

In trace analysis involving tautomers the solvent and pH would affect the tautomeric equilibrium. Thus in the analysis of the insecticide carbaryl the analyzed chromogen is 2-nitroso-1-naphthol (λ_{max} 425 mμ, mϵ 2.6)(173–175). However, in neutral and alkaline aqueous solutions the pure chromogen gives λ_{max} 425 and 433 mμ(mϵ 5.0 and 7.9), respectively (168).

Comparson of the spectra in methanol of 1,4-naphthoquinone oxime, 1,4-naphthoquinone methyloxime, and 1-nitroso-4-methoxynaphthalene definitely indicates that the first-named compound is present mainly as the oxime (Fig. 4.3)(155). Infrared-spectroscopy studies also indicate that 1,4-naphthoquinone oxime is present mainly as the oxime tautomer in the solid form and in chloroform solution(162).

4. Nitrosoanthrol⇄Anthraquinone Oxime

Since the oxidation–reduction potential values of the quinone/quinol systems derived from 1,4-benzoquinone, 1,4-naphthoquinone, and 9,10-anthraquinone are 0.7, 0.5, and 0.2 volt, respectively, with reference to the normal hydrogen electrode, there is a decreasing tendency in the

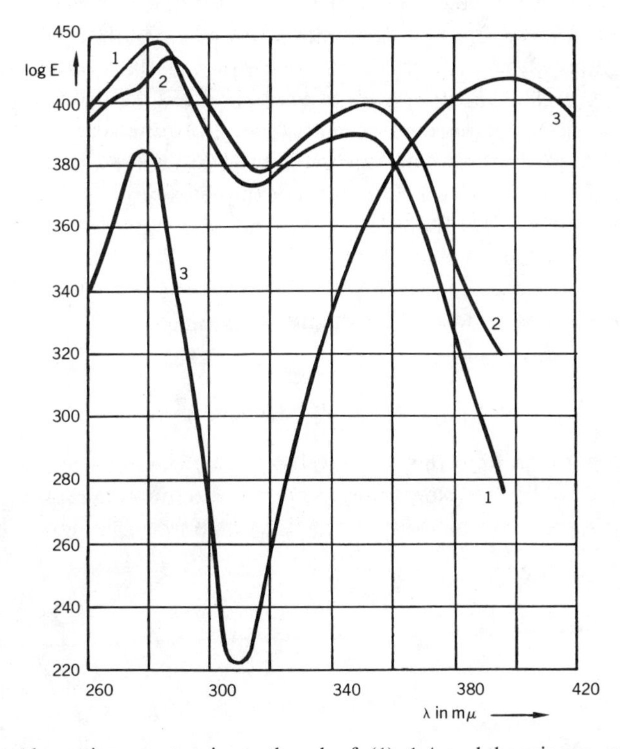

Fig. 4.3. Absorption spectra in methanol of (1) 1,4-naphthoquinone oxime, (2) 1,4-naphthoquinone methyloxime, and (3) 1-nitroso-4-methoxynaphthalene(155).

series to form a completely benzenoid structure. Thus it is not surprising that 9,10-anthraquinone oxime (λ_{max} 229, 261, 325 mμ; mϵ 17, 16, 4.8, respectively) is spectrally similar in methanol to 9,10-anthraquinone methyloxime (λ_{max} 231, 266, 325 mμ; mϵ 17, 16, 6, respectively)(155). Since a 10-nitroso-9-anthrol tautomer would absorb at longer wavelength with greater intensity, the compound is in the oxime form.

I. Azophenol ⇆ Quinone Hydrazone

Azophenols have been used extensively in the identification and determination of phenols. The intense color of these chromogens facilitates their separation by chromatography and their quantitative determination by colorimetry, as shown by the large number of references available in the literature.

In spite of their extensive analytical use, azophenols have many spectral properties that have not been completely exploited. For example, maximum use has not been made of the tautomerism of azophenols and azonaphthols formed in the determination of asphaltic phenols with *p*-diazobenzenesulfonic acid(176); 4-amino-2-hydroxynaphthalenesulfonic acid with *p*-diazobenzenesulfonic acid(177); 1,4-dihydroxynaphthalene, hydroquinone, or pyrogallol with 2,4-dinitrophenylhydrazine(178); aromatic esterase through cleavage of 2-naphthyl esters and coupling of the resultant 2-naphthol with diazotized *p*-phenetidine(179); and similarly hydrolyzing esterases through determination of 1-naphthol with diazotized 4-amino-4′-methoxydiphenylamine(180).

In this chapter we discuss some of the interesting tautomeric properties of phenols that could be useful in organic trace analysis.

1. Azo Enol ⇄ Hydrazone Keto

Compounds containing the azo enol ⇄ hydrazone keto structure could be considered to be simpler analogs of the azophenol tautomers. For example, consider 2-(4′-methoxyphenylazo)dimedone (**50**) and its possible tautomers.

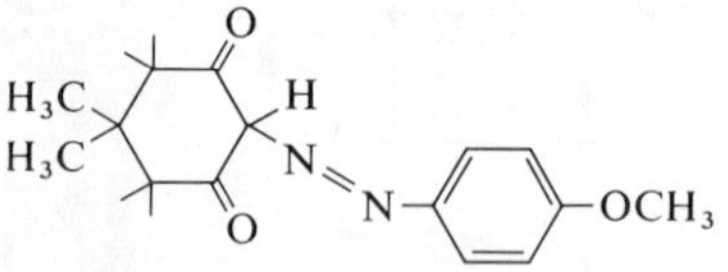

51
Azo-diketone

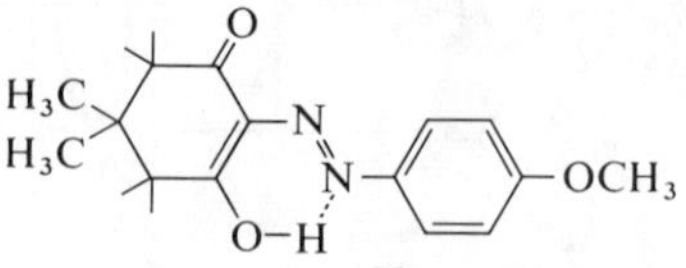

52
Azo-enol

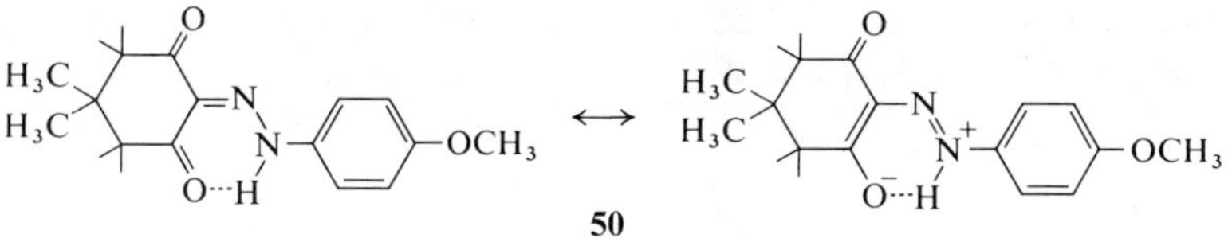

50

Hydrazone

Eistert and Geiss(44) have shown that the azo–diketone tautomer is not present in methanolic solution; neither is it present in the analogous azo dye obtained with Meldrum's acid. This is shown by a comparison of the spectrum of **50** in methanol (λ_{max} 235, ~ 400 mμ; mϵ 10.5 and 22.0, respectively) with that of the analogous azo dye obtained from 2-methyldimedone (**53**), λ_{max} 235, 318, 390 mμ; mϵ 9.5, 14.0, 0.59, respectively. The long-wavelength band is derived from the azo group $n \rightarrow \pi^*$ transition.

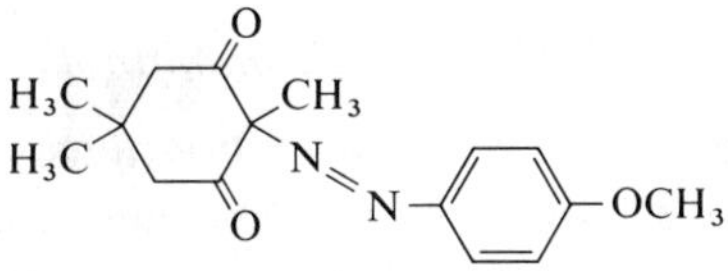

53

Although this equilibrium has not been studied thoroughly, other analogous equilibria have. The drastic effect of solvent and structural changes on the tautomeric equilibrium of compounds analogous to **50**, as reflected in their absorption spectra, is a problem an analyst should be aware of when analyzing any material or solution in which such an equilibrium could be present.

The bis(phenylhydrazone) of xylo-4,5,6-trihydroxycyclohexane-1,2,3-trione has been shown to exist in two forms, whose tautomeric isomerization (**54** ⇄ **55**) has been confirmed by a study of the visible, ultraviolet, infrared, and nuclear-magnetic-resonance spectra of the tautomers and appropriate model compounds(181).

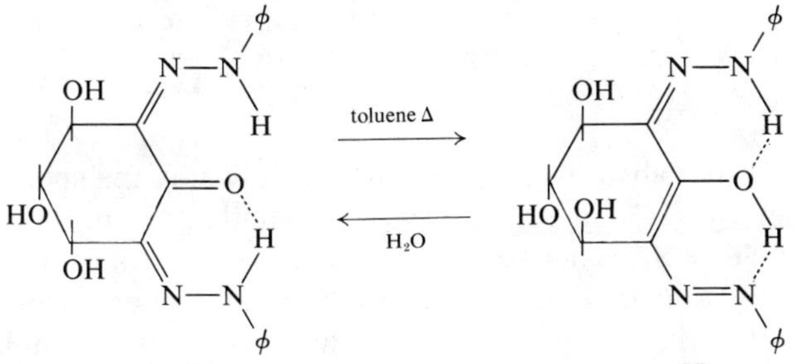

toluene Δ → ; ← H_2O (4.17)

54 Yellow

55 Red

The absorption spectrum of **55** in ethanol (30 min after dissolution) gave λ_{max} value of 251, 286*s*, and 487 mμ (mϵ 18.8, 6.0 and 34.3, respectively).

2. *Azophenols*

Examination of the ultraviolet(182–184) and infrared(185, 186) spectra of 2- and 4-phenylazophenols indicates that these compounds are predominantly present in azo form either as solids or in solution.

It has been suggested that the ability of 4-(2,4-dinitrophenylazo)-phenol to undergo a Diels–Alder reaction indicates that this compound is the hydrazone tautomer(187). However, spectral evidence would be necessary to confirm this. It is possible that with the proper substituents (e.g., a 2′-carbonyl group) some 2-phenylazophenols could exist as the hydrazone tautomer.

3. *1-Phenylazo-2-naphthols*

1-Phenylazo-2-naphthols have seen fairly extensive investigation. The parent compound in ethanol exists in a tautomeric equilibrium(**56** ⇄ **57**) (183, 188–192).

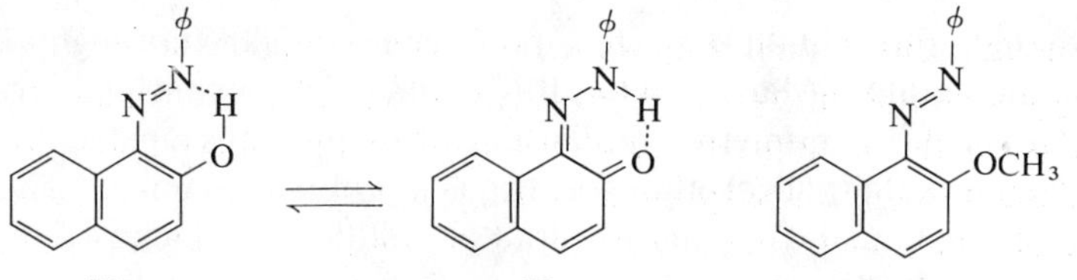

φ N N H O   ⇌   φ N N H O   φ N N OCH$_3$ (4.18)

56   **57**   **58**

No.	Solvent	λ_{max}(mμ)	mϵ
56 ⇄ **57**	EtOH	410*s*	10.0
		480	12.9
	EtOH–H_2O (1 : 1)	410*s*	8.9
		480	18.2
58	EtOH	390	20.0
		469	2.5

The azo tautomer absorbs near 400 mμ, as shown by the spectrum of the methoxy derivative **58** in ethanol(193); the band at 469 mμ, hidden in the spectrum of **56** ⇄ **57**, is derived from the azo $n \rightarrow \pi^*$ transition.

An investigation of numerous derivatives of 1-phenylazo-2-naphthol has shown that these compounds exist in a tautomeric equilibrium that is sensitive to substituents and solvent changes(188). The following equilibrium is postulated:

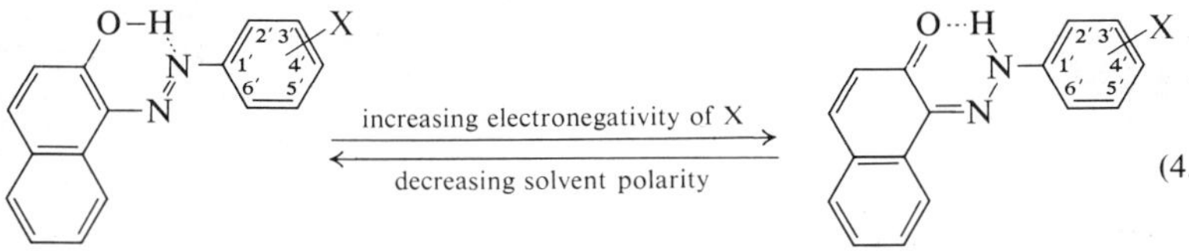

increasing electronegativity of X

decreasing solvent polarity

(4.19)

The concentration of the phenylhydrazone tautomer increases with increased solvent polarity (Table 4.18)(188) or increased electron-acceptor strength of X(188), for example, in the series

$$X = 4\text{-}NH_2 = 4\text{-}OCH_3 < 4\text{-}CH_3 < H$$

and in the series

$$X = 3\text{-}CH_3 < 4\text{-}Cl < 2\text{-}CH_3 < 3\text{-}Cl < 2\text{-}Cl < 2\text{-}CH_3 < 3\text{-}NO_2 < 4\text{-}NO_2 < 2\text{-}NO_2$$

In Table 4.18 the intensities of the hydrazone and azo bands (or shoulders) at their approximate wavelength maxima of 505 and 420 mμ, respectively, and the $\epsilon_{azo}/\epsilon_{hydrazone}$ ratios of 1-(4′-methoxy-3′,5′-dimethyl-phenylazo)-2-naphthol in different solvents are compared with the equilibrium constants, C_{enol}/C_{keto}, of the keto–enol tautomers of ethyl acetoacetate and 2,4-pentanedione(194). In the three compounds the more polar ketonic structures increase with increased solvent polarity. This change correlates fairly well with the X_R value of Brooker et al. (195).

On the basis of infrared-spectroscopy studies Morgan(196) has confirmed that the position of tautomeric equilibrium in 1-arylazo-2-naphthols shifts toward the hydrazone in polar solvents or under the influence of electron-withdrawing substituents in the aryl ring.

Another factor that shifts the equilibrium toward the hydrazone is the substitution of a group in the 2′-position as compared with substitution in the 3′- or 4′-position(188). Thus alcoholic solutions of the 2′-methyl, 2′-chloro and 2′-methoxy derivatives of 1-phenylazo-2-naphthol contain greater amounts of the hydrazone tautomer than do alcoholic solutions of their 3′- and 4′-isomers.

Steric hindrance can also have an effect on the equilibrium. Thus in equation 4.19 the percentage of hydrazone is decreased in sterically unhindered compounds, in which X = H is replaced by X = 4-OH, 3′,5′-diMe-4′-OH, or 4′-OMe. However, in the sterically hindered compound in which X = 3′,5′-diMe-4′-OMe the percentage of hydrazone is decreased relative to X = H but definitely increased in relation to the other compounds(197). The methoxy group is pushed out of the plane of the molecule and thus has less effect on the equilibrium than would the unhindered 4′-OH and 4′-OMe groups.

TABLE 4.18

Effect of Solvent on the Tautomeric Equilibrium of 1-(4-Methoxy-3,5-dimethylphenylazo)-2-naphthol, Ethyl Acetoacetate, and 2,4-Pentanedione (188, 194)

Solvent	X_R[a]	Hydrazone $m\epsilon_{505\,m\mu}$	Azo $m\epsilon_{420\,m\mu}$	$\frac{m\epsilon_{420\,m\mu}}{m\epsilon_{505\,m\mu}}$	$CH_3COCH_2COOC_2H_5$ $\frac{C_{enol}}{C_{keto}}$	$CH_3COCH_2COCH_3$ $\frac{C_{enol}}{C_{keto}}$
Water	< 35				0.004	0.24
H_2O–AcOH (1 : 1)		20.9	9.3	0.45	—	—
H_2O–EtOH (1 : 1)		18.5	10.3	0.56	—	—
AcOH		17.9	10.2	0.57	0.061	2.8
H_2O–EtOH (1 : 3)		16.5	11.3	0.68	—	—
Chloroform	44.2	14.0	10.7	0.76	0.089	3.8
H_2O–EtOH (1 : 7)		15.1	12.0	0.80	—	—
Nitrobenzene	42.6	11.0	11.5	1.04	—	—
95% EtOH	43.9	11.6	13.1	1.13	0.15	5.3
Pyridine	43.9	11.1	12.8	1.15	—	—
Benzene	46.9	11.3	13.1	1.16	0.22	5.7
Carbon tetrachloride	48.7	9.6	13.0	1.35	—	—
n-Hexane	50.9	6.9	14.6	2.12	0.9	12.0

[a] Data from reference 195.

Other compounds of this type reported as existing as hydrazones in solution are 1-(naphthyl-1-azo)-2-naphthol, 1-phenylazo-8-cyano-2-naphthol, 1-phenylazo-2-naphthol-8-carboxylic acid, and 1-phenylazo-2-naphthol-3-carbanilide(190).

The equilibrium is also shifted toward the hydrazone form if the 1-arylazo-2-naphthol contains a 2′-carbonyl group. The increase in hydrazone formation is explained in terms of the NH proton being sandwiched between the two oxygens(183), as in **59**.

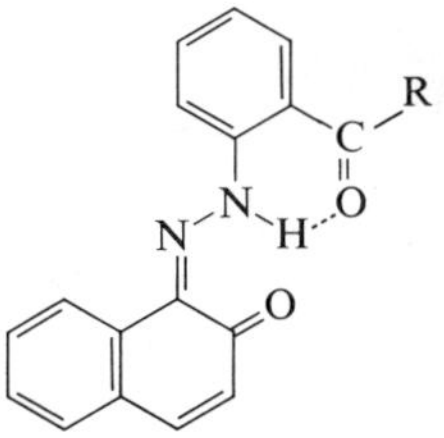

59

Further evidence for the hydrazone structures is derived from the fact that in solutions in which the hydrazone is present the fluorescence emission spectra approximately mirror the hydrazone absorption peaks (198).

4. *2-Phenylazo-1-naphthols*

The reputed quinonoid structure of the parent compound is based on spectral data indicating that 2-phenylazo-1-naphthol is spectrally similar to 1,2-naphthoquinone-2-*N*,*N*-diphenylhydrazone and different from 2-phenylazo-1-methoxynaphthalene(182, 199). Reported absorption spectra of 2-phenylazo-1-naphthol in various solvents indicate that the compound exists predominantly as the hydrazone, for example, in 50% aqueous acetone(200), in a potassium bromide pellet(185), in ethanol(201, 202), and in methylcyclohexane(198). Spectra of 5-amino-2-phenylazo-1-naphthol-3-sulfonic acid and 5-amino-2-phenylazo-1-naphthol-4-sulfonic acid in aqueous solution indicate that these compounds also exist predominantly in the hydrazone form(201). All of these spectra show the presence of one strong band in the visible region near 500 mμ. A more thorough study of these types of compounds is warranted.

5. *3-Phenylazo-2-naphthol*

The azo tautomer would probably be more stable since the hydrazone form would involve high-energy *o*-benzoquinonoid structures. Spectral data seem to bear this out since in ethanol **60** spectrally resembles 2-phenylazonaphthalene (**61**)(203). As one would expect from the single-bond properties of the 2,3-bond, the hydroxy group in **60** shows little conjugative interaction with the phenylazo groups.

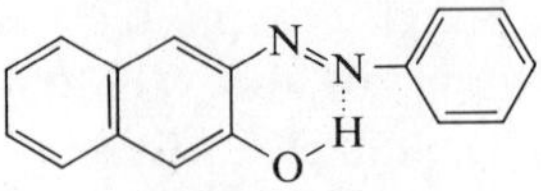

60

λ_{max} 220, 285, 365, 440 mμ
mε 31.6, 15.9, 40.0, 3.2.

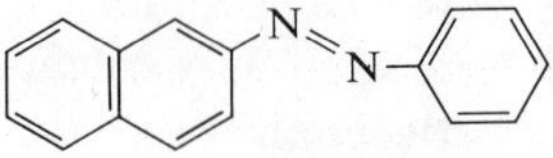

61

λ_{max} 220, 260, 280, 330, 430 mμ
mε 25.1, 12.6, 12.6, 25.1, 1.6.

Other evidence for the azo structure of **60** is its lack of fluorescence and its close spectral resemblance to 3-methoxy-2-phenylazonaphthalene, both dissolved in methylcyclohexane(198).

6. 4-Phenylazo-1-naphthols

The 4-phenylazo-1-naphthol family of azophenols has been the most thoroughly studied spectrally. The parent compound, 4-phenylazo-1-naphthol, exists in solution as an equilibrium mixture of the azo form **62a** and the quinone phenylhydrazone tautomer **63a**, each of which has a characteristic spectrum 182, 188, 191, 204–206.

4-Phenylazo-1-methoxynaphthalene (**62b**), a model for the azo form, and 1,4-naphthoquinone methylphenylhydrazone (**63b**), a model for the phenylhydrazone tautomer, each gives a simple absorption curve with a maximum at 392 mμ for the azo dye and 460 mμ for the hydrazone(189, 193, 205, 206). The spectra of **62** and **63** were measured in ethanol.

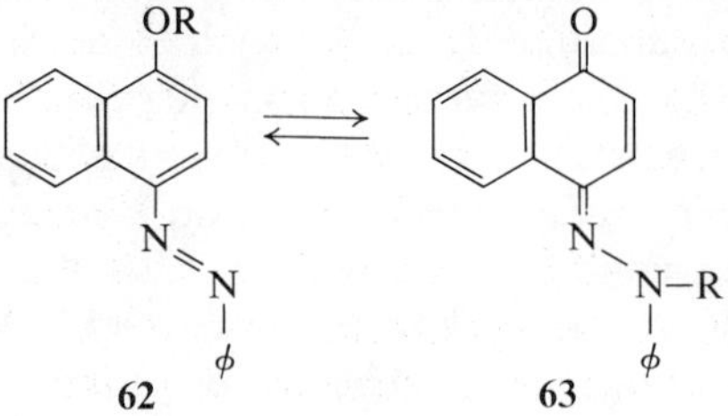

62 ⇄ **63** (4.20)

No.	R	λ_{max} (mμ)	mε
62a ⇄ **63a**	H	408	16.6
		462*s*	10.0
62b	CH_3	392	15.0
63b	CH_3	460	22.0

The tautomerism of the 4′-sulfonic acid analogs of **62** and **63** in neutral aqueous solution has also been investigated(207). The methyl ether absorbed at λ_{max} 402 mμ, mϵ 15.9; the *N*-methyl derivative at λ_{max} 466 mμ, mϵ 25·1; and the unsubstituted sulfonic acid analogous to **62a** $\rightleftarrows$ **63a** at λ_{max} 470 mμ, mϵ 34.4. Thus the latter derivative exists primarily as the hydrazone.

In the 4-phenylazo-1-naphthols the phenylhydrazone concentration increases in the order of substituents 4-CH_3O < 4-CH_3 < 4-Cl < H $\simeq$ 3-CH_3 < 3-CH_3O < 3-Cl and in the order of solvents hexane < ethanol < benzene < chloroform $\simeq$ 50% aqueous ethanol < acetic acid(191).

Substituents in the *ortho* position sterically inhibit the formation of intermolecular hydrogen bonds in the hydrazone tautomers. Both 4-(2′-chlorophenylazo)-1-naphthol and 4-(2′-methoxyphenylazo)-1-naphthol exist almost exclusively as phenylhydrazones in hexane, benzene, and chloroform, and to a slightly smaller degree in acetic acid; however, in ethanol and 50% aqueous ethanol, which form the strongest hydrogen bonds with the phenolic hydroxy group, a considerable amount of the azo tautomer is present(191). The behavior of 4-(2′-tolylazo)-1-naphthol is similar, but the concentration of the azo tautomer is appreciably higher. The phenylhydrazone concentration increases in the order ethanol < 50% ethanol $\simeq$ hexane < benzene < acetic acid < chloroform.

Burawoy and Thomson(191) have emphasized the importance of intermolecular hydrogen bonds between the solvent and 4-arylazo-1-naphthols in the azo $\rightleftarrows$ hydrazone tautomeric equilibrium. The strength of the hydrogen bond increases in the order of solvents chloroform < acetic acid < ethanol < water.

Let us consider the azo (**64**) and hydrazone (**65**) tautomers of 4-(3′-tolylazo)-1-naphthol (see also Fig. 4.4):

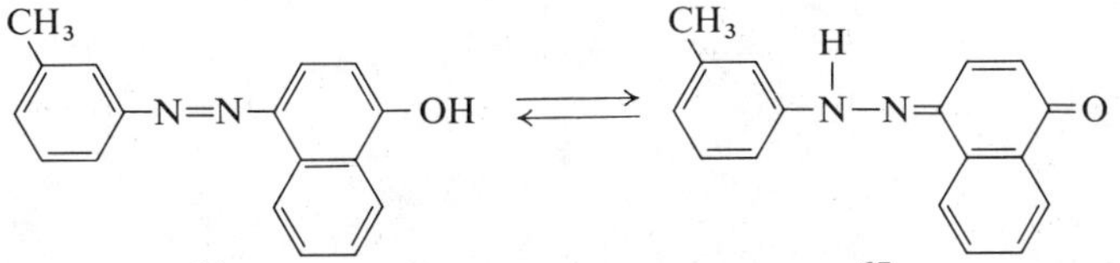

$$\text{CH}_3\text{-C}_6\text{H}_4\text{-N=N-C}_{10}\text{H}_6\text{-OH} \;(\mathbf{64}) \rightleftarrows \text{CH}_3\text{-C}_6\text{H}_4\text{-NH-N=C}_{10}\text{H}_6\text{=O} \;(\mathbf{65}) \qquad (4.21)$$

In 50% ethanol the compound is present mainly as the hydrazone; in ethanol it is present mainly as the azo tautomer. Burawoy and Thompson (191) believe that two strong, but opposing, effects come into play. The azo tautomer forms stronger hydrogen bonds with the hydroxylic solvents; the more polar phenylhydrazone tautomer is favored by increased dielectric constant of the solvent.

The spectrum of 5-phenylazo-8-hydroxyquinoline (Fig. 4.5) consists of

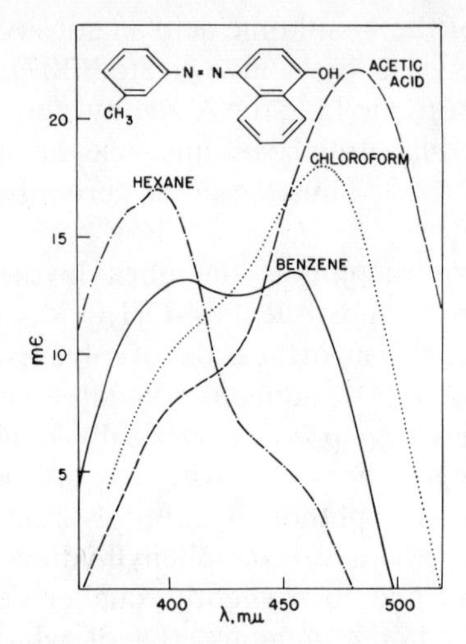

Fig. 4.4. Visible absorption spectra of 4-(3-tolylazo)-1-naphthol in hexane (-·-), benzene (–), chloroform (· · ·), and acetic acid (–––)(191).

two bands, with a maximum at 387 mμ (associated with the azo form, **66**) and a shoulder at 462 mμ (associated with the hydrazone tautomer), the relative intensities of the bands indicating that the azo form occurs to a greater extent in ethanolic solution(205, 208).

The preponderance of the azo tautomer results from the presence of a weak intramolecular hydrogen bond involving a five-membered ring. However, the azo tautomer(**67**) of 5-phenylazo-8-hydroxyquinoline-*N*-oxide has a strong intramolecular hydrogen bond involving a six-membered

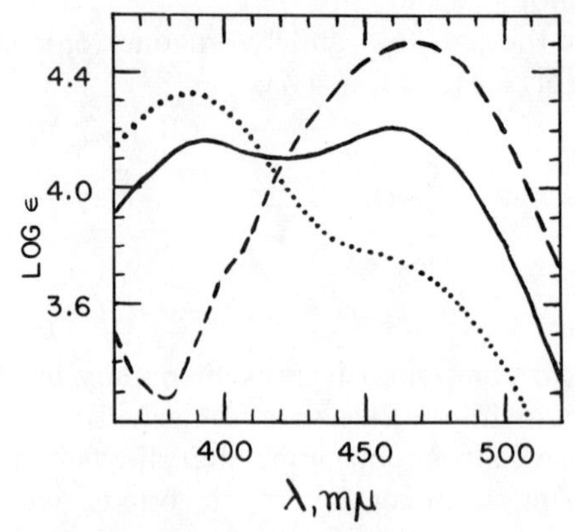

Fig. 4.5. Absorption spectra in 95% ethanol of 5-phenylazo-8-hydroxyquinoline (· · ·), 5-(2-methoxycarbonylphenylazo)-8-hydroxyquinoline (–), and 4-(2-methoxycarbonylphenylazo)-1-naphthol (–––)(208).

ring, which stabilizes this form over that of the hydrazone to an increased extent (205) (see Table 4.19). On the other hand, in 4-(2-methoxycarbonylphenylazo)-1-naphthol, the hydrazone tautomer (**68**) is overwhelmingly present (Fig. 4.5) (208), mainly as a result of a hydrogen bond's stabilizing the hydrazone tautomer. In 5-(2′-methoxycarbonylphenylazo)-8-hydroxyquinoline the intramolecular hydrogen bond between the methoxycarbonyl ketonic oxygen and the azonium hydrogen involved in the six-membered ring of the phenylhydrazone tautomer (**69**) had greater stability than the intramolecular hydrogen bond between the heterocyclic nitrogen and the hydroxyl hydrogen in the five-membered ring of the azo tautomer (**70**) (208).

The importance of intramolecular hydrogen bonding can be seen from the data in Table 4.19, where the ϵ_P/ϵ_A ratio gives an approximate idea of the relative amounts of these tautomers in solution.

The weak intramolecular hydrogen bonding in 8-hydroxy-5-phenylazoquinoline (**66**) can also be neutralized by intermolecular hydrogen bonding with a hydroxylic solvent, since absorption resulting from the hydrazone tautomer increases in the following order: carbon tetrachloride < chloroform < ethanol < acetic acid (205). With an increase in the

TABLE 4.19
Effect of Some Substituents on the Tautomeric equilibrium of 1-Phenylazo-4-naphthols (208)

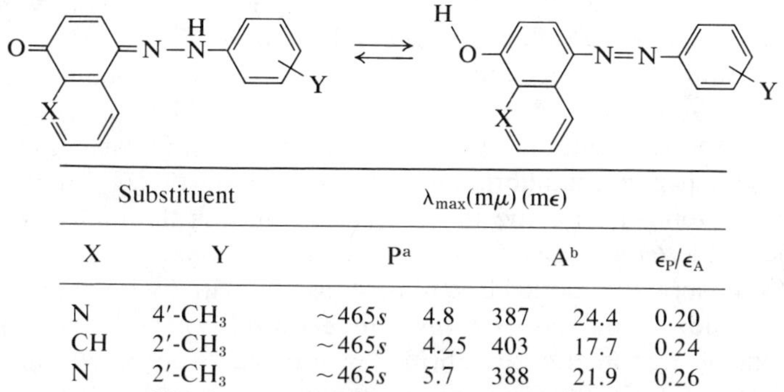

Substituent		λ_{max}(mμ) (mϵ)				
X	Y	P[a]		A[b]		ϵ_P/ϵ_A
N	4′-CH_3	~465*s*	4.8	387	24.4	0.20
CH	2′-CH_3	~465*s*	4.25	403	17.7	0.24
N	2′-CH_3	~465*s*	5.7	388	21.9	0.26
	H	462*s*	5.5	387	21.1	0.26
	4′-$COOC_2H_5$	~465*s*	8.4	402	23.7	0.33
CH	4′-CH_3	~465*s*	10.5	407	23.9	0.44
	H	464	11.5	409	15.4	0.74
N	2′-$COOCH_3$	460	16.1	395	14.4	1.1
CH	4′-$COOC_2H_5$	465	33.6	~400	8.45	~4
	2′-$COOCH_3$	466	31.9	~400	5.0	~6

[a]P = Phenylhydrazone tautomer; *s* = Shoulder.
[b]A = Azo tautomer. All spectra in ethanol.

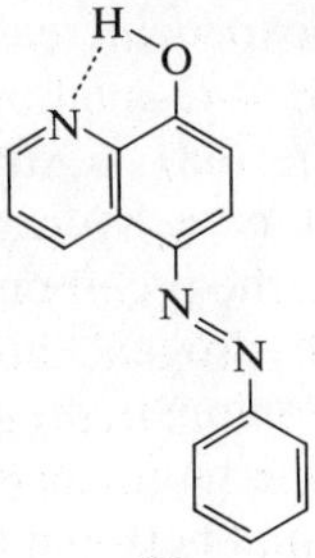

66 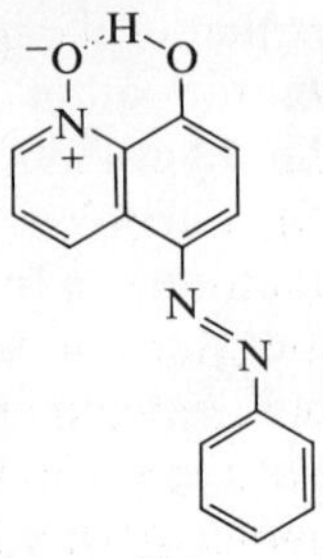 67 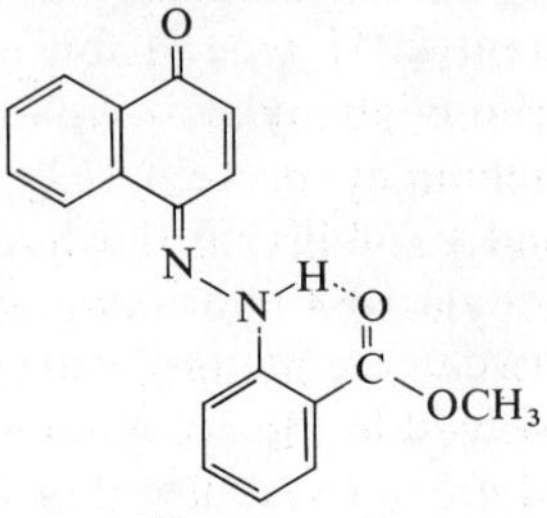 68

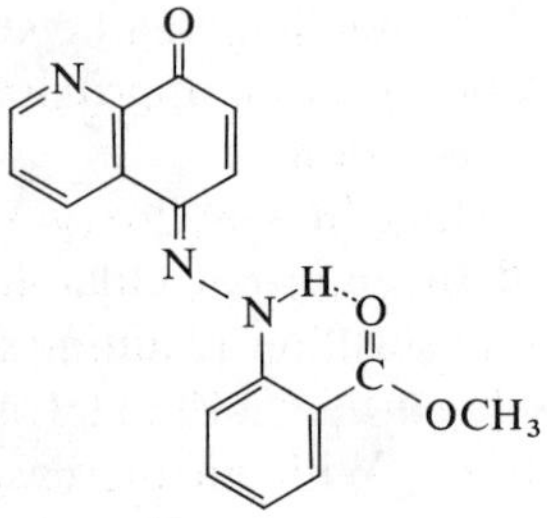

69 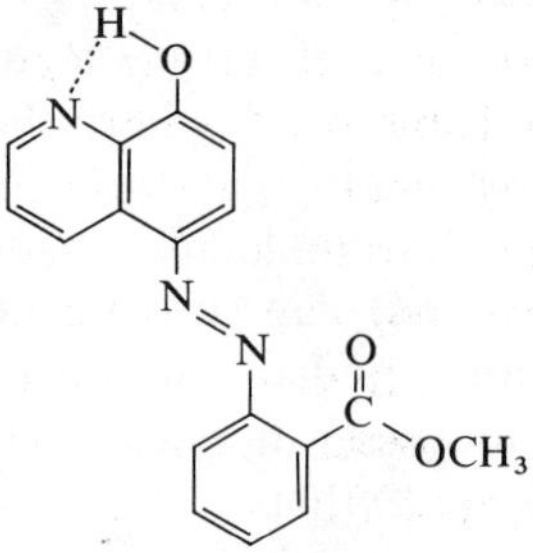 70

strength of the intramolecular hydrogen bond, as in 8-hydroxy-5-phenylazoquinoline *N*-oxide (**67**) intermolecular hydrogen bonding with a hydroxylic solvent has no effect on the equilibrium.

Temperature is another factor that affects tautomeric equilibria. An alcoholic solution of 4-phenylazo-1-naphthol at −140°C consists almost entirely of the azo tautomer, which on irradiation with light at appropriate wavelengths is photoequilibriated with its *cis* isomer(209). With gradual increase in temperature the unstable *cis* isomer is thermally converted into the stable *trans* isomer via the phenylhydrazone. Each step has an activation energy of about 11 kcal/mole, as compared with about 23 kcal/mole for the normal *cis* → *trans* conversion(209). Cooling has little effect on the tautomeric equilibrium in toluene or tetrahydrofuran, but shifts it strongly in favor of the hydrazone in methylcyclohexane, and in favor of the azo compound in ethanol–methanol.

7. Polynuclear Azophenols

1-Phenylazo-2-anthrol(189), 4-phenylazo-1-anthrol(189, 204), and 10-phenylazo-9-anthrol(204) (λ_{max} 511, 506, and 473 mμ, respectively) are believed to exist almost completely in the hydrazone form since the spectra of these compounds in ethanol are entirely different from the spectra of the analogous methoxy derivatives (λ_{max} 435 and 442 mμ for the 1,2- and 1,4-derivatives, respectively).

Examination of the absorption spectra in chloroform solution of a large number of phenylazo-9,10-diphenyl anthrols has led to the conclusion (210) that such compounds as 9,10-diphenyl-1-phenylazo-2,3-anthracenediol, 9,10-diphenyl-1-phenylazo-3-methoxy-2-anthrol, 9,10-diphenyl-1-phenylazo-2,6-anthracenediol, and 9,10-diphenyl-2-phenylazo-1-anthrol exist predominantly as hydrazones ($\lambda_{max} \sim$ 540 mμ, mϵ 17–20).

As shown by the long-wavelength maxima and the millimolar absorptivities structural changes gradually cause the tautomeric equilibrium of the following compounds to shift toward the pure azo compound: 9,10-diphenyl-2-phenylazo-1-anthrol, λ_{max} 525, 550 mμ, mϵ 17.8, 16.6, respectively; 9,10-diphenyl-2-phenylazo-3-methoxy-1-anthrol, λ_{max} 553 mμ, mϵ 11.3; 9,10-diphenyl-2-phenylazo-1,3-anthracenediol, λ_{max} 580 mμ, mϵ 8.7; and 9,10-diphenyl-2-phenylazo-3-anthrol, λ_{max} 450–500 mμ, mϵ 4.0. The latter band is derived from the azo $n \rightarrow \pi^*$ transition.

It is apparent from the data collected in this section that tautomeric equilibria and the derived absorption spectra are affected by changes in structure, solvent, temperature, and so on. Consequently in the analysis of any azophenol its tautomeric and spectral properties should be understood before any analysis is attempted.

8. Hydrazone Phenols

Examples of tautomerism arising from photochromism are found in some of the quinolyl hydrazones, as shown for salicylaldehyde 2-quinolylhydrazone (**71**)(211):

$$\mathbf{71} \underset{\text{dark}}{\overset{h\nu}{\rightleftharpoons}} \mathbf{72} \qquad (4.22)$$

71: λ_{max} 237, 310*s*, 358 mμ; mϵ 40.0, 27.3, 38.5

72: λ_{max} 247, 289, 298, 400 mμ; mϵ 37.9, 27.4, 26.5, 25.0

The spectra were obtained in ethanol. The equation seems reasonable since the *N*-methyl and *O*-methyl derivatives of **71** and the 2-pyridylhydrazone and phenylhydrazone analogs of **71**, as well as benzaldehyde 2-quinolylhydrazone and 4-hydroxybenzaldehyde 2-quinolylhydrazone, show no photochromic properties in 95% ethanol.

References

1. F. Mashio and Y. Kimura, *Nippon Kagaku Zasshi*, **81**, 434 (1960).
2. P. R. Hills and F. J. McQuillin, *J. Chem. Soc.*, **1953**, 4060.
3. B. Eistert and F. Haupter, *Ber.*, **91**, 1921 (1958).
4. G. Hesse, F. Ramisch, and K. Renner, *Ber.*, **89**, 2137 (1956).
5. Grossmann, *Z. Physik. Chem.*, **109**, 305 (1924).
6. G. S. Hammond, W. G. Borduin, and G. A. Guter, *J. Am. Chem. Soc.*, **81**, 4682 (1959).
7. A. Seher, *Arch. Pharm.*, **292**, 519 (1959).
8. B. Eistert and W. Reiss, *Ber.*, **87**, 108 (1954).
9. S. T. Yoffe, E. M. Popov, K. V. Vatsuro, E. K. Tulikova, and M. I. Kabachnik, *Tetrahedron*, **18**, 923 (1962).
10. G. Hesse and G. Krehbiel, *Ann.*, **593**, 35 (1955).
11. M. Cordon, J. D. Knight, and D. J. Cram, *J. Am. Chem. Soc.*, **76**, 1643 (1954).
12. K. Bernauer, *Ann.*, **588**, 230 (1954).
13. K. Alder, *Ann.*, **593**, 23 (1955).
14. N. J. Leonard, H. A. Laitinen, and E. H. Mottus, *J. Am. Chem. Soc.*, **75**, 3300 (1953).
15. B. Eistert, E. Merkel, and W. Reiss, *Ber.*, **87**, 1513 (1954).
16. J. H. Boothe, R. G. Wilkinson, S. Kushner, and J. H. Williams, *J. Am. Chem. Soc.*, **75**, 1732 (1953).
17. C. H. DePuy and P. R. Wells, *J. Am. Chem. Soc.*, **82**, 2909 (1960).
18. A. Stoll, D. Stauffacher, and E. Seebeck, *Helv. Chim. Acta*, **36**, 2027 (1953).
19. N. J. Leonard and P. M. Mader, *J. Am. Chem. Soc.*, **70**, 2707 (1948).
20. E. G. Meek, J. H. Turnbull, and W. Wilson, *J. Chem. Soc.*, 2891 (1953).
21. E. G. Meek, J. H. Turnbull, and W. Wilson, *J. Chem. Soc.*, 811 (1953).
22. B. Eistert and W. Reiss, *Ber.*, **87**, 92 (1954).
23. G. Briegleb and W. Strohmeier, *Z. Naturforsch.*, **6b**, 6 (1951).
24. P. G. Dayton, *Compt. Rend.*, **238**, 2316 (1954).
25. F. Korte and F. Wusten, *Ann. Chem.*, **647**, 18 (1961).
26. L. N. Ferguson, *Modern Structural Theory of Organic Chemistry*, Prentice-Hall, Englewood Cliffs, N.J., 1963, p. 375.
27. B. A. Arbuzov and V. S. Vinogradova, *Dokl. Akad. Nauk SSSR*, **106**, 465 (1956).
28. L. N. Owen and H. M. B. Somade, *J. Chem. Soc.*, 1030 (1947).
29. E. Gelles and R. W. Hay, *J. Chem. Soc.*, 3673 (1957).
30. W. D. Kumler, E. Kun, and J. N. Shoolery, *J. Org. Chem.*, **27**, 1165 (1962).
31. Levin, *Bull. Acad. Sci. USSR, Phys. Ser.*, **11**, 413 (1947).
32. R. A. Morton, A. Hassan, and T. C. Calloway, *J. Chem. Soc.*, 883 (1934).
33. B. Eistert and E. Merkel, *Ber.*, **86**, 895 (1953).
34. B. Eistert, F. Weygand, and E. Csendes, *Ber.*, **84**, 745 (1951).
35. G. Hammond and F. Schultz, *J. Am. Chem. Soc.*, **74**, 329 (1952).
36. H. H. Szmant and H. J. Planinsek, *J. Am. Chem. Soc.*, **76**, 1193 (1954).
37. R. E. Lutz and C. Dien, *J. Org. Chem.*, **21**, 551 (1956).
38. R. D. Campbell and H. M. Gilow, *J. Am. Chem. Soc.*, **82**, 5426 (1960).
39. P. B. Russell and J. Mentha, *J. Am. Chem. Soc.*, **77**, 4245 (1955).
40. W. R. Chan and C. H. Hassall, *J. Chem. Soc.* 3495 (1956).
41. American Petroleum Institute Research Project No. 44, I, 365 (1949).
42. American Petroleum Institute Research Project No. 44, I, 366 (1949).
43. N. M. Chopra, *J. Chem. Soc.*, 588 (1955).
44. B. Eistert and F. Geiss, *Ber.*, **94**, 929 (1961).
45. A. Gero, *J. Org. Chem.*, **19**, 1960 (1954).

46. G. Schwarzenbach and C. Wittwer, *Helv. Chim. Acta*, **30**, 669 (1947).
47. J. W. Sidman, *Chem. Rev.*, **58**, 689 (1958).
48. N. S. Bayliss and E. G. McRae, *J. Phys. Chem.*, **58**, 1006 (1954).
49. G. D. Johnson, *J. Am. Chem. Soc.*, **75**, 2720 (1953).
50. F. Arndt, L. Loewe, and L. Capuano, *Rev., Fac. Sci. Univ. Istanbul*, **8A**, 122 (1943); through *Chem. Abstr.*, **39**, 1624 (1945).
51. R. Filler and S. M. Naqvi, *J. Org. Chem.*, **26**, 2571 (1961).
52. E. H. Holst and W. C. Fernelius, *J. Org. Chem.*, **23**, 1881 (1958).
53. P. Karrer, J. Kebrle, and V. Albers-Schonberg, *Helv. Chim. Acta*, **34**, 1014 (1951).
54. J. Kenner and G. N. Richards, *J. Chem. Soc.*, 2240 (1953).
55. K. H. Meyer and P. Wertheimer, *Ber.*, **47**, 2374 (1914). See also R. D. Campbell and F. J. Schultz, *J. Org. Chem.*, **25**, 1877 (1960).
56. E. M. Kosower, *J. Am. Chem. Soc.*, **80**, 3267 (1958).
57. P. B. Russell, *J. Am. Chem. Soc.*, **74**, 2654 (1952).
58. W. Wislicenus and A. Ruthing, *Ann.*, **379**, 229 (1912).
59. P. B. Russell and E. Csendes, *J. Am. Chem. Soc.*, **76**, 5714 (1954).
60. G. Dimroth, *Ann.*, **399**, 110 (1913).
61. M. I. Kabachnik, S. T. Yoffe, and K. V. Vatsuro, *Tetrahedron*, **1**, 317 (1957).
62. M. I. Kabachnik, S. T. Yoffe, E. M. Popov, and K. V. Vatsuro, *Tetrahedron*, **12**, 76 (1961).
63. G. Schwarzenbach, M. Zimmerman, and V. Prelog, *Helv. Chim. Acta*, **34**, 1954 (1951).
64. R. D. Campbell and H. M. Gilow, *J. Am. Chem. Soc.*, **84**, 1440 (1962).
65. K. H. Meyer and P. Kapelmeier, *Ber.*, **44**, 2724 (1911).
66. W. Dieckmann, *Ber.*, **55**, 2470 (1922).
67. L. F. Fieser and M. Fieser, *J. Am. Chem. Soc.*, **70**, 3215 (1948).
68. R. S. Rasmussen, D. D. Tunnicliff, and R. R. Brattain, *J. Am. Chem. Soc.*, **71**, 1068 (1949).
69. P. Ramart-Lucas and J. Hoch, *Bull. Soc. Chim. France*, **19**, 220 (1952).
70. Y. Hirshberg and F. Bergmann, *J. Am. Chem. Soc.*, **72**, 5118 (1950).
71. R. N. Jones, *J. Am. Chem. Soc.*, **67**, 1956 (1945).
72. B. Eistert, H. Schneider, and R. Wollheim, *Ber.*, **92**, 2061 (1959).
73. R. C. Fuson and Q. Soper, *J. Am. Chem. Soc.*, **65**, 915 (1943).
74. B. Eistert and H. Munder, *Ber.*, **91**, 1415 (1958).
75. B. Eistert and H. Munder, *Ber.*, **88**, 215 (1955).
76. C. A. Buehler and J. O. Harris, *J. Am. Chem. Soc.*, **72**, 5015 (1950).
77. T. Ukai, S. Kanahara, and S. Kanetomo, *J. Pharm. Soc. Japan*, **74**, 43 (1954).
78. H. Beyer and U. Hess, *Ber.*, **90**, 2435 (1957).
79. G. Henseke and W. Lemke, *Ber.*, **91**, 101 (1958).
80. T. Ukai and S. Kanahara, *J. Pharm. Soc. Japan*, **74**, 45 (1954).
81. T. Ukai and S. Kanahara, *J. Pharm. Soc. Japan*, **75**, 31 (1955).
82. A. Burawoy, *J. Chem. Soc.*, 1181 (1939).
83. Z. Reyes and R. M. Silverstein, *J. Am. Chem. Soc.*, **80**, 6373 (1958).
84. Z. Reyes and R. M. Silverstein, *J. Am. Chem. Soc.*, **80**, 6367 (1958).
85. S. K. Mitra, *J. Indian Chem. Soc.*, **15**, 205 (1938).
86. E. Campaigne and R. E. Cline, *J. Org. Chem.*, **21**, 32 (1956).
87. G. Klose, P. Thomas, E. Uhlemann, and J. Marki, *Tetrahedron*, **22**, 2695 (1966).
88. V. V. Ershov and G. A. Nikiforov, *Russ. Chem. Rev.*, **35**, 817 (1966).
89. R. H. Thomson, *Quart. Rev.*, **10**, 27 (1956).
90. G. W. Wheland, *Resonance in Organic Chemistry*, Wiley, New York, 1955, p. 405.

91. W. Mayer and R. Weiss, *Angew. Chem.*, **68**, 680 (1956).
92. R. H. Thomson, *J. Chem. Soc.*, 1737 (1950).
93. C. Daglish, *J. Am. Chem. Soc.*, **72**, 4859 (1950).
94. R. Livingstone and M. C. Whiting, *J. Chem. Soc.*, 3631 (1955).
95. R. G. Cooke and H. Dowd, *Australian J. Sci. Res.*, **A5**, 760 (1952).
96. D. B. Bruce and R. H. Thomson, *J. Chem. Soc.*, 2759 (1952).
97. D. B. Bruce and R. H. Thomson, *J. Chem. Soc.*, 1089 (1955).
98. C. J. P. Spruit, *Rec. Trav. Chim.*, **68**, 309 (1949).
99. C. A. Grob and J. Voltz, *Helv. Chim. Acta*, **33**, 1796 (1950).
100. Y. Bansho and K. Nukada, *Kogyo Kagaku Zasshi*, **63**, 620 (1960); through *Chem. Abstr.*, **57**, 8516 (1963).
101. J. A. Barltrop and K. J. Morgan, *J. Chem. Soc.*, 4245 (1956).
102. R. N. Jones, *J. Am. Chem. Soc.*, **67**, 2127 (1945).
103. E. Faffitte and R. Lalande, *Bull. Soc. Chim. France* [5], **24**, 761 (1957).
104. H. Baba and T. Takemura, *Tetrahedron*, **24**, 4779 (1968).
105. T. Takemura and H. Baba, *Tetrahedron*, **24**, 5311 (1968).
106. L. F. Fieser, *J. Am. Chem. Soc.*, **53**, 2329 (1931).
107. C. Marschalk, *Bull. Soc. Chim. France* [5], **4**, 1547 (1937).
108. A. Etienne and B. Rutimeyer, *Bull. Soc. Chim. France*, 1595 (1956).
109. K. Bredereck and E. F. Sommermann, *Tetrahedron Letters*, 5009 (1966).
110. G. Bichet, *Ann. Chim. France*, **7**, 234, (1952).
111. Meyer and Sander, *Ann.*, **420**, 113 (1920).
112. L. Nedelec and J. Rigaudy, *Bull. Soc. Chim. France*, 1204 (1960).
113. L. F. Fieser and Y. Hershberg, *J. Am. Chem. Soc.*, **59**, 1028 (1937).
114. L. I. Smith, H. R. Davis, Jr., and A. W. Sogn, *J. Am. Chem. Soc.*, **72**, 3651 (1950).
115. L. F. Fieser and M. Fieser, *J. Am. Chem. Soc.*, **61**, 596, (1939).
116. M. G. Ettlinger, *J. Am. Chem. Soc.*, **76**, 2769 (1954).
117. R. A. Durie and J. S. Shannon, *Australian J. Chem.*, **11**, 168 (1958).
118. V. C. Farmer, *Spectrochim. Acta*, 870 (1959).
119. A. B. Hornfeldt, *Arkiv Kemi*, **22**, 211 (1964).
120. C. D. Hurd and K. L. Kreuz, *J. Am. Chem. Soc.*, **72**, 5543 (1950).
121. J. Green and S. Marcinkiewicz, *J. Chromatog.*, **10**, 354 (1963).
122. M. C. Ford and D. Mackay, *J. Chem. Soc.*, 4985 (1956).
123. A. Hornfeldt, *Acta Chem. Scand.*, **19**, 1249 (1965).
124. P. Karrer and F. Kehrer, *Helv. Chim. Acta*, **27**, 142 (1944).
125. S. Gronowitz and A. Bugge, *Acta Chem. Scand.*, **20**, 261 (1966).
126. H. J. Jakobsen and S. O. Lawesson, *Tetrahedron*, **23**, 871 (1967).
127. H. J. Jakobsen and S. O. Lawesson, *Tetrahedron*, **21**, 3331 (1965).
128. S. Gronowitz, P. Moses, and A. B. Hornfeldt, *Arkiv Kemi*, **17**, 273 (1961).
129. R. A. Hoffman and S. Gronowitz, *Arkiv Kemi*, **16**, 563 (1960).
130. S. Gronowitz and P. Moses, *Acta Chem. Scand.*, **16**, 105 (1962).
131. S. Gronowitz, P. Moses, and A. B. Hornfeldt, *Arkiv Kemi*, **17**, 237 (1961); through *Chem. Abstr.*, **56**, 4277 (1962).
132. A. I. Kosak, R. J. F. Palchak, W. A. Steele, and C. M. Selwitz, *J. Am. Chem. Soc.*, **76**, 4450 (1954).
133. O. Dimroth, *Ann.*, **377**, 127 (1910).
134. K. H. Meyer, *Ann.*, **380**, 212 (1911).
135. G. W. Stacy, F. W. Villaescusa, and T. E. Wollner, *J. Org. Chem.*, **30**, 4074 (1965).
136. G. W. Stacy and T. E. Wollner, *J. Org. Chem.*, **32**, 3028 (1967).
137. G. M. Okengendler and M. A. Mostoslavskii, *Ukrain. Khim. Zh.*, **26**, 69 (1960).

138. A. I. Kiss and B. R. Muth, *Magyar Tudomanyos Akad. Kozponti Fiz. Kutato Intezetenek Koslemenyei*, **3**, 213 (1955); through *Chem. Abstr.*, **53**, 3876 (1959).
139. R. Chong and P. S. Clezy, *Australian J. Chem.*, **20**, 935 (1967).
140. G. O. Dudek and R. H. Holm, *J. Am. Chem. Soc.*, **83**, 2099, 3914 (1961).
141. J. W. Ledbetter, Jr., *J. Phys. Chem.*, **70**, 2245 (1966).
142. Z. Holzbecher, *Collection Czech. Chem. Commun.*, **20**, 1292 (1955).
143. L. A. Kazitsyna, L. L. Polstyanko, N. B. Kupletskaya, T. N. Ignatovich, and A. P. Terent'ev, *Dokl. Akad. Nauk SSSR*, **125**, 807 (1959); through *Chem. Abstr.*, **53**, 16689 (1959).
144. G. O. Dudek and E. P. Dudek, *Chem. Commun.* 464 (1965).
145. L. N. Ferguson and I. Kelly, *J. Am. Chem. Soc.*, **73**, 3707 (1951).
146. H. J. Bielig and E. Bayer, *Ann.*, **580**, 135 (1953).
147. L. A. Kazitsyna, N. B. Kupletskaya, L. L. Polstyanko, B. S. Kikot, Y. A. Kolesnik, and A. P. Terent'ev, *Zh. Obshch. Khim.*, **31**, 313 (1961); through *Chem. Abstr.*, **55**, 22247 (1961).
148. R. S. Becker and W. F. Richey, *J. Am. Chem. Soc.*, **89**, 1298 (1967).
149. G. Dudek, *J. Org. Chem.*, **32**, 2016, (1967).
150. M. D. Cohen and S. Flavian, *J. Chem. Soc.*, **B**, 317 (1967).
151. M. D. Cohen and G. M. J. Schmidt, *J. Phys. Chem.*, **66**, 2442 (1962).
152. D. Heinert and A. E. Martell, *J. Am. Chem. Soc.*, **85**, 183, 188 (1963).
153. Y. Matsushima and A. E. Martell, *J. Am. Chem. Soc.*, **89**, 1322 (1967).
154. L. Schirch and R. A. Slotter, *Biochemistry*, **5**, 3175 (1966).
155. A. Schors, A. Kraaijeveld, and E. Havinga, *Rec. Trav. Chim.*, **74**, 1243 (1955).
156. D. P. Johnson and F. E. Critchfield, *Anal. Chem.*, **33**, 910 (1961).
157. S. Osaki, *Bunseki Kagaku*, **7**, 275 (1958).
158. H. H. Hodgson, *J. Chem. Soc.*, 520 (1937).
159. P. Ramart-Lucas and M. Martynoff, *Bull. Soc. Chim. France*, **15**, 571 (1948); ibid., **16**, 53 (1949).
160. E. Hertel and F. Lebok, *Z. Physik. Chem.*, **47**, 315 (1940).
161. S. E. Hoekstra, *The Properties of Alkylated* p-*Benzoquinone Oximes*, Ph.D. Thesis, University of Leiden, The Netherlands, 1960.
162. D. Hadzi, *J. Chem. Soc.*, 2725 (1956).
163. E. Havinga and A. Schors, *Rec. Trav. Chim.*, **69**, 457 (1950).
164. H. Uffmann, *Tetrahedron Letters*, 4631 (1966).
165. R. K. Norris and S. Sternhell, *Tetrahedron Letters*, 97 (1967).
166. A. Burawoy, M. Cais, J. Chamberlain, F. Liversedge, and A. Thompson, *J. Chem. Soc.*, 3727 (1955).
167. A. Burawoy, *J. Chem. Soc.*, 3721 (1955).
168. D. Dyrssen and E. Johonsson, *Acta Chem. Scand.*, **9**, 763 (1955).
169. H. Shimura, *J. Chem. Soc. Japan*, **76**, 867 (1955).
170. E. Baltazzi, *Compt. Rend.*, **232**, 986 (1951).
171. K. S. Tewari, *Naturwissenschaften*, **46**, 445 (1959).
172. W. R. Edwards, Jr., and C. W. Tate, *Anal. Chem.*, **23**, 826 (1951).
173. D. O. Eberle and F. A. Gunther, *J. Assoc. Offic. Agr. Chemists*, **48**, 927 (1965).
174. R. J. Gajan, W. R. Benson, and J. M. Finocchiaro, *J. Assoc. Offic. Agr. Chemists*, **48**, 958 (1965).
175. W. R. Benson and R. J. Gajan, *J. Org. Chem.*, **31**, 2498 (1966).
176. H. Collatz and H. J. Rummlen, *Bitumen, Teere*, **13**, 582 (1962).
177. V. Sandulescu, S. Traian, and E. Tibrea, *Standardizarea*, **17**, 128 (1965).
178. S. Yaroslavsky, *Chem. Ind.* (*London*), 554 (1961).

179. W. Pilz, *Mikrochim. Acta*, 614 (1961).
180. W. Pilz and I. Johann, *Z. Anal. Chem.*, **210**, 113 (1965).
181. H. S. Isbell and A. J. Fatiadi, *Carbohydrate Res.*, **2**, 204 (1966).
182. R. Kuhn and F. Bar, *Ann.*, **516**, 143 (1935).
183. J. Ospenson, *Acta Chem. Scand.*, **4**, 1351 (1950).
184. A. Burawoy and J. T. Chamberlain, *J. Chem. Soc.*, 3734 (1952).
185. D. Hadzi, *J. Chem. Soc.*, 2143 (1956).
186. B. L. Kaul, R. Srinivasan, and K. Venkataraman, *Chima*, **19**, 213 (1965).
187. W. Lauer and S. Miller, *J. Am. Chem. Soc.*, **57**, 520 (1935).
188. A. Burawoy, A. G. Salem, and A. R. Thompson, *J. Chem. Soc.*, 4793 (1952).
189. J. N. Ospenson, *Acta Chem. Scand.*, **5**, 491 (1951).
190. A. Burawoy, *J. Oil Colour Chemists Assoc.*, 298 (1952).
191. A. Burawoy and A. R. Thompson, *J. Chem. Soc.*, 1443 (1953).
192. P. Ramart-Lucas, *Bull. Soc. Chim. France*, **11**, 75 (1944).
193. P. Ramart-Lucas and J. Hoch, *Bull. Soc. Chim. France*, **14**, 986 (1947).
194. K. H. Meyer, *Ber.*, **45**, 2843 (1912).
195. L. G. S. Brooker, A. C. Craig, D. W. Heseltine, P. W. Jenkins, and L. L. Lincoln, *J. Am. Chem. Soc.*, **87**, 2443 (1965).
196. K. J. Morgan, *J. Chem. Soc.*, 2151 (1961).
197. A. Burawoy, A. Salem and A. Thompson, *Chem. Ind.*, (*London*), 668 (1952).
198. G. Gabor, Y. Frei, D. Gegiou, M. Kaganowitch, and E. Fischer, *Israel J. Chem.*, **5**, 193 (1967).
199. A. Burawoy and I. Markowitsch-Burawoy, *Ann.*, **503**, 180 (1933); ibid., **504**, 71 (1933).
200. J. Epstein, M. Demek, and V. C. Wolff, *Anal. Chem.*, **29**, 1050 (1957).
201. V. V. Perekalin and M. V. Savostyanova, *Zh. Obshch. Khim.*, **21**, 1329 (1951).
202. G. M. Badger and R. G. Buttery, *J. Chem. Soc.*, 2243 (1954).
203. H. E. Fierz-David, L. Blangey, and E. Merian, *Helv. Chim. Acta*, **34**, 846 (1951).
204. H. Shingu, *Sci. Papers Inst. Phys. Chem. Research (Tokyo)*, **35**, 78 (1938).
205. G. M. Badger and R. G. Buttery, *J. Chem. Soc.*, 614 (1956).
206. M. M. Shemyakin, F. A. Mendelvich, A. M. Simonov, and V. B. Petrinska, *Dokl. Akad. Nauk SSSR*, **115**, 526 (1957); through *Chem. Abstr.*, **52**, 5360 (1958).
207. J. B. Müller, L. Blangey, and H. E. Fierz-David, *Helv. Chim. Acta*, **35**, 2579 (1952).
208. E. Sawicki, *J. Org. Chem.*, **22**, 743 (1957).
209. E. Fischer and Y. F. Frei, *J. Chem. Soc.*, 3159 (1959).
210. A. Etienne and J. Bourdon, *Bull. Soc. Chim. France*, 396 (1955).
211. J. L. Wong and F. N. Bruscato, *Tetrahedron Letters*, 4593 (1968).

CHAPTER 5

ABSORPTION SPECTRA OF ZWITTERIONIC RESONANCE STRUCTURES

III. TAUTOMERISM INVOLVING METHENE, IMINE, OR AMINE GROUPS

The most important tools in the investigation of the tautomerism of heterocyclic and other compounds have been discussed by Katritsky and Lagowski(1). The physical methods that have been used to study tautomerism include ultraviolet and visible spectroscopy(2,3), basicity measurements(4–8), infrared spectroscopy(9, 10), fluorescence spectroscopy(11), X-ray crystallography(12), dipole moments(13), polarography(14), the Hammett equation(15), nuclear-magnetic-resonance spectroscopy(16), refractive index(17), and molar refractivity(18).

Of these methods this treatise is mainly concerned with ultraviolet and visible spectroscopy, basicity measurements, and fluorescence measurements. The first two methods have seen a tremendous amount of use, whereas the latter, which would be concerned with the excited singlet state, has seen very little use. These methods are most reliable when the absorption spectra, the basicity, or the fluorescence spectra of the comparison standards are widely different. Each of the standards (with the mobile hydrogen replaced by an alkyl group) has a conjugation system that closely resembles one of the tautomeric forms and thus is usually similar absorptiometrically and fluorimetrically and has a pK_a within approximately 0.3 pK_a unit of the appropriate tautomer. If $—NH—\dot{C}{=}O \leftrightarrows —N{=}\dot{C}—OH$ represents the tautomeric equilibrium, then $CH_3—\dot{N}—\dot{C}{=}O$ and $—N{=}\dot{C}—O—CH_3$ would be appropriate comparison standards. When the standards have closely similar physical properties, photometric methods are useless by themselves.

I. Lactam ⇄ Lactim Tautomers

The prototropic tautomerism of heteroaromatic compounds containing five- and six-membered rings substituted with hydroxy, sulfhydryl, or amino groups has been thoroughly discussed(19). For most compounds of the aza type the lactam form predominates in solution or in the solid state.

A. Simple Amides

These compounds are present predominantly in the amide form. Since *N*,*N*-disubstituted amides may show abnormal light absorption due to steric inhibition of resonance involving the amide group, sterically unhindered *N*,*N*-disubstituted amides of α,β-acetylenic acids have been used as comparator standards (20). For example, the amide **1** spectrally resembles the dimethylamide **2** and differs from the methoxy standard **3**.

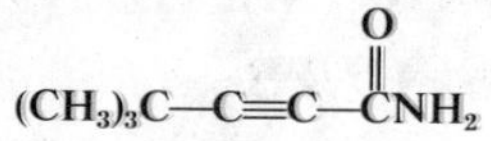

$$(CH_3)_3C{-}C{\equiv}C{-}\overset{O}{\overset{\|}{C}}NH_2$$

1

λ_{max} 209 mμ, mϵ 11.5

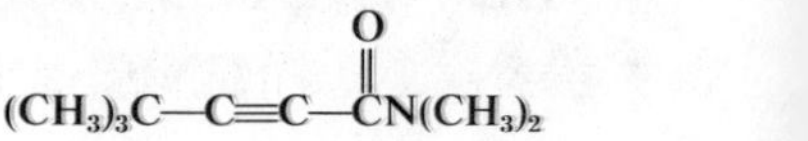

$$(CH_3)_3C{-}C{\equiv}C{-}\overset{O}{\overset{\|}{C}}N(CH_3)_2$$

2

λ_{max} 209 mμ, mϵ 11.2

$$(CH_3)_3C{-}C{\equiv}C{-}\overset{OCH_3}{\overset{|}{C}}{=}NH$$

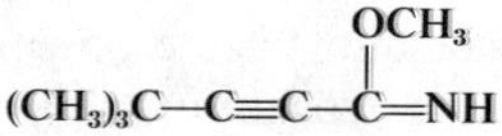

3

λ_{max} < 205 mμ, mϵ 20.0 at 205 mμ

In addition compound **1** shows a doublet at 2.83 and 2.92 μ characteristic of the N—H_2 stretching frequencies and a carbonyl band at 5.96 μ.

It should be possible to stabilize the iminol form through intramolecular hydrogen bonding. An example of this is the postulated equilibrium in equation 5.1 (21). Photometric investigation would be necessary to confirm such an equilibrium.

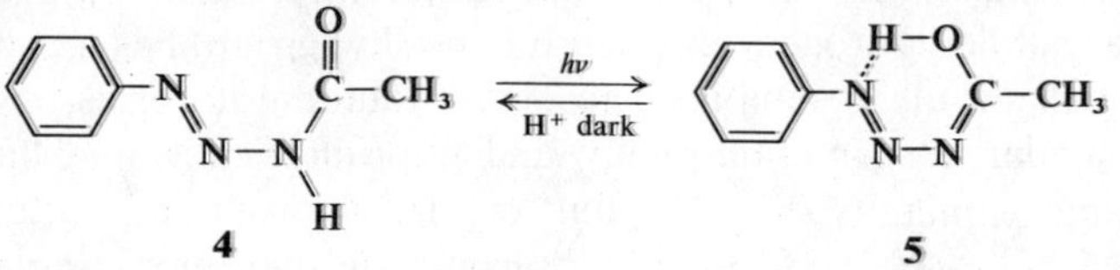

$$C_6H_5{-}N{=}N{-}NH{-}C({=}O){-}CH_3 \underset{H^+\ \text{dark}}{\overset{h\nu}{\rightleftharpoons}} C_6H_5{-}N{=}N{-}N{=}C(OH){-}CH_3 \quad (5.1)$$

4 Colourless **5** Red

Another type of amide in which the iminol form is reported to be present involves strong conjugation between positive and negative resonance terminals through the C=N chain (22), as in the following:

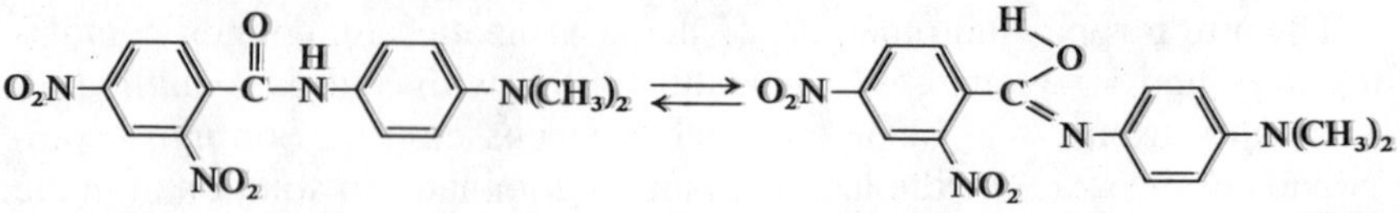

$$O_2N{-}C_6H_3(NO_2){-}C({=}O){-}NH{-}C_6H_4{-}N(CH_3)_2 \rightleftharpoons O_2N{-}C_6H_3(NO_2){-}C(OH){=}N{-}C_6H_4{-}N(CH_3)_2 \quad (5.2)$$

6 **7**

Compound **7** is a deep-red solid that absorbs at λ_{max} 430 mμ, mϵ 2.0. The *m*-dimethylamino isomer of **6** has a long-wavelength shoulder at 300 mμ, mϵ 6.3, and the dinitrophenylacetyl homolog of **6** (i.e., CH_2 between the ring and the C—O group) also absorbs at short wavelengths. A close analog of **7** is the compound *N*,*N*-dimethyl-2′,4′-dinitro-4-stilbeneamine, which absorbs at 477 mμ, mϵ 15.1, in ethanol(23).

Thus this type of tautomerism certainly bears more thorough investigation for its possible usefulness in trace analysis.

B. Five-membered Rings

With the help of the usual ultraviolet and infrared spectral examination of 2-thiazolone and its methylated comparative standards, 2-thiazolone has been shown to exist in the lactam form (**8**)(24, 25).

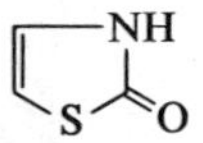

8

λ_{max} 216, 234 mμ
mϵ 3.3, 5.6

In the 5-isoxazolones three tautomers can be obtained:

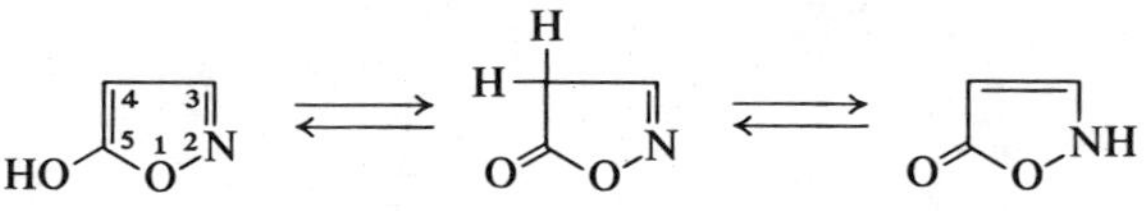

9 ⇄ **10** ⇄ **11** (5.3)

Both solvents and substituents play a role in this equilibrium. Thus infrared and ultraviolet spectral studies show that 3,4-dimethyl-5-hydroxyisoxazole exists in aqueous and chloroform solution, and in the solid phase in the OH form; 3-phenyl-4-bromo-3-phenyl- and 4-methyl-3-phenyl-isoxazol-5-ones exist as mixtures of the CH and NH forms in solution, the proportion of the NH form increasing with the polarity of the solvent (Table 5.1)(26).

In part the assignments are based on ultraviolet spectra comparison, as shown for **12** as compared with **13**, **14**, and **15**; and for **16** as compared with **17**, **18**, and **19**, in cyclohexane solution.

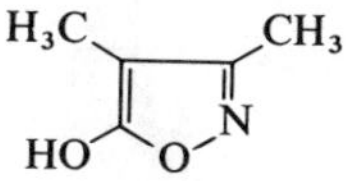

12 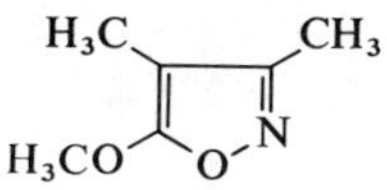 **13** 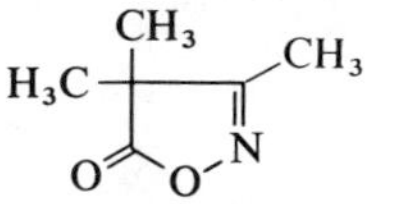 **14** 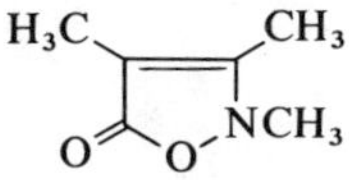 **15**

TABLE 5.1
The Tautomeric Nature of Isoxazol-5-ones (26)

Substituent at positions		Cyclohexane solution (UV)	Tetrachloro-ethylene solution (IR)	Chloroform solution (IR)	Aqueous solution (UV)	Solid state (IR)
3	4					
CH_3	CH_3	>80% OH	–	>98% OH	>80% OH	OH
C_6H_5	H	>90% CH	–	>98% CH	70% CH + 30% NH	CH
C_6H_5	CH_3	>90% CH	90% CH + 10% NH	70% CH + 30% NH	>90% NH	NH
C_6H_5	Br	>90% CH	>95% CH	90% CH + 10% NH	20% CH + 80% NH	NH

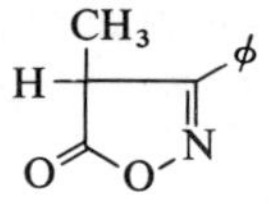

16 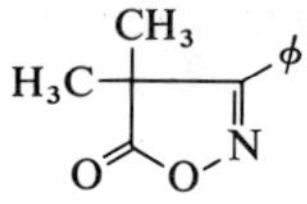 17 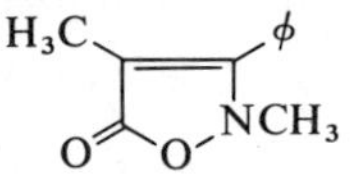 18 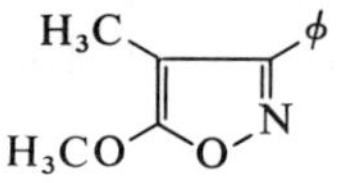 19

No.	λ_{max} (mμ)	mε
12	213	6.6
13	209	5.6
14	<201	
15	262	7.9
16	253	13.2
17	254	10.1
18	231, 280	10.6, 7.0
19	228	12.6

On the other hand, De Sarlo (27) examined the spectra of 3,4-dimethylisoxazolin-5-one and the appropriate reference compounds and reached the conclusion that in dioxane and hydroxylic solvents the NH (or 2-H) form predominates, whereas in nonpolar solvents, such as cyclohexane, the CH (or 4-H) form predominates. Seemingly good spectral evidence is presented in this proof also.

Spectral data in ethanol of the lactam vinylogs of 2-thiazolinone [e.g., **20a** and **b** (28) and **20c** (29)] would seem to indicate that these compounds exist mainly in the NH form. However, a more thorough investigation of solvent effects, etc., would be necessary to establish the position of equilibrium of the various possible tautomers.

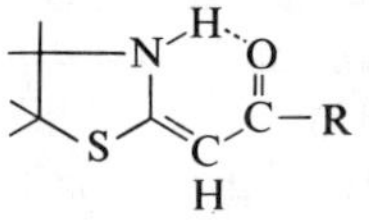

20

	R	λ_{max} (mμ)	mε
a	OC_2H_5	290	33.0
b	$\overset{O}{\overset{\|\|}{C}}OC_2H_5$	342	45.4
c	CH_3	250	6.3
		312	50.0

C. Pyridones

Aza heterocyclic compounds with a hydroxy group α to the nitrogen are cyclic amides; those with a hydroxy group γ to the nitrogen are vinylogs of amides. The amide form predominates by a factor of about 10^3. Thus for

$$K_t = \frac{\%\text{ of lactam}}{\%\text{ of lactim}}$$

the tautomeric equilibrium constant K_t for 2-pyridone has been reported to be 320(30) and 910(8). Both values have been estimated by pK_a methods involving the use of the basic ionization constants of the hydroxy compound (K_1) and that of its *O*-methyl and/or *N*-methyl derivatives (8, 19, 30); for example,

$$K_t = \frac{[\text{NH form}]}{[\text{OH form}]} = \frac{K_{\text{NMe}}}{K_{\text{OMe}}} = \frac{K_{\text{OMe}}}{K_1 - 1} = \frac{K_1 - 1}{K_{\text{NMe}}}$$

For 4-pyridone K_t was approximately 3000(8, 30). The predominance of the oxo forms of 2- and 4-hydroxypyridine was proved spectrally in 1942 by Specker and Gawrosch(31).

Mason(3) in a study of the pK_a values and absorption spectra in several solutions of 68 hydroxylated aza compounds has shown that compounds with a hydroxy group α or γ to a ring-nitrogen atom are spectrally similar to their *N*-methyl derivatives and different from their *O*-methyl derivatives in both organic and aqueous solvents. He has estimated K_t values from the spectra and has found these values to increase with conjugation between the oxygen and the nitrogen atom, and with the addition of fused benzene rings—and to decrease with aza substitution, increase in temperature, and decrease in the dielectric constant of the solvent.

The evidence based on ultraviolet, infrared, and Raman spectra, ionization constants, and X-ray crystallography studies for the predominance of the lactone form of 2- and 4-hydroxypyridines and pyrimidines has been summarized(32).

In compounds with a hydroxy group neither α nor γ to a ring-nitrogen atom tautomerism of the hydroxy form to a zwitterionic form is repressed by an increase in temperature or a decrease in the dielectric constant of the solvent. An example of this type of equilibrium and the spectral proof of it is given in Table 5.2. At pH 6.8 in water 3-hydroxypyridine has a value of 1.27 for

$$K_t = \frac{[\text{zwitterion}]}{[\text{enol}]}$$

The relatively low K_t value for 3-hydroxypyridine in comparison with those of its 2- and 4-isomers has been explained by the stabilization of the zwitterions of α- and γ-hydroxy (α or γ to the nitrogen atom) compounds through mesomerism with uncharged canonical forms (e.g.,

TABLE 5.2
Absorption Spectra of 3-Hydroxypyridine and Some Comparison Standards (3)

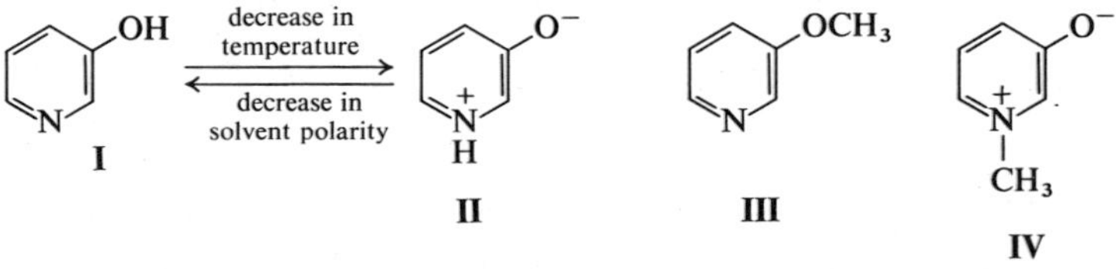

I ⇄ II (decrease in temperature →; ← decrease in solvent polarity) III IV

λ_{max} ($m\mu$) ($m\epsilon$)			
I (in EtOH)	**II** (pH 6.8)	**III** (pH 7)	**IV** (pH 7)
	246 (5.1)		249 (8.1)
279 (4.4)	278 (2.3)	276 (4.0)	
325 (0.025)	315 (3.1)		320 (5.8)

equation 5.4), whereas β-hydroxy compounds cannot be so stabilized (19).

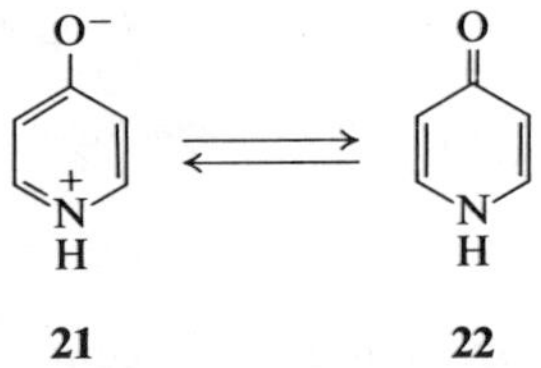

$$\mathbf{21} \rightleftharpoons \mathbf{22} \qquad (5.4)$$

As to be expected, the Z_b-band of 3-hydroxypyridine at 315 $m\mu$ in aqueous solution shifts to longer wavelength, 325 $m\mu$ in ethanol, with a decrease in the polarity of the solvent (Table 5.2). Additional confirmation of the presence of the zwitterionic structure in aqueous solution is provided by the violet shift of the long-wavelength band on the addition of acid to the neutral solution; for example, at pH 6.8, λ_{max} 315 $m\mu$, $m\epsilon$ 3.1; at pH 2, λ_{max} 283 $m\mu$, $m\epsilon$ 5.84 (33). The *N*-methylated zwitterion in Table 5.2 shows the same phenomenon.

An example of the effect of solvent polarity on the tautomeric equilibrium of a 2-pyridol type of molecule is shown in Table 5.3. As has been shown by Mason (3), the lactam form absorbs at longer wavelength than the lactim. Although both tautomers have the same conjugation-chain length, the lactam form has a stronger electron donor at the positive resonance terminal and a stronger electron acceptor at the negative resonance terminal. Consequently the excited state of the lactam form

TABLE 5.3
Effect of Increasing Solvent Polarity on the Tautomeric Equilibrium of 4-Methyl-6-hydroxy-7-azaindoline (34)

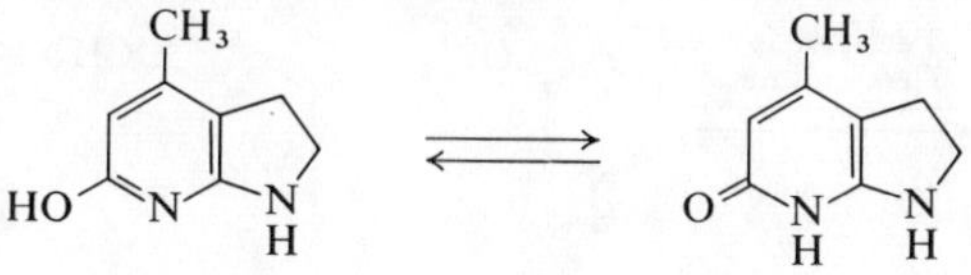

Solvent	$m\epsilon$[a] Lactim band λ_{max} 310mμ	$m\epsilon$[a] Lactam band λ_{max} 343mμ
EtOH (abs.)	—	12.0
EtOH (abs.)–dioxane:		
1:1	3.6*i*[b]	10.2
1:3	4.1*s*[c]	7.9
3:17	5.0	5.6
1:9	6.0	4.4
1:19	6.8	2.1
Dioxane	7.2	0.79*s*[c]
4-Methyl-6-methoxy-7-azaindoline	8.3	—

[a] All $m\epsilon$ values estimated from spectral curves.
[b] Inflection.
[c] Shoulder.

is at relatively lower energy than is the excited state of the lactim form, and so the lactam absorbs at longer wavelengths. As shown with other lactams (3), the lactams of 6-hydroxy-7-azaindolines are stabilized by increased polarity of the solvent (34). In these same molecules simple structural changes can also have a major effect on the equilibrium (Table 5.4).

Aqueous solutions of glutaconimide (λ_{max} 224 and 305 mμ; $m\epsilon$ 6.8 and 9.95, respectively) contain 60% of **23**, 25% of **24**, and 15% of **25** (35). These results are based on spectral and ionization studies.

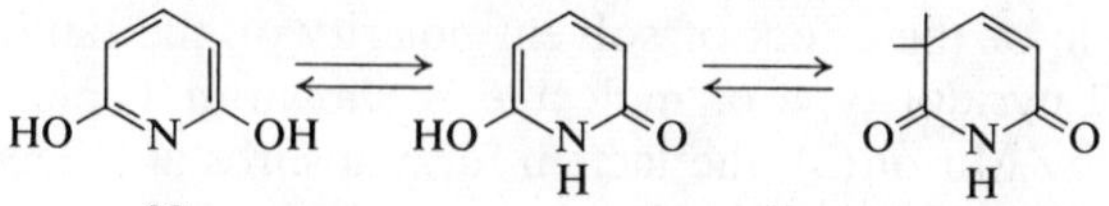

$$\mathbf{23} \rightleftharpoons \mathbf{24} \rightleftharpoons \mathbf{25} \qquad (5.5)$$

The presence of appreciable amounts of the dihydroxy tautomers is in marked contrast to the simple α- and γ-pyridols, which exist to the extent

TABLE 5.4
Effect of Structural and Solvent Changes on the Lactam ⇄ Lactim Equilibrium(34)

	H_2O–95% EtOH				95% EtOH–dioxane					Dioxane
R	98:2	50:50	25:75	EtOH (abs.)	50:50	25:75	15:85	10:90	5:95	
% Lactim				< 1	15	29	42	60	73	88
H K_t				> 99	5.9	2.4	1.4	0.68	0.38	0.14
% Lactim			< 1	25	36	61	86	88		96
n-C_4H_9 K_t			> 99	3.0	1.8	0.64	0.2	0.13		0.04
% Lactim	< 1	25	52	79	89	90				
C_6H_5 K_t	> 99	3.1	0.91	0.26	0.13	0.11				

of about 0.1% as the hydroxy form. The increased stability of **25** is believed to be due to the decrease in basicity of the pyridine nitrogen flanked by two oxygen functions(35).

Investigation of the absorption spectra of 2,4-dihydroxypyridine and six methylated standards in aqueous alcohol has disclosed that the compound exists predominantly as 4-hydroxy-2-pyridone (λ_{max} 277mμ, mϵ 5.0)(36).

Pyridoxine, pyridoxal, and pyridoxamine are present in aqueous solution mainly as the zwitterion(37). The following equilibrium involving lactam ⇄ lactim and ring ⇄ chain tautomerism is possible:

In addition the aldehyde hydrate forms are probably present in solutions containing water.

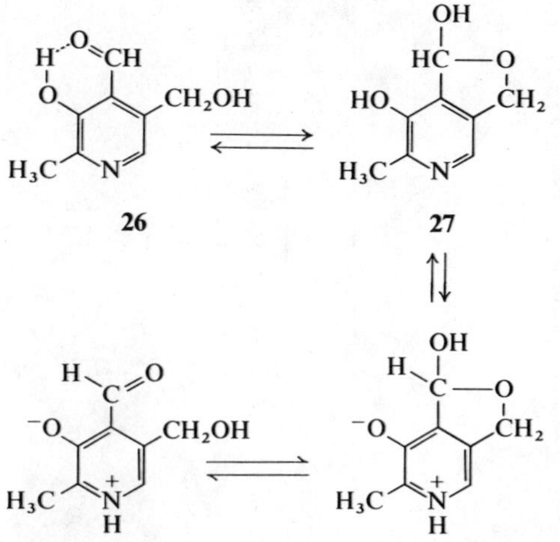

(5.6)

26 ⇄ **27** ⇅ **28** ⇄ **29**

29: λ_{max} 390 mμ, mϵ 0.11

28: λ_{max} 317 mμ, mϵ 8.9

The hydration of the 5-desoxy derivative of pyridoxal(**30**) shows the following equilibrium in aqueous solution at pH 6.88:

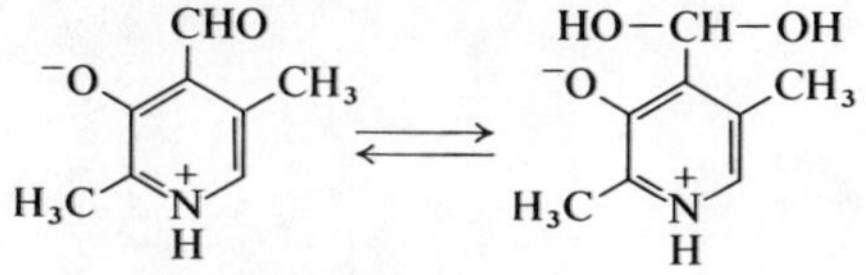

(5.7)

30 ⇄ **31**

30: λ_{max} 381 mμ, mϵ 4.3

31: λ_{max} 324 mμ, mϵ 2.8

In this case the short-wavelength band is derived from the addition of water to the carbonyl bond of the aldehyde derivative and not from any tautomeric phenomena.

Nonzwitterionic forms were reported to be present in aqueous solution in the following percentages: 3-hydroxypyridine, 46%; pyridoxine, 12%; pyridoxal, 8%; and pyridoxamine hydrochloride, 3%(37). For all these compounds the ratio of dipolar ion to uncharged form decreased with an increase in the percentage of dioxane in an aqueous dioxane solution.

Examination of the ionization constants (Table 5.5) and absorption spectra (Table 5.6) of some vinylogs of the pyridols and appropriate alkylated standards has disclosed that these compounds exist, predominantly as hydroxy compounds(38). In carbon tetrachloride solution the aldoximes absorb near 3593 cm^{-1}, owing to the O–H stretching vibration, and no N–H vibration is observed. Consequently in nonpolar solvents these compounds exist predominantly in the enolic form. The 2-aldoximes have the *trans* or *cis* configuration since the infrared absorption corresponds to a free O–H group (Table 5.6). Since the nuclear *N*-methyl derivatives of pyridine-2-aldoxime and pyridine-4-aldoxime cannot be extracted from alkaline solution with chloroform or ether and show no spectral evidence of a nitroso $n \rightarrow \pi^*$ transition in alkaline solution, these compounds must have the Z_z resonance structure (e.g., **32**).

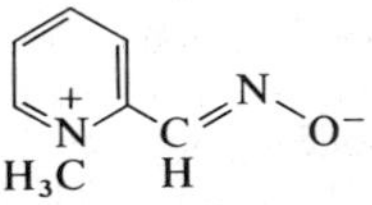

32

λ_{max} 220, 285*s*, 335 mμ
mϵ 5.1, 5.2, 18.2

On this basis the following equilibrium can be postulated for 2-pyridine aldoxime(38):

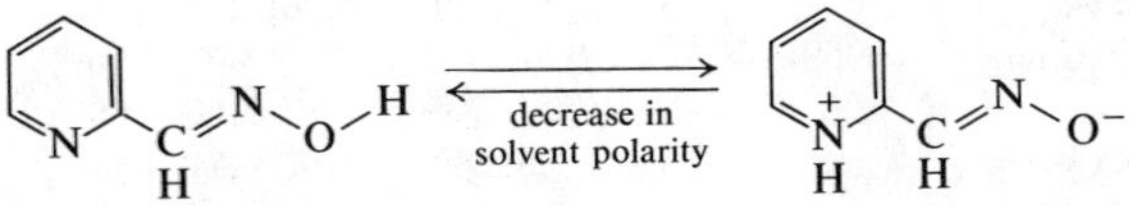

33 ⇄ (decrease in solvent polarity) **34** (5.8)

33: λ_{max} 240, 280 mμ

34: λ_{max} 350 mμ

D. Pyrimidinones

The pyrimidinones have been extensively investigated because of the important role some of them play in the biological properties of the nucleic acids. One hypothesis that has been advanced suggests that the

TABLE 5.5
Ionization Constants of Some *N*-Heteroaromatic Aldoximes and of Their *O*-Alkyl and Nuclear *N*-Alkyl Derivatives in Water at 20°C. The Tautomeric Equilibrium Constants (K_t = [NH form]/[OH form]) of the Aldoximes and of the Corresponding Hydroxy Compounds (38)

Compound	pK_1	log K_t	
		aldoxime	hydroxy
Pyridine-2-aldoxime	3.63 ± 0.02	−4.37	2.96
O-*n*-Propyl ether	3.58 ± 0.05		
1-Methiodide	8.00 ± 0.01		
Pyridine-3-aldoxime	4.10 ± 0.01	−5.12	−0.08
O-*n*-Propyl ether	4.12 ± 0.04		
1-Methiodide	9.22 ± 0.03		
Pyridine-4-aldoxime	4.77 ± 0.01	−3.80	3.29
O-*n*-Propyl ether	4.81 ± 0.04		
1-Methiodide	8.57 ± 0.02		
Quinoline-4-aldoxime	4.55 ± 0.04	−3.77	3.88
1-Methiodide	8.32 ± 0.04		

TABLE 5.6
Frequency of the O–H Stretching Vibration Absorption of Some *N*-Heteroaromatic Aldoximes in Carbon Tetrachloride Solution (ν_{OH}), Their Ultraviolet-Absorption-Band Maxima and Those of the Corresponding Nuclear *N*-Methyl Derivatives, and the Tautomeric Equilibrium Constants (K_t = [NH form]/[OH form]) Calculated from the Spectra (38)

Compound	ν_{OH} (cm^{-1})	λ_{max} ($m\mu$)[a]	log ϵ_{max}	pH or solvent	log K_t
Pyridine-2-aldoxime	3596	355*s*, 277, 239	0.15, 3.88, 4.02	6.9	−4.1
		279, 240	3.89, 4.01	EtOH	
1-Methochloride		335, 285*s*, 220	4.26, 3.72, 3.71	11	
Pyridine-3-aldoxime	3591	275*s*, 243	3.67, 4.06	7.2	
		275*s*, 245	3.68, 4.06	EtOH	
1-Methochloride		335*s*, 288, 242	3.62, 4.17, 4.11	12	
Pyridine-4-aldoxime	3590	340*s*, 246	0.50, 4.10	7.4	−3.8
1-Methochloride		336, 239	4.34, 3.78	11	
Quinoline-4-aldoxime	3591	312, 234	3.81, 4.29	7	
		327, 241	3.95, 4.35	1	
		323, 235	4.26, 4.36	13	
1-Methochloride		387, 246	4.31, 4.33	12	
Quinoline-2-aldoxime	3593				

[a] *s* = Shoulder.

tautomerism of the heterocyclic bases contained in the nucleic acids may play a role in the spontaneous mutation of genes (39, 40). The structures reported in Table 5.7 are based on spectral and ionization comparisons with appropriate *N*-alkyl and *O*-alkyl derivatives.

In the majority of cases the compounds are present in the lactam form. However, there are enough exceptions to necessitate basing assignment on thorough investigation. The assignment of a structure to a compound with a mobile hydrogen is most reliable when the possible tautomers differ widely in physical properties. The advantages of electronic spectral studies of tautomerism are that the solvents and conditions used in the study can be applied directly to trace analysis. Where more than one mobile hydrogen and more than one ring-nitrogen atom are present in the compounds under study, many of the assignments have to be considered as tentative. Before the relative amounts of the various tautomers of such a compound can be determined, a thorough study with all available tools is necessary.

Some of the compounds in Table 5.7 for which fairly reasonable assignments have been made are 2-pyrimidone, 4-pyrimidone, and 5-hydroxypyrimidine (Nos. 1, 2, and 3); uracil (No. 4); 4,6-dihydroxypyrimidine (No. 6); cytosine, 5-methylcytosine, and 5-hydroxymethylcytosine (Nos. 7, 8, and 9); 2-thiouracil (No. 10); thymine (No. 13); orotic and barbituric acids (Nos. 14 and 15); alloxan (No. 16); and dialuric acid (No. 17).

Fox and Shugar (50) have concluded that tautomeric structures of the pyrimidine nucleosides closely resemble those of the corresponding alkyl pyrimidinones. However, since the environment can affect the tautomeric equilibrium, the structures of some of the bases in the nucleic acids could be affected by their environment.

The large number of tautomers possible for some of these bases can be seen in cytosine (No. 7 in Table 5.7), for which the following seven tautomers are possible:

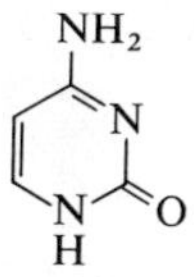

35a 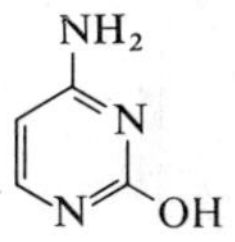 **35b** 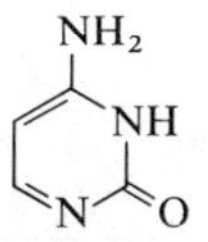 **35c** 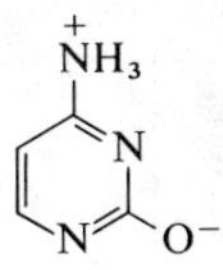 **35d**

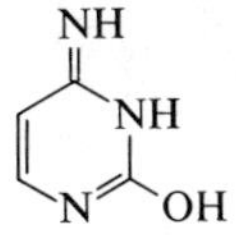

35e 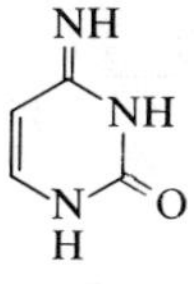 **35f** 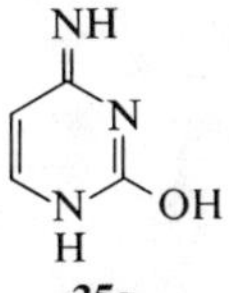 **35g**

TABLE 5.7

Structure, Spectra, and Ionization Constants of Some Pyrimidinones and Their Alkylated Derivatives

Predominant structures[a]	Ref.	pH or solvent	λ_{max} (mμ)	mϵ	pK$_1$[b]	pK$_2$[c]	Other data
1 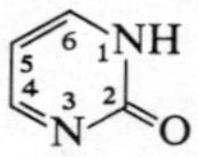 (NH, N, O; positions 1–6)	41, 42	6.2	<215 298	>10.0 4.7	2.24	9.17	See also (43, 44) $\nu_{C=O}$ 1644 cm^{-1}, very strong (32)[d]
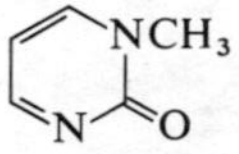 (NCH$_3$, N, O)	42	6.0	215 302	10.0 5.4	2.50	—	$\nu_{C=O}$ 1670 cm^{-1}, very strong
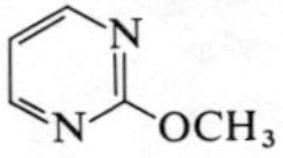 (N, N, OCH$_3$)	41	6.98	264	4.8		—	
2 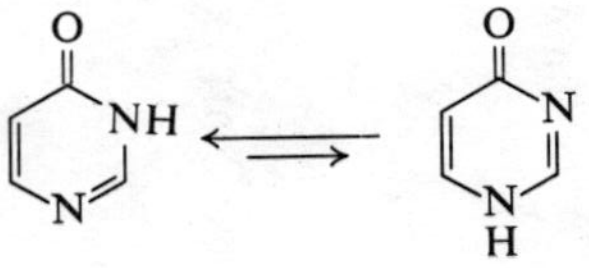 (O, NH, N) ⇌ (O, N, NH)	41, 42	6.2	223 260	7.3 3.74	1.69	8.60	See also (43, 44) $\nu_{C=O}$ 1687 cm^{-1}, (32)[d] $K_{o/p} = 1.5$ (19) $K_{o/p} = 2.5$ (3)
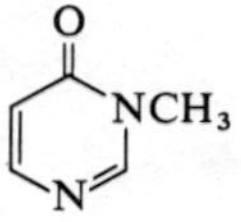 (O, NCH$_3$, N)	42	5.0	221 269	6.8 3.9	1.84	—	
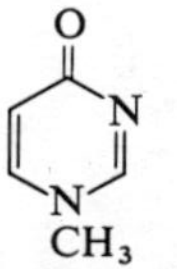 (O, N, NCH$_3$)	42	6.0	240	14.6	2.02	—	

	Compound	Ref.	pH	λ	log ε			Remarks
	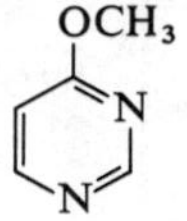 OCH₃, N, N (4-methoxypyrimidine)	42	6.95	247–248	3.35	2.5		
3	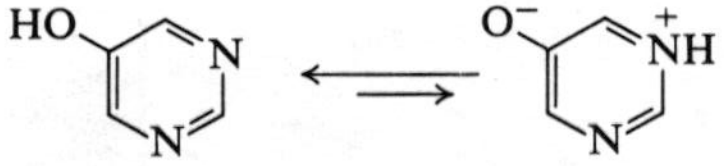 HO, N, N ⇌ O⁻, $\overset{+}{N}H$, N	3	4.32	214 271 325*s*	9.7 4.75 0.11	1.87	6.78	$K_{OH/NH} \geqslant 0.02$
		3	95% EtOH	218 276	9.97 5.3			
4	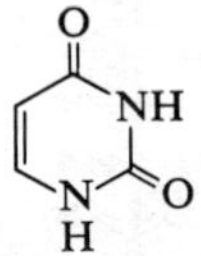 O, NH, N, O, H	45	4.4–7.2	260	8.2	−3.38[e]	9.45	See also (44, 47–49) $pK_3 = >13$
	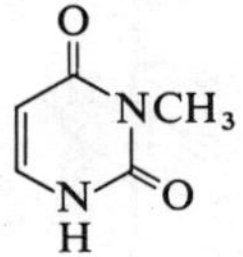 O, NCH₃, N, O, H	45	3–7.2	258.5	7.3	9.97	—	See also (50, 51)
	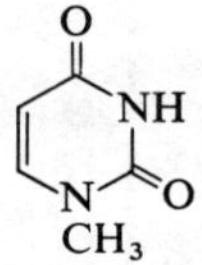 O, NH, N, O, CH₃	45	5.4–7.2	207.5 267.5	8.8 9.75	−3.4[e]	9.73	See also (50)
	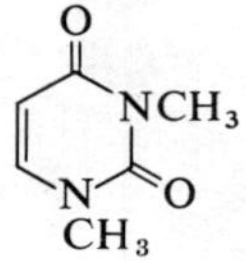 O, NCH₃, N, O, CH₃	45	1–14	266	8.9	−3.25[e]	—	

TABLE 5.7 (*cont.*)

Predominant structures[a]	Ref.	pH or solvent	λ_{max} (mμ)	mϵ	pK_1[b]	pK_2[c]	Other data
O, NH, N, OC_2H_5	45	4.4–7.2	217.5 259.5	6.65 6.05		8.3	
OC_2H_5, N, N, O, H	45	4.4–7.2	269	5.08		1.0[e]	10.7
OC_2H_5, N, N, O, CH_3	45	7.2	274.5	6.10		~0.65[e]	
OC_2H_5, N, N, OC_2H_5	45	7.2	210 259	7.3 6.12			
5 O_2N, O, NH, N, O, H	45, 46	7.7	229 258 341	6.0 4.8 13.5		5.3	$pK_3 = 11.7$
O_2N, O, NH, N, O	42	4.0	240 310	8.24 11.4		7.2	

	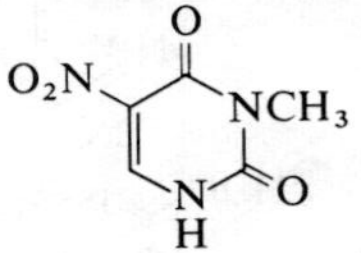	42	4.0	235 298	8.16 9.44			
6	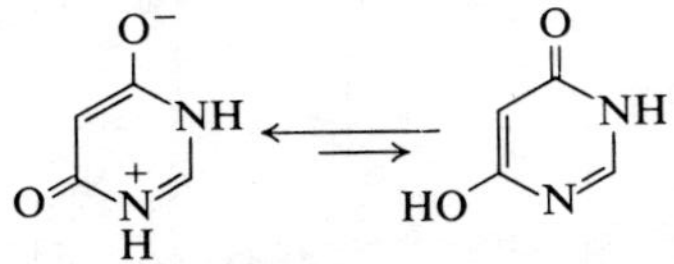	52	0.01 *M* H_2SO_4	252	7.2	−0.3	5.4	See also (53, 54); further evidence from NMR spectra and absorption spectra of anion and cation
	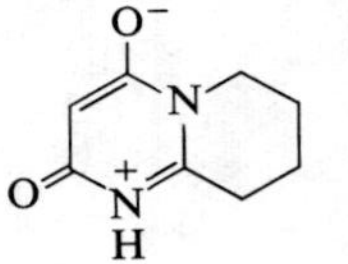	52	4.8	255	9.33	−0.24	7.27	
	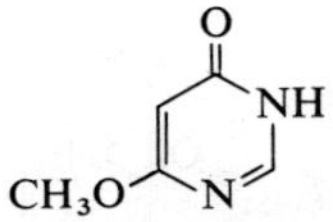	52	7.0	233	3.59	−0.1	8.47	
	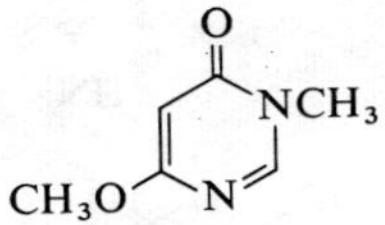	52	7.0	259	3.98	−0.4	—	
	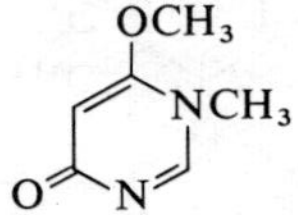	52	7.0	253 285	2.19 1.14	∼−2.7	—	
	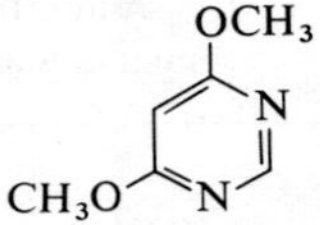	52	7.0	240	3.03	1.5	—	

TABLE 5.7 (*cont.*)

Predominant structures[a]	Ref.	pH or solvent	λ_{max} (mμ)	mϵ	pK_1[b]	pK_2[c]	Other data
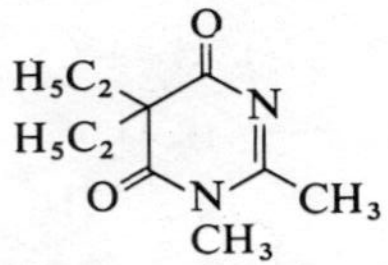 O, H_5C_2, H_5C_2, N, O, N, CH_3, CH_3	53	7.0	255	9.33	—	—	
7 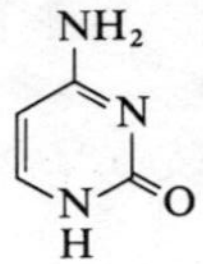 NH_2, N, N, O, H	55	7.0	267	6.17	4.45	12.2	See also (45, 56, 57)
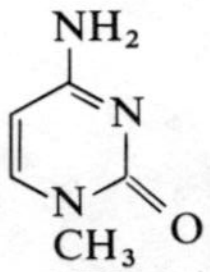 NH_2, N, N, O, CH_3	55	7.0	230*s* 274	7.76 8.3	4.57	—	
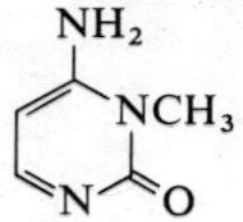 NH_2, NCH_3, N, O	55	12.0	294	12.0	7.4	—	
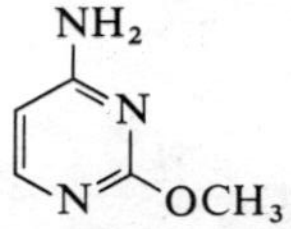 NH_2, N, N, OCH_3	45	7.2	225 271	8.0 7.2	5.3	—	
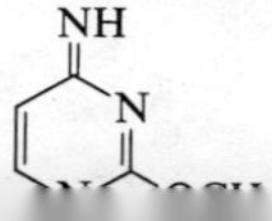 NH, N, N, OCH_3	55	14.1	274	8.3	< 12.0	—	

$N(CH_3)_2$, N, O, N, CH_3	55	7.0	219*s* 282	11.0 12.0	4.20	—	
$N(CH_3)_2$, N, O, N, H	55	7.0	222*s* 242 275	9.55 6.76 10.2	4.25	12.3	
$N(CH_3)_2$, N, N, OCH_3	55	10.0	239 284	10.7 7.59	~6.17	—	
N, N, O^-	55	13.0	220 292	11.5 4.57		9.17	
NH, NCH_3, O, N, CH_3	55	11.5	225 273	10.0 8.7	9.3	—	
8 NH_2, H_3C, N, O, N, H	45	7–10.0	210.5 273.5	14.2 6.23	4.6	12.4	See also (58)
9 NH_2, HOH_2C, N, O, N, H	59	7.4	269.5	5.75			See also (60)

TABLE 5.7 (*cont.*)

Predominant structures[a]	Ref.	pH or solvent	λ_{max} (mμ)	mϵ	pK_1[b]	pK_2[c]	Other data
10 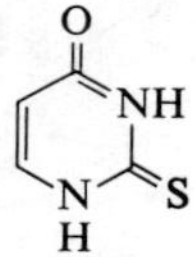	61, 62	7.0	274	10.7	7.75	12.7	See also (63); dipole moment (64); X-ray studies (65); NMR (66)
11 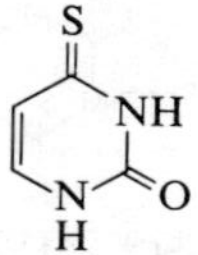	61	7.0	327	16.6			
12 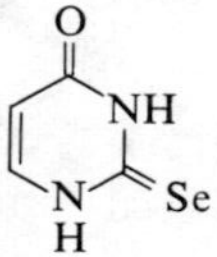	67		315	12.6			
13 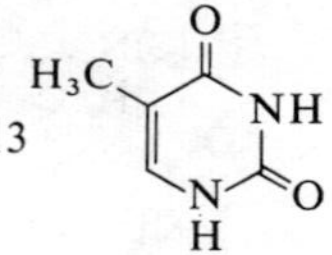	45	4.4–7.2	207 264.5	9.5 7.9		9.9	See also (68) $pK_3 = >13$
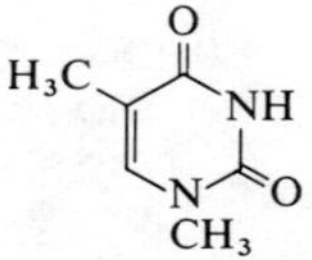	68	5.9	273	11.2		10.09	
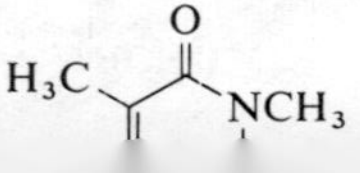	68	5.9	264.5	8.13		10.52	

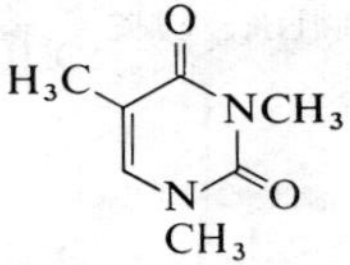	68	5–12	272	9.77		—	
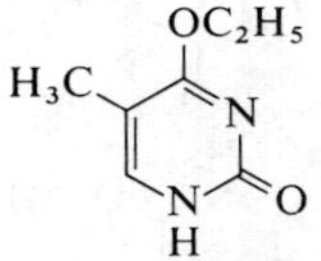	68	5.9	275	5.62		11.1	
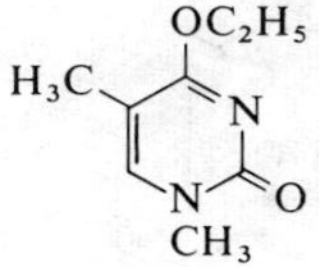	68	5–12	280	6.46			
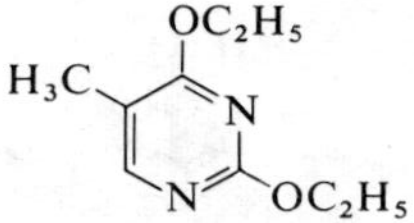	68	5–12	217 265	9.55 7.08			
14	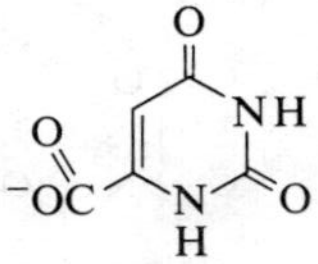 45	4.4–7.2	207 278.5	11.6 7.68	2.6[c]	9.45	$pK_3 = >13$
15	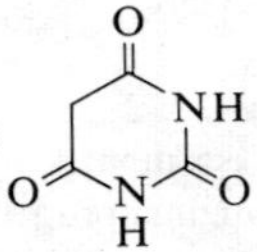 50	1–2	205 258	10.5 0.60		4.12	See also (69, 70) X-ray (71, 72) IR (73)

TABLE 5.7 (*cont.*)

Predominant structures[a]	Ref.	pH or solvent	λ_{max} (mμ)	mϵ	pK_1[b]	pK_2[c]	Other data
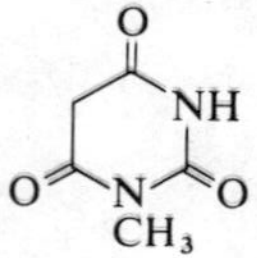 O, NH, O, N, O, CH_3	50	1–2	220	8.7		4.35	
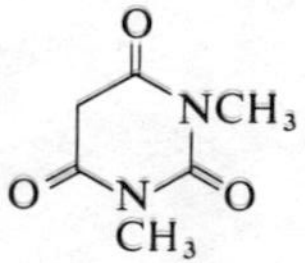 O, NCH_3, O, N, O, CH_3	50	1–2	226	7.9		4.68	
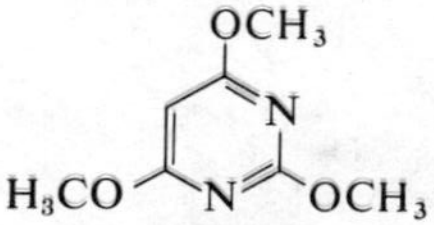 OCH_3, N, H_3CO, N, OCH_3	50	5–13	248	7.76			
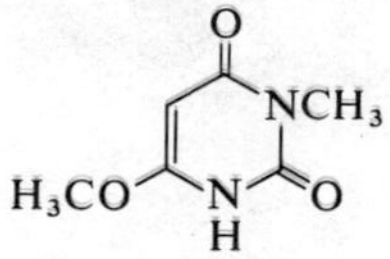 O, NCH_3, H_3CO, N, O, H	50	5–13	257	12.0			
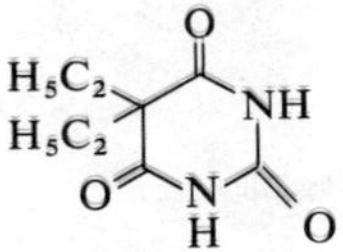 O, H_5C_2, H_5C_2, NH, O, N, H, O	50	1–5	212.5	9.1		7.97	
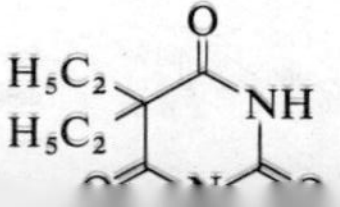 O, H_5C_2, H_5C_2, NH	50	1–6	222.5	7.9			

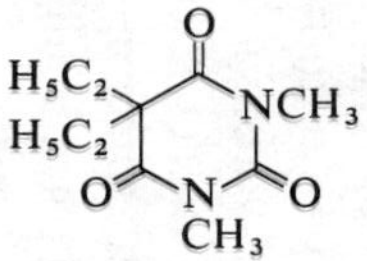

H_5C_2, H_5C_2, O, NCH_3, O, N, O, CH_3	50	Water	228	6.3		
16 O, O, NH, O, N, O, H	74	4.0	265*s*	0.079		See also (75, 76)
17 H, HO, O, NH, O, N, O, H	74	1% HCl	270	0.32		See also (75, 77)

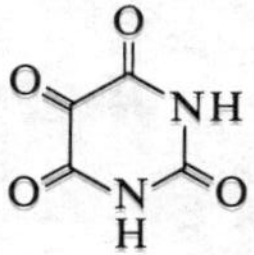

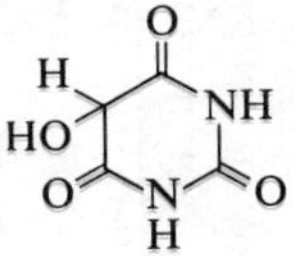

[a]Unnumbered compounds are comparison standards. Where several tautomers have been identified, predominant one is shown first.

[b]Basic ionization constant.

[c]Acidic ionization constant.

[d]In water.

[e]Data from reference 46.

Spectra and ionization constants of cytosine and the alkylated derivatives of these tautomers are compared in Table 5.7. The predominant tautomer is reported to be **35a.** In the same way the structure of 4,6-dihydroxypyrimidine (No. 6 in Table 5.7), which has six possible tautomers, has been assigned.

In the triazenes also the carbonyl form prevails; for example, in ammeline (**36**), ammelide (**37**), and cyanuric acid (**38**) (78).

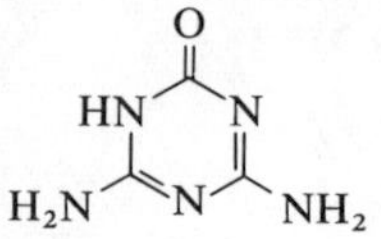

36 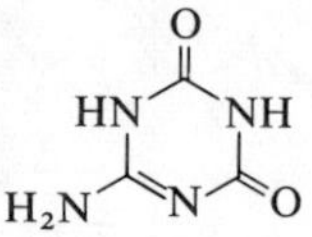 **37** 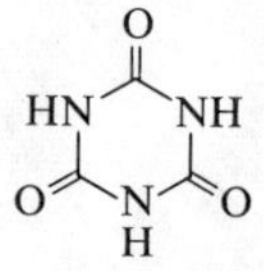 **38**

In line with these assignments are the lack of spectral bands in the ultraviolet region of **38** and of the cation of **37** (proton to aza nitrogen).

E. Quinolone-Type Compounds

The absorption spectra, ionization constants, and tautomeric equilibrium constants of many quinolone-type compounds are compared with the data for their *N*- and *O*-methylated derivatives in Table 5.8.

Where the hydroxy group is conjugated with the aza group, both in the same ring, the lactam form usually predominates. Whether a Z_z or a Z_n-structure is a major contributor to the ground state of the various molecules could be ascertained by spectral, dipole-moment, and ionization-constant studies of the ground and excited states. These properties could be of value in trace analysis studies.

Where the hydroxy group is β to the aza nitrogen, the enolic form can predominate, as in 3-hydroxyquinoline, or the zwitterionic form can predominate, as in 4-hydroxyisoquinoline.

Where the hydroxy group is in a different ring, an enol or lactam form may predominate, depending on the position of the hydroxy and ring-nitrogen groups. In the hydroxycinnolines and hydroxyquinolines of this type the enol form predominates. By ultraviolet spectroscopy 3-hydroxy-, 3-hydroxy-1-methyl-, and 3-hydroxy-6,7-dimethoxy-1-methylisoquinoline have been shown to exist as lactim tautomers in nonhydroxylic solvents, such as diethyl ether, and as lactam tautomers in water (102). 1-Chloro-3-hydroxyisoquinoline is in the hydroxy, or lactim, form in all common solvents, and 3-hydroxycinnoline is in the lactam form in all solvents tried. A group of methyl, chloro, and bromo derivatives of 6,8-dihydroxyisoquinoline exist as hydroxy tautomers with no detectable contribution from quinolone forms (103).

TABLE 5.8

Lactam ⇄ Lactim Tautomerism of Quinolone-Type Compounds

Principal structures[a]	Ref.	pH or solvent	λ_{max} (mμ)[b]	mϵ	pK$_1$	pK$_2$	log K_t[c]	Other data
1	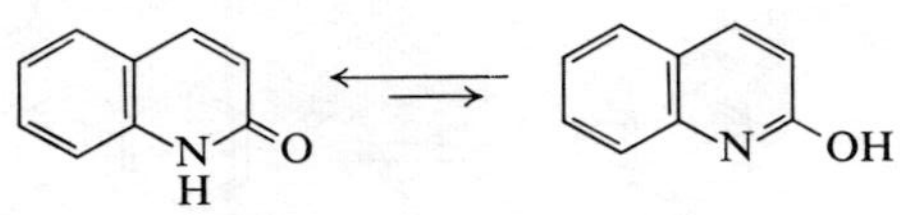 3	5.5	224	26.7	−0.31[d]	11.7[d]	4.85	See also (79–81); IR (10)
			245	8.5				
			270	6.6				
			324	6.3				
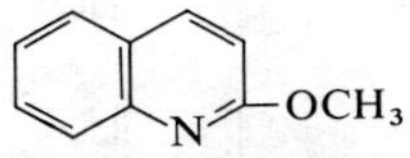	3	6.8	260*s*	2.0	3.17[d]	—	—	
			267	2.0				
			307	2.07		—	—	
			318	2.2				
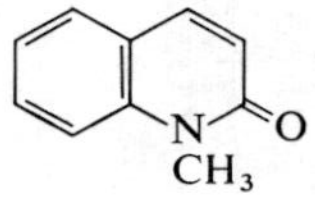	3	7.0	228	34.1	−0.71[d]	—	—	
			245	10.3				
			272	6.9				
			325	6.5				
2	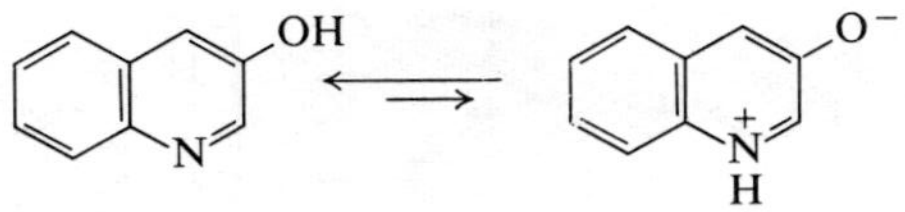 3	6.18[e]	270*s*	3.06	4.30[d]	8.26[d]	−1.1	IR (10)
			321	4.1				
			330	4.08				
			365	0.43				
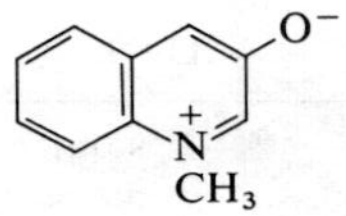	3	10.0	255	14.6	5.42	—	—	
			308	1.68				
			320	1.7				
			384	7.1				

TABLE 5.8 (*cont.*)

Principal structures	Ref.	pH or solvent	λ_{max} (mμ)[b]	mϵ	pK_1	pK_2	log K_t[c]	Other data
3 O^-, $\overset{+}{N}$, H ⇌ OH, N	82	7.0	303*s* 315 328	7.9 13.0 10.0	2.27[f]	11.3[f]	4.2[f]	See also (81, 83–85); IR (10)
O^-, $\overset{+}{N}$, CH_3	82	13.0	279*s* 309*s* 322 335	2.0 7.9 12.6 12.6	2.46[f]	—	—	See also (81, 83–85)
OCH_3, N	82	13.0	276*s* 283	6.3 7.9	6.65[f]	—	—	
4 OH, N ⇌ O^-, $\overset{+}{N}$, H	3	6.78[e]	240 268*s* 312 450	37.4 3.3 2.87 0.19	5.20	8.54	−1.1[g]	See also (86); IR (10)
O^-, $\overset{+}{N}$, CH_3	3	8.5	273 318 332 462	33.1 1.24 1.15 3.8	6.12	—	—	See also (86)

5	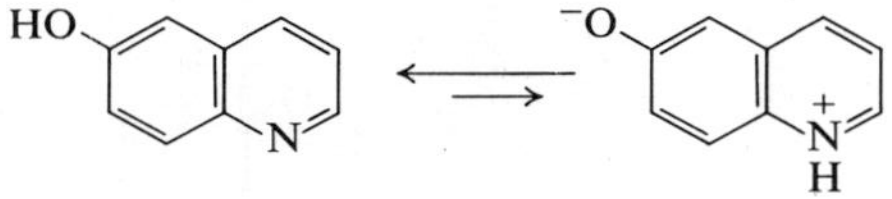	3	7.02[e]	225 272 325 400*s*	30.5 3.06 3.86 0.049	5.17	8.88	−2.0[g]	IR (10)
	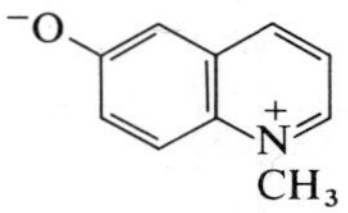	3	10.0	270 316 325 408	26.8 4.0 3.95 3.5	7.15	—	—	
	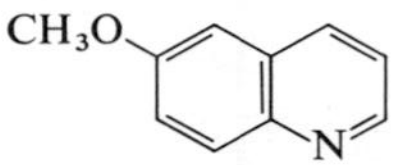	3	7.5	223 268 325	32.4 2.94 3.9	5.06			
6	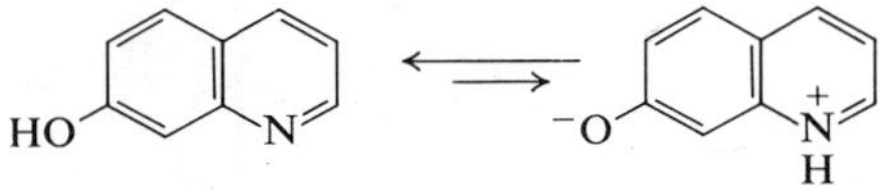	3	7.16[e]	225 258 402	26.7 12.8 3.03	5.48	8.85	0.69[h] 0.37[i]	IR (10)
	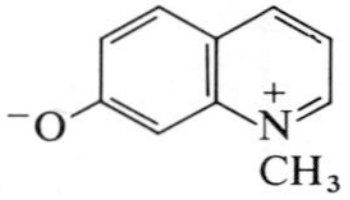	3	8.0	261 311 406	32.5 1.55 10.0	5.56	—	—	See also (86)
7	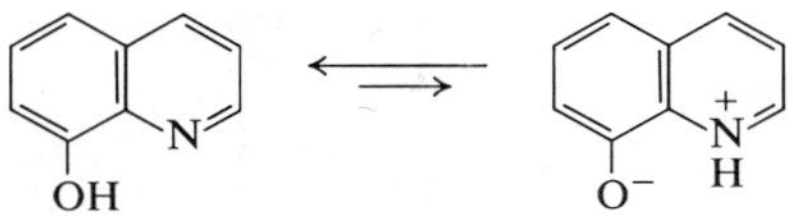	3, 8	7.51[e]	240 270*s* 303 430	17.2 2.3 2.1 0.051	5.13	9.89	−1.5[g]	See also (87, 88); IR (10)
	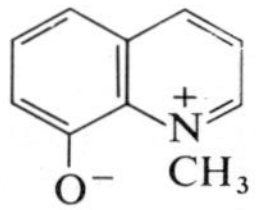	3, 8	10.0	273 334 346 442	37.0 1.2 1.3 1.52	6.81	—	—	See also (89)

TABLE 5.8 (*cont.*)

Principal structures[a]	Ref.	pH or solvent	λ_{max} (mμ)[b]	mϵ	pK_1	pK_2	log K_t[c]	Other data
8 (OH, $\overset{+}{N}$, O^- ⇌ O^-, $\overset{+}{N}$, OH)	83	95% EtOH	240 340 360	12.6 7.9 6.3				
(OCH_3, $\overset{+}{N}$, O^-)	83	95% EtOH	250 340	15.9 7.9				
(O^-, $\overset{+}{N}$, OCH_3)	83	95% EtOH	325 335	12.6 15.9				
(OH, N-H, O ⇌ O, N-H, OH)	90	Methanol	230 270 280 316	15.5 9.1 8.7 6.76	0.76	5.86		
(OCH_3, N-H, O)	90	Methanol	230 265 275 315	14.1 7.4 7.08 6.3				

9	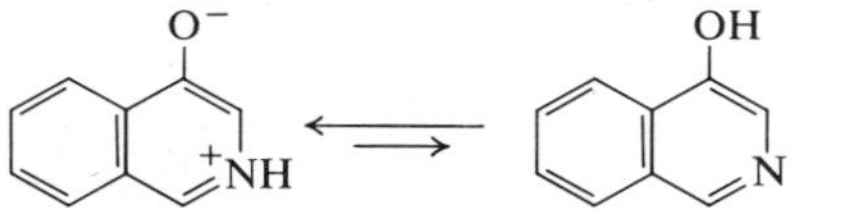 O^-, ^+NH ⇌ OH, N	3	6.74[e]	240 320*s* 359	8.8 4.3 7.6	4.8	8.68	4.19	IR (10)
	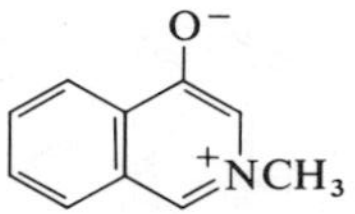 O^-, $^+NCH_3$	3	8.5	248 320*s* 364	9.18 3.7 9.7	4.93	—	—	
10	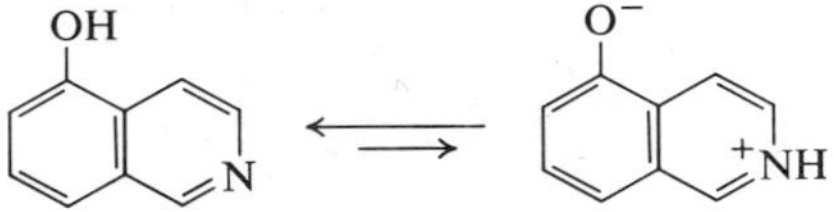 OH, N ⇌ O^-, ^+NH	3	6.92[e]	230 296 330 400*s*	22.8 3.57 4.4 0.14	5.40	8.45	−1.44[g]	IR (10)
11	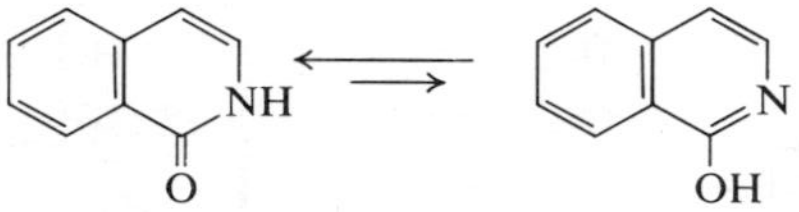 NH, O ⇌ N, OH	91	95% EtOH	280 325	9.3 4.9	−1.2[d,j]	—	4.85[f]	
	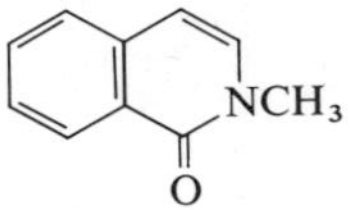 NCH_3, O	91	95% EtOH	228 280 285	10.0 7.9 7.9	−1.8[d]			
	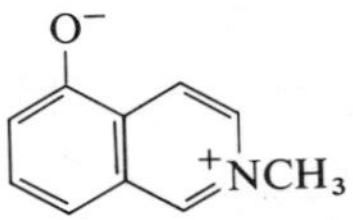 O^-, $^+NCH_3$	3	10.0	233 269 349 408	16.8 21.0 6.12 3.72	6.90	—	—	
12	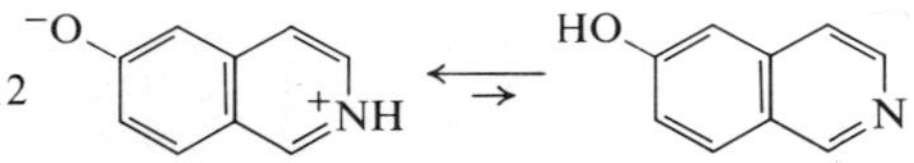 ^-O, ^+NH ⇌ HO, N	3,86	7.5[k]	229 263 353	43.6 17.2 8.04	5.85	9.15	0.30[g]	IR (10)
	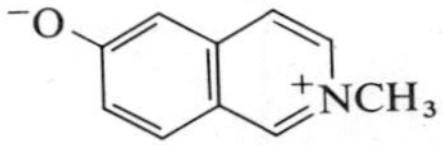 ^-O, $^+NCH_3$	3, 86	10.0	230 267 358	34.4 22.9 12.2	6.02	—	—	

TABLE 5.8 (*cont.*)

Principal structures[a]	Ref.	pH or solvent	λ_{max} (mμ)[b]	mϵ	pK_1	pK_2	log K_t[c]	Other data
13 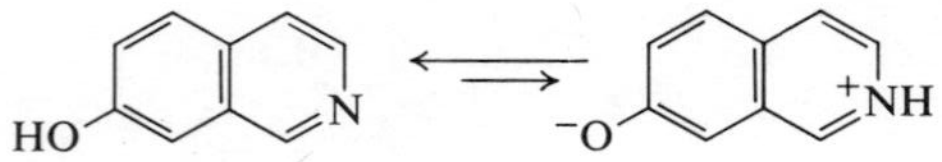 (HO-isoquinoline ⇌ $^-$O-isoquinolinium NH$^+$)	3	7.29[e]	221 257 338 400	49.0 6.55 3.06 0.11	5.70	8.88	−1.4[g]	IR(10)
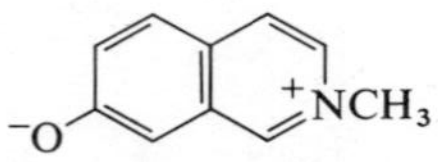 $^-$O … $^+$NCH$_3$	3	10.0	261 298 408	50.4 8.2 2.97	7.09	—	—	
14 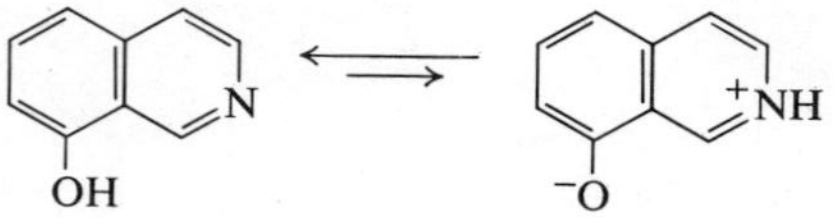 (OH … N ⇌ $^-$O … $^+$NH)	3	7.03[e]	247*s* 328 430	15.5 4.34 2.7	5.66	8.40	0.38[h] 0.06[i]	IR(10)
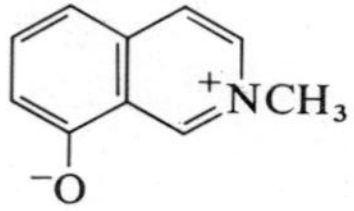 $^-$O … $^+$NCH$_3$	3	10.0	257 334 400	20.0 5.0 5.8	5.81	—	—	
15 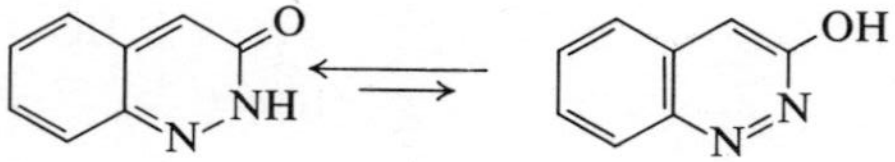 (O, NH, N ⇌ OH, N=N)	92	Methanol	300 312 400	0.69 0.58 3.0	0.21	8.61		
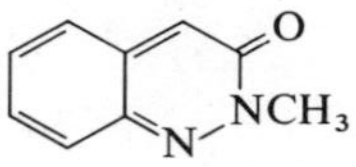 O, NCH$_3$, N	92	Methanol	305 400	1.3 3.2				See also (93)
O$^-$ ⇌ OH	93	95% EtOH	236*fs* 282 292	11.0 2.8 3.0	−0.35	9.27		See also (84)

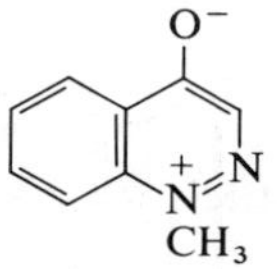	93	95% EtOH	240*fs* 284 296 341 360	10.3 2.1 2.3 12.6 10.9				See also (84)
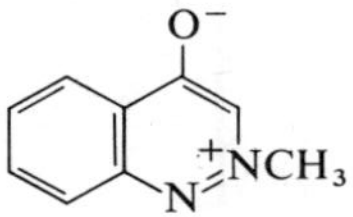	93	95% EtOH	254*fs* 352 369	8.1 12.4 12.8				
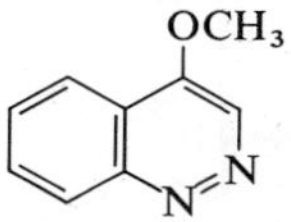	93	95% EtOH	226 284*s* 292 317	39.7 5.2 5.6 4.5	3.21	—		See also (84)
17 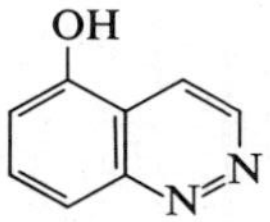	3	4.66 or 95% EtOH	244 305*s* 356	32.5 1.19 2.48	1.92	7.40		See also (94); IR (10)
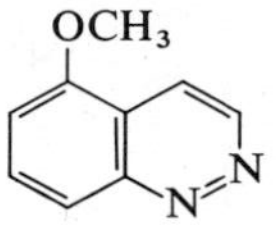	94	7.0	245 305s 353	33.9 1.4 2.75				
18 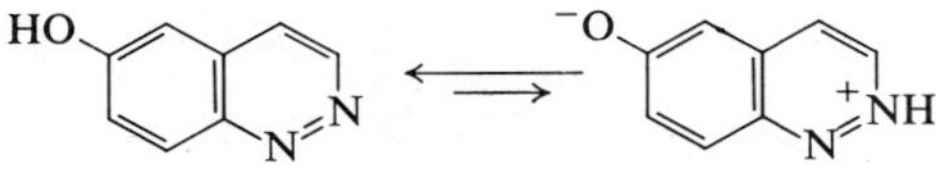	3	5.58[e]	238 281 320 402	24.0 3.75 4.7 1.2	3.65	7.52	~—1.0	See also (94); IR (10)
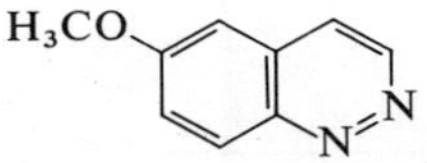	94	7.00	240 316	35.5 5.25				

TABLE 5.8 (*cont.*)

Principal structures[a]	Ref.	pH or solvent	λ_{max} (mμ)[b]	mϵ	pK$_1$	pK$_2$	log K_t[c]	Other data
19 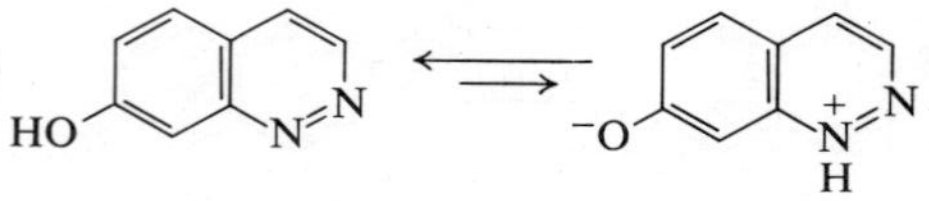	3	5.44[e]	235	42.6	3.31	7.56	~1.52	See also (94); IR (10)
			267	5.14				
			354	2.97				
			448	0.28				
	94	7.0	237	45.7				
			269	2.88				
			346	3.16				
20 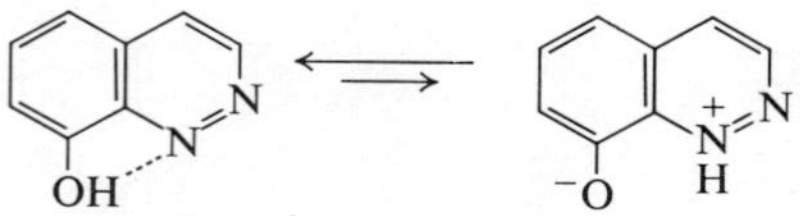 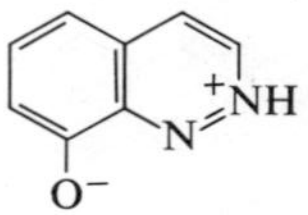	3	5.47[e]	242	32.2	2.74	8.20	~−2.4	See also (95); IR (10)
			296	1.24				
			359	2.61				
			525	0.021				
	95	Methanol	242	80.6				
			295	5.05				
			350	8.06				
21 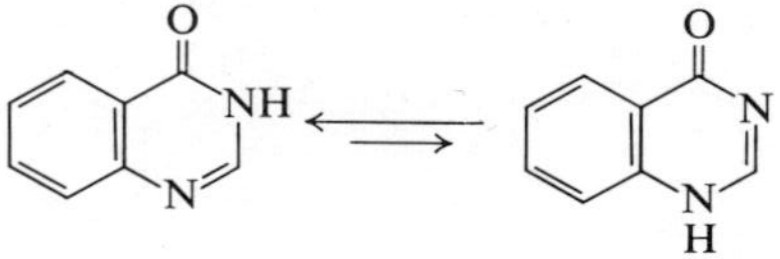	84	95% EtOH	224	22.9	2.12	9.81	0.85[l]	IR (96, 97)
			265	6.46				
			301	3.98				
			313	3.47				

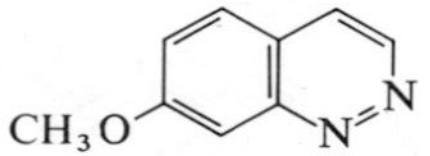

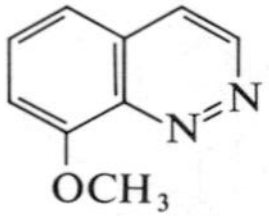

	O, NCH$_3$, N	84	95% EtOH	230	22.4			
				267	7.08			
				276	6.6			
				302	3.6			
				314	2.88			
	O, N, N, CH$_3$	84	95% EtOH	230	12.0			
				269	3.98			
				278	4.27			
				307	6.9			
				317	6.17			
	OCH$_3$, N, N	84	95% EtOH	225	26.3	3.13		
				261	5.13			
				298	3.02			
				309	3.47			
22	N, N, OH	3	6.03	239	28.1	3.41[m]	8.65	−3.85
				324	2.55			
	+NCH$_3$, N, O$^-$	3	10.5	275	6.2	7.26		
				317	5.5			
				328	5.15			
				443	0.17			
23[n]	H, N, O, N ⇌ N, OH, N	98	4.0	228	20.9	−1.38	9.12	
				250	6.17			
				254	6.03			
				287	5.0			
				343	5.5			

TABLE 5.8 (*cont.*)

Principal structures[a]	Ref.	pH or solvent	λ_{max} (mμ)[b]	mϵ	pK_1	pK_2	log K_t[c]	Other data
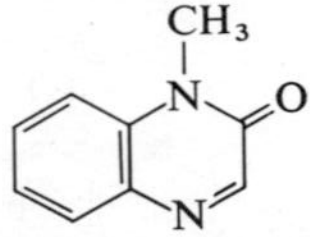	98	95% EtOH	231	25.1	−1.15			
			252*s*	4.37				
			282	5.4				
			336	5.13				
			346	5.37				
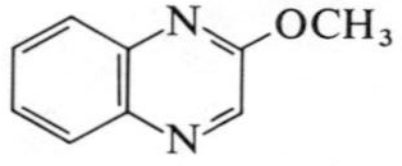	98	4.5	223	15.5	0.28			See also (99)
			241	18.2				
			245	17.8				
			326	5.6				
			333*s*	4.9				
24 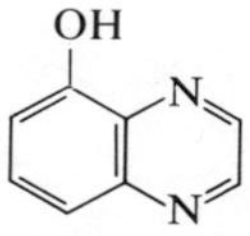	3	95% EtOH	252	30.0	0.9	8.65	−4.8	IR (10)
			326	2.9				
			363*s*	1.52				
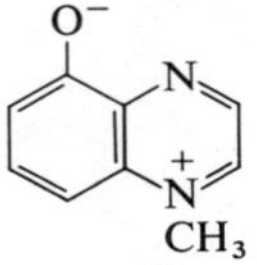	3	10.0	237	9.1	5.74			
			288	28.4				
			335	1.78				
			550	2.44				
25 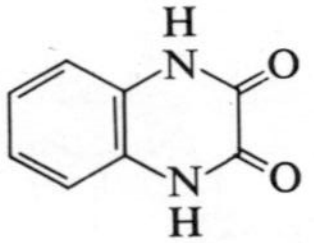	98	5.1	236	7.9	9.52[d]			See also (99)
			258	4.5				
			262*s*	4.3				
			312	11.8				
			326	10.0				
			342*s*	4.79				

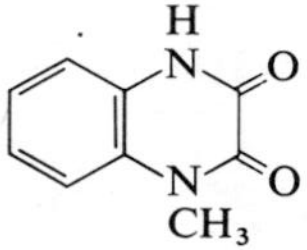	98	5.2	230	10.5	9.74			
			236*s*	8.7				
			256	5.0				
			261*s*	4.6				
			312	1.12				
			325	9.5				
			341*s*	4.5				
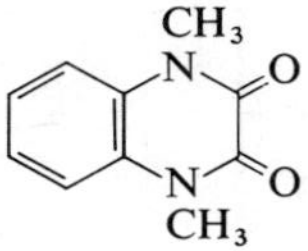	98	5.05	232	11.5				
			238	10.3				
			254	5.82				
			259*s*	5.4				
			313	11.2				
			325	9.8				
			340*s*	4.7				
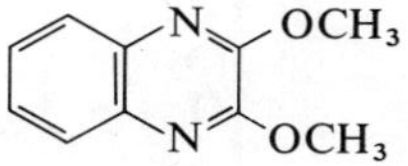	98	5.1	222*s*	16.3	−1.15			
			244	15.2				
			301*s*	6.4				
			311	9.8				
			325	8.2				
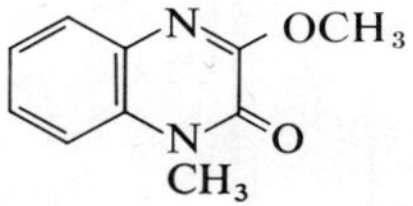	98	5.1	227	16.0				
			248	7.6				
			252*s*	7.4				
			312	9.3				
			319	9.8				
26 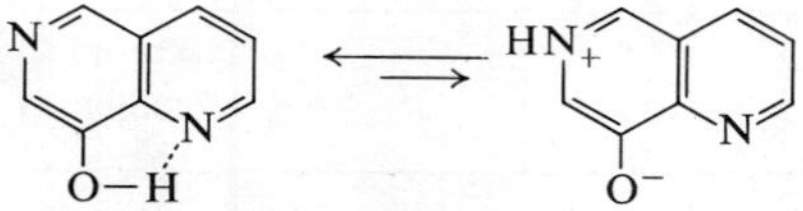	3	6.2	239	14.5	4.08	8.33	0.09[h]	IR (10)
			257	12.0			−0.05[i]	
			325	3.1				
			368	2.16				

TABLE 5.8 (*cont.*)

	Principal structures[a]	Ref.	pH or solvent	λ_{max} (mμ)[b]	mϵ	pK_1	pK_2	log K_t[c]	Other data
	H_3CN^+ … O^-	3	7.0	260 362	9.6 4.55	4.34			
27 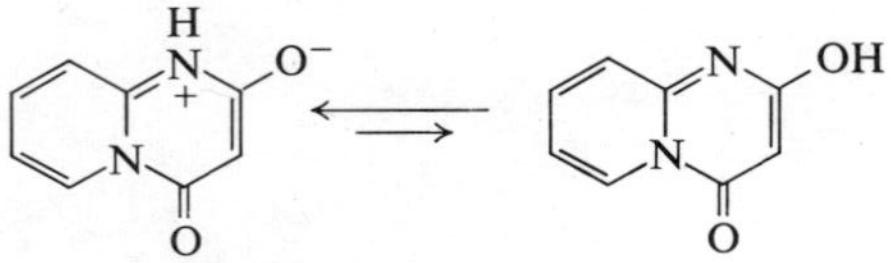	H, N^+, O^-, N, O ⇄ N, OH, N, O	100	0.01 N H_2SO_4	230 252 312	30.4 12.7 4.3	0.07	7.09	1.3	
	CH_3, N^+, O^-, N, O	100	6.99	230 257 322	32.4 12.0 4.75	0.18			
	N, OCH_3, N, O	100	6.99	222.5 254 325	23.8 12.0 7.23	1.36			
28 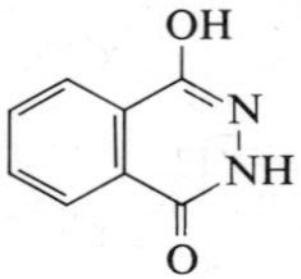	OH, N, NH, O	101	95% EtOH	263 298	2.93 3.6				IR(101)
	OCH_3, N	101	95% EtOH	227 252 262	11.1 3.7 4.2				

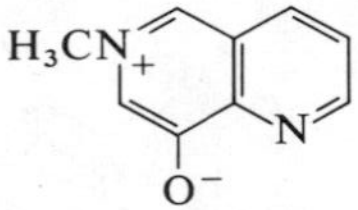

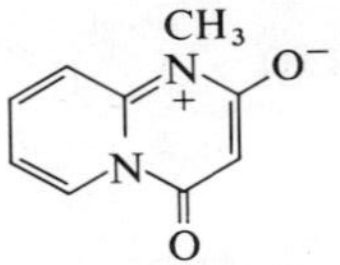

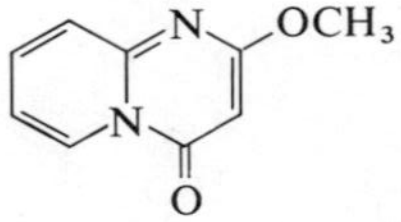

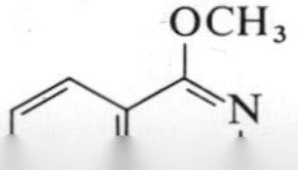

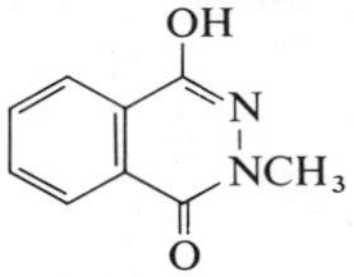 OH, N, NCH_3, O	101	95% EtOH	228	11.1
			255	3.2
			264	3.5
			304	6.0
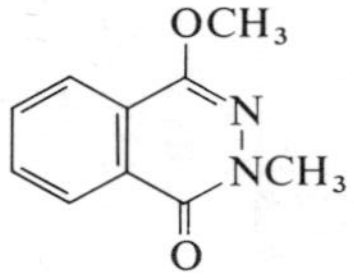 OCH_3, N, NCH_3, O	101	95% EtOH	227*i*	16.6
			234*i*	12.8
			252	7.4
			263	7.4
			298	9.3
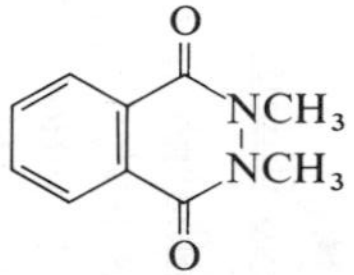 O, NCH_3, NCH_3, O	101	95% EtOH	214	57.0
			230*s*	11.0
			236	11.5
			242*s*	8.85
			302	4.84
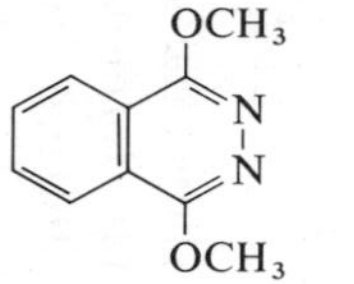 OCH_3, N, N, OCH_3	101	95% EtOH	245	5.32
			280	4.75

[a]Structures that are not numbered are reference compounds.

[b]*s* = Shoulder, *fs* = fine structure, *i* = inflection.

[c]K_t = [Imine]/[enol].

[d]Data from reference 30.

[e]In alcohol increased contribution from enol tautomer.

[f]Data from reference 8.

[g]Average value from calculations of K_t spectrally and by ionization constant.

[h]Calculated from ionization constants.

[i]Calculated from spectra.

[j]The *O*-methyl derivative has p*Ka* 3.05 as compared with −1.8 for *N*-methyl derivative, thus indicating preponderance of the NH form.

[k]In cyclohexane increased contribution from enol form.

[l]$K_{o/p}$ value.

[m]*O*-Methyl derivative, pK_a 3.51.

[n]The cationic salt of compound 23 is closely similar spectrally to the cationic salt of the *N*-methyl derivative and different from that of the *O*-methyl derivative.

On the basis of its absorption spectra compound 3 of Table 5.8 appears to have a zwitterionic structure as an important contributor to the ground state since the long-wavelength maximum at pH 7 of 328 mμ shifts to 302 mμ at pH 1.08(82). The assumption here is that the proton has added to the oxygen atom. The other possibility is that the proton has added to the NH group of the non-dipolar lactam form. This could also cause a violet shift. The *N*-methyl derivative shows the same violet shift on protonation. The *N*-methyl derivatives of 5-hydroxyquinoline and 6-hydroxyisoquinoline show the same violet shift on salt formation(86).

Study of the electronic spectra of 4-hydroxy-1,5-naphthyridine has disclosed that it exists as the fluorescent keto form in such polar solvents as ethanol, acetonitrile, and water (nonfluorescent in this solvent) and as the nonfluorescent enol form in such nonpolar solvents as dioxane and chloroform(104).

The wide spectral differences between two tautomers, of which the lactim has a Z_n-structure and the lactam a Z_b-structure, is apparent from the following equation(3, 87, 88):

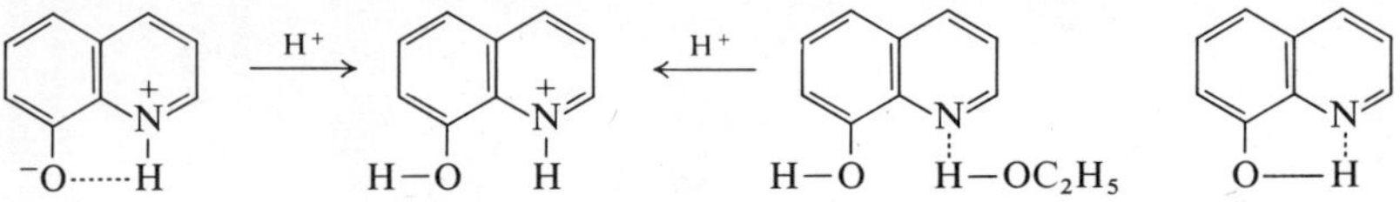

$^{-}$O······H (N$^+$) ; H-O (N$^+$-H) ; H-O (N···H-OC$_2$H$_5$) ; O—H (···N)

39 λ_{max} 430 mμ ⟶(H$^+$) **40** λ_{max} 358 mμ ⟵(H$^+$) **41** λ_{max} 308 mμ (EtOH) **42** λ_{max} 317 mμ (toluene) (5.9)

As to be expected, on protonation the Z_n-structure shows a red spectral shift, whereas the Z_z-structure shows a violet shift. The Z_n-structure shows a red spectral shift with decreasing polarity of the solvent, probably due to the increased importance of the intramolecular hydrogen bond in the nonpolar solvent. The Z_b-band of the *N*-methylated derivative shows the expected pH and solvent changes(89):

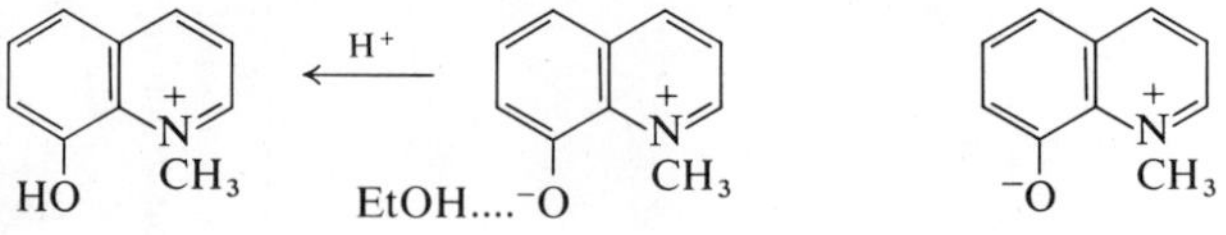

HO (N$^+$ CH$_3$) ; EtOH....$^{-}$O (N$^+$ CH$_3$) ; $^{-}$O (N$^+$ CH$_3$)

43 λ_{max} 370 mμ ⟵(H$^+$) **44** λ_{max} 484 mμ (EtOH) **45** λ_{max} 554 mμ ($CHCl_3$) (5.10)

8-Hydroxycinnoline shows the same phenonenon as demonstrated in equation 5.9. The lactam form has a Z_z-band at 525 mμ, whereas the lactim form absorbs at 360 mμ; protonation gives a salt that absorbs at 430 mμ(95). Thus the Z_z-band shifts to the violet and the Z_n-band shifts to the red on protonation of the negative resonance terminals.

F. Purinones

The tautomeric equilibria of the hydroxypurines are much more involved than those of ring systems that contain one or two nitrogen atoms. For each hydroxypurine there are two possible enol tautomers. However, infrared and ultraviolet spectra indicate that the compounds exist predominantly in the lactone form(105–107). For each hydroxypurine several alternative amide forms are possible. There are five such forms for 2-hydroxypurine, four for the 6-isomer, and three for the 8-isomer. Tentatively assigned structures for these isomers based on infrared and ultraviolet spectral investigations are **46**, **47**, and **48**(106, 107).

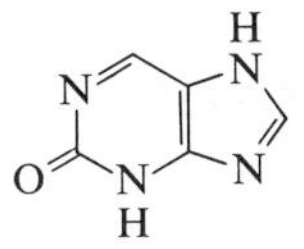

46
λ_{max} 238, 315 mμ
mε 2.88, 4.9

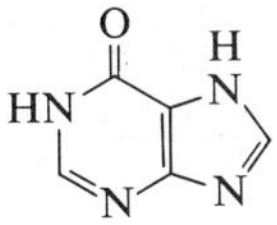

47
λ_{max} 249 mμ
mε 10.5

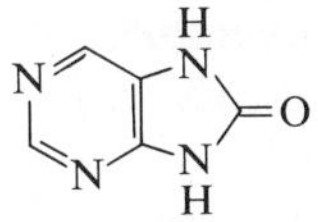

48
λ_{max} 235, 277 mμ
mε 3.24, 11.2

By means of ultraviolet-spectra and ionization-constant investigations of xanthine(108–111), guanine(112), uric acid(113, 114), and their methyl derivatives, these compounds have also been assigned lactam structures.

G. Pteridinones

The monohydroxypteridines exist predominantly in the oxo form (86, 115), but complex equilibria can be present since more tautomers are possible than in monohydroxy compounds with fewer aza nitrogen atoms. Thus in 4-hydroxypteridine the following equilibrium has been postulated(8, 86):
The *ortho*-lactam structure predominates, but appreciable amounts of the enolic tautomer could be present due to its stabilization by the intramolecular hydrogen bond.

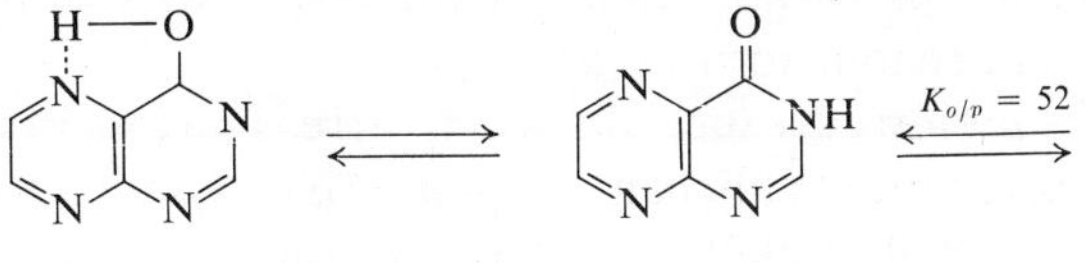

49 ⇄ **50** $\underset{K_{o/p} = 52}{\rightleftarrows}$ **51** (5.11)

λ_{max} 230, 265, 310 mμ (pH 5.6)
mε 9.55, 3,47, 6.61

Tautomerism in these molecules is complicated also by the ability of some of them to add water to a double bond. Thus Brown and Mason (86) suggest that the strongly bound molecule of water in pteridin-2-one and pteridin-6-one and their *N*-methyl derivatives is water of constitution which is added across the 3,4- and 7,8-double bonds, respectively, as shown for 6-pteridinone below:

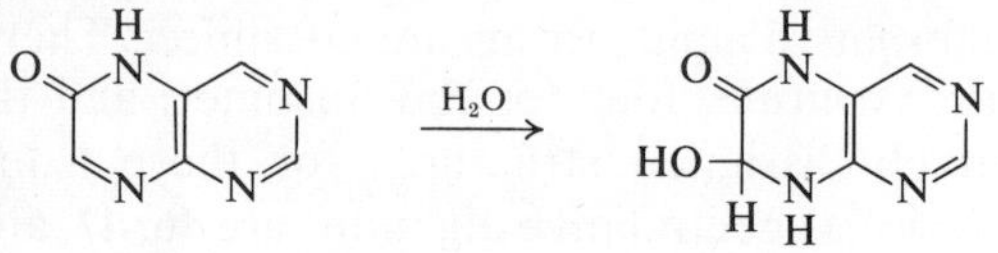

H_2O (5.12)

52 **53**

This reversible water addition to hydroxypteridines (116) and other heterocyclic compounds (5) has been discussed. For this type of covalent hydration to take place the aromatic ring must have a high proportion of —CH=N— groupings so that aromatic stability is undermined and the carbon atom has a low electron density while the nitrogen atom acquires a high electron density through the attraction of π-electrons. Then, if the nitrogen atoms are in the 1,3-position to each other and the hydrated structure represents a resonance gain, covalent hydration is favored (5).

The various structures assigned to the polyhydroxypteridines are shown in Table 5.9. The assignments have been based on spectral comparison with *N*- and *O*-methylated derivatives. For example, some of the evidence for the structure of 6-carboxy-2,4,7-pteridintrione (No. 5 in Table 5.9) is based on the spectral comparisons in Figure 5.1. The compound is mainly present in aqueous solution as *B*, but the shoulder near 370 mμ is apparently derived from the tautomer *C*, which absorbs at as long a wavelength as its comparison standard *D*.

Since these equilibria are complicated, much more needs to be done before they are understood completely.

H. Acridinols and Similar Compounds

The electronic absorption spectra of 9-acridanone, 10-methylacridanone, and 9-methoxyacridine show that 9-acridanone exists in the carbonyl form and not in the enolic form (124).

The lactam form is preferred, especially when there is no loss in Kekulé resonance, as shown for 6(5*H*)-phenanthridinone (No. 1 in Table 5.10) or where there is a gain in Kekulé resonance at the expense of high-energy quinonoid structures, as in compounds 5 and 16 in Table 5.10.

The phenol form predominates in 3-hydroxyphenanthridine (No. 2 in Table 5.10), since the carbonyl form contains high-energy *o*- and *p*-

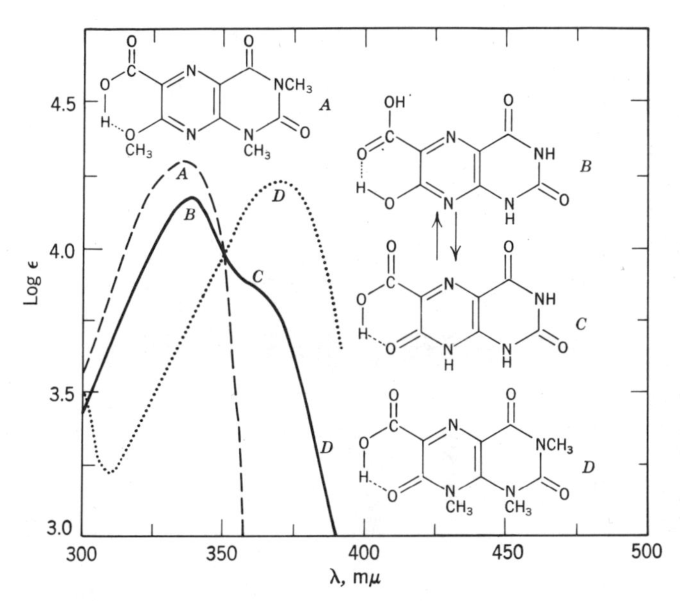

Fig. 5.1. Ultraviolet absorption spectrum of 7-hydroxy-2,4-dioxotetrahydropteridine-6-carboxylic acid ($B \rightarrow C$) and its trimethyl derivatives at pH 0.0(122).

benzoquinonoid structures. The carbonyl forms of this and the other compounds absorb at longer wavelengths than their enolic isomers (Table 5.10) because the positive resonance terminal is a powerful electron-donor —NH— group, whereas the negative resonance terminal is a strong electron-acceptor —C=O group, as shown in the C=O resonance forms **54**, **55**, and **56** of Table 5.10 compounds 3, 8, and 6, respectively.

One of the factors conducive to the shift toward longer wavelengths of the long-wavelength band of **54**, **55**, and **56** would appear to be the increase in the length of the chain of conjugation between the resonance terminals.

In the enolic tautomers the resonance terminals are weaker electron donors and electron acceptors; hence these compounds absorb at shorter wavelengths than do the carbonyl forms.

Resonance is one of the important factors in tautomeric equilibrium. Thus the increased resonance of the carbonyl form of compound 3 in Table 5.10 overcomes the high energy of the *p*-benzoquinonoid structure so that the carbonyl form is favored. Where a non-dipolar carbonyl form is impossible, the zwitterionic resonance forms are apparently of such high energy that the enolic form is preferred (e.g., compounds 4, 7, and 9 in Table 5.10).

TABLE 5.9

Tautomerism of Polyhydroxypteridines

Principal structures	Ref.	pH or solvent	λ_{max} (mμ)	mϵ	pK_a Proton gained	pK_a Proton lost
1 (pK_t = 1·88)	117, 118	0.1 N H_2SO_4	228 325	10.2 7.0	− 3.00	7.91
2	119	3.5	223 251 365	11.5 10.5 5.5		5.85
3	120	1.0	269 324	8.32 11.8		3.43 9.80
4	121	1.8	~ 295 ~ 325	8.3 7.08		3.49 9.60

5		122	0.0	271	8.7	~2.0
				399	14.8	5.98
						9.90
6		122	0.0	246	12.0	~1.7
				385	7.6	7.20
						9.63

TABLE 5.10
Tautomeric Equilibrium of Tricyclic and Tetracyclic Aza Phenols

Principal structures[a]	Ref.	pH or solvent	$\lambda_{max}(m\mu)$[b]	$m\epsilon$	pK_1	pK_2	Log K_t[c]
1 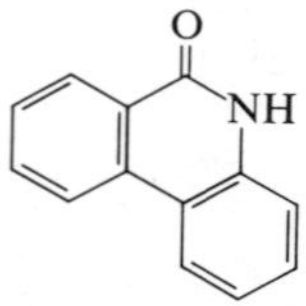 O, NH	123	95% EtOH	230	50.0	>-1.5[d]		3.9
			250	12.6			
			260	20.0			
			320	10.0			
			338	7.9			
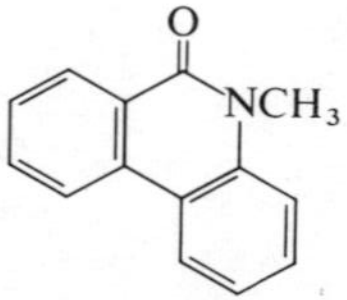 O, NCH_3	123	95% EtOH	238	63.0			
			260	20.0			
			320	10.0			
			338	7.9			
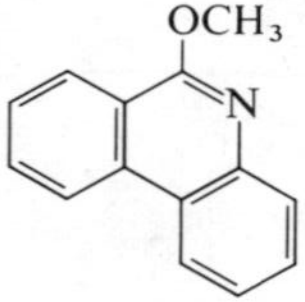 OCH_3, N	123	95% EtOH	220	31.6	2.38[d]		
			235	39.8			
			245	39.8			
			252	31.6			
			305	31.6			
2 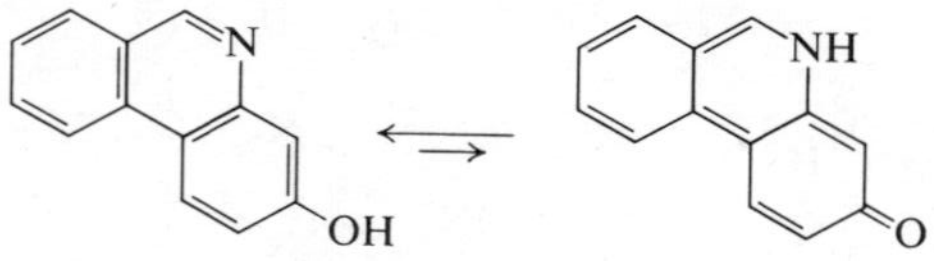 N, OH ⇌ NH, O	3	6.80	253	45.2	4.89	8.79	-3.0
			347	2.24			
			360	2.27			
			450*s*	0.017			
		95% EtOH	254	43.6			
			351	2.46			
			365	2.4			

3	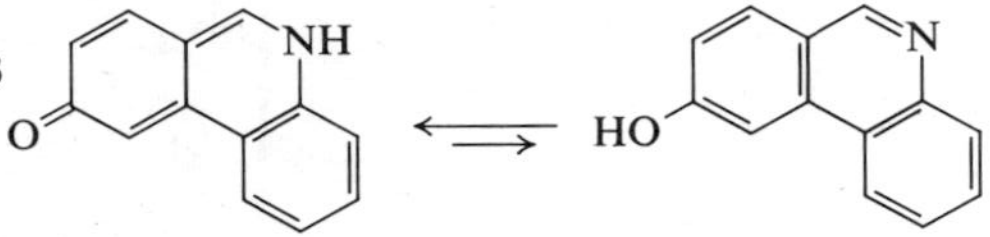 54	3	6.89	251	43.6	5.35	8.43	0.7
				299	8.0			
				328	3.5			
				377	10.4			
			Dioxane	254	50.0			
				297	13.6			
				328	1.4			
4	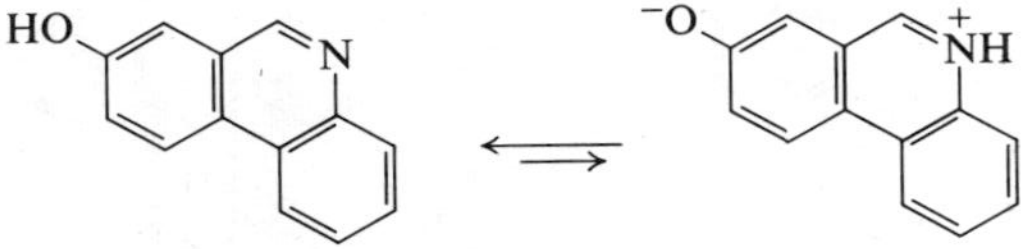	3	H_2O–EtOH (9:1)	301	4.68	4.39	8.68	< −1
				347	2.3			
				361	2.3			
				395*s*	0.315			
			95% EtOH	253	44.1			
				303	4.06			
				349	2.51			
				364	2.55			
5	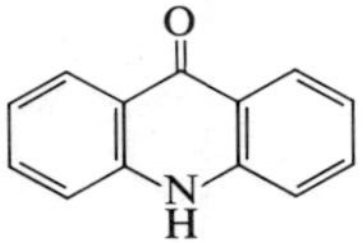	125[e]	95% EtOH	251	57.5	−0.32 (29)		7
				255	60.3			
				295	2.14			
				308	1.12			
				381	8.9			
				399	8.9			
	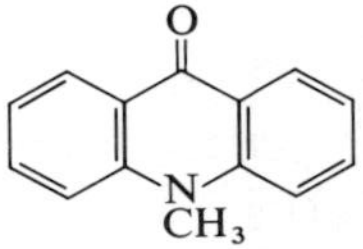	125	95% EtOH	256	57.5			
				295	2.7			
				306*s*	1.2			
				387	8.3			
				404	9.6			
6	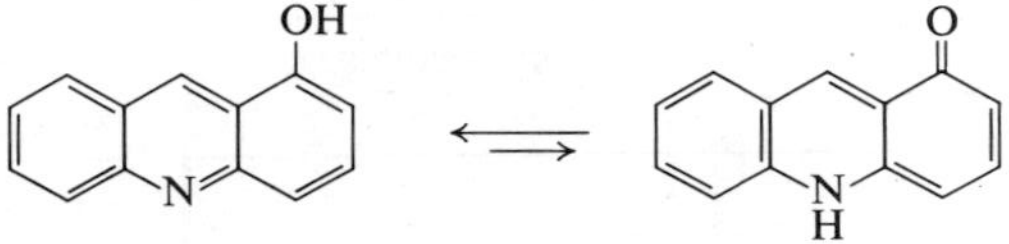 56	3[f]	H_2O–EtOH (4:1)	360	5.0	4.45	9.4	−0.52
				398	2.8			
				570	1.0			

TABLE 5.10 (*cont.*)

Principal structures[a]	Ref.	pH or solvent	$\lambda_{max}(m\mu)$[b]	mε	pK$_1$	pK$_2$	Log K_t[c]
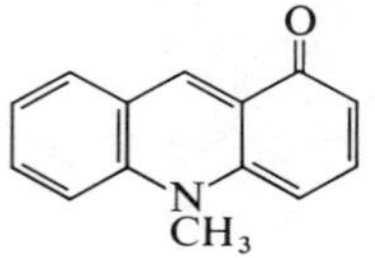 (N-CH_3, C=O structure)	3	9–10	360	4.5			
			570	4.0			
7 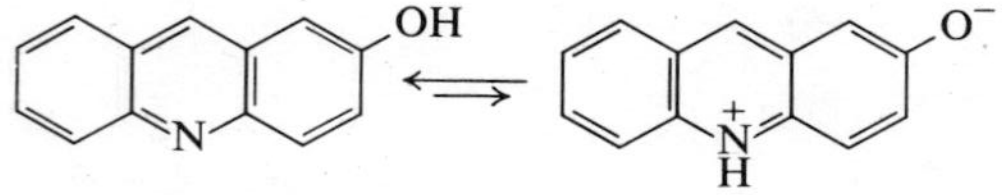 (OH ⇌ O^-, $\overset{+}{N}H$)	3[f]	H_2O–EtOH	355	6.3	5.52	8.81	−1.82
		(9:1)	380	4.5			
			500*s*	0.06			
		95% EtOH	354	7.0			
			396	5.0			
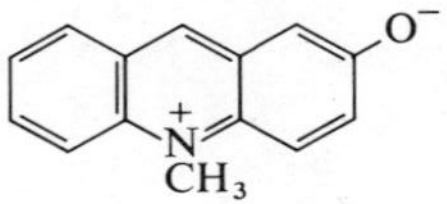 (O^-, $\overset{+}{N}$-CH_3)	3	9–10	370	10.0			
			500	4.0			
8 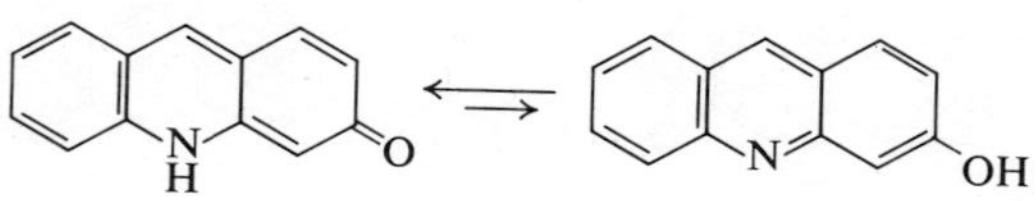 (NH, =O ⇌ N, OH) 55	3[g]	H_2O–EtOH	351	12.5	4.86	9.9	0.40
		(4:1)	363	16.0			
			466	10.0			
		H_2O–EtOH	351	8.0			
		(1:9)	391	5.6			
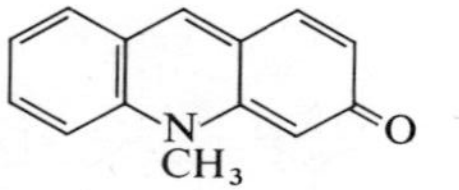 (N-CH_3, =O)	3	9–10	350	12.0			
			470	14.0			
9 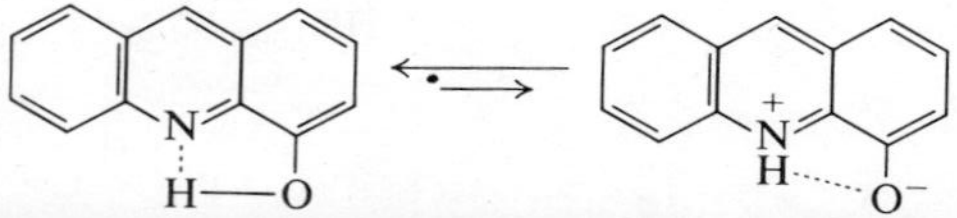 (N···H—O ⇌ $\overset{+}{N}$-H···O^-)	3[f]	H_2O–EtOH	358	3.67	4.18	10.7	−1.4
		(9:1)	386	3.45			
			550 *s*	0.0042			
		95% EtOH	360	3.55			
			392	3.55			

No.	Compound	Ref.	Solvent	λ (nm)	ε × 10⁻³
	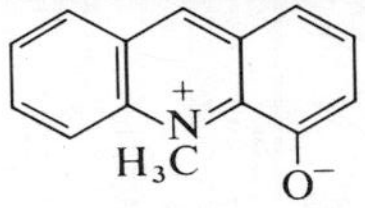 H_3C, N^+, O^-	3	9–10	380	3.6
				550	1.2
10	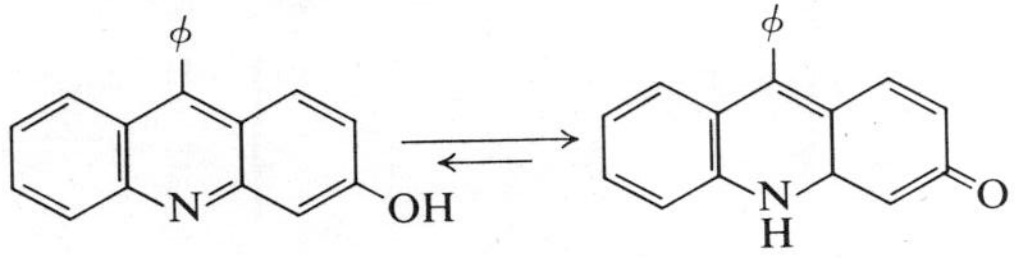 ϕ, N, OH ⇌ ϕ, N–H, O	129	Cyclohexane	258	
				295 *s*	
				342	
				356	
				381	
				402	
				428 *s*	
			EtOH (abs.)	236	30.2
				259	77.6
				286	16.2
				322 *s*	2.46
				338 *s*	5.4
				352	9.77
				394	4.8
				410 *s*	3.8
				483	2.6
				500 *s*	2.24
			H_2O–EtOH (4 : 1)	236	36.3
				259 *s*	26.3
				279	49.0
				355 *s*	11.2
				365	12.6
				468	9.55
	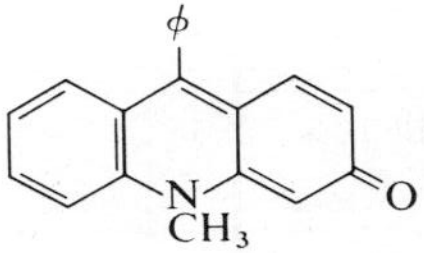 ϕ, N–CH_3, O	129	EtOH (abs.)	238	38.0
				288	49.0
				352	11.5
				363 *s*	10.7
				450 *s*	8.1
				478	11.8
				504	10.2

TABLE 5.10 (*cont.*)

Principal structures[a]	Ref.	pH or solvent	$\lambda_{max}(m\mu)^{b}$	mϵ	pK_1	pK_2	Log K_t^c
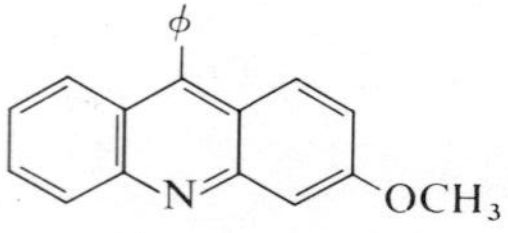 (9-phenyl-3-methoxyacridine; ϕ, N, OCH_3)	129	EtOH (abs.)	236	26.3			
			258	117.5			
			286 *s*	5.25			
			320	2.95			
			335	6.17			
			351	10.5			
			380	6.6			
			396 *s*	4.7			
11[h,i] 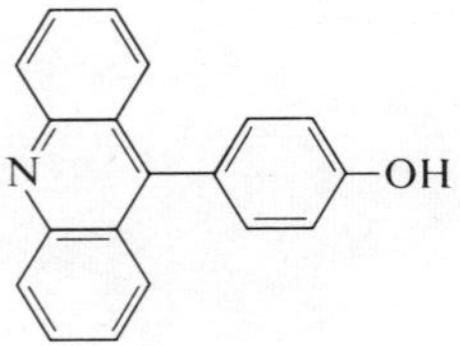 (N, OH)	130[j]	EtOH–ether	255 *fs*	158.5			
		(2 : 1)	290 *fs*	3.16			
			360 *fs*	15.9			
12[h] 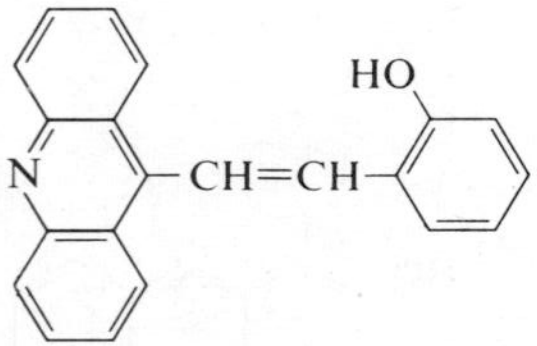 (N, CH=CH, HO)	130[j]	EtOH–ether	255 *s*	100.0			
		(2 : 1)	365 *fs*	12.6			
			410 *fs*	12.6			
13 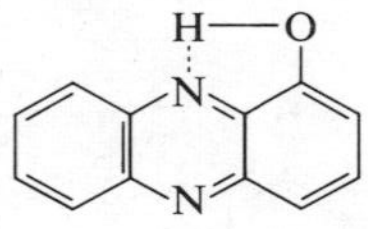 (H—O, N, N)	133	95% EtOH	240	15.9			
			265	50.0			
			360	6.3			
			370	7.9			
			420	2.0			

14	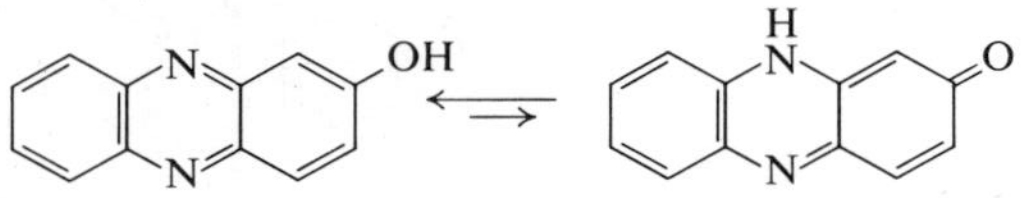 133	Water	260	50.0	2.6	7.5	−0.8
			280	20.0			
			370	10.0			
			415	5.0			
			490	2.5			
		H_2O–EtOH (1 : 1)	260	63.0			
			280	15.9			
			365	10.0			
			415	6.3			
			500 *s*	0.79			
		95% EtOH	260	63.0			
			350	7.9			
			410	6.3			
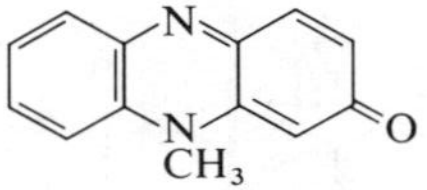	134	95 % EtOH	231	39.8			
			292	100.0			
			369	7.9			
			531	10.0			
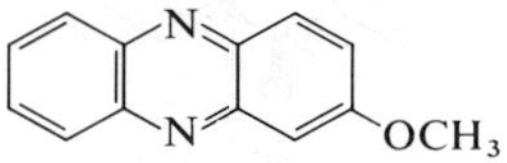	133	95% EtOH	255	79.4			
			360	10.0			
			395 *s*	7.9			
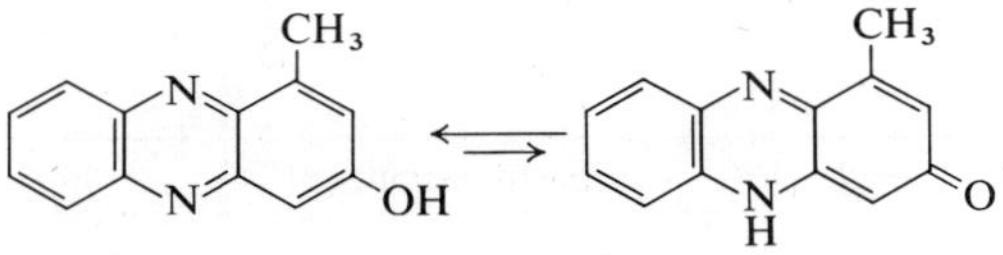	135	Methanol[k]	218	22.9			
			261	77.6			
			361	8.9			
			412	6.3			
			520–530	0.10			
15	136	Dioxane	216	30.9			
			260	91.2			
			359	10.0			
			401	6.9			

TABLE 5.10 (*cont.*)

	Principal structures[a]	Ref.	pH or solvent	$\lambda_{max}(m\mu)$[b]	mϵ	pK_1	pK_2	Log K_t[c]
16	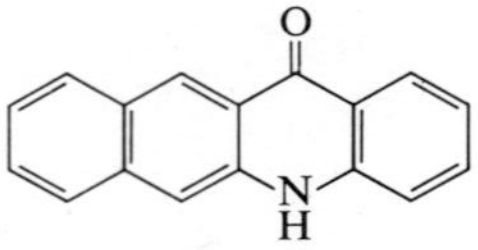 O, N, H	137	Methanol	265	39.8			
				300	15.9			
				470	3.16			
				480	3.16			
	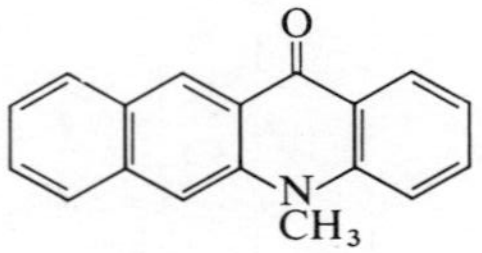 O, N, CH_3	137	Methanol	260	31.6			
				330	6.3			
				470	6.3			
				480	6.3			
17	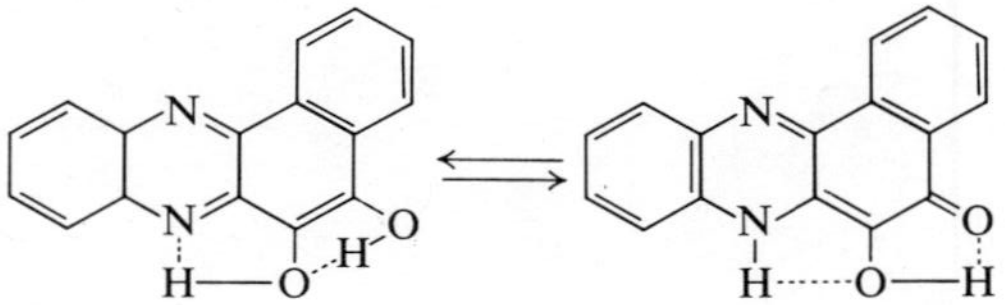 N, N, H—O, H, O ⇌ N, N, H⋯O—H, O	133	Dioxane	315	39.8			
				350 *s*	6.3			
				377	3.9			
				460	4.1			
			Dioxane–water(4 : 1)[l]	495 *br*	2.8			
			Dioxane–water (1 : 1)[l]	550	4.0			
	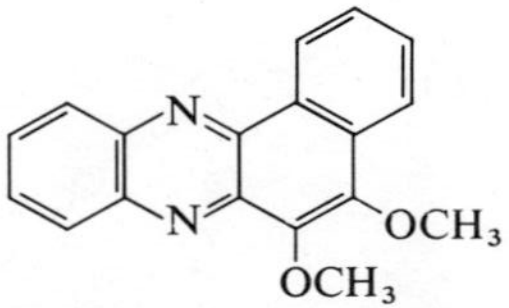 N, N, OCH_3, OCH_3	133	95% EtOH	230	38.0			
				284	39.8			
				300	39.8			
				370	6.3			
				410	8.1			

[a]Structures that are not numbered are reference compounds.
[b]S = Shoulder; fs = fine structure.
[c]Tautomeric constant where K_t = [NH form]/[OH form].
[d]Data from reference 30.
[e]See also reference 126.
[f]See also references 126–128.
[g]See also references 126 and 127.
[h]2′-, 3′-, and 4′-Hydroxy derivatives have enolic form.
[i]Spectrum similar to that of methyl ether.
[j]See also references 131 and 132.
[k]Contains 0.1% acetic acid.
[l]Contains 1% hydrazine as antioxidant.

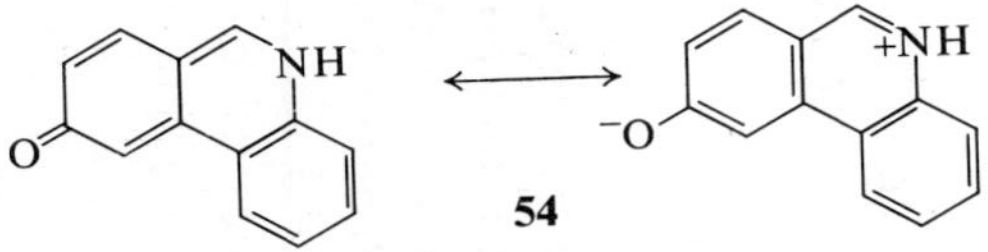

NH ⟷ $^{-}$O, $^{+}$NH

54

λ_{max} 377 mμ

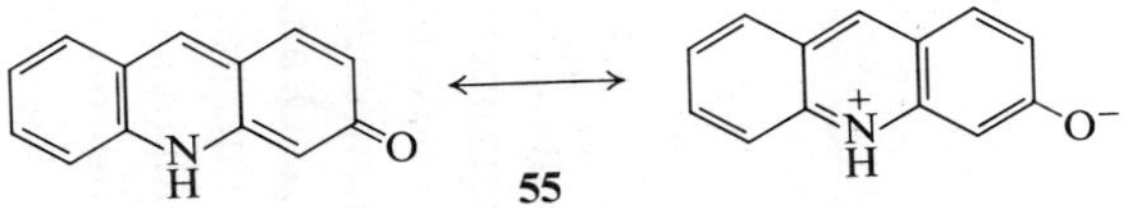

N H, O ⟷ $\overset{+}{N}$ H, O^{-}

55

λ_{max} 466 mμ

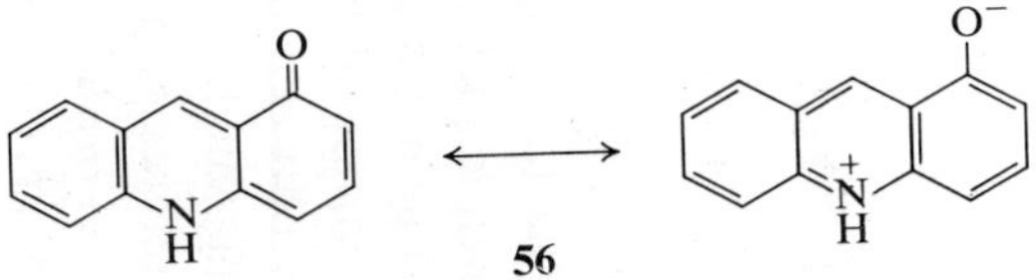

O, N H ⟷ O^{-}, $\overset{+}{N}$ H

56

λ_{max} 570 mμ

The importance of intramolecular hydrogen bonding is shown in 1-hydroxyphenazine (No. 13 in Table 5.10), where the enolic form is preferred and in 5,6-dihydroxybenzo[*a*]phenazine (No. 17 in Table 5.10), where both forms contain intramolecular hydrogen bonds so balanced that the polarity of the solvent determines whether the carbonyl form (λ_{max} 550 mμ) or the dihydroxy form (λ_{max} 460 mμ) predominates.

The conclusions of Badger, Pearce, and Pettit (133) concerning the tautomeric equilibrium of the phenazinols are reinforced by the report that 10-ethyl-2(10*H*)-phenazinone and 5-ethyl-1(5*H*)-phenazinone are red and blue, respectively (138).

As can be seen from the data in Table 5.10, increasing polarity of the solvent shifts the equilibrium toward the carbonyl form. Examples of this phenomena are 1-hydroxyacridine (Fig. 5.2) and 2-phenazinol (Fig. 5.3). The spectrum of the carbonyl form of the former compound shows bands at 296 and 570 mμ derived from that form. The spectrum of the carbonyl form of 2-phenazinol resembles the spectrum of 3*H*-phenothiazin-3-one.

The tautomeric equilibria of the *N*-oxides of acridinols and other analogous large heterocyclic compounds have not, as yet, been thoroughly studied. Ionescu et al. (139) report that the spectra of the methylated comparison standards were fairly similar. On the basis of ionization constants the tautomers **57** and **58** are believed to exist in aqueous solution in comparable amounts.

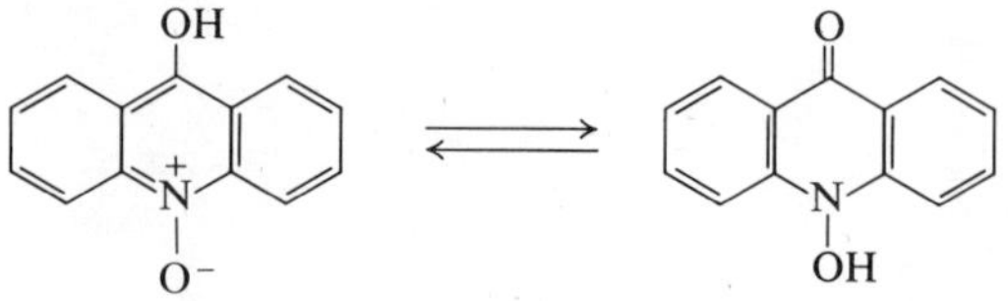

OH, N^{+}, O^{-} ⇄ O, N, OH (5.13)

57 **58**

The tautomeric equilibria can be summarized by the thermodynamic principle that the tautomeric form that predominates will always be the least acidic of all the possible forms giving rise to a common anion. The proportion of NH tautomers will depend on the possibilities of conjugation between the C=O and N—H resonance terminals. Other things being equal, this conjugation is most effective for "*para*, *para*" quinonoid structures (e.g., compound 12 in Table 5.8) and decreasingly less for "*para*, *ortho*", "*ortho*, *ortho*", (e.g., compounds 13 and 4 in Table 5.8, respectively), when no uncharged form can be written for the NH form (e.g., No. 5 in Table 5.8). Another factor that needs emphasis is the

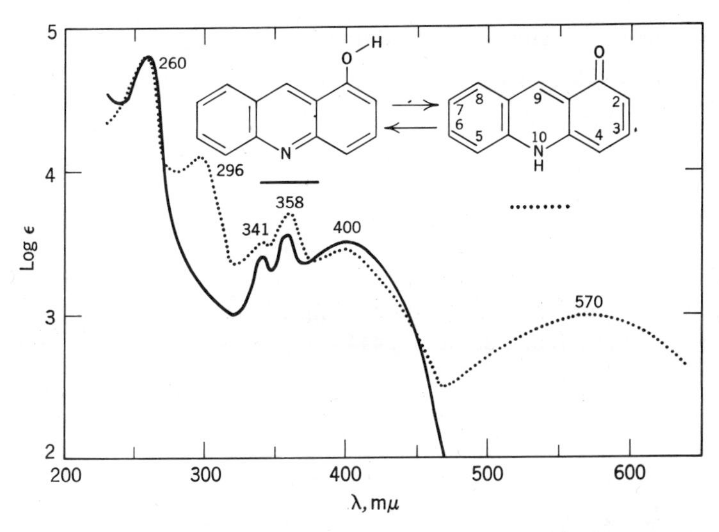

Fig. 5.2. Ultraviolet–visible absorption spectra of 1-hydroxyacridine in absolute ethanol (——) and 20% aqueous ethanol (· · · ·)(126).

annulation effect that tends to stabilize the NH form (e.g., pyrid-4-one, pK_t 3.3; quinol-4-one, pK_t 4.3; and acridan-9-one, pK_t 7.0). With an increase in the polarity of the solvent the proportion of NH tautomer increases.

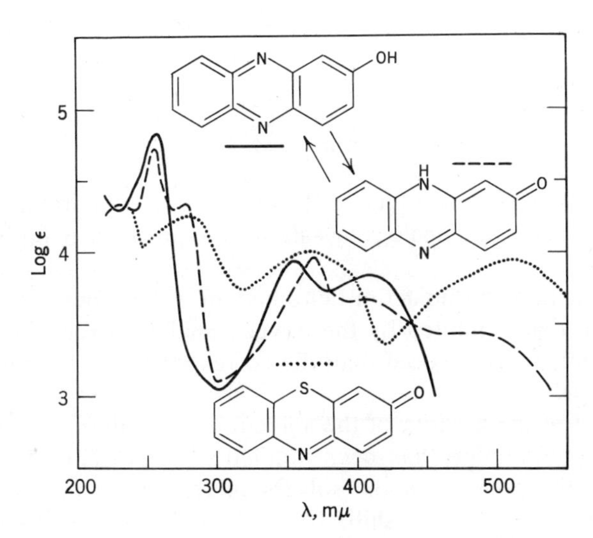

Fig. 5.3. Ultraviolet–visible absorption spectra of 2-phenazinol in absolute ethanol (——) and water (– – –), and of 2-phenothiazone in ethanol (· · · ·)(133).

II. Thiolactam ⇄ Thiolactim Tautomerism

The absorption spectra and the tautomeric equilibria of the thiolactam ⇄ thiolactim heterocyclic compounds have been discussed(19, 140). In both the cyclic and noncyclic compounds of this type the thiolactam is usually the preferred form.

Thiourea(105) and its *N*-alkyl(141) and *N*-phenyl(142) derivatives exist in the thiocarbonyl form, as shown by ultraviolet, infrared, and Raman spectral studies. The preponderance of the thiocarbonyl form (**59**) can be clearly shown by comparing it spectrally with its *S*-methyl derivative (**60**), both in methanol(141).

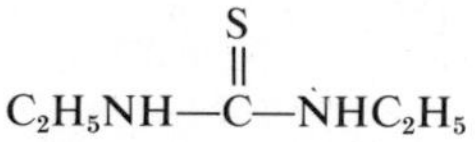

$$C_2H_5NH{-}C({=}S){-}NHC_2H_5$$

59

λ_{max} 234, 265 mμ
mε 6.3, 7.3

$$C_2H_5N{=}C(SCH_3){-}NHC_2H_5$$

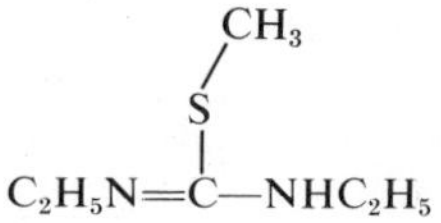

60

Tail absorption < 225 mμ

For some thiocarbazones both tautomers can be obtained, and the equilibrium in solution involves measureable amounts of tautomers. Thus the absorption spectra in carbon tetrachloride of both tautomers of dithizone (**61** and **62**) have been reported(143).

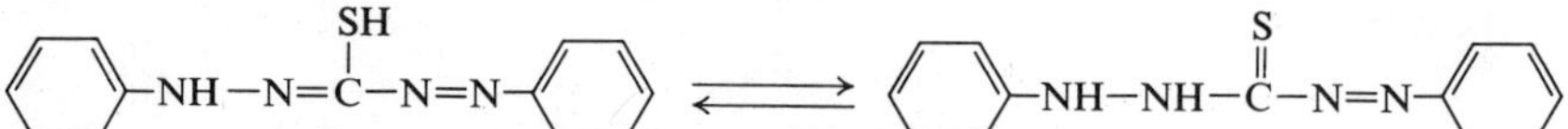

$$C_6H_5{-}NH{-}N{=}C(SH){-}N{=}N{-}C_6H_5 \rightleftarrows C_6H_5{-}NH{-}NH{-}C({=}S){-}N{=}N{-}C_6H_5 \quad (5.14)$$

61: λ_{max} 265, 465 mμ; mε 9.55, 24.0

62: λ_{max} 300, 400, 624 mμ; mε 7.08, 12.6, 30.0

A study of the solvent effect on the tautomeric equilibrium of dithizone and its derivatives and analogs has shown the presence of both tautomers (144, 145).

The thiolactam ⇄ thiolactim tautomerism of heterocyclic compounds is similar to that exhibited by the oxygen analogs, except that the thiolactam form is more favored than the corresponding lactam form (Table 5.11)(146).

The effect of the polarity of the solvent on the tautomeric equilibrium of 8-mercaptoquinoline has been studied (Table 5.12)(147, 148). An increase in the polarity of the solvent increases the proportion of the zwitterionic form and also shifts the Z_b-band to shorter wavelength, as is to be expected.

A study of the ultraviolet and infrared spectra and the ionization con-

TABLE 5.11
Comparison of K_t Values, [NH]/[SH] and [NH]/[OH] Forms, in water at 20°C (146)

Compound	K_t	Compound	K_t
	Pyridines		
2-Mercapto	49,000	2-Hydroxy	340
3-Mercapto	150	3-Hydroxy	1
4-Mercapto	35,000	4-Hydroxy	2,200
	Isoquinolines		
1-Mercapto	680,000	1-Hydroxy	18,000
3-Mercapto	1,000		
	Quinolines		
2-Mercapto	240,000	2-Hydroxy	3,000
3-Mercapto	34	3-Hydroxy	0.06
4-Mercapto	110,000	4-Hydroxy	24,000
5-Mercapto	15	5-Hydroxy	0.05
6-Mercapto	5	6-Hydroxy	0.01
8-Mercapto	27	8-Hydroxy	0.04

TABLE 5.12
Solvent Effect on the Z_b-Band and the Tautomeric Equilibrium of 8-Mercaptoquinoline

Solvent	Ref.	$\lambda_{max}(m\mu)$	$m\epsilon$	Z_b-form ($\sim$%)
Water	147	446	$\sim$1.600	97
	148	448	$\sim$2.032	
Methanol	148	490	0.133	4
Ethanol	147	503	0.022	1.3
	148	503	0.043	
n-Butanol	148	509	0.027	0.8
tert-Butanol	148	528	0.017	0.5
Acetonitrile	147	550	0.008	0.5
Acetone	147	565	0.002	0.14
Chloroform	147	562	0.0015	0.09
Ethyl ether	147	–	$< 10^{-7}$	$< 10^{-5}$
Isooctane	147	–	$< 10^{-7}$	$< 10^{-5}$

stants of methyl derivatives of 8-mercaptoquinoline has disclosed that these compounds exist predominantly as the zwitterion in the solid state and in aqueous solution(149). In chloroform solution the amount of zwitterion is considerably decreased, and in addition the long-wavelength Z_b-band is shifted toward much longer wavelengths, as, for example, in **64**. Acid causes the expected violet-shift effect (equation 5.15).

$$\mathbf{63} \xleftarrow{H^+} \mathbf{64} \qquad (5.15)$$

No.	Solvent	λ_{max}(mμ)	mϵ
63	1 *M* HCl	317	7.4
64	Water	421	2.4
	Chloroform	523	0.17

On the other hand, 2-phenyl-8-mercaptoquinoline is present in water and chloroform solution in the thiol form, as shown by the ultraviolet absorption spectra of the compound and its 5-methyl derivative in aqueous and chloroform solution.

As is to be expected, the annulation effect can further increase the stability of the imino form. Thus 9-thioacridone is overwhelmingly present in the imino form, as shown by a comparison of its absorption spectrum in methanol with the spectra of the methylated standards (Fig. 5.4)(124).

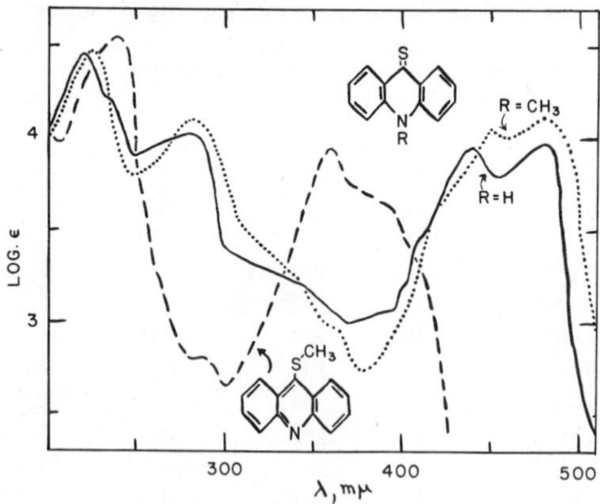

Fig. 5.4. Electronic absorption spectra in methanol of 9-methylthioacridine (– – –), 10-methyl-9-thioacridone (· · · ·), and 9-thioacridone (——)(124).

III. Amino ⇄ Imino Forms

A. Primary Amines

Hydroxy groups are much less acidic than mercapto groups, whereas amino groups are immensely less acidic than hydroxy groups. Thus the following values of pK_a at 25°C have been reported: aniline, ~27 (150); phenol, 9.98 (151); and thiophenol, 6.5 (152). Since the tautomeric form that predominates is always the least acidic of all the possible forms giving rise to a common anion, the amino tautomer will be considerably more stabilized than its imino form. This phenomenon has been demonstrated with many compounds; an example is 2-aminopyrimidine (**65**) and its alkylated derivatives (**66**, **67**, and **68**) (153).

NH_2 **65** $NHCH_3$ **66** $N(CH_3)_2$ **67** NH CH_3 **68**

No.	pK_a	λ_{max}(mμ)	mε
65	3.54	224	13.5
		292	3.16
66	3.82	234	17.0
		307	2.7
67	3.96	243	18.2
		318	2.2
68	10.8	236	15.5
		345	2.9

The definite difference between **65** and **68** in ionization constants and absorption spectra and the similarities between **65**, **66**, and **67** show the predominance of the amino form in 2-aminopyrimidine.

It has been shown by ultraviolet–visible absorption spectra and other physical constants that monocyclic, dicyclic, and tricyclic *N*-heteroaromatic amines exist predominantly in the amino, and not the tautomeric imino, form (19, 153–161).

In this manner some of the more complicated heterocyclic compounds that exist as the carbonyl amine tautomer include 2-amino- and 2-alkylaminothiazolin-4-ones (162), 5-aryl-2-amino-4-oxazolinones (163), and 5-phenyl-2-amino-4-selenazolinones (164).

Although the amino tautomer predominates in a large number of compounds, some structural changes—for example, annulation—favor the imino form. If a polynuclear amine contains high-energy *o*-benzoquinonoid bonds that can be rearranged and relieved by tautomerism, then the imino form may be favored. There is some evidence that certain 5-aminonaphthacenes are predominantly in the imino form(160, 165).

B. Acylamines

Acylamino groups have properties that are intermediate between those of free amino and hydroxy groups, as shown by their acidity, basicity, electron-donor properties in absorption spectroscopy, and in their electrophilic substitution reactions on the benzene ring. Depending on the electron-donor or electron-acceptor strength of the acyl substituent, the acylamino group can more closely resemble either the amino or the hydroxy group.

When an acetyl or a benzoyl group is substituted on the amino group of an aminopyridine, the acidity is increased, but not enough to stabilize the imino form. Thus from the basicity and ultraviolet absorption spectra studies it has been shown that 2-, 3-, and 4-acetamidopyridines and benzamidopyridines exist in aqueous solution predominantly in the acylamino form(166).

As the electronegativity of the acyl group increases, the acidity and the K_t value ([imino]/[amino]) increase. Thus the substitution of a methanesulfonyl group into 2-, 3-, and 4-aminopyridine has considerably enhanced the acidity of these compounds to pK_a values of 8.02, 7.02, and 9.07 in water and has also resulted in fairly high K_t values of ~10, 0.1, and 30, respectively(167). The low value of the 3-substituted derivative is probably due to the lack of a non-dipolar form with which the zwitterion could undergo resonance stabilization.

In the investigation of the tautomeric equilibrium of a group of 2-sulfonamidopyridines it has been shown that, as the electronegativity of the acyl group increases and/or as the polarity of the solvent increases, the acidity of the compound (decreasing pK_a) and its K_t value increase (Table 5.13).

The absorption spectra and ionization constants of the amine and imine tautomers are widely different, as shown by the spectral data for 4-methanesulfonamidopyridine (**69**) and its methylated standards (**70** and **71**) in aqueous solution(167).

The effect of solvent polarity on the tautomeric equilibrium of α-(p-

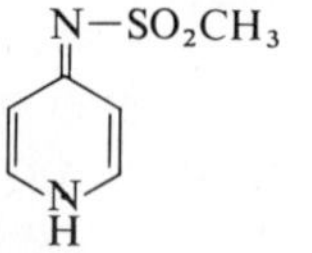

$N{-}SO_2CH_3$ / N-H — **69** $N{-}SO_2CH_3$ / $N{-}CH_3$ — **70** $H_3C{-}N{-}SO_2CH_3$ / N — **71**

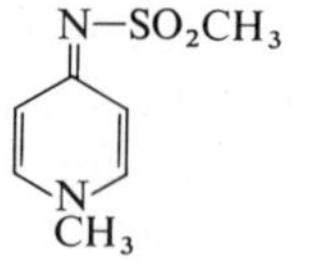

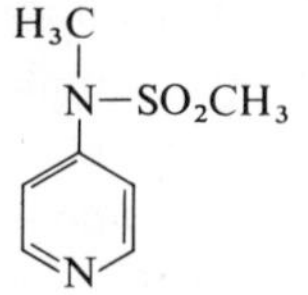

No.	pK_a	$\lambda_{max}(m\mu)$	$m\epsilon$
69	3.64	281	25.2
70	3.42	287	27.3
71	5.14	236	7.9

cyanophenylsulfonamido)pyridine, as demonstrated by the change in absorption spectra, is shown in Figure 5.5.

C. Secondary Amines

Where an amino group connects two separate conjugated systems and a fairly basic proton acceptor (e.g., an aza nitrogen group) is present in

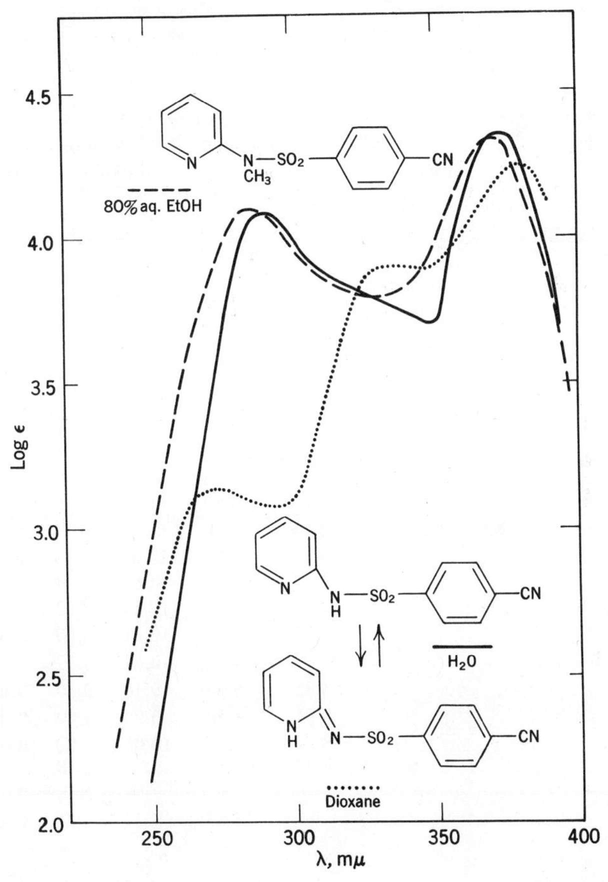

Fig. 5.5. Ultraviolet absorption spectra of α-(*p*-cyanophenylsulfonamido)pyridine in water (——) and dioxane (· · · ·) and of α-(*N*-methyl-*p*-cyanophenylsulfonamido)pyridine in 80% aqueous ethanol (– – –)(169).

TABLE 5.13

Change in Ionization and Tautomeric Equilibrium Constants of 2-Acylaminopyridines with Change in the Electronegativity of the Acyl Group and the Polarity of the Solvent

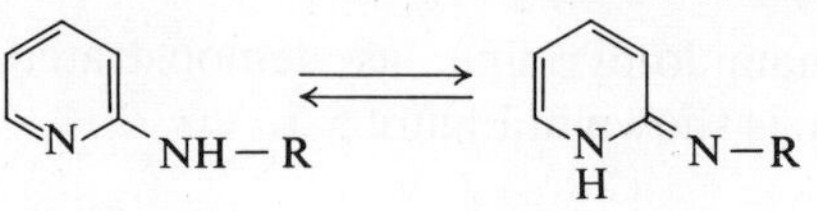

I ⇌ II

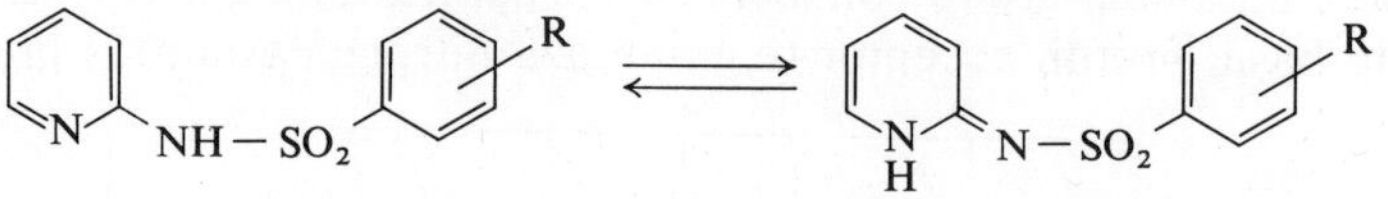

III ⇌ IV

R	Water pK$_a$	Water K_t[a]	50% Aqueous ethanol pK$_a$	50% Aqueous ethanol K_t	80% Aqueous dioxane pK$_a$	80% Aqueous dioxane K_t	100% Dioxane K_t	Ref.
				I ⇌ II				
H		0.000005						166
$COCH_3$		0.001						166
$COCHCl_2$		0.013					0.00016	168
$COCCl_3$		0.11					0.002[b]	168
$COCF_3$		>1.0						
SO_2CH_3	8.02							
				III ⇌ IV				
p-$N(CH_3)_2$				1.82	11.25	0.18	0.01	169
p-NH_2	~8.43	9.54	9.54	1.60	11.23	0.16	0.01	169
p-CH_3			9.16	3.55	10.60	0.54	0.03	169
p-OCH_3			9.25	3.18	10.64	0.41	0.03	169
p-$NHCOCH_3$			9.25	4.61	10.54	0.53	—	169
H			9.05	4.72	10.41	0.69	0.04[c]	169
p-OC_6H_5			—	4.28	10.37	0.57	—	169
p-F			8.88	6.06	10.18	0.80	0.05	169
p-Br			8.75	8.50	10.04	1.00	0.07	169
p-Cl			8.74	7.25	10.05	1.02	0.08	169
p-CN			8.37	15.80	9.59	2.58	0.16	169
m-NO_2			—	—	9.58	2.48	0.21	169
p-NO_2			—	23.83	9.52	2.86	0.26	169

[a] K_t = [imine]/[amine]
[b] In cyclohexane $K_t = 0.00003$.
[c] In cyclohexane $K_t = 0.01$.

at least one of the systems, tautomerism becomes possible. Preliminary investigation indicates that in di-2-pyridylamine such a tautomerism does not take place. However, when benzo groups are fused onto the pyridine rings, both tautomers are found in various proportions in solutions (Table 5.14). With an increase in annulation, a greater amount of conjugation is possible in tautomer **II**, and the tautomeric equilibrium shifts toward preponderance of this tautomer. Thus solvent polarity, annulation, and the, at times, competitive effects of intermolecular and intramolecular hydrogen bonding play an important role in this tautomeric equilibrium.

IV. Methene Aza ⇄ Methinimine Tautomers

Generally a compound that contains a methyl group attached to an electronegative group does not tautomerize readily. Where a CH or CH_2 group connects two aza heterocyclic rings or an aza heterocyclic ring and an electronegative group, the acidity of the CH or CH_2 group is increased and tautomerism becomes possible.

TABLE 5.14

Annulation Effects on the Absorption Spectra and Tautomeric Equilibria of Di-2-pyridylamines in Ethanol (170)

I ⇄ II

Compound	λ_{max} (mμ) (mε)			
	I		II	
	0 → 0 Band	0 → 1 Band	0 → 0 Band	0 → 1 Band
Di-2-pyridylamine	315 (15.0)	267 (20.0)		
Di-2-quinolylamine[a]	~375 (~32)	~360 (~21)	419 (1.2)	400 (1.5)
2-Pyridyl-9-phenanthridylamine[b]	357 (15.4)	339 (15.4)	394 (6.5)	374 (8.3)
2-Quinolyl-9-phenanthridylamine[b]	365 (17)	348 (12)	405 (17.8)	388 (20)
Di-9-phenanthridylamine[c]	364 (10.7)	348 (10)	422 (30.9)	395 (34.7)

[a] Due to increasing strength of the intramolecular hydrogen bond, the bands of **II** tautomer shift toward longer wavelength with decreasing polarity: for example, in *n*-heptane the 0 → 0 band is at 436 mμ. The concentration of tautomer **II** increases at the two extremes of solvent polarity.

[b] Concentration of tautomer **II** increases and the corresponding bands shift toward longer wavelength in the solvents ethanol < carbon tetrachloride < carbon disulfide.

[c] In dimethylformamide.

A. Hydrazone ⇄ Azo Type

It has been reported that freshly prepared phenylhydrazones of aliphatic aldehydes and ketones exist as hydrazone tautomers that tautomerize rapidly in solution to benzeneazoalkanes (171). This rearrangement has been denied by other workers (172, 173).

When diethyl malonate, Meldrum's acid, or dimedone is coupled with diazotized aniline, the products dissolved in methanol have λ_{max} of 348, 373, and 384 mμ (mε 15.0, 25.0, and 22.0), respectively (174). These compounds are stated to be the phenylhydrazone tautomers. For example, the product from the reaction of Meldrum's acid with diazotized *p*-anisidine is believed to be present as the hydrazone **72** in methanolic solution. Spectral comparison with the analogous diazotization product (**73**) from ethyl Meldrum's acid bears this out (174).

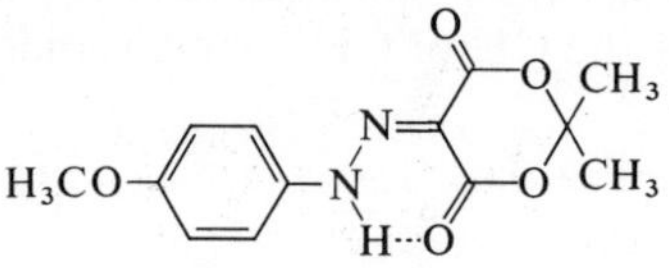

72

λ_{max} 232, 399 mμ
mε 11.0, 23.6

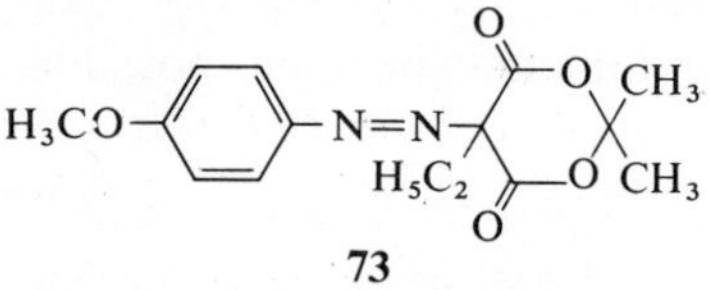

73

λ_{max} 236, 318, ~400 mμ
mε 8.5, 22.0, 0.4

There is a possibility of **72** being in equilibrium with a small amount of azoenol, especially in the analogous dimedone derivative.

B. Heterocyclic imines

In competition between a nitrogen and a carbon atom for a proton, the nitrogen atom usually gets the proton. Thus indole, which could conceivably have its most acidic hydrogen either in the 1-, 2-, or 3-position, has it attached to the nitrogen of the 1-position. Even in indazole the 1-position is preferred, as shown by the close spectral resemblance of indazole to 1-methylindazole and its spectral difference from 2-methylindazole (175). In the type of tautomer analogous to the latter compound a high-energy *o*-benzoquinonoid structure would be involved. The third possible tautomer containing the acidic hydrogen on the carbon atom in the 3-position (e.g., **74**) is not present in appreciable amounts since the spectrum of indazole shows the absence of a long-wavelength, low-intensity azo $n \rightarrow \pi^*$ band.

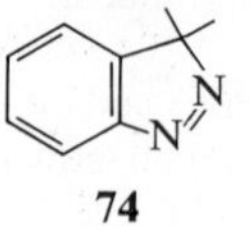

74

Some evidence that has been obtained indicates that tautomerism is possible in these types of molecules. Thus 5*H*-1-pyrindine (**75**) could have the tautomers **76a** and **77**.

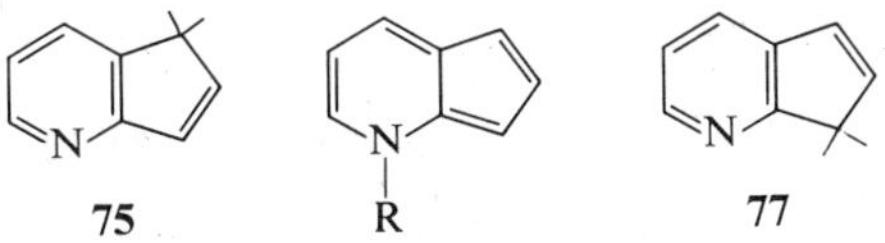

75 **76** **77**

a: R=H
b: R=CH_3

Compound **76b** has a long-wavelength band at 456 mμ, mε 0.69(176). Neat 5*H*-1-pyrindine shows the presence of a band at 470mμ mε 0.00076. From pK_a measurements and spectral data Reese estimates the presence of 0.1% of **76a**. The presence of equilibrium between **75** and **77** was not investigated.

C. R—CH_2—X=Y

An electronegative group can increase the acidity of a CH_2 group enough for ionization or tautometism to take place, depending on the solvent, the temperature, and the presence and strength of a proton-acceptor group in the molecule.

Tschitschibabin and co-workers(177) showed that solid 2-(2′,4′-dinitrobenzyl)pyridine (**78**) forms the blue tautomer **79a** when treated with sunlight. Irradiation at −60°C(178) and flash photolysis(179) have also been found to give the same sequence of reactions:

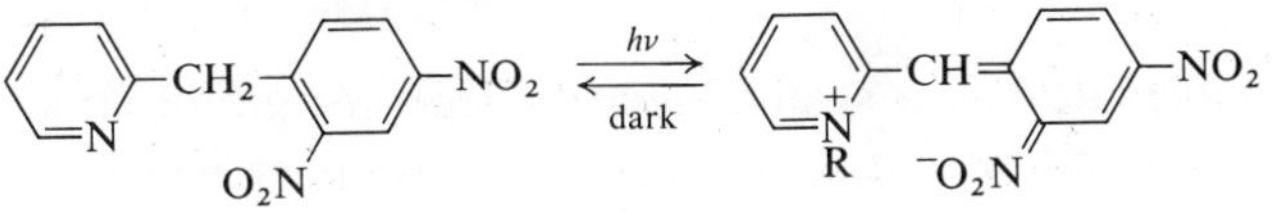

$$\textbf{78} \underset{\text{dark}}{\overset{h\nu}{\rightleftharpoons}} \textbf{79} \qquad (5.16)$$

No.	R	Solvent	λ_{max} (mμ)
78			254
79a	H	EtOH at −60°C	567.5
		Solid at −60°C	610
79b	CH_3	EtOH or acetone	565

The structure of the blue compound would fit in with the report that **79b** absorbs at 565 mμ in acetone or ethanol(178).

Electronegative substituents, such as the diphenylphosphine oxide and the phenylsulfone groups, are not potent enough to tautomerize a hydrogen on a methylene group attached to a pyridine ring. The picolyldiphenylphosphine oxides exist as such, the imino tautomer (**80**) being even less favored ($K_t > \sim 10^8$)(180) than in the phenylsulfone analog (**81**)(181).

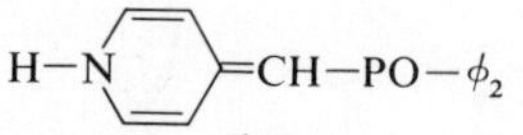

H—N⟨ ⟩=CH—PO—ϕ_2 **80**   H—N⟨ ⟩=CH—SO_2—ϕ **81**

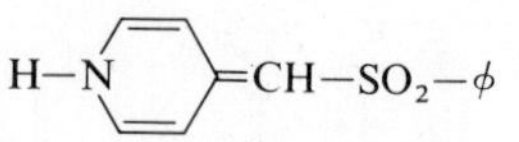

On the other hand, a carbonyl and an aza heterocyclic group connected through a methylene group will favor tautomerism to the imino form (182). This has been shown with a large number of quinoxaline and quinoline derivatives. The imino form is stated to be in highest concentration in chloroform solution and in lower concentrations in dimethyl sulfoxide, as shown for **82** and **83**.

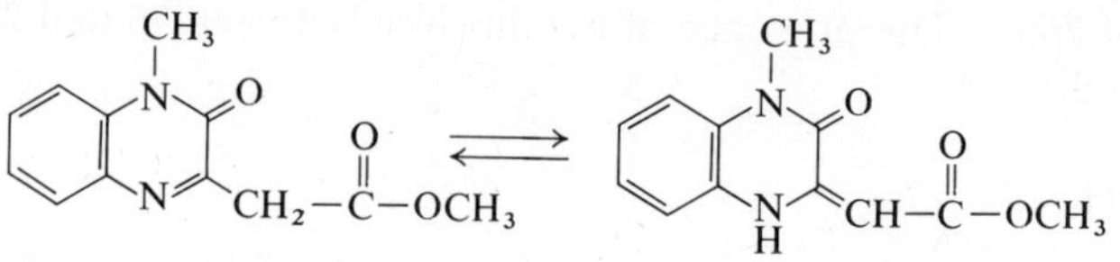

$$\mathbf{82}\ (\text{N-}CH_3\text{ quinoxalinone–}CH_2\text{–C(=O)–}OCH_3) \rightleftharpoons \mathbf{83}\ (\text{N-}CH_3\text{, N–H quinoxalinone=}CH\text{–C(=O)–}OCH_3) \quad (5.17)$$

No.	Solvent	λ_{max} (mμ)	mϵ
82	DMSO	360	8.95
		377	8.96
		400	5.15
83	Chloroform	360	17.4
		379	20.2
		400	12.7

When a molecule has a strong electron-acceptor group attached to the methylene but no strong proton-acceptor group, it will ionize readily, but any tautomer that is formed will have a short lifetime, as in the following example (183):

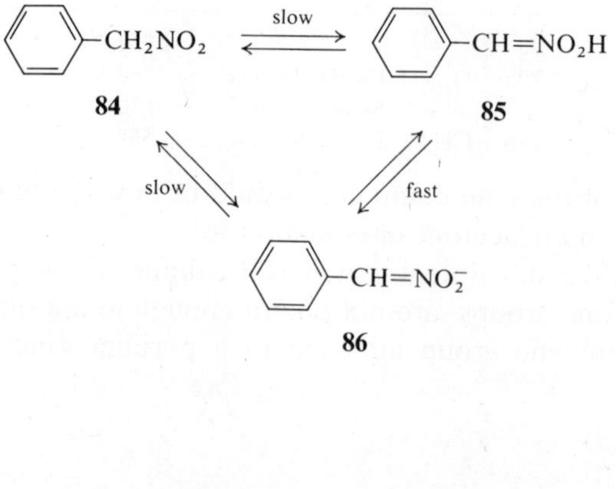

$$\underset{\mathbf{84}}{C_6H_5\text{—}CH_2NO_2} \overset{\text{slow}}{\rightleftharpoons} \underset{\mathbf{85}}{C_6H_5\text{—}CH{=}NO_2H}; \quad \mathbf{84} \overset{\text{slow}}{\rightleftharpoons} \underset{\mathbf{86}}{C_6H_5\text{—}CH{=}NO_2^-} \overset{\text{fast}}{\rightleftharpoons} \mathbf{85} \quad (5.18)$$

No.	λ_{max} (mμ)	mϵ
84	250*s*	0.5
	280*s*	0.2
85	283	~26
86	295	~20

The various spectra were obtained in 10% ethanol. The transient spectrum of the *aci* form of phenylnitromethane was obtained by extrapolation.

D. Dipyridylmethane-Type Compound

Di-2-pyridylmethane exists in solution as such, its absorption spectrum closely resembling that of 2-picoline(184). On the other hand, di-2-thiazolinylmethane exists in methanolic solution as the imino tautomer **87**, as shown by comparison of its spectrum with that of 2-methylthiazoline (**88**)(28).

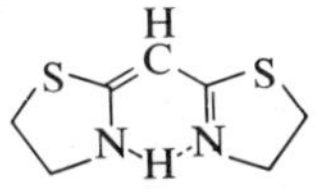

87

λ_{max} 250, 320, 330, 360 mμ
mϵ 33, 21, 16, 6

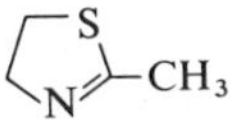

88

λ_{max} 230, 243 *s*, 266 mμ
mϵ 5.2, 4.0, 2.0

With an increase in CH_2 acidity the proportion of imine tautomer increases; for example, di-2-pyridylmethane < 2-pyridyl-2-quinolylmethane < di-2-quinolylmethane < di-2-quinolylcyanomethane(170, 185).

The spectra obtained for the tautomers of di-2-quinolylmethane are shown in Figure 5.6. The colorless tautomer is obviously the methylene form.

Intramolecular hydrogen bonding is not needed for tautomerism to the imine form to take place. In 2-quinolyl-4-quinolylmethane the acidic hydrogen is attached to the 4-substituted quinoline nitrogen atom since this atom is more basic than the 2-substituted nitrogen atom(186). However, this type of bonding does tend to increase the stability of the chelated imino form. In di-6-phenanthridylmethane there is only one non-sterically hindered configuration for the methane form (**89**) and one for the imino form (**90**)(187).

In any other disposition of the rings the molecules are sterically hindered. Intermolecular hydrogen bonds are difficult to form with these molecules.

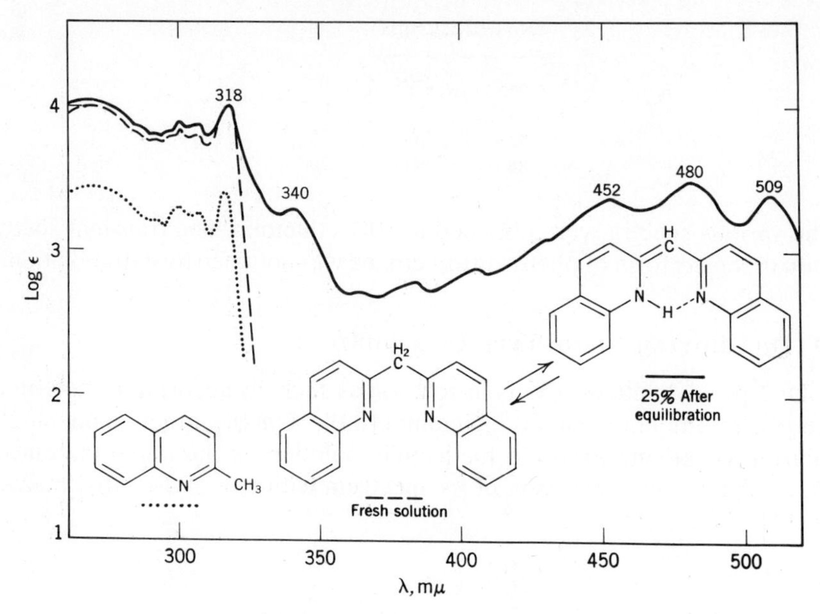

Fig. 5.6. Ultraviolet–visible absorption spectra in dimethylformamide of 2-methylquinoline (· · · ·) and of di-2-quinolinylmethane, fresh solution (– – –) and after equilibration (——)(185).

89 ⇌ 90 (5.19)

For maximum resonance the imine is forced into the configuration of **90**. The colorless form (**89**) could not be isolated in the solid state. Because of the strong hydrogen bond in **90**, the molecule could not form a salt with acid. It has a much greater proportion of the red imino tautomer than any of the other analogous derivatives studied.

The importance of minimal steric hindrance in the red imino form is shown by the inability of tri-6-phenanthridylmethane to form the highly hindered imine(187). In tri-2-quinolylmethane there is much less steric hindrance in both forms so that the absorption spectrum of this compound in ethanol at −180°C or in benzene at 20°C shows the presence of both tautomers(188).

Some of the compounds that have shown this type of tautomerism are di-2-quinolylmethane(188), tri-2-quinolylmethane, 6-phenanthridyl-2-quinolylmethane(187), 2-quinolyl-4-quinolylmethane, di-6-phenanthridyl-methane(187), di-(2-benzo[*f*]quinolyl)methane(189), tri-(2-benzo[*f*]-quinolyl)methane(189), and di-2-quinolylcyanomethane.

The tautomeric equilibrium of these aza compounds is displaced toward the red imine form with a decrease in solvent polarity(189). The spectrally determined results for di- and tri-2-quinolylmethane are listed in Table 5.15. Other factors besides solvent polarity affect the tautomeric equilibrium (Table 5.16).

TABLE 5.15
Percentage of the Red Imino Form of Di-2-quinolylmethane (*A*) and Tri-2-quinolylmethane (*B*) at 20°C in Various Solvents (189)

	Red form (%)	
Solvent	*A*	*B*
Ethanol	0.6	21.9
Methanol	0.7	16.6
n-Propanol	1.0	26.3
Isopropanol	1.6	
Benzene	7.8	21.4
Pyridine	9.1	21.9
Dioxane	9.6	
Carbon tetrachloride	10.0	
Dimethylformamide	13.4	
Heptane	14.8	
Carbon disulfide	19.0	52.5

TABLE 5.16
Percentage of the Red Imino Form of Some Heterocyclic Methanes at 20°C in Dimethylformamide and Ethanol (189)

	Red form (%)	
Compound	Ethanol	Dimethylformamide
Di-(2-benzo[*f*]quinolyl)methane	–	0.9
Tri-(2-benzo[*f*]quinolyl)methane	–	1.4
Di-2-quinolylmethane	0.6	13.4
6-Phenanthridyl-2-quinolylmethane	7.7	6.7
Tri-2-quinolylmethane	21.9	–
Di-6-phenanthridylmethane	–	50.2

V. Aromatic Hydrocarbon Methene Tautomers

1,2-Benzocyclohepta-1,3,5-triene (**91**) and 1,2-benzocyclohepta-1,3,6-triene (**92**) have been prepared in pure form (190). A possible tautomeric equilibrium has not been studied.

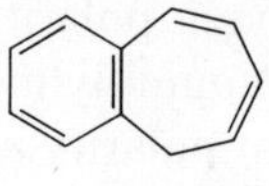

91

λ_{max} 275 mμ
mϵ 7.4

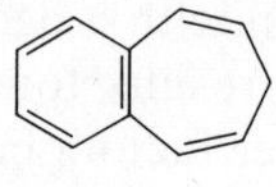

92

λ_{max} 258*s*, 228 mμ
mϵ 5.0, 44.7

Benzanthrene has been isolated in two tautomeric forms, namely, 1*H*-benzanthrene (**93**) and 7*H*-benzanthrene (**95**) (191). Compound **93** has been converted into **95**, but not vice versa. 7*H*-Benzanthrene is more fully aromatic and more energetically stable. Both compounds give the same spectrum in sulfuric acid, but only **95** is obtained on dilution with water.

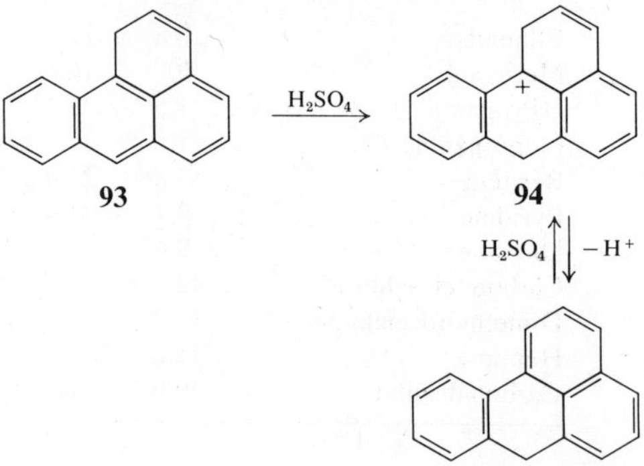

93 $\xrightarrow{H_2SO_4}$ **94** ; **94** $\underset{H_2SO_4}{\overset{-H^+}{\rightleftarrows}}$ **95** (5.20)

93

λ_{max}(mμ)	mϵ
233	23.4
251	87.1
324	6.5
337	9.3
354	7.9
374	3.0
389	0.44

94

λ_{max}(mμ)	mϵ
216	~25
255	47
405	11
475	6.2
580	~4

95

λ_{max}(mμ)	mϵ
222	44.7
229	46.8
247.5	21.4
329	17.4
344	14.5

As shown above, the absorption spectra (in ethanol) of the two tautomers are dissimilar. Compound **93** would spectrally resemble 1-vinylanthracene, whereas **95** would resemble 1-phenylnaphthalene.

The tendency of an aromatic molecule that contains many high-energy *o*-quinonoid rings to rearrange to a more stable configuration by tautomerizing a hydrogen atom from a methyl group has been demonstrated in 6-methylpentacene (**96**)(192).

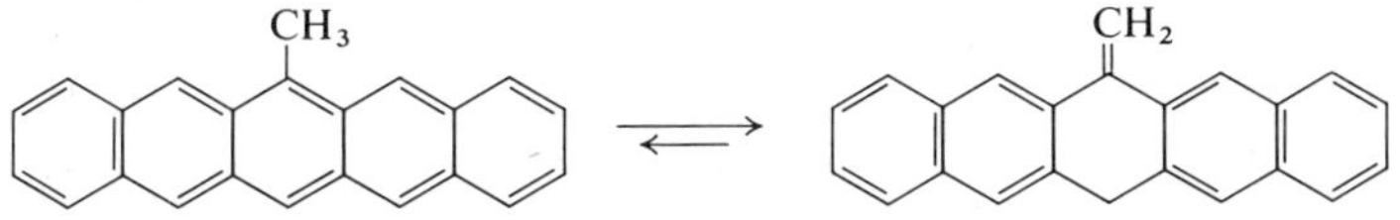

CH_3 CH_2

96 **97** (5.21)

Unstable **Stable**

λ_{max} 585, 541, mμ, etc. λ_{max} 378, 359, mμ, etc.

On the other hand, 5-methyltetracene, 9-methylanthracene, 1-methylnaphthalene, and toluene are present only as the methyl tautomers.

The tautomerism of the dihydroacenes has been discussed(193). The blue 5,18-dihydroheptacene (**98**) is isomerized to the more energetically stable 6,17-dihydropentacene (**99**) and 7,16-dihydropentacene (**100**) by sublimation or in boiling nitrobenzene, but the reaction is not reversible (194). The spectra of these compounds were obtained in trichlorobenzene. Because of the insulation properties of the CH_2 groups the absorption spectra of **98** and pentacene are closely similar, as are the spectra of **99** and tetracene, and of **100** and anthracene.

Indene derivatives also show tautomeric properties. The tautomers 4- and 7-indenol are stable at room temperature, but above 200°C either isomer will form a mixture of the two(195). Treatment of either of the isomers with triethylamine in pyridine at room temperature for 2.5 hr gave an equilibrium mixture of the two isomers.

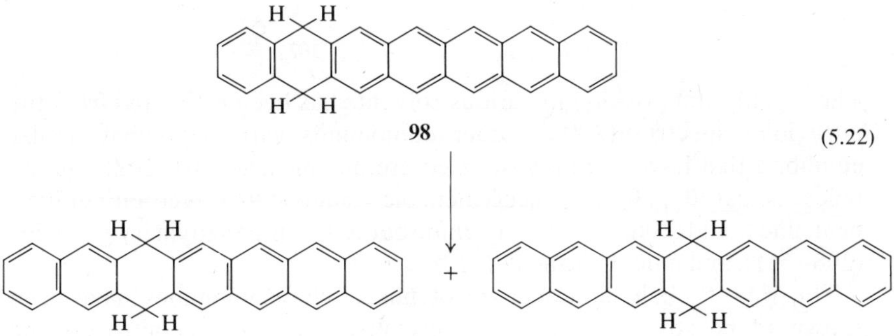

H H H H

98 (5.22)

H H H H

H H H H

99 **100**

No.	$\lambda_{max}(m\mu)$	$m\epsilon$
98	478	1.9
	501	3.0
	543	10.0
	587	12.0
99	409	1.6
	424	2.1
	452	3.2
	482	3.5
100	330	5.2
	347	6.0
	364	6.6
	385	5.2

Neutralization of a red alkaline solution of the anion of 2,5-diacetylindene gave 2,6-diacetylindene (196).

VI. Ring ⇄ Chain Tautomerism

The organic chemistry of ring ⇄ chain tautomerism has been reviewed by Jones (197). Since for many of these compounds the tautomeric change in structure caused by a new substituent or a change in solvent or temperature causes a drastic change in absorption spectra, the analyst has to be aware of these properties so as to understand and make maximal use of this phenomenon whenever he is investigating this type of molecule.

An example is the tautomeric equilibrium of phthalaldehydic acid (198, 199):

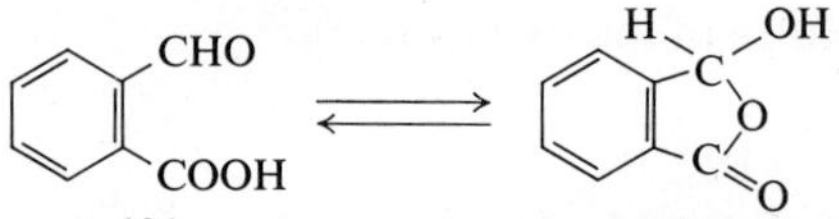

CHO COOH **101** ⇄ H OH C O C O **102** (5.23)

The equilibrium position in various solvents has been estimated from the ultraviolet spectrum (200). Other compounds with somewhat similar equilibria that have been investigated are opianic acid (201, 202), phthalonic acids (203, 204), *o*-acetylbenzoic acid (199), lycorenine (205), penicillic acid (206), phthalyl chloride (207), β-benzoylacrylic acids (208–211), and mucohalic acids (212).

The ring ⇄ chain tautomerism of phenacylethanolamines have been shown to be strongly dependent on substituents on the nitrogen, the aromatic ring, and on the carbon atoms adjacent to the nitrogen (213–217).

The steric and electronic effects of various substituents on the tautomerism of α-(β-hydroxyethylamino) desoxybenzoin has been investigated spectrally (Table 5.17)(214). Substitution of an α-hydrogen of the β-hydroxyethyl group by a methyl, ethyl, phenyl, or benzyl group increases steric hindrance and shifts the equilibrium toward predominance of the cyclic tautomer. In the α-methyl compounds, where the equilibrium is evenly balanced, substitution of a *para*-chloro or *para*-methoxy in the benzoyl group shifted the equilibrium toward or away from the cyclic forms, respectively. These shifts are caused primarily by electronic effects on the activity of the carbonyl group since *para*- substitutions in the adjacent α-phenyl group were inconsequential.

In the sugars 2-deoxyhexoses have a much greater tendency to change from the cyclic forms to the open-chain forms than have ordinary aldohexoses(218).

TABLE 5.17

Long-Wavelength Maxima and Ring ⇄ Chain Equilibrium Constants of α-(β-Hydroxyethylamino) desoxybenzoin Hydrochloride 5×10^{-5} M in 95% Ethanol(213)

$$X-\phi-HC(-N^{+}H_2-CHR-CH_2-OH)-C(=O)-\phi-X' \rightleftarrows X-\phi-HC(-N^{+}H_2-CHR-CH_2-O-)-C(OH)-\phi-X'$$

Substituent					
R	*p*-X	*p*-X′	λ_{max} (mμ)	mϵ	~K_t[a]
CH_3	OCH_3	OCH_3	284	17.1	0.04
C_2H_5	H	OCH_3	285	17.3	0.05
C_2H_5	OCH_3	OCH_3	285	17.3	0.05
CH_3	H	H	245	8.46	0.11
C_2H_5	H	H	246	3.98	1.0
C_2H_5	OCH_3	H	245	4.02	1.0
C_6H_5	OCH_3	OCH_3	285	7.69	1.2
C_6H_5	H	H	246	2.30	5.7
$CH_2C_6H_5$	H	H	245	1.68	9.0
CH_3	Cl	Cl	263	1.86	~12.0
C_2H_5	Cl	Cl	261	1.70	19.0
$CH_2C_6H_5$	OCH_3	OCH_3	285	3.24	24.0
C_6H_5	Cl	Cl	260	1.43	~32.0
$CH_2C_6H_5$	Cl	Cl	261	1.31	~99.0

[a] K_t = [Ring]/[chain].

Solvents play an important role in these equilibria also. D-Glucose is present mainly as the ring form (**103**) in aqueous solution.

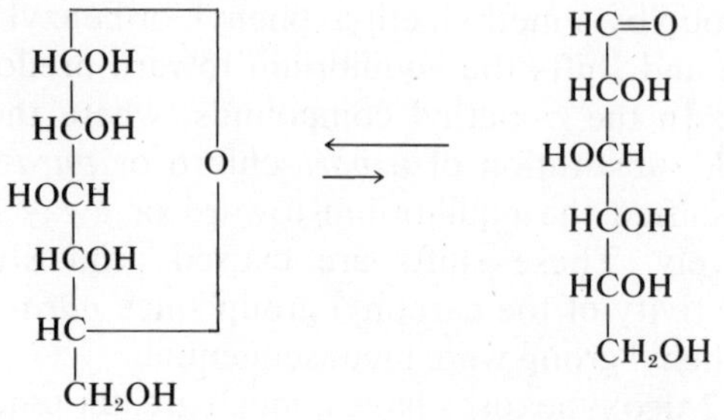

$$\text{HCOH–HCOH–HOCH–HCOH–HC(–O–ring)–}CH_2OH \;(\mathbf{103}) \rightleftharpoons HC{=}O\text{–HCOH–HOCH–HCOH–HCOH–}CH_2OH \;(\mathbf{104}) \quad (5.24)$$

Less than 0.01% of the open chain form (**104**) is present in neutral solution, but up to 0.1% of this form is present in strongly acid solution. It reverts gradually to the ring structure on neutralization(219).

2-(2-Hydroxyanilino)-1,4-naphthoquinone and some of its derivatives undergo striking changes in tautomeric equilibrium with change in solvent, as shown by their absorption spectra(220). The parent compound is present in ethanol primarily as the "chain" tautomer (**105**) and in 1 *N* hydrochloric acid as the ring form (**106**) with some chain tautomer, as shown by the shoulder at 480 mμ.

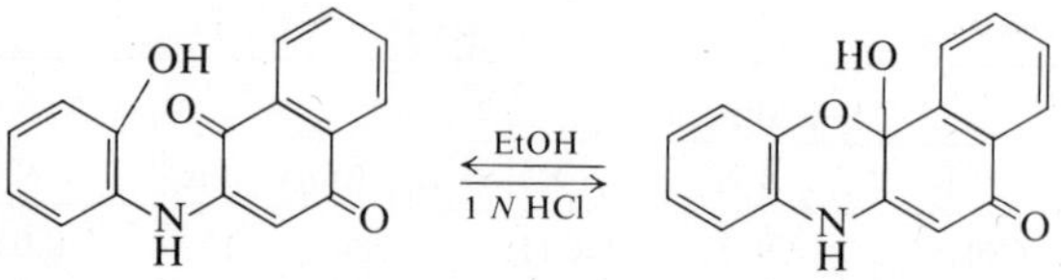

(5.25) **105** $\rightleftharpoons$ **106** (EtOH / 1 *N* HCl)

105: λ_{max} 280, 480 mμ; mε 22.4, 4.57

106: λ_{max} 260, 395, 480*s* mμ; mε 20.4, 7.4, 2.7

The spectrum of **105** is closely similar to that of 2-(2-methoxyanilino)-1,4-naphthoquinone in ethanol (λ_{max} 280, 480 mμ; mε 26.9, 5.75, respectively). The substitution in the aniline ring of a 5-methyl, a 5-chloro, or 3,5-dichloro groups has little effect on the equilibrium in ethanol or 1 *N* hydrochloric acid. However, substitution in the aniline ring of a 4-nitro, 6-acetyl, or 3,5-dimethyl groups shifts the equilibrium in either ethanol or 1 *N* hydrochloric acid predominantly toward the ring form analogous to **106**, as shown by the absence of the ~480-mμ band. Similarly substitution of a methyl group on the nitrogen, as in 2-(*N*-methyl-2-hydroxyanilino)-1,4-naphthoquinone, shifts the equilibrium to the ring form.

Solvent effects on azobenzene-2-sulfenyl derivatives have been studied with the help of equivalent conductivities, molecular-weight determinations and the electronic spectra(221). The results show that, with the exception of the nonionic cyanide, all azobenzene-2-sulfenyl derivatives exist in water and ethanol as true salts, and in benzene and chloroform as equilibria between ionic and nonionic tautomers (**107** and **108**). For example, azobenzene-2-sulfenyl iodide dissolves in water with a pale greenish-yellow color as the salt, and in chloroform and benzene with a violet-red and blue color, respectively(221). The increasing absorption at longer wavelengths (Fig. 5.7) is stated to be due to a displacement of the $n \rightarrow \pi^*$ band to longer wavelengths with the increased polarizability of the iodine atom and possibly to an independent contribution of the SI group.

107 (R⁻) $\underset{\longrightarrow}{\xleftarrow{\text{increased solvent polarity}}}$ **108** (5.26)

Changes in temperature can have a marked effect on ring ⇄ chain tautomerism. Some examples of this phenomenon have been shown in Chapter 3. Photochromism has been reported for over 25 compounds

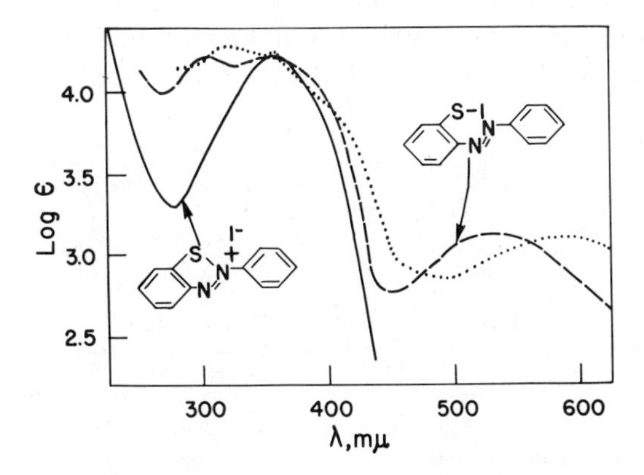

Fig. 5.7. Ultraviolet–visible absorption spectra of azobenzene-2-sulfenyl iodide in water (——), chloroform (– – –), and benzene (· · · ·)(221).

containing a 2*H*-chromene unit (222). The following reaction has been postulated:

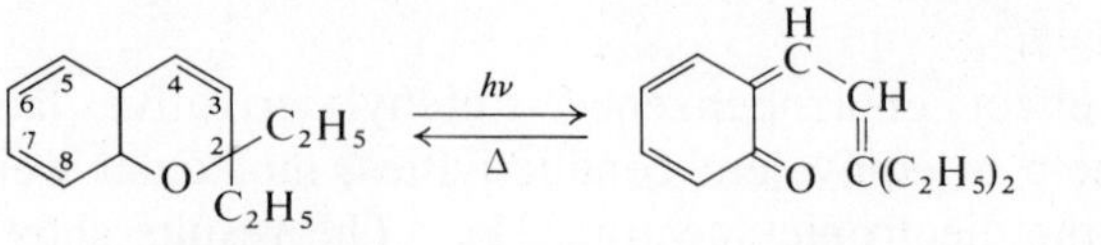

$$\text{109} \underset{\Delta}{\overset{h\nu}{\rightleftarrows}} \text{110} \quad (5.27)$$

109 $\lambda_{max} \sim 310$ mμ (major band), 320, 323, 330 mμ (0 → 0 band)

110 λ_{max} 400–600 mμ (broad band)

Benzo annulation in positions 5,6 or 7,8 increases the stability of the colored form; benzo annulation in position 6,7 decreases it. Conjugative substituents in position 2 increase the stability of the chain form, whereas bulky substituents in position 4 decrease it. The efficiency of chain formation seems to be much higher for excitation into vibrational components higher than the 0 → 0 band. The results suggest that this photochemical reaction may require at least the addition of some vibrational energy in the first excited singlet state.

An example of a pseudobase ring ⇄ chain prototropic system where both tautomers have been isolated is the following (223):

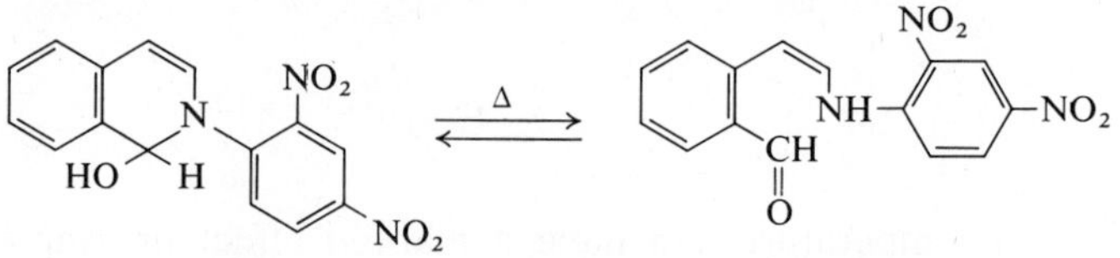

$$\text{111} \underset{}{\overset{\Delta}{\rightleftarrows}} \text{112} \quad (5.28)$$

111 Yellow

112 Violet-red

A remarkably facile equilibrium between 2,3-diphenylindenone oxide (**113**) and its valence tautomer, 1,3-diphenyl-2-benzopyrylium-4-oxide (**114**) has been reported (224). The absorption spectra in ethanol are given.

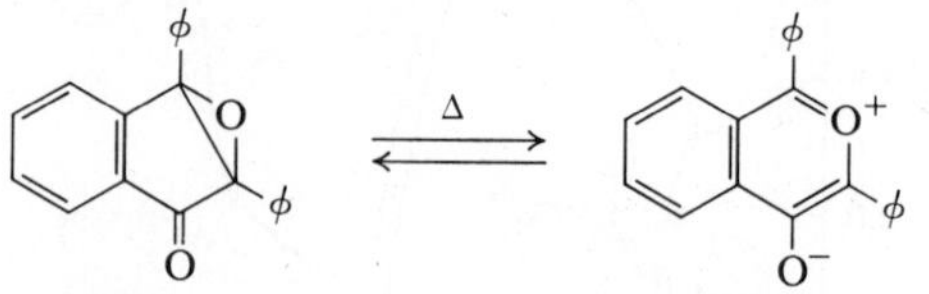

$$\text{113} \underset{}{\overset{\Delta}{\rightleftarrows}} \text{114} \quad (5.29)$$

113

λ_{max} (mμ)	mε
220	25.0
250*s*	10.0
300	0.72
365*fs*	0.1

114

λ_{max} (mμ)	mε
236	21.1
285	29.4
392	12.9
406	14.2
544	25.3

The red color is destroyed instantaneously with bicyclo[2.2.1]heptadiene or dimethyl acetylenedicarboxylate, reagents known to be highly reactive toward dipolar ions (225, 226). Ullman and Henderson (227) believe that both the forward and reverse reactions in equation 5.29 proceed through vibrationally excited ground states of the two tautomers. It is believed that the vibrationally excited ground-state molecules arise during crossing from electronically excited states and that the vibrational energy is not randomly distributed immediately after crossing but is concentrated at the reaction site.

The related triphenyl- and tetraphenyl-2,3-epoxycyclopent-4-en-1-ones (**115a** and **b**) undergo reversible photoisomerization to analogous dipolar species, the deep-red compounds **116a** and **b** (228, 229).

115 Colourless ⇄ **116** Red (5.30)

a: R = H
b: R = C_6H_5

Another example of a photochromic reaction is a reversible cyclohexadiene-hexatriene isomerization involving 1,2-dihydro-1,4-diphenyl-1-hydroxy-2,2,3-tricyano-9-xanthenone (**117**) (230, 231):

117 ⇄ **118** (hν or Δ / dark); **117** → **119** (polar solvent); **118** → **119** (OH⁻) (5.31)

No.	Solvent	λ_{max} (mμ)	mε
117	*n*-Hexane	312	14.5
118	*n*-Hexane	532	
119	Methanol	489	12.0

References

1. A. R. Katritzky and J. M. Lagowski, in A. R. Katritzky, Ed., *Advances in Heterocyclic Chemistry*, Vol. 1, Academic Press, New York, 1963, pp. 311–338.
2. S. F. Mason, in A. R. Katritzky, Ed., *Physical Methods in Heterocyclic Chemistry*, Vol. 2, Academic Press, New York, 1962.
3. S. F. Mason, *J. Chem. Soc.*, 5010 (1957).
4. A. Albert and E. Serjeant, *Ionization Constants of Acids and Bases*, Wiley, New York, 1962.
5. A. Albert, in A. R. Katritzky, Ed., *Physical Methods in Heterocyclic Chemistry*, Vol. 1, Academic Press, New York, 1962.
6. A. Albert, R. Goldacre, and J. Phillips, *J. Chem. Soc.*, 2240 (1948).
7. L. A. Flexser, L. P. Hammett, and A. Dingwall, *J. Am. Chem. Soc.*, **57**, 2103 (1935).
8. S. F. Mason, *J. Chem. Soc.*, 674 (1958).
9. A. R. Katritzky and A. P. Ambler, in A. R. Katritzky, Ed., *Physical Methods in Heterocyclic Chemistry*, Vol. 2., Academic Press, New York, 1962.
10. S. F. Mason, *J. Chem. Soc.*, 4874 (1957).
11. R. Gompper and H. Herlinger, *Ber.*, **89**, 2816 (1956).
12. W. Cochran, in A. R. Katritzky, Ed., *Physical Methods in Heterocyclic Chemistry*, Vol. 1, Academic Press, New York, 1962.
13. D. G. Leis and B. C. Curran, *J. Am. Chem. Soc.*, **67**, 79 (1945).
14. F. von Sturm and W. Hans, *Angew. Chem.*, **67**, 743 (1955).
15. H. H. Jaffé, *J. Am. Chem. Soc.*, **77**, 4445 (1955).
16. R. A. Hoffman and S. Gronowitz, *Arkiv Kemi*, **16**, 515, 563 (1961).
17. L. Knorr, *Ber.*, **44**, 1138 (1911).
18. K. von Auwers, *Ber.*, **63**, 2111 (1930).
19. A. R. Katritzky and J. M. Lagowski, in A. R. Katritzky, Ed., *Advances in Heterocyclic Chemistry*, Vols. 1 and 2, Academic Press, New York, 1963.
20. C. A. Grob and B. Fischer, *Helv. Chim. Acta*, **38**, 1704 (1955).
21. G. Oddo and A. Algerino, *Ber.*, **69**, 279 (1936).
22. E. A. Smirnov, *Zh. Obshch. Khim.*, **27**, 1922 (1957); through *Chem. Abstr.*, **52**, 2537 (1958).
23. V. A. Ismailskii, A. N. Guseva, and E. S. Solov'eva, *Zh. Obshch. Khim.*, **26**, 1766 (1956).
24. G. Klein and B. Prijs, *Helv. Chim. Acta*, **37**, 2057 (1954).
25. G. DeStevens, A. Frutchey, A. Halamandaris, and V. A. Luts, *J. Am. Chem. Soc.*, **79**, 5263 (1957).
26. A. J. Boulton and A. R. Katritzky, *Tetrahedron*, **12**, 41 (1961).
27. F. DeSarlo, *Tetrahedron*, **23**, 831 (1967).
28. R. Kuhn and F. Drawert, *Ann.*, **590**, 55 (1954).
29. R. Kuhn, G. Quadbeck, and E. Rohm, *Ber.*, **86**, 468 (1953).
30. A. Albert and J. N. Phillips, *J. Chem. Soc.*, 1294 (1956).
31. H. Specker and H. Gawrosch, *Ber.*, **75**, 1338 (1942).
32. A. Albert and E. Spinner, *J. Chem. Soc.*, 1221 (1960).
33. S. F. Mason, *J. Chem. Soc.*, 1253 (1959).
34. L. N. Jakhontov, D. M. Krasnokutskaya, E. M. Peresleni, J. N. Sheinker, and M. V. Rubtsov, *Tetrahedron*, **22**, 3233 (1966).
35. A. R. Katritzky, F. D. Popp, and J. D. Rowe, *J. Chem. Soc.*, 562 (1966).
36. H. J. Den Hertog and D. J. Buurman, *Rec. Trav. Chim.*, **75**, 257 (1956).
37. D. Metzler and E. Snell, *J. Am. Chem. Soc.*, **77**, 2431 (1955).

38. S. F. Mason, *J. Chem. Soc.*, 22 (1960).
39. J. D. Watson and F. H. C. Crick, *Cold Spring Harbor Symposia Quant. Biol.*, **18**, 123 (1953).
40. E. Freese, *J. Mol. Biol.*, **1**, 87 (1959).
41. M. P. V. Boarland and J. F. W. McOmie, *J. Chem. Soc.*, 3716 (1952).
42. D. J. Brown, E. Hoerger, and S. F. Mason, *J. Chem. Soc.*, 211 (1955).
43. D. J. Brown, and L. N. Short, *J. Chem. Soc.*, 331 (1953).
44. J. R. Marshall and J. Walker, *J. Chem. Soc.*, 1004 (1951).
45. O. Shugar and J. Fox, *Biochim. Biophys. Acta*, **9**, 199 (1952).
46. A. R. Katritzky and A. J. Waring, *J. Chem. Soc.*, 1540 (1962).
47. J. R. Loofbourow, M. M. Stimson, and M. J. Hart, *J. Am. Chem. Soc.*, **65**, 148 (1943).
48. D. J. Brown, *J. Chem. Soc.*, 3647 (1959).
49. A. Giner-Sorolla and A. Bendich, *J. Am. Chem. Soc.*, **80**, 5744 (1958).
50. J. J. Fox and D. Shugar, *Bull. Soc. Chim. Belges*, **61**, 44 (1952).
51. F. A. Kuehl, Jr., M. N. Bishop, L. Chaiet, and K. Folkers, *J. Am. Chem. Soc.*, **73**, 1770 (1951).
52. A. R. Katritzky, F. D. Popp, and A. J. Waring, *J. Chem. Soc.*, Ser. B, 565 (1966).
53. D. J. Brown and T. Teitei, *Australian J. Chem.*, **17**, 567 (1964).
54. G. M. Kheifets and N. V. Khromov-Borisov, *Zh. Obshch. Khim.*, **34**, 3134 (1964).
55. D. J. Brown and J. M. Lyall, *Australian J. Chem.*, **15**, 851 (1962).
56. G. W. Kenner, C. B. Reese, and A. R. Todd, *J. Chem. Soc.*, 855 (1955).
57. A. R. Katritzky and A. J. Waring, *J. Chem. Soc.*, 3046 (1963).
58. G. R. Wyatt, *Biochem. J.*, **48**, 581 (1951).
59. G. R. Wyatt and S. S. Cohen, *Biochem. J.*, **55**, 774 (1953).
60. C. S. Miller, *J. Am. Chem. Soc.*, **77**, 752 (1955).
61. G. B. Elion, W. S. Ide, and G. H. Hitchings, *J. Am. Chem. Soc.*, **68**, 2137 (1946).
62. D. Shugar and J. J. Fox, *Bull. Soc. Chim. Belges*, **61**, 293 (1952).
63. R. Andrisano and G. Modena, *Gazz. Chim. Ital.*, **81**, 405 (1951).
64. W. C. Schneider and I. F. Halverstadt, *J. Am. Chem. Soc.*, **70**, 2626 (1948).
65. G. S. Parry and F. Strachan, *Acta Cryst.*, **13**, 1035 (1960).
66. G. J. Pitt, *Acta Cryst.*, **1**, 168 (1948).
67. H. G. Mautner, *J. Am. Chem. Soc.*, **78**, 5292 (1956).
68. E. Wittenberg, *Ber.*, **99**, 2391 (1966).
69. M. M. Stimson, *J. Am. Chem. Soc.*, **71**, 1470 (1949).
70. O. Rosen and F. Sandberg, *Acta Chem. Scand.*, **4**, 666 (1950).
71. S. Ghose, G. A. Jeffrey, B. M. Craven, and J. O. Warwicker, *Angew. Chem.*, **72**, 754 (1960).
72. S. Ghose, G. A. Jeffrey, B. M. Craven, and J. O. Warwicker, *Acta Cryst.*, **13**, 1034 (1960).
73. Y. Sata, K. Kotera, T. Takabashi, and T. Meshi, *Yakugaku Zasshi*, **80**, 976 (1960).
74. J. W. Patterson, A. Lazarow, F. J. Lemm, and S. Levey, *J. Biol. Chem.*, **177**, 187 (1949).
75. A. Lazarow, J. W. Patterson, and S. Levey, *Science*, **108**, 308 (1948).
76. R. S. Tipson and L. H. Cretcher, *J. Am. Pharm. Assoc., Sci. Ed.*, **41**, 440 (1951).
77. J. Davoll and D. H. Laney, *J. Chem. Soc.*, 2124 (1956).
78. R. C. Hirt and R. H. Schmitt, *Spectrochim. Acta*, **12**, 127 (1958).
79. R. A. Morton and E. Rogers, *J. Chem. Soc.*, **127**, 2698 (1925).
80. H. Ley and H. Specker, *Ber.*, **72**, 192 (1939).
81. R. D. Brown and F. N. Lahey, *Australian J. Sci. Res.* **3**, 615 (1950).
82. G. F. Tucker, Jr., and J. L. Irvin, Jr., *J. Am. Chem. Soc.*, **73**, 1923 (1951).

83. M. Colonna, *Boll. Sci. Fac. Chim. Ind. Bologna*, **15**, 1 (1957).
84. J. M. Hearn, R. A. Morton, and J. C. E. Simpson, *J. Chem. Soc.*, 3318 (1951).
85. J. K. Landquist, *J. Chem. Soc.*, 1038 (1951).
86. D. J. Brown and S. F. Mason, *J. Chem. Soc.*, 3443 (1956).
87. Q. Gernando and T. Thirunamachandran, *Anal. Chim. Acta*, **17**, 447 (1957).
88. F. Umland and H. Puchelt, *Anal. Chim. Acta*, **16**, 334 (1957).
89. J. P. Saxena, W. H. Stafford, and W. L. Stafford, *J. Chem. Soc.*, 1579 (1959).
90. R. F. C. Brown, C. K. Hughes, and E. Ritchie, *Australian J. Chem.*, **9**, 277 (1956).
91. G. W. Ewing and E. A. Steck, *J. Am. Chem. Soc.*, **68**, 2181, (1946).
92. E. J. Alford and K. Schofield, *J. Chem. Soc.*, 1811 (1953).
93. D. E. Ames, R. F. Chapman, and H. Z. Kucharska, *J. Chem. Soc.*, 5391 (1965).
94. A. R. Osborn and K. Schofield, *J. Chem. Soc.*, 4207 (1956).
95. E. J. Alford, H. Irving, H. S. Marsh, and K. Schofield, *J. Chem. Soc.*, 3009 (1952).
96. H. Culbertson, J. C. Decius, and B. E. Christensen, *J. Am. Chem. Soc.*, **74**, 4834 (1952).
97. S. F. Mason, *J. Chem. Soc.*, 4874 (1960).
98. G. W. H. Cheeseman, *J. Chem. Soc.*, 108 (1958).
99. J. W. Clark-Lewis, *J. Chem. Soc.*, 422 (1957).
100. A. R. Katritzky and A. J. Waring, *J. Chem. Soc.*, 1544 (1962).
101. J. A. Elvidge and A. P. Redman, *J. Chem. Soc.*, 1710 (1960).
102. D. A. Evans, G. F. Smith, and M. A. Wahid, *J. Chem. Soc.* Ser. B, 590 (1967).
103. J. D. White and D. S. Straus, *J. Org. Chem.*, **32**, 2689 (1967).
104. D. N. Bailey, D. M. Hercules, and T. D. Eck, *Anal. Chem.*, **39**, 877 (1967).
105. S. F. Mason, *J. Chem. Soc.*, 2071 (1954).
106. S. F. Mason, *The Chemistry and Biology of Purines*, Ciba Foundation Symposium, 1957, p. 60.
107. D. J. Brown and S. F. Mason, *J. Chem. Soc.*, 682 (1957).
108. W. Pfleiderer and G. Nuebel, *Ann.*, **647**, 155 (1961).
109. W. Pfleiderer, *Ann.*, **647**, 161 (1961).
110. A. G. Ogston, *J. Chem. Soc.*, 1376 (1935).
111. H. F. W. Taylor, *J. Chem. Soc.*, 765 (1948).
112. W. Pfleiderer, *Ann.*, **647**, 167 (1961).
113. E. Johnson, *Biochem. J.*, **51**, 133 (1952).
114. F. Bergmann and S. Dikstein, *J. Am. Chem. Soc.*, **77**, 691 (1955).
115. A. Albert, D. J. Brown, and G. Cheeseman, *J. Chem. Soc.*, 1620, 4219 (1952).
116. Y. Inoue and D. D. Perrin, *J. Chem. Soc.*, 2600 (1962).
117. E. Lippert and H. Prigge, *Z. Elektrochem.*, **64**, 662 (1960).
118. W. Pfleiderer, *Ber.*, **90**, 2582 (1957).
119. W. Pfleiderer, *Ber.*, **90**, 2604 (1957).
120. W. Pfleiderer, *Ber.*, **90**, 2588 (1957).
121. W. Pfleiderer, *Ber.*, **90**, 2631 (1957).
122. W. Pfleiderer, *Ber.*, **90**, 2617 (1957).
123. C. B. Reese, *J. Chem. Soc.*, 895 (1958).
124. R. M. Acheson, M. Burstall, C. Jefford, and B. Sansom, *J. Chem. Soc.*, 3742 (1954).
125. R. D. Brown and F. N. Lahey, *Australian J. Sci. Res.*, **A3**, 593 (1950).
126. A. Albert and L. Short, *J. Chem. Soc.*, 760 (1945).
127. H. H. Perkampus and T. Rossel, *Z. Elektrochem.*, **62**, 94 (1958).
128. A. Gureirch and Y. Sheinker, *Khim. Nauka Prom.*, **3**, 129 (1958); through *Chem. Abstr.*, **52**, 11566 (1958).
129. N. Campbell and A. G. Cairns-Smith, *J. Chem. Soc.*, 1191 (1961).
130. V. Zanker and A. Reichel, *Z. Elektrochem.*, **63**, 1133 (1959).

131. N. F. Kazarinova and I. Y. Postovskii, *Zh. Obshch. Khim.*, **27**, 3325 (1957); through *Chem. Abstr.*, **52**, 9126 (1958).
132. Y. N. Sheinker and I. Y. Postovskii, *Zh. Fiz. Khim.*, **32**, 394 (1958); through *Chem. Abstr.*, **52**, 18406*i* (1958).
133. G. M. Badger, R. S. Pearce, and R. Pettit, *J. Chem. Soc.* 3204 (1951).
134. Y. Fellion, *Ann. Chim.*, **12**, 426 (1957).
135. H. Musso and H. Kramer, *Ber.*, **91**, 2001 (1958).
136. H. Musso, *Ber.*, **91**, 349 (1958).
137. R. M. Acheson and C. W. Jefford, *J. Chem. Soc.*, 2676 (1956).
138. Y. S. Rozum, *Sbornik Statei Obshch. Khim., Akad. Nauk SSSR*, **1**, 600 (1953); through *Chem. Abstr.*, **49**, 1065 (1955).
139. M. Ionescu, A. R. Katritzky, and B. Ternai, *Tetrahedron*, **22**, 3227 (1966).
140. A. Albert and G. B. Barlin, in A. Albert, G. M. Badger, and C. W. Shoppee, Eds., *Current Trends in Heterocyclic Chemistry*, Academic Press, New York, 1958.
141. R. A. Jeffreys, *J. Chem. Soc.*, 2221 (1954).
142. A. K. Chibisov and Y. A. Pentin, *Zh. Obshch. Khim.*, **31**, 11, 359 (1961); through *Chem. Abstr.*, **55**, 20621 (1961).
143. T. Uemura and S. Miyakawa, *Bull. Chem. Soc. Japan*, **22**, 115 (1949).
144. P. S. Pel'kis, *Dokl. Akad. Nauk SSSR*, **88**, 999 (1953); through *Chem. Abstr.*, **48**, 9943 (1954).
145. P. S. Pel'kis and R. G. Dubenko, *Ukrain. Khim. Zh.*, **23**, 748 (1957); through *Chem. Abstr.*, **52**, 14551 (1958).
146. A. Albert and G. B. Barlin, *J. Chem. Soc.*, 2384 (1959).
147. P. D. Anderson and D. M. Hercules, *Anal. Chem.*, **38**, 1702 (1966).
148. H. S. Lee and H. Freiser, *J. Org. Chem.*, **25**, 1277 (1960).
149. A. Kawase and H. Freiser, *Anal. Chem.*, **39**, 22 (1967).
150. J. B. Conant and G. W. Wheland, *J. Am. Chem. Soc.*, **54**, 1212 (1932).
151. F. G. Bordwell and G. D. Cooper, *J. Am. Chem. Soc.*, **74**, 1058 (1952).
152. M. M. Kreevoy, E. T. Harper, R. E. Duvall, H. S. Wilgus, and L. T. Ditsch, *J. Am. Chem. Soc.*, **82**, 4899 (1960).
153. D. J. Brown, E. Hoerger, and S. F. Mason, *J. Chem. Soc.*, 4035 (1955).
154. C. L. Angyal and R. L. Werner, *J. Chem. Soc.*, 2911 (1952).
155. J. D. S. Goulden, *J. Chem. Soc.*, 2939 (1952).
156. L. N. Short, *J. Chem. Soc.*, 4584 (1952).
157. A. R. Osborne, K. Schofield, and L. N. Short, *J. Chem. Soc.*, 4191 (1956).
158. L. C. Anderson and N. V. Seeger, *J. Am. Chem. Soc.*, **71**, 340 (1949).
159. S. J. Angyal and C. L. Angyal, *J. Chem. Soc.*, 1461 (1952).
160. S. F. Mason, *J. Chem. Soc.*, 1281 (1959).
161. C. B. Reese, *J. Chem. Soc.*, 895 (1958).
162. E. Akerblom, *Acta Chem. Scand.*, **21**, 1437 (1967).
163. H. Najer, R. Giudicelli, J. Menin, and N. Voronine, *Bull. Soc. Chim. France*, 207 (1967).
164. J. Giudicelli, J. Menin, and H. Najer, *Bull. Soc. Chim. France*, 1099 (1968).
165. A. Etienne and B. Rutimeyer, *Bull. Soc. Chim. France*, 1588 (1956).
166. R. A. Jones and A. R. Katritzky, *J. Chem. Soc.*, 1317 (1959).
167. R. A. Jones and A. R. Katritzky, *J. Chem. Soc.*, 378 (1961).
168. Y. N. Sheinker, E. M. Peresleni, N. P. Zosimova, and Y. I. Pomerantsev, *Zh. Fiz. Khim.*, *33*, 2096 (1959).
169. T. A. Mastrukova, Y. N. Sheinker, I. K. Kuznetsova, E. M. Peresleni, T. B. Sakharova, and M. I. Kabachnik, *Tetrahedron*, **19**, 357 (1963).
170. H. H. Credner, H. J. Friedrich, and G. Scheibe, *Ber.*, **95**, 1881 (1962).

171. R. O'Connor, *J. Org. Chem.*, **26**, 4375 (1961).
172. A. J. Bellamy and R. D. Guthrie, *J. Chem. Soc.*, 2788 (1965).
173. A. V. Chernova, R. R. Shagidullin, and Y. P. Kitaev, *Bull. Acad. Sci. USSR*, 1470 (1964).
174. B. Eistert and F. Geiss, *Ber.*, **94**, 929 (1961).
175. V. Rousseau and H. G. Lindwall, *J. Am. Chem. Soc.*, **72**, 3047 (1950).
176. C. B. Reese, *J. Am. Chem. Soc.*, **84**, 3979 (1962).
177. A. E. Tschitschibabin, B. M. Kuindshi, and S. W. Benewolenskaja, *Ber.*, **58**, 1580 (1925).
178. R. Hardwick, H. S. Mosher, and P. Passailaigue, *Trans. Faraday Soc.*, **56**, 44 (1960).
179. G. Wettermark, *J. Am. Chem. Soc.*, **84**, 3658 (1962).
180. A. R. Katritzky and B. Ternai, *J. Chem. Soc.*, Ser. B, 631 (1966).
181. S. Golding, A. R. Katritzky, and H. Z. Kucharska, *J. Chem. Soc.*, 3090 (1965).
182. R. Mondelli and L. Merlini, *Tetrahedron*, **22**, 3253 (1966).
183. W. Kemula, *Roczniki Chemii*, **35**, 1169 (1961).
184. H. J. Friedrich, W. Gückel, and G. Scheibe, *Ber.*, **95**, 1378 (1962).
185. E. Daltrozzo, G. Hohlneicher, and G. Scheibe, *Ber. Bunsenges. Physik. Chem.*, **69**, 190 (1965).
186. G. Scheibe and H. J. Friedrich, *Ber.*, **94**, 1336 (1961).
187. H. J. Friedrich and G. Scheibe, *Ber. Bunsenges. Physik. Chem.*, **65**, 767 (1961).
188. G. Scheibe and W. Riess, *Ber.*, **92**, 2189 (1959).
189. H. Friedrich and G. Scheibe, *Z. Elektrochem.*, **65**, 851 (1961).
190. G. Wittig, H. Eggers, and P. Duffner, *Ann.*, **619**, 10 (1958).
191. H. Dannenberg and H. Kessler, *Ann.*, **606**, 184 (1957).
192. E. Clar, *Ber.*, **82**, 495 (1949).
193. E. Clar, *Polycyclic Hydrocarbons*, Academic Press, New York, 1964, P. 62.
194. E. Clar and C. Marschalk, *Bull. Soc. Chim. France*, **17**, 444 (1950).
195. S. Friedman, M. L. Kaufman, B. D. Blaustein, R. E. Dean, and I. Wender, *Tetrahedron*, **21**, 485 (1965).
196. G. Baddeley and W. Pickles, *J. Chem. Soc.*, 2855 (1957).
197. P. R. Jones, *Chem. Rev.*, **63**, 461 (1963).
198. M. Carmack, M. B. Moore, and M. E. Balis, *J. Am. Chem. Soc.*, **72**, 844 (1950).
199. O. H. Wheeler, *Can. J. Chem.*, **39**, 2603 (1961).
200. N. P. Buu-Hoi and C. K. Lin, *Compt. Rend.*, **209**, 221 (1939).
201. N. P. Buu-Hoi, *Compt. Rend.*, **212**, 242 (1941).
202. N. P. Buu-Hoi and P. Cagniant, *Compt. Rend*, **212**, 268 (1941).
203. N. P. Buu-Hoi and C. K. Lin, *Compt. Rend.*, **209**, 346 (1939).
204. N. P. Buu-Hoi and P. Cagniant, *Compt. Rend.*, **212**, 908 (1948).
205. T. Kitagawa, W. I. Taylor, S. Uyeo, and H. Yajima, *J. Chem. Soc.*, 1066 (1955).
206. E. Shaw, *J. Am. Chem. Soc.*, **68**, 2510 (1946).
207. J. Scheiber, *Ann.*, **389**, 121 (1912).
208. C. L. Browne and R. E. Lutz, *J. Org. Chem.*, **18**, 1638 (1953).
209. R. E. Lutz, P. S. Bailey, C. K. Dien, and J. W. Rinker, *J. Am. Chem. Soc.*, **75**, 5039 (1953).
210. R. E. Lutz, C. T. Clark, and J. P. Feifer, *J. Org. Chem.*, **25**, 346 (1960).
211. R. E. Lutz and H. Moncure, Jr., *J. Org. Chem.*, **26**, 746 (1961).
212. H. H. Wasserman and F. M. Precopio, *J. Am. Chem. Soc.*, **74**, 326 (1952).
213. N. H. Cromwell and K. C. Tsou, *J. Am. Chem. Soc.*, **71**, 993 (1949).
214. C. E. Griffin and R. E. Lutz, *J. Org. Chem.*, **21**, 1131 (1956).
215. R. E. Lutz and R. H. Jordan, *J. Am. Chem. Soc.*, **71**, 996 (1949).

216. R. E. Lutz and J. W. Baker, *J. Org, Chem.*, **21**, 49 (1956).
217. R. E. Lutz and C. E. Griffin, *J. Org. Chem.*, **25**, 928 (1960).
218. A. Gottschalk, Ed., *Glycoproteins*, Elsevier, New York, 1966, P. 107.
219. E. Pacsu and L. Hiller, *J. Am. Chem. Soc.*, **70**, 523 (1948).
220. A. Butenandt, E. Biekert, and W. Schäfer, *Ann.*, **632**, 143 (1960).
221. A. Burawoy, F. Leversedge, and C. Vellins, *J. Chem. Soc.*, 4481 (1954).
222. R. S. Becker and J. Michl, *J. Am. Chem. Soc.*, **88**, 5931 (1966).
223. T. Zincke, *Ann.*, **396**, 103 (1913).
224. E. F. Ullman and J. E. Milks, *J. Am. Chem. Soc.*, **84**, 1315 (1962); ibid., **86**, 3814 (1964).
225. R. Huisgen, R. Grashey, P. Lauer, and H. Leitermann, *Angew. Chem.*, **72**, 416 (1960).
226. R. Huisgen, H. Stangl, H. J. Sturm, and H. Wagenhofer, *Angew. Chem.*, **73**, 170 (1961).
227. E. F. Ullman and W. A. Henderson, Jr., *J. Am. Chem. Soc.*, **88**, 4942 (1966).
228. E. F. Ullman, *J. Am. Chem. Soc.*, **85**, 3529 (1963).
229. J. M. Dunston and P. Yates, *Tetrahedron Letters*, 505 (1964).
230. K. R. Huffman, M. Loy, W. A. Henderson, Jr., and E. F. Ullman, *J. Org. Chem.*, **33**, 3469 (1968).
231. K. R. Huffman, M. Loy, W. A. Henderson, Jr., and E. F. Ullman, *Tetrahedron Letters*, 931 (1967).

CHAPTER 6

ABSORPTION SPECTRA OF ZWITTERIONIC RESONANCE STRUCTURES

IV. SOME STRUCTURAL WAVELENGTH DETERMINANTS

Some of the structural factors that can affect the position of the wavelength maximum of an unsaturated compound are intramolecular hydrogen bonding, the electron-donor and electron-acceptor strengths of the resonance nodes and terminals, the length of the conjugation chain, the annulation effect of fused benzo rings, extraconjugation, cross-conjugation, competitive conjugation at a resonance terminal, and other coupling effects.

I. Intramolecular hydrogen bonding

Many of the important spectral aspects of intramolecular hydrogen bonding have been and will be discussed in appropriate sections of this treatise. Theoretical and other aspects of hydrogen bonding have been covered by experts(1). The intramolecular hydrogen bond can drastically affect physical properties – such as melting and boiling points, solubility in alkaline solutions and organic solvents, volatility in steam, partition coefficients, acid or base strength, infrared and ultraviolet absorption spectra, tautomerism, and hydrolysis(2) – all of value to an organic analyst.

As has been shown in preceding chapters, solvent effects and the intermolecular hydrogen bond can compete with and weaken the intramolecular hydrogen bonds. The importance of this type of bond to an analyst stems from its property of causing a violet shift of an $n \rightarrow \pi^*$ transition or of a $\pi \rightarrow \pi^*$ transition of the Z_b or Z_z type (Table 6.1) while causing a red shift in a $\pi \rightarrow \pi^*$ transition of the Z_n type. The violet shift of an $n \rightarrow \pi^*$ band and the red shift of a $\pi \rightarrow \pi^*$ band as caused by intramolecular hydrogen bonding can be seen in Table 6.1 for 2-nitrosophenol and 2-phenylazo-3-naphthol when compared with their methyl ethers.

Unlike the *p*-hydroxyphenylazo isomers, most *o*-hydroxyazo compounds are insoluble in alkali and cannot be methylated with diazo-

TABLE 6.1

The Effect of Intramolecular Hydrogen Bonding on the Wavelength Position of Some $\pi \rightarrow \pi^*$ Transitions of the Z_n Type

Compound	Solvent	λ_{max} (mμ) (mϵ)		Ref.
		OH	OCH_3	
Acetylacetone, enol	*n*-Heptane	270 (12.0)[a]	258 (13.5)	3
Salicylaldehyde	*n*-Hexane	328.5 (3.23)	310 (5.6)	5
2-Hydroxybenzophenone	95% EtOH	335 (5.0)	250[b] (10.0)	6
2-Hydroxy-3-naphthoic acid	Dioxane	366 (2.8)	341[c] (1.78)	7
Tropolone	Isooctane	370*fs*[d] (3.98)	340*fs*[d] (3.2)	8
1-Hydroxyanthraquinone	Methanol	402[e] (5.5)	378 (5.25)	9
N-[4-(Diethylamino)-*o*-tolyl]-8-hydroxy-1, 4-naphthoquinone imine	Cyclohexane	628[f]	570	10
2-Nitrosophenol	CCl_4	400[g] (2.6)	369[h] (4.85)	11
2-Nitrophenol	95% EtOH	343.5 (3.6)	317 (2.85)	4
4-Nitroresorcinol	95% EtOH	345 (14.0)	329.5[i] (9.5)	4
4-Methyl-2-phenylazophenol	*n*-Hexane	393.5[j] (9.8)	359 (9.0)	4
1-(4-Nitrophenylazo)-2-amino-8-naphthol	Benzene	547 (21.1)	496[k] (18.6)	12
1-Phenylazo-2-amino-8-naphthol[l]	Benzene	509 (24.8)	439[k] (13.6)	12
2-Phenylazo-3-naphthol	Methylcyclohexane–isohexane (1:1)	355,370*s*[m]	350[n]	13

[a]Presence of intramolecularly-hydrogen-bonded C=O group shown by presence of strong C=O ɔsorption at 5.85 μ and very strongly hydrogen-bonded C=O at 6.1–6.3 μ (4). The latter band was ɔsent in the methoxy derivative.

[b]Also shoulder at 300 mμ with mϵ = 1.6. Steric hindrance would be important here, especially in .e methoxy derivative. The long-wavelength $\pi \rightarrow \pi^*$ transition is found at 290 mμ in 4-hydroxy-ɛnzophenone and its methyl ether (6).

[c]Methyl ether.

[d]Fine structure.

[e]The 2-isomer absorbs at 368 mμ, whereas its methyl ether absorbs at 363 mμ.

[f]Compound without hydroxy groups absorbs at 582 mμ. A large number of derivatives with ıbstituents other than the 8-hydroxy group absorbed at shorter wavelengths.

[g]$n \rightarrow \pi^*$ Transition at 698 mμ, mϵ 0.065.

[h]$n \rightarrow \pi^*$ Transition at 777 mμ, mϵ 0.053.

[i]3-Methoxy derivative.

[j]4-Phenylazophenol absorbs at 336.5 mμ in hexane; its methyl ether absorbs at 338 mμ.

[k]Compound without hydroxy group.

[l]The absorption spectra of 40 phenyl-substituted derivatives and their deoxygenated analogs reported . benzene and methylcyclohexane. All showed red shift due to intramolecular hydrogen bond.

[m]Shoulder; $n \rightarrow \pi^*$ Band at 433 mμ.

[n]$n \rightarrow \pi^*$ Band at 450 mμ.

methane (14). They acylate with difficulty and react slowly with phenyl isocyanate. The hydroxy group in 1-hydroxy-8-phenylazonaphthalene dyes is also inactive. These dyes are insoluble in aqueous alkali, and their absorption spectra are insensitive to pH changes (12).

Consider compounds **1** and **2**, whose spectra are listed in Table 6.1.

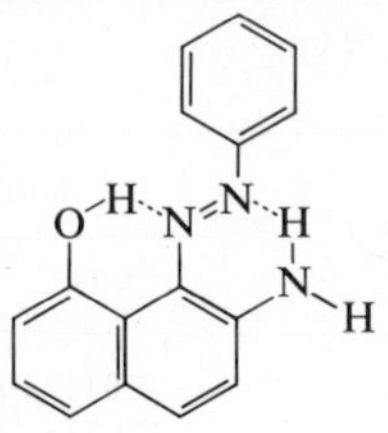

The hydrogen bonds in **1** lock the molecule into a configuration in which maximum coplanarity is achieved. According to Ross and Reissner(12), this serves to lower the energy of excitation of the molecule and to increase the transition probability as compared with **2**, an effect reflected in absorbance at longer wavelengths and with greater intensity by the 8-hydroxy system as compared with the analogous 1-arylazo-2-naphthylamines. The authors note that the average increase in wavelength of 58 mμ on going from the compounds of type **2** to compounds of type **1** corresponds to a net decrease in the transition energy of approximately 7 kcal/mole, a representative value to be expected for a hydrogen bond.

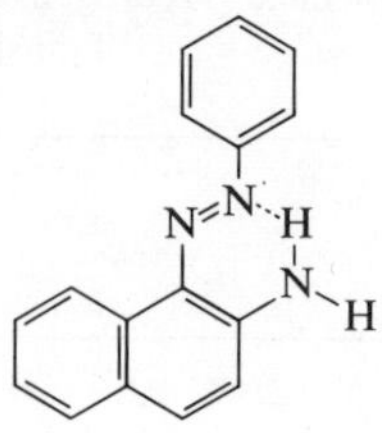

II. Resonance Nodes

Colorimetric analysis is based mainly on the formation of a chromogen whose color and intensity are derived from a $\pi \rightarrow \pi^*$ transition involving a polymethine chain with two atoms of elements from Groups IV, V, or VI at its extremities. Usually the two extremities, or resonance terminals, are carbon, nitrogen, oxygen, or sulfur atoms. This particular electronic structure leads to equalization of the bond lengths in the polymethine chain with simultaneous alternating π-electron distribution. The alternating charge distribution follows from quantum-mechanical calculations, either by the free-electron (FE) method(14) or the linear-combination-of-atomic-orbitals molecular-orbital (LCAO-MO) method (15, 16) and has been experimentally confirmed by proton-magnetic-resonance measurements(17, 18).

The alternation of π-electron density occurs irrespective of whether or not the molecule contains a zwitterionic, cationic, or anionic resonance structure.

In this section the discussion is centered on the effect of substituents in positions of high or low electron density in the polymethine chain on the position of the long-wavelength band. The Forster–Knott rule has been formulated to explain these wavelength shifts(19).

According to Forster(20), Lewis(21), and Lewis and Calvin(22), as amplified and modified by Knott(19), the wavelength maximum of a compound will increase as the contributions by any of the intermediate ionic excited structures decrease. Since contributions by the various structures depend largely on their relative energies, the long-wavelength maximum can be shifted to the red by raising the energy of any one intermediate excited ionic structure.

In any structure of this type all positions (except the resonance terminals) that will accept a positive charge are called positive resonance nodes; positions that will accept a negative charge are called negative resonance nodes, as shown in the following example:

$$(CH_3)_2N{-}\overset{1}{C}H{=}\overset{2}{C}H{-}\overset{3}{C}H{=}\overset{4}{C}H{-}NO_2 \leftrightarrow (CH_3)_2\overset{+}{N}{=}CH{-}CH{=}CH{-}CH{=}\overset{-}{N}O_2$$

$$\updownarrow \qquad\qquad\qquad\qquad\qquad\qquad \updownarrow$$

$$(CH_3)_2\overset{+}{N}{=}CH{-}CH{=}CH{-}\overset{-}{C}H{-}NO_2 \leftrightarrow (CH_3)_2N{-}\overset{+}{C}H{-}CH{=}CH{-}CH{=}\overset{-}{N}O_2$$

$$\updownarrow \qquad\qquad\qquad\qquad\qquad\qquad \updownarrow$$

$$(CH_3)_2\overset{+}{N}{=}CH{-}\overset{-}{C}H{-}CH{=}CH{-}NO_2 \leftrightarrow (CH_3)_2N{-}CH{=}CH{-}\overset{+}{C}H{-}CH{=}\overset{-}{N}O_2$$

The 1- and 3-positions are positive nodes; the 2- and 4-positions are negative nodes. The dimethylamino nitrogen and the nitro oxygen are the positive and negative resonance terminals, respectively. The top two structures are the limiting resonance structures; the remaining formulas represent intermediate resonance structures.

A. Positivizing a Positive Node

We shall discuss first the effects of substitution on a positive CH node or replacement of the CH group by a nitrogen atom or some other group. Positivizing the positive node by means of appropriate substituents or replacements at the node that will cause this node to more readily accept the positive charge in the excited state will shift the long-wavelength band to the violet. It is possible for the electron-donor strength of a positive node to be increased sufficiently for it to compete with the positive resonance terminal as such a terminal. The system then acts as if its chain of conjugation had been shortened, and the compound absorbs at very much shorter wavelengths, with considerably decreased intensity.

In the zwitterionic resonance structures of the Z_n type an increase in the number of methine groups increases the amount of energy necessary for the separation of charge involved in the excited state. An increase in methine groups would make more positive nodes available and thus more lower energy intermediate excited-state structures. These phenomena would cause wavelength-maxima convergence with increasing chain length; that is, the red shift in wavelength maxima would become more gradual. Lewis and Calvin(22) gave an equation of the form $\lambda_{max} = kn^{1/2}$ to denote the shift of the wavelength maximum with n, the number of CH =CH groups in the conjugation chain of compounds of this type. Examples of other compounds of this type that show this phenomenon, as well as compounds of the cationic or anionic resonance nonconvergent type, whose long-wavelength maxima obey the equation $\lambda_{max} = kn$, have been given (23).

B. Negativizing a Positive Node

The data in Table 6.2 show the effect of negativizing a positive node on the long-wavelength $\pi \rightarrow \pi^*$ transition. Replacement of a =CH— of a positive node with a more electronegative =N— causes the expected red shift. Some more examples of the red-shift effect of negativizing a positive node are shown in the comparison of the long-wavelength bands of compounds **3a**(36), **3b**(36), and **3c**(37), of which one of the intermediate resonance structures is shown.

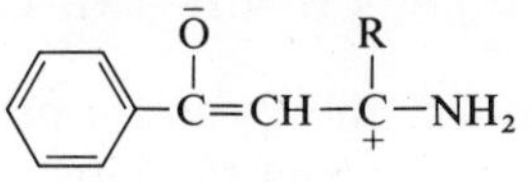

$C_6H_5-C(\bar{O})=CH-C^+(R)-NH_2$ **3**

$R-(furan^+)=CH-CH=C(\bar{O})-CH_3$ 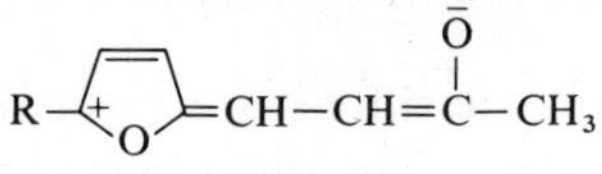 **4**

No.	R	λ_{max} (mμ)	mϵ
3a	H	324	22.0
3b	$COOCH_3$	352	17.0
3c	COC_6H_5	360	11.4
4a	H	316	22.9
4b	NO_2	346	18.6

A similar example(38) is seen in the furan derivatives **4a** and **b**, for which one of the intermediate resonance structures is shown.

A somewhat more complicated example of this phenomenon can be found in comparing the spectra in chloroform of tetrahydro-α, γ-dioxo-octaethylporphinogen (**5a**) and octaethylxanthoporphinogen (**5b**)(39). One of the intermediate resonance structures is shown.

H_5C_2 O^- C_2H_5 H_5C_2 C_2H_5 N N H H X X H H N N H_5C_2 C_2H_5 H_5C_2 O C_2H_5

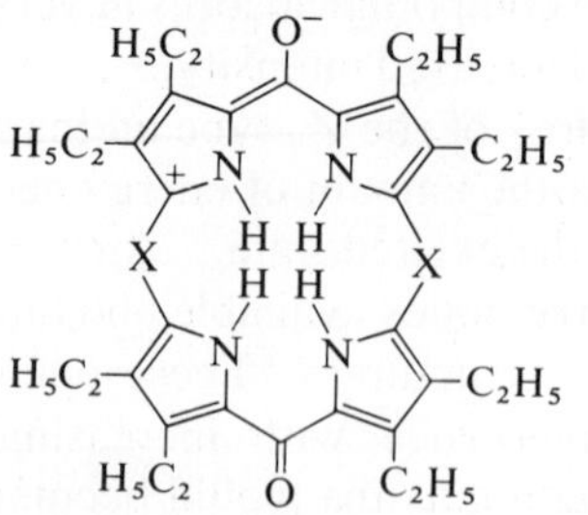

5

	X	λ_{max} (mμ)	mϵ
a	CH_2	326	36.4
b	C=O	340	50.1

Intermediate resonance structure	X = CH				X = N			
	Solvent	λ_{max} (mμ)	(mϵ)	Ref.	Solvent	λ_{max} (mμ)	(mϵ)	Ref.
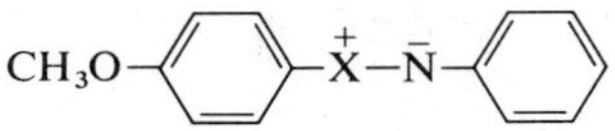 $CH_3O-C_6H_4-\overset{+}{X}-\overset{-}{N}-C_6H_5$	95% EtOH	314	(16.6)	24	95% EtOH	343	(36.3)	25
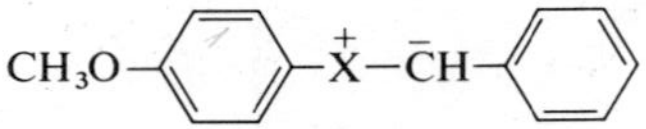 $CH_3O-C_6H_4-\overset{+}{X}-\overset{-}{C}H-C_6H_5$	95% EtOH	318[a]	(31.6)	26	Methanol	331	(15.1)	27
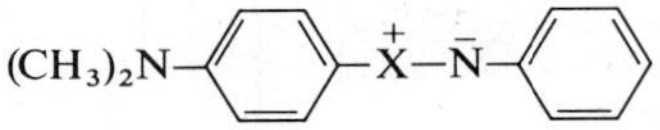 $(CH_3)_2N-C_6H_4-\overset{+}{X}-\overset{-}{N}-C_6H_5$	95% EtOH	334	(34.7)	28	95% EtOH	409	(31.6)	29
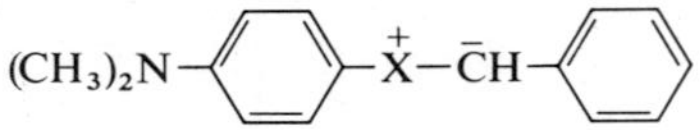 $(CH_3)_2N-C_6H_4-\overset{+}{X}-\overset{-}{C}H-C_6H_5$	95% EtOH	346	(30.8)	30	95% EtOH	376	(6.3)	31
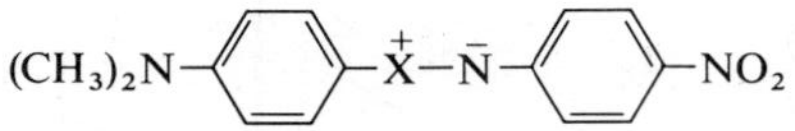 $(CH_3)_2N-C_6H_4-\overset{+}{X}-\overset{-}{N}-C_6H_4-NO_2$	Ethylene chloride	405	(25.0)	31	95% EtOH	470	(32.0)	32
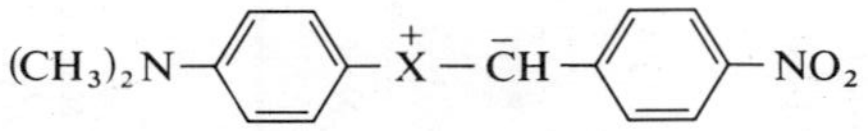 $(CH_3)_2N-C_6H_4-\overset{+}{X}-\overset{-}{C}H-C_6H_4-NO_2$	95% EtOH	430	(30.9)	33	95% EtOH	445	(12.0)	28
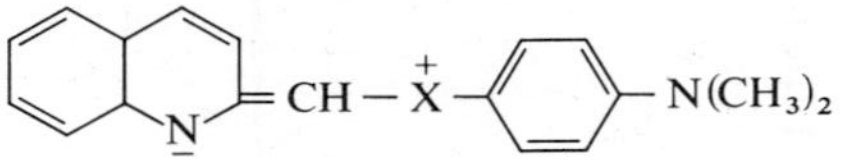 (quinolinide, $\underset{-}{N}$)$=CH-\overset{+}{X}-C_6H_4-N(CH_3)_2$	Methanol	396	(40.2)	34	Methanol	418		35
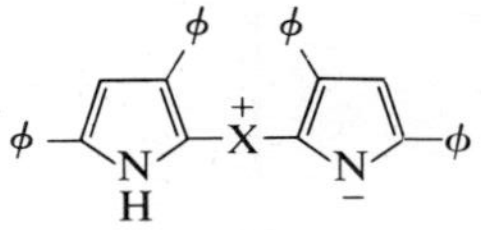 ϕ, ϕ-pyrrole (NH)$-\overset{+}{X}-$pyrrole ($\underset{-}{N}$) with ϕ substituents	Pyridine	550		19	95% EtOH	614		19
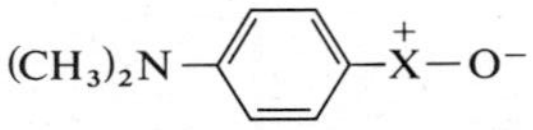 $(CH_3)_2N-C_6H_4-\overset{+}{X}-O^-$	95% EtOH	342		19	95% EtOH	423		19

[a]Fine structure.

The enhanced resonance effect of the tricyanovinyl group over the dicyanovinyl group can be explained through the resonance-node effect. The effect of these groups as compared with that of the nitro group is shown in the following compounds of the general formula 4-X—C_6H_4—$N(CH_3)_2$, whose long-wavelength maxima increase from 390 mμ (mε 19.1) for X = NO_2 in ethanol (40) to 442 and 535 mμ for X = —CH=$C(CN)_2$ and —C(CN)=$C(CN)_2$, respectively, both in dioxane–water (45:55)(41). Thus in 4-$(CH_3)_2N$—C_6H_4—CX—$C(CN)_2$ substitution of X = CN for X = H negativizes the positive node and results in a strong red shift.

The effect of negativizing a positive node by means of a benzoyl group can be seen in the comparison of the position of the long-wavelength band of 2-(*N*-methyl-*p*-anisidino)-1,4-naphthoquinone (λ_{max} 278, 476 mμ; mε 24.6, 4.7, respectively, in ethanol) and its irradiated product, 7-methyl-10-methoxy-12*a*-hydroxy-5-benzo[*c*]phenoxazone (λ_{max} 256, 398 mμ; mε 22.9, 18.2, respectively, in ethanol) (42). In the naphthoquinone dye coupling of the benzoyl and O=$\dot{C}$—CH=C—$\dot{N}$— resonance systems probably also plays a role in the red shift of the long-wavelength maximum.

C. The Negative Node

Negativizing the negative node will result in a violet shift unless other factors intervene. If X is considered a node in the resonance structures **6**, negativizing it should cause a violet shift. This does not occur, and hence is believed that X = N may play an important role as a resonance terminal since in the structure where the NO_2 oxygen is the resonance terminal the excited state contains two high-energy *p*-benzoquinonoid rings. Because of this the nitrophenyl group has a mainly extraconjugative effect. This type of phenomenon is discussed in Section VII.

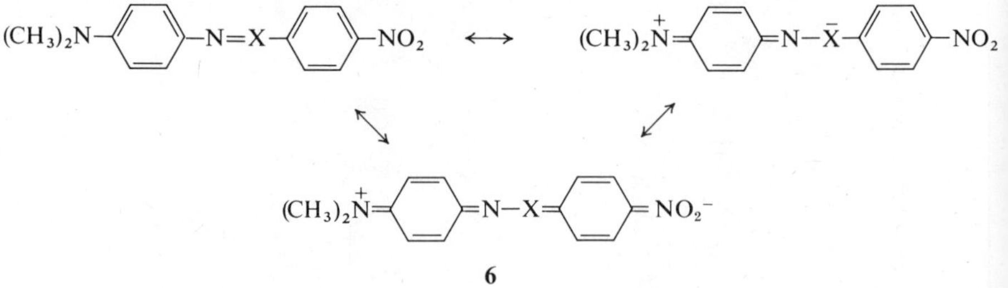

$(CH_3)_2N$—C_6H_4—N=X—C_6H_4—NO_2 ⟷ $(CH_3)_2\overset{+}{N}$=C_6H_4=N—$\bar{X}$—C_6H_4—NO_2

$(CH_3)_2\overset{+}{N}$=C_6H_4=N—X=C_6H_4=NO_2^-

6

X	λ_{max} (mμ)	mε	Ref.
CH	445	12	28
N	470	32	32

In the intermediate resonance structure **7** replacement of X = CH by N gives a lower energy intermediate structure, because the nitrogen accepts the negative charge more readily. Since the contribution of this intermediate ionic excited structure to the hybrid structure is greater than that where X = CH or CCH_3, a violet shift results. When this node is positivized by the substitution of a methyl group, a red shift results. However, the steric strain due to the CH and CCH_3 groups must also be an important factor here.

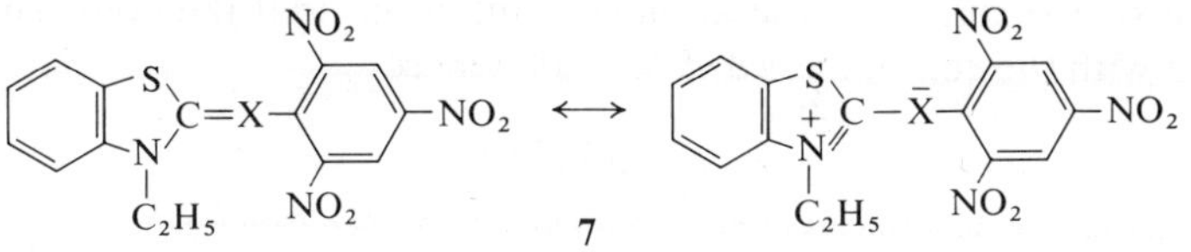

NO_2 NO_2 NO_2 C_2H_5 ⟷ NO_2 NO_2 NO_2 C_2H_5

7

X	λ_{max} (mμ) (43)
N	430
CH	552
CCH_3	605

In Table 6.3 the expected spectral shifts on negativizing or positivizing a negative node are demonstrated in several solvents.

TABLE 6.3
Effect of Resonance-Node Changes on the Long-Wavelength Maximum (44)

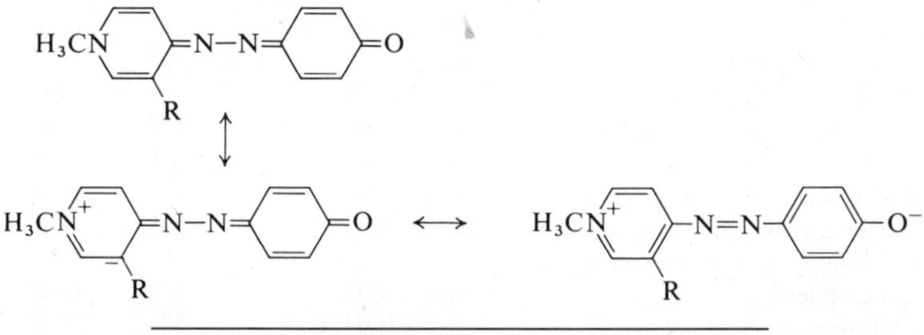

H_3CN =N—N= =O, R ↕ H_3CN^+ =N—N= =O, R ⟷ H_3CN^+ —N=N— —O^-, R

R	Benzyl alcohol λ_{max} (mμ)	(mε)	Acetone λ_{max} (mμ)	(mε)	Chloroform λ_{max} (mμ)	(mε)
NH_2	583	(57.8)	573	(77.5)	570	(66.0)
OCH_3	572	(63.5)	555	(62.5)	537	(49.5)
H	555	(74.0)	530	(67.7)	520	(57.0)
NO_2	520	(30.9)	485	(35.9)	460	(33.7)

Many more compounds with cationic or anionic resonance structures that show these nodal effects are discussed in other chapters.

III. Ortho–Para Effect

Wizinger (45) investigated the *ortho–para* spectral effect with zwitterionic, cationic, and anionic resonance structures. Some of his results are given in Table 6.4 (45–83). (Kauffmann (84, 85) investigated this effect between the years 1906 and 1920 and called it the law of the distribution of auxochromes). Essentially these authors pointed out that the color shifted to the red with the compounds **8** < **9** < **10**, where

TABLE 6.4

Spectral Effect of Substituents *Ortho* or *Para* to a Resonance Terminal on a Benzene Ring

	Compound[a]	Solvent	λ_{max} (mμ)	mϵ	Ref.
1.	Salicylaldehyde	95% EtOH	325	3.0	46
	3-Methoxysalicylaldehyde	0.1 *N* HCl in H_2O–EtOH (1 : 1)	350	2.5	47
	3-Methyl-6-isopropylgentisaldehyde	Methanol	390	4.07	48
	Salicylaldehyde	Aqueous alkali	378	5.5	46
	3-Methoxysalicylaldehyde	Aqueous alkali	395	6.03	47
2.	4-Hydroxybenzaldehyde	95% EtOH	284	17.4	46
	Vanillin	95% EtOH	310	10.7	49
	Syringaldehyde	95% EtOH	308	12.6	49
	4-Hydroxybenzaldehyde	Aqueous alkali	330	26.3	46
	Vanillin	95% EtOH, KOH	353	29.5	49
	Syringaldehyde	95% EtOH, KOH	370	27.5	49
3.	*p*-Anisaldehyde	95% EtOH	277	15.5	50
	Veratraldehyde	Water	308	9.3	51
4.	*o*-Anisaldehyde	95% EtOH	321	4.6	50
	2,5-Dimethoxybenzaldehyde	95% EtOH	352	–	45
5.	4′-Hydroxyacetophenone	95% EtOH	278	13.8	49
	3′,4′-Dihydroxyacetophenone	95% EtOH	307	7.9	52
	3′-Methoxy-4′-hydroxyacetophenone	95% EtOH	303	8.3	49
	3′,5′-Dimethoxy-4′-hydroxy-acetophenone	95% EtOH	302	11.0	49
	4′-Hydroxyacetophenone	95% EtOH, KOH	328	24.6	49
	3′,4′-Dihydroxyacetophenone	Aqueous alkali	342	16.2	53
	3′-Methoxy-4′-hydroxyacetophenone	95% EtOH, KOH	348	24.0	49
	3′,5′-Dimethoxy-4′-hydroxy-acetophenone	95% EtOH, KOH	362	23.4	49
6.	Salicylic acid	Water	296	3.39	54
	Gentisic acid	Water	320	4.27	54
7.	Coumarin	95% EtOH	308	5.5	55
	6-Methoxycoumarin	95% EtOH	344	4.4	56
	6-Hydroxycoumarin	95% EtOH	349	4.27	56

TABLE 6.4 (*cont'd*)

	Compound[a]	Solvent	λ_{max} (mμ)	mϵ	Ref.
8.	2-Hydroxybenzophenone	95% EtOH	335	5.0	6
	1-Hydroxyanthraquinone	Methanol	402	5.5	9
	1,4-Dihydroxyanthraquinone	Methanol	470	19.0	57
9.	4-Hydroxybenzophenone	95% EtOH	290	12.6	6
	2-Hydroxyanthraquinone	Dioxane	360	3.2	58
10.	3,4-Dihydro-8-hydroxy-1(2*H*)-naphthalenone	95% EtOH–HCl	335	3.09	59
	3,4-Dihydro-5,8-dihydroxy-1(2*H*)-naphthalenone	95% EtOH	372	3.5	60
11.	2-Aminoacetophenone	95% EtOH	359	4.4	61
	2-Amino-3-hydroxyacetophenone	95% EtOH	375	5.0	62
		0.1 *N* NaOH	400	4.0	62
12.	4-Aminoacetophenone	95% EtOH	316	20.0	61
	4-Amino-3-hydroxyacetophenone	95% EtOH	335	20.0	62
	4-Amino-3-hydroxyacetophenone	0.1 *N* NaOH	365	10.0	62
13.	Methyl anthranilate	95% EtOH	340	5.0	63
	Methyl 5,10-dihydro-1-phenazine-carboxylate	95% EtOH	450	~ 13.0	64
	Tetramethyl diaminopyromellitate	95% EtOH	445	5.0	65
14.	2-Aminobenzamide	95% EtOH	335	3.98	66
	5,10-Dihydro-1-phenazine-carboxamide	95% EtOH	430	~ 12.0	64
15.	2-Nitrophenol[b]	Aqueous acid	350	3.07	67
	2-Nitro-4-chlorophenol	Aqueous acid	362	2.93	67
	2-Nitro-4-cresol	Aqueous acid	368	2.98	67
	2-Nitro-4-acetylaminophenol	Aqueous acid	375	2.25	67
	2-Nitro-4-phenylphenol	Aqueous acid	378	2.36	67
	2-Nitro-4-methoxyphenol	Aqueous acid	391	3.11	67
	2-Nitrohydroquinone	Aqueous acid	395	2.8	68
	2-Nitro-4-aminophenol	Aqueous acid	421	2.36	67
16.	4-Nitrophenol	Aqueous acid	318	9.8	68
	4-Nitrocatechol	95% EtOH	345	6.0	68
	4-Nitro-2-aminophenol	Aqueous acid	370	4.7	53
17.	2-Nitrodiphenylsulfide	95% EtOH	368	4.5	69
	2-Nitro-4-chlorodiphenylsulfide	95% EtOH	380	4.3	69
	2-Nitro-4-acetylaminodiphenyl-sulfide	95% EtOH	394	3.7	69
	2-Nitro-4-aminodiphenylsulfide	95% EtOH	428	2.2	69
18.	2-Phenylazophenol	Acetone	368*s*[c]		45
	2-Phenylazo-4-cresol	Acetone	390		45
	2-Phenylazohydroquinone	Acetone	425		45
19.	4-Nitroaniline	95% EtOH	375	15.5	40
	4-Nitro-1,2-phenylenediamine	95% EtOH	408	9.1	
20.	4-Nitrodiphenylamine	EtOH (abs.)	390	21.2	70
	3-Nitrophenoxazine	95% EtOH	450		45
	3-Nitrophenothiazine	95% EtOH	454		45

TABLE 6.4 (*cont'd*)

	Compound[a]	Solvent	λ_{max} (mμ)	mϵ	Ref.
21.	2-Nitroaniline	EtOH (abs.)	404	5.3	70
	1-Nitrocarbazole	EtOH (abs.)	403	7.6	70
	3-Nitro-5-chloro-1,2-phenylene-diamine	6 *N* HCl in H_2O–EtOH (1:1)	406	4.7	
	4-Chloro-2-nitroaniline	Acetone	410	5.0	71
	2,3-Dinitroaniline	95% EtOH	410	5.1	
	2-Nitro-*p*-toluidine	95% EtOH	418	5.7	72
	3-Nitro-4-aminobiphenyl	95% EtOH	427	4.4	73
	3-Nitro-4-amino-*p*-terphenyl	95% EtOH	430	4.07	73
	2-Nitro-3-dibenzofuranamine	95% EtOH	430	4.0	74
	2-Nitro-4-methylthioaniline	95% EtOH	432	4.0	
	3-Nitro-9-methyl-2-carbazolamine	95% EtOH	440*s*[c]	3.7	75
	4-Methoxy-2-nitroaniline	95% EtOH	442	5.5	
	1-Nitro-2-dibenzothiophenamine	95% EtOH	446	2.8	76
	4-Nitro-3-dibenzothiophenamine	95% EtOH	446	4.4	77
	3-Nitro-2-fluorenamine	95% EtOH	447	5.4	78
	3-Nitro-5-chloro-1,2-phenylene-diamine	95% EtOH	448	3.4	
	1-Nitro-2-dibenzoselenophenamine	95% EtOH	450	2.0	79
	3-Nitro-7-methyl-2-fluorenamine	95% EtOH	451	5.1	78
	3-Nitro-7-ethyl-2-fluorenamine	95% EtOH	451	5.1	78
	4-Nitro-3-dibenzoselenophenamine	95% EtOH	452	4.3	77
	4-Nitro-3-dibenzothiophenamine-5,5-dioxide	95% EtOH	452	4.0	
	1-Nitro-2-fluorenamine	95% EtOH	466	5.0	78
	3-Nitro-2-dibenzothiophenamine	95% EtOH	468	3.5	76
	3-Nitro-2-dibenzoselenophenamine	95% EtOH	470	2.8	79
	Nitro-1,4-phenylenediamine	95% EtOH	475	4.1	
	4-Nitro-9-methyl-3-carbazolamine	95% EtOH	481	4.6	79
	N[4]-(β-Hydroxyethyl)-2-nitro-1,4-phenylenediamine	95% EtOH	492		45
	2-Nitro-3-carbazolamine	95% EtOH	500	3.2	79
	N[1],*N*[4]-Bis(β-hydroxyethyl)-2-nitro-1,4-phenylenediamine	95% EtOH	512		45
	N[1],*N*[4],*N*[4]-Tris(β-hydroxyethyl)-2-nitro-1,4-phenylenediamine	95% EtOH	530		45
	1-Nitro-5-methyl-5,10-dihydro-phenazine	95% EtOH	Violet		80
	1-Nitro-5-phenyl-5,10-dihydro-phenazine	95% EtOH	Blue		80
22.	Isatin	95% EtOH	416	0.71	81
	7-Methylisatin	95% EtOH	422	0.85	81
	5-Methylisatin	95% EtOH	427	0.72	81
	7-Methoxyisatin	95% EtOH	437	1.07	81
	5-Methoxyisatin	95% EtOH	472	0.78	81

TABLE 6.4 (*cont'd*)

	Compound[a]	Solvent	λ_{max} (mμ)	mϵ	Ref.
23.	Indigo	$Cl_2C{=}CCl_2$[d]	605	16.6	82, 83
	6,6′-Dimethoxyindigo	$Cl_2C{=}CCl_2$	570	10.5	82, 83
	6,6′-Dichloroindigo	$Cl_2C{=}CCl_2$	590	14.8	82, 83
	6,6′-Dimethylindigo	$Cl_2C{=}CCl_2$	595	8.7	82, 83
	6,6′-Dinitroindigo	$Cl_2C{=}CCl_2$	635		82, 83
	5,5′-Dimethoxyindigo	$Cl_2C{=}CCl_2$	645		82, 83
	5,5′-Dichloroindigo	$Cl_2C{=}CCl_2$	620	17.8	82, 83
	5,5′-Dimethylindigo	$Cl_2C{=}CCl_2$	620	14.5	82, 83
	5,5′-Dinitroindigo	$Cl_2C{=}CCl_2$	580		82, 83
24.	Thioindigo	DMF	543		45
	6,6′-Dichlorothioindigo	DMF	539		45
	6,6′-Diethylthio-thioindigo	DMF	531		45
	6,6′-Diethoxythioindigo	DMF	515		45
	6,6′-Diaminothioindigo	DMF	490		45
	5,5′-Dichlorothioindigo	DMF	536		45
	5,5′-Diethylthio-thioindigo	DMF	573		45
	5,5′-Diethoxythioindigo	DMF	584		45
	5-5′-Diaminothioindigo	DMF	638		45

[a]Unnumbered compounds are reference compounds.
[b]With electronegative substituents, such as COOH, COO^-, $COOCH_3$, NH_3^+, NMe_3^+, and $COCH_3$ in the 4-position, λ_{max} ranges from 335 to 342 mμ (67).
[c]Shoulder.
[d]*sym*-Tetrachloroethylene.

X=Y is an electron-acceptor and Z is an electron-donor group.

X=Y Z: Z: **8** X=Y Z: Z: **9** X=Y Z: :Z **10**

When a benzene compound contains an electron-donor and an electron-acceptor group in conjugative positions, substitution of another electron-donor group *ortho* or *para* to the positive resonance terminal causes the long-wavelength band to shift to the red, to lose some of its intensity usually, and to increase its half-band width. The effect on the wavelength position is especially striking (Table 6.4). These phenomena appear to be derived from an electronic nodal effect. Thus in the quinolone derivative **11A**, which is in resonance with the intermediate resonance structure **11B**, and the limiting resonance structure **11C**, substitution of an electron-donor group at the negative node causes a strong red shift, as shown by the spectra in ethanol of **11**, **12**, and **13** (86).

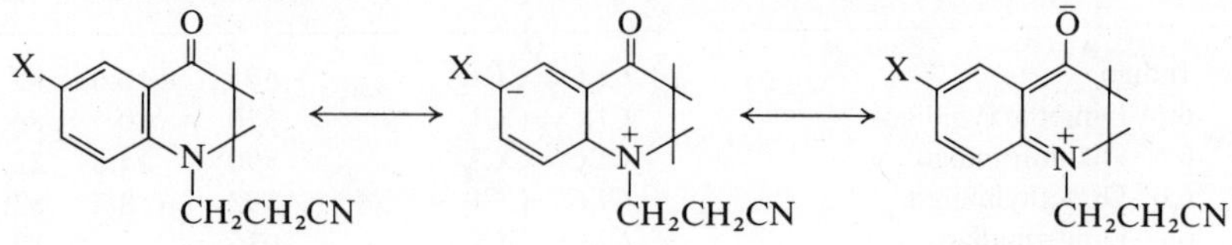

A ⟷ B ⟷ C

11

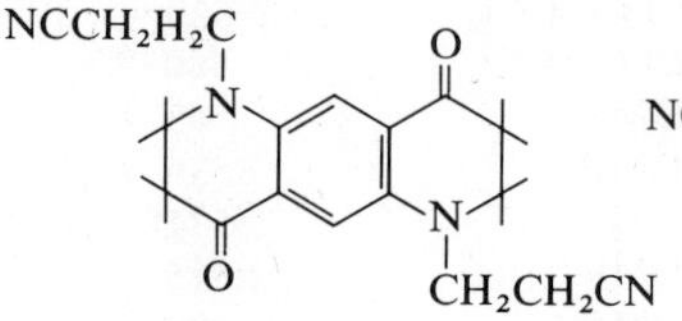

12

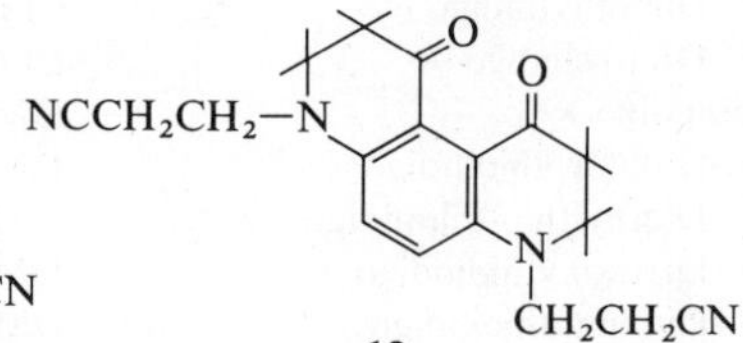

13

No.	X	λ_{max}(mμ)	mε
11	Cl	393	4.5
	OH	410	5.07
	$N(CH_2CH_2CN)_2$	424–431	3.3
12		478	4.5
13		492–496	5.5

Thus increasing positivization of the negative node shifts the wavelength maximum of **11** from 393 mμ when X = Cl to 410 mμ when X = OH, and 424–431 mμ when $X = N(CH_2CH_2CN)_2$. In **12** and **13** there is a further shift to the red due to the additive red-shift effect of negativizing a positive node and positivizing a negative node.

The red color of dihydroxanthommatin (**14**), as opposed to the yellow color of the longer conjugated xanthommatin (**15**) (87), is explained readily by the *ortho–para* effect.

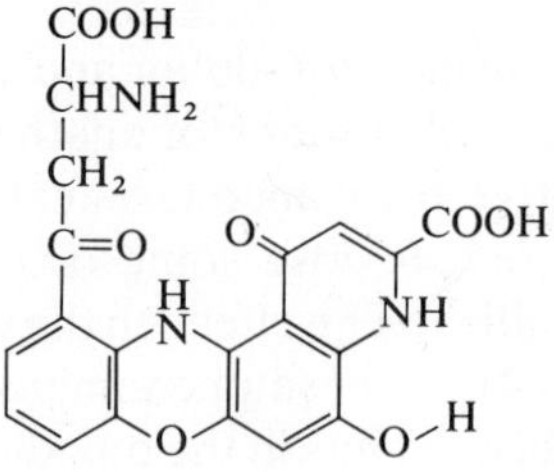

14
Red

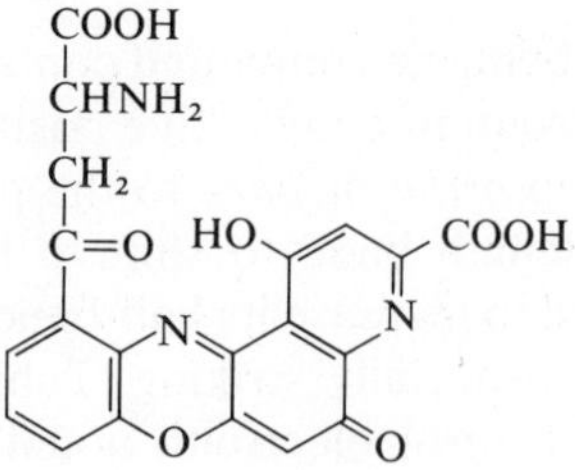

15
Yellow

In Table 6.4 a large number of compounds containing the *o*-nitroaniline group are compared spectrally. The compared band is derived from a transition involving the nitro group as the negative resonance terminal and the amino group as the positive resonance terminal. As can be seen in the table, the wavelength maximum shifts from 404 mμ for 2-nitroaniline to somewhere near 600 mμ for 1-nitro-5-phenyl-5,10-dihydrophenazine, mainly due to the *ortho–para* effect. The *para* effect of substituents with increasing donor strength is seen in 2-nitroaniline (**16**, X = H) and a few of its derivatives. The same effect can be seen in the reported spectra of fourteen 4-substituted 2-nitroanilines dissolved in aqueous alkali(88) and of fifteen 4-substituted 2-nitrophenols in acidic and alkaline solution(67).

NH_2 NO_2 X

16

X	λ_{max}(mμ)	X	λ_{max}(mμ)
H	404	CH_3O	442
CH_3	418	NH_2	475
C_6H_5	427	$NHCH_2CH_2OH$	492
CH_3S	432		

On the other hand, negativizing a negative node (e.g., replacement of a ring CH with an aza nitrogen) causes a violet shift, as shown, for example, by the spectra in ethanol of **17a** and **b**.

X NH_2 H_3C NO_2

17

	X	λ_{max}(mμ)	mε
a	CH	418	5.7
b	N	349	33.3

In a similar manner the spectra of 5-substituted 2-nitrophenols in acidic and alkaline solution(89) and 5-substituted 2-nitroanilines in weakly basic solution(90) show weaker phenomena, though opposite to the 4-substituted isomers. Electron-donor groups in the 5-position cause a slight violet shift or have no effect on the long-wavelength band. Electron-acceptor groups cause a slight red shift, with the nitro group in the

5-position causing the largest change, as shown by comparison of 2-nitroaniline (λ_{max} 413 mμ, mϵ 24.2) with 2,5-dinitroaniline (λ_{max} 440 mμ, mϵ 22.7)(90).

The *ortho–para* effect can be found in the isatin and indigoid compounds. However, where a group is substituted in the molecule so that resonance can take place with the C=O or the NH (or S) resonance terminals, this interfering conjugation shifts the long-wavelength absorption to the violet. This phenomenon is discussed in another section.

IV. Resonance Terminals

A knowledge of the electron-donor and electron-acceptor strengths of resonance terminals is often valuable in choosing the appropriate reagent for the colorimetric analysis of a trace constituent in a complicated mixture. With knowledge of such factors as solvent polarity and pH, background absorption, interferences, the reactivity of the reagent, and the desired wavelength maximum of the chromogen, one can select a reagent that will overcome many of these factors and will make available the necessary resonance terminal.

A. Positive Resonance Terminal

Groups that are *ortho-para* directing are usually electron donating. When these various electron-donor groups are placed in a benzene ring *para* to a strong electron-acceptor group, they interact in the excited state, and the strength of interaction and electron shift is shown by the displacement of the $\pi \rightarrow \pi^*$ band involving the entire conjugation system. Thus in Table 6.5 the groups are placed in order of increasing wavelength displacement or, assuming a strict relationship, in order of increasing electron-donor strength.

The greater the electron-donor and electron-acceptor strengths of the terminals, the further in the visible will a compound absorb, other factors being equal. It has been established that electron-donor or electron-acceptor substituents in benzene produce red shifts, because the benzene ring is amphoteric in that it can act either as an electron donor or acceptor, or as a conjugation chain. The magnitude of these spectral shifts has been studied with a large number of mono-, di-, and tri-substituted benzene derivatives(40, 107, 118, 119). In most of these polysubstituted compounds the benzene ring acts mainly as the conjugation chain.

Where the electronegative group is substituted on the positive resonance terminal, the electron-donor strength of this terminal is decreased due to a competing resonance, as shown for **18**.

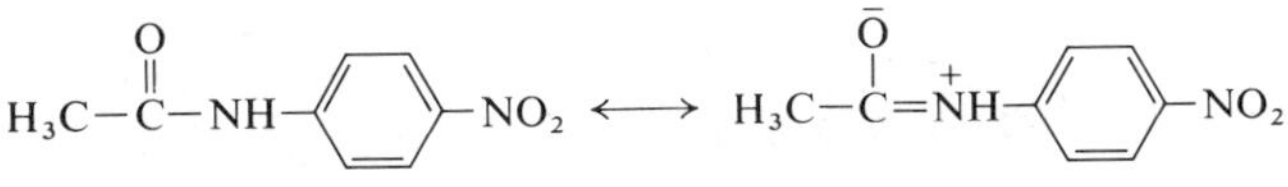

$$H_3C-\overset{\overset{O}{\|}}{C}-NH-C_6H_4-NO_2 \longleftrightarrow H_3C-\overset{\overset{\bar{O}}{|}}{C}=\overset{+}{N}H-C_6H_4-NO_2$$

18

The effect of methyl and substituted methyl groups at the ends of conjugated systems on the position of the long-wavelength band has been studied (101, 120). The effect of alkyl groups (substituted in the amine group of a positive resonance terminal) on the position of the long-wavelength band is demonstrated by the data in Table 6.5 and has been shown for 4-tricyanovinylanilines (121).

Many factors of value in an understanding of a methoxy group as a resonance terminal have been covered in a comprehensive review (122). The properties of the acetoxy group as a resonance terminal in simple benzenoid systems have been discussed from an ultraviolet-spectral viewpoint (123). The effect of terminal nitro and amino groups on the electronic spectra of conjugated hydrocarbon systems has been reported (124). On the other hand, the R—As: group has little, if any, electron-donor property in such a compound as 1,2,3,4-tetrahydro-7-methoxy-1-methyl-4-oxoarsinoline (**19**) when comparison is made with **20** (125).

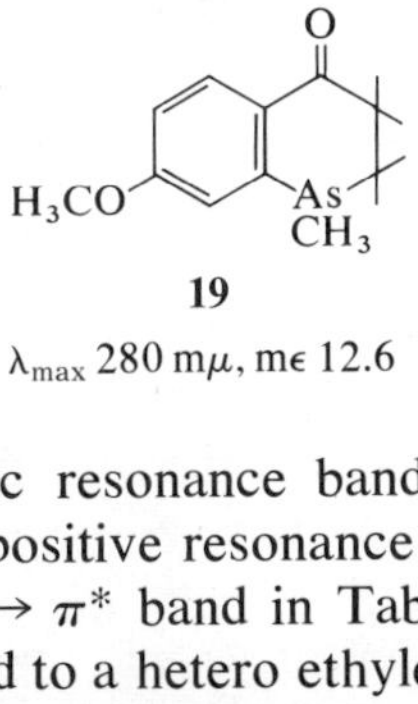

H_3CO … As–CH_3 (**19**): λ_{max} 280 mμ, mε 12.6

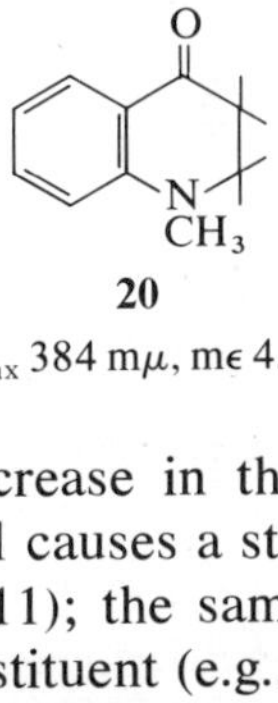

N–CH_3 (**20**): λ_{max} 384 mμ, mε 4.0

In zwitterionic resonance bands an increase in the electron-donor strength of the positive resonance terminal causes a strong red shift (as seen for the $\pi \rightarrow \pi^*$ band in Table 6.6)(11); the same electron-donor group conjugated to a hetero ethylenic substituent (e.g., $C{=}\ddot{X}$ or $N{=}\ddot{X}$, where $X = O$, S, or N) has the opposite effect on the $n \rightarrow \pi^*$ band derived from the hetero ethylenic substituent.

A very strong electron-donor group not presented in Table 6.5 is the ferrocenyl group (126). Although the diethylamino group is the strongest electron donor in the compounds listed in Table 6.5, the ferrocenyl group may be much stronger. Thus 4-diethylaminoazobenzene absorbs at 415 mμ, mε 29.5, whereas 4-ferrocenylazobenzene absorbs at 488 mμ, mε 4.64. Since the transitions involved in the latter band are uncertain, this interesting group will need much more thorough study for possible use in trace-analysis research.

TABLE 6.5

Effect of Increasing Electron-Donor Strength of the Positive Resonance Terminal on the Z_n-Band[a]

X–⟨C₆H₄⟩–Ac (I) X–⟨C₆H₄⟩–CHO (II) X–⟨C₆H₄⟩–NO_2 (III) X–⟨C₆H₄⟩–N=N–⟨C₆H₅⟩ (IV)

X	I			II			III			IV		
	λ_{max} (mμ)	(mε)	Ref.	λ_{max} (mμ)	(mε)	Ref.	λ_{max} (mμ)	(mε)	Ref.	λ_{max} (mμ)	(mε)	Ref.
H	242	(12.6)	91	245	(12.9)	92	260	(8.3)	93	318	(20.9)	94
F	242[b]	(11.5)	95	246	(11.5)	92	262	(9.0)	92	323	(20.4)	96
$OCOCH_3$	248	(14.1)	97	249	(14.5)	98	270	(10.0)	99	325	(21.9)	98
Cl	249[b]	(16.0)	95	254	(16.0)	92	270	(11.0)	92	328	(23.4)	96
CH_3	252	(15.1)	97	257[c]	(12.6)	100	273[d]	(10.0)	101	333	(23.4)	102
Br	253[b]	(16.0)	95	259	(16.5)	92	276	(11.5)	92	329	(26.9)	96
Cyclopropyl							280[c]	(11.0)	103			
$CH{=}NCH_3$							280	(17.0)	104			
$C{\equiv}CH$							286	(15.0)	104			
SCN							287	(11.2)	105			
I	262[b]	(16.0)	95	271	(13.5)	92	294	(12.5)	92	331	(28.2)	96
$CH{=}CH_2$							301	(14.5)	104			
$NHCOCF_3$							301			337		
OC_6H_5							302	(11.8)	106			
OCH_3	275	(17.8)	91	277	(15.5)	50	305	(13.8)	107	343	(36.3)	25
C_6H_5	283	(24.6)	108				307	(15.4)				
NHCN							313	(13.5)	109			
$N(NO)C_2H_5$							313	(15.1)	70			
OH	279	(14.1)	98	282	(14.0)	50	314	(13.0)	50	350	(26.9)	98
$NHCOCH_3$	285	(20.0)	98	297	(20.4)	98	316	(11.5)	110	347	(23.4)	111
SH							318	(12.0)	107			
$CH_2COO_2H_5$							320			352	(26.3)	102

SC_6H_5	304	(17.4)	112				337	(13.5)	113			
SCH_3							340	(12.6)	114	362	(23.7)	102
SeC_6H_5							345	(11.5)	115			
$SeCH_3$							350	(12.3)	115			
$CH{=}N^+(O^-){-}C_6H_5$							352[e]	(20.0)	116			
NH_2	317	(20.1)	40	332	(26.3)	104	375	(15.5)	40	385	(24.6)	102
$NHCH_3$							386	(18.4)	40	402	(25.7)	102
NHC_2H_5	332	(25.9)	40				390	(19.0)	40	405	(26.3)	102
$N(CH_3)_2$	337	(25.6)	40	342	(29.5)	40	390	(19.0)	40	408	(27.5)	102
NHC_6H_5							390	(21.1)	70	420[f]	(28.2)	117
$N(C_2H_5)_2$				348	(33.1)	40	400	(21.6)	40	415	(29.5)	102

[a]All spectra in 95% ethanol except when stated otherwise.

[b]In absolute ethanol.

[c]In hexane.

[d]Substituents on methyl group of *p*-nitrotoluene caused a violet shift; for example, NH_2, 271.5 mμ, mε 10.0; Br, 270.5 mμ, mε 12.0; OH, 270.5 mμ, mε 10.0; Cl, 265 mμ, mε 12.0; and NH_3^+, 262 mμ, mε 11.0(101).

[e]In methanol.

[f]In 50% aqueous ethanol.

TABLE 6.6
The Effect of Electron-Donor Groups on $n \rightarrow \pi^*$ and $\pi \rightarrow \pi^*$ Bands(11)

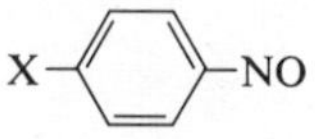

X—⟨C₆H₄⟩—NO (I)   X—⟨C₆H₄⟩—C(=S)—⟨C₆H₄⟩—X (II)

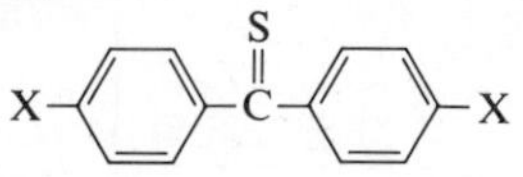

	λ_{max}(mμ) (mϵ)			
	I		II	
X	$n \rightarrow \pi^*$	$\pi \rightarrow \pi^*$	$n \rightarrow \pi^*$	$\pi \rightarrow \pi^*$
H	756 (0.045)	282 (9.5) 306 (7.0)	605 (0.066)[a]	315 (17.0)[a]
RO[b]	728 (0.047)	346 (11.6)	593 (0.36)[a]	344 (34.3)
$(CH_3)_2N$	674 (0.055)	420 (30.4)	573 (0.85)[a]	434 (43.8)

[a]In ether. Remainder of compounds in 95% ethanol.
[b]$R = CH_3$ for the nitroso derivative and $R = C_2H_5$ for the thioketone.

Many other compounds show the expected red shift with the increasing electron-donor strength of the substituents(40, 118). 2,1,3-Benzoselenadiazole derivatives show the same phenomenon(127) as do the nitro compounds **21** (79).

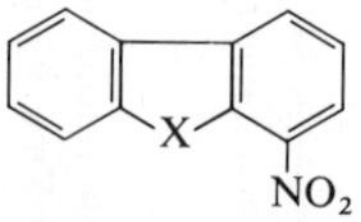

NO_2

21

$X = CH_2$, O, S, Se, or NH

B. Negative Resonance Terminal

Electron-acceptor groups are listed in Table 6.7 in the order of their red-shift effect when substituted in a benzene ring or in the *para* position in aniline or *N*, *N*-dimethylaniline. On the whole, most of the compounds show the same order in the three types of compounds. There are some exceptions in the monosubstituted benzene derivatives since some of the substituents have both electron-donor and electron-acceptor properties, the former being the stronger. This is shown by their absorption at longer wavelength than one would expect on the basis of their electron-acceptor properties (e.g., styrene and phenylacetylene).

TABLE 6.7

'ect of Increasing Electron-Acceptor Strength of the Negative Resonance Terminal on the Z_n-Band

X–C$_6$H$_5$ (I) X–C$_6$H$_4$–NH$_2$ (II) X–C$_6$H$_4$–N(CH$_3$)$_2$ (III)

	I			II			III		
X	λ_{max}(mμ)	(mϵ)	Ref.	λ_{max}(mμ)	(mϵ)	Ref.	λ_{max}(mμ)	(mϵ)	Ref.
	204*fs*[a,b]	(6.3)	128	234[a]	(11.5)	129	251[a]	(12.9)	130
)$_2$NH$_2$	219[a]	(8.9)	131	262	(17.8)	40	276	(24.6)	40
)$_2$CH$_3$	217[a]	(6.8)	132	268	(19.5)	132	275[c]	(26.9)	133
≡CH	244[d]	(13.5)	104	277	(16.0)	104			
I=CH$_2$	245	(15.0)	104	277	(16.0)	104			
N	230[e]	(11.0)	104	278	(24.0)	104	294	(26.9)	134
)NH$_2$	225[a]	(10.0)	135	285	(20.0)	66	305	(20.0)	66
)OH	227[a,f]	(10.5)	136	288	(17.4)	40	308	(25.4)	40
)OCH$_3$	228[a,f]	(11.8)	136	294	(20.0)	63	312	(25.0)	63
)$_2$C$_6$H$_5$	235[a]	(15.5)	137	291	(19.1)	138	307	(25.1)	137
I=N—CH$_3$	245	(15.0)	104	299	(21.0)	104			
)CH$_3$	243[a]	(13.2)	139	316	(20.0)	61	337	(25.7)	40
IO	244[a]	(12.9)	140	332	(26.3)	104	342	(29.8)	40
)C$_6$H$_5$	253	(17.8)	141	335	(20.0)	142	353	(24.0)	143
I=NC$_6$H$_5$	263	(16.8)	144				356	(31.6)	31
)$_2$	260	(7.9)	145	375	(15.5)	40	390	(19.0)	40
=NCH$_3$	261	(8.0)	146						
=NC$_6$H$_5$	318	(21.4)	102	385	(24.6)	102	408	(27.5)	102
SO	321	(9.3)	147				408	(28.2)	147
)	306	(7.0)	11				423	(29.4)	40
SC$_6$H$_4$N(CH$_3$)$_2$[g]							434	(43.8)	143

Forbidden $\pi \rightarrow \pi^*$ bands at longer wavelengths. Spectra in 95% ethanol unless otherwise stated.
Fine structure.
In hexane.
Band also at 235 mμ, mϵ 15.0.
Band also at 222 mμ, mϵ 13.0.
n absolute ethanol.
Thiobenzophenone has long-wavelength $\pi \rightarrow \pi^*$ band at 315 mμ, mϵ 17.0, in ether(146).

The effect of various electron-acceptor groups substituted in a benzene ring *ortho*, *meta*, or *para* to an electron-donor group has been reported(118). The conjugative ability of a few electron-acceptor groups has been correlated with their effect on the primary 203.5-mμ band of benzene(148). The results are similar to those reported in Table 6.7.

The relative electron-acceptor strength of other groups can be deduced from their absorption spectra. It has been reported that the oxirane and thiirane groups are weakly electron withdrawing relative to benzene, as shown by the absorption spectra in hexane of the *p*-methoxybenzene derivatives **22** (103). Structure **23** is believed to contribute to the singlet excited state.

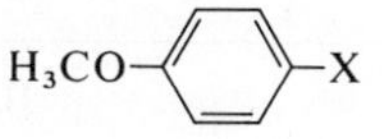

$H_3CO-C_6H_4-X$

22

X	λ_{max}(mμ)	mϵ
CH_3	223	7.7
oxiranyl (O)	230	12.4
thiiranyl (S)	238	13.3

$H_3C\overset{+}{O}=C_6H_4=CHCH_2\overset{|}{}{}^-O$

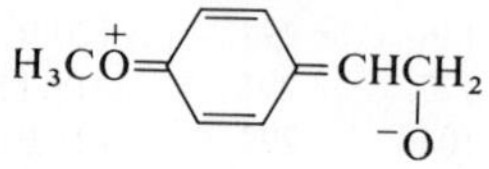

23

The electron-acceptor strengths of the tetrazole and pentazole groups are approximately equivalent to those of the carboxamide and acetyl groups, respectively, as shown by the absorption spectra in ethanol of their 4-dimethylaminophenyl derivatives (**24**) (149).

$(CH_3)_2N-C_6H_4-N(N{=}N)(X{=}N) \longleftrightarrow (CH_3)_2\overset{+}{N}=C_6H_4=\overset{+}{N}(\bar{N}-N)(\underline{X}-N)$

24

X	λ_{max}(mμ)	mϵ
CH	295	15.5
N	333	15.0

The much greater electron-acceptor strength of the thiocarbonyl group as compared with the carbonyl group is apparent from the data in Table 6.7. Other compounds, such as **25** (spectra in ethanol) confirm this (150).

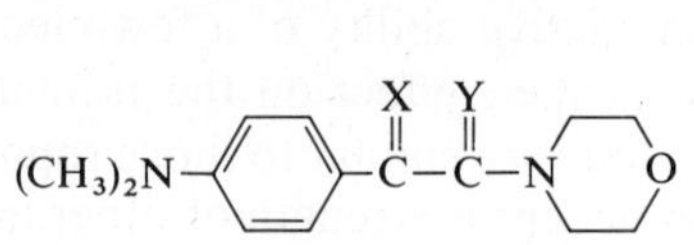

$(CH_3)_2N-C_6H_4-\overset{X}{\overset{\|}{C}}-\overset{Y}{\overset{\|}{C}}-N(CH_2CH_2)_2O$

25

X	Y	$\lambda_{max}(m\mu)$	$m\epsilon$
O	O	350	28.7
O	S	354	27.0
S	S	458	47.7

The 2,1,3-selenadiazole group appears to have a high order of electron-acceptor strength since the amino derivatives of 2,1,3-benzoselenadiazole (**26a** and **b**), absorb at much longer wavelengths than do the comparable *p*-nitroanilines(127). The dipolar structure believed to contribute strongly to the excited state is shown. The spectra were obtained in ethanol.

26

	R	λ_{max} $(m\mu)$	$m\epsilon$
a	H	426	6.3
b	CH_3	449	6.76

Another highly electronegative group is the tricyanovinyl group. The *p*-aminophenyl derivatives **27a** and **b**(121), absorb at even longer wavelengths than any of the compounds that have been mentioned. These compounds absorb at a wavelength approximately 80 $m\mu$ longer than the analogous $p\text{-}R_2NC_6H_4CH{=}C(CN)_2$ compounds(41). The dicyanovinyl group is approximately as strong an electron-acceptor group as the nitro group. The spectra listed below were measured in acetone.

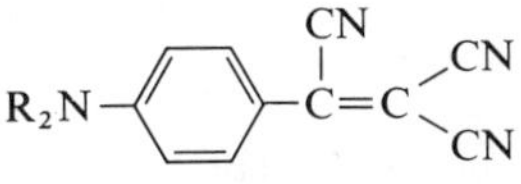

27

	R	λ_{max} $(m\mu)$	$m\epsilon$
a	H	480	37.2
b	CH_3	514	41.5

Substitution of the strongly electronegative 2,1,3-selenodiazole and tricyanovinyl groups for the nitro group in the various nitro-substituted reagents used in trace analysis should make available some useful test reagents.

C. Amphoteric Resonance Terminals

The term "amphoteric resonance terminals" refers to terminal groups that can be either electron donors or electron acceptors. The phenyl group has this property as has the 2-thienyl group, which is a stronger electron acceptor and electron donor than the benzene ring(151).

The nitroso and nitrone groups appear to be able to behave as amphoteric resonance terminals. Some of these compounds are present in the dipolar form in the ground state, as shown, for example, in **28**(152).

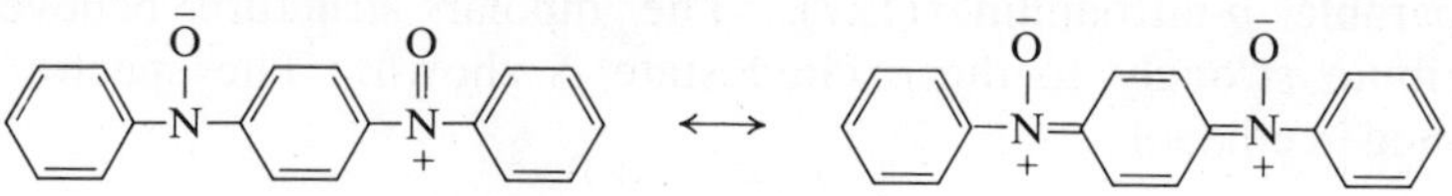

28

Solvent	λ_{max} (mμ)	mϵ
Methanol	417	47.0
Chloroform	423	43.0

In *p*-dinitrosobenzene the dipolar form **29** is believed to predominate in the ground state(153).

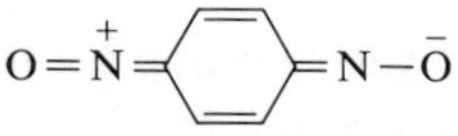

29

As shown for **30**, the electron-acceptor properties of the nitroso group are much more powerful than its electron-donor properties. The spectra were measured in ethanol.

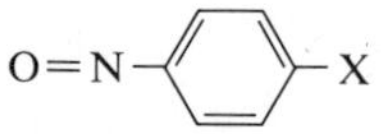

30

X	λ_{max} (mμ)	mϵ	Ref.
NO	405	3.58	153
$N(CH_3)_2$	423	29.4	40
NO_2	282	45.7	154

The electron-donor and electron-acceptor strengths of resonance terminals are also affected by electronic effects involving groups outside the main conjugation chain, as will be shown in other sections of this chapter.

V. Conjugation-Chain Length

A. Polyenes

The importance of the length of a conjugation chain to the position and intensity of a long-wavelength band derived from a $\pi \rightarrow \pi^*$ transition is emphasized by the data in Table 6.8. Before exploring this facet of the absorption spectra of organic compounds, we shall discuss the minimum conjugation necessary for a compound to absorb in the ultraviolet region.

For organic compounds to absorb light in the ultraviolet–visible region

TABLE 6.8
Effect of Resonance Chain Length on the Long-Wavelength Band

R—(CH=CH)$_n$—R′

R	R′	n	Solvent	λ_{max} (mμ)	mϵ	Ref.
CH_3	CH_3	1	Gas	187*fs*[a]	6.3	155
		2		227[b]		156
		3		263		
		4		299[c]		
		5		326[d]		
		6		352		
		8		396		
		9		413		
C_6H_5	CH_2OH	1	Methanol	251	16.4	160
		2		286	32.5	
		3		317	53.5	
C_6H_5[e]	C_6H_5	0	Chloroform	242	18.3	164
		1	Cyclohexane	303	24.6	165
		2		327	39.8	
		3		350	74.1	
		4		375	85.1	
		5		392	93.3	
		6		408	112.0	
		7		424	135.0	
C_6H_5	CHO	0	AcOH	250	1.0	166
		1		287	2.3	
		2		325	4.0	
		3		360	5.1	
		4		387	6.1	
		5		410		
		6		430		

TABLE 6.8 (*cont'd*)

R	R′	n	Solvent	λ_{max} (mμ)	mϵ	Ref.
4-$CH_3OC_6H_4$	CHO	0	95% EtOH	277	15.5	50
		1	Chloroform	322	30.5	167
		2		352	42.0	
		3		382	52.0	
		4		403	61.5	
		5		425	68.7	
		6		440	78.0	
		7		457	90.2	
$(CH_3)_2N$	CHO	0	Gas	197	8.7	168
		1	CH_2Cl_2	283	37.0	23
		2		362	51.0	
		3		422	56.0	
		4		463	65.0	
		5		492	68.0	
		5		513	72.0	
5-Nitro-2-furyl	CHO	0	Water	310	11.5	169
		1	95% EtOH	347	23.2	170
		2		375	24.0	
		3		401	45.2	
		4		425	59.0	
(1,3-indandione-2-ylidene)C=CH	C_6H_5	0	95% EtOH	343		171[f]
		1		389		
		2		428		
		3		465		
		4	AcOH	497		171
		5		508		
(barbituric acid-5-ylidene)=CH–	C_6H_5	0	AcOH	331		171
		1		375		
		3		451		
		5		503		
C_6H_5CO	C_6H_5	0	AcOH	254	1.8	166
		1		310	2.3	
		2		343	3.6	
		3		376	3.1	
		4		402	5.7	
		5		418	6.4	
		6		445		

Table 6.8 (*cont'd*)

R	R′	n	Solvent	λ_{max} (mμ)	mϵ	Ref.
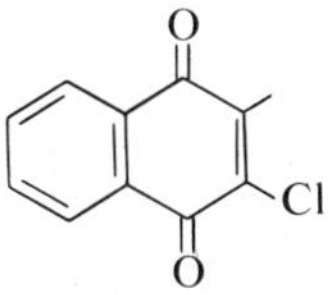 O, O, Cl	$N(C_2H_5)_2$	0	Dioxane	494	3.1	172
		1		564	13.3	
		2		625	28.9	
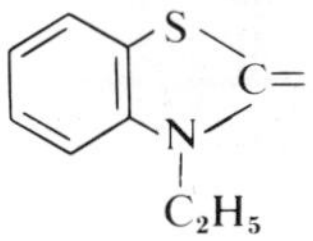 S, C=, N, C_2H_5	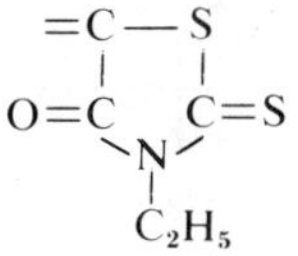 =C—S, O=C, C=S, N, C_2H_5	0[g]		426		173
		1[g]		524		
		2[g]		607		
		3[g]		636		
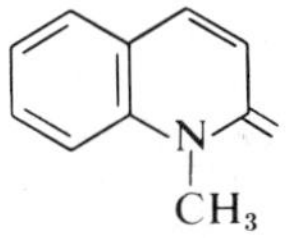 N, CH_3	$=C(CN)_2$	1[g]	95% EtOH	Orange		174[h]
		2[g]		Violet,	588	
		3[g]		Blue,	689	
		4[g]		Blue green		

[a]*fs* = Fine structure.
[b]In ethanol λ_{max} 227 mμ, mϵ 22.4(157).
[c]In ethanol λ_{max} 320 mμ, mϵ 58.9(158).
[d]In methanol λ_{max} 343 mμ (fine structure), mϵ 83.2(159).
[e]Wavelength maxima for this series also reported in benzene(161), and more polar solvents in which $n = 7$, 11, and 15 gave λ_{max} of 465, 530, and 570 mμ, respectively(162, 163).
[f]Wavelength maxima given for many other polymethine series.
[g]Where chain is $(=CH—CH=)_n$.
[h]Wavelength maxima for many other ω, ω-dicyano polymethines listed.

of the spectrum the presence of π- or nonbonding n-electrons in a molecule is necessary. Usually a chain of at least two conjugated double and/or triple bonds is necessary for an organic compound to absorb with some intensity (m$\epsilon > 5$) in the ultraviolet–visible region. A hetero atom—for example, nitrogen or sulfur—can take the place of one unsaturated bond. Examples are listed in Table 6.9.

In all cases it can be seen that these transitions involve a dipolar form as an important contributor to the excited state structure, with the resonance terminals separated by one or two atoms. Examples are acetamide (**31**), 1,3-butadiene (**32**), and butatriene (**33**).

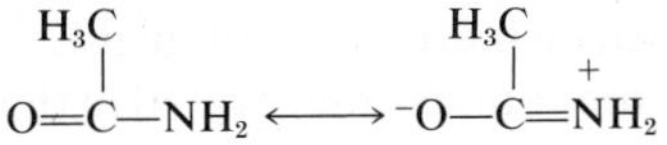

$$\underset{O=C—NH_2}{\overset{H_3C}{|}} \longleftrightarrow \underset{^{-}O—C=\overset{+}{N}H_2}{\overset{H_3C}{|}}$$

31

TABLE 6.9

$\pi \rightarrow \pi^*$ Transitions of the Simplest Conjugated Systems

Compound	Solvent	λ_{max} (mμ)	mϵ	Ref.
CH_3CONH_2	Water	<200		175
$CH_2{=}CHCN$	95% EtOH	203	6.0	176
$CH_2{=}CHCHO$	*n*-Hexane	208	>10.0	177
$CH_2{=}CHCH{=}CH_2$	*n*-Hexane	217	20.9	178
CH_3COSH	*n*-Hexane	219	2.2	179
$CH_3C{\equiv}CCHO$	95% EtOH	225	2.8	180
$CH_2{=}CHC{\equiv}CH$	Methanol	228	7.8	181
$C_2H_5CH{=}CHN$ (ring with S)	*n*-Hexane	228	10.0	182
$CH_3CH{=}CHNO_2$	*n*-Hexane	229	9.5	183
$CH_2{=}CHSCH_3$	95% EtOH	230	25.0	184
$(CH_3)_2NNO_2$	Dioxane	240	6.3	185
$CH_2{=}C{=}C{=}CH_2$[a]	95% EtOH	241	20.4	186
CH_3CSNH_2	Water	262	12.0	188
CH_3CSSCH_3	95% EtOH	305	12.0	189

[a] $CH_2{=}C{=}CHCH_3$ in the gaseous state has a long-wavelength band at 178 mμ, mϵ 20.0, and a shoulder at 186 mμ, mϵ 3.98 (187).

$$H_2C{=}CH{-}CH{=}CH_2 \longleftrightarrow H_2\overset{\pm}{C}{-}CH{=}CH{-}\overset{\mp}{C}H_2$$

32

$$H_2C{=}C{=}C{=}CH_2 \longleftrightarrow H_2\overset{\pm}{C}{-}C{\equiv}C{-}\overset{\mp}{C}H_2$$

33

Besides the compounds listed in Table 6.8 many other families of polymethine compounds with varying chain lengths have been studied spectrally—for example, α-(4-pyrimidyl)-ω-phenylpolyenes (190), α-(α-quinolyl)-ω-phenylpolyenes (191), dialkylpolyenes (192), polyenal anils (193), polyenal phenylhydrazones (194), methylpolyene aldehydes (195), methylpolyene acids (195), α-furylpolyene aldehydes (195), α-furylpolyene acids (195), 4-(*N*-methylaniline) polyene aldehydes (23), methoxypolyenyne aldehydes (23), α,α-diphenyl-ω,ω-diphenylpolyenes (161, 165), 2,3-bis(ω-phenylpolyenyl) quinoxalines (196), and 3-methyl-2-(ω-phenylpolyenyl) quinoxalines (196).

As can be seen from the data in Table 6.8, increasing the chain of conjugation between resonance terminals shifts the long-wavelength maximum toward the red, with an increase in intensity. These shifts also take place in the families of compounds enumerated in the preceding paragraph.

This shift can be seen in the chromogens obtained in the determination of malonaldehyde (197), isonicotinic acid (198), or hydrogen cyanide (199) with barbituric acid. With the aldehyde a trimethine oxonol is obtained that absorbs at 485 mμ. In the determination of isonicotinic or hydrocyanic acid a pentamethine oxonol is obtained that absorbs at 574 or 580 mμ, respectively.

Polyenes in which one or more CH groups in a chain have been replaced by N groups are known—for example, phenylpolyene azines (144), 4-*p*-nitrophenylpolyene azines (144), phenylpolyenal anils (144), polyenal anils (193), and polyenal phenylhydrazones (194).

B. Poly-ynes

The absorption spectra of a few families of poly-ynes have been reported, including those of bis-α,ω-(1-propenyl)poly-yne (200), methyl poly-ynecarboxylic acids (200), di-*tert*-butyl poly-ynes (201), dimethylpoly-ynes (202), diphenylpoly-ynes (203, 204), and a series of diarylpoly-ynes in which the aryl group is 1-anthryl (205, 206), 9-anthryl (206), 9-phenanthryl (206), 1-naphthyl (206), and 2-naphthyl (206).

The poly-ynes are readily recognized by their well-defined spectral fine structure, the maxima showing a spacing of approximately 1850–2300 cm^{-1} as compared with the sparser and typical polyene-fine-structure spacing of approximately 1600 cm^{-1} (200, 202). These poly-yne fine-structure properties provide an easy method of characterizing polyacetylenic groupings—for example, among the extracts of Compositae (207–209), in the seedfat of *Onguekoa klaineana* Pierre (210), and in the antibiotics nemotin, nemotinic acid, nemotin A, and agrocybin (211, 212).

As in the polyenes, the intense bands of the poly-ynes increase in intensity as n increases. However, unlike those of the polyenes, the long-wavelength bands of the few poly-ynes that have been studied are of only moderate intensity and decrease in intensity as n increases (202). In all cases there is a red spectral shift of either set of bands.

C. Cumulenes

Another type of conjugated linear chain that can be present in a molecule is the cumulene type, for example,

$$>C{=}(C{=}C)_n{=}C<$$

Compounds that contain this chain usually absorb at longer wavelength than those that contain the $(\text{—C=C—})_n$ or $(\text{—C}\equiv\text{C—})_n$ groupings of comparable length, as shown by the absorption spectra of 1,1,6,6-tetraphenyl-1,5-hexadien-3-in (**34**), 1,1,6,6-tetraphenyl-1,3,5-hexatriene (**35**),

1,1,6,6-tetraphenyl-1,2,3,5-hexatetraene (**36**)(213), and 1,1,6,6-tetraphenylhexapentaene (**37**)(214).

$\phi_2C{=}CH{-}C{\equiv}C{-}CH{=}C\phi_2$ (**34**) $\phi_2C{=}CH{-}CH{=}CH{-}CH{=}C\phi_2$ (**35**)

$\phi_2C{=}C{=}C{=}CH{-}CH{=}C\phi_2$ (**36**) $\phi_2C{=}C{=}C{=}C{=}C{=}C\phi_2$ (**37**)

No.	Solvent	$\lambda_{max}(m\mu)$
34	Chloroform	370
35	Chloroform	375
36	Chloroform	435
37	Benzene	489

The powerful conjugative effect of the cumulene group is shown in a comparison of several types of linear chains attached to the 9- and 10-positions of anthracene (Fig. 6.1).

Increasing the chain of conjugation in the cumulenes also shifts the wavelength maximum to the red, as shown for **38** and **39**.

$\phi_2C{=}(C{=}C)_n{=}C\phi_2$ (**38**)

39 [two tricyclic X-bridged ring systems joined by $=(C{=}C)_n=$]

No.	X	Solvent	$\lambda_{max}(m\mu)$ $n=1$	$n=2$	$n=3$	Ref.
38		Benzene	420	489	530, 557	214
39a	–	Benzene	484	543	540, 597	214
39b[a]	$C(CH_3)_2$	Chloroform	481	555	610	215

[a] $m\epsilon$ 47.9, 109.6, and 158.5, respectively.

The compounds **39a** are fluorenonylidene derivatives. Especially noteworthy is the great increase in intensity with the increasing chain length of the anthronylidene derivatives (**39b**), the $m\epsilon$ values rising from 47.9 for $n = 1$ to 109.6 for $n = 2$ and 158.5 for $n = 3$.

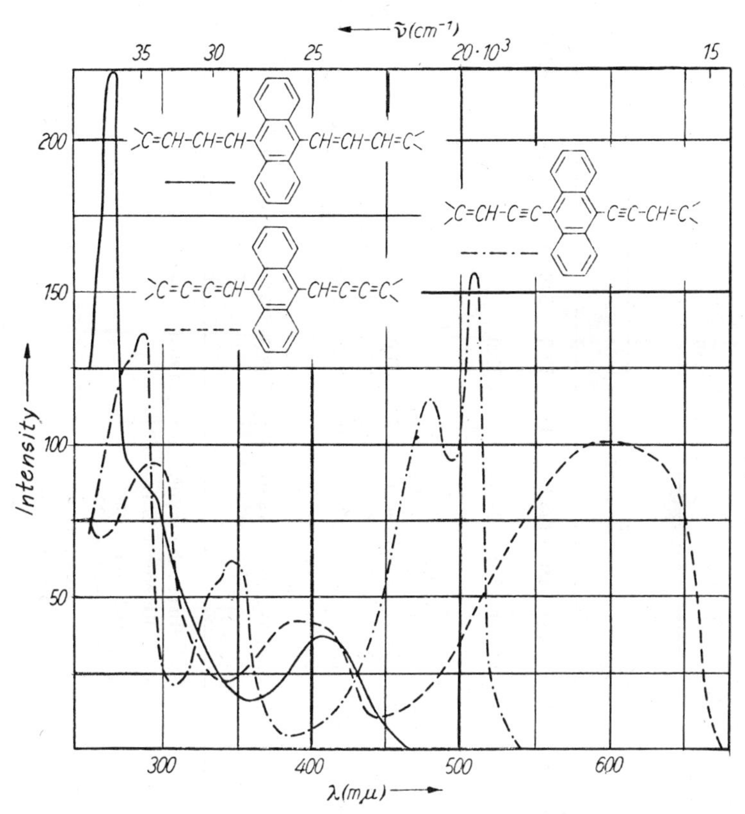

Fig. 6.1. Ultraviolet–visible absorption spectra in chloroform of 9,10-bis[1,1-diphenylbutadien-(1,3)-yl-(4)]anthracene (——), 9,10-bis[1,1-diphenylbutatrienyl-(4)]anthracene (---), and 9,10-bis[1,1-diphenylbuten-(1)-in-(3)-yl-(4)]anthracene (—·—)(213).

D. Miscellaneous Types

A conjugation chain can contain five- or six-membered aromatic or heterocyclic rings. The rings can be connected in a chain through various lengths and numbers of linear or circular (as in porphyrins) chains or fused, as in the polynuclear compounds. Examples of a few of these types are listed in Table 6.10. In all families of compounds (except benzene in the acene series) as n increases the wavelength maximum shifts toward longer wavelength and the molar absorptivity increases.

E. Convergence and Nonconvergence

In most of the families of compounds listed in Tables 6.8 and 6.10 as n increases there is a greater separation of charge in the dipolar

TABLE 6.10
Effect of Increasing Conjugation-Chain Length on the Long-Wavelength $\pi \rightarrow \pi^*$ Band

Structure	Solvent	λ_{max}(mμ) (mϵ)										Ref.
		n										
		0		1		2		3		4		
1,4-C_6H_5—$(C_6H_4)_n$—C_6H_5	Chloroform	251.5	(18.3)	280	(25.0)	300	(39.0)	310	(62.5)	317.5	(>56.0)	216
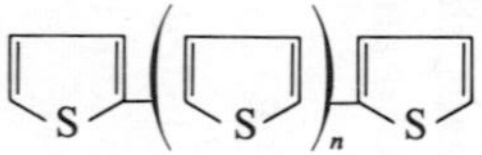	Isooctane	301	(12.8)	350	(24.0)	385	(32.0)					217
1,4-C_6H_5—(CH=CH—$C_6H_4)_n$—CH=CH—C_6H_5	*n*-Hexane	294*fs*[a]	(28.0)	350*fs*	(60.0)	378*fs*[b]	—					218
1,4-C_6H_5—(C≡C—$C_6H_4)_n$—C≡C—C_6H_5	Chloroform	279*fs*[c]	(38.0)	320*fs*	(54.0)	343*fs*	(83.0)					219
1,4-C_6H_5—(N=N—$C_6H_4)_n$—N=N—C_6H_5	95% EtOH	328	(21.9)	359	(43.7)	380	(57.5)	400[d]	(63.0)			220
1,4-H_2N—C_6H_4—(N=N—$C_6H_4)_n$—N=N—C_6H_4—NH_2[e]	Benzene	397	(28.5)	431[f]	(45.6)	448[g]	(51.6)	500[h]	(70.8)			222
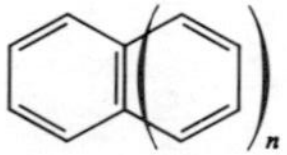		202[i]	(6.6)	285[j]	(4.0)	375[k]	(8.3)	471[l]	(11.2)	580[m]	(12.6)	

[a]*fs* = Fine structure.
[b]In chloroform λ_{max} 380 mμ, mϵ 79 (219).
[c]In ethanol (220).
[d]In tetrahydrofuran.
[e]Spectra of other families reported, for example, 4-nitroazobenzene, 4,4′-dinitroazobenzene, 4-aminoazobenzene, 4-hydroxyazobenzene, and 4-amino-4′-hydroxyazobenzene.
[f]In ethanol at λ_{max} 474 mμ, mϵ 60.3 (221).
[g]In tetrahydrofuran at λ_{max} 490 mμ, mϵ 66.1 (221).
[h]Reference 221.
[i]In hexane (104).
[j]In hexane (223).
[k]In isooctane.
[l]In hexane (224).
[m]In benzene (225).

structures contributing to the excited state; hence the wavelength shifts to longer wavelength at a slower rate. The increasing separation of charge involves increasing amounts of energy so that the wavelength maxima approach a limiting value. This is called convergence. For example, in the *p*-polyphenyls sexiphenyl in hexane absorbs at 308 mμ, and the series approaches a limiting wavelength calculated to be 313 mμ(226).

There are two ways of decreasing the energy involved in the separation of charge and shifting the wavelength maxima further into the red. A cationic salt can be formed, as shown in equation 6.1 (227), or a favorable environment can be created for the separation of charge, as shown in Table 6.11 (228). A typical example of a convergent series is the zwitterionic structures **40**; addition of a proton forms the nonconvergent series of cationic resonance structures.

$$\text{(benzothiazole, S, N–}C_2H_5\text{)}\,C{=}(CHCH)_n{=}NC_6H_5 \;\xrightarrow[\text{MeOH}]{H^+}\; \text{(benzothiazole, S, N–}C_2H_5\text{)}\,C{=}(CHCH)_n{=}\overset{+}{N}HC_6H_5 \tag{6.1}$$

40 **41**

	40		**41**	
n	λ_{max} (mμ)	mε	λ_{max} (mμ)	mε
0	302	11	299	14
1	394	38	414	55
2	448	59	516	107
3	485	68	613	76

According to Brooker(229), rapid convergence of absorption maxima in a vinylogous series can be attributed to a considerable difference in the energies of the extreme resonance structures involved and to an increase in this difference as the chain is lengthened.

Nonconvergence in a series (with increase in n the wavelength maximum shifts to the red at a steady rate and does not approach a limiting value) normally indicates approximate energetic equivalence of the limiting resonance structures(230). In the majority of the series shown in Tables 6.8 and 6.10 convergence is the rule. Examples of zwitterionic resonance nonconvergent series have been presented(230). In the strongly polar Z_z-series the structures are closest to energetic equivalence in pyridine and furthest from energetic equivalence in water. Thus in water and in aqueous pyridine the series is convergent, whereas in pyridine it is nonconvergent (Table 6.12).

An empirical relation between the wavelength maxima and n has been found for many of these series(231, 232). For many members of the convergent series $\lambda^2_{max} = kn$(22). This relation holds for the diphenyl-

TABLE 6.11
Effect of Decreasing the Energy Involved in the Separation of Charge (228)

$$C_6H_5-(CH=CH)_nCHO \longleftrightarrow {}^+C_6H_5=(CHCH=)_nCHO^-$$

$$\xrightarrow{AcCl,\ SbCl_5} (SbCl_6)^- \; {}^+C_6H_5=(CH-CH=)_nCHO^-Ac^+$$

	In $CHCl_3$		+ $SbCl_5$ + AcCl	
n	λ_{max} (mμ)	mϵ	λ_{max} (mμ)	mϵ
0	245	10.5	—	—
1	289	24.0	318	35.0
2	324	34.0	455	55.0
3	353	40.0	545	67.5
4	380	50.0	612	86.0
5	407	63.5	670	110.0
6	430	67.5	730	124.0

TABLE 6.12
Effect of Solvent on Convergence of a Highly Polar Zwitterionic Series (229)

$$H_3C\overset{+}{N}C_5H_4-(CH=CH)_n-C(=C(O^-)-O-N=C-\phi) \longleftrightarrow H_3CNC_5H_4=(CH-CH)_n=C(C(=O)-O-N=C-\phi)$$

	λ_{max} (mμ) (mϵ)							
	n							
Solvent	0		1		2		3	
Water	364	28.0	435	34.8	470	37.7	488	30.0
Pyridine–water:								
30:70	375	29.7	463	39.7	515	44.0	538	34.6
60:40	381	32.4	475	45.4	538	52.5	565	35.9
90:10	392	34.4	498	59.0	580	75.8	639	44.2
Pyridine	400	35.0		–	610	118.0	710	93.7

polyenes (22), α-(α-quinolyl)-ω-phenylpolyenes (191), α-furyl polyenecarboxylic acids (232), α-furylpolyene aldehydes (232), methylpolyenecarboxylic acids (232), methylpolyene aldehydes (232), α-(4-pyrimidyl)-ω-phenylpolyenes (190), ω-*p*-methoxyphenylpolyene aldehydes (167), 5-nitrofurylpolyene aldehydes (170), dimethylpoly-ynes (200), and diphenylpoly-ynes (203, 204). It is reasonable to believe that as n increases in the series to much higher values the linear relation with λ_{max}^2 would fall off. A few exceptions to this rule are known where a plot of λ_{max}^2 against the number of ethylenic groups in the chain (i.e., n) gives a curve that is convex upward—for example, *p*-dimethylaminophenylpolyenals [$(CH_3)_2NC_6H_4(CH{=}CH)_nCHO$] (233), the *p*-phenylenediamine Schiff bases of phenylpolyenals [$C_6H_5(CH{=}CH)_nCH{=}NC_6H_4N{=}CH(CH{=}CH)_nC_6H_5$] (144), and the *p*-polyphenyls (164). On the other hand, in the 1,1′-dianthrylpoly-ynes (205) and 9,9′-dianthrylpoly-ynes (206) a linear relation is found between the position of the long-wavelength maximum and n^2. In the analogous series of 2-naphthyl, 9-phenanthryl, and 1-naphthyl poly-ynes the linear relation is between the wavelength maximum and $n^{1.3}$, $n^{1.4}$, and $n^{1.5}$, respectively (206).

In the members of the nonconvergent series the wavelength maximum is a linear function of n, $\lambda_{max} = kn$. The series dissolved in pyridine (Table 6.12) is one example. Another series of moderate polarity that is nonconvergent in pyridine or aqueous pyridine is **42** (230).

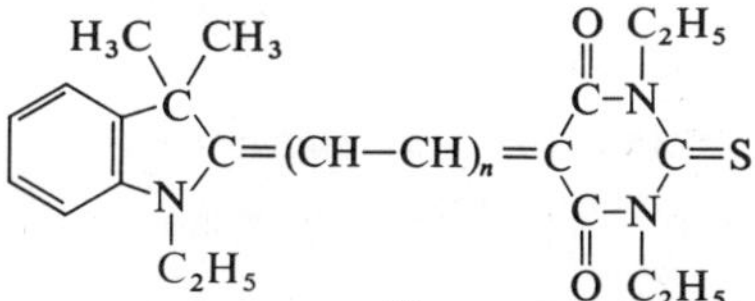

42

Solvent	λ_{max} (mμ)		
	$n = 1$	$n = 2$	$n = 3$
Pyridine	495	595	698
Water–pyridine (2:3, v/v)	498	594	693

This series is relatively insensitive to changes of solvent. An example of a weakly polar series, Z_n in structure, that is affected differently by solvent polarity is **43** (230).

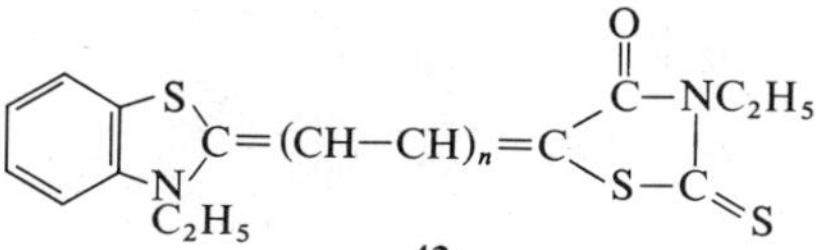

43

Solvent	λ_{max} (mμ) $n=0$	$n=1$	$n=2$	$n=3$
Pyridine	432	528	605	635
Water–pyridine (1 : 1)	432	540	633	733

In anhydrous pyridine the values show marked convergence, whereas in aqueous pyridine they are nonconvergent.

Thus in the Z_z structure depicted in Table 6.12 the dipolar structure contributes much more strongly to the ground state. Highly polar aqueous solvents tend to stabilize or decrease the energy of the dipolar form; consequently there is a greater energetic dyssymetry between the limiting resonance structures, and the series converges. In the series **42** the limiting resonance structures of these dyes are near the isoenergetic point, especially in pyridine; hence the series is nonconvergent. In series **43**, where the dipolar resonance structure contributes mainly to the excited state, highly polar aqueous solutions lower the energy of this state; hence an approximate energetic equivalence of the limiting resonance structures is obtained, and the series is nonconvergent. Convergence of a series is thus partly a function of the intrinsic polarity of the vinylogs and partly of the polarity of the solvent.

F. Violet Shifts

Examples of violet shifts with chain lengthening are known. Structures **44**(234) and **45a**(235) show this phenomenon.

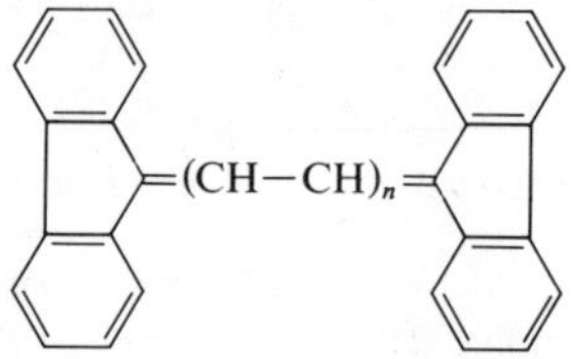

=(CH—CH)$_n$=

44

No.	X	λ_{max} (mμ) $n=0$	$n=1$	$n=2$
44		460	446	555
45a	S	545	508	528
45b	NH	595[a]	600[a]	

[a]In xylene

The longer wavelength absorption of **44**, $n = 0$ as compared with $n = 1$ is apparently due to the decreased steric hindrance in the excited state of the former compound. Thus the excited state is of lower energy, and the compound absorbs at relatively longer wavelength.

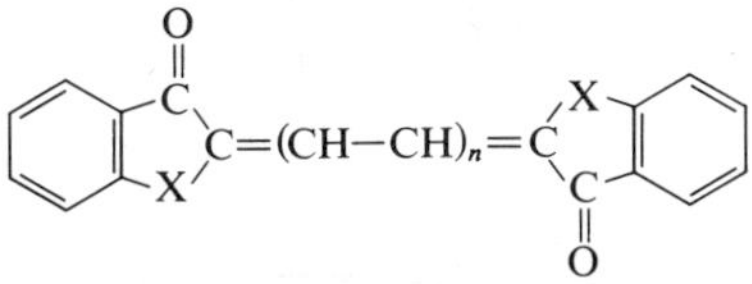

C=(CH—CH)$_n$=C

45

In thioindigo and indigo lengthening the chain means the addition of a positive resonance node, as in **46**, which would have a strong violet-shift effect balancing out the bathochromic effect of the lengthening of the conjugation chain.

C—$\overset{+}{C}$H—CH=C

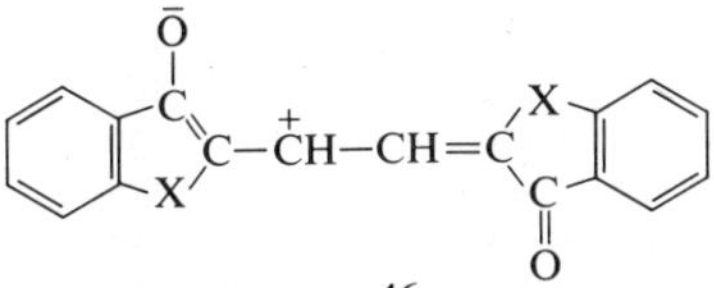

46

If the conjugation chain in **45** is lengthened by inserting a benzene ring, as in **47**, then a further violet shift is effected (235). This shift is derived from the loss of Kekulé resonance of the benzene ring in the excited state of the molecule. The change in energy between the ground and excited states is increased, and the molecule consequently absorbs at shorter wavelength (compare with **45**, $n = 1$).

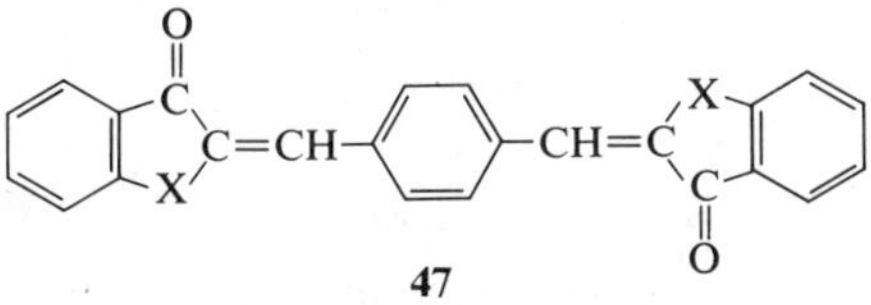

47

X	λ_{max} (mμ)
S	488
NH	519

G. Acidity and Basicity

The effect of chain lengthening on the acidity or basicity of resonance terminals has been little studied. In one series of polyenic diphenylketones with a lengthening of the chain the basicity of the carbonyl negative resonance terminal increases (Table 6.13)(166). Apparently with increasing length of the chain the electron density on the carbonyl oxygen increases.

TABLE 6.13
Effect of Increasing Separation of Charge and Increasing Length of Conjugation Chain on Basicity, Wavelength, and Intensity (166)

$$C_6H_5(CH{=}CH)_n{-}\overset{O}{\overset{\|}{C}}{-}C_6H_5 \xrightarrow[H_2SO_4]{20\%} C_6H_5(CH{=}CH)_n{-}\overset{^+OH}{\overset{\|}{C}}{-}C_6H_5$$

		In AcOH		In 20% H_2SO_4	
n	pK_a	λ_{max} (mμ)	mϵ	λ_{max} (mμ)	mϵ
0	−6.2	254	1.8	338	
1	−4.2	310	2.3	390	0.74
2	−3.5	343	3.6	460	2.9
3	−3.2	376	3.1	530	4.0
4	−3.0	402	5.7	585	5.2
5	−2.8	418	6.4	650	5.2
6	−2.7	445	–	735	–

H. Application

Some of the polyene series are found in nature. The spectra of these compounds shift to the red with the increasing chain of conjugation. Examples are available in the carotenoid (Table 6.14)(236) and piperine series (**48**)(237).

$$H_2C\langle O_2\rangle C_6H_3{-}(CH{=}CH)_n{-}\overset{O}{\overset{\|}{C}}{-}N\langle C_5H_{10}\rangle$$

48

	n	λ_{max} (mμ)	mϵ
a	0	290	48.6
b	1	325	19.0
c	2	343	37.1
d	3	364	48.9

Piperine and piperettine (**48c** and **d**) were characterized and determined in black and white peppers.

A reagent or its vinylog can be picked for an analysis of a test substance, depending on whether excess reagent will make the blank color

TABLE 6.14
Effect of Conjugation-Chain Length on the Wavelength Maxima of Some Carotenoids (236)

Compound	Number of conjugated double bonds	λ_{max} (mμ)[a]
Phytofluene	5	367
retro-Dihydro-C_{19}-aldehyde	5[b]	402
ζ-Carotene	7	425
5,6-Dihydro-α-carotene	9	469
α-Carotene	10	474
β-Carotene	11	480
16,16′-Homo-β-carotene	12	496
Rhodoviolascin	13	527

[a] In hexane or petroleum ether.
[b] And one C=O bond.

analytically prohibitive and whether the absorbance of the vinylogous chromogen at longer wavelengths is advantageous, other things being equal. Thus 4-dimethylaminobenzaldehyde or 4-dimethylaminocinnamaldehyde can be used for the colorimetric analysis of 4-aminoantipyrine, a urinary metabolite of aminopyrine (238). The chromogens **49** are formed.

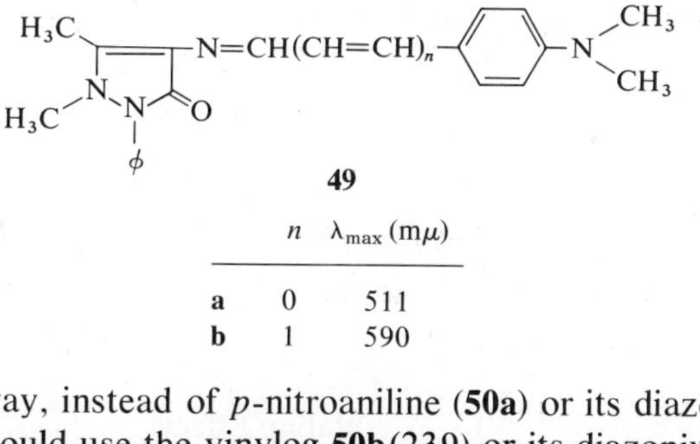

H_3C, H_3C, N, N, O, ϕ — N=CH(CH=CH)$_n$— C$_6$H$_4$ —N(CH$_3$)$_2$

49

	n	λ_{max} (mμ)
a	0	511
b	1	590

In the same way, instead of *p*-nitroaniline (**50a**) or its diazonium salt as reagents one could use the vinylog **50b** (239) or its diazonium salt as the reagent. The wavelength shift may be favorable in analysis. The spectra of **50** were obtained in ethanol.

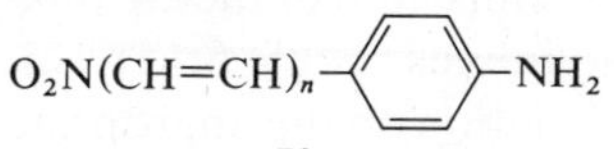

$O_2N(CH=CH)_n$—C$_6$H$_4$—NH_2

50

	n	λ_{max} (mμ)	mϵ
a	0	375	15.5
b	1	420	22.9

Some reagents can be used in the analysis of one-, three-, and five-carbon fragments. The chromogens thus obtained belong to a vinylogous series. Thus in the analysis for formic acid, malonaldehyde, and glutaconaldehyde with thiobarbituric acid the series **51** is obtained (240).

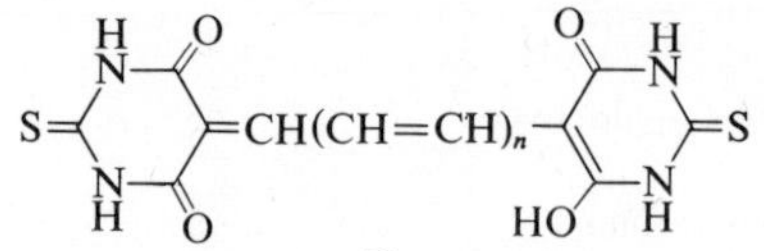

$$S{=}\text{(thiobarbituric ring)}{=}CH(CH{=}CH)_n{-}\text{(thiobarbituric ring, HO)}{=}S$$

51

Test substance	n	λ_{max} (mμ)
Formic acid	0	452
Malonaldehyde	1	532
Glutaconaldehyde	2	622

In the determination of shikimic acid with thiobarbituric acid a carboxy derivative of **51**, $n = 2$, λ_{max} 660 mμ, is believed to be the final chromogen (241).

A polyene test substance containing a CH_2 group at one end and a CH_2OH at the other can be assayed through dehydration to a compound containing a longer chain of conjugation and absorbing at longer wavelength. An example of this type of assay is the determination of vitamin A (**52**) in cod liver oil through dehydration to **53**(242).

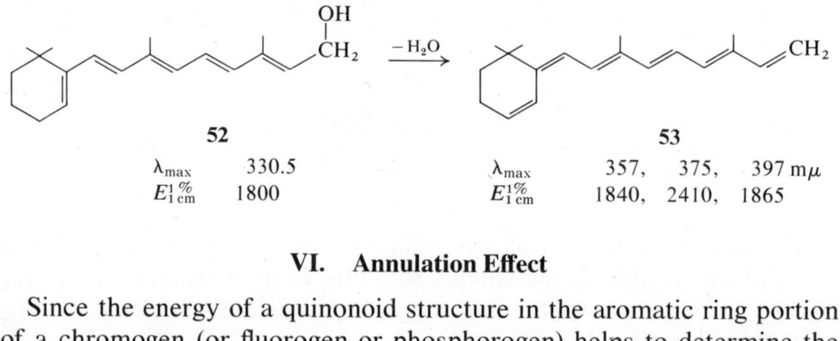

OH
CH$_2$ $\xrightarrow{-H_2O}$ CH$_2$

52 **53**

	52		**53**		
λ_{max}	330.5	λ_{max}	357,	375,	397 mμ
$E^{1\%}_{1\,cm}$	1800	$E^{1\%}_{1\,cm}$	1840,	2410,	1865

VI. Annulation Effect

Since the energy of a quinonoid structure in the aromatic ring portion of a chromogen (or fluorogen or phosphorogen) helps to determine the position of absorption (or emission) of the chromogen, etc., the oxidation–reduction potentials of quinones (Table 6.15) can be of some use in understanding the effect of annulation on the appropriate wavelength bands.

A thorough study of the annulation effect has never been attempted. Probably the most complete study of this phenomenon has been reported by Clar(247) in his study of the absorption spectra of the polycyclic hydrocarbons. Linear annulation results in a red spectral shift,

TABLE 6.15

Oxidation–Reduction Potentials of Quinones in Alcohol at 20–30°C (243–246)

Parent compound	Carbonyl positions	E_q (volt)
Naphthalene	1,4,5,8	0.972
Diphenyl	4,4′	0.954
Benzene	1,2	0.792
Naphthalene	2,6	0.76
Benzene	1,4	0.715
Diphenyl	2,5	0.694
Phenanthrene	1,2	0.660
Phenanthrene	3,4	0.621
Pyrene	1,6	0.612
2-Tosylnaphthalene	1,4	0.605
Naphthalene	1,2	0.576
Sodium 2-naphthalenesulfonate	1,4	0.553
3-Nitrophenanthrene	9,10	0.551
3-Cyanophenanthrene	9,10	0.553
Phenanthrene	1,4	0.523
3-Benzoylphenanthrene	9,10	0.519
Pyrene	1,8	0.514
Picene	13,14	0.503
Benzo[*c*]phenanthrene	5,6	0.492
3-Phenanthrenesulfonic acid	9,10	0.490
Anthracene	1,2	0.490
Naphthalene	1,4	0.484
2,5-Dimethoxybenzene	1,4	0.476
Pyrene	4,5	0.474
Chrysene	5,6	0.465
Phenanthrene	9,10	0.460
Dibenz[*a*,*j*]anthracene	5,6	0.455
2-Phenylnaphthalene	1,4[a]	0.452
Picene	5,6	0.451
8-Hydroxynaphthalene	1,4	0.447
Dibenz[*a*,*h*]anthracene	5,6	0.446
Benzo[*a*]pyrene	6,12	0.443
	4,5	0.442
	3,6	0.441
	1,6	0.438
4-Methoxynaphthalene	1,2	0.433
Benz[*a*]anthracene	5,6	0.430
2-Acetamidonaphthalene	1,4	0.417
Picene	5,6,7,8	0.416
2-Methylnaphthalene	1,4	0.408
3-Hydroxyphenanthrene	9,10	0.407
Dibenz[*a*,*j*]anthracene	5,6,8,9	0.406
Anthracene	1,4	0.401

TABLE 6.15 (*cont'd*)

Parent compound	positions	E_q (volt)
Chrysene	6,12	0.392
3-Aminophenanthrene	9,10	0.362
2-Naphthol	1,4	0.356
2-Methoxynaphthalene	1,4	0.353
Dibenz[*a*,*j*]anthracene	7,14	0.302
3-Methyl-2-naphthol	1,4	0.299
Dibenz[*a*,*h*]anthracene	7,14	0.292
2-Anilinonaphthalene	1,4	0.286
2-Naphthylamine	1,4	0.274
N-Methyl-2-naphthylamine	1,4	0.232
Benz[*a*]anthracene	7,12	0.228
2,5,8-Trihydroxy-3-methylnaphthalene	1,4	0.200
Anthracene	9,10	0.154

[a]Naphthalene.

especially of the *p*-bands. The maximum annulation effect is obtained by linear fusion of benzene rings in the acene series, as shown by the strong red shift of these compounds going from benzene (~205 mμ) to pentacene (~587 mμ). Angular annulation results in fewer quinonoid rings (as compared to linear annulation) and causes a violet shift. Thus the bands shift toward shorter wavelengths in the series perylene, benzo-[g,h,i]perylene, and coronene. Annulation effects on the β-, *p*-, and α-bands (Clar's nomenclature) of arenes have also been discussed by Altiparmakian and Braithwaite (248).

The absorption spectra of polynuclear cinnolines show the presence of β-, *p*-, and α-bands, with the latter two bands showing a slight red shift as compared with the analogous hydrocarbon bands. In addition, the α-bands of the cinnolines as compared with the analogous hydrocarbons show an increase in intensity, a loss in fine structure, and the presence of nonbonding $n \rightarrow \pi^*$ absorption at long wavelengths. The effect of annulation, with and without alkylation and *N*-oxidation, on these various bands has been discussed (248). These effects usually cause a red shift of the various $\pi \rightarrow \pi^*$ bands.

Another effect of annulation can be seen in Table 6.16. In the simpler molecules of this series the dipolar Z_z-structure is most important in the ground state. When the molecule contains four or more rings, the Z_n-structure predominates. Thus on annulation the Z_z-structure changes to a Z_n one, and the spectrum shifts to the red.

On the other hand, in the comparable diazomerocyanine dyes all

TABLE 6.16
Absorption Maxima of Some Phenol Betaines in Benzyl Alcohol and Chloroform (249)

$H_3C\overset{+}{N}$—CH=CH—O^-

Benzo rings present	Structure	In benzyl alcohol		In chloroform	
		λ_{max} (mμ)	mϵ	λ_{max} (mμ)	mϵ
None[a]	Z_z	523	41	620	
A	Z_z	610	49	677	38
C	Z_z	615	66	643	38
AB	Z_n	700	55	575	28
CD	Z_n	670	45	600	17

[a] 4-(4-Oxystyryl)*N*-methylpyridyl betaine.

nine had Z_n-structures, as shown by the red spectral shift with increasing polarity and hydrogen-bonding strength of the solvent (250).

The striking extreme results for these compounds are shown in Table 6.17. The intensity for most compounds decreased with an increase in the number of fused rings. The compound with no fused rings had the

TABLE 6.17
Absorption Maxima of Some Diazomerocyanines (250)

H_3CN=N—N=O

Benzo rings present	In benzene		In benzyl alcohol	
	λ_{max} (mμ)	mϵ	λ_{max} (mμ)	mϵ
None	512	47.4	558	77.0
AB	558	26.4	625	29.0
CD	498	30.0	530	26.0
ABCD	517	15.1	542	14.2

highest spectral intensities in all examined solvents, the six-ring compound had the lowest intensities. The compound in Table 6.17, with rings A and B, absorbed at the longest wavelength in all solvents, whereas the compound with rings C and D absorbed at the shortest wavelengths. Some of the aspects of the structures in Tables 6.16 and 6.17 have been discussed from the viewpoint of hydrogen-bonding and solvent effects (251). The increasing energy of Kekulé resonance in a benzene or pyridine ring on annulation, the decreasingly lower energy of the *p*-benzoquinone > 1,4-naphthoquinone > 9,10-anthraquinone type of structures, and the high energy involved in the separation of charge are some of the factors involved in the reported spectral changes.

The effect of annulation on the spectra of 10 quinonemethides and quinodimethanes, where the negative resonance terminal is either a quinonoid oxygen or a dicyanomethene carbon, was a violet shift (252). The quinonoid-type structures involved in this study were *p*-benzoquinonoid, 1,4-naphthoquinonoid, and 9,10-anthraquinonoid. Protonation of the benzoquinonoid derivatives resulted in a violet shift, of the naphthoquinonoid derivatives mostly a violet shift, and of the anthroquinonoid derivatives a red shift. An example is the compound *p*-$(NC)_2C{=}C_6H_4{=}C({-}S{-}CH_2)_2$, λ_{max} 595 mμ, mϵ 32.4, in chloroform. Annulation to the naphthoquinone and anthraquinone analogs gives λ_{max} 515 and 450 mμ, mϵ 37.2 and 7.6, respectively. Protonation of these compounds at the dicyanomethene carbon atom of the monocyclic dicyclic, and tricyclic compounds gives λ_{max} 329, 429, and 562 mμ, mϵ 27.5, 5.6, and 1.3, respectively. In the last salt a carbethoxy group was present instead of a cyano group. In the salts annulation causes a strong red shift.

In the azobenzene type of dye annulation usually shifts the long-wavelength maximum to the red, as shown for **54** (253).

$$H_3C{-}\overset{O}{\overset{\|}{C}}{-}C_6H_4{-}N{=}N{-}C_6H_4{-}N(CH_3)_2$$

A A

54

Benzo rings	λ_{max} (mμ)	mϵ
None	445	32.0
AA	467	20.5

A similar red-shift effect and intensity decrease has been shown for 4-(4-dimethylaminostyryl)quinoline (λ_{max} 430 mμ, mϵ 12.6) on fusing a benzo ring to the 2,3-position to give the acridine analog (λ_{max} 470 mμ, mϵ 8.9)(254). The spectra of both compounds were measured in ether–alcohol (1:2, v/v) at −183°C. A red-shift effect is also shown by 1,1′-dianthrylpenta-acetylene (long-wavelength λ_{max} 473 mμ, mϵ 40.7, in tetrahydrofuran)(205) as compared with diphenylpenta-acetylene (λ_{max} 437 mμ, mϵ 13.0, in benzene)(204).

When annulation takes place on a benzene ring in the middle of a conjugation chain, the spectral changes are a little more complicated, as shown for **55**(255).

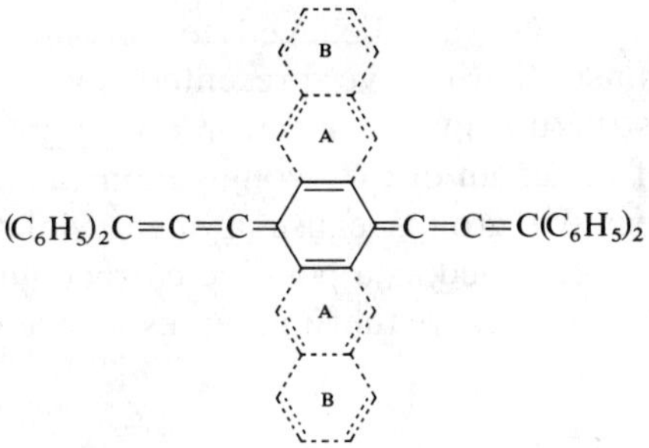

$(C_6H_5)_2C{=}C{=}C{=}$[ring]${=}C{=}C{=}C(C_6H_5)_2$

55

Benzo rings	λ_{max} (mμ)
None	540
AA	583
AABB	565

Annulation can also cause violet shifts(234), as shown for fulvene, benzofulvene, and dibenzofulvene, and for the quinonoid derivatives **56**, **57**, and **58**.

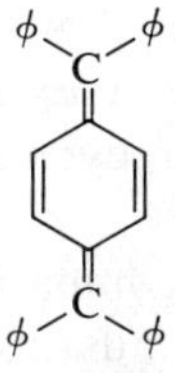

56 Orange

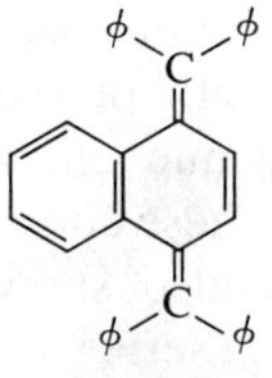

57 Yellow

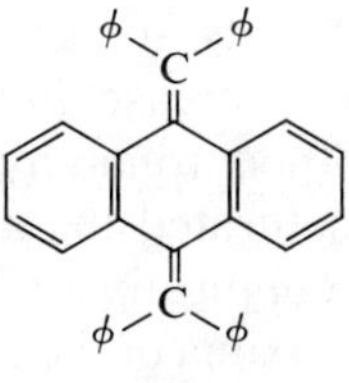

58 Colourless

The higher relative energy of the ground state and the lower relative energy of the excited state of **56**, and the smaller amount of steric hindrance in **56**, as compared with the other molecules probably account for the violet shift.

VII. Extraconjugation

By extraconjugation is meant any additional conjugation involving one or both resonance terminals with a conjugated system outside of the main system of conjugation and resulting in a red hyperchromic spectral shift.

It has been pointed out (256) that in the dinitrophenylhydrazone of a simple ketone, such as acetone the wavelength maximum of 2,4-dinitrophenylhydrazine (λ_{max} 352 mμ) (193) is displaced toward the visible, 360 mμ (257).

Such structures as **59C** are believed to account for this red shift. Szmant and Planinsek (256) have presented some evidence for this extraconjugative structure in that electron-donor groups substituted at the positive end of the chain of extraconjugation cause a slight red shift, whereas electronegative groups cause a slight violet shift. Structure **59D** has also been mentioned as a possible extraconjugative contributor (258). Structures **59A** ↔ **B** are the limiting resonance structures.

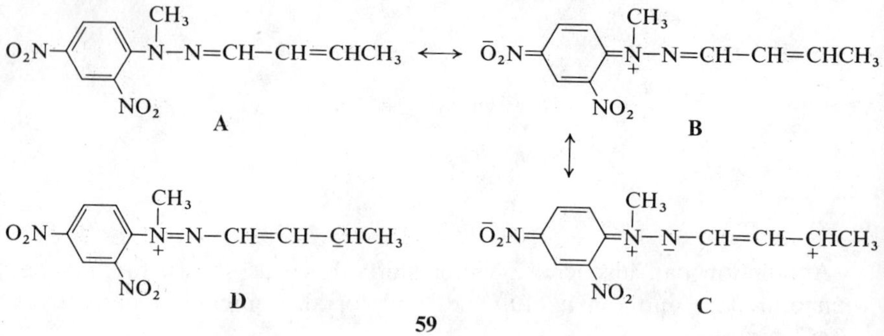

$O_2N-C_6H_3(NO_2)-N(CH_3)-N=CH-CH=CHCH_3$ (**A**) ⟷ $\bar{O}_2N=C_6H_3(NO_2)=\overset{+}{N}(CH_3)-N=CH-CH=CHCH_3$ (**B**)

↕

$O_2N-C_6H_3(NO_2)-\overset{+}{N}(CH_3)=N-CH=CH-\underline{C}HCH_3$ (**D**) $\quad \bar{O}_2N=C_6H_3(NO_2)=\overset{+}{N}(CH_3)-\overset{-}{N}-CH=CH-\overset{+}{C}HCH_3$ (**C**)

59

Table 6.18 lists some more definite examples of this type of extraconjugation. Increase in the length of the chain of extraconjugation causes a bathochromic-hyperchromic effect. The cyclopropyl group shows its unsaturated character in these compounds.

The extraconjugative effect is also shown by comparing the long-wavelength maxima of the following series of compounds (261):

$CH_2{=}CH{-}CH{=}CH_2$	$CH_3S{-}CH{=}CH_2$	$CH_2{=}CH{-}S{-}CH{=}CH_2$
60	**61**	**62**
λ_{max} 217 mμ, mε 20	λ_{max} 240s mμ, mε 10	λ_{max} 275 mμ, mε 5.0

A similar phenomenon is shown in the aniline derivatives (262) (all spectra in methanol).

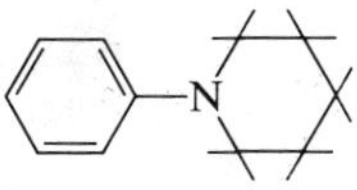

$CH_3CH_2H_2C$ C_2H_5 C_2H_5 $CH_3CH_2H_2C$ C_2H_5 C_2H_5

63 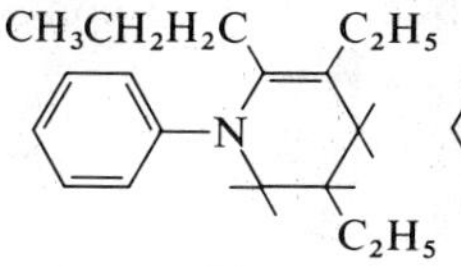 **64** 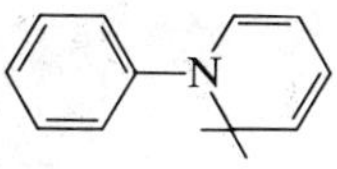 **65** 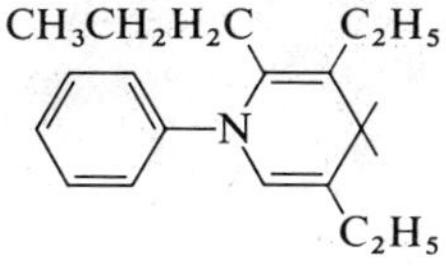 **66**

No.	λ_{max} (mμ)	mϵ
63	252	8.9
64	285	21.4
65	288	19.2
66	340	12.5

TABLE 6.18
Extraconjugation

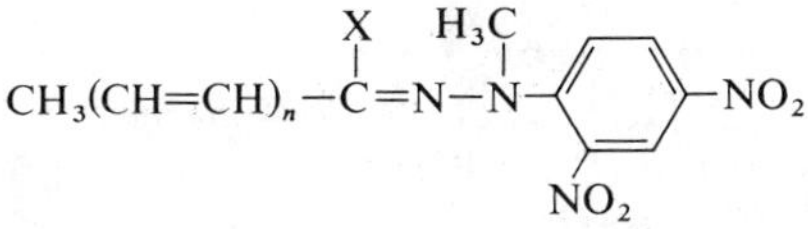

$CH_3(CH{=}CH)_n{-}C(X){=}N{-}N(CH_3){-}C_6H_3(NO_2)_2$

n[a]	X	λ_{max} (mμ)	mϵ
0	CH_3	375	19.1
1	H	397	22.3
2	H	410	25.4
3	H	424	28.9

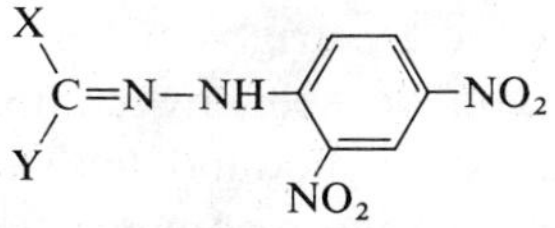

$XYC{=}N{-}NH{-}C_6H_3(NO_2)_2$

X[b]	Y	λ_{max} (mμ)	mϵ
CH_3	C_2H_5	368	24.0
CH_3	i-C_3H_7	368	22.9
i-C_3H_7	i-C_3H_7	369	24.1
CH_3	$CH_2{=}CH$	374	27.4
CH_3	C_3H_5[c]	375	26.8
C_3H_5	C_3H_5	381	25.6

[a]This series in methanol(259).
[b]In methylene chloride(260).
[c]Cyclopropyl.

Extraconjugation has also been postulated for some *p*-nitrophenylsulfides (**67**)(263):

$O_2N-C_6H_4-S-R$

67

Thus the long-wavelength bands of **67a**(264) or **b** (in dioxane) are at shorter wavelength than those of **67c** and **d**, probably due to an extraconjugative structure, such as **68**, contributing to **67c** and **d**.

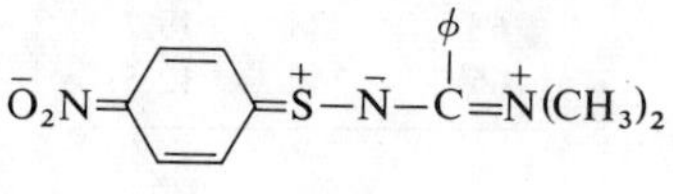

$\bar{O}_2N{=}C_6H_4{=}\overset{+}{S}-\overset{-}{N}-C(\phi){=}\overset{+}{N}(CH_3)_2$

68

No.	R	Solvent	λ_{max} (mμ)	mϵ
67a	CH_3	Ethanol	340	12.6
67b	$N(CH_3)C(C_6H_5){=}NCH_3$	Ethanol	333	12.6
67c	$N{=}C(C_6H_5)NH_2$	Dioxane	390	17.8
67d	$N{=}C(C_6H_5)N(CH_3)_2$	Dioxane	400	15.9

Extraconjugation plays a part in the long-wavelength absorption of 3,5-diacetyl-2,6-dimethyl-1,4-dihydropyridine (λ_{max} 412, mϵ 8.0, in ethanol), obtained in the colorimetric determination of formaldehyde with 2,4-pentanedione(265). The method has been used to determine formaldehyde precursors [e.g., triglycerides(266)] and atmospheric formaldehyde(267). As can be seen by comparing the diacetyl compound with the dicarbethoxy compound (**69**), decreasing the electron-acceptor strength of both negative resonance terminals shifts the wavelength maximum to the violet approximately 43 mμ.

The role of extraconjugation can be seen in the spectral comparison of the analogous carbethoxy derivative (**69**) with appropriate compounds lacking the extraconjugation.

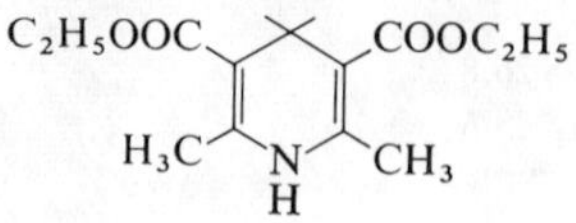

C_2H_5OOC, $COOC_2H_5$, H_3C, N, H, CH_3

69

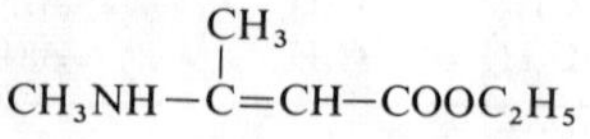

$CH_3NH-C(CH_3){=}CH-COOC_2H_5$

70

CH_3, $COOC_2H_5$, N, H, CH_3

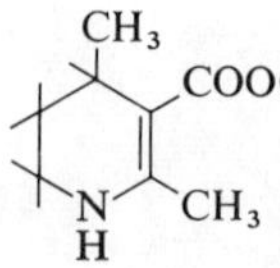

71

No.	Solvent	λ_{max} (mμ)	mϵ	Ref.
69	Methanol	369	6.1	268
70	Methanol	274	14.4	268
71	Ethanol	285	17.8	269

An interesting extraconjugative effect is shown by *o*-nitroanilinoboron halides, such as the series shown in Table 6.19(270). The transition causing the purple color in the anilinoborons is believed to involve the boron and the nitro groups. The colors are observed only in systems, such as **72**, in which boron is linked to the aromatic ring through nitrogen (271) or oxygen(272). These compounds are Z_n-structures since their long-wavelength bands shift to the red with increasing solvent polarity.

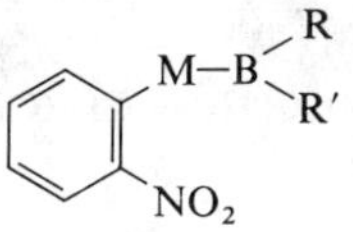

M—B(R)(R′)
NO_2
72

M=O or NR

Other examples of extraconjugation are discussed in Section X.B.

VIII. Cross-Conjugation

Cross-conjugation is found in a compound that has more than two resonance terminals and more than two limiting resonance structures.

TABLE 6.19
Extraconjugation in Some *o*-Nitroaniline Derivatives Dissolved in Benzene(269)

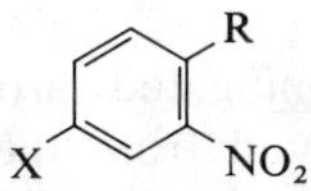

R
X
NO_2

	R=NH_2		R=NH—B(Cl)C_6H_5	
X	λ_{max} (mμ)	mϵ	λ_{max} (mμ)	mϵ
NO_2	382	4.95	510	3.60
H	392	5.02	516	2.70
CH_3	404	4.88	524	2.80
Cl	408	5.00	530	2.90
CH_3O	430	5.38	542	3.20

The conjugation may be competitive or additive and usually involves interaction of one of the main resonance terminals with a third resonance terminal; hence part of the main conjugation system is common to at least two competitive or additive conjugation systems. Cross-conjugation can be extraconjugative and result in a red shift—or it can be competitive and result in a violet shift or in no wavelength shift with doubling, tripling, etc., in intensity.

A. Cross-Conjugation with a Carbonyl Group

Red shifts derived from extraconjugation at the positive resonance terminal have been discussed in the preceding section. Additional conjugation attached to a carbonyl carbon can also cause a red shift. As an example, consider the limiting resonance structures **73** (182):

$$CH_3(CH{=}CH)_n{-}\overset{\overset{\large O}{\|}}{C}{-}CH{=}CH{-}NHC_6H_5 \leftrightarrow CH_3(CH{=}CH)_n{-}\overset{\overset{\large O^-}{|}}{C}{=}CH{-}CH{=}\overset{+}{N}HC_6H_5$$

73

Extraconjugation in **73**, $n = 1$, probably involves such resonance structures as

$$CH_3\overset{+}{C}H{-}CH{=}\overset{\overset{\large O^-}{|}}{C}{-}CH{=}CH{-}NHC_6H_5 \leftrightarrow CH_3CH{-}\overset{-}{C}H{-}\overset{\overset{\large \bar{O}}{|}}{C}{=}CH{-}CH{=}\overset{+}{N}HC_6H_5$$

$$\swarrow$$

$$CH_3CH{=}CH{-}\overset{\overset{\large O}{\|}}{C}{-}CH{=}CH{-}\overset{+}{N}H{=}C_6H_5^-$$

As can be seen from the data in Table 6.20, extraconjugation is available at both ends of the O=C—CH=CH—NR_2 chromophore in compound **73**, $n = 1$.

In some of the cross-conjugated aromatic carbonyl compounds two $\pi \rightarrow \pi^*$ transitions derived from the two chromophores can be found (Table 6.21). For some of the chromophores an extraconjugation effect can be seen as the conjugative power of the other group is increased. This is shown for the *p*-$(CH_3)_2NC_6H_4CH{=}CHC{=}O$ chromophore in Table 6.21 and for 10 *p*-dimethylaminobenzylidene dyes containing this chromphore. The absorption long-wavelength band for these latter dyes in methanol ranges from 436 mμ for 3-*p*-dimethylaminobenzylidene-1-ethyloxindole (**74**), to 490 mμ for 5-*p*-dimethylaminobenzylidene-1,3-diethyl-2-thiobarbituric acid (**75**) (230).

74

75

In some of the benzophenones the relative intensities of the two main bands are dependent on the number and type of chromophores present; for example, in cyclohexane solution benzophenone, two $C_6H_5C{=}O$ chromophores, λ_{max} 248 mμ, mϵ 20.0; 4-methoxybenzophenone, $C_6H_5C{=}O$ chromophore, λ_{max} 247 mμ, mϵ 10.5, and 4-$CH_3OC_6H_4C{=}O$ chromophore, λ_{max} 274 mμ, mϵ 17.0; and bis-4,4′-dimethoxybenzophenone, two 4-$CH_3OC_6H_4C{=}O$ chromophores, λ_{max} 278 mμ, mϵ 27.0(279).

Cross-conjugation of an *exo* and an *endo* double bond with a carbonyl group has been investigated and found to have an extraconjugative effect(280). The additive effect of two polyenic conjugation chains attached to a carbonyl group can be seen in Table 6.22.

Cross-conjugation can have a violet-shift effect. A resonance structure contributing to this violet shift is shown in Table 6.23. Decreasing the basicity of the nitrogen attached to the carbonyl group and thus decreasing the strength of cross-conjugation shifts the wavelength maximum to the red. Another way of looking at this effect is to consider the carbonyl carbon as a positive node; thus the decrease in the electron-donor strength of the attached nitrogen would cause a red shift.

TABLE 6.20
Extraconjugation at Both Resonance Terminals(182)

$$R{-}(CH{=}CH)_n{-}\overset{O}{\overset{\|}{C}}{-}CH{=}CH{-}N(R^1)(R^2)$$

R	R^1	R^2	n	λ_{max} (mμ)[a]	mϵ
n-C_3H_7	C_2H_5	C_2H_5	0	307	28.0
CH_3	H	C_6H_5	0	338	27.5
CH_3	C_2H_5	C_2H_5	1	339	24.0
C_6H_5	H	C_6H_5	0	374[b]	31.0
CH_3	H	C_6H_5	1	377	29.5

[a]Spectra in ethanol.
[b]$C_6H_5C{=}O$ band at 254 mμ, mϵ 17.5.

TABLE 6.21
Cross-Conjugation and the Absorption Spectra of Aromatic Carbonyl Compounds

$$X-\overset{\overset{\displaystyle O}{\|}}{C}-Y$$

			λ_{max}(mμ) (mϵ) of appropriate chromophore				
X	Y	Solvent	$C_6H_5C{=}O$	$C_6H_5CH{=}CHC{=}O$	$Me_2NC_6H_4C{=}O$	$Me_2NC_6H_4CH{=}CHC{=}O$	Ref.
p-$Me_2NC_6H_4$	CH_3	95% EtOH			337 (25.6)		40
p-$Me_2NC_6H_4$	C_6H_5	Methanol	248 (15.9)		356 (27.0)		273
p-$Me_2NC_6H_4$	*p*-$Me_2NC_6H_4$	95% EtOH			390[a] (23.0)		274
CH_3	C_6H_5	95% EtOH	241 (13.2)				275
C_6H_5	C_6H_5	95% EtOH	253 (18.6)				274
p-$Me_2NC_6H_4$	$C_6H_5CH{=}CH$	Methanol		303 (20.0)	387 (26.0)		273
CH_3	$C_6H_5CH{=}CH$	95% EtOH		286 (23.4)			276
C_6H_5	$C_6H_5CH{=}CH$	95% EtOH	253 (5.4)	310 (22.4)			277
CH_3	*p*-$Me_2NC_6H_4{-}CH{=}CH$	*n*-Heptane				360 (29.5)	278
C_6H_5	*p*-$Me_2NC_6H_4{-}CH{=}CH$	Methanol	264 (17.8)			419 (32.4)	273

[a] Another band at 370 mμ, mϵ 19.1.

TABLE 6.22
Additive Effect of Polyenic Cross-Conjugation with a Carbonyl Group (281)

$C_6H_5(CH{=}CH)_nCO(CH{=}CH)_mC_6H_5$			$C_6H_5(CH{=}CH)_{n+m}C_6H_5$	
n	m	λ_{max} (mμ)	$n+m$	λ_{max} (mμ)
1	0	310	1	295
1	1	330	2	328
2	1	357	3	349
2	2	375	4	375
3	3	430	6	420
$CH_3(CH{=}CH)_4CO(CH{=}CH)_4CH_3$		415	$CH_3(CH{=}CH)_8CH_3$	405

TABLE 6.23
Violet-Shift Effect of Cross-Conjugation (282)

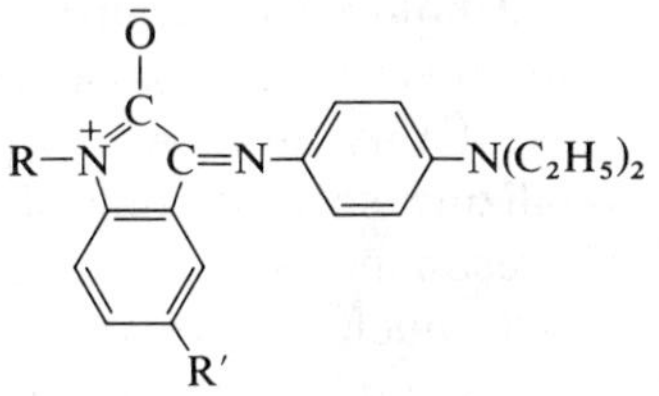

R′	R	λ_{max} (mμ)
H	H	487
Br	C_2H_5	498
Br	CH_3	502
Br	H	506
Br	C_6H_5	518
NO_2	H	524
Br	Ac	530

B. Cross-Conjugation with Ring Carbonyl

Examples of cross-conjugation with ring carbonyl show that a band is derived from each of two different chromophores oriented at approximately right angles to each other. Thus the absorption spectra of the tetracyclones **76** shows that substitution of electron-donor groups in the *para* position of either the 2- and/or 5-phenyl ring affects mainly the absorption band near 512 mμ, whereas substitution in the *para* position of the 3- and/or 4-phenyl ring affects primarily the absorption band near 342 mμ (283).

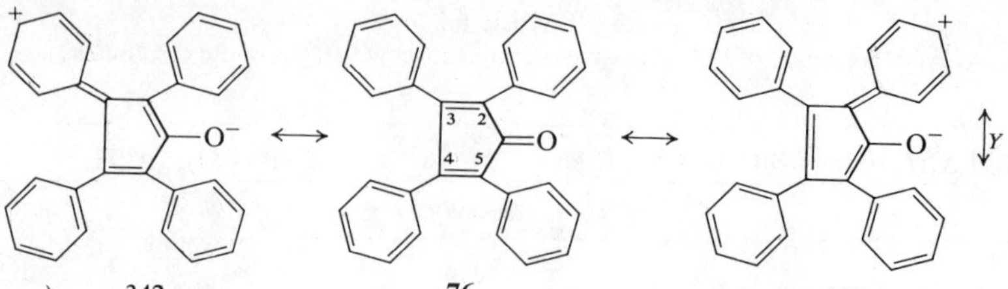

$\lambda_{max} \sim 342$ mμ ⟷ **76** ⟷ $\lambda_{max} \sim 512$ mμ

Thus the longer chain of conjugation is associated with the band near 512 mμ, whereas the shorter chain is associated with the band near 342 mμ.

In some compounds in which the cationic resonance terminals and an anionic resonance terminal are each part of a different bulky ring, all three terminals cannot occupy the same plane. Thus, when the acidic nucleus is twisted out of the plane, a cationic resonance structure is formed that has been called holopolar because of the complete charge separation(284). Thus in the thiacarbocyanine **77**, the resonance structures **77A** ↔ **B** represent this form, which predominates in methölic solution. In nonstabilizing solvents (e.g., lutidine) the compound realigns into a zwitterionic resonance structure, as represented by **77A** ↔ **C**, in which one of the basic nuclei is twisted out of the plane. This meropolar substance (the name implies less than complete charge separation) is formed by a change that has been called allopolar isomerism(284).

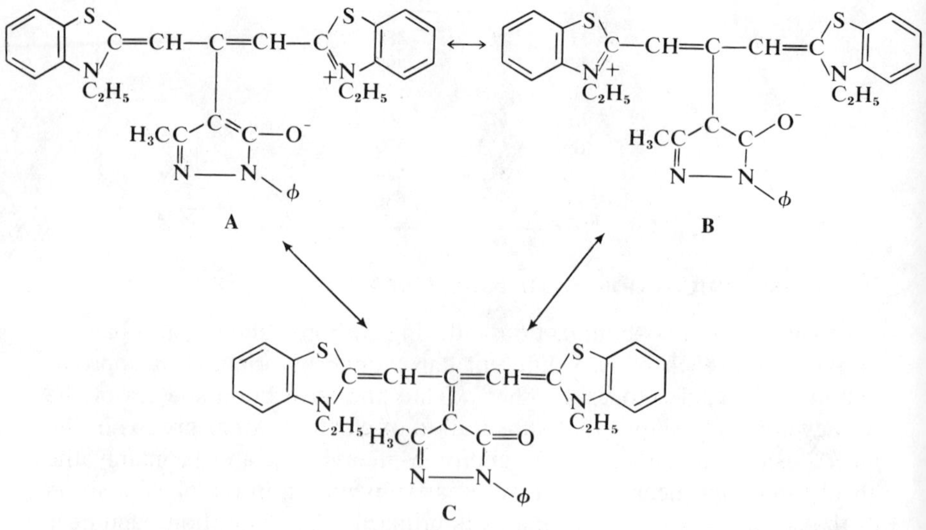

A ⟷ **B**

C

77

Thus in methanolic solution the acidic nucleus is twisted out of the plane of the molecule, and the cationic resonance structures **77A** ↔ **B** are the limiting resonance structures from which the band at about 565 mμ (mϵ 140) is derived. In lutidine solution one of the basic nuclei is twisted out of the plane of the molecule, and the structures **77A** ↔ **C** and **77B** ↔ **C** are the limiting resonance structures from which the band at 510 mμ (mϵ 50) is derived.

It is possible to prevent this isomerization by forcing the positive resonance terminals into planarity, as in **78**, where the structure is firmly held in a holopolar arrangement (284).

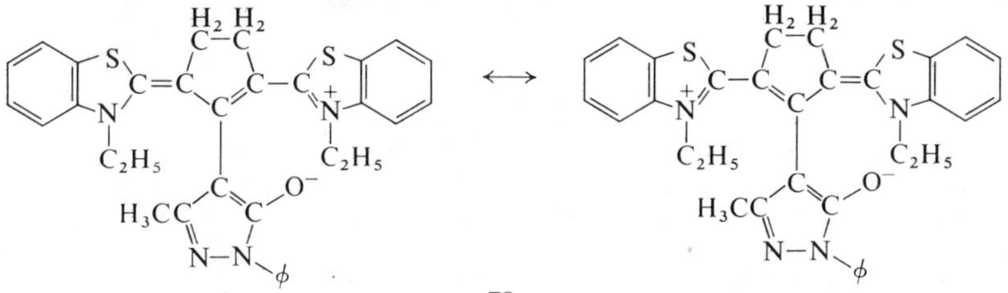

78

An example of allopolar isomerism as affected by solvent polarity is shown in Table 6.24. The isomerism of the compound in the table is also affected by temperature since decreasing the temperature of a solution in *n*-butanol to 4°C increases the relative amount of the holopolar isomer (286).

The bisdimethylaminofuchsone **79** shows a solvent effect similar to that shown in Table 6.24 (285). Thus in methanol the compound is present as the holopolar isomer (λ_{max} 555 mμ, mϵ 74) and in benzene as the meropolar isomer (λ_{max} 430, 480 mμ; mϵ 32, 36, respectively).

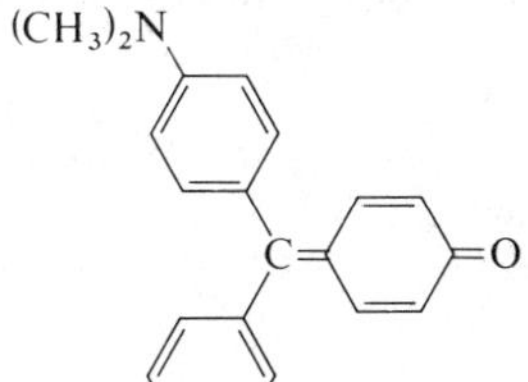

79

C. Meta-Substituted Benzene Derivatives

Examples of *meta*-substituted benzene derivatives are shown in Tables 6.25 and 6.26. Since the substituents in the benzene ring are in the

TABLE 6.24

Effect of Hydrogen Bonding and Protonation in Changing a Zwitterionic (**A** ↔ **B**) Resonance Band to a Cationic (**B** ↔ **C**) One (285, 286)

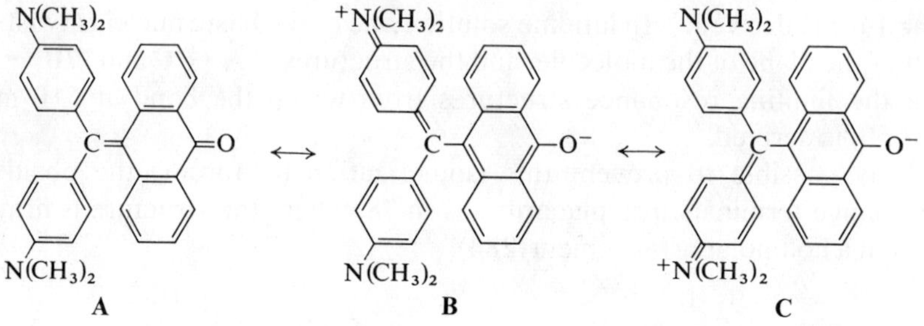

$N(CH_3)_2$ / C=O / $N(CH_3)_2$ — **A** ⟷ $^+N(CH_3)_2$ / C / O^- / $N(CH_3)_2$ — **B** ⟷ $N(CH_3)_2$ / C / O^- / $^+N(CH_3)_2$ — **C**

Solvent	A ↔ B[a]		B ↔ C[b]	
	λ_{max} (mμ)	mϵ	λ_{max} (mμ)	mϵ
Cyclohexane	420	12.3	—	
Benzene	445	13.0	~550[c]	6.0
Methanol	475	13.8	~600[c]	9.3
1 *M* $HClO_4$	480	15.0	620[d]	13.0
12 *M* $HClO_4$	—	—	625[d]	115.0

[a]Where the lower $(CH_3)_2N$-ϕ group is twisted out of the plane of the molecule – that is, the meropolar isomer.

[b]Where the anthracene ring is twisted out of the plane of the molecule – that is, the holopolar isomer.

[c]Maximum remains in the presence of alkali.

[d]Where **B** and **C** are protonated on the oxygen atom.

TABLE 6.25

Spectral Effect of Cross-Conjugation in *Meta*-Disubstituted Benzene Derivatives

Compound	λ_{max} (mμ)	mϵ	Ref.
$C_6H_5NO_2$	260	8.1	287
1,3-$C_6H_4(NO_2)_2$	234	17.0	287
1,3,5-$C_6H_3(NO_2)_3$	224	26.9	287
$C_6H_5CH{=}CHC_6H_5$	295	26.3	288
1,3-$(C_6H_5CH{=}CH)_2C_6H_4$	298	53.7	288
$C_6H_5N{=}NC_6H_5$[a]	318	21.9	221
1,3-$(C_6H_5N{=}N)_2C_6H_4$[a]	320	34.0	221
m,m'-$C_6H_5N{=}NC_6H_4N{=}NC_6H_4N{=}NC_6H_5$[a]	320	46.0	221

[a]In 95% ethanol.

TABLE 6.26
Spectral Effect of Cross-Conjugation in the *m*-Polyphenyls (164, 289)

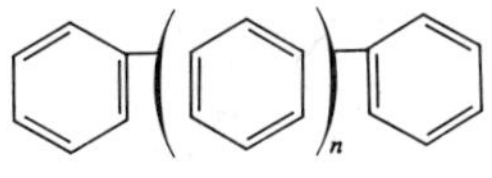

Compound	n	λ_{max} (mμ)[a]	mϵ	mϵ / number of biphenyl chromophores
Biphenyl	0	252	18.3	18.2
Terphenyl	1	252	44.0	22.0
Quatrophenyl	2	249[b]	59.0	19.7
Quinquaphenyl	3	249[b]	79.5	19.9
Sexaphenyl	4	249[b]	102.0	20.4
Octaphenyl	6	249[b]	146.0	20.9
Noviphenyl	7	253	184.0	23.0
Deciphenyl	8	253	213.0	23.7
Undeciphenyl	9	253	215.0	21.5
Duodeciphenyl	10	253	237.0	21.5
Tredeciphenyl	11	253	252.0	21.0
Quatrodeciphenyl	12	253	283.0	21.8
Quindeciphenyl	13	254	309.0	22.1
Sedeciphenyl	14	255	320.0	21.3

[a]Spectra in chloroform unless otherwise specified.
[b]In cyclohexane.

meta-position, they are insulated from each other. Thus with several substituents in the ring, the compound contains several chromophoric systems having in common part of the benzene conjugation chain. For most of the compounds there is little effect on the position of the wavelength maximum. The violet shift in the *m*-polynitrobenzenes is probably due to the effect of negativizing a negative node.

Using 1,3,5-trinitrobenzene as an example, the threefold increase in intensity can be explained by the presence of three chromophores involving each of the nitro groups as negative resonance terminals and a positive resonance terminal on the benzene ring. This effect can be seen more dramatically in the *m*-polyphenyls, whose wavelength maxima are derived from the biphenyl chromophore and whose molar absorptivities depend on the number of this chromophore. Thus in *m*-terphenyl (**80**) the resonance in each of the two chromophores could be depicted in the following fashion:

80

Cross-conjugation involving a benzene ring can also result in two separate transitions. This can occur when a benzene ring is substituted with either a strong electron-donor group and two electron-acceptor groups of different strengths or a strong electron-acceptor group and two electron-donor groups of different strengths. An example of the latter is shown in Figure 6.2. The two chromophores in the dibenzothiophene compound are the *o*-nitrophenylsulfide and the *o*-nitroaniline groups.

D. Miscellaneous Types

The pyrazolone azomethine dyes are examples of compounds that contain two negative resonance terminals and a positive terminal. The nitrogen at position 2 is postulated as the negative resonance terminal in the electronic transition from which the *X*-band is derived (Table 6.27) (290), and the carbonyl oxygen is postulated as the negative resonance terminal in the transition from which the *Y*-band is derived. The smaller shift of the *X*-band when the 3-position is occupied by a substituent other

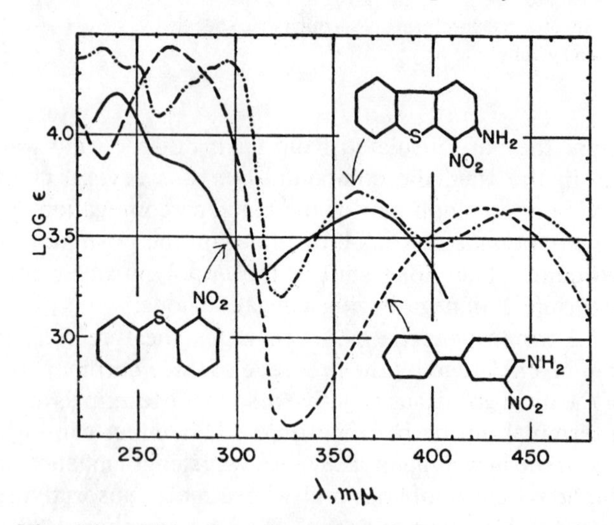

Fig. 6.2. Ultraviolet–visible absorption spectra in 95% ethanol of 2-nitrodiphenylsulfide (——), 3-amino-4-nitrodibenzothiophene (—··—), and 3-nitro-4-aminobiphenyl (---)(77).

TABLE 6.27

Spectral Effect of Cross-Conjugation in Some Pyrazolone Azomethine Dyes (290)

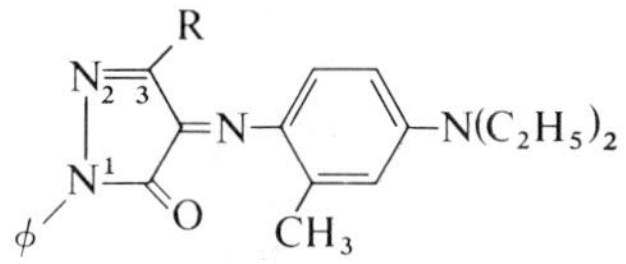

R	λ_{max} (mμ) (mϵ)			
	X-Band		*Y*-Band	
	Cyclohexane	Methanol	Cyclohexane	Methanol
H	520 (22)	573 (41)	447 (13)	422 (6)
CH_3	499 (29)	539 (38)	444 (22)	444 (13)
C_6H_5	514 (22)	554 (34)	450 (18)	452 (14)
$C_2H_5CO_2$	535 (29)	575 (41)	460 (19)	462 (12)
C_6H_5NH	504 (40)	533 (44)	434 (11)	426 (7)

than hydrogen is ascribed to a shielding effect; the larger groups reduce the accessibility of the 2-nitrogen atom to the oriented polar-solvent molecules. However, the λ_{max} of the *Y*-band is nearly independent of the polarity of the solvent for all 3-substituents other than hydrogen. Thus, when substituents other than hydrogen are attached to the 3-position, the benzene ring is forced to occupy a position wherein it effectively shields the oxygen atom from the approach of the solvent molecules.

IX. Electronic Interactions between Nonconjugated Groups

In this section we discuss electronic interactions that are explainable in terms of nonclassical resonance and molecular-orbital theories. Many examples have been given by Ferguson and Nnadi (291). In the first two portions we consider transannular interactions of simple unsaturated aliphatic ketones.

A. Transannular Interaction with a Nonconjugated Double Bond

Acetone exhibits in hexane a $\pi \rightarrow \pi^*$ band at 188 mμ, mϵ 0.9, and an $n \rightarrow \pi^*$ band at 279 mμ, mϵ 0.015. This spectrum is similar to that of other aliphatic carbonyl compounds whose wavelength position as affected by substituents can be calculated (292). Conjugation with a double bond, as in a hexane solution of acrolein, shifts the $\pi \rightarrow \pi^*$ band to 210 mμ, mϵ 7.0, and the $n \rightarrow \pi^*$ band to 328 mμ, mϵ 0.013 (177). The $\pi \rightarrow \pi^*$

band is found between 210 and 250 mμ in simple conjugated enones. Extensive work on the effect of substituents on the wavelength position of the enone chromophore and the consequent prediction of this position bears out this conjugation effect(293–297). Where the double bond is not conjugated with a carbonyl group, only an $n \rightarrow \pi^*$ band is found in the ultraviolet; for example, 4-methyl-4-pentene-2-one in isooctane absorbs at 290 mμ, mϵ 0.079(298).

On the other hand, where transannular conjugation takes place in the excited state, as in **81**, absorption in the ultraviolet is found(299). Here the molecular π-orbitals which are parallel end to end show some overlapping. Other examples are given where the unconjugated carbonyl and olefinic double-bond molecular π-orbitals are parallel face to face, and the $\pi \rightarrow \pi^*$ bands are found at 210–244 mμ and the $n \rightarrow \pi^*$ bands near 300 mμ(300).

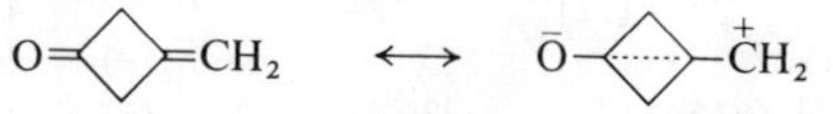

O=◇=CH_2 ⟷ $\bar{O}$–◇–$\overset{+}{C}H_2$

81

Homoconjugation, where the orbital overlap of the carbonyl carbon and an olefinic carbon is crosswise, results in a shift of the $\pi \rightarrow \pi^*$ band to the 200 to 230-mμ region and an enhancement of the $n \rightarrow \pi^*$ carbonyl band – for example, as in **82** compared with **83**(298, 301).

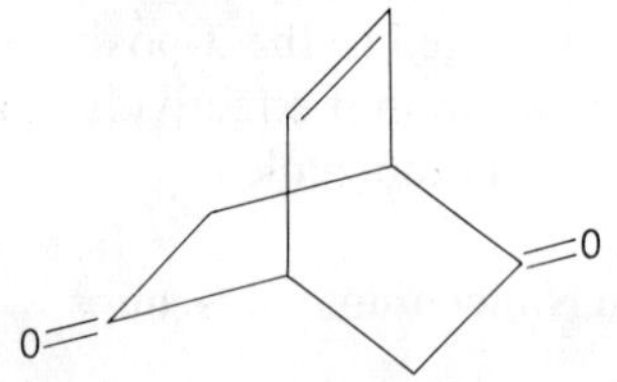

82

λ_{max} 223, 296, 307mμ
mϵ 2.29, 0.27, 0.27

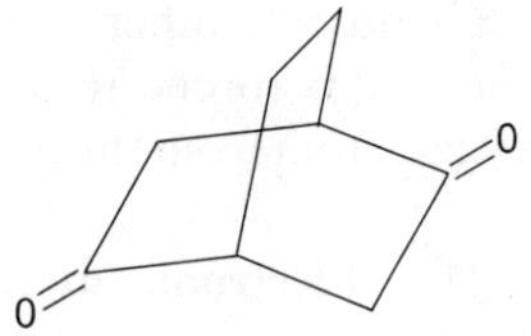

83

λ_{max} 296 mμ
mϵ 0.032

Thus in these compounds (e.g., **84**) homoconjugation takes place in the excited state, as shown by the ultraviolet absorption spectra and the normal infrared frequency of the isolated C=O group, the latter indicating the lack of conjugation of the two unsaturated groupings in the ground state(301, 302).

These various $\pi \rightarrow \pi^*$ bands have been attributed to charge transfer from the olefinic double bond to the carbonyl group, occurring only when the geometry is favorable for overlap of the orbitals concerned (301, 303). This effect is demonstrated in a comparison of the spectra of cyclodecanone (**85**) *trans*-5-cyclodecenone (**86**), and *cis*-5-cyclodecenone (**87**)(304).

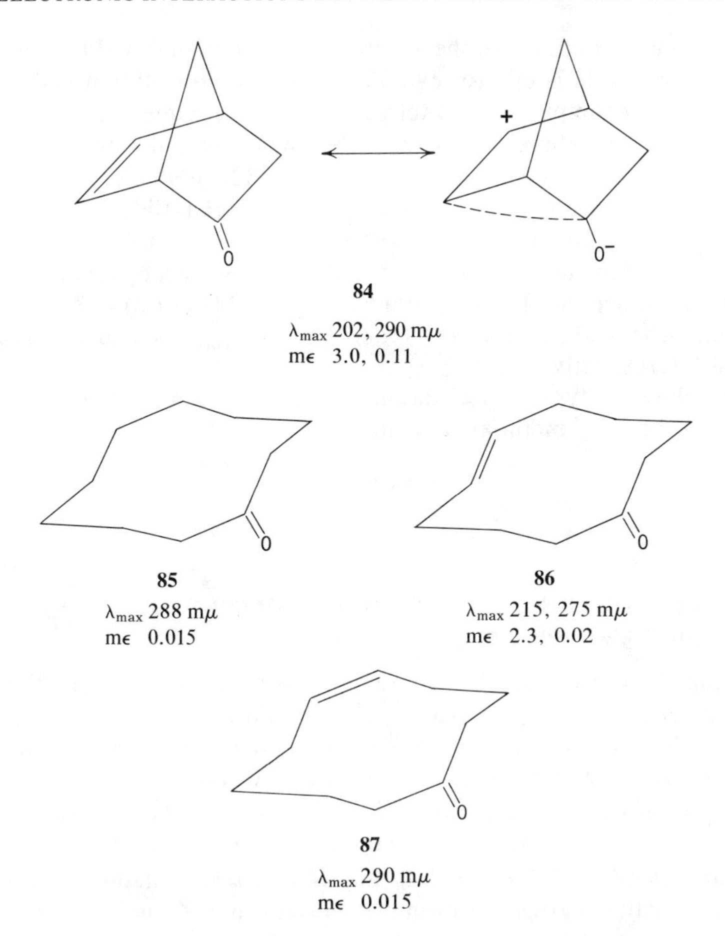

The lack of π-type overlap in the *cis* compound (**87**) is shown by its spectral resemblance to **85**.

B. SPIROCONJUGATION

A special type of homoconjugation results when two perpendicular π-systems are joined by a common tetrahedral atom. Spiroconjugation occurs whenever four *p*-centers are located by *D2d* symmetry on atoms bonded to the tetrahedral center(305). The resultant electronic delocalization will affect the electronic spectra and chemical reactivity of these compounds. On the basis of theories developed from these concepts the properties of the spirenes $[(CH{=}CH)_mC(CH{=}CH)_n]$ and the polyene ketals $[(CH{=}CH)_qC(OR)_2]$ have been predicted and discussed (305). It was predicted and shown in the spirenes that, where *m* and *n* are both odd or both even, the spectra show a red shift and an intensity decrease; where *m* is odd (or even) and *n* is even (or odd), a violet shift results. For example, 1,2,3,4-tetrachlorocyclopentadiene has a λ_{max} of 276 mμ, mϵ 4.8, whereas its spirene derivative (*m* and *n* are even) with a 5,5-$[C(C_2H_5)]_2]_4$ substituent has a λ_{max} of 325 mμ, mϵ 0.50, and its spiro derivative (*m* even and *n* odd) with a 5,5-$[C(C_2H_5)_2]_2$ substituent has a λ_{max} of 284 mμ, mϵ 3.40 (all compounds in ethanol).

The ketals demonstrate the red-shift effect, as shown by comparing the spectra in isooctane of cyclopentadiene (λ_{max} 241 mμ, mϵ 3.2) and its 5,5-dimethoxy and 5,5-$(OCH_2—)_2$ derivatives (λ_{max} 270 and 280 mμ, mϵ 1.2 and 1.1, respectively)(305).

The following type of zwitterionic resonance can be postulated, as shown for the 5,5-dimethoxy derivative:

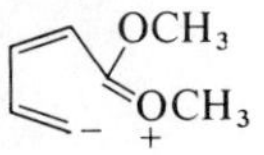

OCH_3
$\overset{+}{O}CH_3$

88

C. Transannular Interaction with Nonconjugated Electron-Donor or Electron-Acceptor Groups

Examples of transannular interaction with nonconjugated electron-donor or electron-acceptor groups are presented in Table 6.28. In the compounds **I** the strong $\pi \rightarrow \pi^*$ absorption of the α, β-unsaturated ketone chromophore at 238 mμ shifts to the violet, whereas the weak $n \rightarrow \pi^*$ absorption of the chromophore at 310 mμ shifts to the red, as group X becomes more electronegative. Examination of the absorption spectra of compounds **I**, X = CH_2, NCH_3, and $\overset{+}{N}(CH_3)_2$ indicated that there is a good correlation between transition energies and *Z*, and that the shift in the $\pi \rightarrow \pi^*$ transition is primarily electrostatic(306).

Interaction between nonclassically conjugated C═C bonds has also been surmised from wavelength-maximum red shifts and intensity changes(291).

The higher intensity bands of compounds **II** in Table 6.28, X = S and NCH_3, show the expected red shift of a Z_n-band with an increase in the polarity of the solvent; this suggests that the excited state of the transition is more polar than the ground state. These results are consistent with the assignment of the new band to the transition of a lone-pair electron from the ring hetero atom to the antibonding π-orbital of the carbonyl group, as depicted by **89**.

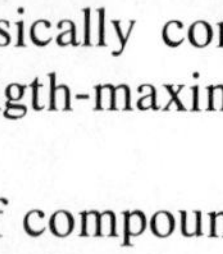

O^-
S^+

89

TABLE 6.28
Spectral Effect of Nonconjugated Hetero Atoms with Ring Carbonyl

I   II

	I			II		
X	λ_{max} (mμ)[a]	mϵ	Ref.	λ_{max} (mμ)[a]	mϵ	Ref.
CH_2	238	15.4	306	283	0.014	308
O				272	0.016	308
NCH_3	233	14.1	306	225[b]	6.3	309
$\overset{+}{N}(CH_3)_2$	229	15.6	306			
S	230	16.0	307	226[c]	2.45	308
				242[d]	2.78	308
SO_2	229	14.8	307			
$\overset{+}{S}CH_3$	220	20.0	307			

[a] In 95% ethanol unless otherwise specified.
[b] In ether.
[c] In *n*-hexane.
[d] In water.

D. Spectral Effect of Nonconjugated Charged Group on a Nearby Resonance Terminal

Where a positive resonance terminal is near a positively charged group or a negative resonance terminal is near a negatively charged group, the spectra of Z_n-structures will show a violet shift and the spectra of Z_z-structures will show a red shift. The opposite shifts will be found where a positive resonance terminal is near a negatively charged group or a negative resonance terminal is near a positively charged group. This is because dipolar structures contribute more strongly to the excited state of the Z_n-structure and to the ground state of the Z_z-structure.

The violet-shift effect of a positive charge adjacent to a positive resonance terminal is shown in Table 6.29. This shift is explained by the higher transition energy from the ground state to an excited state in which a high energy form containing adjacent positive charges is a major contributor, as, for example, in **90**.

$^{-}O_2N{=}C_6H_4{=}\overset{+}{N}(H){-}\overset{+}{N}H_3$

90

TABLE 6.29

Spectral Effect of a Positive Charge Adjacent to a Positive Resonance Terminal

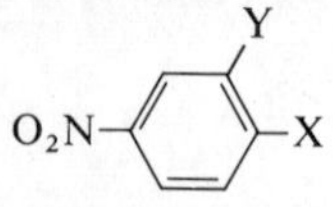

X	Y	Solvent	λ_{max} (mμ)	mϵ	Ref.
$NHNH_2$	H	Aqueous alkali	405	10.5	310
NH_2	H	Water	380	13.5	311
NH_2	NH_3^+	1.2 *N* HCl in H_2O–EtOH (1 : 1)	365	13.6	*a*
$NHNH_3^+$	H	0.02 *N* HCl in H_2O–EtOH (1 : 1)	325	8.9	310
H	H	95% EtOH	260	9.6	101

[a]Present work.

The violet shifts demonstrated by the compounds listed in Table 6.30 can be explained on the basis of the repulsion of positive charges in the excited state (e.g., **91B**). The bias in the bases toward dipolar resonance forms with a positive charge at C-3 would be more pronounced in 3-methyl derivatives because the nonprotonated form of **91B** would be stabilized by the methyl substituent. Because of this the destabilizing influence

TABLE 6.30

Spectral Effect in Ethanol of Nonconjugated Positive Charge (312–314)

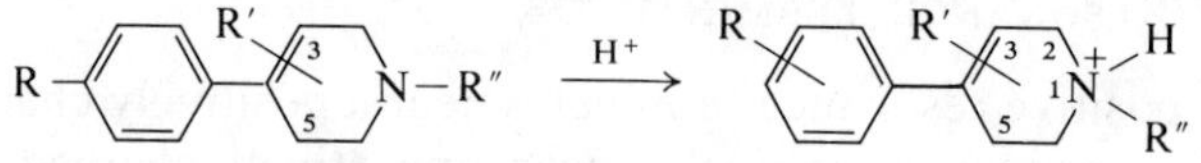

R	R′	R″	λ_{max} (mμ) Base	Hydrochloride	$\Delta\lambda$ (mμ)[a]
H	3-CH_3	H	237 (7.4)	230 (7.8)	−7
H	3-CH_3	C_2H_5	235 (7.6)	230 (8.1)	−5
CH_3	3-CH_3	C_2H_5	237 (10.0)	234 (10.5)	−3
H	5-CH_3	H	240 (11.2)	237 (11.2)	−3
H	5-CH_3	C_2H_5	240 (10.7)	238 (10.7)	−2
CH_3	5-CH_3	C_2H_5	243 (13.5)	243 (13.5)	0
CH_3O	3-CH_3	$C_6H_5CH_2$	239 (13.2)	241 (12.9)	+2
$(CH_3)_2N$	H	$(CH_3)_2N$	287 (20.0)	294 (20.4)	+7
$(CH_3)_2N$	3-CH_3	C_2H_5	269 (15.9)	278 (15.1)	+9
$(CH_3)_2N$	5-CH_3	C_2H_5	281 (19.1)	290 (18.2)	+10
$(CH_3)_2N$	H	CH_3	287 (17.8)	298 (19.1)	+11

[a]$\Delta\lambda = \lambda_{max}$ (HCl) $- \lambda_{max}$ (base); plus values represent a red shift, minus values a violet shift.

of the protonated nitrogen on this form (e.g., **91B**) would be expected to have a correspondingly greater effect in 3-methyl derivatives than in the corresponding 5-methyl or unsubstituted analogs.

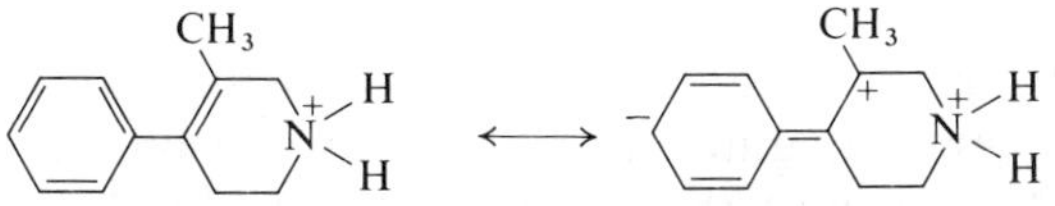

A **91** B

On the other hand, in the methoxy and dimethylamino derivatives resonance structures, such as **92A** ↔ **B**, are of great importance. In these compounds the 3-position has become the negative resonance terminal. In these compounds the stabilization of the excited-state contributory structure **92B** is due to the stabilization of the negative charge by the nearby positive charge. The decrease in transition energy causes the red shifts (Table 6.30).

A **92** B

These wavelength shifts are not due to solvent effects since **93** and its hydrochloride have similar spectral characteristics in ethanol (hydrochloride is hydrolyzed since base is too weak for protonation under these conditions). Since position 3 is the negative resonance terminal, substitution of an electron-donor group at position 5 causes a red shift, and substitution of a positively charged group causes an even larger red shift.

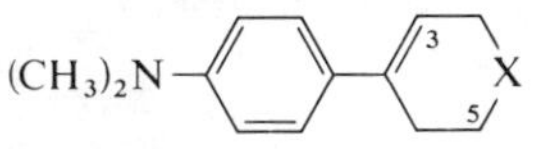

93

X	λ_{max} (mμ)	mϵ
CH_2 (base)	283	20.4
CH_2 (HCl)	283	24.0
NCH_3	287	17.8
$\overset{+}{N}HCH_3$	298	19.1

On the other hand, where the positive resonance terminal is close to an electron-donor or a positively charged group, violet shifts result, as shown for **94** in methanol (315).

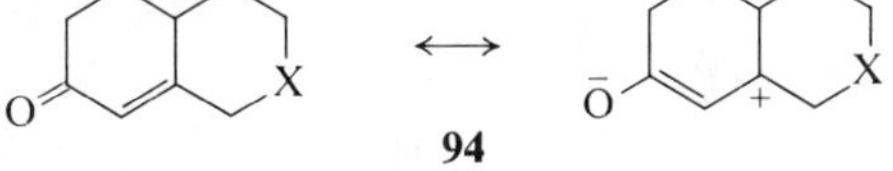

94

X	λ_{max} (mμ)
CH_2	244
NCH_3	222
$\overset{+}{N}HCH_3$	218

Kamlet(116) has presented many other examples to show that, when the nitrogen atom is separated from the positive resonance terminal of the chromophore by a single carbon atom, a strong violet shift is in most cases observed (6–22 mμ). Quaternization of the nitrogen atom enhances the effect by a further 4–5 mμ. When two carbon atoms separate the groups, the shift is smaller (2-11 mμ) and is again enhanced by quaternization or conversion into the amine oxide but reduced by a reduction of the basicity of the nitrogen group through acetylation.

E. Resonance Involving Three-Membered Rings

The spectra of a group of *para*-substituted stilbene oxides demonstrate that oxirane transmits electronic effects via second-order conjugation from one phenyl group to the other(316). References were the spectra of the corresponding bibenzyls and *para*-substituted styrene oxides. This effect can be seen for 4-methoxy-4′-nitro-*trans*-stilbene oxide (**95**), for which some of the resonance forms are shown.

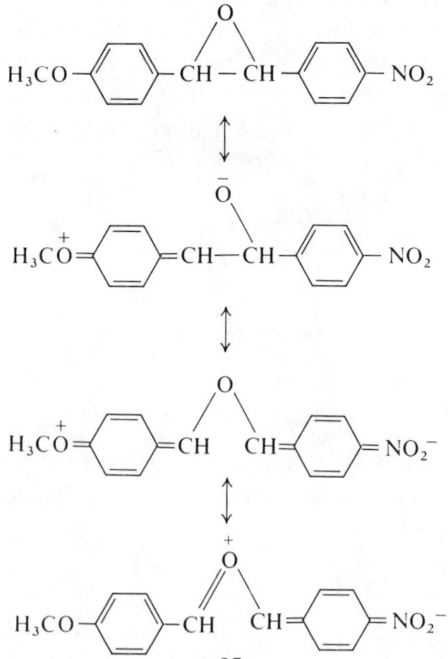

O
H₃CO—C₆H₄—CH—CH—C₆H₄—NO₂

↕

Ō
H₃CÓ⁺=C₆H₄=CH—CH—C₆H₄—NO₂

↕

O
H₃CÓ⁺=C₆H₄=CH CH=C₆H₄=NO₂⁻

↕

O⁺
H₃CO—C₆H₄—CH CH=C₆H₄=NO₂⁻

95

In every case the absorption bands of the *trans* isomers of the oxirane derivatives are markedly enhanced in comparison with those of the *cis* isomers. These differences have been attributed to the conjugative properties of the oxirane ring and have been interpreted as indicating a steric requirement for delocalization (316–320).

Electronic interaction has been postulated as taking place between aryl groups separated by a methylene group through the formation of a three-membered ring in the excited state; an example is **96.**

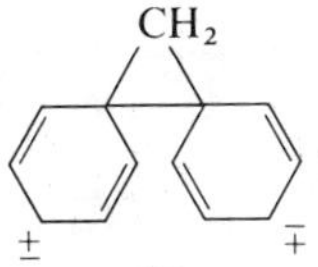

CH_2

96

For instance, resonance structures **96** have been proposed for diphenylmethane to account for its spectral differences from toluene (321). The effect is magnified by substituting dissimilar electron-attracting groups in the two rings, as shown by comparing the ethanolic solutions of **97**, *p*-nitrotoluene (**98**), and *p*-methylanisole (**99**) (291). It is obvious that the spectrum of **97** is more than the sum of its parts.

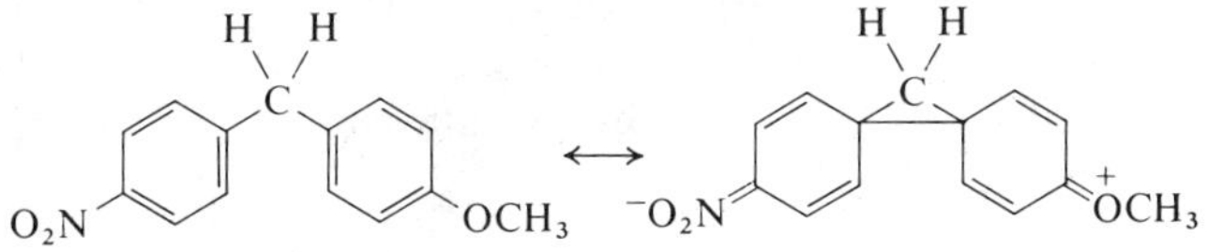

97

λ_{max} 280, 287 mμ
mε 24.4, 16.9

98

λ_{max} 274 mμ
mε 9.49

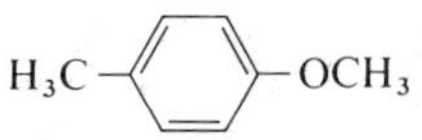

99

λ_{max} 277, 286 mμ
mε 2.19, 1.79

X. Coupling Effects

The coupling concept has been developed by Dahne and Leupold (18). Where a molecule contains polymethine and/or polyene chromophores capable of interacting across a π-band, the interaction can lead to coupling effects, which can determine the position of the ultraviolet–visible absorption maximum.

A. Polymethines Coupled through a Single Bond(s)

One particular type of coupling is that of two trimethinemerocyanines across two separate single bonds to give a quadrupole merocyanine—for example, 2,5-diamino-1,4-benzoquinone (**100**)

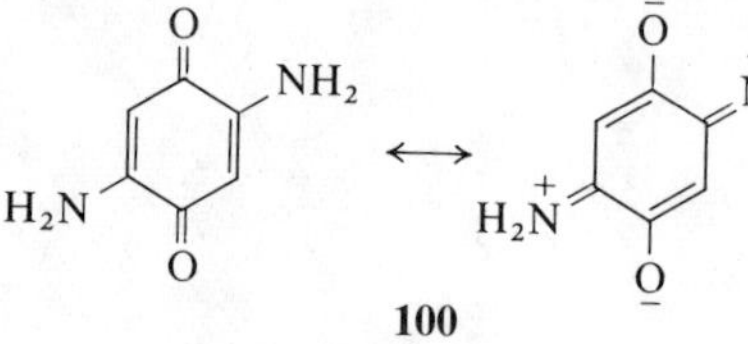

100

λ_{max} 336, 488 mμ

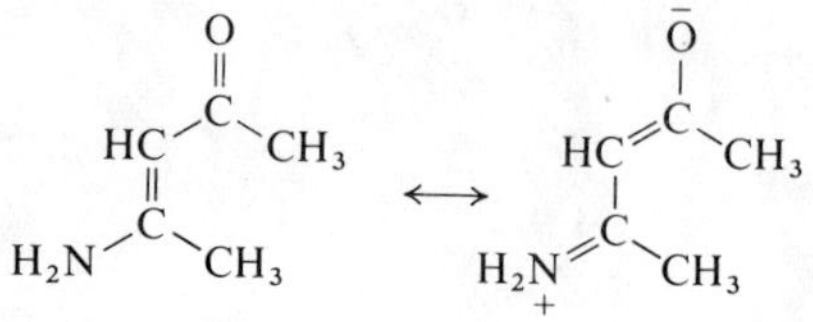

101

λ_{max} 285 mμ, mε 6.12

The separation of charge in the quadrupole is aided by the adjacency of charges of opposite sign. The extraordinarily powerful red-shift effect of coupling can be seen in comparing the long-wavelength $\pi \rightarrow \pi^*$ maxima of **100** in dimethyl sulfoxide and **101** in heptane (278).

Coupling of two pentamethinemerocyanines occurs in 1,5-diamino-2,6-naphthoquinone (**102**). The coupling single bonds are at *a* in **102.** The red-shift effect of coupling can be seen by comparing the long-wavelength maxima of **102** in dimethyl sulfoxide (18) and the pentamethinemerocyanine **103** (182).

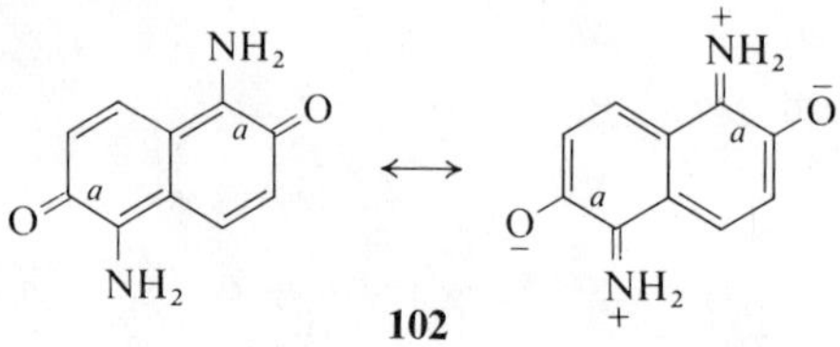

102

λ_{max} 311, 580 mμ

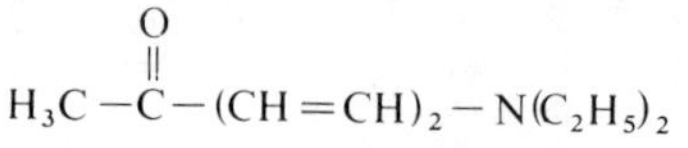

$H_3C-C(=O)-(CH=CH)_2-N(C_2H_5)_2$

103

λ_{max} 378 mμ, mε 46.5

Many other compounds show the coupling phenomenon—for example, indanthrene dyes and the quinone **104** (172), in which the coupling single bonds are at *a*.

B. Polymethines Coupled End to End

In polymethines coupled end to end the chain of conjugation is lengthened in the excited state. A simple example is the series of compounds

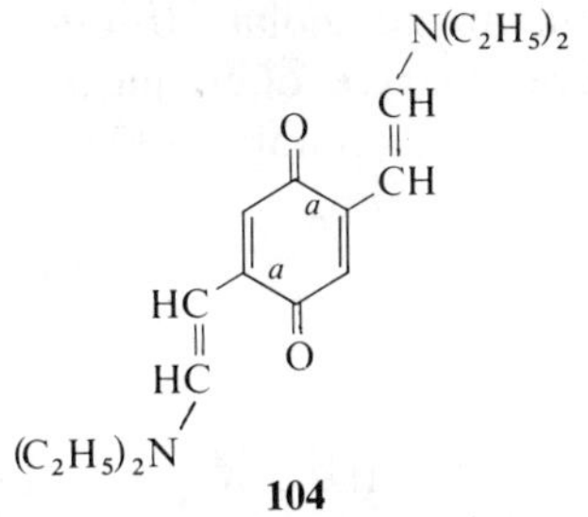

$N(C_2H_5)_2$
CH
O ‖
CH
a
a
HC
‖ O
HC
$(C_2H_5)_2N$

104

λ_{max} 537, 645 mμ in dioxane

– such as **105** (spectrum in heptane) and **106** – that contain the chromophore 4-$R_2NC_6H_4CH{=}CHC{=}O$,

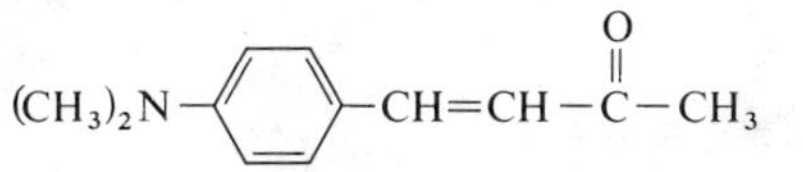

$(CH_3)_2N-C_6H_4-CH{=}CH-C(=O)-CH_3$

105

$(CH_3)_2N-C_6H_4-CH{=}$[pyrimidine ring: O, $N-C_2H_5$, $=X$, $N-C_2H_5$, O] $\longleftrightarrow$ $(CH_3)_2\overset{+}{N}{=}C_6H_4{=}CH-$[pyrimidine ring: $\bar{O}$, $\overset{+}{N}-C_2H_5$, $-X^-$, $N-C_2H_5$, O]

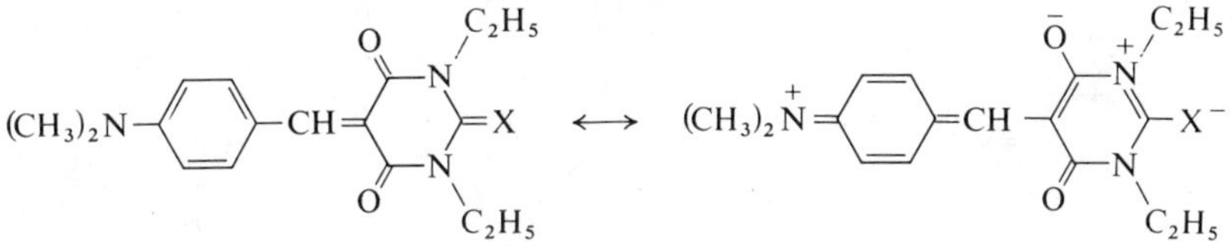

106

No.	X	λ_{max}(mμ)	mε	Ref.
105		360	29.4	278
106a	O	466		230
106b	S	490		230

A hexapolar excited-state structure could be postulated for **106**.

The long-wavelength absorption of **107** in ethanolic solution(322) may be due to the coupling effect and the formation of a quadrupole in the excited state.

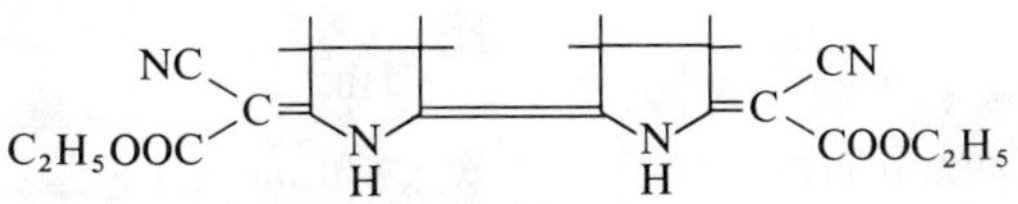

NC, C_2H_5OOC $>C{=}$ [N–H ring] = [N–H ring] $={C}<$ CN, $COOC_2H_5$

107

λ_{max} 524, 563 mμ,
mε 51.4, 78.6

The unusually long-wavelength absorption of **108** [λ_{max} 832 mμ in pyridine(323)] could be due only to extended conjugation in the excited state. For this to occur an octapolar structure is postulated. Many other molecules that have this type of conjugation and absorb at surprisingly long wavelengths have been reported(323, 324).

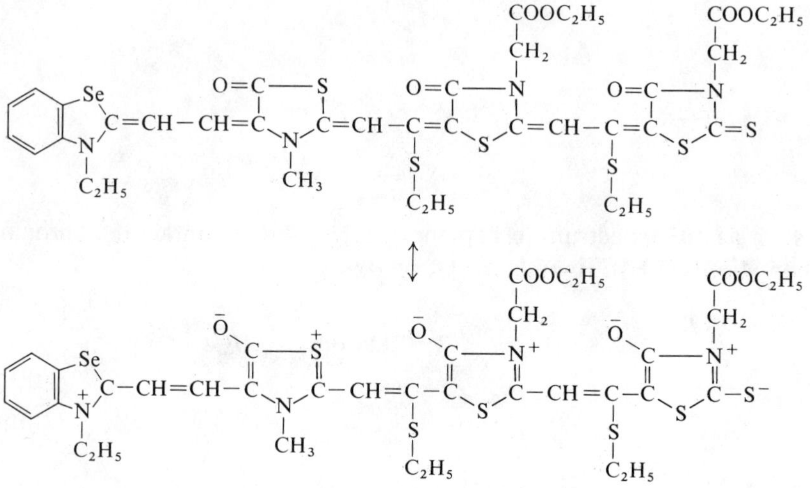

108

C. Polyene–Polymethine Coupling

The polyene–polymethine type of coupling has been discussed by Dahne and Leupold(18). These compounds show a red-shift effect in comparison with the analogous polymethine. Examples are the quinones **109**(18) and **110**(325).

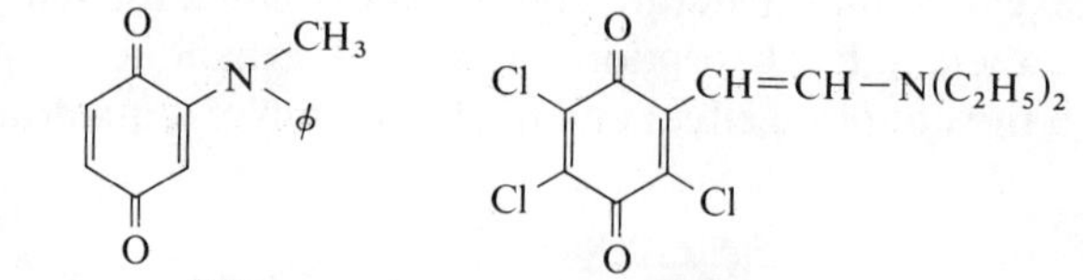

109

λ_{max} 497 mμ in DMSO

110

λ_{max} 640 mμ, mϵ 8.8

Excited-state quadrupolar structures, such as **111**, probably contribute to the red shift. The coupling bonds are at *a*. Similar coupling is postulated in the 4-amino-1,2-benzoquinones, in which the polyene structural unit is linked in the opposite direction with the trimethine structural unit(18).

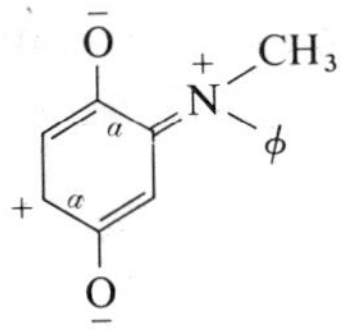

111

Many other chromogens with polyene–polymethine coupling are found among the monosubstituted naphthoquinones, anthraquinones, and higher condensed quinones (326).

In all the examples of coupling that have been presented the polymethine has had a Z_n-structure; that is, the dipole moment is greater in the excited state than in the ground state. Coupling effects involving Z_b- and Z_z-structures in various combinations with themselves and with a Z_n-structure or a polyene would be worthwhile examining spectrally.

D. Polyene–Polyene Coupling

The absorption spectra of the quinones would seem to indicate some form of coupling (327, 328); for example, compare the data for **112** (in water) and **113** (328).

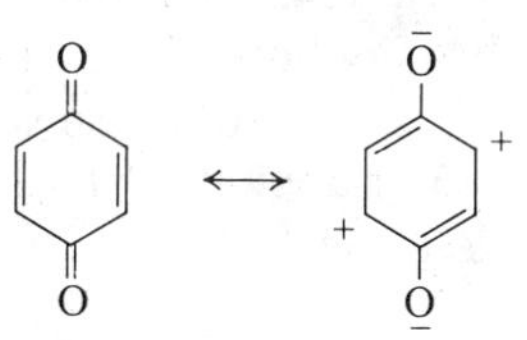

112

λ_{max} 247, 292, 436 mμ
mϵ 21.5, 0.33, 0.020

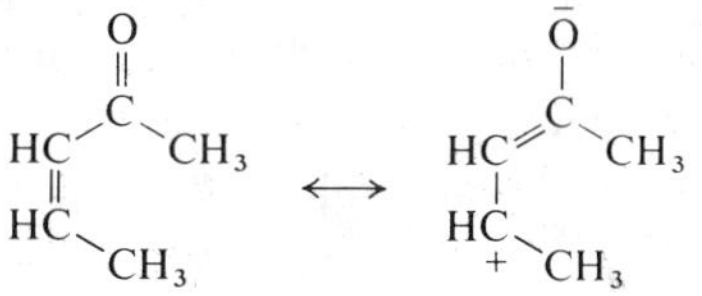

113

λ_{max} 224, 314 mμ
mϵ 10.0, 0.046

In both **112** and **113** the long-wavelength bands at 436 and 314 mμ, respectively, are $n \rightarrow \pi^*$ bands, as shown by their low intensities and violet shifts with increasing solvent polarity. The longer wavelength absorption of the $\pi \rightarrow \pi^*$ bands of **112** as compared with those of **113** could be explained by the coupling effect. Many other quinones show this phenomenon.

E. H-Chromophores

The term "H-chromophore," meaning a linkage of conjugated systems in the form of a H, was suggested by Klessinger and Luttke (329) for the indigoid dyes. Dahne and Leupold (18) have postulated that in indigo (**114A**, X = NH) there is a coupling of two high-energy heptamethine structural units through a vinylene bridge, as in **114B**, X = NH.

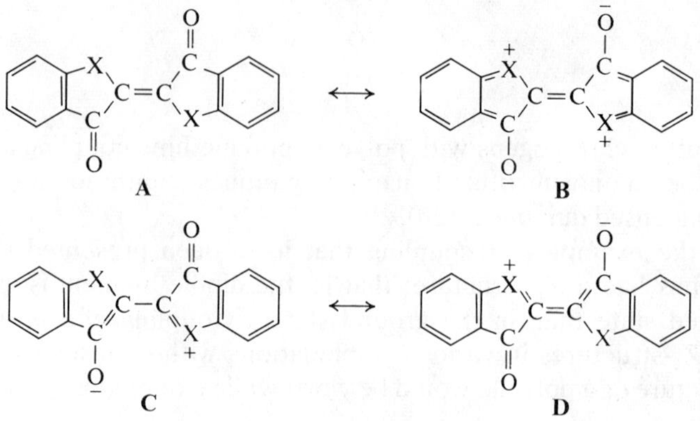

114

In **114C** and **D** two intersecting trimethines have a common pair of π-electrons. Thus the indigoid structure can be considered as a resonance hybrid of the four polymethine structures **114**. The quadrupolar structure cannot be written for all the H-chromophores listed in Table 6.31. However, whether the postulation is right or wrong, the structure common to these compounds is essentially **115** or a vinylog of this structure, where X = NR, Se, S, or O and Y = O, S, or Se.

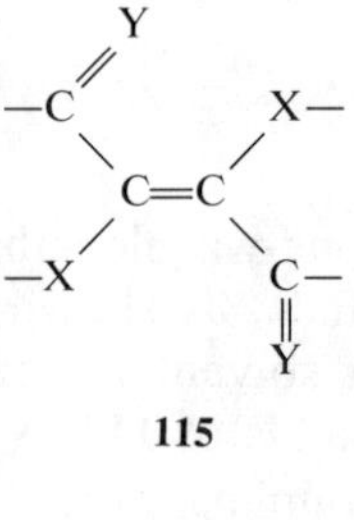

115

Where this type of coupling is absent, as in **116**, β-anilinovinylphenylketone, and isoindigo (**117** and **118**, respectively), the compounds absorb at shorter wavelength.

Many compounds that contain the H-chromophore (**115**) are shown in Table 6.31. The extremely long wavelength absorption of these compounds as compared with those of the analogous simple polymethines of the same chain length emphasizes the importance of this type of coupling effect.

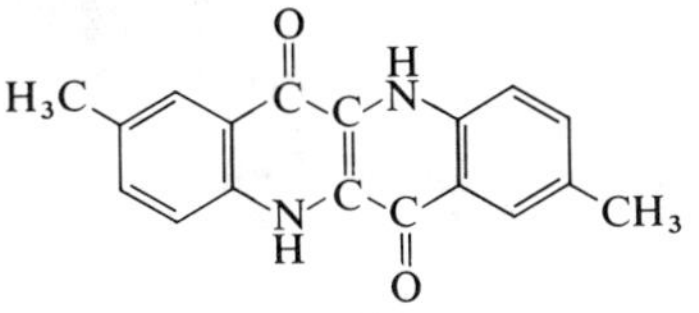

116

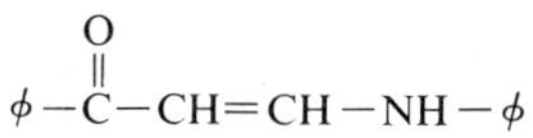

$\phi-\overset{O}{\overset{\|}{C}}-CH{=}CH-NH-\phi$

117

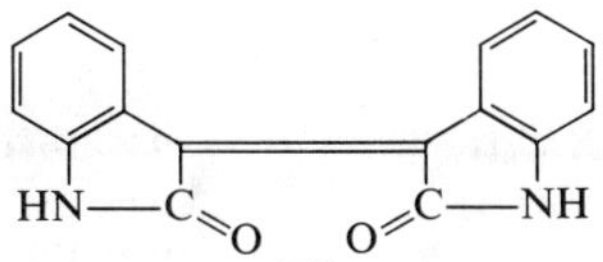

118

No.	Solvent	λ_{max} (mμ) or color	Ref.
116	Pyridine	464	329
117	Methanol	371	235
118		Yellow	235

The long-wavelength absorption of some of the squaric acid dyes indicates that the coupling effect is important here also; an example is **119** (spectrum in dimethylformamide)(339).

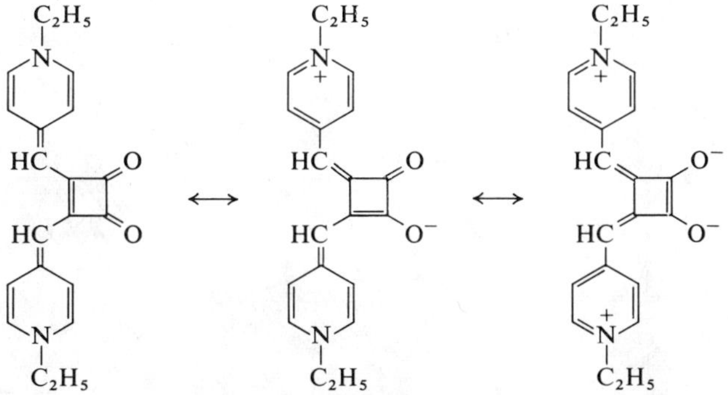

119

λ_{max} 338, 369, 458, 495, 555, 600 mμ
mϵ 13.5, 17.3, 10.3, 15.9, 67.0, 126.9

Substitution of the 4-*N*-ethyl pyridinyl group by a 4-*N*-ethyl quinolinyl group shifts the long-wavelength maximum to 670 mμ, mϵ 87.6. These cyclobutene dyes could be considered as having a H-chromophore.

TABLE 6.31
Long-Wavelength $\pi \rightarrow \pi^*$ Bands of Some H-Chromophores and Comparison Compounds

No.	Compound	Solvent	λ_{max} (mμ)	mϵ	Ref.
1[a]	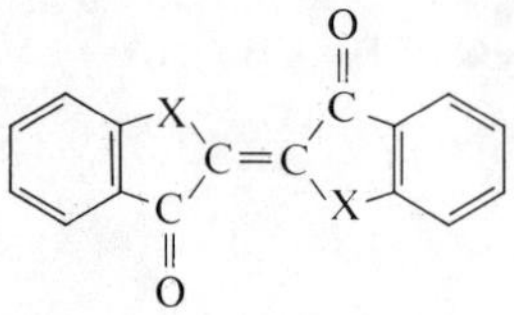 X = NH	95% EtOH	610		330
	X = Se		562		
	X = S		543		
	X = O		416		
	$C_6H_5NH—CH=CH—COC_6H_5$	95% EtOH	374	31.0	36
	$C_6H_5S—CH=CH—COC_6H_5$	95% EtOH	335	19.0	331
	$C_6H_5O—CH=CH—COC_6H_5$	95% EtOH	291	18.0	331
2	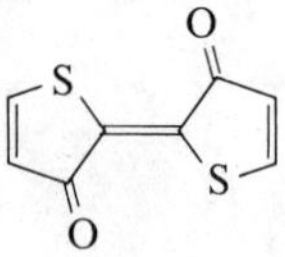	Cyclohexane	502		332
3	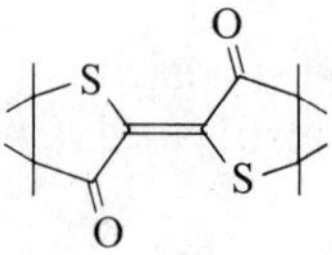	Cyclohexane	450	13.5	332
4[b]	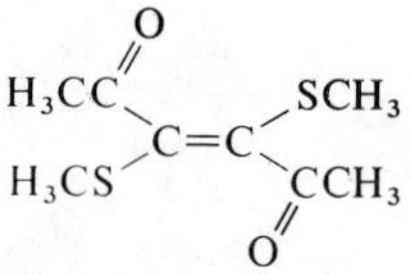	Cyclohexane	315	8.9	332
5	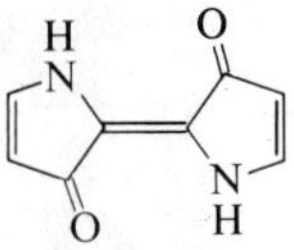		486		333
6	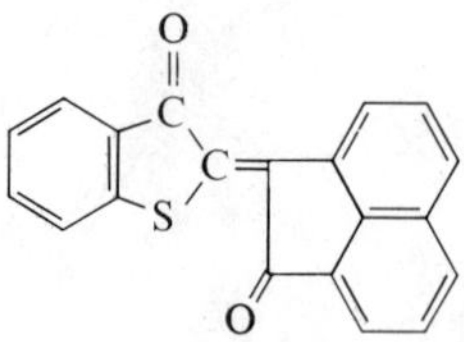		478		334
7[c]	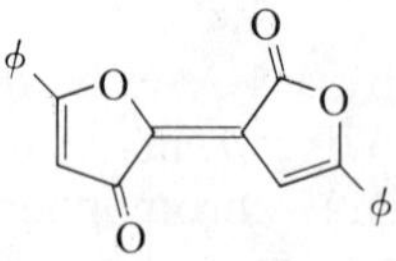	DMF	500, 535	31.6	335, 336

TABLE 6.31 (*cont'd*)

No.	Compound	Solvent	λ_{max} (mμ)	mϵ	Ref.
8	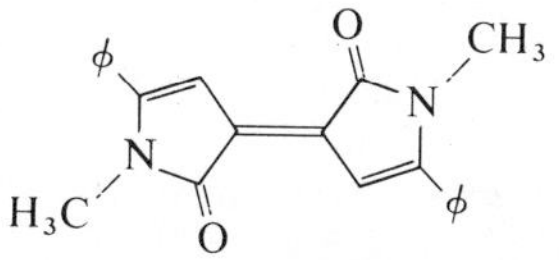	95% EtOH	~570	17.8	335
9	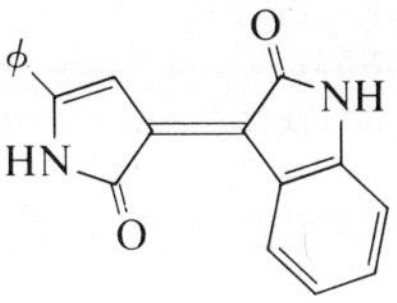	95% EtOH	~510	14.0	335
10	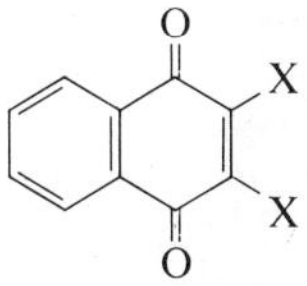 $X = NHC_6H_5$	Cyclohexane	434	10.5	332
	$X = SC_6H_5$		374	5.25	
	$X = SCH_3$		386	6.8	
11	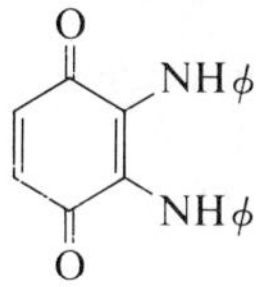	Cyclohexane	535		332
12	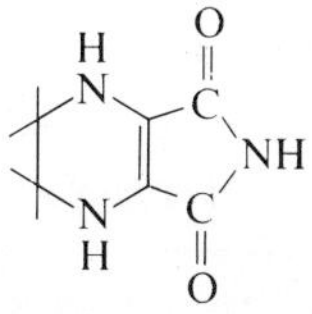	Cyclohexane	Red		332
13	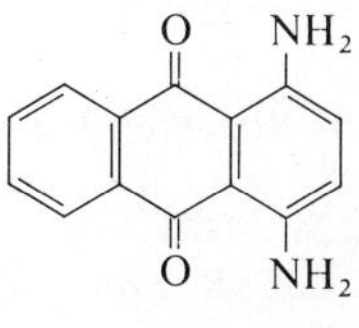	Methanol	550	13.0	337
			590	12.4	
14	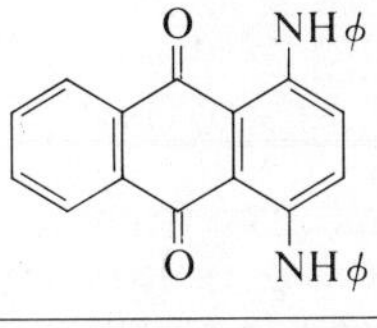	DMF	600	16.0	338
			650	16.0	

[a]Compound **114A** in text.
[b]Strong deviation from planarity so that indigoid character is completely lost.
[c]Pechmann dye.

References

1. D. Hadzi and H. W. Thompson, Eds., *Hydrogen Bonding*, Pergamon, New York, 1959.
2. G. M. Badger, *Rev. Pure Appl. Chem.*, **7**, 55 (1957).
3. B. Eistert, E. Merkel, and W. Reiss, *Ber.*, **87**, 1513 (1954).
4. R. Rasmussen, D. Tunnicliff, and D. Brattain, *J. Am. Chem. Soc.*, **71**, 1068 (1949).
5. A. Burawoy and J. T. Chamberlain, *J. Chem. Soc.*, 3734 (1952).
6. P. Grammaticakis, *Bull. Soc. Chim. France*, **53**, 865 (1953).
7. M. R. Padhye, N. R. Rao, and K. Venkataraman, *Proc. Indian Acad. Sci.*, **37**, 297 (1953).
8. W. E. Doering and L. H. Knox, *J. Am. Chem. Soc.*, **73**, 828 (1951).
9. R. H. Peters and H. H. Sumner, *J. Chem. Soc.*, 2101 (1953).
10. A. P. Lurie, G. H. Brown, J. R. Thirtle, and A. Weissberger, *J. Am. Chem. Soc.*, **83**, 5015 (1961).
11. A. Burawoy, M. Cais, J. Chamberlain, F. Liversedge, and A. Thompson, *J. Chem. Soc.*, 3721, 3727 (1955).
12. D. L. Ross and E. Reissner, *J. Org. Chem.*, **31**, 2571 (1966).
13. G. Gabor and E. Fischer, *J. Phys. Chem.*, **66**, 2478 (1962).
14. H. Kuhn, *Chimia*, **9**, 237 (1955); *Angew. Chem.*, **71**, 93 (1959).
15. S. Dahne and D. Leupold, *Ber. Bunsenges. Physik. Chem.*, **70**, 618 (1966).
16. D. Leupold, *Z. Physik. Chem.*, **223**, 405 (1963).
17. G. Scheibe, W. Seiffert, H. Wengenmayr, and C. Jutz, *Ber. Bunsenges. Physik. Chem.*, **67**, 560 (1963).
18. S. Dahne and D. Leupold, *Angew. Chem., Intern. Ed.*, **5**, 984 (1966). See discussion and references therein.
19. E. B. Knott, *J. Chem. Soc.*, 1024 (1951).
20. D. Forster, *Z. Elektrochem.*, **45**, 548 (1939).
21. G. N. Lewis, *J. Am. Chem. Soc.*, **67**, 770 (1945).
22. G. N. Lewis and M. Calvin, *Chem. Rev.*, **25**, 273 (1939).
23. S. S. Malhotra and M. C. Whiting, *J. Chem. Soc.*, 3812 (1960).
24. B. Witkop, J. Patrick, and H. Kissman, *Ber.*, **85**, 949 (1952).
25. R. P. Zelinski and W. A. Bonner, *J. Am. Chem. Soc.*, **71**, 1791 (1949).
26. M. Calvin and H. W. Alter, *J. Chem. Phys.*, **19**, 765 (1951).
27. H. H. Keasling and F. W. Schueler, *J. Am. Pharm. Assoc., Sci. Ed.*, **39**, 87 (1950).
28. G. Smets and A. Delvaux, *Bull. Soc. Chim. Belges*, **56**, 106 (1947).
29. W. Anderson, *J. Chem. Soc.*, 1722 (1952).
30. A. Haddow, R. J. C. Harris, G. A. R. Kon, and E. M. F. Roe, *Phil. Trans. Roy. Soc. (London)*, **A241**, 147 (1948).
31. V. A. Izmailskii and E. A. Smirnov, *Zh. Obshch. Khim.*, **26**, 3042 (1956).
32. A. I. Kiprianov and I. K. Ushenko, *Izvest. Akad. Nauk SSSR., Otdel. Khim. Nauk*, 492 (1950).
33. R. Merckx, *Bull. Soc. Chim. Belges*, **58**, 460 (1949).
34. L. G. S. Brooker and R. H. Sprague, *J. Am. Chem. Soc.*, **63**, 3203 (1941).
35. F. Hamer, *J. Chem. Soc.*, 3197 (1952).
36. E. R. H. Jones, T. Shen, and M. C. Whiting, *J. Chem. Soc.*, 236 (1950).
37. R. Lutz, T. Amacker, S. King, and N. Shearer, *J. Org. Chem.*, **15**, 181 (1950).
38. R. Andrisano and G. Pappalardo, *Atti. Accad. Nazl. Lincei*, **15**, 64 (1953).
39. H. H. Inhoffen, J. H. Fuhrhop, and F. von der Haar, *Ann.*, **700**, 92 (1966).
40. W. D. Kumler, *J. Am. Chem. Soc.*, **68**, 1184 (1946).

41. W. A. Sheppard and R. M. Henderson, *J. Am. Chem. Soc.*, **89**, 4446 (1967).
42. M. Ogata and H. Kano, *Tetrahedron*, **24**, 3725 (1968).
43. L. Yagupolski and M. Marenets, *Zh. Obshch. Khim.*, **23**, 481 (1953); through *Chem. Abstr.*, **48**, 3964 (1954).
44. S. Hunig and G. Kobrich, *Ann.*, **617**, 210 (1958).
45. R. Wizinger, *Chimia*, **19**, 339 (1965).
46. R. Morton and A. Stubbs, *J. Chem. Soc.*, 1347 (1940).
47. R. A. Robinson and A. K. Kiang, *Trans. Faraday Soc.*, **51**, 1398 (1955).
48. A. Quilico and C. Cardani, *Gazz. Chim. Ital.*, **83**, 1088 (1953).
49. H. Lemon, *J. Am. Chem. Soc.*, **69**, 2998 (1947).
50. A. Burawoy and J. T. Chamberlain, *J. Chem. Soc.*, 2310 (1952).
51. O. Goldschmid, *J. Am. Chem. Soc.*, **75**, 3780 (1953).
52. N. A. Valyashko and N. N. Valyashko, *Zh. Obshch. Khim.*, **26**, 146 (1956).
53. L. Doub and J. M. Vandenbelt, *J. Am. Chem. Soc.*, **77**, 4535 (1955).
54. L. J. Kleckner and A. Osol, *J. Am. Pharm. Assoc., Sci. Ed.*, **41**, 103 (1952).
55. B. K. Ganguly and P. Bagchi, *J. Org. Chem.*, **21**, 1415 (1956).
56. S. Mangini and R. Passerini, *Gazz. Chim. Ital.*, **87**, 243 (1957).
57. C. J. P. Spruit, *Rec. Trav. Chim.*, **68**, 309 (1949).
58. H. Hartmann and E. Lorenz, *Z. Naturforsch.*, **7a**, 360 (1952).
59. F. A. Hochstein et al., *J. Am. Chem. Soc.*, **75**, 5455 (1953).
60. T. Momose, Y. Ohkura, and S. Goya, *Pharm. Bull. Tokyo*, **3**, 401 (1955).
61. W. F. Forbes and I. R. Leckie, *Can. J. Chem.*, **36**, 1371 (1958).
62. A. Butenandt, E. Bierkert, M. Dauble, and K. H. Kohrmann, *Ber.*, **92**, 2172 (1959).
63. P. Grammaticakis, *Bull. Soc. Chim. France* (5), **18**, 220 (1951).
64. L. Birkofer, *Ber.*, **85**, 1023 (1952).
65. L. I. Smith and R. L. Abler, *J. Org. Chem.*, **22**, 811 (1957).
66. P. Grammaticakis, *Bull. Soc. Chim. France* (5), **20**, 207 (1953).
67. M. Rapoport, C. K. Hancock, and E. A. Meyers, *J. Am. Chem. Soc.*, **83**, 3489 (1961).
68. J. Smith, *J. Chem. Soc.*, 2861 (1951).
69. A. Mangini, R. Passerini, and S. Serra, *Gazz. Chim. Ital.*, **84**, 47 (1954).
70. W. Schroeder, P. Wilcox, K. Trueblood, and A. Dekker, *Anal. Chem.*, **23**, 1740 (1951).
71. V. Gold and B. W. V. Hawes, *J. Chem. Soc.*, 2102 (1951).
72. F. Smith and L. M. Turton, *J. Chem. Soc.*, 1701 (1951).
73. E. Sawicki and F. E. Ray, *J. Org. Chem.*, **19**, 1903 (1954).
74. A. Cerniani, R. Passerini, and G. Righi, *Boll. Sci. Fac. Chim. Ind. Bologna*, **12**, 75 (1954).
75. E. Sawicki, *J. Am. Chem. Soc.*, **76**, 664 (1954).
76. E. Sawicki, *J. Org. Chem.*, **19**, 608 (1954).
77. E. Sawicki, *J. Org. Chem.*, **19**, 1163 (1954).
78. E. Sawicki, B. Chastain, and H. Bryant, *J. Org. Chem.*, **21**, 754 (1956).
79. E. Sawicki, *J. Am. Chem. Soc.*, **77**, 957 (1955).
80. F. Kehrmann and J. Effront, *Helv. Chim. Acta*, **4**, 519 (1904).
81. A. Mangini and R. Passerini, *Gazz. Chim. Ital.*, **85**, 840 (1955).
82. P. W. Sadler and R. L. Warren, *J. Am. Chem. Soc.*, **78**, 1251 (1956).
83. P. W. Sadler, *J. Org. Chem.*, **21**, 316 (1956).
84. H. Kauffmann and W. Franck, *Ber.*, **39**, 2722 (1906).
85. H. Kauffmann, *Ber.*, **52**, 1422 (1919).
86. J. Braunholtz and F. Mann, *J. Chem. Soc.*, 1817 (1953).
87. M. Osanai, *J. Insect Physiol.*, **12**, 1295 (1966).

88. J. O. Schrenk, C. K. Hancock, and R. M. Hedges, *J. Org. Chem.*, **30**, 3504 (1965).
89. C. K. Hancock and A. D. H. Clague, *J. Am. Chem. Soc.*, **86**, 4942 (1964).
90. C. K. Hancock, R. A. Brown, and J. P. Idoux, *J. Org. Chem.*, **33**, 1947 (1968).
91. N. A. Valyashko and Y. S. Rozum, *Zh. Obshch. Khim.*, **17**, 755 (1947).
92. A. Burawoy and A. R. Thompson, *J. Chem. Soc.*, 4314 (1956).
93. J. Burgers et al., *Rec. Trav. Chim.*, **77**, 491 (1958).
94. L. Pentimalli, *Tetrahedron*, **5**, 27 (1959).
95. W. F. Forbes and A. S. Ralph, *Can. J. Chem.*, **34**, 1447 (1956).
96. P. P. Birnbaum, J. H. Linford, and D. W. G. Style, *Trans. Faraday Soc.*, **49**, 735 (1953).
97. O. C. Musgrave, *J. Chem. Soc.*, 1104 (1957).
98. L. Skulski, *Bull. Acad. Polon. Sci., Ser. Sci. Chim.*, **10**, 207 (1962).
99. P. Grammaticakis, *Bull. Soc. Chim. France* (5), **18**, 534 (1951).
100. E. A. Braude and F. Sondheimer, *J. Chem. Soc.*, 3754 (1955).
101. A. Burawoy and E. Spinner, *J. Chem. Soc.*, 2557 (1955).
102. E. Sawicki, *J. Org. Chem.*, **22**, 915 (1957).
103. L. A. Strait, R. Ketcham, D. Jambotkar, and V. P. Shah, *J. Am. Chem. Soc.*, **86**, 4628 (1964).
104. A. Burawoy and J. P. Critchley, *Tetrahedron*, **5**, 340 (1959).
105. V. Bellairta and A. Ricci, *Boll. Sci. Fac. Chim. Ind. Bologna*, **13**, 76 (1955).
106. R. Passerini and G. Righi, *Boll. Sci. Fac. Chim. Ind. Bologna*, **10**, 163 (1952).
107. A. Burawoy, J. P. Critchley, and A. R. Thompson, *Tetrahedron*, **4**, 403 (1958).
108. W. Baker, M. P. V. Boarland, and J. F. W. McOmie, *J. Chem. Soc.*, 1476 (1954).
109. R. Huisgen and H. Koch, *Ann.*, **591**, 200 (1955).
110. H. E. Ungnade, *J. Am. Chem. Soc.*, **76**, 5133 (1954).
111. H. Henbest and T. Owen, *J. Chem. Soc.*, 2968 (1955).
112. F. Montanari, *Boll. Sci. Fac. Chim. Ind. Bologna*, **17**, 33 (1959).
113. A. Mangini and R. Passerini, *Boll. Sci. Fac. Chim. Ind. Bologna*, **7**, 63 (1949).
114. A. Mangini and R. Passerini, *J. Chem. Soc.*, 1168 (1952).
115. L. Chierici and R. Passerini, *Atti. Accad. Nazl. Lincei*, **15**, 69 (1953).
116. M. J. Kamlet, *J. Org. Chem.*, **22**, 576 (1957).
117. G. M. Badger, R. G. Buttery, and G. E. Lewis, *J. Chem. Soc.*, 1888 (1954).
118. L. Doub and J. M. Vandenbelt, *J. Am. Chem. Soc.*, **69**, 2714 (1947); ibid., **71**, 2414 (1949).
119. G. M. Badger and R. Pearce, *J. Chem. Soc.*, 3072 (1950).
120. W. A. Sweeney and W. M. Schubert, *J. Am. Chem. Soc.*, **76**, 4625 (1954).
121. B. C. McKusick, R. E. Heckert, T. L. Cairns, D. D. Coffman, and H. F. Mower, *J. Am. Chem. Soc.*, **80**, 2806 (1958).
122. L. A. Wiles, *Chem. Rev.*, **56**, 329 (1956).
123. J. M. Vandenbelt, *Appl. Spectry.*, **17**, 120 (1963).
124. R. W. H. Berry, P. Brocklehurst, and A. Burawoy, *Tetrahedron*, **10**, 109 (1960).
125. F. G. Mann and A. J. Wilkinson, *J. Chem. Soc.*, 3336 (1957).
126. W. F. Little and A. K. Clark, *J. Org. Chem.*, **25**, 1979 (1960).
127. E. Sawicki and A. Carr, *J. Org. Chem.*, **22**, 503 (1957).
128. N. S. Bayliss and L. Hulme, *Australian J. Chem.*, **6**, 257 (1953).
129. G. Leandri and A. Tundo, *Boll. Sci. Fac. Chim. Ind. Bologna*, **10**, 160 (1952).
130. V. Baliah and T. Rangarajan, *Naturwissenschaften*, **46**, 107 (1959).
131. P. Mesnard and R. Crockett, *Compt. Rend.*, **248**, 872 (1959).
132. V. Baliah and S. Shanmuganathan, *J. Indian Chem. Soc.*, **35**, 31 (1958).
133. A. Mangini, L. Ruzzier, and A. Tundo, *Boll. Sci. Fac. Chim. Ind. Bologna*, **14**, 81 (1956).

134. R. T. Arnold, V. J. Webers, and R. M. Dodson, *J. Am. Chem. Soc.*, **74**, 368 (1952).
135. K. Miescher, A. Marxer, and E. Urech, *Helv. Chim. Acta*, **34**, 1 (1951).
136. E. Fehnel, *J. Am. Chem. Soc.*, **72**, 1404 (1950).
137. G. Leandri, A. Mangini, and R. Passerini, *Gazz. Chim. Ital.*, **84**, 73 (1954).
138. H. H. Szmant and J. J. McIntosh, *J. Am. Chem. Soc.*, **73**, 4356 (1951).
139. V. Baliah and S. Shanmuganathan, *J. Phys. Chem.*, **62**, 255 (1958).
140. W. F. Forbes and W. A. Mueller, *Can. J. Chem.*, **33**, 1145 (1955).
141. R. Elderfield and V. Meyer, *J. Am. Chem. Soc.*, **76**, 1887 (1954).
142. P. Grammaticakis, *Bull. Soc. Chim. France*, **20**, 93 (1953).
143. A. Burawoy, *J. Chem. Soc.*, 1181 (1939).
144. L. Ferguson and E. Branch, *J. Am. Chem. Soc.*, **66**, 1467 (1944).
145. H. Ungnade and I. Ortega, *J. Org. Chem.*, **17**, 1475 (1952).
146. A. Burawoy, *J. Chem. Soc.*, 1177 (1939).
147. G. Leandri and A. Mangini, *Spectrochim. Acta*, 421 (1959).
148. L. Freedman and G. Doak, *J. Org. Chem.*, **21**, 811 (1956).
149. I. Ugi, H. Perlinger, and L. Behringer, *Ber.*, **91**, 2324 (1958).
150. P. Barrett, *J. Chem. Soc.*, 2056 (1957).
151. H. H. Szmant and H. J. Planinsek, *J. Am. Chem. Soc.*, **76**, 1193 (1954).
152. C. J. Pedersen, *J. Am. Chem. Soc.*, **79**, 2295 (1957).
153. J. H. Boyer, V. Toggweiler, and G. A. Stoner, *J. Am. Chem. Soc.*, **79**, 1748 (1957).
154. K. Nakamoto and R. E. Rundle, *J. Am. Chem. Soc.*, **78**, 1113 (1956).
155. American Petroleum Institute Research Project No. 44, II, 558 (1953).
156. F. Bohlmann and H. J. Mannhardt, *Ber.*, **89**, 1307 (1956).
157. E. A. Braude and C. J. Timmons, *J. Chem. Soc.*, 2007 (1950).
158. J. Cymerman, I. Heilbron, E. R. H. Jones, and R. N. Lacey, *J. Chem. Soc.*, 500 (1946).
159. F. Bohlmann, *Ber.*, **85**, 386 (1952).
160. F. Bohlmann, *Ber.*, **85**, 1144 (1952).
161. J. J. Panouse, *Bull. Soc. Chim. France*, 1568 (1956).
162. R. Kuhn, *J. Chem. Soc.*, 605 (1938).
163. K. W. Hausser and R. Kuhn, *Z. Physik. Chem.*, **B29**, 371, 378, 384, 391, 417 (1935).
164. A. E. Gillam and D. H. Hey, *J. Chem. Soc.*, 1170 (1939).
165. G. Kortüm and G. Dreesen, *Ber.*, **84**, 182 (1951).
166. J. F. Thomas and G. Branch, *J. Am. Chem. Soc.*, **75**, 4793 (1953).
167. D. Marshall and M. C. Whiting, *J. Chem. Soc.*, 4082 (1956).
168. H. D. Hunt and W. T. Simpson, *J. Am. Chem. Soc.*, **75**, 4540 (1953).
169. R. F. Raffauf, *J. Am. Chem. Soc.*, **72**, 753 (1950).
170. H. Saikachi and H. Ogawa, *J. Am. Chem. Soc.*, **80**, 3642, (1958).
171. R. Wizinger and H. Sontag, *Helv. Chim. Acta*, **38**, 363 (1955).
172. D. Buckley, H. B. Henbest, and P. Slade, *J. Chem. Soc.*, 4891 (1957).
173. A. Van Dormael, *Ind. Chim. Belg.*, **24**, 463 (1959).
174. M. Strell, W. B. Braunbruck, and L. Reithmayr, *Ann.*, **587**, 195 (1954).
175. L. J. Saidel, *Nature*, **172**, 955 (1953).
176. R. Heilmann, J. Bonnier, and G. Gaudemaris, *Compt. Rend.*, **244**, 1787 (1957).
177. E. A. Braude and E. A. Evans, *J. Chem. Soc.*, 3334 (1955).
178. H. Booker, L. Evans, and A. Gillam, *J. Chem. Soc.*, 1453 (1940).
179. J. I. Cunneen, *J. Chem. Soc.*, 134 (1947).
180. F. Gunstone and W. Russell, *J. Chem. Soc.*, 3782 (1955).
181. K. Georgieff, W. Cave, and K. Blaikie, *J. Am. Chem. Soc.*, **76**, 5494 (1954).
182. K. Bowden, E. A. Braude, E. R. H. Jones, and B. Weedon, *J. Chem. Soc.*, 45 (1946).
183. E. A. Braude, E. R. H. Jones, and G. G. Rose, *J. Chem. Soc.*, 1104 (1947).

184. C. C. Price and J. Zomlefer, *J. Am. Chem. Soc.*, **72**, 14 (1950).
185. R. N. Jones and G. D. Thorn, *Can. J. Res.*, **27B**, 828 (1949).
186. W. M. Schubert, T. H. Liddicoet, and W. A. Lanka, *J. Am. Chem. Soc.*, **74**, 569 (1952).
187. L. C. Jones, Jr., and L. W. Taylor, *Anal. Chem.*, **27**, 228 (1955).
188. D. Rosenthal and T. I. Taylor, *J. Am. Chem. Soc.*, **79**, 2684 (1957).
189. L. W. C. Miles and L. N. Owen, *J. Chem. Soc.*, 2934 (1950).
190. M. Korach and W. Bergmann, *J. Org. Chem.*, **14**, 1118 (1949).
191. C. Compton and W. Bergmann, *J. Org. Chem.*, **12**, 363 (1947).
192. R. Kuhn and C. Grundmann, *Ber.*, **71**, 442 (1938).
193. E. Hertle and M. Schinzel, *Z. Physik. Chem.*, **B48**, 289 (1941).
194. E. A. Braude and E. R. H. Jones, *J. Chem. Soc.*, 498 (1945).
195. K. W. Hauser et al., *Z. Physik. Chem.*, **B29**, 363 (1935).
196. F. Bohlmann, *Ber.*, **84**, 860 (1951).
197. K. Taufel and R. Zimmermann, *Naturwissenschaften*, **47**, 133 (1960).
198. H. D. Eilhauer, P. Fahrig, and G. Krautschick, *Z. Anal. Chem.*, **228**, 276 (1967).
199. R. Bonnichsen and A. C. Maehly, *J. Forensic Sci.*, **11**, 516 (1966).
200. E. R. H. Jones, M. C. Whiting, J. B. Armitage, C. L. Cook, and N. E. Entwistle, *Nature*, **168**, 900 (1951).
201. F. Bohlmann, *Ber.*, **86**, 657 (1953).
202. C. L. Cook, E. R. H. Jones, and M. C. Whiting, *J. Chem. Soc.*, 147 (1954).
203. J. B. Armitage, N. Entwistle, E. R. H. Jones, and M. C. Whiting, *J. Chem. Soc.*, 147 (1954).
204. H. H. Schlubach and V. Franzen, *Ann.*, **573**, 110 (1951).
205. S. Akiyama and M. Nakagawa, *Bull. Chem. Soc. Japan*, **40**, 340 (1967).
206. S. Akiyama, K. Nakasuji, K. Akashi, and M. Nakagawa, *Tetrahedron Letters*, 1121 (1968).
207. N. Sörensen and J. Stene, *Ann.*, **549**, 80 (1941).
208. T. Bruun, C. M. Haug, and N. Sörensen, *Acta Chem. Scand.*, **4**, 850 (1950).
209. K. Stavholt and N. Sörensen, *Acta Chem. Scand.*, **4**, 1567, 1575 (1950).
210. A. Castille, *Ann.*, **543**, 104 (1939).
211. M. Anchel, J. Polatnick, and F. Kavanagh, *Arch. Biochem. Biophys.*, **25**, 208 (1950).
212. F. Kavanagh, A. Hervey, and W. J. Robbins, *Proc. Natl. Acad. Sci. U.S.*, **36**, 1, 102 (1950).
213. R. Kuhn and H. Fischer, *Ber.*, **94**, 3060 (1961).
214. R. Kuhn and H. Zahn, *Ber.*, **84**, 566 (1951).
215. H. Fischer and W. D. Hell, *Angew. Chem.*, **79**, 931 (1967).
216. A. E. Gillam and D. H. Hey, *J. Chem. Soc.*, 1174 (1939).
217. J. H. Uhlenbroek and J. D. Bijloo, *Rec. Trav. Chim.*, **79**, 1181 (1960).
218. J. Dale, *Acta Chem. Scand.*, **11**, 971 (1957).
219. G. Drefahl and G. Plötner, *Ber.*, **93**, 990 (1960).
220. L. Skattebol and N. A. Sörensen, *Acta Chem. Scand.*, **12**, 2101 (1958).
221. H. Dahn and H. Castelmur, *Helv. Chim. Acta*, **36**, 638 (1953).
222. K. Ueno and S. Akiyoshi, *J. Am. Chem. Soc.*, **76**, 3667 (1954).
223. G. Favini and M. Simonetta, *Gazz. Chim. Ital.*, **89**, 2111 (1959).
224. A. J. Lindsey, *Anal. Chim. Acta*, **20**, 175 (1959).
225. E. Clar, *Ber.*, **69**, 608 (1936).
226. E. Clar and R. Sandke, *Ber.*, **81**, 52 (1948).
227. L. G. S. Brooker, F. L. White, G. H. Keyes, C. P. Smyth, and P. F. Oesper, *J. Am. Chem. Soc.*, **63**, 3192 (1941).

228. W. Krauss and H. Grund, *Z. Elektrochem.*, **58**, 767 (1954).
229. L. G. S. Brooker, *Nuclear and Theoretical Organic Chemistry*, Interscience, New York, 1945, p. 132.
230. L. G. S. Brooker, G. H. Keyes, R. H. Sprague, R. H. Van Dyke, E. Van Lare, G. Van Zandt, F. L. White, H. W. J. Cressman, and S. G. Dent, Jr., *J. Am. Chem. Soc.*, **73**, 5332 (1951).
231. K. Venkataraman, *The Chemistry of Synthetic Dyes*, Vol. I, Academic Press, New York, 1952, p. 351.
232. L. N. Ferguson, *Chem. Rev.*, **43**, 385 (1948).
233. W. Konig, W. Schramek, and B. Rosch, *Ber.*, **61**, 2074 (1928).
234. A. Pullman and B. Pullman, *Discussions Faraday Soc.*, 46 (1950).
235. E. B. Knott, *J. Soc. Dyers Colourists*, **67**, 302 (1951).
236. L. Zechmeister, Cis-Trans *Isomeric Carotenoids*, *Vitamins A* and *Arylpolyenes*, Academic Press, New York, 1962, p. 35.
237. C. Genest, D. M. Smith, and D. G. Chapman, *J. Agr. Food Chem.*, **11**, 508 (1963).
238. M. Strell and S. Reindl, *Arzneimittel-Forsch.*, **11**, 552 (1961).
239. L. Skulski and J. Plenkiewicz, *Roczniki Chem.*, **37**, 45 (1963).
240. H. Schmidt, *Naturwissenschaften*, **46**, 379 (1959).
241. L. D. Saslaw and V. S. Waravdekar, *Biochim. Biophys. Acta*, **37**, 367 (1960).
242. H. Thies and M. Steinigen, *Deut. Apotheker-Ztg.*, **106**, 1451 (1966).
243. M. G. Evans and J. DeHeer, *Quart. Rev.*, **4**, 94 (1950).
244. G. Fritz and H. Hartmann, *Z. Elektrochem. Angew. Physik. Chem.*, **55**, 184 (1951).
245. E. J. Moriconi, B. Rakoczy, and W. F. O'Connor, *J. Org. Chem.*, **27**, 2772 (1962).
246. L. F. Fieser and M. Fieser, *Organic Chemistry*, Heath, Boston, 1944, p. 726.
247. E. Clar, *Polycyclic Hydrocarbons*, Vol. I, Academic Press, New York, 1964, pp. 40–69.
248. R. H. Altiparmakian and R. S. Braithwaite, *J. Chem. Soc., Sect. B*, 1113 (1967).
249. S. Hunig and O. Rosenthal, *Ann.*, **592**, 161 (1955).
250. S. Hunig and H. Herrmann, *Ann.*, **636**, 32 (1960)
251. S. Hunig, G. Bernhard, W. Liptay, and W. Brenninger, *Ann.*, **690**, 9 (1965).
252. R. Gompper and H. U. Wagner, *Tetrahedron Letters*, 165 (1968).
253. F. Gerson, E. Heilbronner, O. Neunhoeffer, and H. Paul, *J. Prakt. Chem.* (4), **9**, 160 (1959).
254. V. Zanker and G. Schiefele, *Z. Elektrochem.*, **62**, 86 (1958).
255. W. Ried and G. Dankert, *Ber.*, **92**, 1223 (1959).
256. H. H. Szmant and H. J. Planinsek, *J. Am. Chem. Soc.*, **72**, 4042 (1950).
257. J. J. Wren, *J. Chem. Soc.*, 2208 (1956).
258. F. Ramirez and A. Kirby, *J. Am. Chem. Soc.*, **76**, 1037 (1954).
259. F. Bohlmann, *Ber.*, **84**, 490 (1951).
260. M. Hawthorne, *J. Org. Chem.*, **21**, 1523 (1956).
261. C. Price and H. Morita, *J. Am. Chem. Soc.*, **75**, 4747 (1953).
262. D. Craig, L. Schaefgen, and W. Tyler, *J. Am. Chem. Soc.*, **70**, 1624 (1948).
263. J. Goerdler, D. Drause-Loevenich, and B. Wedekind, *Chem. Ber.*, **90**, 1638 (1957).
264. A. Mangini and R. Passerini, *Boll. Sci. Fac. Chim. Ind. Bologna*, **7**, 63, 64, 65 (1949).
265. T. Nash, *Biochem. J.*, **55**, 416 (1953).
266. F. Dunsbach, *Z. Klin. Chem.*, **4**, 262 (1966).
267. V. D. Yablochkin, *Lab. Delo*, 719 (1966).
268. D. Hofmann, E. M. Kosower, and K. Wallenfels, *J. Am. Chem. Soc.*, **83**, 3314 (1961).
269. K. Tsuda, Y. Satch, N. Ikekawa, and M. Mishima, *J. Org. Chem.*, **21**, 800 (1956).
270. J. C. Lockhart, *J. Chem. Soc.*, 3737 (1962).

271. J. C. Lockhart, *Chem. Ind (London)*, 2006 (1961).
272. T. Colclough, W. Gerrard, and M. F. Lappert, *J. Chem. Soc.* 3006 (1956).
273. E. Katzenellenbogen and G. Branch, *J. Am. Chem. Soc.*, **69**, 1615 (1947).
274. H. H. Szmant and C. McGinnis, *J. Am. Chem. Soc.*, **74**, 240 (1952).
275. T. W. Campbell, S. Linden, S. Godshalk, and W. G. Young, *J. Am. Chem. Soc.*, **69**, 880 (1947).
276. A. L. Wilds, *J. Am. Chem. Soc.*, **69**, 1985 (1947).
277. M. K. Seikel and T. A. Geissman, *J. Am. Chem. Soc.*, **72**, 5720 (1950).
278. N. H. Cromwell and W. R. Watson, *J. Org. Chem.*, **14**, 411 (1949).
279. E. J. Moriconi, W. F. O'Connor, and W. F. Forbes, *J. Am. Chem. Soc.*, **82**, 5454 (1960).
280. H. S. French, *J. Am. Chem. Soc.*, **74**, 514 (1952).
281. F. Bohlmann, *Ber.*, **89**, 2191 (1956).
282. M. Kisteneva, *Zh. Obshch. Khim.*, **26**, 2019 (1956).
283. S. P. Coan, D. E. Trucker, and E. I. Becker, *J. Am. Chem. Soc.*, **75**, 900 (1953).
284. L. G. S. Brooker and P. W. Vittum, *J. Photographic Sci.*, **5**, 71 (1957).
285. S. Hunig, H. Schweeberg, and H. Schwarz, *Ann.*, **587**, 132 (1954).
286. S. Hunig and H. Schwarz, *Ann.*, **599**, 131 (1956).
287. M. J. Kamlet, J. C. Hoffsommer, and H. G. Adolph, *J. Am. Chem. Soc.*, **84**, 3925 (1962).
288. E. R. Blout and V. W. Eager, *J. Am. Chem. Soc.*, **67**, 1315 (1945).
289. R. Alexander, Jr., *J. Org. Chem.*, **21**, 1464 (1957).
290. G. H. Brown, B. Graham, P. W. Vittum, and A. Weissberger, *J. Am. Chem. Soc.*, **73**, 919 (1951).
291. L. N. Ferguson and J. C. Nnadi, *J. Chem. Educ.* **42**, 529 (1965).
292. P. Maroni and J. Dubois, *J. Chim. Phys.*, **51**, 402 (1954). P. Maroni, *Ann. Chim. (Paris)*, **2**, 757 (1957).
293. A. I. Scott, *Interpretation of the Ultraviolet Spectra of Natural Products*, Macmillan, New York, 1964, pp. 55–68.
294. L. F. Fieser and M. Fieser, *Steroids*, Reinhold, New York, 1959, pp. 15–24.
295. H. French and L. Wiley, *J. Am. Chem. Soc.* **71**, 3702 (1949).
296. L. Evans and A. Gillam, *J. Chem. Soc.*, 815 (1941).
297. R. B. Woodward, *J. Am. Chem. Soc.*, **63**, 1123 (1941); *ibid.*, **64**, 72, 76 (1942).
298. R. C. Cookson and N. S. Wariyar, *J. Chem. Soc.*, 2302 (1956).
299. F. F. Caserio and J. D. Roberts, *J. Am. Chem. Soc.*, **80**, 5837 (1958).
300. S. Winstein, L. Devries, and R. Orloski, *J. Am. Chem. Soc.*, **83**, 2020 (1961).
301. H. Labhart and G. Wagniere, *Helv. Chim. Acta*, **42**, 2219 (1959).
302. P. D. Bartlett and B. E. Tate, *J. Am. Chem. Soc.*, **78**, 2473 (1956).
303. W. Bradley and M. C. Clark, *Chem. Ind. (London)*, 589 (1961).
304. E. M. Kosower, W. D. Closson, H. L. Goering, and J. C. Gross, *J. Am. Chem. Soc.*, **83**, 2013 (1961).
305. H. E. Simmons and T. Fukunga, *J. Am. Chem. Soc.*, **89**, 5208 (1967).
306. E. M. Kosower and D. C. Remy, *Tetrahedron*, **5**, 281 (1959).
307. V. Georgian, *Chem. Ind. (London)*, 930, (1954); ibid., 1480 (1957).
308. N. J. Leonard, T. W. Milligan, and T. L. Brown, *J. Am. Chem. Soc.*, **82**, 4077 (1960).
309. N. J. Leonard and M. Oki, *J. Am. Chem. Soc.*, **77**, 6239 (1955).
310. J. T. Clarke and E. R. Blout, *J. Polymer Sci.*, **1**, 419 (1946).
311. M. Gillois and P. Rumpf, *Bull. Soc. Chim. France*, **21**, 112 (1954).
312. A. F. Casy, A. H. Beckett, M. A. Iorio, and H. Z. Youssef, *Tetrahedron*, **21**, 3387 (1965).

313. A. H. Beckett, A. F. Casy, and M. A. Iorio, *Tetrahedron*, **22**, 2745 (1966).
314. A. H. Beckett, A. F. Casy, R. G. Lingard, and M. A. Iorio, and K. Hewitson, *Tetrahedron*, **22**, 2735 (1966).
315. C. B. Clarke and A. R. Pinder, *J. Chem. Soc.*, 1967 (1958).
316. L. A. Strait, D. Jambotkar, R. Ketcham, and M. Hrenoff, *J. Org. Chem.*, **31**, 3976 (1966).
317. A. D. Walsh, *Trans. Faraday Soc.*, **45**, 179 (1949).
318. J. F. Music and F. A. Matsen, *J. Am. Chem. Soc.*, **72**, 5256 (1950).
319. N. H. Cromwell, F. H. Schumacher, and J. L. Adelfang, *J. Am. Chem. Soc.*, **83**, 974 (1961).
320. N. H. Cromwell, *Record Chem. Progr.*, **19**, 215 (1958).
321. F. H. C. Stewart, *J. Org. Chem.*, **27**, 3374 (1962).
322. J. Elvidge, J. Fitt, and R. Linstead, *J. Chem. Soc.*, 244 (1956).
323. E. B. Knott, *J. Chem. Soc.*, 1490 (1954).
324. E. B. Knott, *J. Chem. Soc.*, 1482 (1954).
325. D. Buckley, S. Dunsten, and H. B. Henbest, *J. Chem. Soc.*, 4880 (1957).
326. A. I. Kiprianov and A. V. Stetsenko, *Zh. Obshch. Khim.*, **23**, 1912 (1953); through *Chem. Abstr.*, **49**, 260 (1955).
327. W. Flaig, J. C. Salfeld, and E. Baume, *Ann.*, **618**, 117 (1958).
328. E. A. Braude, *J. Chem. Soc.*, 490 (1945).
329. M. Klessinger and W. Luttke, *Tetrahedron*, **19**, Suppl. 2, 315 (1963).
330. M. Klessinger and W. Luttke, *Ber.*, **99**, 2136 (1966).
331. K. Bowden, E. A. Braude, and E. R. H. Jones, *J. Chem. Soc.*, 948 (1946).
332. W. Luttke and M. Klessinger, *Angew. Chem., Intern. Ed.*, **5**, 598 (1966).
333. M. Klessinger, *Tetrahedron*, **22**, 3355 (1966).
334. S. K. Guha and J. N. Chatterjea, *Ber.*, **92**, 2768 (1959).
335. A. Treibs, K. Jacob, and A. Dietl, *Ann.*, **702**, 112 (1967).
336. E. Klingsberg, *Chem. Rev.*, **54**, 59 (1954).
337. S. E. Sheppard and P. T. Newsome, *J. Am. Chem. Soc.*, **64**, 2937 (1942).
338. E. Merian, *Chimia*, **13**, 181 (1959).
339. H. E. Sprenger and W. Ziegenbein, *Angew. Chem. Intern. Ed.*, **6**, 553 (1967).

CHAPTER 7

ABSORPTION SPECTRA OF ZWITTERIONIC RESONANCE STRUCTURES V. INTENSITY DETERMINANTS

The absorption intensity of a compound in solution is one important factor in determining the ease and sensitivity of any analytical assay for the compound. As shown in Table 7.1, the range of intensities is very wide for the $S_0 \rightarrow S_1$ transitions of neutral molecules (e.g., mϵ from 0.0035 to 700).

Whereas the difference in energy between the excited and the ground states determines the wavelength maximum of a band, the intensity of absorption is a measure of the probability with which the particular transition occurs. According to Braude and Sondheimer(24), the transition probability is governed by three factors – namely, the transition moment, the chromophore area, and selection rules.

The transition moment is related to the change in dipole moment in passing from the ground to the excited state. The intensity of an absorption band is proportional to the square of the transition moment. The relationship between the chromophore area and absorption intensity has been derived and applied to the electronic spectra of polyenes and polycyclic aromatic hydrocarbons(25).

In reference to selection rules and analogous matters Hercules(26) has emphasized three major factors that influence the intensity of an electronic transition:

1. The multiplicities of the ground and excited states (i.e., whether they are singlet, triplet, etc.).
2. The degree of overlap between the orbitals involved in electron promotion (i.e., the orbital from which the excited electron came and the one to which it was promoted).
3. The symmetries of the wavefunctions of the ground and excited states.

Each of these factors can introduce some degree of forbiddenness into an electronic transition, thus reducing the probability of a transition and, other things being equal, decreasing the molar absorptivity of a band. The ramifications of these various factors will be discussed from the organic analysis viewpoint.

TABLE 7.1
Absorption Intensities of Some $S_0 \rightarrow S_1$ Transitions

Compound	Transition	Solvent	λ_{max} (mμ)	mϵ	Ref.
Diazomethane	$n \rightarrow \pi^*$	Vapor	450	0.0035	1
Acetone	$n \rightarrow \pi^*$	Water	265	0.019	2
Propylene sulfide	$n \rightarrow \sigma^*$	Octane	275[a]	0.030	3
Nitrosobenzene	$n \rightarrow \pi^*$	95% EtOH	756	0.045	5
Thiobenzophenone	$n \rightarrow \pi^*$	Ether	605	0.066	6
Amyl nitrite	$n \rightarrow \pi^*$	95% EtOH	357[b]	0.075	6
Dimethyl nitrosamine	$n \rightarrow \pi^*$	95% EtOH	344	0.097	6
Trimethylene disulfide	$n \rightarrow \sigma^*$	95% EtOH	333	0.147	7
Benzene[c]	$\pi \rightarrow \pi^*$	Water	254	0.204	8
Methyl iodide	$n \rightarrow \sigma^*$	95% EtOH	255	0.360	9
Trimethylamine	$n \rightarrow \sigma^*$	Vapor	227	0.900	10
Chrysene[c]	$\pi \rightarrow \pi^*$	95% EtOH	360[b]	1.00	11
Aniline[c]	$\pi \rightarrow \pi^*$	Water	280	1.43	8
Methyl ether	$n \rightarrow \sigma^*$	Vapor	184	2.52	12
4-Acetyldibenzoselenophene	$\pi \rightarrow \pi^*$	95% EtOH	262	4.2	
Piperazine	$n \rightarrow \sigma^*$	Vapor	196	5.0	13
1,4-Dioxane	$n \rightarrow \sigma^*$	Vapor	180	6.0	14
N,N'-Diethylthiourea	$\pi \rightarrow \pi^*$	Methanol	265	7.3	15
$(CH_3)_3C—CH_2—C(CH_3)_3$	$\sigma \rightarrow \sigma^*$	Vapor	154	~10.0	16
Thioacetamide	$\pi \rightarrow \pi^*$	95% EtOH	265	12.9	17
Chrysene[d]	$\pi \rightarrow \pi^*$	95% EtOH	319[b]	15.9	11
3,6-Dinitro-9-methylcarbazole[e]	$\pi \rightarrow \pi^*$	95% EtOH	368	18.0	
3-(2-Furyl)acrolein	$\pi \rightarrow \pi^*$	Dioxane	312	26.3	18
(benzothiazoline ring, S, N–C_2H_5) C=$(CHCH)_2$=Nϕ	$\pi \rightarrow \pi^*$	Methanol	448	59.0	19
Chrysene[f]	$\pi \rightarrow \pi^*$	95% EtOH	267	138.0	11
m-Sedeciphenyl[g]	$\pi \rightarrow \pi^*$	Chloroform	255	320.0	20
$\alpha,\beta,\gamma,\delta$-Tetraphenylporphin[h]	$\pi \rightarrow \pi^*$	Benzene	420	395.0	21
$CH_3(C\equiv C)_6CH_3$[i]	$\pi \rightarrow \pi^*$	95% EtOH	284	445.0	22
$(CH_3)_3C(C\equiv C)_7C(CH_3)_3$[i]	$\pi \rightarrow \pi^*$	Ether	311	527.0	23
Li_2 $\alpha,\beta,\gamma,\delta$-Tetraphenylporphin[h]	$\pi \rightarrow \pi^*$	Pyridine–methanol (3%)	445	610.0	21
Mg $\alpha,\beta,\gamma,\delta$-Tetraphenylporphin[h]	$\pi \rightarrow \pi$	Benzene	425	700.0	21

[a]Also band at 218 mμ, mϵ 0.60. Saturated oxygen and sulfur heterocyclics, in which there are two pairs of lone-pair electrons on the hetero atom, give two $n \rightarrow \sigma^*$ absorption bands (4).

[b]Fine structure.

[c]α-Band (Clar's nomenclature).

[d]*p*-Band (Clar's nomenclature).

[e]Long-wavelength band from *p*-nitroaniline-type chromophore.

[f]β-Band (Clar's nomenclature).

[g]High intensity due to cross-conjugation in structure consisting of 15 biphenyl chromophores.

[h]Soret band.

[i]Not long-wavelength band. Most intense band.

I. Effect of Multiplicities of Ground and Excited States on Intensity

The $S_0 \rightarrow T_1$ transition always occurs at a longer wavelength than the corresponding $S_0 \rightarrow S_1$ transition, though the absorption intensity is much weaker as a change of electronic spin on promotion is highly forbidden. Thus mϵ for a $(S_0 \rightarrow T_1)n \rightarrow \pi^*$ transition is on the order of 10^{-5} as compared with a $(S_0 \rightarrow S_1)n \rightarrow \pi^*$ transition with an mϵ of 10^{-1} for the azine compounds(4). The $(S_0 \rightarrow S_1)n \rightarrow \pi^*$ transitions are forbidden and hence of low intensity since the transition from a ground-state to an excited-state orbital is fairly difficult. Transitions of the $n \rightarrow \sigma^*$ type have moderate intensities, whereas $\sigma \rightarrow \sigma^*$ and $\pi \rightarrow \pi^*$ transitions usually have high intensities, since they are allowed(4). The order of intensities among the various transitions is usually(4, 27)

$$(S_0 \rightarrow T_1)\pi \rightarrow \pi^* < (S_0 \rightarrow T_1)n \rightarrow \pi^*$$
$$<<< (S_0 \rightarrow S_1)n \rightarrow \pi^* < (S_0 \rightarrow S_1)n \rightarrow \sigma^*$$
$$< (S_0 \rightarrow S_1)\sigma \rightarrow \sigma^* < (S_0 \rightarrow S_1)\pi \rightarrow \pi^*$$

The data in Table 7.1 bear this out.

Mixing of transitions can also cause an intensification of the absorption. For example, the $n \rightarrow \pi^*$ absorption of unsaturated ketones is intensified due to the mixing of the forbidden $n \rightarrow \pi^*$ transition (which lacks an electric dipole moment) with the allowed $\pi \rightarrow \pi^*$ transition, in which an electron is transferred from the C=C double bond (π) to the antibonding orbital of the carbonyl group (π^*)(28).

II. Colorimetric Analysis and Intensity

In colorimetric analysis the main concern is to end up with a chromogen that has a $(S_0 \rightarrow S_1)\pi \rightarrow \pi^*$ band of high intensity and long wavelength. This can be done fairly readily. If a compound absorbs at fairly short wavelengths or with low intensity, there is a much greater probability of interference from the increasingly larger number of compounds that absorb at shorter wavelengths or with greater intensity at the wavelength maximum of the test substance. In such a case the compound is reacted with a reagent to give a chromogen that absorbs at longer wavelength with greater intensity (Table 7.2).

III. Annulation Effects in Polynuclear Compounds

Clar(44) in a lifetime study has shown that the spectra of polynuclear aromatic hydrocarbons demonstrate the presence of three main types of

TABLE 7.2
Spectral Properties of Test Substances and Their Derived Chromogens

Test substance	λ_{max} ($m\mu$)	$m\epsilon$	Ref.	Chromogen	λ_{max} ($m\mu$)	$m\epsilon$	$\Delta\lambda$ ($m\mu$)	$\frac{m\epsilon_{chrom}}{m\epsilon_{ts}}$ [a]	Ref.
CH_2O	288	0.0135	29	Ac, Ac, N, H, H_3C, CH_3	412	7.7	124	570	
H_2NCONH_2	< 220		30	$(p\text{-}(CH_3)_2N\phi CH{=}N)_2C{=}O$	425	~7.2	> 205	–	31
Phenol	268	1.5	32	O=, N=, =O, ϕ–N, N, CH_3, CH_3	455	17.0	187	11.3	33
Sulfadiazine	243[b]	14.2	34	H, O, N, S=, N, H, O, =CHCH=CH–, HO, H, N, =S, N, O, H	533	156.0	290	11.0	35
Glycine	< 200	–	36	O, HO, =N–, O, O	570	21.1	> 370		37
2-Deoxyribose	< 200			H, O, N, S=, N, H, O, =CHCH=CH, HO, H, N, =S, N, O, H	532	151.0	> 332		38

TABLE 7.2 (*cont'd*)

Test substance	λ_{max} (mμ)	mϵ	Ref.	Chromogen	λ_{max} (mμ)	mϵ	$\Delta\lambda$ (mμ)	$\frac{m\epsilon_{chrom}}{m\epsilon_{ts}}$ [a]	Ref.
Pentachloro-nitrobenzene (Terrachlor)	~314[c]	~1.5		$H_2N-C_{10}H_6-N{=}N-C_6H_4-C(=O)OC_2H_5$	525	21.0	~211		40
Ethylamine	213[d]	0.79	10	$O_2N-C_6H_3(NO_2)-NHC_2H_5$	353	18.0	140	22.8	41
Pyruvic acid	330	0.02	42	$O_2N-C_6H_3(NO_2)-NH-N{=}C(COOH)(CH_3)$	365	20.5	35	1025.0	43

[a]At wavelength maxima of chromogen and test substance. Actually if mϵ were taken for test substance and chromogen at λ_{max} of chromogen, values would be much higher.

[b]Also band of about one-fourth intensity at 308 mμ.

[c]Values for 2,5-dichloronitrobenzene (39).

[d]Fine structure.

absorption band, termed the α-, *para*, and β-bands. The α-bands are weak (mϵ 0.1–1.0). The *para*-bands are of moderate intensity (mϵ ~ 10) and shift strongly to longer wavelengths with linear annulation in the polyacene series (e.g., naphthalene, anthracene, naphthacene, pentacene, and hexacene). The β-bands are of high intensity (mϵ ~ 100) and move moderately to the red with both linear and angular annulation. The α- and β-bands are related since the ratio of the frequencies of the β- and α-bands is a constant 1.35 for the aromatic hydrocarbons. These various bands are shown in the spectrum of phenanthrene and benzo[*c*]cinnoline in Figure 7.1 (45).

IV. Spectroscopic Moments

According to the theory of Sklar(46) and Forster(47), the added intensity produced in the benzene 260 mμ transition (α-band) by monosubstitution is proportional to the square of a transition, or "spectroscopic," moment induced by the presence of any substituent that destroys the symmetry. With polysubstituted nuclei, the intensity increases may be calculated by adding vectorially the spectroscopic moments of the individual substituents, the spectroscopic moment being the square root of the intensity increment conferred by each substituent singly(48). In

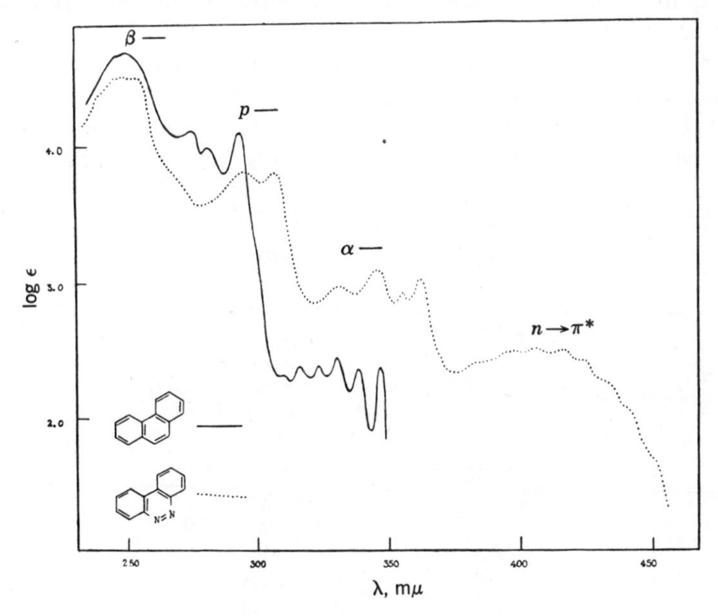

Fig. 7.1. Absorption spectra of phenanthrene (——) and benzo[*c*]cinnoline (· · · · ·) in hexane(45).

the case of the six-membered-ring systems the intensity changes in the α-band consequent on the replacement of a CH group by a nitrogen atom or by substitution in the nucleus generally may be accounted for quantitatively according to the theory of Sklar and Forster, as developed by Platt(48). An example of the intensity change is seen in the α-band of benzo[*c*]cinnoline as compared with that of phenanthrene (Fig. 7.1). The same intensification of the α-band is seen in quinoline and isoquinoline as compared with naphthalene and in many other aza compounds as compared with their parent hydrocarbons.

V. Distance between Resonance Terminals

Since the intensity of an absorption band is proportional to the square of the change in dipole moment on going from the ground to the excited state – and since the larger the separation of charge, the greater the change in dipole moment – an increase in the straight-line distance between resonance terminals will result in a higher intensity. The intensity of a band is stated to be proportional to the square of the distance between the ends of a conjugated system(49).

Other things being equal, increasing the length of conjugation of a molecule increases the molar absorptivity (see Section V in Chapter 6). Even an increase in the chain of extraconjugation can cause an increase in intensity (Table 7.3).

In aromatic derivatives that contain an electron-donor and an electron-acceptor group in the *ortho* and *para* positions, the *ortho* isomer has a longer chain of conjugation and a lower molar absorptivity than the *para* derivative. The intensity is decreased because the straight-line distance

TABLE 7.3
Increase in Extraconjugation Resulting in Increased Intensity(50)

$H_3C(CH{=}CH)_n{-}C(R){=}N{-}N(CH_3){-}C_6H_3(NO_2)_2$ (2,4-dinitrophenyl; O_2N, NO_2)

n	R	λ_{max} (mμ)	mϵ
0	CH_3	375	19.1
1	H	397	22.3
2	H	410	25.4
3	H	424	28.9

between resonance terminals is much shorter in the *ortho* isomer. For example, in ethanol *o*-nitroaniline absorbs at 404 mμ, mε 5.3, whereas *p*-nitroaniline absorbs at 372 mμ, mε 15.2(51); 2-phenylazoaniline absorbs at 413 mμ, mε 6.3, and 4-phenylazoaniline absorbs at 380 mμ, mε 25.0(52). The same explanation would suffice for the difference in intensity between cyclopentadiene (λ_{max} 239 mμ, mε 3.4), cyclohexadiene (λ_{max} 260 mμ, mε 9.9), and 1,3-butadiene (λ_{max} 217 mμ, mε 21.0).

The integrated intensity (measured by the area under a band envelope) is approximately proportional to the square of the distance between the resonance terminals of the chromophore. This relation has been used successfully in correlating spectral intensities of linear polyenes (53–55), α, β-unsaturated ketones (56), and α- and γ-pyrones and pyridones (57). The relation for the latter compounds is seen in Table 7.4.

TABLE 7.4
Integrated Intensity and Distance between Hetero Atom and Carbonyl Oxygen (57)

Compound	λ_{max} (mμ)	Relative intensity	r^2(Å^2)	Ratios Compounds	Intensities	r^2/r^2
1. γ-Pyrone	246	0.91	14.98	1:3	2.5	2.9[a]
2. 2,6-Dimethyl-γ-pyrone	247	0.98	14.98	2:3	2.7	2.9
3. 5-Methyl-α-pyrone	295	0.36	5.20			
4. γ-Pyridone	256	1.00	15.37			
5. *N*-Methyl-γ-pyridone	262	0.99	15.37	4:6	2.4	2.9
6. α-Pyridone	297	0.42	5.38	5:7	2.4	2.9
7. *N*-Methyl-α-pyridone	300	0.41	5.38			

[a] r^2 of $1/r^2$ of $3 = 14.98/5.20 = 2.9$.

In the carotenoids and arylpolyenes the all-*trans* forms have the longest distance between resonance terminals, whereas the central-mono-*cis* form would have the shortest distance. Consequently the intensities of the long-wavelength bands of all mono-*cis* isomers should lie between the extremes of the high intensity of the all-*trans* and the low intensity of the central-mono-*cis* forms (58). As expected, the all-*trans* forms of lycopene, diphenylhexatriene, and diphenyloctatetraene absorb at longer wavelength and with greater intensity than any of the known *cis* isomers. In the case of lycopene over 40 isomers are known.

Some zwitterionic molecules act as if they were cationic resonance structures absorbing at fairly long wavelengths with very high intensity. Thus **1**(59) absorbs in the same region as the cationic resonance structure **2**(60) but has three times the intensity.

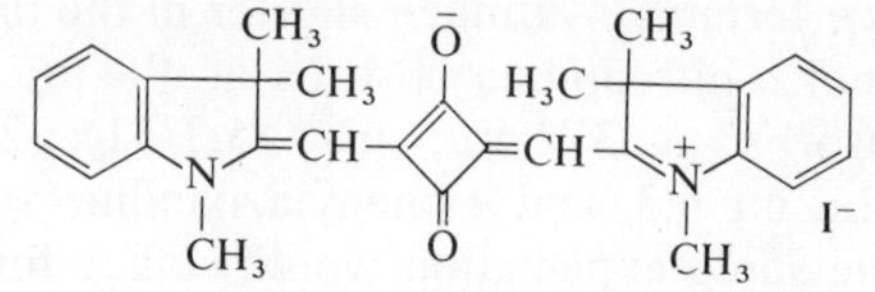

1

λ_{max} 625 mμ, mϵ 460 in ethanol

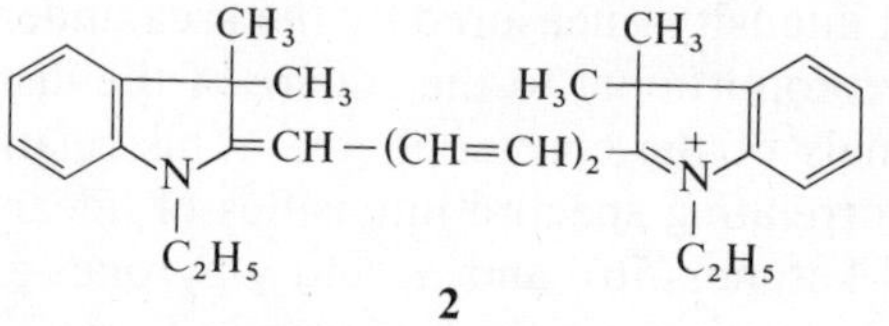

2

λ_{max} 638 mμ, mϵ 144 in methanol

In some betaines the positive and negative charges can diffuse over the whole molecule, but, as shown for **3** and **4**(61), one could consider the positive charge as residing mainly on the methylated nitrogen, whereas the negative charge alternates to a large extent between the other two nitrogen atoms(61, 62). The low intensity of the visible-region-band system is probably partially explainable on the basis of the short distance between the resonance terminals. However, it is obvious that the position of the long-wavelength bands depends on the entire conjugation system, as shown by comparing **4** with **3** (spectra in methanol).

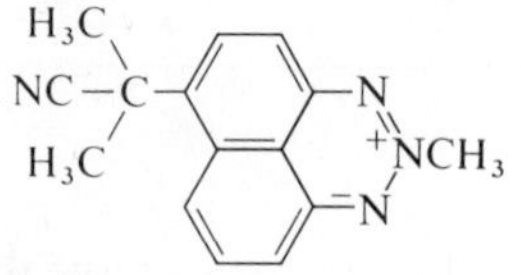

3

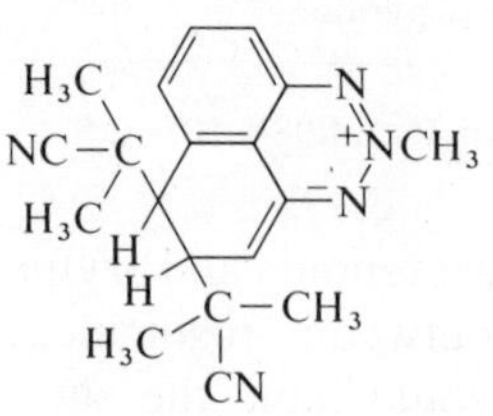

λ_{max} (mμ)	mϵ
233	35.5
355	12.9
545, 585	0.83, 0.83
635	0.59

4

λ_{max} (mμ)	mϵ
200	21.4
300	7.9
367	3.7
490	1.8

VI. Cross-Conjugation and Intensity

Where a chromophore is present several times in a cross-conjugated form in a molecule, the intensity will be increased. For example, 1,3,5-

trinitrobenzene (λ_{max} 224 mμ, mϵ 26.9) has three nitrobenzene (λ_{max} 259.5 mμ, mϵ 8.1) chromophores and so has three times the intensity(63). The molar absorptivity of the *m*-polyphenyls depends on the number of cross-conjugated biphenyls present. Many more examples are given in Chapter 6 in the section on cross-conjugation.

VII. Molecular Conformation and Intensity

The geometry of a molecule can have a decided effect on the intensity of a band. From the Franck–Condon principle that the absorption of light occurs in a fraction of time much shorter than that required for atom vibrations, it follows that, other factors being equal, the more completely the shape (i.e., the bond distances and angles) of the molecule in the lower state corresponds to the shape of its upper state, the higher is the probability of light absorption(64).

The molar absorptivity is greater the higher the absorption probability. The data in Table 7.5 indicate that, in addition to the importance of the straight-line distance between resonance terminals, a conformation factor is necessary to explain the changes in intensity.

TABLE 7.5
Effect of Absorption Probability on Band Intensity(64)

Compound	λ_{max} (mμ) (mϵ) Hexane	λ_{max} (mμ) (mϵ) 0.1 *N* NaOH	Double-bond character of parent hydrocarbon
3-Hydroxy-2-naphthaldehyde	390 (1.9)	442 (5.4)	$\frac{1}{3}$[a]
Salicylaldehyde	328 (3.6)	382 (7.08)	$\frac{1}{2}$
2-Hydroxy-1-naphthaldehyde	370 (4.7)	395 (7.9)	$\frac{2}{3}$[b]
1-Hydroxy-2-naphthaldehyde	381 (4.7)	405 (10.5)	$\frac{2}{3}$
Acetylacetone enol	273 (7.9)	290 (24.0)	~1

[a]The double-bond character of the 2,3-position of napthalene.
[b]The double-bond character of the 1,2-position.

VIII. Steric Hindrance and Intensity

The effect of steric hindrance on the absorption spectra of some dienones has been reported(65). It was found that, as hindrance is increased, the ring double bond is gradually pushed out of the plane of the dienone chain. The intensity of the dienone band is then decreased and a lower wavelength "enone" band is formed, which gradually increases in intensity with increasing steric hindrance (Table 7.6).

Many examples of the effect of steric hindrance on intensity are shown in Chapter 6.

TABLE 7.6
Effect of Steric Hindrance on Spectra(65)

Compound	Enone band		Dienone band	
	λ_{max} (mμ)	mϵ	λ_{max} (mμ)	mϵ
Cyclohexenyl–$CH{=}CHC(O)CH_3$			281	20.8
6,6-Dimethylcyclohexenyl (H_3C CH_3)–$CH{=}CHC(O)CH_3$	228	4.1	281	13.0
2,6,6-Trimethylcyclohexenyl (H_3C CH_3, CH_3)–$CH{=}CHC(O)CH_3$	223	6.5	296	10.7
2,6,6-Trimethylcyclohexenyl (H_3C CH_3, CH_3)–$CH{=}CHC(O)CH_2CH_3$	220	6.5	295	9.4
2,6,6-Trimethylcyclohexenyl (H_3C CH_3, CH_3)–$CH{=}C(CH_3)C(O)CH_3$	228	11.6	278	4.5

IX. Resonance Nodes and Terminals

Such effects as positivizing a positive node or negativizing a negative node cause sharp decreases in the intensity of a band and move it toward shorter wavelength. Negativizing a positive node or positivizing a negative node moves the long-wavelength band to the red and increases its intensity. In the former case the overall effect is analogous to a shortening of the chain of conjugation, whereas in the latter case the effect is analogous to an increase in the chain of conjugation.

Examples of this phenomenon have been presented in Section II of Chapter 6.

X. Solvent, Temperature, and Other Effects

Much of this material has been discussed in Chapters 3, 4, and 6. In the Z_n type of structure an increase in solvent polarity shifts the wavelength maximum to the red and usually increases the intensity. In the Z_z- and Z_b-structures an increase in solvent polarity shifts the wavelength maximum to the violet and usually decreases the intensity. An example of this shift in intensity can be seen in the betaine **5**(66) and in the zwitterion **6**(67).

5

$C_2H_5N^+$—CH=CH—$C^-(NO_2)(CHO)$

6

No.	Solvent	λ_{max} (mμ)	mϵ
5	Benzene	588	9.76
	Ethanol	538	2.74
6	Benzene	519	~95
	Acetone	480	~60
	Methanol	456	40.0
	Water	424	~28

Many examples of absorption intensification with a decrease or an increase in temperature are presented in Sections II and V.F of Chapter 3.

The sharpening and intensification of the absorption spectra of heme compounds at liquid-air temperatures was accomplished by making use of light scattering by ice crystals or a kaolin suspension(68). A greater than fortyfold intensification of the absorption spectra of cytochrome *c* reduced by cysteine has been achieved by freezing the solution, homogenizing it to a snow, and then obtaining the spectrum of the snow (Fig. 7.2). The procedure should be most useful in characterization of trace amounts of organic compounds in dilute solutions.

One last point that needs emphasis is the poor reproducibility of the literature data on molar absorptivity. A study of 20,000 spectra reported in 1958 and 1959 indicated that wavelength reproducibility was good. The range for four out of five maxima was 2 mμ or less. However, 20% of the intensity values had an ϵ ratio equal to or greater than 1.26 : 1 (70).

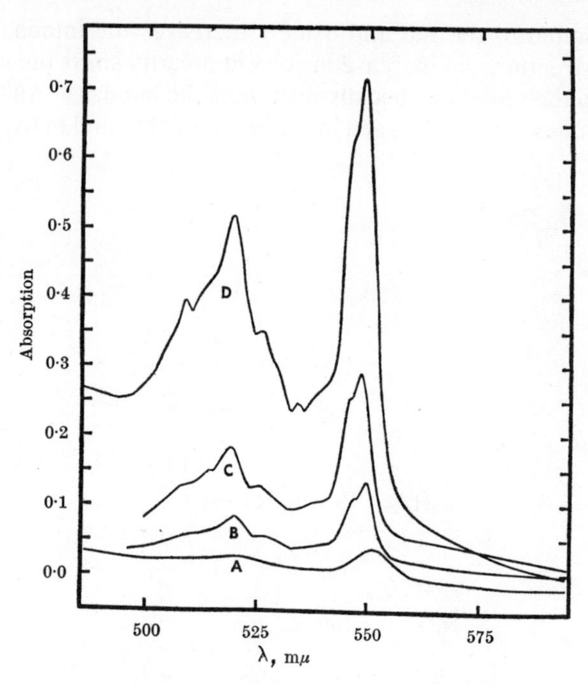

Fig. 7.2. Absorption spectrum of cytochrome *c* reduced by cysteine. *A*, at 25°C; *B*, frozen pellets as colected at 81°K; *C*, less than 1-mm diameter pellets; *D*, tamped "snow" from homogenization. All spectra were made on aliquots of the same solution and in the same cuvette (69).

References

1. R. Brinton and D. Volman, *J. Chem. Phys.*, **19**, 1394 (1951).
2. R. Mariella, R. Raube, J. Budde, and C. Moore, *J. Org. Chem.*, **19**, 678 (1954).
3. W. A. Haines, R. V. Helm, C. W. Bailey, and J. S. Bell, *J. Phys. Chem.*, **58**, 270 (1954).
4. S. F. Mason, in A. R. Katritzky, Ed., *Physical Methods in Heterocyclic Chemistry*, Academic Press, New York, 1963, pp. 1–88.
5. A. Burawoy, M. Cais, J. Chamberlain, F. Liversedge, and A. T. Thompson, *J. Chem. Soc.*, 3721 (1955).
6. A. Burawoy, *J. Chem. Soc.*, 1177 (1939).
7. J. A. Barltrop, P. M. Hayes, and M. Calvin, *J. Am. Chem. Soc.*, **76**, 4348 (1954).
8. L. Doub and J. M. Vandenbelt, *J. Am. Chem. Soc.*, **69**, 2714 (1947).
9. R. Haszeldine, *J. Chem. Soc.*, 1764 (1953).
10. E. Tannenbaum, E. M. Coffin, and A. J. Harrison, *J. Chem. Phys.*, **21**, 311 (1953).
11. E. Clar and D. G. Stewart, *J. Am. Chem. Soc.*, **74**, 6235 (1952).
12. A. J. Harrison and D. R. W. Price, *J. Chem. Phys.*, **30**, 357 (1959).
13. L. W. Pickett, M. E. Corning, G. M. Wieder, D. A. Semenov, and J. M. Buckley, *J. Am. Chem. Soc.*, **75**, 1618 (1953).
14. L. W. Pickett, N. J. Hoeflich, and T. C. Liu, *J. Am. Chem. Soc.*, **73**, 4865 (1951).
15. R. A. Jeffreys, *J. Chem. Soc.*, 2221 (1954).

16. D. W. Turner, *Chem. Ind.* (*London*), 626 (1958).
17. A. Burawoy, *J. Chem. Soc.*, 1180 (1939).
18. E. Blout and M. Fields, *J. Am. Chem. Soc.*, **70**, 189 (1948).
19. L. G. S. Brooker, F. L. White, G. H. Keyes, C. P. Smyth, and P. F. Oesper, *J. Am. Chem. Soc.*, **63**, 3192 (1941).
20. A. E. Gillam and D. H. Hey, *J. Chem. Soc.*, 1170 (1939).
21. G. D. Dorough, J. R. Miller, and F. M. Huennekens, *J. Am. Chem. Soc.*, **73**, 4315 (1951).
22. E. R. H. Jones, M. C. Whiting, J. B. Armitage, C. L. Cook, and N. E. Entwhistle, *Nature*, **168**, 900 (1951).
23. F. Bohlmann, *Ber.*, **86**, 657 (1953).
24. E. A. Braude and F. Sondheimer, *J. Chem. Soc.*, 3754 (1955).
25. E. A. Braude, *J. Chem. Soc.*, 379 (1950).
26. D. M. Hercules, Ed., *Fluorescence and Phosphorescence Analysis*, Interscience, New York, 1966, p. 12.
27. D. R. Kearns and W. A. Case, *J. Am. Chem. Soc.*, **88**, 5087 (1966).
28. R. C. Cookson and S. MacKenzie, *Proc. Chem. Soc.* (*London*), 423 (1961).
29. R. Bieber and G. Trumpler, *Helv. Chim. Acta*, **30**, 1860 (1947).
30. I. M. Klotz and T. Askounts, *J. Am. Chem. Soc.*, **69**, 801 (1947).
31. R. C. Hoseney and K. F. Finney, *Anal. Chem.*, **36**, 2145 (1964).
32. J. C. Dearden and W. F. Forbes, *Can. J. Chem.*, **37**, 1294 (1959).
33. T. W. Stanley, E. Sawicki, H. Johnson, and J. D. Pfaff, *Mikrochim. Acta*, 48 (1965).
34. A. E. O. Marzys, *Analyst*, **86**, 460 (1961).
35. R. O. Sinnhuber and T. C. Yu, *Food Technology*, **12**, 9 (1958).
36. R. B. Setlow and W. R. Guild, *Arch. Biochem. Biophys.*, **34**, 233 (1951).
37. W. Troll and R. K. Cannan, *J. Biol. Chem.*, **200**, 803 (1953).
38. V. S. Waravdekar and L. D. Saslaw, *J. Biol. Chem.*, **234**, 1945 (1959).
39. L. Doub and J. M. Vandenbelt, *J. Am. Chem. Soc.*, **77**, 4535 (1955).
40. H. J. Ackermann, L. J. Carbone, and E. J. Kuchar, *J. Agr. Food Chem.*, **11**, 297 (1963).
41. M. Richardson, *Nature*, **197**, 290 (1963).
42. M. Errera and J. P. Greenstein, *Arch. Biochem. Biophys.*, **14**, 477 (1947).
43. S. Markees and F. Gey, *Helv. Physiol. Acta*, **11**, 49 (1953).
44. E. Clar, *Polycyclic Hydrocarbons*, Vols. I and II, Academic Press, London and New York, 1964.
45. G. M. Badger, *Proc. Roy. Soc.* (*N.S. Wales*), **40**, 87 (1956).
46. A. Sklar, *J. Chem. Phys.*, **10**, 135 (1942).
47. T. Forster, *Z. Naturforsch*, **2a**, 149 (1947).
48. J. Platt, *J. Chem. Phys.*, **19**, 263 (1951).
49. L. Zechmeister, A. L. LeRosen, W. A. Schroeder, A. Polgár, and L. Pauling, *J. Am. Chem. Soc.*, **65**, 1940 (1943).
50. F. Bohlmann, *Ber.*, **84**, 490 (1951).
51. W. Schroeder, P. Wilcox, K. Trueblood, and A. Dekker, *Anal. Chem.*, **23**, 1740 (1951).
52. M. Martynoff, *Compt. Rend.*, **235**, 54 (1954).
53. L. Pauling, *Fortschr. Chem. Org. Naturstoffe*, **3**, 203 (1939); *Helv. Chim. Acta*, **32**, 2241 (1949).
54. L. Zechmeister, A. L. LeRosen, W. A. Schroeder, A. Polgár, and L. Pauling, *J. Am. Chem. Soc.*, **65**, 1941 (1943).
55. L. Zechmeister and R. B. Escue, *J. Am. Chem. Soc.*, **66**, 322 (1944).
56. R. B. Turner and D. Voitle, *J. Am. Chem. Soc.*, **73**, 1403 (1951).

57. J. A. Berson, *J. Am. Chem. Soc.*, **75**, 3521 (1953).
58. L. Zechmeister, Cis-Trans *Isomeric Carotenoids, Vitamins A, and Arylpolyenes*, Academic Press, New York, (1962).
59. A. Treibs and K. Jacob, *Ann.*, **699**, 153 (1966).
60. A. Leifer, D. Bonis, M. Boedner, P. Dougherty, A. J. Fusco, M. Koral, and J. E. LuValle, *Appl. Spectry.*, **21**, 71 (1967).
61. H. Beecken, P. Tavs, and H. Sieper, *Ann.*, **704**, 166 (1967).
62. H. Beecken and P. Tavs, *Ann.*, **704**, 172 (1967).
63. M. J. Kamlet, J. C. Hoffsommer, and H. G. Adolph, *J. Am. Chem. Soc.*, **84**, 3925 (1962).
64. N. Melchior, *J. Am. Chem. Soc.*, **71**, 3647 (1949).
65. E. A. Braude and E. R. H. Jones, *J. Am. Chem. Soc.*, **72**, 1041 (1950).
66. K. Dimroth, G. Arnoldy, S. von Eicken, and G. Schiffler, *Ann.*, **604**, 221 (1957).
67. C. Reichardt, *Ann.*, **715**, 74 (1968).
68. D. Keilin and E. F. Hartree, *Nature*, **164**, 254 (1949).
69. W. B. Elliott and G. F. Doebbler, *Nature*, **198**, 690 (1963).
70. J. P. Phillips, *Anal. Chem.*, **34**, 171 (1962).

CHAPTER 8

ABSORPTION SPECTRA OF CATIONIC RESONANCE STRUCTURES I. POSITION OF PROTON

I. Introduction

A large proportion of colorimetric methods consist of the formation and quantitative measure of a cationic resonance chromogen. These chromogens are usually formed through (*a*) ionization of a compound by extraction of an anion from the compound, (*b*) salt formation at the negative resonance terminal of a zwitterionic resonance structure, and (*c*) condensation of the test substance and 1 or 2 equivalents of the reagent. Examples of these types are shown in equations 8.1 (1), 8.2 (2), and 8.3.

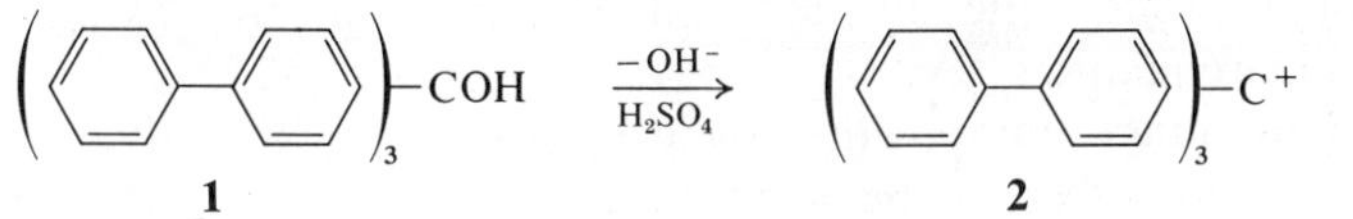

$$(C_6H_5C_6H_4)_3COH \xrightarrow[H_2SO_4]{-OH^-} (C_6H_5C_6H_4)_3C^+ \tag{8.1}$$

1 → **2**

(8.2): H_2N-quinoxaline ↔ $H_2\overset{+}{N}$=…$\overset{-}{N}$ (**3**) $\xrightarrow{H^+}$ H_2N-quinoxalinium ($\overset{+}{N}$H) (**4**)

(8.3): **5** (HO_3S, OH, OH, SO_3H) + CH_2O $\xrightarrow[\Delta]{H_2SO_4}$ **6** (HO_3S, OH, HO, SO_3H, $\overset{+}{O}$, HO_3S, SO_3H)

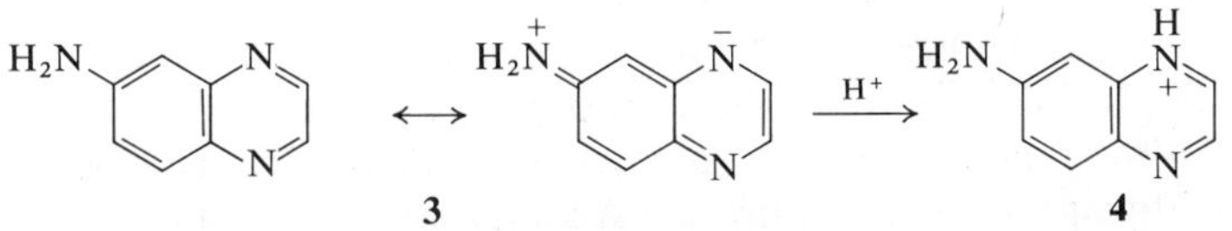

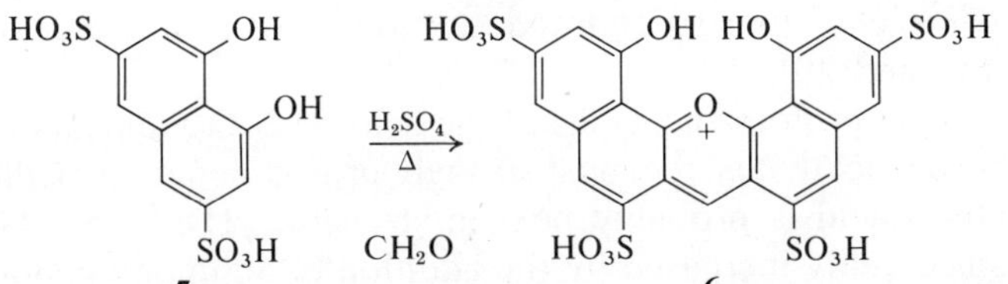

No.	Solvent	λ_{max} (mμ)	mϵ
1		Colorless	
2	H_2SO_4	544	112.0
3	EtOH	407	10.0
4	6 *N* HCl in H_2O–EtOH (1 : 1)	492	10.0
5	Water	286	0.013
6	H_2SO_4	578	15.7

As shown in the three equations, the aim of organic trace colorimetry is to form a chromogen that absorbs at longer wavelengths and with greater intensity than the test substance. Examples of some of the chromogens found useful in trace analysis are presented in Table 8.1. All these analytical methods show favorable wavelength and intensity shifts.

II. Strong Acids in Trace Analysis

Since a very large number of Z_n-structures can add a proton to the negative resonance terminal in appropriate acidic media (thus shifting the long-wavelength maximum to the red, usually with an increase in intensity), the basicity of the resonance terminals will determine the strength of acid necessary for salt formation. Since the negative resonance terminal is weakly basic, a strong acid is necessary for salt formation to take place. Table 8.2 lists such a group of acids. The one most used in organic trace analysis is sulfuric acid and its mixtures with water. One of the simplest methods of obtaining standard concentrations of strong sulfuric acid solutions is to dilute the concentrated sulfuric acid at hand to an appropriate concentration and then to titrate the diluted solution with standard alkali.

The percentage composition by weight of sulfuric acid solutions between 32 and 39% can be accurately computed from specific gravity measurements (17). With quinalizarin the percentage of sulfuric acid in the concentration range 85–99% sulfuric acid can be determined spectrally (18); with 1,1′-dianthrimide the concentration range 80–99% is covered (19). Sulfuric acid solutions of various concentrations can then be prepared by appropriate dilutions. An alternative way of doing this is to start with 100% sulfuric acid. This concentration can be prepared by adding 20% oleum to concentrated sulfuric acid until a $d^{15.5}_{15.5}$ value of 1.8393 is reached (20).

Of the simple protonic acids disulfuric acid ($H_2S_2O_7$) appears to be the most highly acidic, that is, the most strongly proton-donating medium (21). Fluorosulfuric acid is probably next in strength. However, its acidity can be considerably increased by the addition of antimony pentafluoride and antimony pentafluoride–sulfur trioxide; the resulting solutions are the most highly acidic known (22). They have been termed superacids. Even nitrobenzene is completely ionized in these fluorosulfuric acid media.

With the advent of thin-layer chromatography and direct spectral examination the use of strongly acidic adsorbents in the formation of highly colored conjugate acids is worthy of investigation. The acidic strength of surfaces has been investigated (23). In some cases—for

Some Cationic Resonance Chromogens Formed in Organic Trace Analysis

Test substance	λ_{max} (mμ)	mϵ	Chromogen	λ_{max} (mμ)	mϵ	Ref.
Pentachloronitro-benzene	Colorless		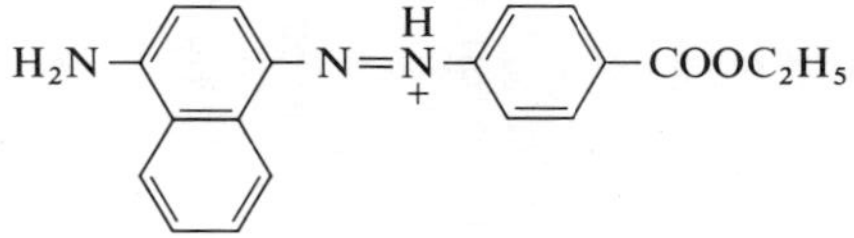 H_2N–N=$\overset{H}{N^+}$–$COOC_2H_5$	525	21.0	3
Furfural (4)	272	13.2	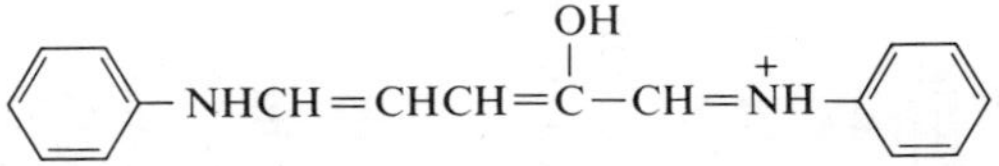 NHCH=CHCH=C(OH)–CH=$\overset{+}{N}$H	509		5
Benzoyl chloride	Colorless		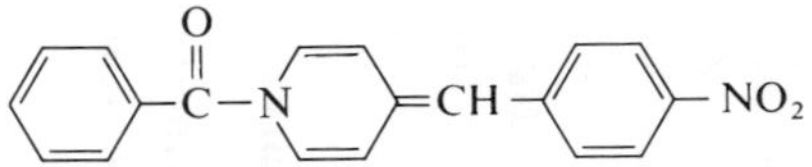 C(=O)–N=CH–NO_2	450	15.0	6
N-Acetyl-D-glucos-amine	Colorless		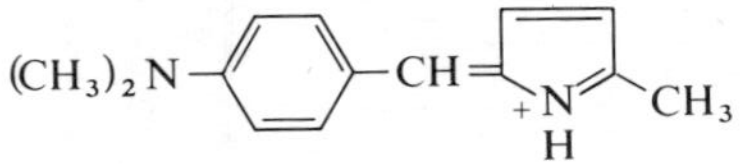 $(CH_3)_2N$–CH=$\overset{+}{N}$H CH_3	545	14.2	7
p-$(CH_3)_2NC_6H_4N{=}NSO_3Na$ (Dexon)	420	16.1	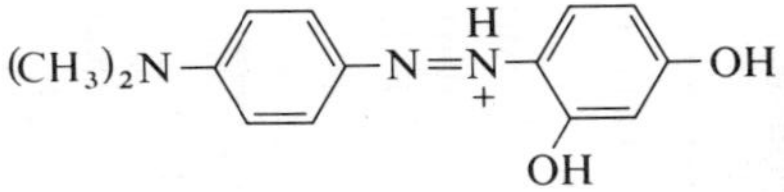 $(CH_3)_2N$–N=$\overset{H}{N^+}$–OH, OH	450	28.6	8
n-Hexanethiol (9)	224*s*	0.13	$H_2N-\phi-\phi-\overset{H}{\underset{+}{N}}=CHCH=CHCH=CH\overset{H}{N}-\phi-\phi-NH_2$	525	56.0	10
Hydroxyproline (11)	<200		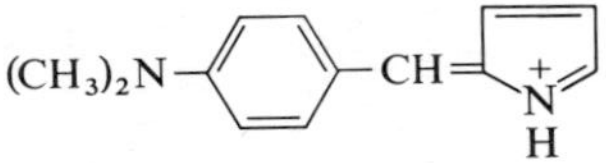 $(CH_3)_2N$–CH=$\overset{+}{N}$H	558	88.4	12
Prednisone (13)	238	15.5	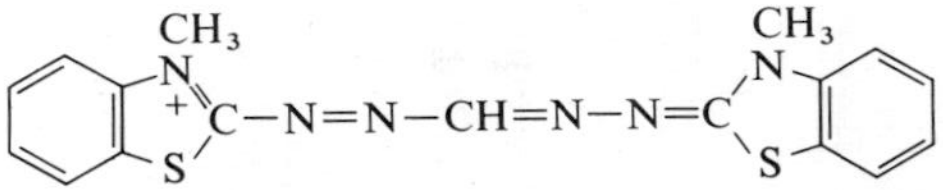 CH_3 $\overset{+}{N}$ S C–N=N–CH=N–N=C N S CH_3	620	28.4	14

TABLE 8.2
The pK_a of Strong Acids (15)

Acid	pK_a	Acid	pK_a
Fluorosulfuric	−12.6[a]	*p*-Toluenesulfonic	−5.48
Chlorosulfuric	−11.27	$(R_2OH \cdots OR_2)^+$	
Sulfuric	−11.0	$(SbCl_6)^-$	−4.70
$(R_2OH)^+(SbCl_6)^-$	−10.50	Trifluoroacetic	−2.51
Perchloric	−10.08	Trichloroacetic	−1.40
Hydrogen fluoride		Picric	+0.52
(anhydrous)	−10.0	Monochloroacetic	+1.54
Methyl hydrogen			
sulfate	−7.76		

[a] H_0 for the neat acid (16).

example, fresh silica–alumina catalysts—the adsorbent is approximately equal in acidity to 90% sulfuric acid (24). Indicators have been used to assess the acid strength (Table 8.3).

III. Basicity

Since a large number of organic compounds have weakly basic properties, a knowledge of their pK_a values or, better yet, the concentration of sulfuric acid at which they are half-ionized would be invaluable in organic trace analysis. For example, the strength of acid necessary to treat a thin-layer chromatogram so as to form a colored cationic salt from a weak base could be quickly ascertained. In the same way the strength of acid necessary to extract a weak base from the solution of a complicated mixture in an organic solvent could be easily determined. Through-

TABLE 8.3
Indicators Used for Acid-Strength Measurements (23)

Indicator	Color		pK_a	H_2SO_4 (% wt)
	Basic	Acidic		
Neutral red	Yellow	Red	+6.8	8×10^{-8}
4-Phenylazo-1-naphthylamine	Yellow	Red	+4.0	5×10^{-5}
4-Phenylazo-*N*,*N*-dimethylaniline	Yellow	Red	+3.3	3×10^{-4}
4-Phenylazodiphenylamine	Yellow	Purple	+1.5	0.02
Dicinnamalacetone	Yellow	Red	−3.0	48.0
Benzalacetophenone	None	Yellow	−5.6	71.0
Anthraquinone	None	Yellow	−8.2	90.0

out this treatise many examples are given of the usefulness of the basicity and the spectra of the neutral compound and its cationic salt in characterization and assay.

The basicity properties of a large number of weak organic bases are presented in Table 8.4 for the use of the organic trace analyst in his investigations.

Any attempt to intercompare pK values obtained by various authors with various systems shows how chaotic this field is. For example, the following pK_a values have been reported for acetone: −7.2(51), −4(53), −2(54), −1.6(55), and −0.66(56). An attempt to explain some of these discrepancies has been made(57). For the analyst it appears that the most useful comparison of the basicities of organic compounds is obtained through the value of the concentration of sulfuric acid at which the compounds are half-dissociated. Paul and Long(58) have reviewed H_0 and related indicator-acidity functions so useful in determining the pK values of weak organic bases through such equations as

$$pK_{BH^+} = H_0 + \log \frac{C_{BH^+}}{C_B} \tag{8.4}$$

IV. Position of Proton

Since a zwitterionic resonance structure dissolved in acidic solution can add a proton to the positive or negative resonance terminal or even to some other part of the molecule, the position of proton addition will help to determine the wavelength position and the band intensity of the cation thus formed.

In a zwitterionic resonance structure the basicity of the positive resonance terminal is considerably decreased when compared with the basicity of the same molecule without the negative resonance terminal. On the other hand, the basicity of the negative resonance terminal is considerably increased when compared with the basicity of the same molecule minus the positive resonance terminal. Thus, depending on their relative basicity, either the positive or the negative resonance terminal can add the proton or a mixture of tautomers can be formed. We shall discuss this type of tautomerism.

A. Nitrogen Compounds

Since amino and aza nitrogen atoms are usually more basic than other types of functional groups, the basicity and spectral properties of a large number of these compounds have been investigated.

TABLE 8.4
Basic Dissociation Constants of Some Weak Bases

Compound	pK_a (aqueous solution)	% H_2SO_4 at half-dissociation	Ref.
4-Phenylazo-*N*,*N*-dimethylaniline	+3.3	3×10^{-4}	25
4-Hydroxyquinoline	+2.27		26
4-Phenylazodiphenylamine	+1.5	0.02	23
Phenazine	+1.23		27
4-Nitroaniline	+1.00		28, 29
Diphenylamine	+0.78		30
Acetanilide	+0.4		31
γ-Pyrone	+0.3		32
Urea	+0.2		33
2-Nitroaniline	−0.29[a]		35
1,2,5-Trimethylpyrrole	−0.2[b]	0.5 *M*[d]	36
	−0.5[c]		
2-Hydroxyquinoline	−0.31		26
Acetamide	−0.5		33
4,4′-Dimethoxytriphenylmethanol	−1.24[e]	~14	37
1-Cyclopropyl-3-methyl-2-cyclopenten-1-ol		12	38
4′-Methoxyflavone	−0.8		39
Tropolone	−0.86		40
Thiourea	−0.96		33
Tropone	−1.02	18	40
Flavone	−1.2	21.6	39
4-Methoxyazobenzene	−1.36[f]		34
N,*N*-Dimethylbenzamide	−1.62	28	41
Hexamethylene oxide	−2.02	33	42
2′,4′,6-Trimethylflavone	−2.05	33.6	39
Tetrahydrofuran	−2.08	35	42
Benzamide	−2.16	35.2	41
Tricyclopropylmethanol	−2.34	22	38
Azobenzene	−2.5[g]	39.6	25
1,3-Dimethylcyclopentadiene		35	38
Tetrahydropyran	−2.79	43	42
Dicinnamalacetone	−3.0	48	24
4-Nitrodiphenylamine	−3.13[f]	5.9 *M*	30
Dioxane	−3.16	48	42
1-Phenyl-3-methylcyclopentadiene		50	38
1,3-Diphenylcyclopentadiene		54.5	38
4,4′,4″-Trimethyltriphenylmethanol	−3.56[e]	32	37
Diethyl ether	−3.59	52	42
p-Ferrocenylazobenzene	−3.63[f]	53	43
1,3,5-Trimethoxybenzene	−3.7	53.2	44

TABLE 8.4 (*contd.*)

Compound	pK_a (aqueous solution)	% H_2SO_4 at half-dissociation	Ref.
4,4′-Dihydroxybenzophenone	−4.19	57.6	45
4-Methoxychalcone	−4.25[h]		46
Pyrrole	−4.4	5.3 *M*	36
4-Nitroazobenzene	−4.70[f]		34
9-Anthraldehyde	−4.81	63	47
Chalcone	−5.0[i]	64	46
Anthrone	−5.02	64.7	45
Dibenzo[*b*,*f*]tropone	−5.25	66.6	45
1-Acetylanthracene	−5.65	70	47
Dibenzo[*b*,*f*]suberone	−5.69	70.2	45
1-Anthraldehyde	−5.71	70.5	47
2-Acetylnaphthalene	−6.04	73	47
Acetophenone	−6.15[j]	73.6	49
Benzophenone	−6.16[k]	73.8	45
4,4′-Dinitrodiphenylamine	−6.21[f]	10.2 *M*	30
4-Nitrochalcone	−6.22	74	46
Azoxybenzene	−6.45[f]		50
Anisole	−6.5	76.5	44
Triphenylmethanol	−6.63[e]	~50	40
Fluorenone	−6.65	77.6	45
Cyclohexanone	−6.8	78.6	51
m-Bromoacetophenone	−6.9	79	49
Acetone	−7.2	81.5	51
Benzaldehyde	−7.47	84	48
Benzoic acid	−7.60	84.8	48
m-Nitroacetophenone	−7.62	85	49
p-Nitroacetophenone	−7.94	87	47
Anthraquinone	−8.2	90	24
Cyclobutanone	−9.5	98.5	51
4,4′-Dinitrobenzophenone	−10.12	99.6	45
Nitrobenzene	−11.26		52

[a] $pK_a = -0.34$ in 20% alcohol solution (34).
[b] α-Protonation.
[c] β-Protonation.
[d] Molarity at half-ionization.
[e] pK_{R^+}
[f] 20% alcohol.
[g] $pK_a = -2.90$ in 20% alcohol (34).
[h] 5% Dioxane.
[i] Also reported $pK_a = -5.6$ and half-ionization = 71% H_2SO_4 (24).
[j] Also reported $pK_a = -6.36$ (48).
[k] Also $pK_a = -6.41$ (48).

1. Aliphatic and Olefinic Amines

The aliphatic amines with pK_a values in water of 9–11(59) are strong organic bases. Since the carbon and hydrogen atoms in the aliphatic amines are fully saturated, the nitrogen atom with its unshared pair of electrons must attract the proton; for example,

$$R_3N\!:+H^+ \rightarrow (R_3\overset{+}{N}H) \tag{8.5}$$

Wherever a nitrogen is in conjugation with an unsaturated group, the possibility exists of proton addition to some other part of the molecule other than the nitrogen. The highly basic(60) α,β-unsaturated amines form an iminium salt in acid solution(61–63), as shown by equation 8.6 (64), unless proton addition to the carbon violates Bredt's rule(61, 63):

$$\underset{\mathbf{7}}{CH_2{=}CH{-}N(C_4H_9)(C_2H_5)} \xrightarrow{H^+} \underset{\mathbf{8}}{CH_3{-}CH{=}\overset{+}{N}(C_4H_9)(C_2H_5)} \tag{8.6}$$

No.	Solvent	λ_{max} (mμ)	mϵ
7	Hexane	231	4.2
8	Acetonitrile	228	19.8

2. Aromatic Amines

In the large majority of cases proton addition takes place on the amino nitrogen of aromatic amines. The salt thus formed, being iso-π-electronic to the parent hydrocarbon, is closely similar spectrally to the parent hydrocarbon. An example of this spectral phenomenon is shown for 2-aminofluorene in Figure 8.1. In this fashion an aromatic amine of unknown structure can be characterized as to the parent structure of aromatic hydrocarbon present in the amine. Many examples are available in the literature – for example, for *o*-, *m*-, and *p*-nitroanilines(65); *o*-, *m*-, and *p*-aminobenzoic acids(65); naphthylamines(66); aminophenanthrenes (67); and amino derivatives of pyrene, benz[*a*]anthracene, and benzo[*a*]-pyrene(68).

This phenomenon is useful in proof of structure work. The position of nitration of some acylamino compounds has been proven spectrally by demonstrating that the cationic salts of 3-amino-4-nitro-dibenzothiophene(69), 2-amino-3-nitro-9-methylcarbazole(70), 3-amino-2-nitro-

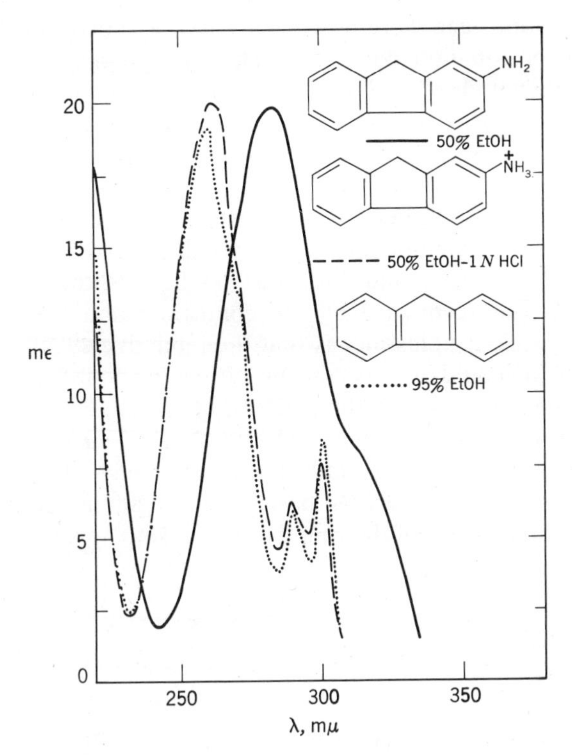

Fig. 8.1. Ultraviolet absorption spectra of 2-aminofluorene (——), its salt (– – –), and fluorene (· · · · ·).

carbazole(70), 2-amino-1-nitrodibenzothiophene(71), 3-amino-4-nitro-9-methylcarbazole(70), 2-amino-1-nitrodibenzoselenophene(72), 2-amino-3-nitrodibenzoselenophene(72), and 2-amino-3-nitrodibenzothiophene (71) are very closely similar spectrally to the deaminated compounds and spectrally different from the isomers of the deaminated compounds.

In the same way it has been shown that the addition of a proton to 2,3-diaminodibenzofuran gives a cationic salt that is isospectral with 3-aminodibenzofuran; the addition of a second proton gives the dicationic salt that is isospectral with dibenzofuran(73). Similarly in its absorption spectrum 4-nitro-1,2-phenylenediamine in 50% ethanol–1.2 *N* hydrochloric acid (λ_{max} 225, 365 mμ; mε 7.93, 13.6 respectively) closely resembles *p*-nitroaniline (λ_{max} 228, 372 mμ; mε 6.3, 15.2, respectively) and differs from *m*-nitroaniline (λ_{max} 235, 374 mμ; mε 15.9, 1.5, respectively). Diprotonation of the nitrophenylenediamine takes place on both amino nitrogens(74). Diprotonation of other 1,2-phenylenediamines takes place similarly, as shown by the absorption spectra.

If the amino nitrogen were made less basic or substituted by a weaker basic group, protonation could take place in the ring. This has been shown for phenol dissolved in fluorosulfuric acid, protonation taking place in the *para* position (22).

However, it is possible for the aromatic ring to compete with an amino group for the proton. Ammonia, with a pK_a of 9.3, is approximately 10^{22} times as strong a base as benzene. Substitution of a phenyl ring for one of the hydrogens of ammonia considerably decreases the basicity of the nitrogen to 4.7. Krogt and Wepster (75) have shown that both the inductive and the resonance effects contribute about equally to this decrease in basicity. This means that the inductive effect as well as the resonance effect should increase the basicity of the benzene ring.

In an aniline derivative (**9**) the electron density in the ring is increased enough to attract cations, such as ArS^+, $(ArN_2)^+$, $(NO_2)^+$ and Br^+. Conceivably a proton could also add to the point of high electron density in the benzene ring. However, the basicity of the amino group is so much greater than the basicity of the ring that there is no apparent competition for the proton.

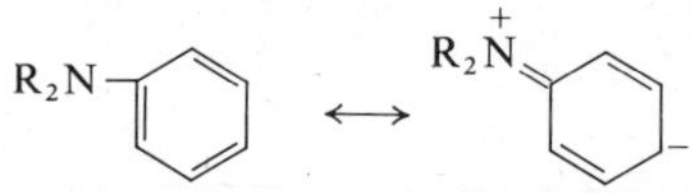

$R_2N-C_6H_5 \longleftrightarrow R_2\overset{+}{N}=C_6H_5^-$

9

Just as *m*-xylene is much more basic than toluene (76), the benzene ring in 1,3-phenylenediamine could be expected to be much more basic than the benzene ring in aniline. In spite of this increase in basicity, the benzene ring in this diamine is apparently still too weakly basic to compete with the amino groups for the proton. This is clearly shown in Table 8.5, where the monocationic salt of the diamine has a close spectral resemblance to aniline.

In 1,3,5-trimethylbenzene the basicity is increased approximately 200 times as compared with *m*-xylene (76). Similarly the basicity of the benzene ring in 1,3,5-triaminobenzene (**10**) has been increased to such an extent that the unsubstituted positions in the ring are of the same order of basicity as the amino nitrogens. The spectral data in Table 8.5 (77)

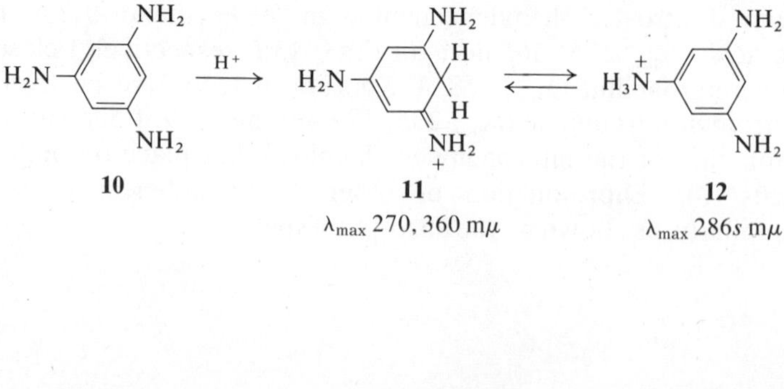

$$\underset{\mathbf{10}}{C_6H_3(NH_2)_3} \xrightarrow{H^+} \underset{\substack{\mathbf{11} \\ \lambda_{max}\ 270,\ 360\ m\mu}}{H_2N\text{-}C_6H_4(NH_2)(\overset{+}{N}H_2)} \rightleftharpoons \underset{\substack{\mathbf{12} \\ \lambda_{max}\ 286s\ m\mu}}{H_3\overset{+}{N}\text{-}C_6H_3(NH_2)_2} \quad (8.7)$$

TABLE 8.5
Ultraviolet Absorption Spectra of Some Benzene Amines and Their Monocationic Salts (77)

	Base		Salt		Deaminated compound	
Compound[a]	λ_{max} (mμ)	mϵ	λ_{max} (mμ)	mϵ	λ_{max} (mμ)	mϵ
Aniline	230	7.9	203	7.9	204	7.9
	280	1.6	254[b]	0.16	254[b]	0.2
1,3-Diaminobenzene	210	40.0	233	7.9	230	7.9
	240[c]	7.9	278	1.6	280	1.6
	289	2.0				
1,3,5-Triaminobenzene	218	20.0[d]	220	32.0[e]	210	40.0
	236	10.0	270	7.9	240[c]	7.9
	280	0.63	286[c]	1.3	289	2.0
			360	3.2		

[a]In water.
[b]Fine structure.
[c]Shoulder.
[d]At pH 7.3.
[e]At pH 4.3.

indicate that both tautomers are present, as shown in Equation 8.7. These results have been confirmed by infrared, ultraviolet, and nuclear magnetic resonance spectral studies (78).

Other 1,3,5-triaminobenzenes that also show the presence of the cationic tautomer involving proton addition to the carbon atom include the 2-methoxy, 2,6-dimethyl, and 2,4,6-trimethyl derivatives with long wavelength absorption bands at 375, 390, and 400 mμ, respectively (77).

Other amines in which protonation takes place on an aromatic carbon atom include 4-aminoazulene (**13**), 5-aminonaphthacene (**15**), and 5-amino-6-phenylnaphthacene (80). The reactions are shown in equations

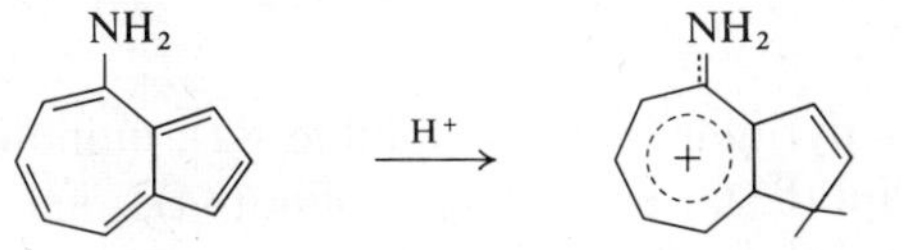

$$\mathbf{13}\ (\text{NH}_2) \xrightarrow{H^+} \mathbf{14}\ (\text{NH}_2) \qquad (8.8)$$

$$\mathbf{15}\ (\text{NH}_2) \xrightarrow{H^+} \mathbf{16}\ (\overset{+}{\text{N}}\text{H}_2,\ \text{H}\ \text{H}) \qquad (8.9)$$

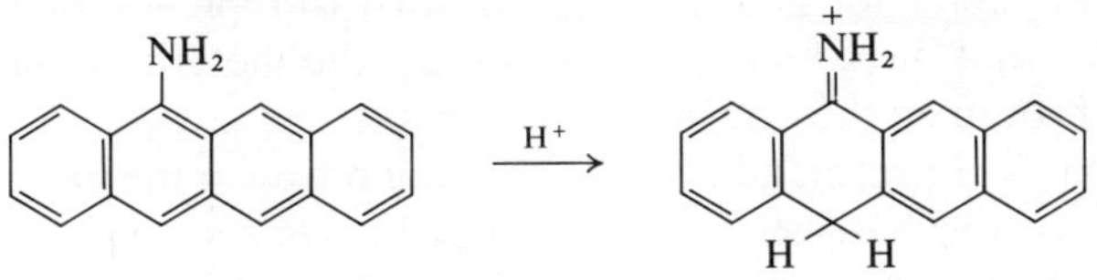

No.	Solvent	$\lambda_{max}(m\mu)$	mϵ	Ref.
13	95% EtOH	500		79
14	2N HCl	350		79
15	95% EtOH	299	97.7	80
		408fs[a]	4.7	
		510	3.7	
16	AcOH	274	22.9	80
		317	9.77	
		402	3.16	

[a]Fine structure.

8.8 and 8.9. If protonation had taken place on the amino nitrogens, the spectra of the salts would not have been so widely different from those of the parent hydrocarbons.

3. *Heterocyclic Amines*

In many heterocyclic monoamines salt formation takes place on the aza nitrogen, causing little change in the absorption spectra of these derivatives. For example, in 2-aminothiazole (**17**) salt formation takes place at the ring nitrogen (81). If proton addition had taken place at the amino nitrogen, the resultant salt would have been isospectral with thiazole (**19**), which it is not.

N S NH$_2$ **17** $\xrightarrow{H^+}$ H N$^+$ S NH$_2$ **18** N S **19** (8.10)

No.	Solvent	$\lambda_{max}(m\mu)$	mϵ
17	EtOH	255	6.3
18	EtOH–HCl	255	8.0
19	EtOH	240	4.0

The amines of pyridine (66), pyrimidine (82), quinoline (66), isoquinoline (83), cinnoline (83), benzo[*c*]cinnoline (84), and acridine (85, 86) show a marked red shift effect when the base is converted into the monocation. This behavior is in marked contrast to typical aromatic amines and proves that in forming monocations the heteroaromatic amines accept the proton on the ring nitrogen atom.

On the basis of proton addition to the ring nitrogen the varying degrees of basicity of the heteroaromatic amines have been explained through the principle of "additional ionic resonance" of the monocationic salts of 120 heterocyclic bases belonging to 30 different ring systems(27). In the heteroaromatic amines the compounds containing an amino group *para* to a ring nitrogen were found to be the most basic. The least basic amines were those that contained an amino group in a position where it could not resonate with the ring nitrogen. For example, 4-aminopyridine (**21**) is more basic than the 2-isomer **22**, which is more basic than the 3-isomer **23**. The pK_a values were measured in water at 20°C.

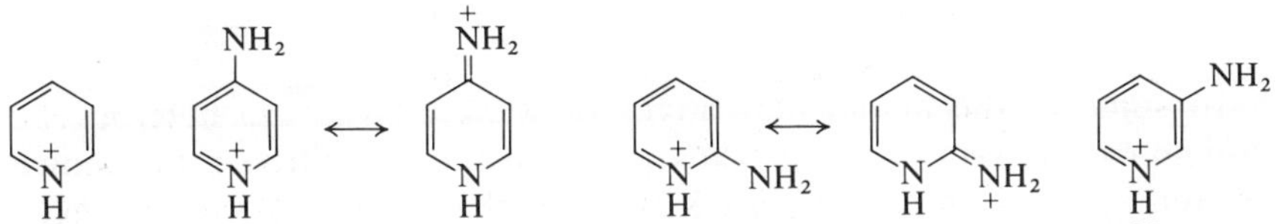

20 pK_a 5.23   **21** pK_a 9.17   **22** pK_a 6.86   **23** pK_a 5.98

In a review of the ultraviolet absorption spectra, basicity, and chemical properties of the amino derivatives of pyridine, quinoline, and isoquinoline, Steck and Ewing(66) have pointed out that 3-aminopyridine readily forms a dihydrochloride, whereas the 2- and 4-aminopyridines form only monohydrochlorides. This is because in **21** and **22** both the ring and amino nitrogens share the positive charge, and only in concentrated sulfuric acid is a second proton able to add to give dicationic salts that are isospectral with the monocationic salt of pyridine(87).

For some aza heterocyclic amines protonation is reported to take place on the aza nitrogen with a slight violet shift. Since the monocationic salts of 2-, 4-, and 5- aminobenzimidazoles do not resemble benzimidazole spectrally, the first proton must add to the ring nitrogen. The dicationic salts of 4- and 5-aminobenzimidazole are closely similar spectrally to the monocationic salt of benzimidazole(88). This means that the second proton must add to the amino nitrogen.

Where an aza arene is weakly basic, protonation of its amino derivatives can take place on the amino group. In the 4-, 5-, 6-, and 7-aminobenzothiazoles protonation takes place on the amino nitrogen since the spectra of the monocationic salts were similar to that of benzothiazole(89). From the available data(90) it would appear that salt formation in 5-amino- and 6-amino-1-methyl benzotriazole takes place first at the amino group. This appears reasonable since the ring nitrogen in benzotriazole, pK_a 1.6, is much more weakly basic than it is in benzimidazole, pK_a 5.53(27). However, examination of the spectral data of 1-methyl-6-aminobenzotriazole in 0.1 *N* hydrochloric acid indicates the possible presence of monocationic tautomers. In 1-methyl-7-aminobenzotriazole (**24**) addition of a proton to the amino group would cause an increase in steric hindrance with the *N*-methyl group. This helps to account for salt formation at the 3-nitrogen in this case.

$$\textbf{24} \xrightarrow{H^+} \textbf{25} \qquad (8.11)$$

In some of the amino derivatives of weakly basic aza heterocyclic compounds proton addition takes place on the aza nitrogen (as shown by red spectral shift in Figure 8.2) when the amino and ring nitrogen groups are involved in *para* resonance and on the amino nitrogen (as shown by the violet shift and close spectral resemblance to the deaminated compound) when the amino and ring nitrogen groups are involved in *ortho* resonance. This phenomenon is shown for the amino-2,1,3-benzoselenadiazoles in Table 8.6. Addition of a second proton then takes place on the amino nitrogen (as shown by the violet shift and the close spectral resemblance to the monocationic salt of the deaminated compound).

In the same way 6-aminoquinoxaline adds its first proton to the 4-aza nitrogen and its second to the amino nitrogen, whereas the 5-isomer adds its first proton to the amino group (92). The effect of amino group position on the change in spectra with increasing acidity can be seen

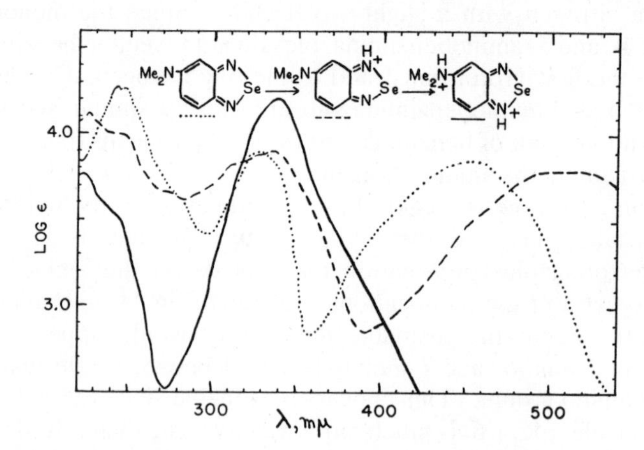

Fig. 8.2. Ultraviolet–visible absorption spectra of 5-dimethylamino-2,1,3-benzoselenadiazole in 95% ethanol (·····), 50% ethanol–6 *N* hydrochloric acid (---), and 95% sulfuric acid (——) (91).

TABLE 8.6
Position of Proton Addition in Amino-2,1,3-benzoselenadiazoles(91)

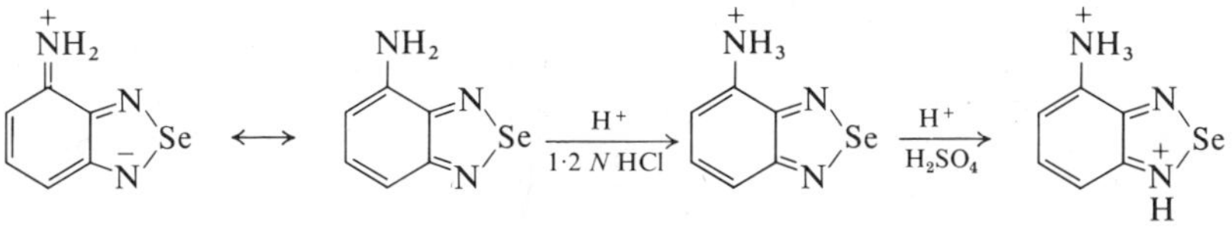

I ⟷ [NH_2 compound] $\xrightarrow[1\cdot 2\ N\ HCl]{H^+}$ II $\xrightarrow[H_2SO_4]{H^+}$ III

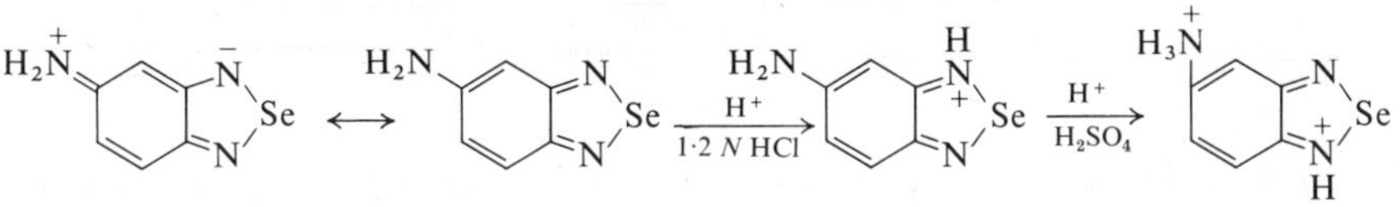

IV ⟷ [H_2N compound] $\xrightarrow[1\cdot 2\ N\ HCl]{H^+}$ V $\xrightarrow[H_2SO_4]{H^+}$ VI

Compound	λ_{max} (mμ)	mϵ	Compound	λ_{max} (mμ)	mϵ
I	242[a]	13.8	**IV**	236[a]	17.8
	330	11.0		324	8.7
	462	1.9		425	6.3
II	231	5.0	**V**	236	9.3
	333	17.0		333	12.6
				459	5.5
III	221	6.6	**VI**	222	6.6
	340[b]	19.1		337	17.8
	380	2.0		380[c]	2.2

[a]In 95% ethanol.
[b]Shoulder;
[c]Shoulder.

with 2,3-diphenyl-6-aminoquinoxaline and 2,3-diphenyl-5-amino-7-chloroquinoxaline(2). The following data (2) from Figure 8.3 show that the 6-amino derivative gives protonation mainly on the 4-aza nitrogen and some on the amino nitrogen, diprotonation, and triprotonation:

	Solvent	λ_{max}(mμ)	mϵ
Parent compound	95% EtOH	407	10.0
Protonation:			
On the 4-aza N	50% EtOH–1.2 *N* HCl	490	10.0
On the amino N	50% EtOH–1.2 *N* HCl	355*s*	6.6
Diprotonation	H_2SO_4–EtOH (1:1)	405	10.7
Triprotonation	H_2SO_4–EtOH (95:5)	480	14.1

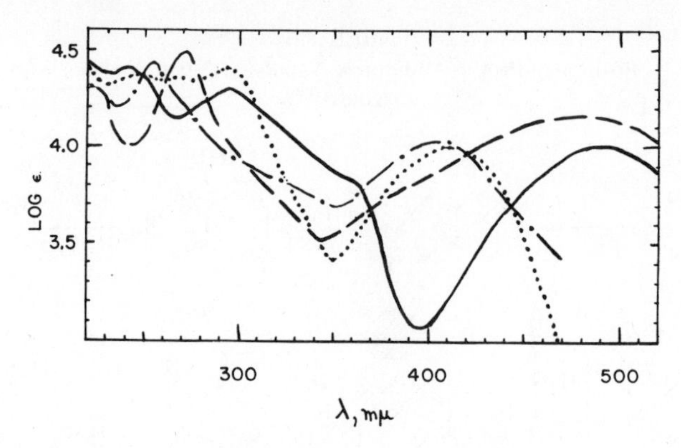

Fig. 8.3. Ultraviolet–visible absorption spectra of 2,3-diphenyl-6-aminoquinoxaline in 95% ethanol (· · · · ·), 50% ethanol–1.2 *N* hydrochloric acid (——), 50% sulfuric acid (—·—), and 95% sulfuric acid (– – –)(2).

In 2,3-diphenyl-5-amino-7-chloroquinoxaline(2) monoprotonation takes place on the amino nitrogen, although a weak broad band at 520 mμ indicates the presence of the tautomer. The trication of this compound forms only in small quantity in strong sulfuric acid, as shown by a shoulder at 520 mμ, mε 2.75.

In 6-amino quinazoline (**26**) and its amino isomers protonation takes place on an aza nitrogen with hydration of the molecule:

H_2N–(quinazoline) **26** $\xrightarrow[H_2O]{H^+}$ H_2N–(3,4-dihydro-4-hydroxy-1H-quinazolinium; H, OH, NH, $\overset{+}{N}$H) **27** (8.12)

6-Aminoquinazoline cation is mostly hydrated, 7-aminoquinazoline cation mostly anhydrous, whereas the 5- and 8-isomers give mixtures containing mostly the hydrated form(93).

The aza nitrogen in a heteroaromatic amine can be made so weakly basic that salt formation will take place on the amino group, for example, in 2,4,7-tribromo- and 2,4,5,7-tetrabromo-9-(4-dimethylaminophenyl) acridine(94).

Where there is steric interference with salt formation at an aza nitrogen atom, protonation will take place on the amino nitrogen; for example, 4,5-diaminoacridine (λ_{max}448mμ, mε 4.27, in absolute ethanol) has diprotonation take place on the amino groups in 5 *N* hydrochloric acid (λ_{max} 353 mμ, mε 10.7)(95). In 4-amino-5-methylacridine (**29**) it is believed that the smaller water molecule can stabilize protonation at the sterically protected aza nitrogen group, but the larger ethanol molecule cannot, as shown by the following(95):

$$\mathbf{28} \xleftarrow[N/4\ HCl]{H_2O} \mathbf{29} \xrightarrow[N/15\ HCl]{EtOH} \mathbf{30} \quad (8.13)$$

28: λ_{max} 460 mμ, mϵ 0.93; 29: λ_{max} 413 mμ, mϵ 3.16; 30: λ_{max} 354 mμ (*fs*), mϵ 8.9

One concentration range was found where the ammonium tautomer was stable in the cold and the azonium tautomer in the hot solution.

4. *Schiff Bases*

The protonation of 4-aminobenzalamines (**31**) has been investigated. Where the azomethine nitrogen is the negative resonance terminal, protonation takes place almost entirely on the negative resonance terminal:

$$(CH_3)_2N{-}C_6H_4{-}CH{=}N{-}R \ (\mathbf{31}) \xrightarrow{H^+} (CH_3)_2N{-}C_6H_4{-}CH{=}\overset{+}{N}H{-}R \ (\mathbf{32}) \quad (8.14)$$

R	λ_{max} (mμ)	mϵ	λ_{max} (mμ)	mϵ	Ref.
H	~ 329	~ 23	~ 386	~ 44.6	96
n-Bu	~ 325	~ 24	~ 390	~ 50	96
C_6H_5	358	34	434	82	97

On the other hand, when the aza and methine groups are interchanged, protonation takes place mainly on the amino nitrogen, as, for example, in the following(97):

$$4\text{-}(CH_3)_2NC_6H_4N{=}CHC_6H_5 \ (\mathbf{33}) \quad (8.15)$$

$$\downarrow H^+$$

$$4\text{-}(CH_3)_2H\overset{+}{N}C_6H_4N{=}CHC_6H_5 \ (\mathbf{34}) \rightleftarrows 4\text{-}(CH_3)_2NC_6H_4\overset{+}{N}H{=}CHC_6H_5 \ (\mathbf{35})$$

34: λ_{max} 260 mμ; 35: λ_{max} 455 mμ

The last type of reaction has been used in the analysis for unsaturated aldehydes, as shown in Table 8.7. The evidence for protonation on the aza nitrogen is based on the fact that $C_6H_5CH{=}CHCH\overset{+}{N}(C_2H_5)C_6H_4N$-

TABLE 8.7
Reaction of Unsaturated Aldehydes with *N*,*N*-Dimethyl-*p*-phenylenediamine (98)

	Color and λ_{max} (mμ)			
n	$CH_3(CH{=}CH)_n{-}CHO$[a]	$C_6H_5(CH{=}CH)_n{-}CHO$	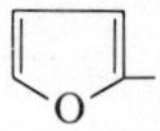 2-Furyl$-(CH{=}CH)_n-CHO$	
0	Colorless	Orange red	Orange red,	480
1	Yellow	Red, 485	Red,	510
2	Red	Red	Violet red,	530
3	Blue red	Blue red	Red violet,	542
4	Violet	Violet	Violet,	553
5	Blue	Blue violet	Blue,	570
6	Blue	Blue	Blue	

[a]See also reference 99 for λ_{max} obtained with other unsaturated carbonyl compounds.

$(CH_2C_6H_5)_2$ absorbs at 505 mμ, mϵ 14.5, in ethylene chloride, whereas $C_6H_5CH{=}CHCH{=}NC_6H_4N(CH_2C_6H_5)_2$ absorbs at 485 mμ, mϵ 31.6, in 1 *M* trichloroacetic acid (100).

In the case of other Schiff bases, where only one basic nitrogen is present in the molecule, the proton would be expected to add to the nitrogen; an example is **36**.

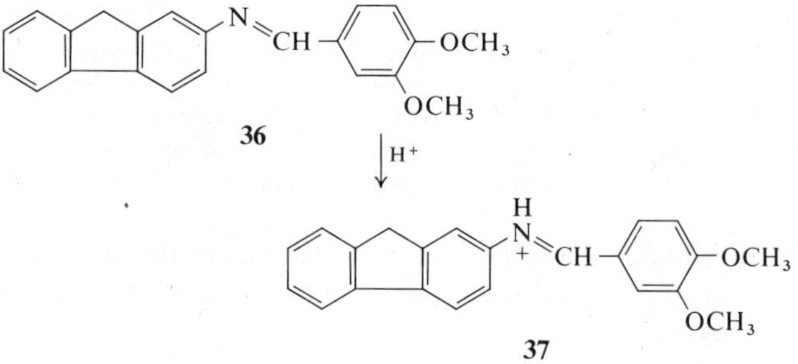

(8.16)

36 $\xrightarrow{H^+}$ **37**

No.	Solvent	λ_{max} (mμ)	mϵ
36	Ethanol	346	28.8
37	Trifluoroacetic acid	420	32.4

5. *Azo Dyes*

In a zwitterionic structure resonance helps to increase the basicity of the negative resonance terminal and decrease the basicity of the positive resonance terminal. This effect is clearly shown in the comparison of

4-phenylazodimethylaniline (**39**) with azobenzene (**38**) and dimethylaniline (**40**), where the pK_a of the aza nitrogen changes from -1.64 for azobenzene (101) to 2.17 for **39** (102) and the pK_a of the dimethylamino nitrogen changes from 4.22 for dimethylaniline (103) to 1.64 for **39**. The

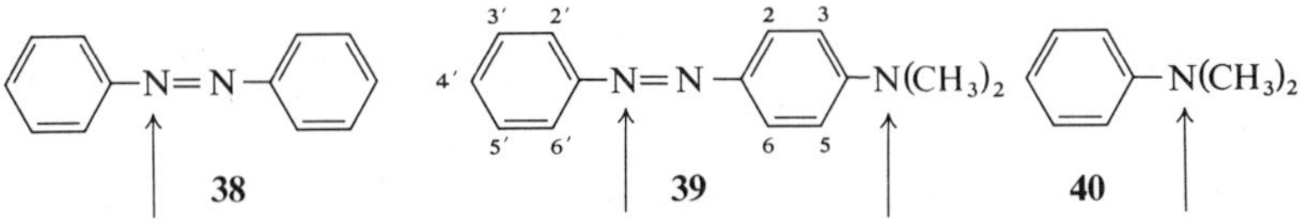

3′ 2′ 2 3

4′ —N=N— —N(CH$_3$)$_2$ —N(CH$_3$)$_2$

5′ 6′ 6 5

38 **39** **40**

pK_a $-1{\cdot}64$ in EtOH–aqueous H_2SO_4 (20 : 80)
pK_a 2·17 pK_a 1·64 in 50% aqueous EtOH
pK_a 4·22 in 50% aqueous EtOH

relative basicities of the β-azo nitrogen and the amino nitrogen in **39** are changed, depending on the substitutions in the ring and on the amino nitrogen. Thus in the 4′-methoxy derivative of **39** the pK_a values of the β-azo and amino nitrogens are both 2.10; in the 3-methyl derivative they are 1.87 and 3.47, respectively; in the 4′-nitro derivative they are 1.76 and 0.82, respectively. Consequently it can be seen that in acidic solution the two basic centers would compete for the proton, and a tautomeric equilibrium would be formed, depending on their relative basicity in the particular solvent system.

In the absence of basic amino groups azobenzene compounds substituted in the 2- or 4-position with an electron-donor group are protonated on the β-azo nitrogen atom, as shown in equation 8.17. Alkoxy, hydroxy, anilino, and aryl groups would be expected to act in this fashion. In all cases the long wavelength zwitterionic resonance band would be shifted to longer wavelength with greater intensity on protonation of the molecule.

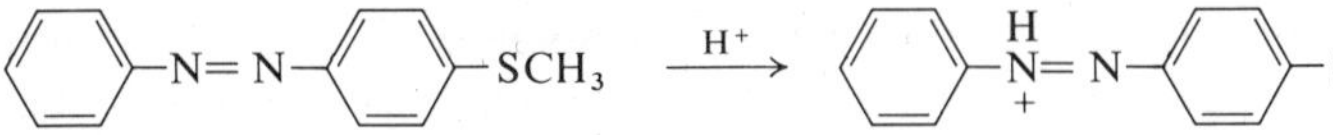

$$C_6H_5-N{=}N-C_6H_4-SCH_3 \xrightarrow{H^+} C_6H_5-\overset{H}{\underset{+}{N}}{=}N-C_6H_4-SCH_3 \quad (8.17)$$

41 **42**

No.	Solvent	$\lambda_{max}(m\mu)$	mϵ	Ref.
41	EtOH	363	24.0	104
42	H_2SO_4–EtOH	532	57.3	105

For the azo dyes substituted with a 4-amino, alkylamino, or dialkylamino group overwhelming evidence has been presented showing that in weak acid solution tautomers are formed containing a proton on the β-azo nitrogen and a proton on the amino nitrogen (102, 106), as shown, for example, in the following (107):

$$C_6H_5-N=N-C_6H_4-N(CH_3)_2 \quad (8.18)$$

43

$\downarrow H^+$

$$C_6H_5-\overset{+}{N}H=N-C_6H_4-N(CH_3)_2 \rightleftarrows C_6H_5-N=N-C_6H_4-\overset{+}{N}H(CH_3)_2$$

44 **45**

No.	Solvent	λ_{max}(mμ)	mε
43	EtOH	408	27.5
44	50% EtOH–0.6 N HCl	517	34.0
45		320	10.1

As expected, protonation at the negative resonance terminal shifts the long wavelength $\pi \rightarrow \pi^*$ band to 517 mμ and increases its intensity. Protonation at the positive resonance terminal causes the expected violet shift, with a decrease in intensity. Increasing the acidity of the solution increases the amount of cationic resonance tautomer at the expense of the dimethylammonium tautomer (107). The latter tautomer is of course, iso-π-electronic and isospectral with azobenzene.

The A_c/A_e ratio,—that is, the ratio of the absorbance of the cationic resonance tautomer to the absorbance of the ammonium tautomer—in equation 8.18 is 3.4 in 50% ethanol–0.6 *N* hydrochloric acid. A methyl group in the 2-position decreases the ratio to 0.025 in 50% ethanol– 1*N* hydrochloric acid; in the 2-position the ratio is 10.0; in the 2′-position it is 0.29; in the 3′-position 4.1; and in the 4′-position 2.6 (107). Thus by a simple change in the position of a methyl group the relative amounts of the tautomers are drastically changed. The A_C/A_E ratios of 73 azobenzene derivatives have been reported and discussed (107).

The relation between the structures of some of these dyes and their basicity and spectra are shown in Figure 8.4. In Table 8.8 the long-wavelength bands of some of the tautomers are identified and characterized.

The colorimetric determination of a very large variety of compounds depends on the formation of 4-aminoarylazo cationic dyes. Some of the compounds that have been determined through the formation of such cationic resonance structures are 3-amino-1,2,4-triazole (109), fenitrothion (110), folic acid (111), metoclopramide (112), nitrazepam (113), sulfamethazine (114), and sulfaquinoxaline (115). Since an increase in the acidity of the solution, short of forming the dicationic salt, can drastically increase the proportion of the intensely colored cationic resonance tautomer, for greatest sensitivity analysis should be at the highest possible pH.

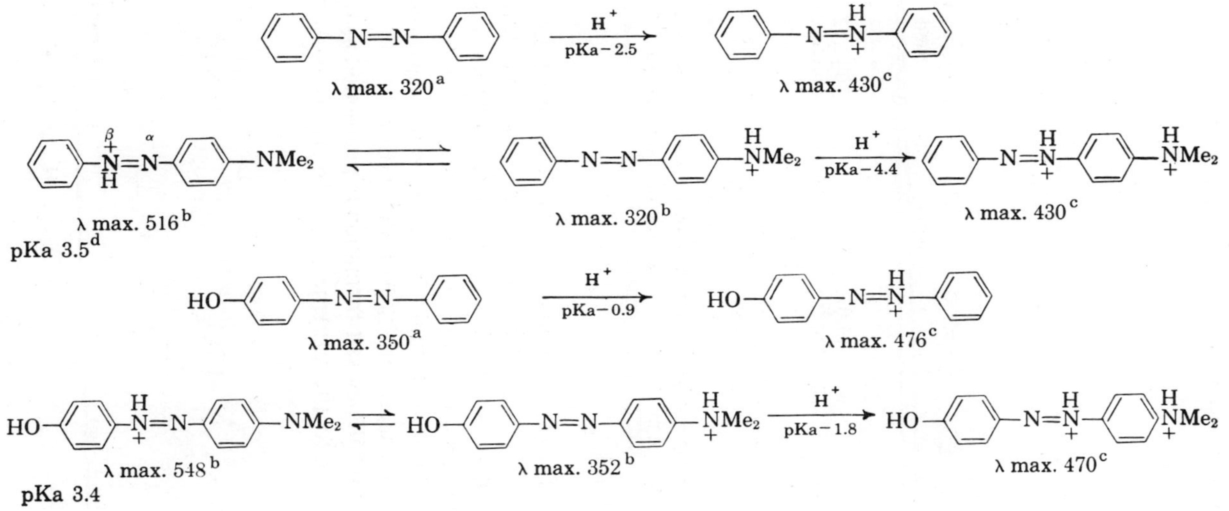

λ max. 320[a] $\xrightarrow[\text{pKa}-2.5]{H^+}$ λ max. 430[c]

λ max. 516[b] $\rightleftharpoons$ λ max. 320[b] $\xrightarrow[\text{pKa}-4.4]{H^+}$ λ max. 430[c]

pKa 3.5[d]

λ max. 350[a] $\xrightarrow[\text{pKa}-0.9]{H^+}$ λ max. 476[c]

λ max. 548[b] $\rightleftharpoons$ λ max. 352[b] $\xrightarrow[\text{pKa}-1.8]{H^+}$ λ max. 470[c]

pKa 3.4

Fig. 8.4. Spectral data and basicity of some azo dyes (108). [a]In 95% ethanol. Azobenzene and 4-hydroxyazobenzene also have low-intensity $n \rightarrow \pi^*$ bands at 443 and 440 mμ, respectively. [b]In 50% ethanol–1 *N* hydrochloric acid. [c]In 95% sulfuric acid.

TABLE 8.8
Spectral Identification of the 4-Aminoazobenzene Tautomers[a] (108)

X	λ_{max} ($m\mu$) ($m\epsilon$)		
	Ammonium tautomer		Azonium tautomer
	$\pi \to \pi^*$ band	$n \to \pi^*$ band	Cationic resonance band
4-$N(CH_3)_2$	320 (9.77)	—[b]	517 (35.5)
4-NH_2	319 (17.0)	—[b]	500 (12.3)
3-CH_3-4-$N(CH_3)_2$	319 (20.0)	445 (0.63)	500*s* (0.5)
3-NH_2	322 (20.0)	430 (0.76)	—[c]
4-$N(CH_3)_3^+I^-$[d]	319 (20.9)	443 (0.54)	—[c]
H[d]	320 (21.4)	443 (0.51)	—[c]

[a]In 1 *N* hydrochloric acid in water–ethanol (1 : 1) unless otherwise specified.
[b]Hidden under the much more intense cationic resonance band.
[c]No cationic resonance band present.
[d]In 95% ethanol.

Other types of azo compounds show interesting properties. The cationic salts of the *cis*- and *trans*-azobenzenes have different spectral and physical properties. In 2- and 3-phenylazoanilines salt formation takes place at the amino group; the *ortho* derivative does show a very weak band at 570 $m\mu$ that could be due to salt formation at the β-azo nitrogen atom or to a decomposition product (108).

In the 1-arylazoazulenes salt formation is postulated as taking place at the azo nitrogen β to the azulene ring (116–121), as in the following example (121):

46
λ_{max} 477, $m\epsilon$ 37.1

47
λ_{max} 671, $m\epsilon$ 42.6

48
λ_{max} 513, $m\epsilon$ 46.8

(8.19)

In the *p*-amino compound a mixture of tautomers could be present, with salt formation taking place at the three nitrogen atoms. The second proton goes to the amino group, as shown by the spectral resemblance to the iso-π-electronic 4-trimethylaminonium dicationic and the phenylazo cationic compounds(121).

In the determination of azulene with oxidized 2-hydrazinobenzothiazole an intensely colored cationic resonance structure (**49a**) is formed which contains its proton on the heterocyclic nitrogen(122), as shown by the close spectral resemblance of this compound to the chromogen **49b** obtained from the determination of azulene with oxidized 3-methyl-2-benzothiazolinone hydrazone(122).

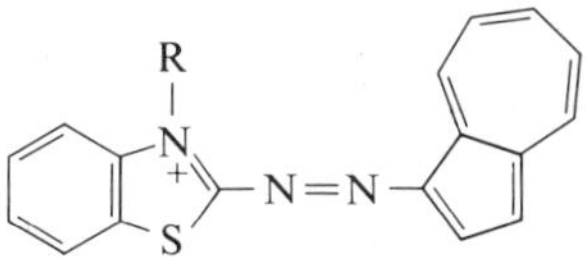

49

	R	λ_{max} (mμ)	mϵ
a	H	583	50.0
b	CH_3	588	52.0

Azulene and its derivatives can also be determined with 4-azobenzenediazonium fluoborate (123). Either the neutral bisazo chromogen or its mono-cationic or dicationic salt can be formed (Fig. 8.5). Protonation is postulated as taking place on the β-azo nitrogen closest to the azulene ring. The second proton then adds to the other β-azo nitrogen. With each proton the wavelength maximum shifts to the red with increasing intensity.

The spectral change with increasing protonation and the tautomeric equilibrium taking place at each level of protonation for the 4-amino-*p*-bisazobenzene dyes are depicted in Figure 8.6.

The absorption spectra of 2,3′,4″-trimethyl-4-dimethylamino-*p*-bisazobenzene and the cationic tautomers with which the long wave length bands are associated are shown in Figure 8.7. As in the 4-phenylazoanilines, the substitution of a methyl group *ortho* to the β-nitrogen atom, as in 2′,2″-dimethyl-4-dimethylamino-*p*-bisazobenzene decreases the A_C/A_E ratio significantly.

p-Bisazobenzene cationic dyes are formed in the determination of anilines, naphthylamines, and anthramines, as well as their *N*-alkyl and *N*,*N*-dialkyl derivatives with 4-azobenzenediazonium fluoborate as the reagent (125). Many of the compounds give mϵ values of about 100.

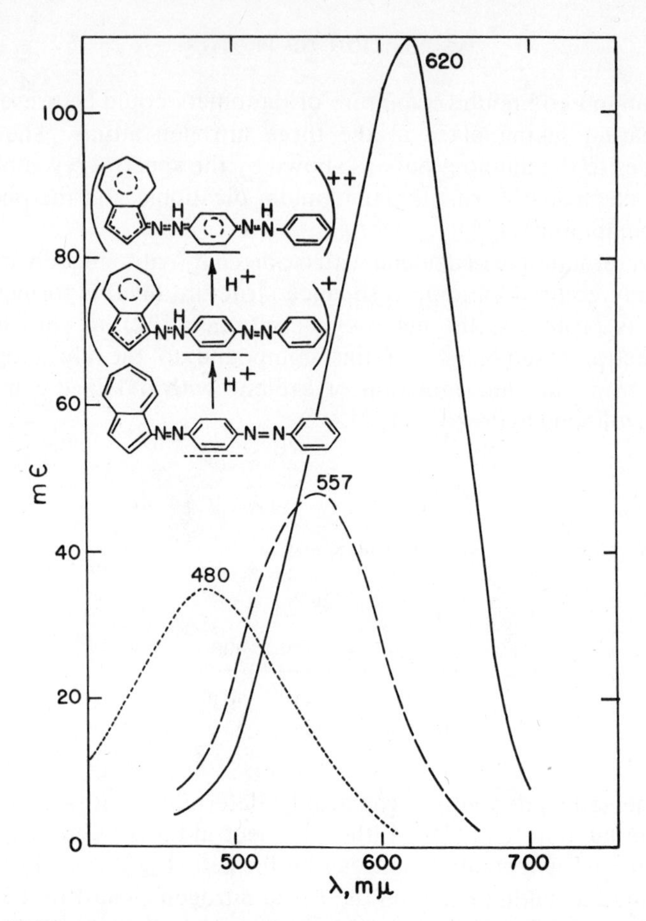

Fig. 8.5. Visible absorption spectra of the chromogens obtained in the determination of azulene with 4-azobenzenediazonium fluoborate in increasingly acidic solutions.

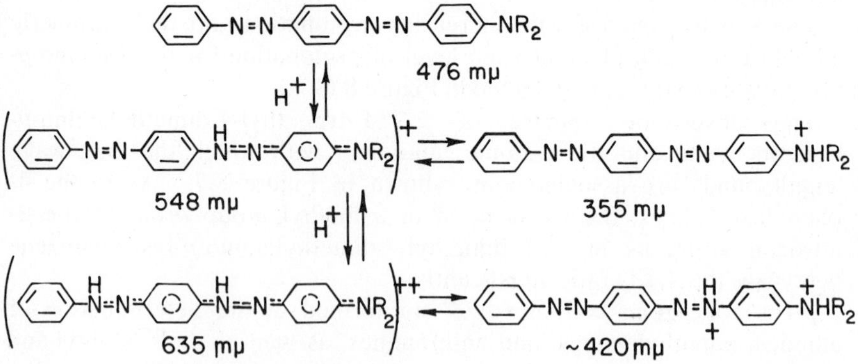

Fig. 8.6. Spectral change with increasing protonation and tautomeric equilibrium at each level of protonation for the 4-amino-*p*-bisazobenzene dyes.

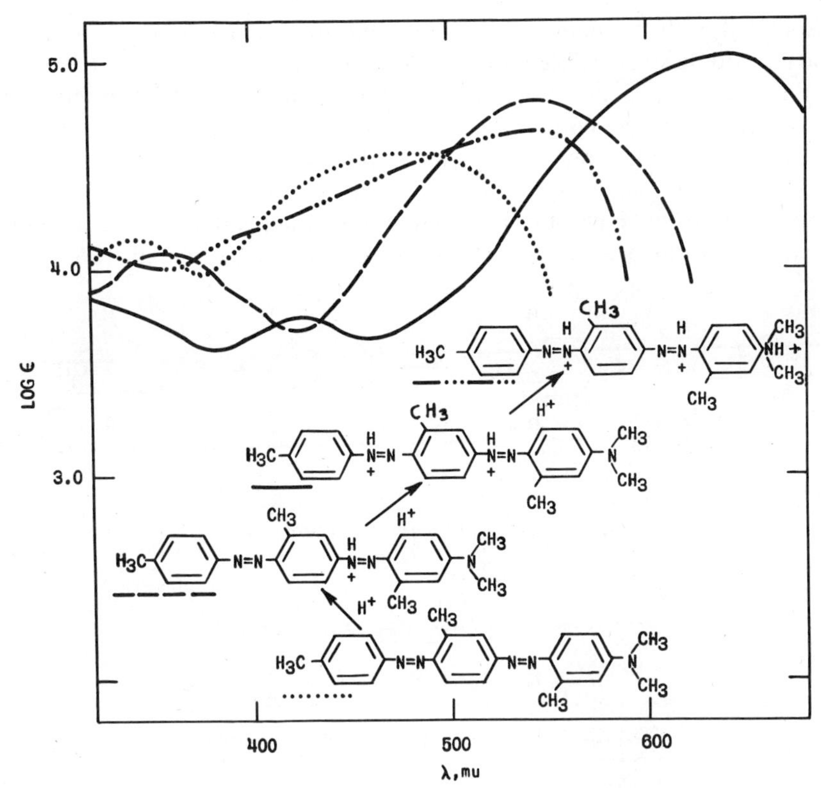

Fig. 8.7. Absorption spectra of *N,N*,2,3′,4″-tetramethyl-4-amino-*p*-bisazobenzene in 95% ethanol containing 0.25% dioxane (· · · · ·), 50% ethanol–1.2 *N* hydrochloric acid (– – –), ethanol–sulfuric acid (——), and 95% sulfuric acid (—· · ·—)(124).

The position of protonation in the chromogen formed in the determination of α-naphthylamine is shown in Figure 8.8.

Salt formation in solvents of increasing acidity by azobenzene dyes that contain two *p*-amino groups has been investigated. The bands

$$C_6H_5\text{-}N{=}N\text{-}C_6H_4\text{-}NH_2 \xrightarrow{HONO} C_6H_5\text{-}N{=}N\text{-}C_6H_4\text{-}N_2^{+}$$

$$\text{I} + \text{(1-naphthylamine, } NH_2\text{)} \xrightarrow{H^+} \left(C_6H_5\text{-}\overset{H}{N}{=}N{=}C_6H_4{=}\overset{H}{N}{=}N{=}C_{10}H_6\text{-}NH_2\right)^{++}$$

λ_{max} 643 mμ, mε 93

Fig. 8.8. Position of protonation in the chromogen formed in the determination of α-naphthylamine.

associated with the neutral compound and its monocationic, dicationic, and tricationic salts are characterized in Figure 8.9.

For some types of compounds proton addition can take place at a negative resonance node, especially when the node is a fairly basic nitrogen atom. Thus phenols reacted with 3-methyl-2-benzothiazolinone hydrazone give azo dyes (spectra in acetonitrile) whose tautomeric equilibrium in 60% perchloric acid solution is dependent on the amount of steric hindrance adjacent to the negative resonance terminal (126):

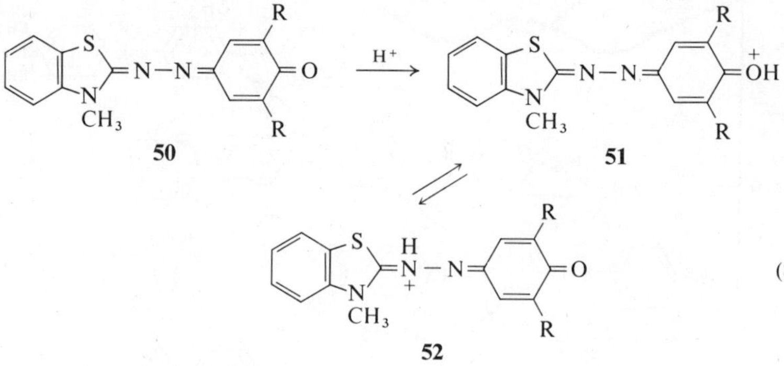

(8.20)

	$R = CH_3$		$R = C_2H_5$		$R = t\text{-}C_4H_9$	
No.	λ_{max} mμ	mε	λ_{max}(mμ)	mε	λ_{max}(mμ)	mε
50	475	46.5	~475	~46.5	471	47.5
51	511	~37.5	513	~37	520	~2.5
52	380s	5.0	380s	6.0	386	32.0

6. *Aza Heterocyclic Compounds*

Salt formation in the monoaza heteroaromatic hydrocarbons takes place on the nitrogen. Protonation of pyridine and its alkylated and halogenated derivatives causes little change in wavelength position and some increase in intensity (127, 128). Phenylpyridines and dipyridyls undergo a red shift of 20–30 mμ on protonation, with little change in intensity (129). The long-wavelength bands of the polynuclear aza hydrocarbons shift to the red 20–50 mμ on acidification (130). For some of the compounds the intensity increases, for others it decreases. The spectra of the cations are much poorer for characterization and analysis since much fine structure is lost and the bands are broadened by salt formation.

Whenever a hydroxy monoaza compound exists as the imino tautomer, protonation takes place on the carbonyl oxygen (26, 131). In the alkoxy and alkylthio (132) derivatives protonation takes place on the ring nitrogen.

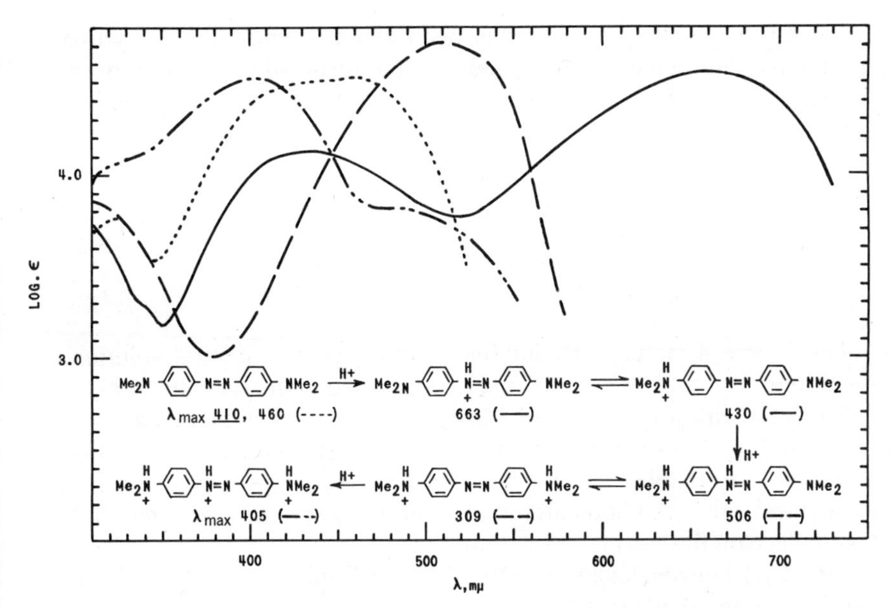

Fig. 8.9. Visible absorption spectra of 4,4′-bis(dimethylamino)azobenzene in 95% ethanol (·····), 95% acetic acid (——), 50% ethanol–1.2 *N* hydrochloric acid (–––), and 95% sulfuric acid (—··—).

Where there are several nitrogen atoms in a ring, protonation usually takes place so as to give maximum resonance (equation 8.21). However, where a more basic ring nitrogen competes with a carbonyl oxygen for a proton, salt formation will take place at the aza nitrogen (equation 8.22).

2-Hydroxypyrimidine (**53a**) and *N*-methylpyrimidone (**53b**) add a proton in acid solution to form isospectral monocationic salts (**54**) that are entirely different spectrally from the 2-methoxypyrimidine monocationic salt (**55**)(133). This can only mean that the proton is bonded to the ring nitrogen.

$$\mathbf{53} \xrightarrow{H^+} \mathbf{54} \qquad \mathbf{55} \qquad (8.21)$$

a: R = H; **b**: R = CH_3

No.	λ_{max}(mμ)	mε
54a	309	3.75
54b	313	3.85
55	274	3.69
	309	2.85

Somewhat similar evidence has been given to show that 4-hydroxypyrimidine (**56**), which exists largely in keto form, adds a proton to a ring nitrogen to form **57**(133). These proton additions have been confirmed (134).

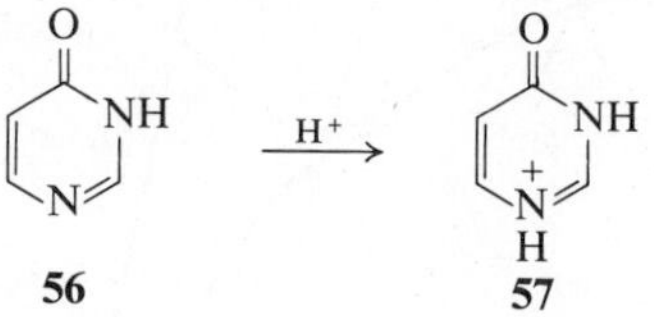

(8.22)

56 **57**

The 2- and 4-mercaptopyrimidines, which exist in neutral solution in the thione structure, appear to bind a proton also to the ring nitrogen.

Proton addition to a benzimidazole derivative takes place on the aza nitrogen(135). The addition of a proton to the spectrally dissimilar 1-methyl-7-azaindole and 7-methyl-7*H*-pyrrolo[2,3-*b*]pyridine forms isospectral salts; this indicates that in both compounds protonation has taken place on the aza nitrogen atom(136).

For 2,1,3-benzoselenadiazole(91, 137) and quinoxaline(2) derivatives salt formation takes place at the negative resonance terminal; the addition of the first proton causes a red shift, followed by an even larger red shift with the addition of the second proton. Examples of this phenomenon are shown in Figures 8.10 and 8.11.

Cinnoline(84), benzo[*c*]cinnoline(84), and phenazine(138, 139) derivatives (including the amines) undergo salt formation at the aza nitrogen atom, which serves as the negative resonance terminal, as shown for 1-methoxyphenazine in equation 8.23(139). The absorption spectra of the cationic resonance structures in equations 8.23 and 8.24 are closely similar to that of **62**.

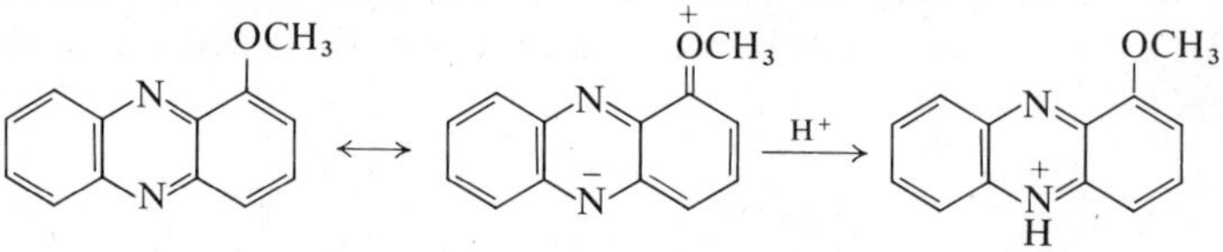

(8.23)

58 **59**

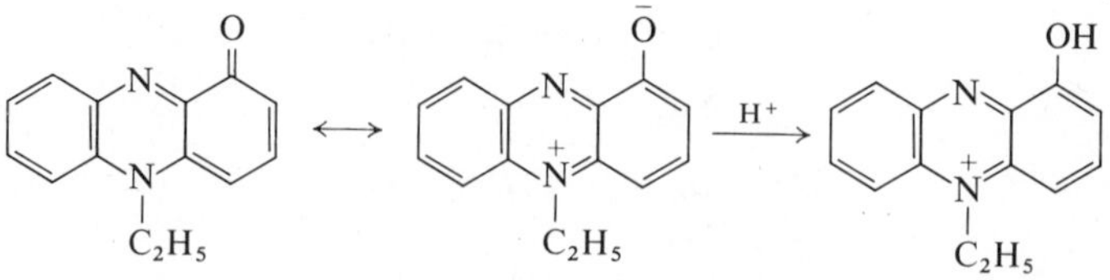

(8.24)

60 **61**

No.	Solvent	$\lambda_{max}(m\mu)$
59	$EtOH-H_2SO_4$	290, 390, 550
61	$EtOH-H_2SO_4$	290, 390, 550
62	95% EtOH	290, 390, 550

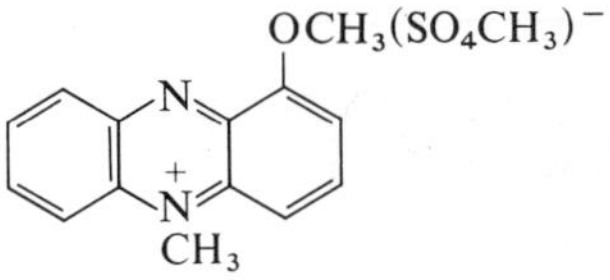

$OCH_3(SO_4CH_3)^-$

CH_3

62

Cryoscopic evidence indicates that polynuclear cinnolines are diprotonated on the nitrogen atoms in concentrated sulfuric acid(140). Intense colors are formed probably because the two adjacent positive charges would repel each other strongly.

For some aza compounds the changes in spectra with tautomeric shifts can be quite dramatic, as shown below(141).

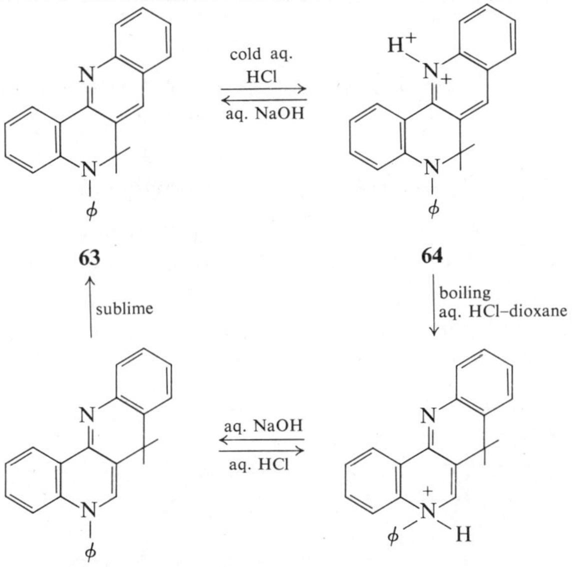

cold aq. HCl

aq. NaOH

63 **64**

sublime

boiling aq. HCl–dioxane

aq. NaOH

aq. HCl

65 **66**

(8.25)

No.	Solvent	$\lambda_{max}(m\mu)$	mϵ
63	EtOH	400	5.0
64	EtOH–1 *N* HCl (1:1)	510	5.1
65	EtOH	420	15.0
66	EtOH–1 *N* HCl (1:1)	410	15.0

The protonation and the hydration of the quinazoline amines have been covered in subsection 3 of this section. This phenomenon has been discussed for a large number of types of quinazoline compounds(142).

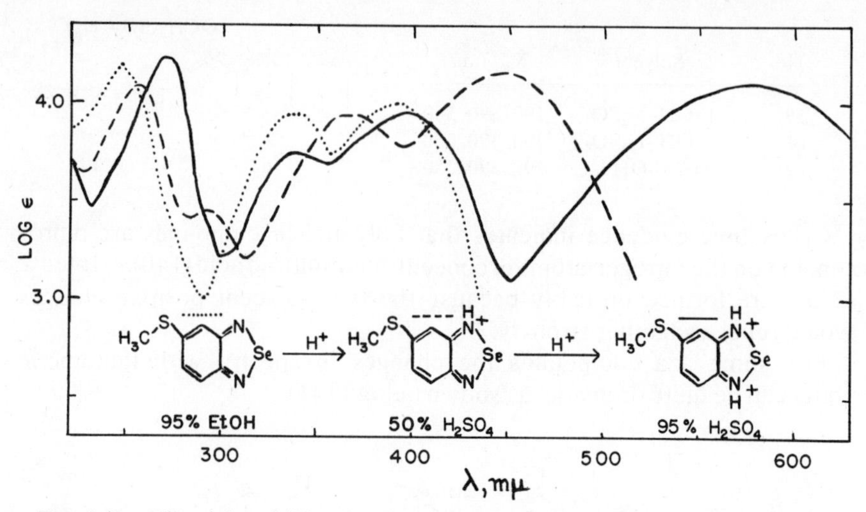

Fig. 8.10. Ultraviolet–visible absorption spectra of 5-methylthio-2,1,3-benzoselenadiazole in 95% ethanol (·····), 50% ethanol–sulfuric acid (---), and 95% sulfuric acid (——)(91).

7. *Imino Heterocyclic Compounds*

Pyrrole (λ_{max}208 mμ, mϵ 7.3, in ethanol) forms a conjugate acid (λ_{max}241 mμ, mϵ 7.9, in aqueous sulfuric acid) by proton addition to the α-position(36). In some of the alkyl derivatives of pyrrole both α- and β-proton addition takes place, as shown, for example, in equation 8.26 (spectra in water)(36).

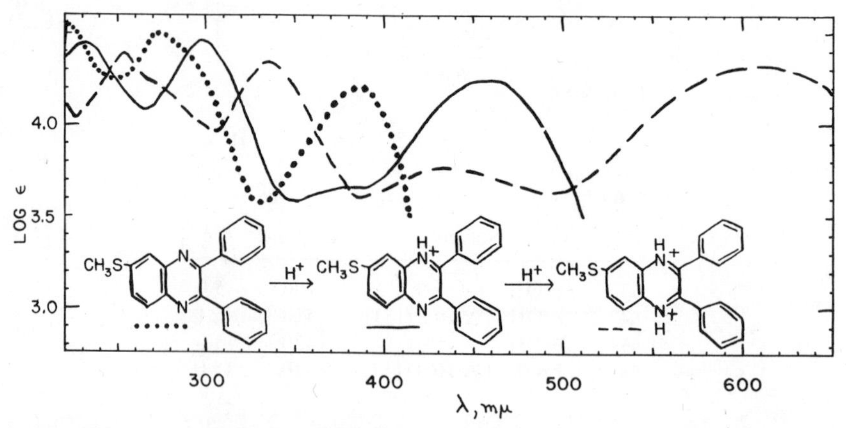

Fig. 8.11. Ultraviolet–visible absorption spectra of 2,3-diphenyl-6-methylmercaptoquinoxaline in 95% ethanol (·····), 50% ethanol–6 *N* hydrochloric acid (——), and 95% sulfuric acid (---)(2).

$$\underset{\textbf{67}}{\text{2,5-dimethylpyrrole}} \xrightarrow{H^+} \underset{\textbf{68}}{} \rightleftarrows \underset{\textbf{69}}{} \qquad (8.26)$$

λ_{max} 209 mμ, mϵ 7.7 $\quad$ λ_{max} 237 mμ, mϵ 4.1 $\quad$ λ_{max} 275 mμ, mϵ 2.5

The basicities, spectra, and position of protonation of many other methylpyrroles(143,144) and *N*-phenylpyrroles(36) have been discussed.

The spectral evidence in Figure 8.12 indicates that the principal conjugate acid of an indole in strong aqueous acids is the 3-protonated isomer(145). This has been confirmed by proton magnetic resonance

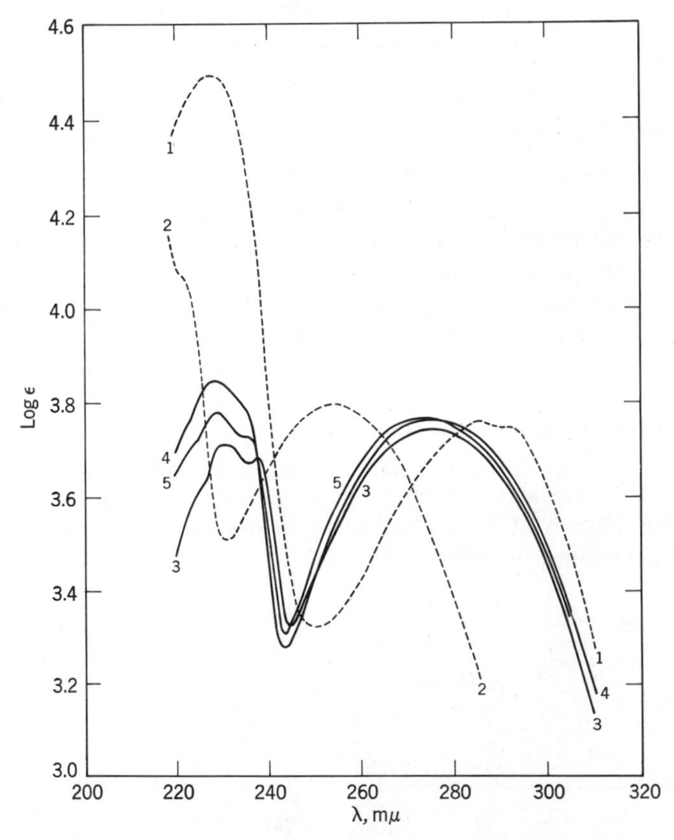

Fig. 8.12. Ultraviolet absorption spectra of 1,2,3-trimethylindole in water (curve 1), 2,3,3-trimethylindolenine in 95% ethanol (curve 2), and their salts: 1,2,3-trimethylindole in 6 *M* sulfuric acid (curve 3), 2,3,3-trimethylindolenine in 0.1 *N* hydrochloric acid (curve 4), and 2,3,3-trimethylindolenine methiodide in water (curve 5).

measurements on a series of methylindoles. Some spectral and chemical evidence that has been obtained indicates that many 2-arylindoles protonate at the 3-position(146), as in the following:

$$\mathbf{70} \xrightarrow{H^+} \mathbf{71} \tag{8.27}$$

No.	Solvent	λ_{max} (mμ)	mϵ
70	Methanol	240	17.0
		309	24.6
71	Phosphoric acid	264	6.76
		343	21.9

B. Carbonyl Compounds

In a large number of organic compounds in which the carbonyl oxygen is the negative resonance terminal the proton adds to the oxygen atom. The only exceptions are those compounds in which a fairly basic amino nitrogen is the positive resonance terminal.

Where a proton adds to the carbonyl oxygen resonance terminal, the spectrum of the compound shifts to the red and the intensity increases.

1. Olefinic Carbonyl Compounds

Saturated carbonyl compounds can form cationic salts in strong acid solution, but the consequent spectra are too weak to be of value in trace analysis. Conjugated carbonyl compounds are believed to form cationic salts that absorb at longer wavelength and greater intensity than the neutral compound, as in **72** and **73**(147).

$$\underset{\mathbf{72}}{(CH_3)_2C{=}CH{-}\overset{O}{\overset{\|}{C}}{-}CH_3} \xrightarrow{H^+} \underset{\mathbf{73}}{(CH_3)_2C{=}CH{-}\overset{^+OH}{\overset{\|}{C}}{-}CH_3} \tag{8.28}$$

No.	Solvent	λ_{max} (mμ)	mϵ
72	EtOH	237	11.5
		313	0.057
73	H_2SO_4	282	17.3

The surprisingly large wavelength shift shown in equation 8.28 may be due to decomposition or condensation. However, other unsaturated carbonyl compounds – such as crotonaldehyde, 3-methyl-3-penten-2-one, and 2-cyclohexen-1-one show a similar phenomenon(148). The pK_{BH^+} values of these compounds in sulfuric acid [based on the H_A function (149)] have also been reported.

As shown in equation 8.29, β-diketones form cationic salts in acid solution(150). A large number of β-diketones showed the same spectral and intensity shifts on formation of the cationic salt.

$$\underset{\mathbf{74}}{O{=}\overset{CH_3}{\overset{|}{C}}{-}CH{=}\overset{CH_3}{\overset{|}{C}}{-}OH} \xrightarrow{H^+} \underset{\mathbf{75}}{H\overset{+}{O}{=}\overset{CH_3}{\overset{|}{C}}{-}CH{-}\overset{CH_3}{\overset{|}{C}}{-}OH} \qquad (8.29)$$

No.	Solvent	λ_{max} (mμ)	mε
74	EtOH	273	7.94
75	H_2SO_4	287	19.8

2. *Carboxylic Acids*

Evidence has been presented that protonation of a carboxyl group takes place on a carbonyl oxygen(151, 152). Consistent with this is the increase in electron density of the carbonyl oxygen and the increase in basicity of benzoic acid with substitution of electron-donor groups in the ring and a decrease in these properties with substitution of an electron-acceptor group in the ring (Table 8.9). These facts and the spectral data

TABLE 8.9
Protonation, Basicity and Spectral Properties of Benzoic Acids(151)

$$\underset{\mathbf{I}}{X{-}C_6H_4{-}C({=}O)OH} \xrightarrow{H^+} \underset{\mathbf{II}}{X{-}C_6H_4{-}C({=}\overset{+}{O}H)OH}$$

	I in 56–63% H_2SO_4		→	II in 95–98% H_2SO_4	
X	λ_{max} (mμ)	mε	pK_a	λ_{max} (mμ)	mε
4-OC_2H_5	260	14.1	−6.61	300	20.4
4-OCH_3	263	15.1	−6.68	300	19.5
4-CH_3	246	14.5	−6.92	277	19.1
H	235	11.0	−7.26	260	18.2
4-F	237	12.3	−7.30	267	18.6
4-Cl	249	19.1	−7.48	279	29.5
3-NO_2	222	22.4	−7.97	234	24.0

are consistent with the carbonyl oxygen's being the negative resonance terminal and the proton acceptor.

2,4,6-Trimethoxybenzoic acid has been reported to be a strong base because of the electron-donor groups and the intramolecular hydrogen bonds present in the cationic resonance structure **76**(153).

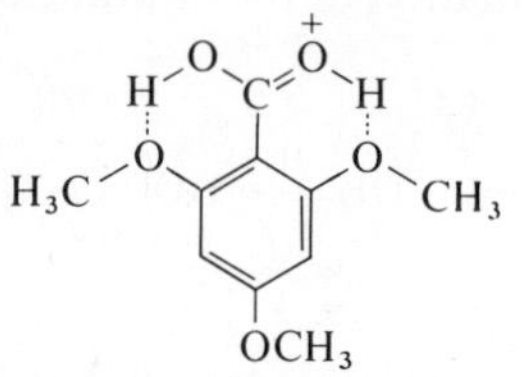

76

On the other hand, in mesitoic acid a structure such as **76** is formed, but to relieve the steric strain water is split off and the mesitoyl cation is formed(152):

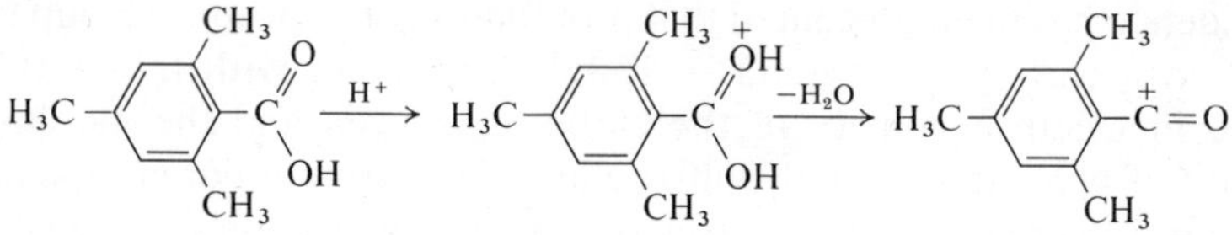

$$\mathbf{77} \xrightarrow{H^+} \mathbf{78} \xrightarrow{-H_2O} \mathbf{79} \quad (8.30)$$

No.	Solvent	λ_{max} (mμ)	mϵ
77	Heptane	243	4.8
		280	0.76
79	H_2SO_4	282	20.0
		338	5.0

Even olefinic acids, such as acrylic and crotonic acids, show a red shift effect on protonation in sulfuric acid(147). Thus crotonic acid absorbs at 209 mμ, mϵ 13.2, whereas its cationic salt absorbs at 235 mμ, mϵ 17.7.

3. *Acid Amides*

In any acid amide one limiting resonance structure to consider would involve the carbonyl oxygen as the negative resonance terminal and the amino nitrogen as the positive resonance terminal. Evidence for protonation at the oxygen or the nitrogen atom has been critically reviewed (154). Katritzky and Jones(154) concluded that protonation predominates at the oxygen atom. In sulfur analogs, such as thioacetamide and thiourea, protonation takes place on the sulfur atom(155). A violet spectral shift results. In the case of thioacetamide the thiocarbonyl $n \rightarrow \pi^*$ band at 327 mμ is lost:

$$CH_3{-}\overset{\overset{\displaystyle S}{\|}}{C}{-}NH_2 \xrightarrow{H^+} CH_3{-}\overset{\overset{\displaystyle \overset{+}{S}H}{\|}}{C}{-}NH_2 \qquad (8.31)$$

80 **81**

No.	Solvent	λ_{max} (mμ)	mϵ
80	EtOH	266	12.6
		327	0.052
81	H_2SO_4	234	14.5

Protonation of the benzamides results in the red spectral shift and intensity increase expected of salt formation at the oxygen atom(41). Consistent with this postulate is the red shift effect and basicity increase caused by electron-donor substituents (equation 8.32). The effect of the 4-nitro group on the basicity and lack of spectral shift with protonation appears to be consistent with this postulate.

$$X{-}C_6H_4{-}\overset{\overset{\displaystyle OH}{\|}}{C}{-}NH_2 \xrightarrow{H^+} X{-}C_6H_4{-}\overset{\overset{\displaystyle \overset{+}{O}H}{\|}}{C}{-}NH_2 \qquad (8.32)$$

82 **83**

X	**82** in water λ_{max} (mμ)	mϵ	pK$_a$	**83** in aqueous H_2SO_4 λ_{max} (mμ)	mϵ
4-OCH_3	253	14.8	−1.80	282	16.9
H	225	9.1	−2.16	245	12.0
4-NO_2	266	11.2	−3.23	264	14.5

The resonance within these molecules appears to be a decisive factor in determining the site of protonation.

4. *Aromatic Carbonyl Compounds*

In aromatic carbonyl compounds salt formation takes place at the carbonyl oxygen except when an amino group is present. The typical spectral changes that result can be seen in the following:

$$X{-}C_6H_4{-}CHO \longrightarrow X{-}C_6H_4{-}\overset{H}{C}{=}\overset{+}{O}H \qquad (8.33)$$

84 **85**

		X = H		X = 3,4-Dimethoxy	
No.	Solvent	λ_{max}(mμ)	mϵ	λ_{max}(mμ)	mϵ
84	EtOH	246	20.0	308	9.2
85	H_2SO_4	298	35.2	343	19.0

Naphthaldehydes, anthraldehydes, and 7-formylcyclopenta[*b*]pyran (156) show similar spectral changes after proton addition to the carbonyl oxygen.

Substitution of electron-donor groups in the ring or halogen atoms in the methyl group shifts the cationic resonance band of the aryl methyl ketone cationic salts to longer wavelength(157). Protonation causes the expected spectral changes (Table 8.10), except for 8,9-diacetylsesquifulvalene, which may have a Z_z-structure (**86**) and thus would have a violet shift on protonation of a carbonyl oxygen.

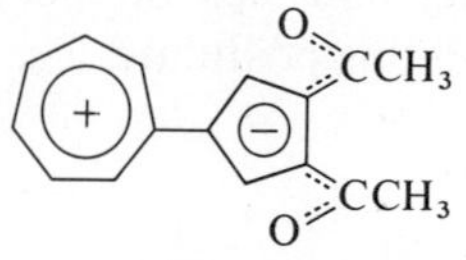

O⋯CCH₃ / O⋯CCH₃

86

ω,ω,ω-Trichloroacetophenone and ω,ω,ω-tribromoacetophenone do not form detectable amounts of oxonium ions in concentrated sulfuric acid (157). The basicities of many aromatic aldehydes and ketones have been reported(47, 48, 148).

Diaryl ketones undergo red spectral shifts and intensity increases on protonation, as shown in equation 8.34 for benzophenone(45). The 4,4′-bisdimethylamino derivative protonates on a nitrogen in aqueous solution. In acidified methanol or ethanol about 1% of the oxonium ion, absorbing at 500 mμ, is formed(45). In aqueous tetrafluoroboric acid both the ammonium (λ_{max}359 mμ) and the oxonium (λ_{max}524 mμ) tautomers are present(160).

$$\underset{\mathbf{87}}{(C_6H_5)_2C{=}O} \xrightarrow{H^+} \underset{\mathbf{88}}{(C_6H_5)_2C{=}\overset{+}{O}H} \qquad (8.34)$$

No.	Solvent	λ_{max}(mμ)	mϵ
87	EtOH–H_2O(1 : 10)	258	17.0
88	H_2SO_4	344	24.4

Interposition of an increasing number of methylene groups between the carbonyl groups of benzil has been investigated(161). Benzil and

TABLE 8.10
Long-Wavelength Bands of Aryl Methyl Ketones and Their Cationic Salts

Compound	Solvent	λ_{max} (mμ)	mϵ	Solvent	λ_{max} (mμ)	mϵ	Ref.
Acetophenone	*n*-Hexane	239	12.5	H_2SO_4 (conc.)	295	26.0	157
p-Methylacetophenone	47% H_2SO_4	260	14.9	H_2SO_4 (conc.)	312	23.8	48
3,4-Dihydroxyacetophenone	95% EtOH	307	8.0	H_2SO_4 (conc.)	340	90.0	158
ω-Chloroacetophenone	*n*-Hexane	245	13.0	H_2SO_4 (conc.)	307	24.5	157
ω,ω-Dichloroacetophenone	*n*-Hexane	251	12.0	H_2SO_4 (conc.)	316	4.2	157
4-Acetylbiphenyl	H_2SO_4 (aqueous)	288	9.2	H_2SO_4 (conc.)	349	27.8	47
2-Acetylfluorene	H_2SO_4 (aqueous)	316	40.0	H_2SO_4 (conc.)	379	72.2	47
8,9-Diacetylsesquifulvalene	Water	498	39.5	Aqueous acid	467	37.0	159

dibenzoylmethane are monoacid bases in concentrated sulfuric acid; 1,3-dibenzoylpropane and higher members are diacid bases.

Diarylcarbonyl compounds in which either or both of the aryl groups are polynuclear would be expected to add a proton to the carbonyl oxygen in concentrated sulfuric acid. These cationic salts would absorb at fairly long wavelength; for example, 1-benzoylpyrene is orange red, whereas 1-α-naphthoylpyrene is violet in sulfuric acid.

One of the few thiocarbonyl compounds that have been studied is the blue thiobenzophenone, which has, in ether, a λ_{max} of 315 and 620 mμ (mϵ 17, and 0.07 respectively) and gives a light-yellow salt (λ_{max}384 mμ, mϵ 25) in sulfuric acid. In this case salt formation is stated to take place at the sulfur atom.

In the diarylcarbonyl compounds that contain a carboxy group *ortho* to the carbonyl the type of cationic resonance structure that is formed depends on the strength of the acidic solvent(162, 163):

O, C, 5, 4′, C, O, OH, X — aq. H_2SO_4, H^+ → $\overset{+}{O}H$, C, C, O, OH, X — conc. H_2SO_4, H^+ → X, C, +O, O·

89 **90** **91**

(8.35)

No.	4′-X	λ_{max}(mμ)	mϵ
90	H	310	
	OCH_3	355	
91	H	410	28.2
	OCH_3	440	48.3

Structures **90** and **91** are in agreement with the spectral data since the hydroxy group on the positive resonance node of **90** would have a violet-shift effect, which would be mostly canceled in **91** by the interaction of the electronegative carbonyl group with the lactone oxygen. Consequently **91** would be expected to absorb at much longer wavelength than **90**. The red shift effect of a 5-methoxy group in *o*-benzoylbenzoic acid is consistent with equation 8.35; that is, in aqueous sulfuric acid its λ_{max} is 375 mμ and in 100% sulfuric acid its λ_{max} is 490 mμ, mϵ 61.5.

Polynuclear carbonyl compounds that contain an *o*-carboxy group should give brilliantly colored carbonium salts in concentrated sulfuric acid, e.g.

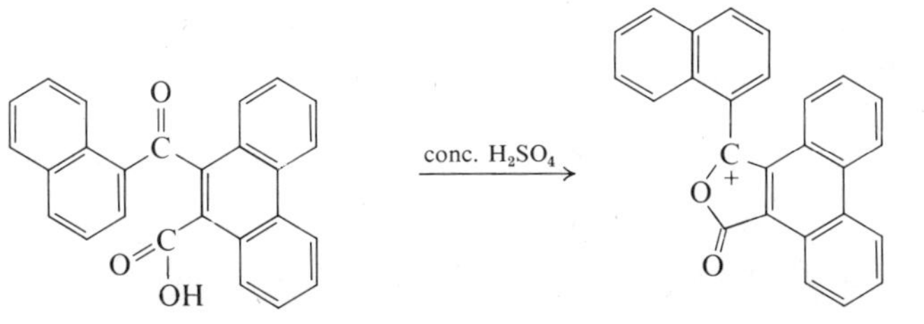

conc. H_2SO_4 (8.36)

92 **93**

Blue (164)

5. *Aralkene Carbonyl Compounds*

In the ω-*p*-methoxyphenylpolyene aldehydes, $CH_3OC_6H_4(CH{=}CH)_n$-CHO, $n = 1$–7, evidence has been presented indicating that protonation takes place at the carbonyl oxygen, with the usual red shift (165, 166).

The chalcones show a strong red shift effect and intensity increase on protonation (Table 8.11). In addition the strong effect of electron-donor groups on the basicity and spectra of the compounds is demonstrated.

A large number of compounds that contain the chalcone chromophore

TABLE 8.11
Absorption Spectra in Neutral and Acid Solutions, and Basicity of Some Chalcones (46)

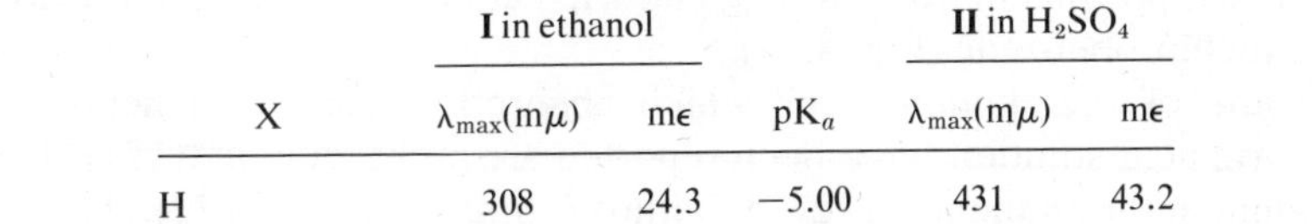

I $\xrightarrow{H^+}$ II

X	I in ethanol			II in H_2SO_4	
	λ_{max}(mμ)	mε	pK$_a$	λ_{max}(mμ)	mε
H	308	24.3	−5.00	431	43.2
4-CH_3	322	25.0	−4.79	439	
4-OCH_3	341	24.0	−4.25	470	—
2,4,6-$(OCH_3)_3$	368	22.3	−2.75	480	—

have brilliant colors in strong sulfuric acid—for example, 2-arylidene-1-indanones and 2-arylidenebenzosuberones (167). This would be expected of proton addition to the carbonyl oxygen.

Chalcones that contain an unsubstituted or alkylated amino group are protonated on the amino nitrogen atom (168):

$$\underset{\substack{\mathbf{94} \\ \lambda_{max}\ 419\ m\mu,\ m\epsilon\ 33.0}}{4\text{-}(CH_3)_2NC_6H_4CH{=}CHCOC_6H_5} \xrightarrow{H^+} \underset{\substack{\mathbf{95} \\ \lambda_{max}\ 294\ m\mu,\ m\epsilon\ 27.0}}{4\text{-}(CH_3)_2H\overset{+}{N}C_6H_4CH{=}CHCOC_6H_5} \qquad (8.37)$$

6. *Ring Carbonyl Compounds*

Compounds of the ring carbonyl type—such as fluorenone, anthrone (45), flavones (39), 1,3,5,7-tetramethyl-6*H*-cyclohepta[*c*]thiophen-6-one (169), perinaphthenone, and benzanthrone show a red shift on protonation of the carbonyl oxygen in strong sulfuric acid solution. The basicity of these and many other monocarbonyl compounds has been reviewed (170).

7. *Quinones*

Quinones are weakly basic compounds that can form cationic salts with strong acids. In aqueous sulfuric acid the following pK_a values have been reported: 1,4-naphthoquinone, −7.2; 9,10-anthraquinone, −7.5; 5,12-naphthacenequinone, −8.4; 3,10-perylenequinone, −5.5; and anthanthrone, −7.9 (171).

Since the carbonyl oxygen is the proton acceptor in carbonyl compounds (excluding amines), and quinones have two carbonyl groups, the assumption in the literature has been that one or two protons can add to these molecules. For example, it would appear that naphthacenequinone can form a monocationic salt with a proton in concentrated sulfuric or with the strong Lewis acid, anhydrous aluminum chloride, in *o*-dichlorobenzene. The same quinone can also form a dicationic salt with two protons in 30% fuming sulfuric acid or with aluminum chloride in *o*-dichlorobenzene (Fig. 8.13).

Some of the quinones for which absorption spectra in neutral and sulfuric acid solution have been reported are anthraquinone (171), 1- and 2-aminoanthraquinones (172), dimethoxyanthraquinones (172), dibenzanthrone and some of its dialkoxy derivatives (172), indanthrone (172), $\Delta^{10,10'}$-bianthrone (172), helianthrone (173), and mesonaphthobianthrone (173). Protonation causes a strong red shift, as exemplified by dibenzanthrone (172):

(8.38)

96 $\xrightarrow{H^+}$ 97

96 in chlorobenzene

λ_{max}(mμ)	mϵ
380	7.7
558	34.8
607	55.7

97 in sulfuric acid

λ_{max}(mμ)	mϵ
335	11.0
570	30.0
760	26.0
845	32.0

Inner ring *o*-quinones give blue to green colors in concentrated sulfuric acid; for example, 9,10-phenanthraquinone, λ_{max} 630 mμ, mϵ 1.7; 5,6-chrysenequinone, λ_{max} 585 mμ, mϵ 12.5; and dibenz[*a*,*h*]anthra-5,6-quinone, λ_{max} 655 mμ, mϵ 10.5 (174). The structure of this chromogen is unknown.

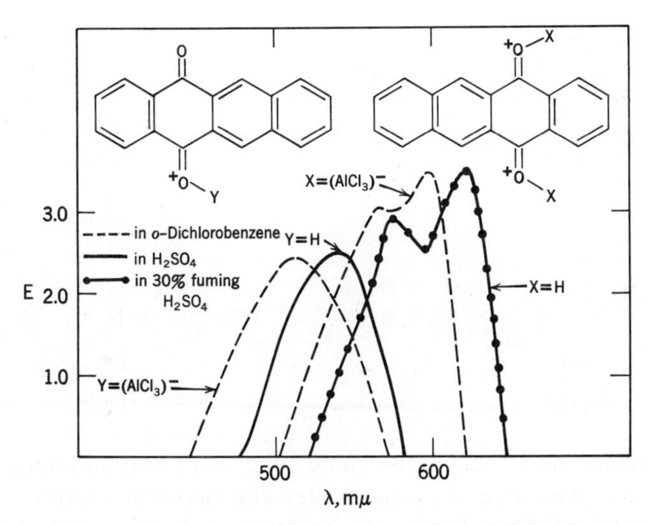

Fig. 8.13. Visible absorption spectra of the monocationic and dicationic salts of tetracenequinone.

A large number of polynuclear quinones have been shown by E. Clar in his publications to have brilliant colors in sulfuric acid.

The formation of a quinonic cationic resonance structure has been used in the trace analysis of 1,4-cyclohexanedione(175). The blue dicationic salt of pentacenequinone is formed in sulfuric acid solution through the reaction of *o*-phthalaldehyde with 1,4-cyclohexanedione.

In the determination of sugars with anthrone in sulfuric acid chromogens containing the chromophore **98** have been isolated(176). The number of protons in the molecule is not known.

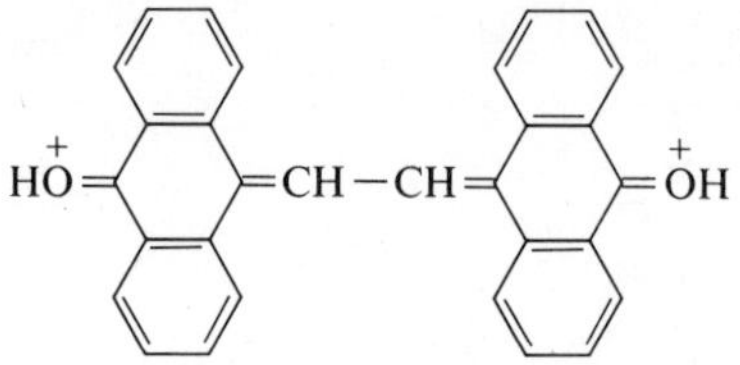

$\overset{+}{H}O=$ ... $=CH-CH=$ ... $=\overset{+}{O}H$

98

λ_{max} 625 mμ, mϵ 39.8, in H_2SO_4

8. Dye Carbonyl Compounds

Dyes that contain a carbonyl oxygen atom as the negative resonance terminal usually protonate on the oxygen, as seen in equation 8.39(177). Notable is the addition of the second proton to the positive resonance node instead of to the amino nitrogen.

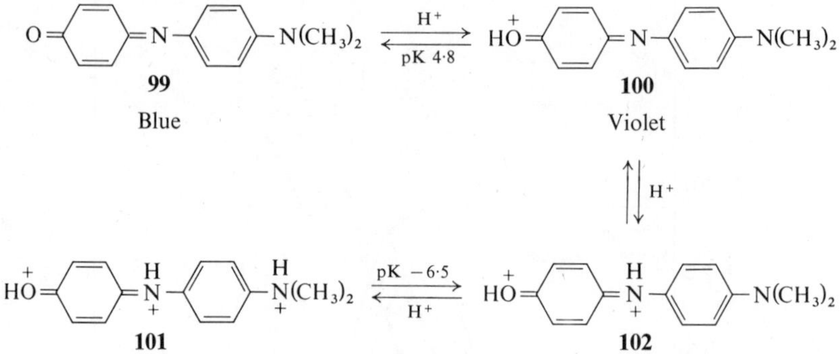

$O=$ ... $=N-$ ... $-N(CH_3)_2$ $\underset{pK\ 4{\cdot}8}{\overset{H^+}{\rightleftarrows}}$ $H\overset{+}{O}=$ ... $=N-$ ... $-N(CH_3)_2$

99 Blue   **100** Violet

$\uparrow\downarrow H^+$

$H\overset{+}{O}=$ ... $=\overset{H}{\underset{+}{N}}-$ ... $-\overset{H}{\underset{+}{N}}(CH_3)_2$ $\underset{H^+}{\overset{pK\ -6{\cdot}5}{\rightleftarrows}}$ $H\overset{+}{O}=$ ... $=\overset{H}{\underset{+}{N}}-$ ... $-N(CH_3)_2$

101 Yellow   **102** Violet   (8.39)

Indophenol anion shows the same type of proton addition, although a blue dicationic dye is formed after the addition of three protons. Since neutral indophenol and its monocationic salt are orange and the blue dication is only obtained in strong acid solution, most methods of analysis involving the formation of indophenol take place in alkaline solution, where the anion predominates – for example, for phenols (178), 4-hydroxyacetanilide (179), urea (180), and other ammonia precursors (181, 182).

Indigo (λ_{max} 610 mμ) forms a cationic salt (λ_{max} 680 mμ) by protonation on the carbonyl oxygen (183). Diprotonation (on the two carbonyl oxygens) shifts the wavelength maximum to 650 mμ. *N,N'*-Dimethylindigo shows similar wavelength shifts.

In a dye with a Z_z-structure protonation at the negative resonance terminal – that is, the carbonyl oxygen – will shift the long-wavelength maximum to the violet; an example is the following:

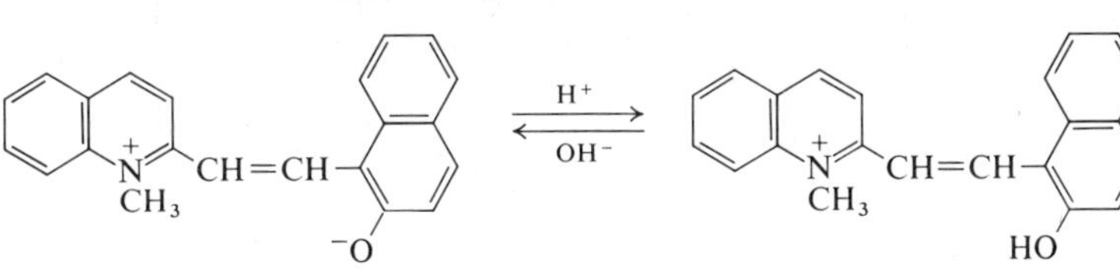

$$\mathbf{103}\ \underset{OH^-}{\overset{H^+}{\rightleftharpoons}}\ \mathbf{104} \qquad (8.40)$$

No.	Solvent	λ_{max}(mμ)	mϵ
103	EtOH–1% tetraethylammonium hydroxide	597	56.8
104	EtOH–1% HCl	460	23.2

Many other examples of this type have been reported (184,185). More examples are available in Section I.B of Chapter 3.

Somewhat similar examples can be found in the spiro derivatives (186,187) that have a ring broken in acid solution to form a highly colored cationic resonance structure (Fig. 8.14).

C. Organic Nitrogen Oxides

N,N-Dimethyl-*p*-nitrosoaniline is a green compound λ_{max} 421, 670 mμ, mϵ 30.5, 0.063, respectively, in ethanol (188)] that forms a colorless solution in alcoholic hydrochloric acid with a band at 357 mμ, mϵ 19.6 (147), and no sign of a $n \rightarrow \pi^*$ band in the 600 to 850 mμ region (189, 190). Consequently protonation takes place on the oxygen atom. In this compound we have the unexplained phenomenon of protonation at the negative resonance terminal of a Z_n-structure resulting in a violet shift of the long-wavelength $\pi \rightarrow \pi^*$ band.

In the azoxybenzenes, which are much weaker bases than the analogous azobenzenes, protonation also takes place at the oxygen atom (50).

The absorption spectra of 2- and 4-nitrodiphenylsulfide in sulfuric acid have long-wavelength bands at 611 and 480 mμ, respectively (191). The

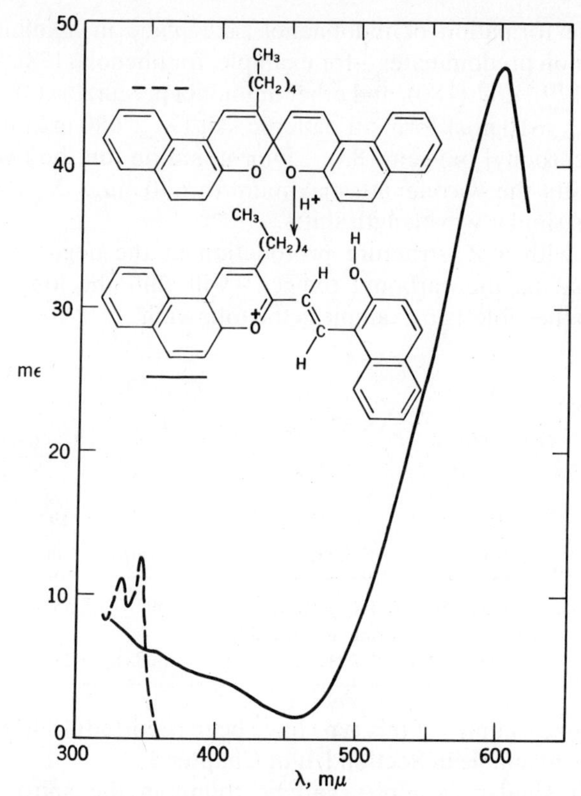

Fig. 8.14. Absorption spectra of the chromogen obtained from the analysis of 2-octanone with 2-hydroxy-1-naphthaldehyde. In 95% ethanol containing 1% dioxane (---) and in acidic ethanol (——).

long wavelength band, as in the 2-derivative, is probably derived from cationic resonance structures, such as **105**.

$S-CH_3$ $\longleftrightarrow$ $\overset{+}{S}-CH_3$

$\overset{+}{N}-OH$ $\overset{+}{N}-OH$

O O

105

The same explanation can be given for the spectra of *p*-nitroanisole (yellow, λ_{max} 419 mμ) and *p*-nitrothioanisole (cherry red, λ_{max} 522 mμ). The latter compound in *o*-dichlorobenzene gives an orange-red color (λ_{max} 500 mμ) with anhydrous aluminum chloride. The data on a representative group of these compounds are presented in Table 8.12.

TABLE 8.12
The Visible Absorption Spectra of Some Nitrophenyl Sulfides and Selenides in Concentrated Sulfuric Acid

Compound	λ_{max} (mμ)	mϵ	Ref.
4-Nitrodiphenylsulfide	480	20.0	191
4′-Phenoxy-4-nitrodiphenylsulfide	520	49.0	192
4′-Phenylthio-4-nitrodiphenylsulfide	505	28.8	192
4-Nitrodiphenylselenide	510	10.5	191
2-Nitrodiphenylsulfide	414	19.1	191
	575	5.4	
	611	5.6	
4′-Phenoxy-2-nitrodiphenylsulfide	457	75.9	192
	570	20.0	
	605	20.0	
4′-Phenylthio-2-nitrodiphenylsulfide	510	30.2	192
	575	15.5	
2-Nitrodiphenylselenide	420	15.5	191
	625	7.4	

3-Nitrodiphenylsulfide, 3-nitrodiphenylselenide(191), and the nitrodiphenyl ethers(192) do not show the halochromic effect.

D. Unsaturated Hydrocarbons

Most work with weak bases and unsaturated hydrocarbons dissolved in concentrated sulfuric acid has assumed that a cation is formed. Since sulfuric acid can act as an oxidizing and condensing agent, complex reactions can take place. Thus the absorption spectrum of a hydrocarbon in sulfuric acid may not be that of its cation.

1. Olefins

Since alkylcarbonium ions in fluorosulfuric acid–antimony pentafluoride solution at −60°C show no absorption maxima above 210 mμ(193), the absorption spectra of alcohols and olefins in sulfuric acid solutions(194) must be due to condensation or other products but not to the simple alkylcarbonium ions. However, in the polyolefin cations in which the positive charge is increasingly delocalized the absorption spectra shift from the ultraviolet to the visible with an increase in the length of the conjugation chain(195), as, for example, in **106**, **107**, and **108**.

$[(CH_3)_2C{=\!=}CH{=\!=}C(CH_3)_2]^+$
106
λ_{max} 305 mμ

$[(CH_3)_2C{=\!=}CH{=\!=}CH{=}CH{=\!=}C(CH_3)_2]^+$
107
λ_{max} 397 mμ

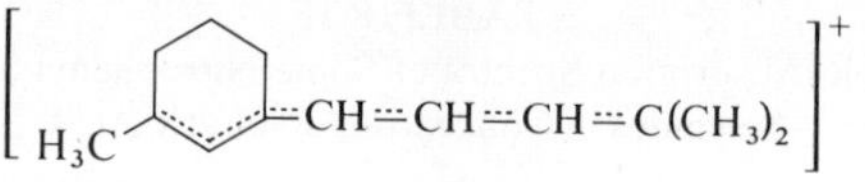

$$\left[H_3C\text{-}C_6H_8 \doteq CH \doteq CH \doteq CH \doteq C(CH_3)_2 \right]^+$$

108

λ_{max} 470 mμ

A further increase in the chain of conjugation of the polyenylic cations shifts the wavelength maximum to even longer wavelengths, as shown by the report that tetraenylic and pentaenylic cations absorb at approximately 550 and 625 mμ, respectively (196). The structure of the retinylic cation was thus shown to be pentaenylic. Somewhat similar cations are probably formed by the reaction of carotenoid 5,6- and 5,8-epoxides with hydrochloric acid (197) or mercuric chloride (198). In each case brilliant blue-green to red-violet colors are obtained.

These cations can be formed either from the polyene or the unsaturated alcohol, as in the following:

$$(CH_3)_2C{=}CHC(OH)(CH_3)_2 \;(\mathbf{109}) \xrightarrow{H^+} [(CH_3)_2C \doteq CH \doteq C(CH_3)_2]^+ \;(\mathbf{111}) \xleftarrow{H^+} (CH_3)_2C{=}CHC(CH_3){=}CH_2 \;(\mathbf{110}) \qquad (8.41)$$

The protonation of 13 polyenes with 2–15 conjugated double bonds has been discussed (165,199,200). Vitamin A acetate, a polyene with five conjugated double bonds, is protonated so that a mesomeric cation with the maximum number of double bonds is formed; that is, protonation takes place at a resonance terminal (201).

Steroids give rise to characteristic and frequently intense spectra in and near the visible region when dissolved in strong sulfuric (202–210), phosphoric (211), or perchloric acid (212). A very large number of steroids on thin-layer chromatograms give brilliant colors with sulfuric acid (213). Innumerable modifications of the strong acid determination of cholesterol are known (214). The green-blue color obtained with cholesterol in the acetic anhydride–sulfuric acid test has been investigated with 60 other steroids in terms of color intensity and rate of reaction (215; see also 216).

The mechanisms of the formation of these brilliant colors are obscure (217). However, many experimental facts have been assembled and should help in unraveling the structures of the chromogens. For example, in the acetic anhydride–sulfuric acid test a blue-green color is obtained only when the steroid has the complete 19-carbon atom steroid skeleton, a side chain of at least 8 carbon atoms, and a free or esterified hydroxy group or a thiol group in the 3-position(215). On the basis of the evidence it has been concluded that some of the chromogens are polyene cations(212,218,219).

2. *Aralkenes*

Since strong acid is usually necessary for protonation and color formation with aralkenes and since strong acids can cause oxidation, condensation, and polymerization, most explanations on color formation based on simple proton addition are tentative.

The aryl polyenes—for example, $C_6H_5(CH{=}CH)_nC_6H_5$, $n = 3–8$—dissolve in concentrated sulfuric acid, the color deepening with the length of the chain(220), as follows:

n	Color
3	Yellow orange
4	Red
5	Violet red
6	Blue
7	Blue green
8	Blue green

The monocation is believed to be the chromogen even with the immediate disulfonation that has been reported for 1,6-diphenylhexatriene(221).

When an ethylene group is attached to an aromatic ring, protonation usually takes place on the β-carbon of the ethylene group, and an arylcarbonium ion is formed. Closer investigation of these reactions has shown that in some cases the color is derived from a polymer. Thus styrene and β-phenylethanol are stated to give phenylmethylcarbonium ion, λ_{max} 300 and 430 mμ in sulfuric acid(222). However, the ion formed from styrene and perchloric acid under the cleanest conditions (low temperature, lowest concentration) is stated to be the dimer **112**(223).

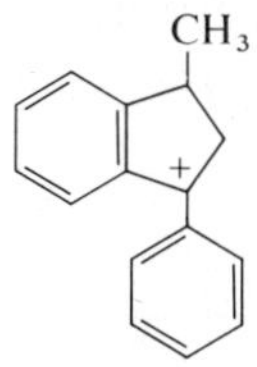

CH_3

+

112

λ_{max} 305, 415 mμ

At room temperature **112** and other diphenylcarbonium ions absorbing at 430 mμ are believed to be formed.

Other compounds of the styrene type that form cations in sulfuric acid supposedly by proton addition to the methylene carbon are 3,3-dimethyl-4-methylene-1,2,3,4-tetrahydro-2-quinolone, scarlet(224); 9-methyl-9-phenyl-10-methylene-9,10-dihydrophenanthrene, λ_{max} 268, 333, 549 mμ; mϵ 22.0, 92.0, and 6.0, respectively(225); 9-benzylidene thioxanthene, red(226); and 1,1-diphenylethylene, yellow(227), and its 4,4′-dimethoxy derivative, orange(227). The 4,4′-bisdimethylamino derivative of 1,1-diphenylethylene (**113**) protonates in acetic acid:

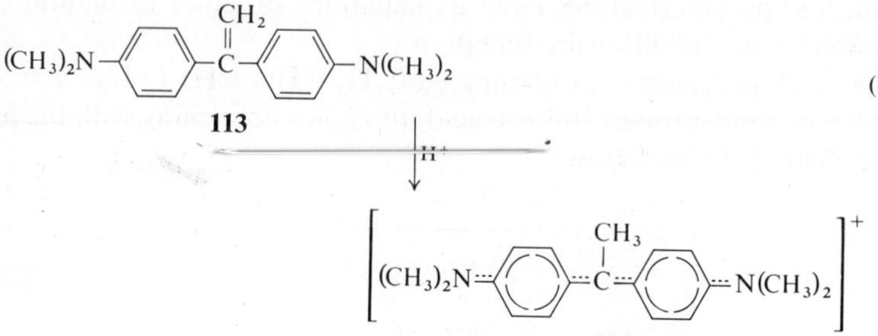

$(CH_3)_2N-C_6H_4-C(=CH_2)-C_6H_4-N(CH_3)_2$ (**113**) $\xrightarrow{H^+}$ $[(CH_3)_2N \cdots C_6H_4 \cdots C(CH_3) \cdots C_6H_4 \cdots N(CH_3)_2]^+$ (**114**, Blue) (8.42)

113

114
Blue

Iodine can also add to the methylene carbon to form the blue cation (228). A somewhat similar protonation is postulated for the more complicated indene derivative **115**(229):

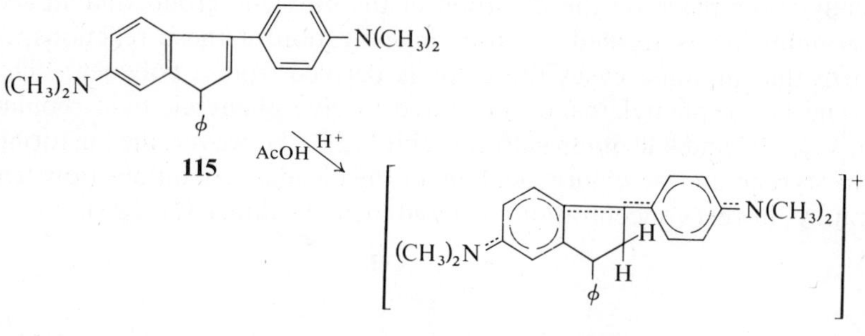

115 $\xrightarrow{AcOH, H^+}$ **116** (8.43)

$(CH_3)_2N$ — N$(CH_3)_2$ — ϕ

115

$[(CH_3)_2N$ ⋯ N$(CH_3)_2]^+$ — H — H — ϕ

116
Blue

On the other hand, in analogous compounds in which a carbonyl oxygen atom is the negative resonance terminal protonation will either take place at the positive resonance terminals when they are amino nitrogens or at the negative resonance terminal when the positive terminals are not amino nitrogens; examples are **117** and **118**(230).

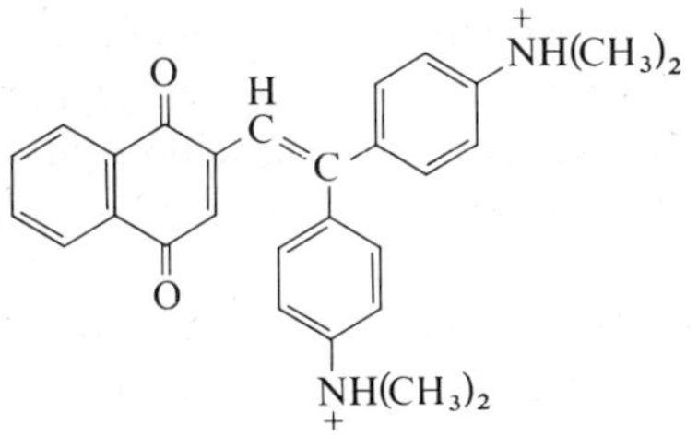

117

Yellow in 3 *N* HCl

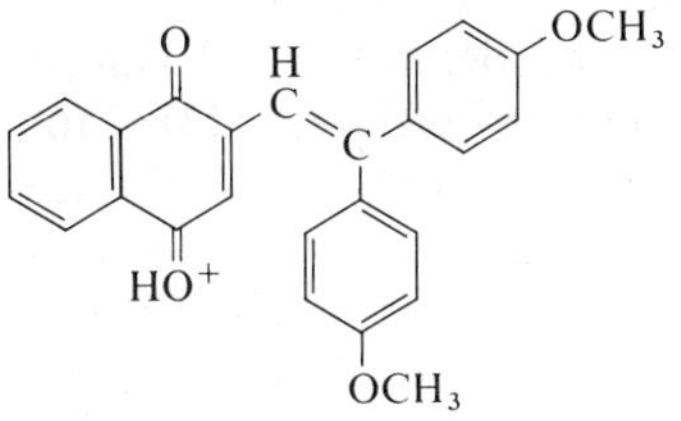

118

Blue in H_2SO_4

Much of the evidence for proton addition to a 1,1-diphenylethylene type of molecule is based on the very close spectral similarity of the monocationic salts of anthracene (**119**) and 1,1-diphenylethylene (**121**) in sulfuric acid (231).

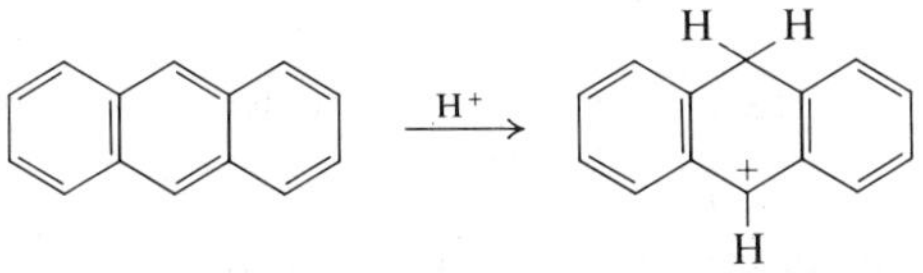

(8.44)

119

λ_{max} 252, 375 mμ
mϵ 200.0, 8.0

120

λ_{max} 315, 425 mμ
mϵ 12.0, 31.0

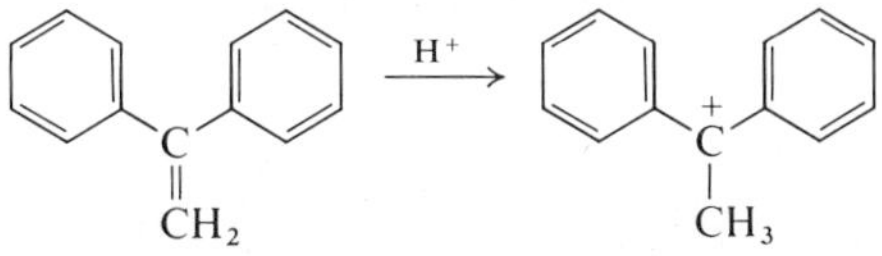

(8.45)

121

λ_{max} 250 mμ
mϵ 11.0

122

λ_{max} 315, 430 mμ
mϵ 8.0, 30.0

However, in the spectrum of 1-(2,6-dimethylphenyl)-1-phenylethylene in 98% sulfuric acid a new band is found at λ_{max} 570 mμ, mϵ 8.0, in place of the 430 mμ band (232). This chromogen is believed to be derived from a reaction between the nonprotonated olefin and the carbonium ion (233,234). In addition to the band near 450 mμ some ethylene derivatives dissolved in solutions in which protonated and nonprotonated compounds are present show a band at longer wavelength; for example, 1,1-diphenylethylene, 607 mμ; 1,1-di-*p*-tolylethylene, 637 mμ; 1,1-di-*p*-anisylethylene, 680 mμ; and tetraphenylethylene, 680 mμ (235). Evans et al. (235) explain these bands as being derived from a π-complex between the acid and the olefinic double bond.

Some pesticides have been determined through protonation of a derived aralkene derivative. Examples are 1,1-dichloro-2,2-bis(*p*-ethylphenyl)-ethane, or Perthane (**123**) (236), and 1,1-dichloro-2,2-bis(*p*-chlorophenyl)-ethane, or TDE or DDD(237). The former compound is dehydrochlorinated, dissolved in 96% sulfuric acid, and measured at 495 mμ. The latter compound is treated similarly and measured at 502 mμ. In both cases the diphenylcarbonium cation is formed, as shown for **123** below.

$HCCl_2$ C H; H_5C_2; C_2H_5 **123** $\xrightarrow{-HCl}$ HCCl ‖ C; H_5C_2; C_2H_5 **124** $\xrightarrow{H^+}$ H_2CCl C+; H_5C_2; C_2H_5 **125** (8.46)

In some of the dyes containing an alkene carbon atom that is a resonance node and not a terminal protonation can take place on this atom; the resulting methylene group will insulate the two parts of the molecule from each other, causing a violet shift in spectra. Examples are shown in equations 8.47(238), 8.48(239), and 8.49(240).

ϕ S S =CH **126** $\xrightarrow{H^+}$ ϕ S^+ S CH_2 **127** (8.47)

No.	Solvent	λ_{max} (mμ)	mε
126	Dioxane	354	21.4
127		345	5.1
128	Methanol	230	5.5
		260	2.0
		410	21.6
129		230	7.0
		260	5.0

H_3CN =CH−C(=O)−SCH_3 **128** $\xrightarrow{H^+}$ H_3CN^+ −CH_2−C(=O)−SCH_3 **129** (8.48)

$$\mathbf{130} \xrightarrow{H^+} \mathbf{131} \qquad (8.49)$$

No.	X	Color
130	O	Red
	NCH_3	Yellow red
131	O	Yellow
	NCH_3	Colorless

3. *Aromatic Hydrocarbons and Their Derivatives*

The relative basicities of some benzene derivatives and a few polycyclic hydrocarbons in anhydrous hydrogen fluoride at 0°C are listed in Table 8.13.

TABLE 8.13
Relative Basicities[a] of Some Aromatic Hydrocarbons (241)

Compound	Log K	Compound	Log K
Benzene	−9.4	Acenaphthene	+1.6
Toluene	−6.3	Hexaethylbenzene	+2.0
p-Xylene	−5.7	Pyrene	+2.1
Diphenyl	−5.5	Dibenz[*a*,*h*]anthracene	+2.2
Triphenylene	−4.6	Benz[*a*]anthracene	+2.3
Naphthalene	−4.0	Anthracene[d]	+3.8
Phenanthrene	—	Perylene	+4.4
m-Xylene	−3.2	2-Methylanthracene	+4.7
Fluorene	−2.4	9-Ethylanthracene	+5.4
Chrysene	−1.7	9-Methylanthracene	+5.7
1-Methylnaphthalene	−1.7	Naphthacene	+5.9
2-Methylnaphthalene	−1.4	9,10-Dimethylanthracene	+6.4
Mesitylene[b]	−0.4	Benzo[*a*]pyrene	+6.5
Hexamethylbenzene[c]	+1.4		

[a]Somewhat similar relative basicities reported for benzene derivatives by Condon (76).
[b]pK_a' is approximately −8.8 (242).
[c]pK_a is −5.0 (171) on Hammett's scale (see Table 8.4).
[d]Calculated from log $K = 3.8$ to be $pK_a = 3.8$ by Hammett's scale.

Phloroglucinol and its monomethyl, dimethyl, and trimethyl ethers (**132**) are believed to protonate at an aromatic carbon atom in strong perchloric acid (243), as shown in equation 8.50 (243–245).

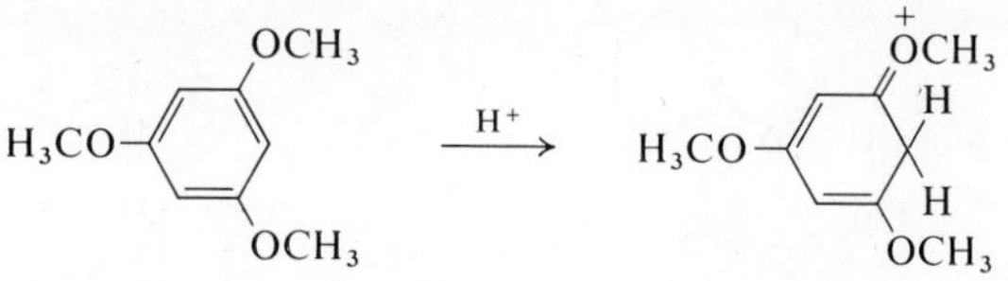

λ_{max} 266 mμ
mε 0.53

λ_{max} 250, 342 mμ
mε 20.0, 13.2

(8.50)

132 **133**

Other polyhydroxy derivatives give brilliant colors in sulfuric acid; this could be due to a protonation similar to that in equation 8.50; for example, 2,8-dihydroxy-3,7-dimethoxydibenzofuran has an intense blue color, and 2,2′, 5,5′-tetrahydroxy-4,4′-dimethoxy-6,6′-dipropylbiphenyl is emerald green(246).

Sesquifulvalene is unstable and exists only in highly dilute solution (247). However, many of its derivatives are much more stable. The tetraphenyl derivative (**134**) has been postulated as protonating on the five-membered ring(247).

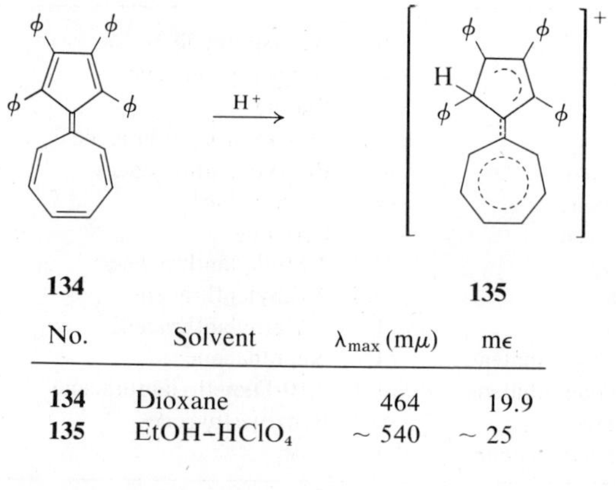

(8.51)

134 **135**

No.	Solvent	λ_{max} (mμ)	mε
134	Dioxane	464	19.9
135	EtOH–$HClO_4$	~ 540	~ 25

The 1- and 3-positions in azulene show considerable nucleophilic activity. The degree of basicity of the azulenes is comparable to that of the nitroanilines and can be explained by the formation of azulenium cations through protonation at the 1- or 3-position(248). Thus azulene, benz[*a*]azulene, benz[*f*]azulene, and cyclohepta[*bc*]acenaphthylene are half-neutralized in 51.0, 53.6, 61.7, and 44.2% sulfuric acid, respectively (249). Protonation takes place with a violet shift, as shown in Figure 8.15(250), as one would expect for a Z_z structure.

The pseudoazulenes (**136**) protonate in a similar fashion (251):

$$\mathbf{136} \xrightarrow{H^+} \mathbf{137} \qquad (8.52)$$

(ϕ, ϕ, ϕ, X; ϕ, H, ϕ, ϕ, +X)

136 Green — **137** Yellow

X = S or NMe

As can be seen from Table 8.13, benzene is very weakly basic, the basicity increasing with substitution of alkyl or other electron donor groups into the ring or through fusion of additional benzene rings. Thus mesitylene (**138**) in anhydrous hydrogen fluoride–boron trifluoride mixtures forms a protonated salt (252):

$$\mathbf{138} \xrightarrow{H^+} \mathbf{139} \qquad (8.53)$$

(CH_3, H_3C, CH_3; CH_3, H, H, +, H_3C, CH_3)

λ_{max} 256, 355 mε 8.8, 11.1

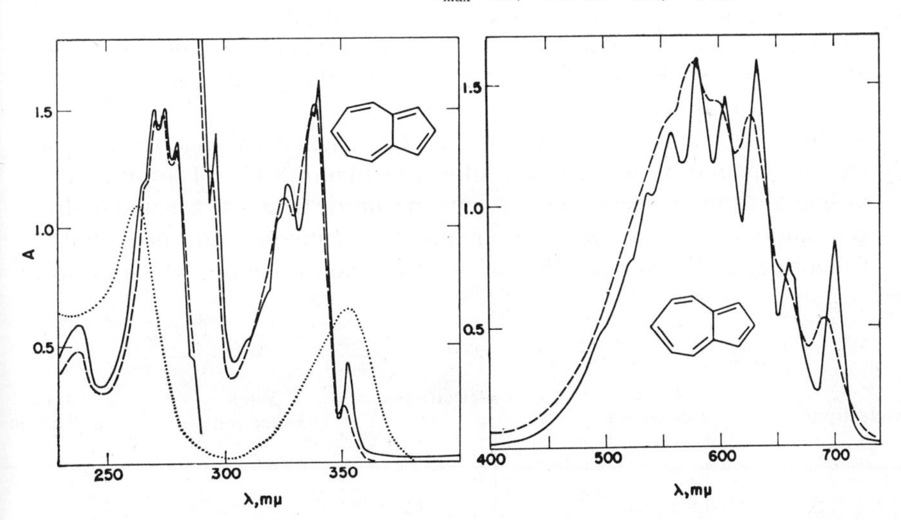

Fig. 8.15. Ultraviolet (left) and visible (right) absorption spectra of azulene in pentane (——) in 95% ethanol (---), and in concentrated hydrochloric acid (· · · · ·)(250). Concentration of azulene in pentane and in ethanol: 230–290 mμ, 3×10^{-5} *M*; 290–400 mμ, 4×10^{-4} *M*; 400–750 mμ, 5×10^{-3} *M*. Concentration in hydrochloric acid, 5×10^{-5} *M*.

Under the same conditions 1,2,3,5-tetramethylbenzene, pentamethylbenzene, naphthalene, anthracene, pyrene, and benzo[*a*]pyrene form cations that absorb at ever increasing wavelengths.

Aside from the more extensive changes taking place with time, polynuclear hydrocarbons in strong acid solution form two kinds of salts: a cation formed by proton addition and a free-radical cation formed by oxidation (electron extraction). The former type can exist as tautomers, each of which is spectrally similar to the structurally analogous monoanion formed by proton addition to the dianion(253); the free-radical cation is spectrally similar to the free-radical anion of the hydrocarbon formed by electron addition to the neutral hydrocarbon(254). The dianion and the dication should also be spectrally similar. The various reactions, with Ar representing the polynuclear aromatic hydrocarbons, are shown in Figure 8.16. These reactions are probably all reversible.

For some compounds the presence of oxygen in the strong acid solvent can result in the formation of the free-radical cation; in the absence of oxygen or an oxidizing agent a covalent complex is formed with a proton or a Lewis acid. Thus in oxygenated mixtures of trifluoroacetic acid, boron trifluoride, and water, perylene, naphthacene, pyrene, and anthracene are present in solution as free-radical cations whose spectra closely resemble those of the corresponding free-radical anions(254). Perylene, naphthacene, and anthracene in concentrated sulfuric acid and oxygenated anhydrous hydrogen fluoride show the presence of the free-radical cation; the solutions of anthracene also show the presence of the protonated compound. Benzo[*a*]pyrene forms only the protonated compound in these various acids(254).

The widely differing spectra of the free-radical cation and the protonated compound of perylene are shown in Figure 8.17. The ultraviolet visible absorption spectra of perylene in sulfuric acid, oxygenated hydrogen fluoride, and oxygenated mixtures of trifluoroacetic acid, boron trifluoride, and water are that of the free radical cation, which closely

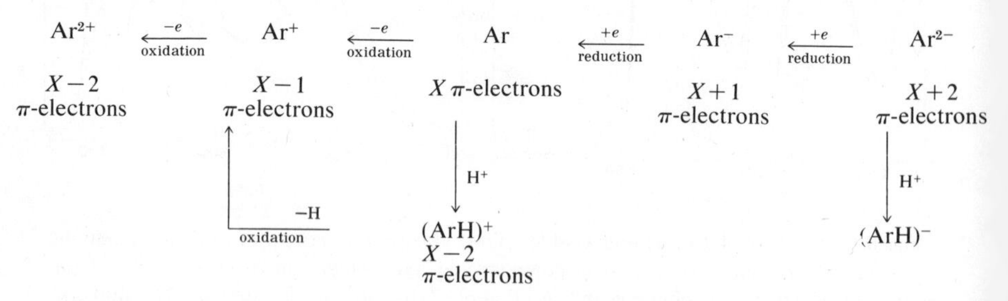

Fig. 8.16. Reactions involved in the formation of salts by polynuclear hydrocarbons (Ar).

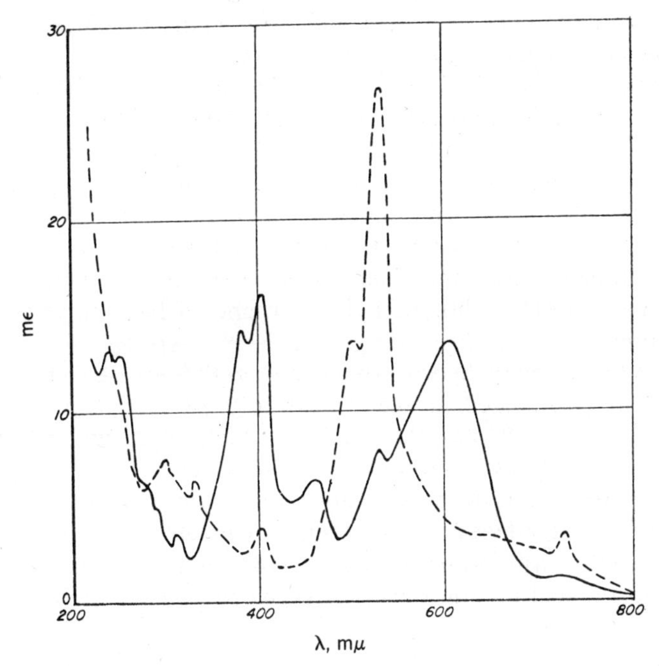

Fig. 8.17. Absorption spectra of perylene in hydrogen fluoride(252). Oxygen excluded (——); contact with oxygen (- - -).

resembles the spectrum of the sodium salt of perylene in tetrahydrofuran (254). Thus for perylene the following different ions are believed to be formed in strong acid solution:

(8.54)

H H + ⟷ H H + $\xrightarrow[\text{oxidation}]{-\text{H}}$ $^{+\cdot}$

140 **141**

Although polynuclear aromatic hydrocarbons form two classes of cation with strong Lewis acids—such as boron trifluoride, phosphorus pentafluoride, antimony pentachloride, and sulfur trioxide(255)— perylene has been reported to give a third type of spectrum with antimony pentachloride in solvents of low dielectric constant(256). The spectrum in solvents of high dielectric constant coincides with the spectrum of perylene in sulfuric acid; in solvents of high dielectric constant perylene reacts with antimony pentachloride to give an entirely changed spectrum, with broad bands near 500, 700, and 1200 mμ.

A detailed spectral investigation has been made of the protonated tautomers of benz[*a*]anthracene and its methylated derivatives (257–259). Two tautomers were formed in which protonation took place on either of the two meso carbon atoms (i.e., 7- or 12-). One would expect the 12-protonated tautomers of the benz[*a*]anthracenes to be closely similar spectrally to the 5-protonated tautomer of naphthacene, and the 7-protonated tautomer of the benz[*a*]anthracenes to be closely similar spectrally to 1-phenyl-1-(1′-naphthyl)ethylene or phenyl-1-naphthylcarbinol in acid solution.

Some of the aromatic hydrocarbons (e.g., zethrene) are so highly basic that they are protonated in acetic acid (260). Other members of this family (e.g., dibenzoheptazethrene) form stable, deep-violet salts with hydrochloric, perchloric, sulfuric, or phosphoric acids.

Some hydrocarbons form strongly basic cations through oxidation. For example, the tendency to form a centrosymmetric system is so strongly pronounced in tribenzoperinaphthene (**142**) that the compound is oxidized in acetic acid by atmospheric oxygen to a blue non-paramagnetic salt (261).

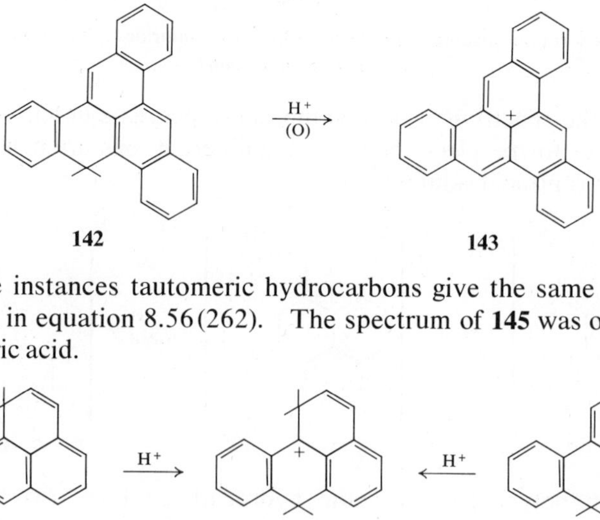

H^+ / (O)

142 **143** (8.55)

In some instances tautomeric hydrocarbons give the same cation, as postulated in equation 8.56 (262). The spectrum of **145** was obtained in 80% sulfuric acid.

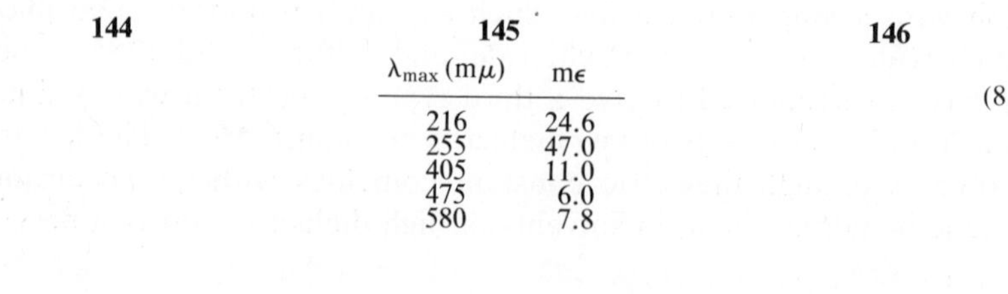

H^+ H^+

144 **145** **146**

λ_{max} (mμ)	mε
216	24.6
255	47.0
405	11.0
475	6.0
580	7.8

(8.56)

4. Carbinols

In any organic compound containing a —CROH— group adjacent to a conjugated system or, better yet, insulating two conjugated systems, solution in a strong acid results in the elimination of water and formation of a lower energy, longer chain cationic resonance system. Some of the older evidence for the formation of these types of cations has been summarized (263).

However, reemphasis is needed to stress that in strong acid solutions a variety of reactions can take place. Thus ascribing ultraviolet absorption to an aliphatic cation formed from solution of an aliphatic alcohol in sulfuric acid (264) is suspect, since such a molecule contains no conjugation and would not be expected to absorb in the ultraviolet. There is evidence to indicate that the spectra of the alkanols in sulfuric acid are actually those of the alkenyl cations (265).

Since many of these compounds are quickly sulfonated in concentrated sulfuric acid, their spectra are derived from the cation of the sulfonic acid. This effect has been demonstrated for 4-biphenylphenylmethanol, which absorbs at λ 508 mμ, mϵ 39.0, in sulfuric acid–acetic acid (5:95, v/v) but at λ474 mμ, mϵ 51.0, in concentrated sulfuric acid (266). The latter cation was shown to be that of a sulfonic acid quickly formed in solution.

The addition of tricyclopropylcarbinol (**147**) to 96% sulfuric acid produces the tricyclopropylmethyl cation (265):

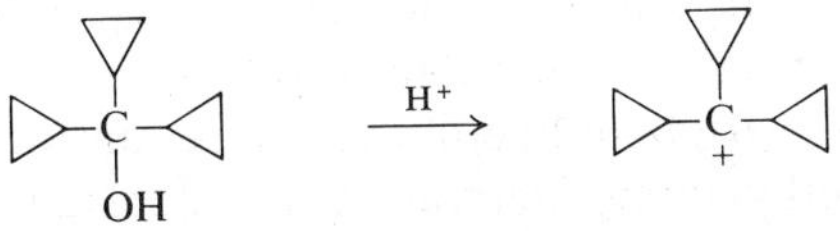

$$(C_3H_5)_3C\text{–}OH \xrightarrow{H^+} (C_3H_5)_3C^+ \qquad (8.57)$$

147 **148**

λ_{max} 270 mμ, mϵ 22.0

This ion is half-formed in 22% sulfuric acid, $pK_{R+} = -2.34$. This means that it is more basic than the triphenylcarbonium ion, which is half-formed at 50% sulfuric acid, $pK_{R+} = -6.63$ (37). As expected, the carbonium ion absorbs at longer wavelength than the protonated dicyclopropyl ketone (265):

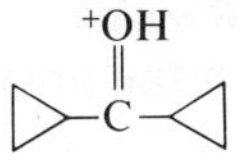

$$(C_3H_5)_2C{=}\overset{+}{O}H$$

λ_{max} 235 (mμ)

The absorption spectra of 31 monoaryl carbinols in concentrated sulfuric acid have been measured(267). For most of the compounds weak absorption was found near 400 mμ, mϵ < 1.0.

Although benzhydrol is unstable in sulfuric acid and undergoes rapid polymerization and sulfonation(268), many benzhydrols ionize in concentrated sulfuric acid to give stable carbonium ions(231, 269–271). For example, 2,2′,4,4′,6,6′-hexamethyl diphenylcarbinol in 100% sulfuric acid has a λ_{max} of 525 mμ, mϵ ~ 200(268).

The basicity constant, pK_{R+}, as defined by Deno et al.(272), has a wide range for many diphenylcarbinols(273, 274). Thus diphenylcarbinol has a pK_{R+}, of -13.3, whereas its much more basic 4,4′-dimethoxy derivative has a pK_{R+} of -5.71, and its much more weakly basic 3,3′-dichloro derivative has a pK_{R+} of -17.3(273). By comparison the 4-methoxydiphenylmethyl cation has a pK_{R+} of -7.9 and absorbs at 466 mμ, mϵ 33.0.

Diarylmethane cations, in which the aromatic groups are polynuclear, should show increased basicity and should absorb at longer wavelengths than does the diphenylmethane cation. This property is made use of in the characterization of polynuclear diaryl ketones(275). These compounds are treated with methyl magnesium bromide, followed by trifluoroacetic acid. 4-Benzoylbiphenyl gives a red carbonium ion absorbing at 510 mμ, whereas 1-naphthoylpyrene gives a blue-green carbonium ion absorbing at 680 mμ. The absorption spectra of approximately 30 arylphenylmethyl carbinols in methanol and in 96% sulfuric acid have been reported(276). The aryl substituent was either a polynuclear aromatic ring system or an oxygen or sulfur heterocyclic ring.

The basicity of the triphenylmethyl cation is considerably increased in the 4,4′,4″-trisdimethylamino derivative and considerably decreased in the 4,4′,4″-trinitro derivative ($pK_{R+} = -6.63$, $+9.36$, and -16.27, respectively)(273). Another substituent effect can be seen in the 4-methoxy, 4,4′-dimethoxy, and 4,4′,4″-trimethoxy derivatives of the triphenylmethyl cation ($pK_{R+} = -3.40$, -1.24, and $+0.82$, respectively). Thus a knowledge of the basicities of these types of compounds would be of value in the selective measurement at the proper acidity of any chromogen of the arylcarbonium type.

The tris(pentafluorophenyl)methyl cation has a λ_{max} of 500 mμ, mϵ 42.7, and is half-ionized in 93% sulfuric acid; hence its pK_{R+} is about -17.5, whereas that of triphenylcarbinol is -6.63(277).

The triphenylmethyl cation absorbs at 430 mμ, mϵ 37–38, in sulfuric acid or methylene chloride when the anion is of the hexachloroantimonate type(278).

When 4-dimethylaminotriphenylcarbinol and 2-dimethylamino-9,10-

diphenyl-9,10-dihydroxy-9,10-dihydroanthracene are dissolved in dilute acid, triphenylmethane dyes are formed that absorb at 410 and 487 mμ respectively (279). On the other hand, 2-dimethylaminotriphenylcarbinol (**150**) and 1-dimethylamino-9,10-diphenyl-9,10-dihydroxy-9,10-dihydroanthracene show no halochromic effect in dilute acid solution. This is believed to be due to an intramolecular hydrogen bond stabilizing the proton on the amino nitrogen.

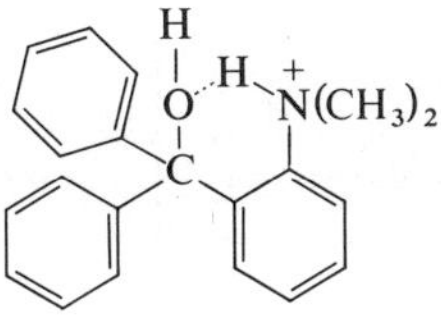

H
O···H—$\overset{+}{N}(CH_3)_2$
C

150

Another aminocarbinol that is reputed not to form a carbonium ion in acid solutions is 2,2′,2″-trisdimethylaminotriphenylcarbinol (280).

Whereas the mononegative and positive radical ions of an aromatic hydrocarbon have very similar electronic absorption spectra, a parallel similarity is not always obtained between the spectra of the anion and the cation of the arylmethyl ions (Table 8.14) (281). Since the transition energies of the arylmethyl anions are much more influenced by ion-pairing and solvation forces than are the aromatic hydrocarbon negative radical anions, the spectra of the arylmethyl cations and anions show a poorer similarity. Phenomena of this type are discussed more fully in Chapter 10, Section IIIB.

Compounds that contain a —CR(OH)— grouping as part of a conjugated ring readily form cations in strong acid solution. An exception would be found in compounds in which this grouping is at a bridgehead. For example, 9-hydroxytryptycene or 9-hydroxy-9,10-*o*-benzeno-9,10-

TABLE 8.14
Long-Wavelength Maxima of Diarylmethyl and Triarylmethyl Ions (281)

Methyl ion	Cation		Anion	
	λ_{max}	mϵ	λ_{max}	mϵ
Tris-*m*-biphenyl	500	10.1	485	45.5
Tris-*p*-biphenyl	510	105.0	580	72.5
Bis-*p*-biphenyl	555	71.4	545	156.0
Bis-*p*-biphenylmethyl	515	64.0	590	145.0

dihydroanthracene gives no bridgehead carbonium ion when it is dissolved in strong acid solution (282). In this hydrocarbon with a rigid propellerlike structure a positive charge on the bridgehead carbon would be out of conjugation with the benzene rings and thus very unstable.

Cycloheptatrienol and its polynuclear derivatives form tropylium cations in strong acid solution, as shown in the formation of the 10-chlorodibenzotropylium ion (spectra in strong acid) (283):

Cl H O-*t*-Bu **151** $\xrightarrow{H^+}$ [Cl]$^+$ **152** (8.58)

No	λ_{max}(mμ)	mϵ	λ_{max}(mμ)	mϵ
151	282	11.0	**152**—*continued*	
152	246	12.6	400	6.5
	255	12.3	513	3.0
	311	69.2	545	3.24
	384	9.1		

In the symmetrical crystal violet (**153a**), there is one visible region band derived from a transition involving resonance between two amino nitrogen terminals. In malachite green (**153b**) two bands are found in the visible region, the one at longer wavelength involving resonance between nitrogen terminals and the one at shorter wavelength involving resonance between a nitrogen terminal and the phenyl group.

$(CH_3)_2N$ $\overset{+}{N}(CH_3)_2$ R

153

	R	Solvent	λ_{max}(mμ)	mϵ	Ref.
a	$N(CH_3)_2$	Water	550*s*	67	284
			590	100	
b	H	AcOH	428	18.8	285
			620	99.1	

If in **153a** and **b** two benzene rings are connected to make a new five-membered ring, **154a**, **b**, and **c** are obtained(286). It would appear that in the fluorene analogs of crystal violet and malachite green the greater rigidity and the presence of the biphenyl linkage has shifted the spectra toward much longer wavelength and has complicated it.

In water the chloride of **154a** absorbs at 590 mμ, mϵ 42.2. This is probably due to polymerization of the dye in aqueous solution. Crystal violet (**153a**) shows a similar phenomenon in water (λ_{max} 550 mμ, mϵ 63.1)(287).

All these various dyes are formed by dissolving the appropriate carbinol in acid solution. However, 3,6-bisdimethylaminofluoren-9-ol hydrochloride (**155**) closely resembles 3,6-bisdimethylaminofluorene hydrochloride spectrally(286), but is widely different from Michler's hydrol blue (**156**)(288).

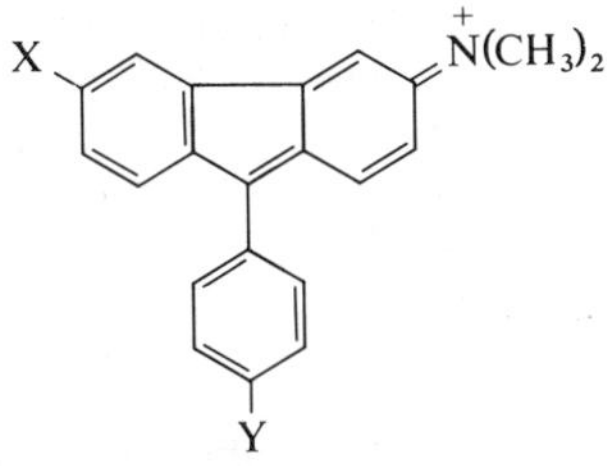

X, $\overset{+}{N}(CH_3)_2$, Y

154

X	Y	λ_{max} (mμ)	mϵ
a $N(CH_3)_2$	$N(CH_3)_2$	470,647,850	~7,75,15
b $N(CH_3)_2$	H	480,505,850,955	~21,22,15,19
c H	$N(CH_3)_2$	470,530,570,728,788	~8,9,10,33,36

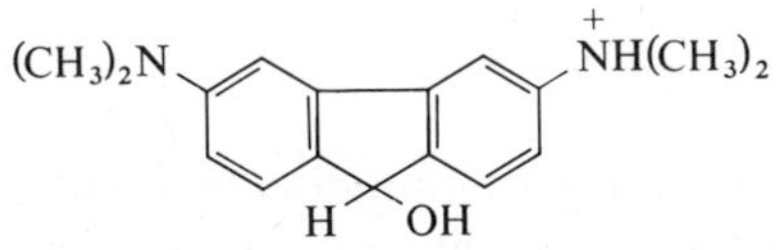

$(CH_3)_2N$, $\overset{+}{N}H(CH_3)_2$, H, OH

155

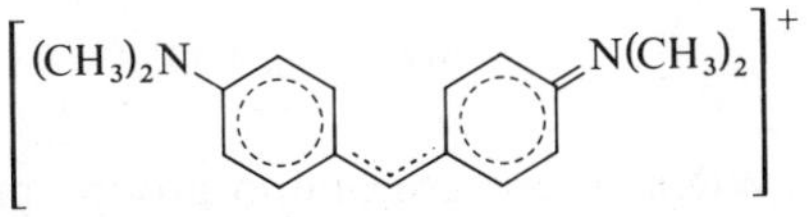

$[(CH_3)_2N \ldots N(CH_3)_2]^+$

156

No.	Solvent	λ_{max} (mμ)	mϵ
155	Ethanol	255	40
		263	39
		350	6
156	Nitromethane	610	131

Heterocyclic compounds containing a ring —CR(OH)— group in potential conjugation with the hetero atom form fairly basic carbonium ions in acid solution. Thus the cation (λ_{max} 370, 438 mμ; mϵ 34.7, 3.39, respectively) formed from 9-isopropyl-9-xanthenol, has a pK_{R+} of −0.84 (289). The strong red shift on cation formation is shown in equation 8.59(290).

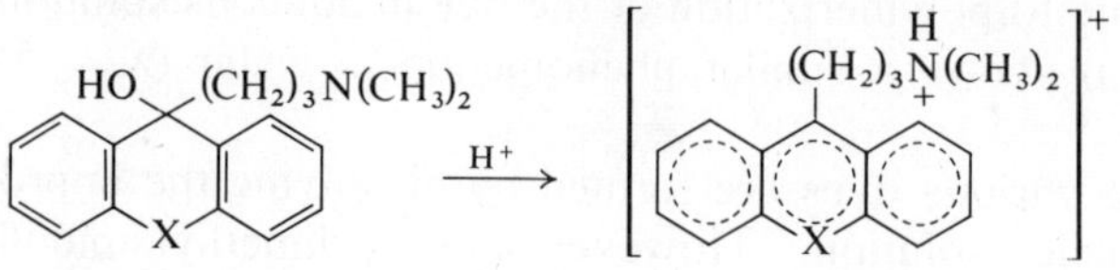

HO (CH$_2$)$_3$N(CH$_3$)$_2$ —H$^+$→ [H (CH$_2$)$_3$N$^+$(CH$_3$)$_2$]$^+$ (8.59)

157 **158**

No.	X	λ_{max} (mμ)	mϵ
157	O	280–290 (broad)	2.5–3.5
	S	268–271	12.2–14.5
158	O	390	17.6
		480	4.8
	S	395	9.7
		520	3.65

The long wavelength maxima of 19 dibenzoxanthylium salts range from the low of **159** to the high of **160** (spectra in glacial acetic acid)(290).

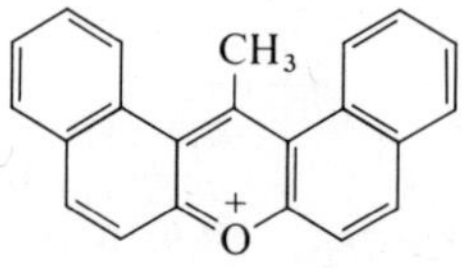

CH$_3$ O$^+$ O$^+$

159 **160**

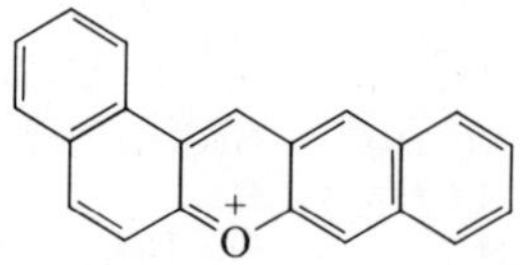

λ_{max} 464 mμ, mϵ 1.29 λ_{max} 580 mμ, mϵ 4.17

Where there are amino groups present in a 9-xanthenol compound, a series of spectral and structural changes take place with change in solvent and pH, as shown for rhodamine B in Figure 8.18(291).

Dicarbonium ions have also been reported. Bisxanthylium (**161a**) (292–296) and acridinium (**161b**)(297) salts and certain of their nonbenzo analogs(298, 299) are known. Substituted tetraarylethylene dications [**162a**(300–303) and **162b**(300, 301, 304)] have been reported.

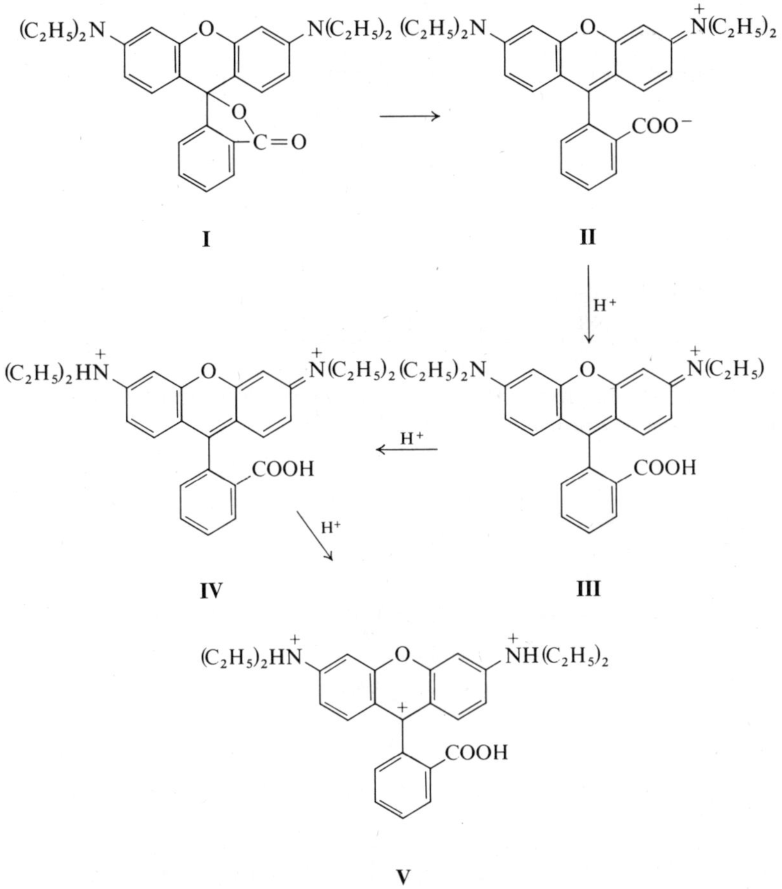

$(C_2H_5)_2N$ O $N(C_2H_5)_2$ $(C_2H_5)_2N$ O $\overset{+}{N}(C_2H_5)_2$

O

C=O

COO^-

I

II

H^+

$(C_2H_5)_2H\overset{+}{N}$ O $\overset{+}{N}(C_2H_5)_2$ $(C_2H_5)_2N$ O $\overset{+}{N}(C_2H_5)$

H^+

COOH

COOH

H^+

IV

III

$(C_2H_5)_2H\overset{+}{N}$ O $\overset{+}{N}H(C_2H_5)_2$

+

COOH

V

No.	Solvent or pH	λ_{max} (mμ)	mϵ	Fluorescence
I	Benzene	316	18	Weak blue
II	> 3	553	110	Strong yellow
III	1 to 3	556	110	Strong yellow
IV	0 to −1	494	15	
V	< −1	366	36	

Fig. 8.18. Spectral and structural changes in rhodamine B with change in solvent and pH (291).

161

a: X = O or S
b: X = NR

162

	R	λ_{max}	mϵ
a	OCH_3	455	50
		530	92(305)
b	$N(CH_3)_2$	860	

The three tetraphenylxylylene glycols have been reported to form dicarbonium ions in sulfuric acid(305):

$$(C_6H_5)_2\underset{OH}{C}-C_6H_4-\underset{OH}{C}(C_6H_5)_2 \longrightarrow (C_6H_5)_2\overset{+}{C}-C_6H_4-\overset{+}{C}(C_6H_5)_2 \quad (8.60)$$

163 **164**

	λ_{max}	mϵ
para	455	59.0
meta	419	73.0
	440	70.0
ortho	373	30.5
	455	44.0

The α,α'-diphenyl-*m*-xylylene glycol dication (λ_{max} 447 mμ, mϵ 35.0) and the α, α'-diphenyl-*p*-xylylene glycol dication (λ_{max} 461 mμ, mϵ 48.0) have been reported(305).

References

1. A. Burawoy, *J. Chem. Soc.*, **64**, 1635 (1931).
2. E. Sawicki, B. Chastain, H. Bryant, and A. Carr, *J. Org. Chem.*, **22**, 625 (1957).
3. H. J. Ackermann, L. J. Carbonne, and E. J. Kuchar, *J. Agr. Food Chem.*, **11**, 297 (1963).
4. R. Andrisano and G. Pappalardo, *Gazz. Chim. Ital.*, **85**, 1430 (1955).
5. G. A. Adams and A. E. Castagne, *Can. J. Res.*, **26B**, 314 (1948).
6. A. M. Agree and R. L. Meeker, *Talanta*, **13**, 1151 (1966).
7. C. Cessi and F. Serafini-Cessi, *Biochem. J.*, **88**, 132 (1963).
8. C. A. Anderson and J. M. Adams, *J. Agr. Food Chem.*, **11**, 474 (1963).
9. American Petroleum Institute Research Project No. 44, II, 593 (1955).
10. J. M. Bakes and P. G. Jeffery, *Talanta*, **8**, 641 (1961).

11. L. J. Saidel, A. R. Goldfarb, and S. Waldman, *J. Biol. Chem.*, **197**, 285 (1952).
12. I. Bergman and R. Loxley, *Anal. Chem.*, **35**, 1961 (1963).
13. H. L. Herzog et al., *J. Am. Chem. Soc.*, **77**, 4781 (1955).
14. J. Bartos, *Ann. Pharm. Franc.*, **20**, 650 (1962).
15. F. Klages, K. Bott, and P. Hegenberg, *Angew. Chem.*, **1**, 563 (1962).
16. R. J. Gillespie in *Friedel–Crafts and Related Reactions*, Vol. I, G. A. Olah, Ed., Interscience, New York, 1963.
17. R. E. Essery, *Anal. Chim. Acta*, **11**, 501 (1954).
18. E. Zimmerman and W. W. Brandt, *Talanta*, **1**, 374 (1958).
19. F. J. Langmyhr and O. B. Skaar, *Anal. Chim. Acta*, **23**, 28 (1960).
20. C. Arcus, M. Coombs, and J. Evans, *J. Chem. Soc.*, 1498 (1956).
21. R. J. Gillespie and K. C. Malhotra, *J. Chem. Soc.*, 1994 (1967).
22. R. J. Gillespie, *Accounts of Chemical Research*, **1**, 202 (1968).
23. C. Walling, *J. Am. Chem. Soc.*, **72**, 1165 (1950).
24. H. Benesi, *J. Am. Chem. Soc.*, **78**, 5490 (1956).
25. I. M. Klotz, H. A. Fiess, J. Y. C. Ho, and M. Mellody, *J. Am. Chem. Soc.*, **76**, 5136 (1954).
26. A. Albert and J. Phillips, *J. Chem. Soc.*, 1294 (1956).
27. A. Albert, R. Goldacre, and J. Phillips, *J. Chem. Soc.*, 2240 (1948).
28. A. I. Biggs and R. A. Robinson, *J. Chem. Soc.*, 388 (1961).
29. M. A. Paul, *J. Am. Chem. Soc.*, **76**, 3236 (1954).
30. D. Dolman and R. Stewart, *Can. J. Chem.*, **45**, 903 (1967).
31. N. F. Hall, *J. Am. Chem. Soc.*, **52**, 5115 (1930).
32. A. Albert and E. P. Serjeant, *Ionization Constants of Acids and Bases*, Wiley, New York, 1962.
33. H. Lunden, *Z. Phys. Chem.*, **54**, 532 (1906).
34. S. Yeh and H. H. Jaffé, *J. Am. Chem. Soc.*, **81**, 3274 (1959).
35. K. N. Bascombe and R. P. Bell, *J. Chem. Soc.*, 1096 (1959).
36. Y. Chiang, R. L. Hinman, S. Theodoropulos, and E. B. Whipple, *Tetrahedron*, **23**, 745 (1967).
37. N. C. Deno, J. J. Jaruzelski, and A. Schriesheim, *J. Am. Chem. Soc.*, **77**, 3044 (1955).
38. N. C. Deno, H. G. Richey, Jr., J. S. Liu, D. N. Lincoln, and J. O. Turner, *J. Am. Chem. Soc.*, **87**, 4533 (1965).
39. C. T. Davis and T. A. Geissman, *J. Am. Chem. Soc.*, **76**, 3507 (1954).
40. H. Hosoya and S. Nagakura, *Bull. Chem. Soc. Japan*, **39**, 1414 (1966).
41. J. T. Edward, H. S. Chang, K. Yates, and R. Stewart, *Can. J. Chem.*, **38**, 1518 (1960).
42. E. M. Arnett and C. Y. Wu, *J. Am. Chem. Soc.*, **84**, 1684 (1962).
43. W. F. Little, R. A. Berry, and P. Kannan, *J. Am. Chem. Soc.*, **84**, 2525 (1962).
44. A. J. Kresge and L. E. Hakka, *J. Am. Chem. Soc.*, **88**, 3868 (1966).
45. R. Stewart, M. R. Granger, R. B. Moodie, and L. J. Muenster, *Can. J. Chem.*, **41**, 1065 (1963).
46. D. S. Noyce and M. J. Jorgenson, *J. Am. Chem. Soc.*, **84**, 4312 (1962).
47. G. Culbertson and R. Petit, *J. Am. Chem. Soc.*, **85**, 741 (1963).
48. K. Yates and B. F. Scott, *Can. J. Chem.*, **41**, 2320 (1963).
49. R. Stewart and K. Yates, *J. Am. Chem. Soc.*, **80**, 6355 (1958).
50. C. Hahn and H. H. Jaffé, *J. Am. Chem. Soc.*, **84**, 949 (1962).
51. H. Campbell and J. T. Edward, *Can. J. Chem.*, **38**, 2109 (1960).
52. J. C. D. Brand, *J. Chem. Soc.*, 997 (1950).
53. W. Gordy and S. Stanford, *J. Chem. Phys.*, **8**, 170 (1940).
54. H. Lemaire and H. J. Lucas, *J. Am. Chem. Soc.*, **73**, 5198 (1951).

55. S. Nagakura, A. Minegishi, and K. Stanfield, *J. Am. Chem. Soc.*, **73**, 5198 (1951).
56. U. L. Haldna and V. A. Palm, *Dokl. Akad. Nauk SSSR*, **135**, 667 (1960).
57. V. A. Palm, U. L. Haldna, and A. J. Talvik, in S. Patai, Ed., *The Chemistry of the Carbonyl Group*, Interscience, New York, 1966.
58. M. A. Paul and F. A. Long, *Chem. Rev.* **57**, 1 (1957).
59. N. Hall and M. Sprinkle, *J. Am. Chem. Soc.*, **54**, 3472 (1932).
60. R. Adams and J. Mahan, *J. Am. Chem. Soc.*, **64**, 2588 (1942).
61. N. Leonard and V. Gash, *J. Am. Chem. Soc.*, **76**, 2781 (1954).
62. J. Johnson, M. Herr, J. Babcock, A. Fonken, J. Stafford, and F. Heyl, *J. Am. Chem. Soc.*, **78**, 430 (1956).
63. J. Elguero, R. Jacquier, and G. Tarrago, *Tetrahedron Letters*, 4719 (1965).
64. G. Opitz, H. Hellmann, and H. Schubert, *Ann.*, **623**, 112 (1959).
65. L. Dede and A. Rosenberg, *Ber.*, **67**, 147 (1934).
66. E. A. Steck and G. W. Ewing, *J. Am. Chem. Soc.*, **70**, 3397 (1948).
67. R. Epsztein, *Mem. Services Chim. Etat* (*Paris*), **36**, 235 (1951).
68. R. N. Jones, *Chem. Rev.*, **41**, 353 (1947).
69. E. Sawicki, *J. Org. Chem.*, **18**, 1492 (1953).
70. E. Sawicki, *J. Am. Chem. Soc.*, **76**, 664 (1954).
71. E. Sawicki, *J. Org. Chem.*, **19**, 608 (1954).
72. E. Sawicki, *J. Am. Chem. Soc.*, **77**, 957 (1955).
73. E. Sawicki and F. E. Ray, *J. Am. Chem. Soc.*, **75**, 2519 (1953).
74. P. Vetesnik, J. Bielavsky, and M. Vecera, *Collection Czech. Chem. Commun.*, **33**, 1687 (1968).
75. S. Krogt and B. Wepster, *Rec. Trav. Chim.*, **74**, 1611 (1955).
76. F. Condon, *J. Am. Chem. Soc.*, **74**, 2528 (1952).
77. H. Kohler and G. Scheibe, *Z. Anorg. Chem.*, **285**, 221 (1956).
78. T. Yamaoka, H. Hosoya, and S. Nagakura, *Tetrahedron*, **24**, 6203 (1968).
79. D. Reid, W. Stafford, and J. Ward, *J. Chem. Soc.*, 1100 (1958).
80. A. Etienne and B. Rutimeyer, *Bull. Soc. Chim. France*, 1588 (1956).
81. W. Wilson and R. Woodger, *J. Chem. Soc.*, 2943 (1955).
82. D. Brown, E. Hoerger, and S. Mason, *J. Chem. Soc.*, 4035 (1955).
83. A. R. Osborn and K. Schofield, *J. Chem. Soc.*, 4191 (1956).
84. J. F. Corbett, P. F. Holt, A. N. Hughes, and M. Vickery, *J. Chem. Soc.*, 1812 (1962).
85. N. Turnbull, *J. Chem. Soc.*, 441 (1945).
86. D. Craig and L. Short, *J. Chem. Soc.*, 419 (1945).
87. J. Ashley, G. Buchanan, and E. Easson, *J. Chem. Soc.*, 60 (1947).
88. G. Leandri, A. Mangini, F. Montanari, and R. Passerini, *Gazz. Chim. Ital.*, **85**, 769 (1955).
89. C. H. Williams, *J. Chem. Soc.*, 2258 (1965).
90. D. Monte, A. Mangini, and R. Passerini, *Boll. Sci. Fac. Chim. Ind. Bologna*, **12**, 168 (1954).
91. E. Sawicki and A. Carr, *J. Org. Chem.*, **22**, 507 (1957).
92. A. Osborn and K. Schofield, *J. Chem. Soc.*, 4200 (1956).
93. W. L. F. Armarego, *J. Chem. Soc.*, 561 (1962).
94. R. M. Acheson, *Acridines*, Interscience, New York, 1956, p. 100.
95. D. P. Craig, *J. Chem. Soc.*, 534 (1946).
96. R. K. McLeod and T. I. Crowell, *J. Org. Chem.*, **26**, 1094 (1961).
97. R. L. Reeves and W. F. Smith, *J. Am. Chem. Soc.*, **85**, 724 (1963).
98. S. Hunig, J. Utermann, and G. Erlemann, *Chem. Ber.*, **88**, 708 (1955).
99. S. Hunig and J. Utermann, *Chem. Ber.*, **88**, 1201 (1955).

100. S. Hunig and J. Utermann, *Chem. Ber.*, **88**, 1485 (1955).
101. H. H. Jaffé and R. W. Gardner, *J. Am. Chem. Soc.*, **80**, 319 (1958).
102. H. Zollinger, *Azo and Diazo Chemistry*, Interscience, New York, 1961, pp. 334 and 335.
103. W. C. Davies and H. W. Addis, *J. Chem. Soc.*, 1622 (1937).
104. E. Sawicki, *J. Org. Chem.*, **22**, 915 (1957).
105. E. Sawicki, *J. Org. Chem.*, **22**, 1084 (1957).
106. G. E. Lewis, *Tetrahedron*, **10**, 129 (1960).
107. E. Sawicki, *J. Org. Chem.*, **22**, 621 (1957).
108. E. Sawicki, *J. Org. Chem.*, **22**, 365 (1957).
109. H. W. Hilton and G. K. Uyehara, *J. Agr. Food Chem.*, **14**, 90 (1966).
110. S. H. Yuen, *Analyst*, **91**, 811 (1966).
111. V. Stoicescu, H. Beral, and C. Ivan, *Pharm. Zentralhalle*, **104**, 776 (1965); through *Anal. Abstr.*, **14**, 2801 (1967).
112. G. Pitel and T. Luce, *Ann. Pharm. Franc*, **23**, 673 (1965).
113. J. Rieder, *Arzneimittel-Forsch.*, **15**, 1134 (1965).
114. T. R. Berg, J. Q. Penrod, and L. G. Blaylock, *J. Assoc. Off. Agr. Chemists*, **48**, 905 (1965).
115. Prophylactics in Animal Feeds Subcommittees, *Analyst*, **90**, 531 (1965).
116. F. Gerson and E. Heilbronner, *Helv. Chim. Acta*, **41**, 1444 (1958).
117. F. Gerson, J. Schulze, and E. Heilbronner, *Helv. Chim. Acta*, **41**, 1463 (1958).
118. F. Gerson, T. Gaumann, and E. Heilbronner, *Helv. Chim. Acta*, **41**, 1482 (1958).
119. F. Gerson and E. Heilbronner, *Helv. Chim. Acta*, **42**, 1877 (1959).
120. A. Morikofer and E. Heilbronner, *Helv. Chim. Acta*, **42**, 1909 (1959).
121. F. Gerson, J. Schulze, and E. Heilbronner, *Helv. Chim. Acta*, **43**, 517 (1960).
122. E. Sawicki, T. W. Stanley, and W. C. Elbert, *Microchem. J.*, **5**, 225 (1961).
123. E. Sawicki, T. W. Stanley, and W. C. Elbert, in N. Cheronis, Ed., *Microchemical Techniques*, Interscience, New York, 1962.
124. E. Sawicki, *J. Org. Chem.*, **23**, 532 (1958).
125. E. Sawicki, J. L. Noe, and F. T. Fox, *Talanta*, **8**, 257 (1961).
126. S. Hunig and G. Kaupp, *Ann.*, **700**, 65 (1966).
127. R. Andon, J. Cox, and E. Herington, *Trans. Faraday Soc.*, **50**, 918 (1954).
128. N. Ikekawa, M. Maruyama, and Y. Sato, *Pharm. Bull. Japan*, **2**, 209 (1954); through *Chem. Abstr.*, **50**, 994 (1956).
129. P. Krumholz, *J. Am. Chem. Soc.*, **73**, 3487 (1951).
130. E. Sawicki, T. W. Stanley, J. D. Pfaff, and W. C. Elbert, *Anal. Chim. Acta*, **31**, 359 (1964).
131. A. R. Katritzky and R. A. Y. Jones, *Proc. Chem. Soc.* (*London*), 313 (1960).
132. A. Mangini and R. Passerini, *Gazz. Chim. Ital.*, **84**, 36 (1954).
133. D. Brown, E. Hoerger, and S. Mason, *J. Chem. Soc.*, 211 (1955).
134. E. Spinner, *J. Chem. Soc.*, 1226 (1960).
135. G. Beaven, E. Holiday, and E. Johnson, *Spectrochim. Acta*, **4**, 338 (1951).
136. M. Robison and B. Robison, *J. Am. Chem. Soc.*, **77**, 6554 (1955).
137. E. Sawicki and A. Carr, *J. Org. Chem.*, **23**, 610 (1958).
138. A. Gray and F. G. Holliman, *Tetrahedron*, **18**, 1095 (1962).
139. Y. Rozum, *Sbornik Statei Obshch. Khim., Akad. Nauk SSSR*, **1**, 600 (1953); through *Chem. Abstr.*, **49**, 1065 (1955).
140. R. H. Altiparmakian and R. S. Braithwaite, *J. Chem. Soc.*, Sect. B, 1112 (1967).
141. J. T. Braunholtz and F. G. Mann, *J. Chem. Soc.*, 3368 (1958).
142. W. L. F. Armarego, in A. R. Katritzky, Ed., *Advances in Heterocyclic Chemistry*, Vol. I, Academic Press, New York, 1963, pp. 253–309.

143. R. J. Abraham, E. Bullock, and S. S. Mitra, *J. Am. Chem. Soc.*, **84**, 2534 (1962).
144. Y. Chiang and E. B. Whipple, *J. Am. Chem. Soc.*, **85**, 2763 (1963).
145. R. L. Hinman and E. B. Whipple, *J. Am. Chem. Soc.*, **84**, 2534 (1962).
146. M. J. Kamlet and J. C. Dacons, *J. Org. Chem.*, **26**, 220 (1961).
147. A. Burawoy, *J. Chem. Soc.*, 1177 (1939).
148. R. I. Zalewski and G. E. Dunn, *Can. J. Chem.*, **46**, 2469 (1968).
149. K. Yates, J. B. Stevens, and A. R. Katritzky, *Can. J. Chem.*, **42**, 1957 (1964).
150. B. Eistert, E. Merkel, and W. Reiss, *Ber.*, **87**, 1513 (1954).
151. R. Stewart and K. Yates, *J. Am. Chem. Soc.*, **82**, 4059 (1960).
152. H. Hosoya and S. Nagakura, *Spectrochim. Acta*, **17**, 324 (1961).
153. L. A. Wiles, *Chem. Rev.*, **56**, 346 (1956).
154. A. R. Katritzky and R. A. Y. Jones, *Chem. Ind. (London)*, 722 (1961).
155. M. J. Janssen, *Spectrochim. Acta*, **17**, 475 (1961).
156. W. Treibs and W. Schroth, *Ann.*, **642**, 82 (1961).
157. E. Spinner and A. Burawoy, *Spectrochim. Acta*, **17**, 558 (1961).
158. N. A. Valyashko and N. V. Valyashko, *Zh. Obshch. Khim.*, **26**, 146 (1956); through *Chem. Abstr.*, **50**, 9152 (1956).
159. E. K. von Gustorf, M. C. Henry, and P. V. Kennedy, *Angew. Chem., Intern. Ed.*, **6**, 627 (1967).
160. M. Meerwein, W. Florian, N. Schön, and G. Stopp, *Ann.*, **641**, 1 (1961).
161. L. A. Wiles and E. C. Baughan, *J. Chem. Soc.*, 933 (1953).
162. D. S. Noyce and P. A. Kittle, *J. Org. Chem.*, **30**, 1896 (1965).
163. D. S. Noyce and P. A. Kittle, *J. Org. Chem.*, **30**, 1899 (1965).
164. P. Lambert and R. Martin, *Bull. Soc. Chim. Belges*, **61**, 513 (1952).
165. A. Wassermann, *J. Chem. Soc.*, 1014, 3228 (1958).
166. A. Wassermann, *J. Chem. Soc.*, 983 (1959).
167. N. Buu-Hoi and N. Xuong, *J. Chem. Soc.*, 2225 (1952).
168. E. Katzenellenbogen and G. Branch, *J. Am. Chem. Soc.*, **69**, 1615 (1947).
169. M. Winn and F. G. Bordwell, *J. Org. Chem.*, **32**, 1610 (1967).
170. V. A. Palm, U. L. Haldna and A. J. Talvik, in S. Patai, Ed., *The Chemistry of the Carbonyl Group*, Interscience, 1966, pp. 421–459.
171. T. Handa, *Bull. Chem. Soc. Japan*, **28**, 483 (1955).
172. R. A. Durie and J. S. Shannon, *Australian J. Chem.*, **11**, 168, 189 (1958).
173. R. Hebert, M. Goren, and A. Veron, *J. Am. Chem. Soc.*, **74**, 5779 (1952).
174. E. Sawicki and W. C. Elbert, *Anal. Chim. Acta*, **22**, 448 (1960).
175. E. Sawicki and H. Johnson, *Anal. Chim. Acta*, **34**, 381 (1966).
176. T. Momose, Y. Ueda, K. Sawada, and A. Sugi, *Pharm. Bull. Tokyo*, **5**, 31 (1957).
177. G. Schwarzenbach, H. Mohler, and J. Sorge, *Helv. Chim. Acta*, **21**, 1636 (1938).
178. H. D. Gibbs, *J. Biol. Chem.*, **72**, 649 (1927).
179. T. Ninomiya, *J. Pharm. Soc. Japan*, **85**, 394 (1965).
180. J. K. Fawcett and J. E. Scott, *J. Clin. Pathol.*, **13**, 156 (1960).
181. G. H. Sloane-Stanley and G. R. N. Jones, *Biochem. J.*, **86**, 16P (1963).
182. F. Liemann, *Z. Ges. Exptl. Med.*, **138**, 191 (1964).
183. R. Pummerer, F. Meininger, G. Schrott, and H. Wagner, *Ann.*, **590**, 195 (1954).
184. R. Wizinger and H. Wenning, *Helv. Chim. Acta*, **23**, 247 (1940).
185. Y. Hirshberg, E. B. Knott, and E. Fischer, *J. Chem. Soc.*, 3313 (1955).
186. W. Dilthey et al., *J. Prakt. Chem.*, **114**, 174 (1926).
187. J. H. Day, *Chem. Rev.*, **63**, 65 (1963).
188. K. Nakamoto and R. E. Rundle, *J. Am. Chem. Soc.*, **78**, 1113 (1956).
189. D. D. MacNicol, A. L. Porte, and R. Wallace, *Nature*, **212**, 1572 (1966).

190. A. Schors, A. Kraaijeveld, and E. Havinga, *Rec. Trav. Chim.*, **74**, 1243 (1955).
191. L. Bartolotti and R. Passerini, *Ricerca Sci.*, **25**, 1095 (1955); through *Chem. Abstr.*, **50**, 224 (1956).
192. A. Compagnini and N. Marziano, *Ricerca Sci.*, **29**, 2339 (1959).
193. G. A. Olah, *Chem. Eng. News*, **45**, 77 (March 27, 1967).
194. J. Rosenbaum and M. C. R. Symons, *Proc. Chem. Soc.* (*London*), 92 (1959).
195. N. C. Deno, *Chem. Eng. News*, **42**, 88 (October 5, 1964).
196. P. E. Blatz and D. L. Pippert, *J. Am. Chem. Soc.*, **90**, 1296 (1968).
197. A. L. Curl and G. F. Bailey, *J. Agr. Food Chem.*, **9**, 403 (1961).
198. H. Y. Yamamoto, C. O. Chichester, and T. O. M. Nakayama, *Anal. Chem.*, **33**, 1792 (1961).
199. A. Wassermann, *J. Chem. Soc.*, 4329 (1954); 979, 983, 986 (1959).
200. A. Wassermann, *Trans. Faraday Soc.*, **53**, 1030 (1957); *Mol. Physics*, **2**, 226 (1959).
201. A. Wassermann, *J. Chem. Soc.*, 979 (1959).
202. L. L. Smith and S. Bernstein, in L. L. Engel, Ed., *Physical Properties of the Steroid Hormones*, Macmillan, New York, 1963, p. 321.
203. S. Bernstein and R. H. Lenhard, *J. Org. Chem.*, **18**, 1146, 1166 (1953); ibid., **19**, 1269 (1954); ibid., **25**, 1405 (1960).
204. A. Zaffaroni, *J. Am. Chem. Soc.*, **72**, 3828 (1950).
205. G. Diaz, A. Zaffaroni, G. Rosenkranz, and C. Djerassi, *J. Org. Chem.*, **17**, 747 (1952).
206. A. Zaffaroni and R. B. Burton, *J. Biol. Chem.*, **193**, 749 (1957).
207. S. Eriksson and J. Sjovall, *Arkiv Kemi*, **8**, 303, 311 (1955).
208. A. M. Bongiovani, *Anal. Biochem.*, **13**, 85 (1965).
209. H. Wilson, *Anal. Biochem.*, **1**, 402 (1960).
210. J. H. Linford and O. J. Fleming, *Can. J. Med. Sci.*, **31**, 182 (1953).
211. W. J. Nowaczynski and P. F. Steyermark, *Arch. Biochem.*, **58**, 453 (1955); *Can. J. Biochem. Physiol.*, **34**, 592 (1956).
212. W. Lange, R. Folzenhogen, and D. Kolp, *J. Am. Chem. Soc.*, **71**, 1733 (1949).
213. E. Heftmann, S. Ko, and R. D. Bennett, *J. Chromatog.*, **21**, 490 (1966).
214. H. Dam in R. P. Cook, Ed., *Cholesterol*, Academic Press, New York, 1958, p. 5.
215. R. P. Cook, *J. Chem. Soc.*, 373 (1961).
216. M. Pesez, *Bull. Soc. Chim. France*, 369 (1958).
217. P. Bladon in R. P. Cook, Ed., *Cholesterol*, Academic Press, New York, 1958, p. 84.
218. H. Kalant, *Biochem. J.*, **69**, 79 (1958).
219. C. Brieskorn and L. Capuano, *Ber.*, **86**, 866 (1953).
220. R. Kuhn and A. Winterstein, *Helv. Chim. Acta*, **11**, 87, 114 (1928).
221. J. A. Leisten and P. R. Walton, *Proc. Chem. Soc.* (*London*), 60 (1963).
222. D. O. Jordon and F. E. Treloar, *J. Chem. Soc.*, 729, 734, 737 (1961).
223. V. Bertoli and P. H. Plesch, *Chem. Commun.*, 625 (1966).
224. A. Searles and R. Kelly, *J. Am. Chem. Soc.*, **78**, 2242 (1956).
225. R. Shriner and L. Geipel, *J. Am. Chem. Soc.*, **79**, 227 (1957).
226. H. Decker and T. Fellenberg, *Ber.*, **38**, 2493 (1905).
227. R. Wizinger, *Organische Farbstoffe*, Dummlers Verlag, Bonn, 1933.
228. R. Wizinger and R. Gross, *Helv. Chim. Acta*, **35**, 411 (1952).
229. C. Dufraisse, A. Etienne, and B. Goffinet, *Compt. Rend.*, **235**, 1349 (1952).
230. M. Gates, *J. Am. Chem. Soc.*, **66**, 124 (1944).
231. V. Gold and F. L. Tye, *J. Chem. Soc.*, 2172 (1952).
232. O. Korver, J. V. Veenland, and T. J. De Boer, *Rec. Trav. Chim.*, **84**, 806 (1965).
233. H. P. Leftin, *J. Phys. Chem.*, **64**, 382, 1714 (1960); ibid., **66**, 1214, 1457 (1962).
234. J. J. Rooney and B. J. Hathaway, *J. Catalysis*, **3**, 447 (1964).

235. A. G. Evans, P. M. S. Jones, and J. H. Thomas, *J. Chem. Soc.*, 104 (1957).
236. C. F. Gordon, L. D. Haines, and I. Rosenthal, *J. Agr. Food Chem.*, **10**, 380 (1962).
237. I. Rosenthal, C. F. Gordon, and E. L. Stanley, *J. Agr. Food Chem.*, **7**, 486 (1959).
238. W. Kirmse and L. Horner, *Ann.*, **614**, 4 (1958).
239. A. R. Katritzky, *J. Chem. Soc.*, 2586 (1955).
240. R. Wizinger, *J. Prakt. Chem.*, **157**, 139 (1941).
241. E. L. Mackor, A. Hofstra, and J. H. Van Der Waals, *Trans. Faraday Soc.*, **54**, 66, 186 (1958).
242. M. Kilpatrick and H. Hyman, *J. Am. Chem. Soc.*, **80**, 77 (1958).
243. W. M. Schubert and R. H. Quacchia, *J. Am. Chem. Soc.*, **84**, 3778 (1962).
244. L. A. Wiles, *Chem. Rev.*, **56**, 329 (1956).
245. A. J. Kresge and Y. Chiang, *Proc. Chem. Soc.*, 81 (1961).
246. F. Dean, A. Osman, and A. Robertson, *J. Chem. Soc.*, 11 (1955).
247. H. Prinzbach, D. Seys, L. Knothe, and W. Faisst, *Ann.*, **698**, 57 (1966).
248. E. Heilbronner and M. Simonetta, *Helv. Chim. Acta*, **35**, 1049 (1952).
249. D. H. Reid, W. H. Stafford, and J. P. Ward, *J. Chem. Soc.*, 1193 (1955).
250. E. Sawicki, T. W. Stanley, and W. C. Elbert, *Anal. Chem.*, **33**, 1183 (1961).
251. G. V. Boyd, *J. Chem. Soc.*, 55 (1959).
252. G. Dallinga, E. L. Mackor, and A. A. Verrijn Stuart, *Mol. Physics*, **1**, 123 (1958).
253. N. H. Velthorst and G. J. Hoijtink, *J. Am. Chem. Soc.*, **87**, 4529 (1965).
254. W. I. Aalbersberg, G. J. Hoijtink, E. L. Mackor, and W. P. Weijland, *J. Chem. Soc.*, 3049 (1959).
255. W. I. Aalbersberg, G. J. Hoijtink, E. L. Mackor, and W. P. Weijland, *J. Chem. Soc.*, 3055 (1959).
256. H. Kuroda, T. Sakurai, and H. Akamatu, *Bull. Chem. Soc. Japan*, **39**, 1893 (1966).
257. E. L. Mackor, G. Dallinga, J. Kruizinga, and A. Hofstra, *Rec. Trav. Chim.*, **75**, 836 (1956).
258. A. A. Verrijn Stuart and E. L. Mackor, *J. Chem. Phys.*, **27**, 826 (1957).
259. C. MacLean, J. H. van der Waals, and E. L. Mackor, *Mol. Physics*, **1**, 247 (1958).
260. E. Clar, G. S. Fell, and M. H. Richmond, *Tetrahedron*, **9**, 96 (1960).
261. E. Clar and D. G. Stewart, *J. Chem. Soc.*, 23 (1958).
262. H. Dannenberg and H. Kessler, *Ann.*, **606**, 184 (1957).
263. H. Burton and P. F. G. Praill, *Quart. Rev.*, **6**, 302 (1952).
264. P. G. Farrell and S. F. Mason, *Proc. Chem. Soc.* (*London*), 958 (1959).
265. N. C. Deno, H. G. Richey, Jr., J. S. Liu, J. D. Hodge, J. J. Houser, and M. J. Wisotsky, *J. Am. Chem. Soc.*, **84**, 2016 (1962).
266. H. Hart and T. Sulzberg, *J. Org. Chem.*, **28**, 1159 (1963).
267. J. F. A. Williams, *Tetrahedron*, **18**, 1487 (1962).
268. H. A. Smith and B. B. Stewart, *J. Am. Chem. Soc.*, **79**, 3693 (1957).
269. C. M. Welch and H. A. Smith, *J. Am. Chem. Soc.*, **72**, 4748 (1950).
270. M. S. Newman and N. C. Deno, *J. Am. Chem. Soc.*, **73**, 3644 (1951).
271. H. A. Smith and R. G. Thompson, *J. Am. Chem. Soc.*, **77**, 1778 (1955).
272. N. C. Deno, J. J. Jaruzelski, and A. Schriesheim, *J. Org. Chem.*, **19**, 155 (1954).
273. N. C. Deno and A. Schriesheim, *J. Am. Chem. Soc.*, **77**, 3051 (1955).
274. N. C. Deno and W. L. Evans, *J. Am. Chem. Soc.*, **79**, 5804 (1957).
275. T. W. Stanley, *Chemist-Analyst*, **49**, 47 (1960).
276. V. Horak, C. Parkanyi, J. Pecka, and R. Zahradnik, *Collection Czech. Chem. Commun.*, **32**, 2272 (1967).
277. R. Filler, C. Wang, M. A. McKinney, and F. N. Miller, *J. Am. Chem. Soc.*, **89**, 1026 (1967).

278. B. A. Timimi, *Chem. Ind.* (*London*), 2148 (1967).
279. J. Robert, *Ann. Chim.* (*Paris*), **10**, 871 (1955); through *Chem. Abstr.*, **50**, 12948 (1956).
280. A. Baeyer, *Ann.*, **354**, 152 (1907).
281. R. Grinter and S. F. Mason, *Proc. Chem. Soc.* (*London*), 386 (1961).
282. P. D. Bartlett and F. D. Greene, *J. Am. Chem. Soc.*, **76**, 1088 (1954).
283. R. W. Murray, *Tetrahedron Letters*, 27 (1960).
284. R. Cigen, *Acta Chem. Scand.*, **12**, 1456 (1958).
285. V. V. Ghaisas, B. J. Kane, and F. F. Nord, *J. Org. Chem.*, **23**, 560 (1958).
286. A. Barker and C. C. Barker, *J. Chem. Soc.*, 1307 (1954).
287. L. Michaelis and S. Granick, *J. Am. Chem. Soc.*, **67**, 1212 (1945).
288. L. G. S. Brooker and R. H. Sprague, *J. Am. Chem. Soc.*, **63**, 3203 (1941).
289. N. C. Deno, P. T. Groves, and G. Saines, *J. Am. Chem. Soc.*, **81**, 5790 (1959).
290. M. Kamel and H. Shoeb, *Ber.*, **99**, 1822 (1966).
291. R. W. Ramette and E. B. Sandell, *J. Am. Chem. Soc.*, **78**, 4872 (1956).
292. A. Werner, *Ber.*, **34**, 3300 (1901).
293. A. Werner, *Ann.*, **322**, 296 (1902).
294. A. Schonberg and S. Nickel, *Ber.*, **67**, 1795 (1934).
295. A. Schonberg and W. Asker, *J. Chem. Soc.*, 272 (1942).
296. R. Wizinger and Y. Al-Atter, *Helv. Chim. Acta*, **30**, 189 (1947).
297. R. M. Acheson, *Acridines*, Interscience, New York, 1956, pp. 280–285.
298. F. Arndt, P. Nachtweg, and E. Scholz, *Ber.*, **57**, 1903 (1924).
299. F. Arndt and C. Lorenz, *Ber.*, **63**, 3121 (1930).
300. R. Wizinger and J. Fontaine, *Ber.*, **60B**, 1377 (1927).
301. R. E. Buckles and N. A. Meinhart, *J. Am. Chem. Soc.*, **74**, 1171 (1952).
302. R. E. Buckles and W. D. Womer, *J. Am. Chem. Soc.*, **80**, 5055 (1958).
303. R. E. Buckles, R. E. Erickson, J. D. Snyder, and W. B. Person, *J. Am. Chem. Soc.*, **82**, 2444 (1960).
304. D. H. Anderson, R. M. Elofson, H. S. Gutowsky, S. Levine, and R. B. Sandin, *J. Am. Chem., Soc.*, **83**, 3157 (1961).
305. H. Hart, T. Sulzberg, and R. R. Rafos, *J. Am. Chem. Soc.*, **85**, 1800 (1963).

CHAPTER 9

ABSORPTION SPECTRA OF CATIONIC RESONANCE STRUCTURES
II. WAVELENGTH AND INTENSITY DETERMINANTS

Cationic resonance structures absorb at longer wavelengths than analogous zwitterionic resonance structures except when the latter structure has a greater dipole moment in the ground state than in the excited state. In this type of molecule containing a Z_b or a Z_z resonance structure protonation at the negative resonance terminal causes a violet shift in spectra.

I. Solvent Effects

Some of the highly polar merocyanines exhibit two absorption bands in alcohols or aqueous alcohol, the long wavelength band being attributed to the dye itself and the shorter wavelength band to a quaternary cation formed between the solvent and solute(1, 2). Addition of hydrochloric acid enhances the shorter wavelength band. An example of the phenomenon is the following(2):

1 $\xrightarrow{H_2O}$ 2 (9.1)

Blue in benzene Yellow

Another area of interest where a changing composition of solvent can affect the wavelength position and intensity of a compound and hence its use as a standard in basicity determination is in the determination of the pK of a weak base(3). The use of the Hammett equation (equation 9.2) with spectral methods depends on two properties of the organic base B: (*a*) that B and its conjugate acid BH^+ have different wavelength maxima or markedly different absorptivities; (*b*) that the wavelength maxima and molar absorptivities of both B and BH^+ are independent of the changing concentration of sulfuric acid in the medium. Though the first of these requisites is generally true, the second is rarely completely correct.

Noyce and Jorgenson(3) have presented a review of methods of correcting this phenomenon. Chalcone and its derivatives are examples of this difficulty. Thus the position of the long-wavelength maximum of chalcone is at 415 mμ, mϵ 37.0, in 80% sulfuric acid ($H_0 - 7.00$) and at 425 mμ, mϵ 43.0, in 88% sulfuric acid ($H_0 - 8.00$). This, then, is the source of the difficulty in using chalcone as a Hammett indicator.

$$\mathrm{pK} = H_0 + \log \frac{C_{\mathrm{BH}^+}}{C_{\mathrm{B}}} \tag{9.2}$$

In a somewhat similar fashion the basicities of the 4-phenylazoanilines are beset by the problem of the tautomeric equilibrium

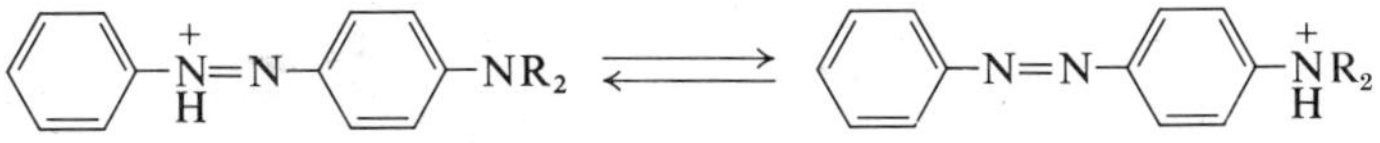

$$C_6H_5\text{-}\overset{+}{N}H\text{=N-}C_6H_4\text{-}NR_2 \rightleftharpoons C_6H_5\text{-N=N-}C_6H_4\text{-}\overset{+}{N}HR_2$$

shifting to the left with increased acidity(4). This shift strongly affects the spectra. Examination of the spectra of cationic resonance structures in aqueous or alcoholic and *o*-dichlorobenzenic solutions has disclosed that these compounds absorb at longer wavelength in the less polar aromatic solvent (Table 9.1).

The effect of acids on the long wavelength maxima of some phenylpolyenealdehydes has been investigated(5), as shown for 5-phenylpentadienal in Table 9.2. In the acetyl chloride solvents the authors believe that the red shift is due to the polarizing effect of $(BF_3Cl)^-$ and $(CH_3CO)^+$, etc. It would appear that the red shift is greater with increasing electronegativity of the inorganic anion.

In the diphenylmethane dye **3** a decreasing polarity of the solvent shifts the visible bands to the red, with an intensity increase(7).

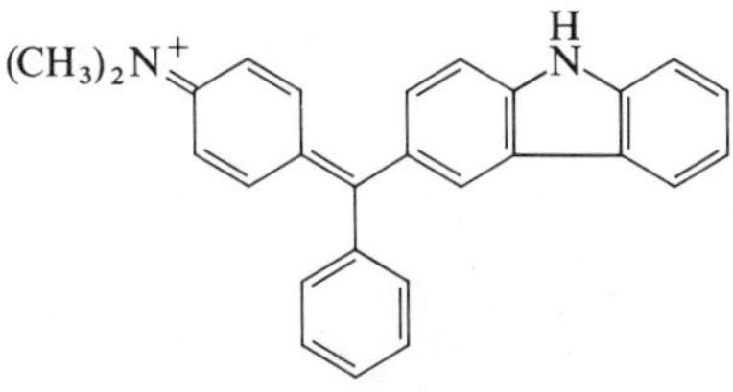

$(CH_3)_2N^+$ H N

3

Solvent	λ_{max}(mμ)	mϵ
Water	408, 562	11.6, 35.0
Methanol	416, 585	15.2, 50.0
Chloroform	440, 629	14.2, 66.0

TABLE 9.1
Solvent Effect on Cationic Resonance Bands

Compound	$\lambda_{max}(m\mu)$[a]	$\lambda_{max}(m\mu)$	Solvent
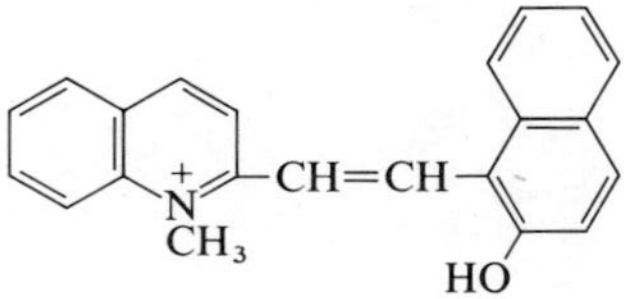 $\overset{+}{N}$ CH_3 —CH=CH— HO	480	452	EtOH–HCl
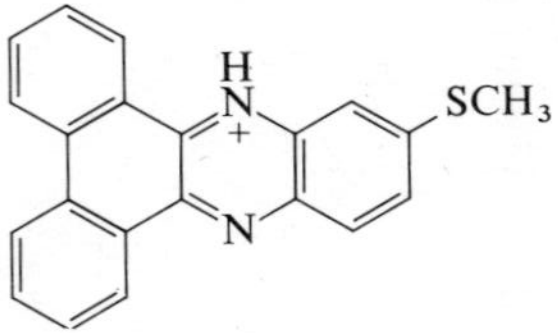 H $\overset{+}{N}$ N SCH_3	521	508	6 *N* HCl in H_2O–EtOH (1 : 1)
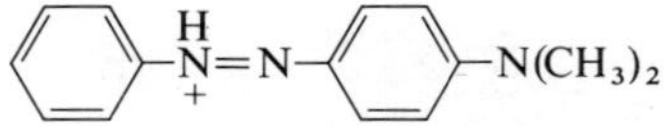 —$\overset{H}{\underset{+}{N}}$=N—$N(CH_3)_2$	530 552	517 530*s*	Aqueous acid
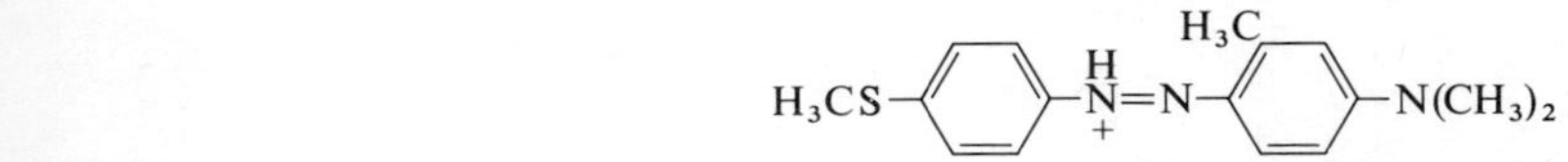 H_3CS—$\overset{H}{\underset{+}{N}}$=N— H_3C —$N(CH_3)_2$	582	553	6 *N* HCl in H_2O–EtOH (1 : 1)
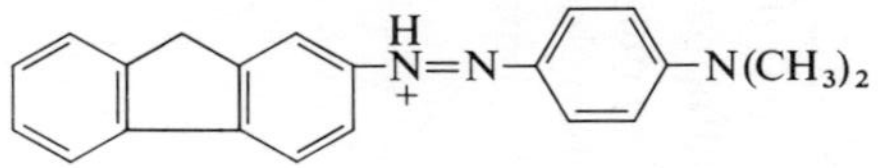 —$\overset{H}{\underset{+}{N}}$=N—$N(CH_3)_2$	587	565	6 *N* HCl in H_2O–EtOH (1 : 1)
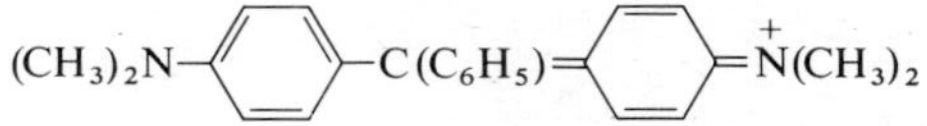 $(CH_3)_2N$—$C(C_6H_5)$=$\overset{+}{N}(CH_3)_2$	628	611	Aqueous acid
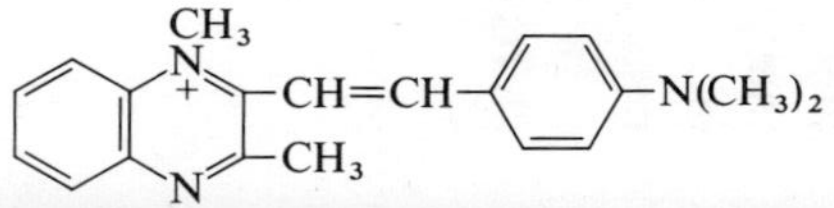 CH_3 $\overset{+}{N}$ —CH=CH—$N(CH_3)_2$ CH_3 N	630[b]	583	Water

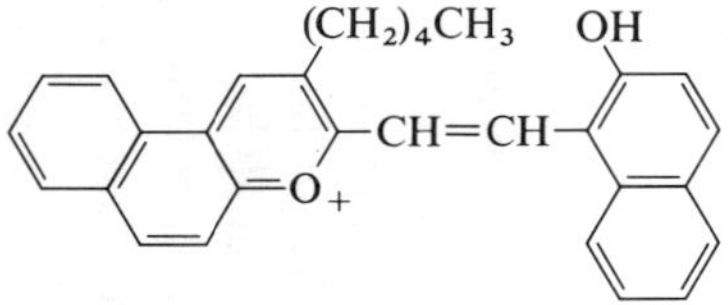 $(CH_2)_4CH_3$, OH; —CH=CH—; O^+	650	600	EtOH–HCl
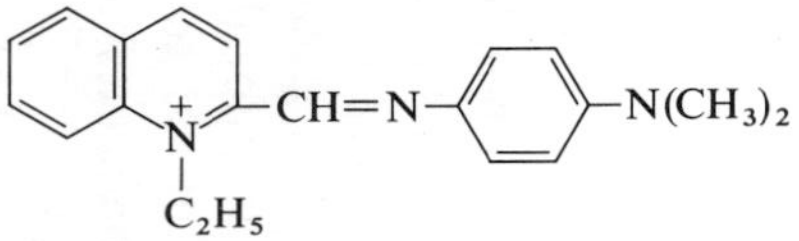 N^+, C_2H_5; —CH=N—; —$N(CH_3)_2$	609[c]	570	95% EtOH

[a]In *o*-dichlorobenzene saturated with hydrogen chloride unless otherwise stated.

[b]In *o*-dichlorobenzene. In this solvent saturated with HCl λ_{max} is 400 mμ.

[c]In *o*-dichlorobenzene. In this solvent or alcohol saturated with hydrogen chloride the longest wavelength absorption is at 322 mμ.

TABLE 9.2
Long Wavelength Maxima of 5-Phenylpentadienal in Various Acid Solutions (5, 6)

Solvent	λ_{max}(mμ)	mϵ
Benzene	320	38
Benzene, 1 *M* CCl_3COOH	350	35
$CHCl_3$, BF_3, CH_3COCl	396	44
$CHCl_3$, $FeCl_3$, CH_3COCl	448	51
$CHCl_3$, $SbCl_5$, CH_3COCl	455	55

Dye **4**, obtained from nicotinic acid through cleavage with alkaline phosphorus oxychloride and condensation with phloroglucinol shows a similar phenomenon (8).

HO–C₆H₂(OH)₂–CH=CH–CH=C(COOH)–CH=C₆H₂(OH)(HO)=$\overset{+}{O}$H

4

Solvent	λ_{max}(mμ)
Methanol	632
Ethanol	639
n-Propanol	641

The chromogen pinacyanol (**5**), formed in the determination of formaldehyde (9), formic acid (10), and chloral (11), shows the same solvent effect but in addition shows a strong intensity decrease with an increase in the polarity of the solvent (Table 9.3) (10). Pure pinacyanol has a λ_{max} of 613 mμ, mϵ 207, in the solvent mixture used for the analysis of formic acid (10) and a λ_{max} of 610 mμ, mϵ 210, in the solvent mixture used for the analysis of formaldehyde. As can be seen from Table 9.3, aqueous solvents would decrease the intensity.

The dipole moment of a compound containing two limiting resonance structures that are equivalent and isoenergetic will be the same in the ground and excited states. Since a solvent will affect each state equally, a change in solvent will have little or no effect on the position of the wavelength maximum. The symmetrical polymethine dyes (12) and pentaenylic cations (13) are two classes of compounds that fit the above description.

TABLE 9.3
Absorption Spectra of Pinacyanol in Various Solvents (10)

Solvent	λ_{max}(mμ)	mϵ
Dimethylformamide	571	95
	618	209
Procedure *A* solvent mixture	569	93
	613	207
2-Methoxyethanol	565	94
	608	210
Methanol	560	90
	606	201
10% Methanol in water (v/v)	552	90
	604	124
Water	550	100
	600	105

II. Temperature Effects

A colored cation formed by protonation of a Z_n-structure can be readily detected. Cooling and heating an *o*-dichlorobenzene solution saturated with hydrogen chloride containing the compound under test will show color changes indicative of the structure and physical properties of the test compound (Table 9.4). The spiro derivative in Table 9.4 has been shown to undergo this reaction (14), giving a green-blue cation (λ_{max} 650 mμ, mϵ 56.0) at room temperature. This thermochromic reaction has been used to detect acetonyl compounds (15).

TABLE 9.4
Thermochromic Changes Involving Cationic Resonance Structures in *o*-Dichlorobenzene Saturated with Hydrochloric Acid

$$(CH_3)_2H\overset{+}{N}-\phi-N{=}N-\phi-\overset{+}{N}H(CH_3)_2 \underset{\text{cool}}{\overset{\Delta}{\rightleftarrows}} (CH_3)_2H\overset{+}{N}-\phi-\overset{+}{N}H{=}N-\phi-N(CH_3)_2$$

0°C Yellow 40°C Red

$$(CH_3)_2N-\phi-N{=}N-\phi-N(CH_3)_2 \underset{\Delta}{\overset{\text{cool}}{\rightleftarrows}} (CH_3)_2N-\phi-\overset{+}{N}H{=}N-\phi-N(CH_3)_2$$

170°C Green yellow 115°C Blue

$$\phi-\overset{+}{N}H{=}N-\phi-N{=}N-\phi-\overset{+}{N}H(CH_3)_2 \overset{\Delta}{\rightleftarrows} \phi-\underset{+}{N}H{=}N-\phi-\overset{+}{N}H{=}N-\phi-N(CH_3)_2$$

−20°C Orange 70°C Purple

$$\phi-N{=}N-\phi-N{=}N-\phi-N(CH_3)_2 \underset{\Delta}{\rightleftarrows} \phi-N{=}N-\phi-\overset{+}{N}H{=}N-\phi-N(CH_3)_2$$

150°C Yellow orange 100°C Red

TABLE 9.4 (*cont'd*)

$$CH_3S-\phi-\overset{H}{\underset{+}{N}}=N-\phi-SCH_3 \underset{\text{cool}}{\overset{\Delta}{\rightleftarrows}} CH_3S-\phi-N=N-\phi-SCH_3$$

Blue — Yellow

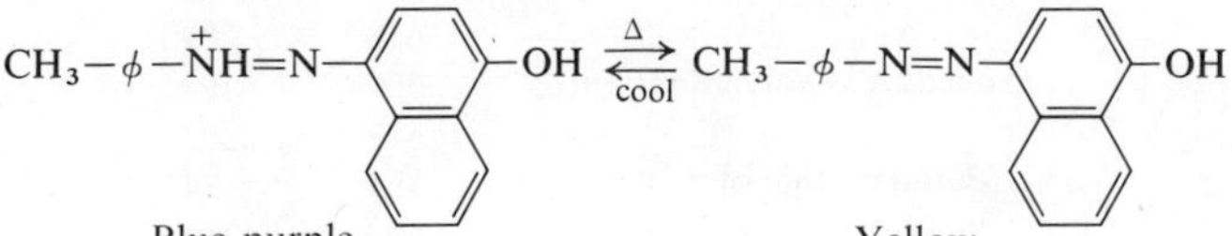

$$CH_3-\phi-\overset{+}{N}H=N-C_{10}H_6-OH \underset{\text{cool}}{\overset{\Delta}{\rightleftarrows}} CH_3-\phi-N=N-C_{10}H_6-OH$$

Blue purple — Yellow

$$\phi-\overset{+}{N}H=N-\phi-NH-\phi \underset{\text{cool}}{\overset{\Delta}{\rightleftarrows}} \phi-N=N-\phi-NH-\phi$$

Violet — Orange

$$(CH_3)_2H\overset{+}{N}-\phi-\overset{OH}{\overset{|}{C}}H-\phi-\overset{+}{N}H(CH_3)_2 \underset{\text{cool}}{\overset{\Delta}{\rightleftarrows}} (CH_3)_2\overset{+}{N}=\phi=CH-\phi-N(CH_3)_2$$

Colorless — Blue

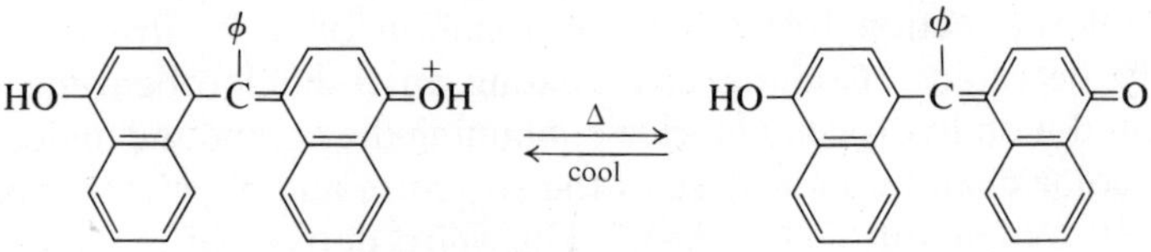

$$HO-C_{10}H_6-\overset{\phi}{\overset{|}{C}}=C_{10}H_6=\overset{+}{O}H \underset{\text{cool}}{\overset{\Delta}{\rightleftarrows}} HO-C_{10}H_6-\overset{\phi}{\overset{|}{C}}=C_{10}H_6=O$$

Green — Yellow

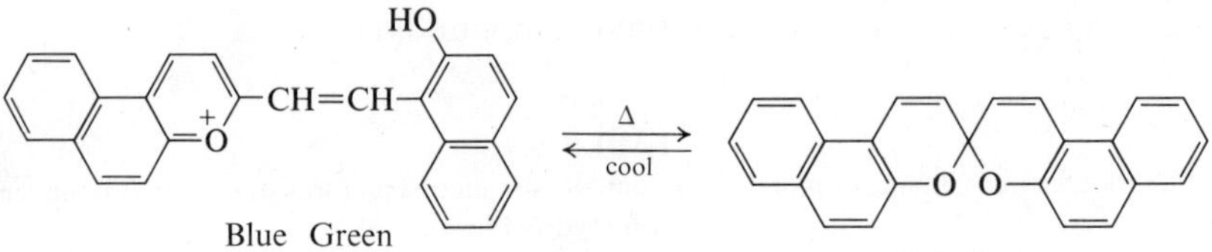

$$(\text{naphthopyrylium})\overset{+}{O}-CH=CH-C_{10}H_6(HO) \underset{\text{cool}}{\overset{\Delta}{\rightleftarrows}} (\text{spiropyran, O O})$$

Blue Green — Colorless

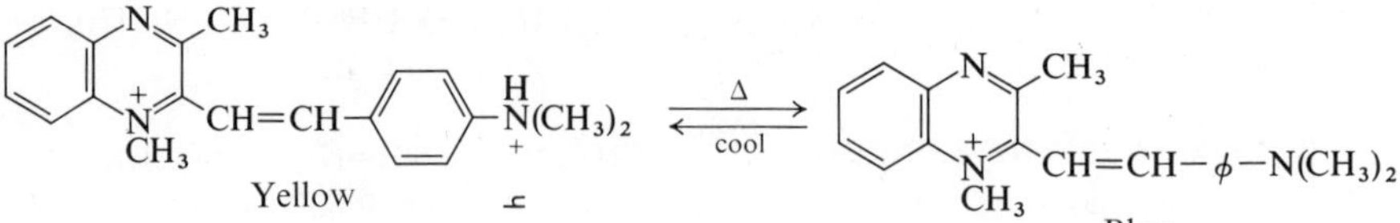

$$(\text{quinoxalinium: } N, CH_3, \overset{+}{N}, CH_3)-CH=CH-\phi-\underset{+}{\overset{H}{N}}(CH_3)_2 \underset{\text{cool}}{\overset{\Delta}{\rightleftarrows}} (\text{quinoxalinium: } N, CH_3, \overset{+}{N}, CH_3)-CH=CH-\phi-N(CH_3)_2$$

Yellow — Blue

$$(\text{quinolinium: } \overset{+}{N}, C_2H_5)-CH=N-\phi-\underset{+}{\overset{H}{N}}(CH_3)_2 \underset{\text{cool}}{\overset{\Delta}{\rightleftarrows}} (\text{quinolinium: } \overset{+}{N}, C_2H_5)-CH=N-\phi-N(CH_3)_2$$

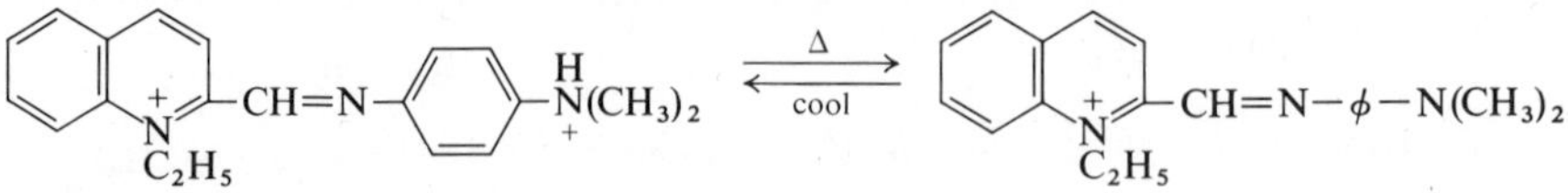

Colorless — Blue

TABLE 9.4 (*cont'd*)

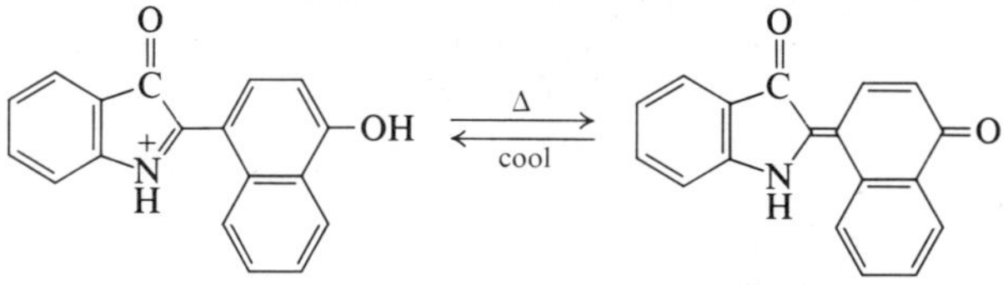

Blue $\underset{\text{cool}}{\overset{\Delta}{\rightleftharpoons}}$ Red

$(CH_3O-\phi-CH=CH)_2C=\overset{+}{O}H \underset{\text{cool}}{\overset{\Delta}{\rightleftharpoons}} (CH_3O-\phi-CH=CH)_2C=O$

Red — Pale yellow

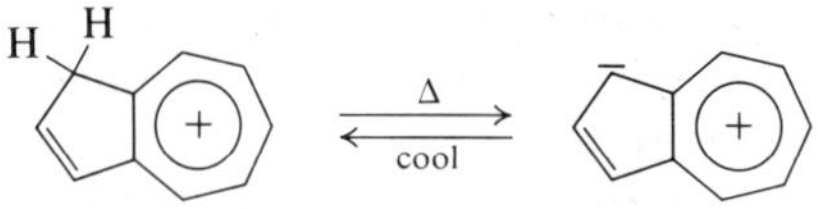

Colorless $\underset{\text{cool}}{\overset{\Delta}{\rightleftharpoons}}$ Blue

The reversible change of color with temperature of numerous organic compounds, a phenomenon known as thermochromism, has been of interest for many years. A colorless solution of the salt of 4-(4′-dimethylaminoxenyl)diphenylcarbinol (**5**) in hot acetic acid gives a deep blue-green color(16). The energy given to the mixture in the form of heat splits off water from the molecule to form the triphenylmethane dye **6**. Structures involving two *p*-quinonoid rings are of high energy, but the extra energy added to the system apparently stabilizes this type of contributing structure.

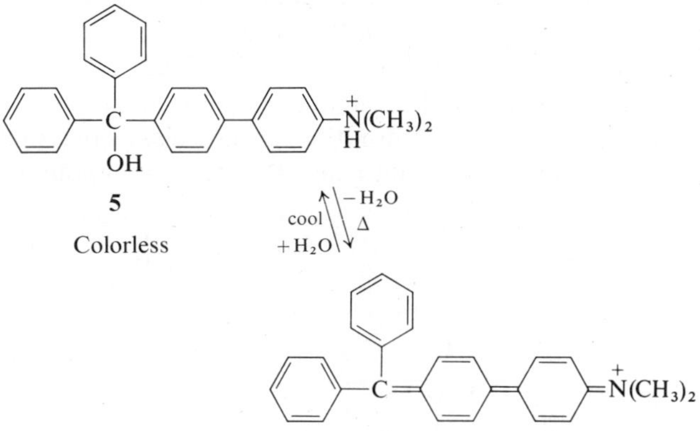

5

Colorless

$\underset{\text{cool}, +H_2O}{\overset{-H_2O, \Delta}{\rightleftharpoons}}$ (9.3)

6

Green, λ_{max} 400, 850 mμ

The same type of phenomenon has been reported for bis-4-(4′-dimethylaminoxenyl)phenylcarbinol(16). A solution of this compound in acetic acid is pale yellow at room temperature and green at 100°C. The hot solution shows bands at 405 and 700 mμ, and probably a band in the infrared region. Tris-4-(4′-dimethylaminoxenyl)carbinol, pale reddish brown in acetic acid, is green in hot solution (λ_{max} 643 mμ and probably a band in the infrared)(16).

On the other hand, crystal violet and malachite green give the same spectra at room temperature and at 117 and 130°C in acetic acid. Bis(4-dimethylaminophenyl)phenylcarbinol dissolved in dilute mineral acids gives a green solution of the triphenylmethane dye malachite green. On the other hand, bis(4-aminophenyl)phenylcarbinol dissolved in dilute mineral acids forms a colorless solution that turns red only on warming (17). Apparently the amino group is a more powerful proton attractor than the dimethylamino group.

4-Aminoazobenzene derivatives show the same relative proton-attractor strength(4). Tris-2-anisylcarbinol is reversibly colorless in acetic acid at room temperature and violet in hot acetic acid, whereas the 3-isomer is colorless in cold or hot acetic acid solution(18). The latter compound cannot form a triphenylmethane dye molecule containing low energy resonance structures involving the methoxy groups. Consequently the carbinol is too weakly basic to ionize. Colorless tris-2-anisylchloromethane contains a covalent C—Cl bond. When a colorless solution of this compound in benzene is cooled to −70°C, a solid purple "ice" is formed that becomes colorless when warmed to room temperature(19). Here again ionization has taken place at the lower temperature. When an ether solution of trixenylcarbinol is heated, violet to red films, which become colorless on cooling, are deposited on the glass surface(20). Many other examples of the ionization of triarylcarbinols and similar compounds at increased temperatures have been given(21–25).

A reversible color reaction that might prove of value in research studies involving pyridine enzyme systems is the thermochromic reaction between a 9-(4-dimethylaminophenyl) xanthene (**7**) type of compound and nicotinic acid:

(9.4)

N(CH$_3$)$_2$ ... H ... O ... **7** + nicotinic acid (N$^+$H, C(=O)O$^-$) $\underset{\text{cool}}{\overset{\Delta}{\rightleftharpoons}}$ $\overset{+}{N}$(CH$_3$)$_2$ ... O ... **8** + dihydronicotinic acid (NH, H, H, C(=O)O$^-$)

It has also been shown that 3-(9-xanthyl)-9-methylcarbazole and the thiaxanthyl analog undergo the same thermochromic oxidation–reduction reaction(26).

III. Steric Effects

A. Introduction

In conjugated molecules the stretching of bonds and the distortion of valency angles have a relatively small effect on the movement of π-electrons, whereas rotation about a bond has a definite effect. On rotation through 90° the atomic *p*-orbitals cease to interact. A fall in molecular absorptivity is a direct consequence of the diminished mutual overlap of *p*-orbitals in the conjugated system and is therefore the first sign of the appearance of steric hindrance in the absorbing molecule.

The displacement of the absorption band is rather more complicated, since it may occur either toward longer or toward shorter wavelengths. Steric hindrance always increases the energy of a molecule in comparison with one that is free from it.

When steric hindrance increases the energy of the excited state more than it does that of the ground state, the absorption spectrum of the hindered molecule as compared with the spectrum of the unhindered comparison compound will shift to the violet. When steric hindrance increases the energy of the ground state more than that of the excited state, the absorption spectrum of the molecule will shift to the red.

The effect of steric hindrance on the absorption spectra of cationic resonance dyes has been reviewed(27). In this section we present examples showing that steric hindrance can cause red or violet shifts and intensity decreases or increases.

B. Diphenylmethane Cations

Protonated benzophenone derivatives can be considered as hydroxylated diphenylcarbonium cations. As shown in Table 9.5, *ortho*-methyl groups in the prime ring of protonated 4-methoxybenzophenone have a steric effect(28). The long wavelength maximum of the 2′, 4′-dimethyl derivative shows a red shift effect, which indicates that steric strain is greater in the ground state for this molecule.

The three main types of diphenylmethane cationic resonance structures are Michler's hydrol blue (**9**), malachite green (**10**), and crystal violet (**11**). The limiting resonance structures of Michler's hydrol blue are shown.

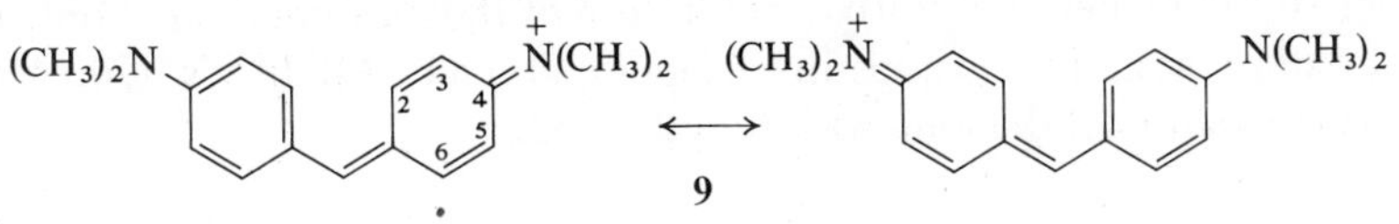

$(CH_3)_2N$–C₆H₄–CH=C₆H₄(2,3,4,5,6)=$\overset{+}{N}(CH_3)_2$ $\longleftrightarrow$ $(CH_3)_2\overset{+}{N}$=C₆H₄=CH–C₆H₄–$N(CH_3)_2$

9

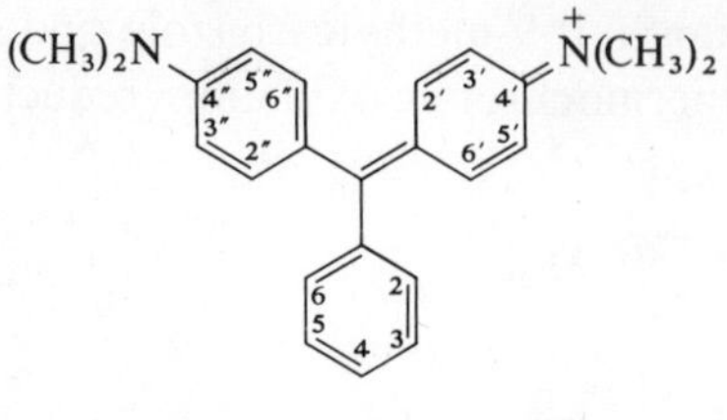

10 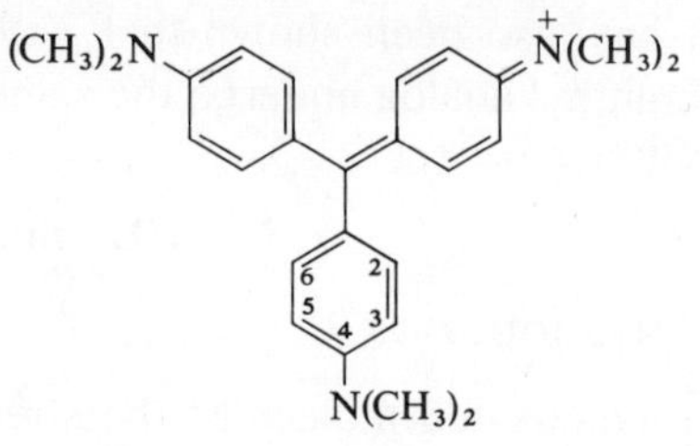 **11**

The shape of a triphenylmethane dye resembles that of a three-bladed propeller. Two types of steric effects predominate in these dyes, that due to the overlap of substituents in any or all of the various 2- and 6-positions, and that due to substituents *ortho* to the amino groups. Methyl

TABLE 9.5
Steric Effect in Protonated 4-Methoxybenzophenone Derivatives (28)

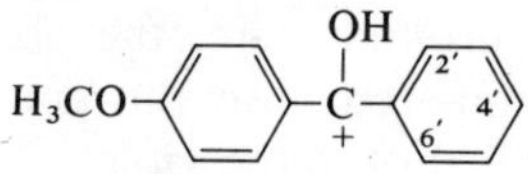

Substituent at position				
2′	4′	6′	λ_{max} (mμ)	mϵ
H	CH_3	H	389	42.0
CH_3	H	H	376	37.4
CH_3	CH_3	H	395	29.4
CH_3	CH_3	CH_3	349	24.0

groups substituted in any of these positions in crystal violet cause a red spectral shift and a decrease in the intensity of the long wavelength band (Table 9.6).

Whereas steric inhibition of the mesomerism of all or any of the dimethylamino groups in crystal violet causes a red shift of the long wavelength band, inhibition of the mesomerism of all of these groups in malachite green and Michler's hydrol blue causes small and moderate violet shifts, respectively (30). Inhibition of the mesomerism of only one of the dimethylamino groups in malachite green or Michler's hydrol blue causes red shifts of the long wavelength band.

Seemingly small changes in the solvent can complicate the visible absorption spectra of the 3-methyl, 3,3′-dimethyl, 3,5-dimethyl, and 3,3′,5,5′-tetramethyl derivatives of crystal violet. Thus in acetic acid containing about 5–25% of water or trace amounts of hydrogen chloride ($\sim 5 \times 10^{-5}\ M$) a new band is found near 430 mμ (mϵ ~ 10) in addition to the long wavelength band. In many of these solutions the bands near 600 and 430 mμ are derived from the dication since steric hindrance of an aromatic dimethylamino group considerably increases its basicity. In the less acidic solutions a mixture of the monocation and dication is present. Increasing the acidity increases the amount of trication at the expense of the monocation.

TABLE 9.6
Long Wavelength Maxima in 98% Acetic Acid of the Alkyl Derivatives of Crystal Violet, Michler's Hydrol Blue, and Malachite Green (29–32)

Substituent	λ_{max} (mμ)	mϵ
	Crystal Violet	
None	589	117
2-Methyl	597	110
2,2′-Dimethyl	605	103
2,2′,2″-Trimethyl	614	100
2,6-Dimethyl	614.5	—
3-Methyl	599.5	88
3,3′-Dimethyl	606.5	77
3,3′,3″-Trimethyl	615	13
3,5-Dimethyl	610.5	96
3,3′,5,5′-Tetramethyl	617.5	50[a]
3,3′,3″,5,5′,5″-Hexamethyl	649.5	1.7[b]
	Michler's Hydrol Blue	
None	607.5	148
2-Methyl	614.5	130
2,2′-Dimethyl	623	121
2,2′,6,6′-Tetramethyl	649	55
α-Methyl	606, 395, 376	11[c], 1.0[c], 0.8
α-Ethyl	620, 403, 381	19[c], 2.1[c], 1.7
α-Isopropyl	610, 402, 383	14.1[c], 2.0[c], 1.7
α-*tert*-Butyl	617.5, 384[d]	0.64[a], 0.77[a]
3-Methyl	602.5	0.6[a]
3,3′-Dimethyl	584	0.3[a]
3,3′,5,5′-Tetramethyl	Colorless[e]	

TABLE 9.6 (*cont'd*)

Substituent	λ_{max} (mμ)	mϵ
	Malachite Green	
None	621[f], 427.5[g]	104, 20
2′-Methyl	635, 437.5	72, 20
2,2′-Dimethyl	634, 430	62, 20
2′,2″-Dimethyl	648, 445	67, 18
2,2′,2″-Trimethyl	647.5, 440	69, 17
3′-Methyl	627.5, 436	63, 16
3′,5′-Dimethyl	630, 452.5	18, 12
3′,3″-Dimethyl	617, 430	11[a], 40[a]
3′,3″,5′,5″-Tetramethyl	616, 446	0.15[a], 0.11[a]
2-Methyl	622.5, 420	123, 15
3-Methyl	618.5, 433	106, 22
2,6-Dimethyl	624, 410	132, 12
2-*tert*-Butyl	623.5, 415	118, 14

[a]Value of mϵ low due to equilibrium of cation with carbinol salt.
[b]In pure acetic acid.
[c]Value of mϵ low due to equilibrium of cation with corresponding olefin.
[d]Dication.
[e]Ultraviolet absorption spectrum is closely similar to that of bis(3,5-dimethyl-4-dimethylaminophenyl)methane hydrochloride in ethanol (30).
[f]These bands in the 620 mμ region are derived from cationic resonance structure in which the amino groups are resonance terminals.
[g]These bands in the 430 mμ region are derived from the cross-conjugated cationic resonance structure in which an amino group and the nonaminated phenyl ring are the resonance terminals.

An α-*tert*-butyl group can cause a somewhat similar effect in Michler's hydrol blue. This group causes rotation of one of the dimethylaminophenyl groups about a central bond. The basicity of the incompletely conjugated nitrogen atom is increased so that it is protonated readily to form a dication. The univalent cation, in equilibrium (equation 9.5) with the dication, would become electronically asymmetrical and would show a red shift, since symmetrical cationic resonance structures usually respond to crowding substituents with a red shift and a decrease in the intensity of the long wavelength band (33).

α-Alkyl derivatives of Michler's hydrol blue give weak absorption bands in acetic acid (Table 9.6) since acetic acid solutions of these compounds consist of an equilibrium mixture of the cation and the olefin (equation 9.6) (31).

$$(CH_3)_2N\text{-}C_6H_4\text{-}C(C(CH_3)_3){=}C_6H_4{=}\overset{+}{N}(CH_3)_2 \underset{-H^+}{\overset{+H^+}{\rightleftharpoons}} (CH_3)_2\overset{+}{N}{=}C_6H_4{=}C(C(CH_3)_3)\text{-}C_6H_4\text{-}\overset{+}{N}H(CH_3)_2 \quad (9.5)$$

12 λ_{max} 617.5 mμ in AcOH

13 λ_{max} 384 mμ in AcOH

Since 1,1-diphenylethylene polymerizes to 1,1,3,3-tetraphenylbut-1-ene in the presence of cationic catalysts (34), and the dimethylamino group in the olefin should facilitate polymerization, it is not surprising that the spectra of many of the α-alkyl derivatives of Michler's hydrol blue change with time. Thus the compounds in equilibrium as shown in equation 9.6 undergo in 24 hr the following changes:

$$[(CH_3)_2NC_6H_4]_2\overset{+}{C}\text{-}CH_2CH_3 \; (\mathbf{14}) \underset{+H^+}{\overset{-H^+}{\rightleftharpoons}} [(CH_3)_2NC_6H_4]_2C{=}CH\text{-}CH_3 \; (\mathbf{15}) \quad (9.6)$$

$$\downarrow H^+$$

$$(CH_3)_2\overset{+}{N}{=}C_6H_4{=}C(C_6H_4N(CH_3)_2)\text{-}CH(CH_3)\text{-}C(C_2H_5)(C_6H_4N(CH_3)_2)_2 \; (\mathbf{16}) \; + \; (CH_3)_2\overset{+}{N}{=}C_6H_4{=}C(C_6H_4\overset{+}{N}H(CH_3)_2)\text{-}CH(CH_3)\text{-}C(C_2H_5)(C_6H_4N(CH_3)_2)_2 \; (\mathbf{17})$$

16 λ_{max} 637.5 mμ, mϵ 8.6

17 λ_{max} 350 mμ, mϵ 14.8

The dimer from the α-isopropyl derivative is exceptionally crowded so that its monocation (λ_{max} 630 mμ, mϵ 0.7) and dication (λ_{max} 353 mμ, mϵ 2.6) have especially weak intensities.

Absorption spectra of derivatives of malachite green containing substituents in the phenyl ring have been examined (32). With the exception of the nitro group, substituents in the 3- or 4-position have little effect on the intensity of the long wavelength band, but, if they are electropositive they cause a violet shift; if electronegative, a red shift. Substituents in the 2-position, however, markedly increase the intensity of this band, and this increase varies directly with the van der Waals radius of the substituent except in the case of the *tert*-butyl derivative; they also show a red shift relative to the 4-substituted isomer. On the other hand, in the *Y*-band, derived from the cross-conjugated structure where the phenyl and amino groups are resonance terminals, a violet shift and a decrease in intensity result.

Comparison of malachite green with its 2-methyl and 2,6-dimethyl derivatives clearly demonstrates the diverse effects of this steric hindrance on the *X*- and *Y*-bands (~ 660 and ~ 420 mμ, respectively) of these molecules (Table 9.6). The red shift and the intensity increase of the *X*-band in these sterically hindered molecules is due to the twisting of the phenyl group out of the plane of the molecule. The conjugated portion of the molecule with the two amino groups as resonance terminals resembles a somewhat sterically hindered Michler's hydrol blue in wavelength maximum and intensity.

C. Spiro Derivatives

Spiro derivatives have been discussed in Chapter 3 in terms of their thermochromism. However, they can also be readily protonated to give brilliantly colored cations (Fig. 9.1). Where substituents are present in the 3- and 3′-positions, steric hindrance in the ring-opened compounds causes a violet shift. From a comparison of thermochromism and protonation in Table 9.7, it can be seen that protonation can open the ring more readily. The necessity for planarity in the ring-opened zwitterionic and cationic resonance structures is shown by the diminishing tendency for thermochromism and the violet shift of the cation as the size of the polymethylene ring increases. The five-membered ring favors a flat structure in the ring-opened compounds (14,35,36).

D. Brunings–Corwin Effect

In some cationic resonance structures and many zwitterionic resonance structures (as has been shown in Chapter 3) an increase in steric hindrance causes a violet shift and a decrease in intensity. An example of this phenomenon is the styryl dye **18a** and its methyl derivative **18b** (37).

In **18b** steric hindrance arises between the methyl group in the chain and the sulfur atom or the *N*-alkyl group. This interference twists the benzothiazole out of the plane of the molecule at its bond to the neighboring carbon atom in the chain, this bond being predominantly single in the ground state. This bond is predominantly a double bond in the excited state, to which the limiting resonance structure containing a positive

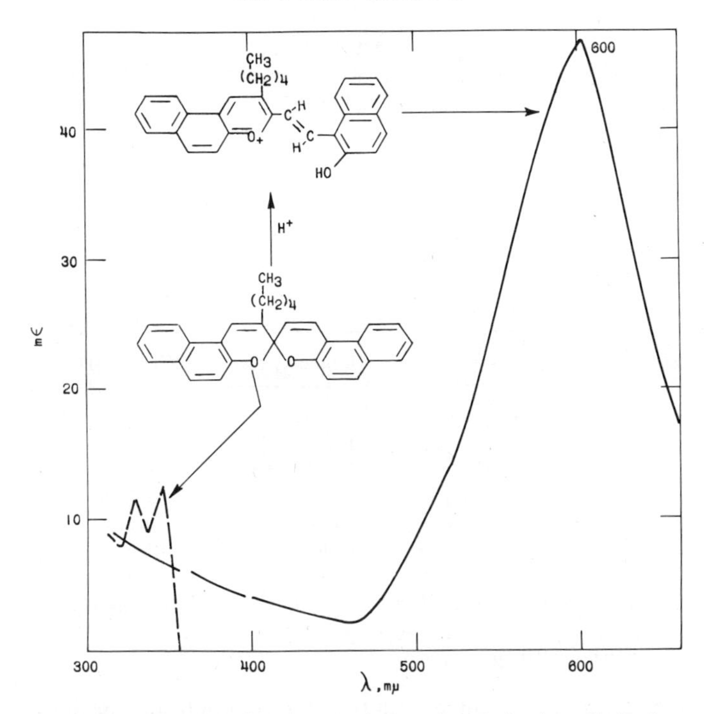

Fig. 9.1. Absorption spectrum in ethanol of the spiro derivative obtained on analysis of 2-octanone with 2-hydroxy-1-naphthaldehyde (– – –) and effect of acid addition (——).

S, $\overset{+}{N}$, C_2H_5, R, $-\overset{|}{C}=CH-$, $-N(CH_3)_2$

18

	R	λ_{max} (mμ)	mϵ
a	H	528	110
b	CH_3	481	62

charge on the dimethylamino group is the main contributor. Consequently steric hindrance will be greater in the excited state than in the ground state. In addition deflection about a double bond consumes more energy than deflection about a single bond. The change in energy in going from the ground to the excited state will be increased, and the molecule will absorb at shorter wavelengths than the nonmethylated analog, as shown by comparing **18a** and **b**.

TABLE 9.7
The Effect of Steric Hindrance on Thermochromism and on the Color of the Salt

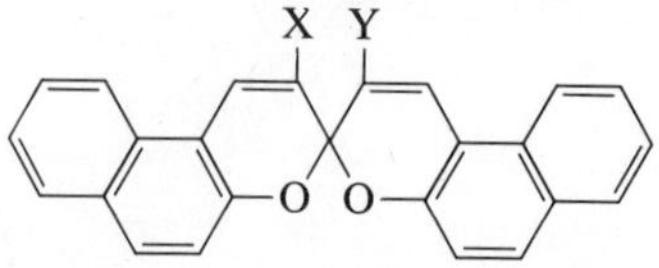

Substituents		Color[a] in	
X	Y	Boiling xylene	CCl_3COOH
H	H	Purple	Blue
$—CH_2—CH_2$		Blue	Blue
$—CH_2—CH_2—CH_2—$		Light violet[b]	Purple
$—CH_2—(CH_2)_2—CH_2—$		Colorless	Red
$—CH_2—(CH_2)_3—CH_2—$		Colorless	Orange
CH_3	CH_3	Colorless	Blue

[a]No ring opening in colorless solutions. In acid solution ring is opened and a cationic resonance structure is formed.
[b]In boiling diphenyl ether.

In some cationic resonance structures steric hindrance primarily affects ground state resonance, and the loss in resonance energy increases E_0 to make $E_1 - E_0$ smaller. This will result in a red shift of the long wavelength band. This type of steric hindrance has been designated the Brunings–Corwin effect(38) and defined in the following manner: *substitution into a symmetrical or nearly symmetrical cationic resonance structure so as to cause a crowding of the two halves of the molecule out of planarity results in a red shift of the long wavelength band.*

Many examples of this effect have been described(27,29). The red shift and intensity decrease of this phenomenon have been attributed to the twisting of adjacent bonds that connect two planar ring systems to a central atom(39).

The example of this steric phenomenon is seen in the comparison of the symmetrical cations **19** and **20**, where spectra were obtained in chloroform(38).

When the hydrogens on the nitrogen atoms are replaced by methyl groups, steric hindrance increases especially in the ground state, and the new molecule, **20**, shows a red shift. An intensity decrease also results since the crowding would cut down the overlap of the π-electronic orbitals. The change in wavelength and intensity can be attributed to steric hindrance because, when the rings are further separated by another

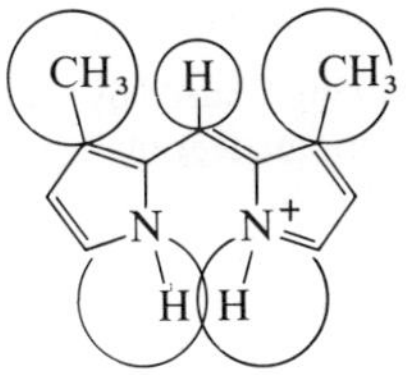

19

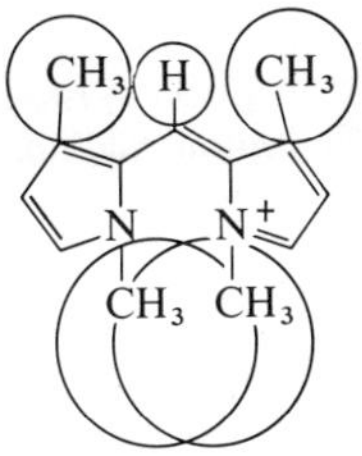

20

λ_{max} 473 mμ, mϵ 135

λ_{max} 510 mμ, mϵ 57.0

double bond, as in **21**, the introduction of the *N*-methyl group produces a violet shift, even though the resultant molecule, **21b**, is coplanar(29). The spectra were measured in chloroform.

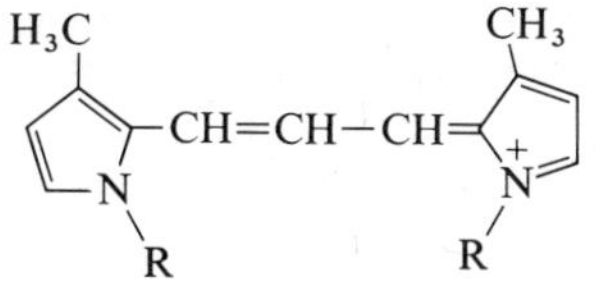

21

	R	λ_{max} (mμ)	mϵ
a	H	595	170
b	CH_3	581	170

Some other examples of this phenomenon are shown in Table 9.8. An apparent exception is **23** as compared with **22**(40). The low energy intermediate resonance structure is shown.

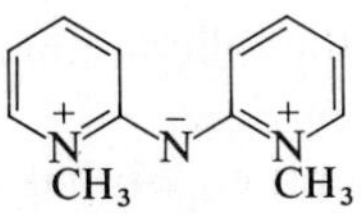

22

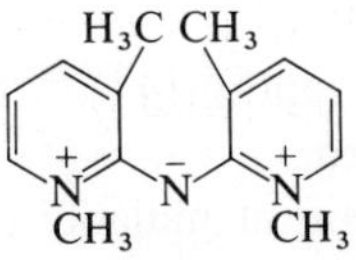

23

λ_{max} 388 mμ, mϵ 30.4

λ_{max} 364 mμ, mϵ 22.6

Overcrowding in **23** would have a red shift effect, but this is balanced by the greater stability of such structures as **23**. This would act similarly to negativizing a negative node and would have a violet shift effect.

TABLE 9.8
Effect of Increased Crowding Caused by Proton Addition to a Zwitterionic Resonance System(39)

	Base[a]		Salt[b]	
R	λ_{max} (mμ)	mε	λ_{max} (mμ)	mε
H	537	40.3	557	87.0
C_6H_5	546	31.5	616	42.0
CH_3	550	7.4	595	37.5
C_2H_5	540	5.0	605	28.2

[a]In carbon tetrachloride. Where R = alkyl; some triethylamine was necessary to stabilize the solution.

[b]In acetic acid.

IV. Insulation

Unlike the triphenylcarbinols and triphenylchloromethanes, the triphenylsilanols and the triphenylchlorosilanes do not form colored compounds in acid solution(41,42). It was postulated that neither a Si═C type of compound nor a silicon cation could be formed under the conditions of the investigation(41). Thus the silicon analog of crystal violet could not be formed.

The boron analog of malachite green (i.e., **24**) has also been prepared and has been found to be light yellow in color(43).

$(CH_3)_2N-C_6H_4-B(\phi)-C_6H_4-N(CH_3)_2$

24

The insulating properties of a saturated carbon atom are seen in a comparison of **25**(44) and **26**(45).

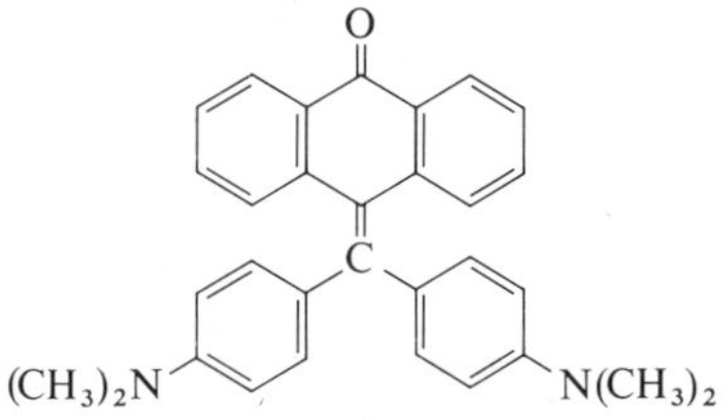

25 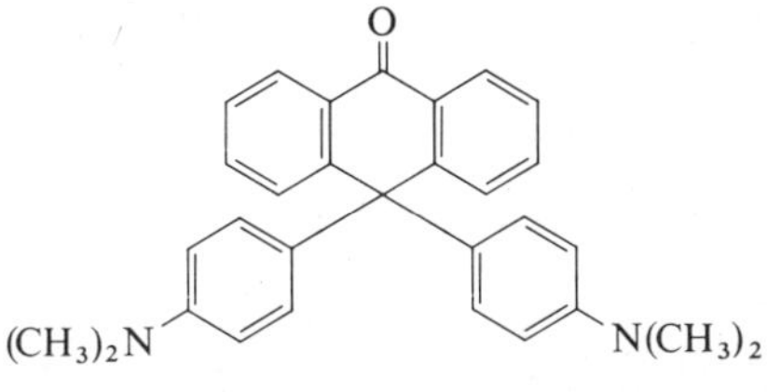 **26**

In a solution of methanolic perchloric acid the proton adds to the oxygen of **25** to give a dye that absorbs at 625 mμ.

On the other hand, in the somewhat analogous structure **26** the salts are colorless compounds. Because of the insulating central carbon atom, the ketone group is not in conjugation with the amino groups. Consequently the basicity of the keto group is much less and the basicities of the amino groups are greater than the basicities of the analogous groups in the anthrafuchsone base. This means that salt formation would first take place at the amino nitrogens.

V. Resonance Nodes

A. Forster–Knott Rule

Forster's rule(46) as applied to ionic dyes essentially states that the long wavelength maximum of a dye will shift to the red with a decreasing tendency of the resonance nodes lying between the resonance terminals to take up the charge. Knott(47) has also applied this rule to nonionic dyes and restated it in the following form: *the wavelength maximum of a compound will increase as the contributions by any of the intermediate ionic excited structures decrease.*

Knott states that, since the contributions by the various structures depend largely on their relative energies, it follows that the wavelength maximum will be increased by raising the energy of any one intermediate ionic excited structure.

Lewis(48) has postulated in an investigation of a large number of 4,4′-diaminodiphenylmethane and 4,4′-diaminodiphenylamine dyes that the position of the long wavelength band depends on the fraction of the characteristic charge that is on the two resonance terminals and that the red shift is larger when this fraction is larger.

These rules have been verified in a study of the absorption maxima of a large number of dyes and their aza analogs(49).

B. Positivizing a Negative Node

The substitution of an electron-donor group at a CH negative node causes a red shift, as shown for **27** (50).

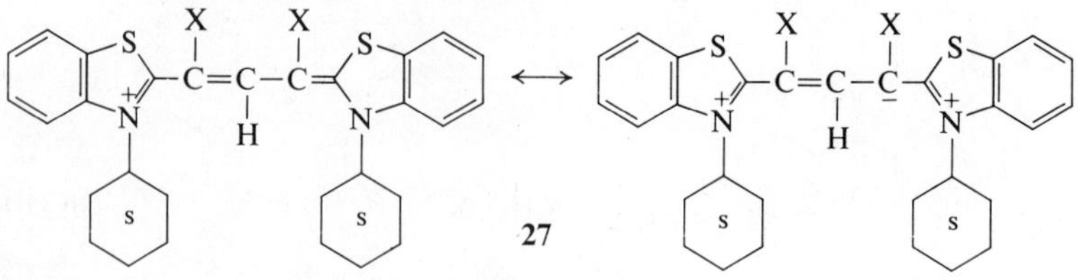

27

X	λ_{max} (mμ)
H	566
CH_3	574
OCH_3	610

C. Negativizing a Positive Node

The substitution at a positive node of an electronegative group for a hydrogen or a stronger electronegative group for a weaker one causes a red shift. An example of this phenomenon is seen in compounds **28** and **29**, for which the intermediate resonance structures are shown.

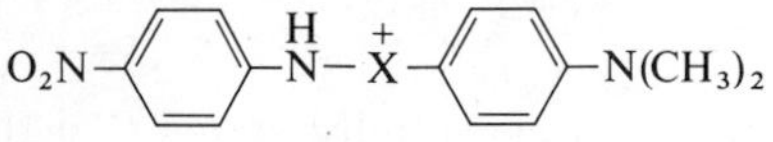

28

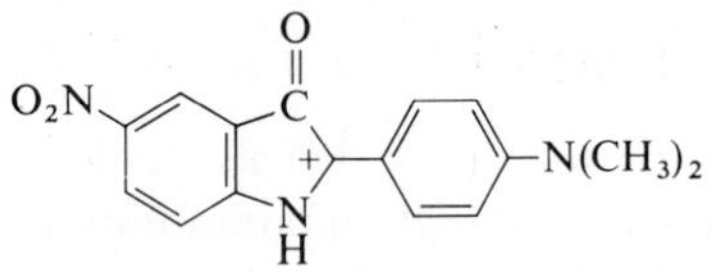

29

No.	X	λ_{max} (mμ)	mϵ	Ref.
28a	CH	469[a]		
28b	N	512[a]	61.6	4
29		665	31.0[b]	51

[a]In 1 *N* HCl in H_2O–EtOH (1 : 1).
[b]mϵ' or mϵ of chromogen obtained analytically.

Thus the chromogen **29**, which is obtained in the determination of dimethylaniline with 5-nitroisatin, absorbs at a very much longer wavelength than structures **28a** and **b**. The same strong red shift is found when the chromogen obtained in the determination of a phenol with an isatin is compared with an analogous azophenol, for example, **30** and **31** (52).

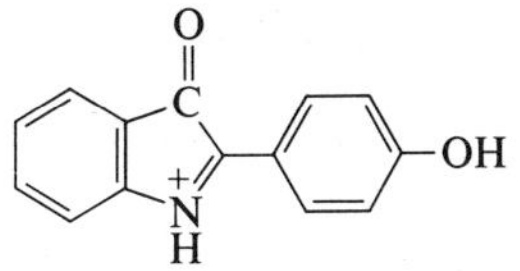

30

λ_{max} 560 mμ

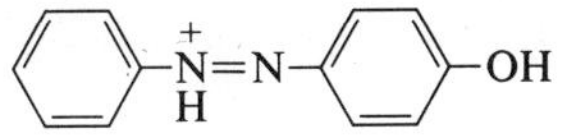

31

λ_{max} 476 mμ in H_2SO_4

Structures **32**(53,54), **33**(55), **34**, and **35**(56) show the red shift phenomenon.

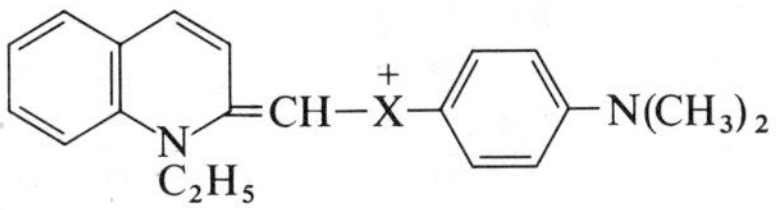

32

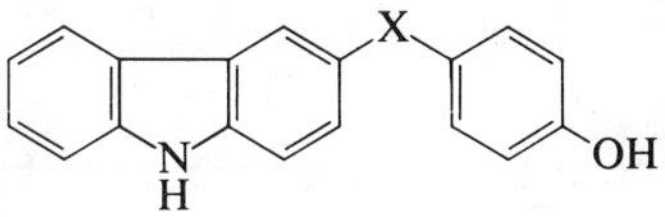

33

No.	X	λ_{max} (mμ)	mε
32	CH	524	151
	N	570	
33	N^+	615	~12[a]
	NH^{2+}	665	40[a]

[a]mε′.

34

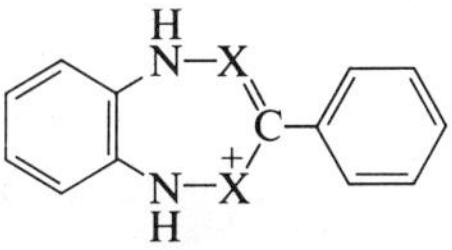

35

No.	X	Color	
34	CH	Yellow	λ_{max} 440 mμ
	N	Violet	
35	CH	Red	
	N	Blue	

The spectral properties of **33** have been used in the characterization and estimation of carbazole in air polluted by fumes of coal-tar pitch(55).

3-Methyl-2-benzothiazolinone hydrazone (MBTH) is a reagent that forms intensely colored chromogens in the trace analysis of a large number of organic compounds. The chromogens obtained in the determination of dimethylaniline (**36b**)(57) and azulene (**36d**)(58) absorb at longer wavelengths than the analogous dimethine derivatives. The intermediate resonance structure depicting the positive and negative resonance nodes is shown.

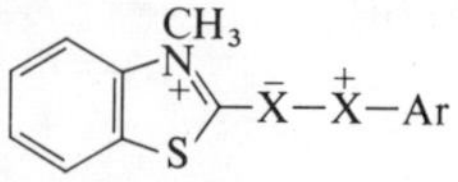

36

	Ar	X	$\lambda_{max}(m\mu)$	$m\epsilon$	Ref.
a	4-Dimethylaminophenyl	CH	524		
b		N	600	79.5	59
c	1-Azulenyl	CH	514	60.3	60
d		N	588	52.5	58

In these compounds the red shift effect of negativizing a positive node overrides the violet shift effect of negativizing the negative node and thus there is an overall red shift.

The negativization of a positive node in the azacarbocyanines can cause a strong red shift, as, for example, in **37** (61).

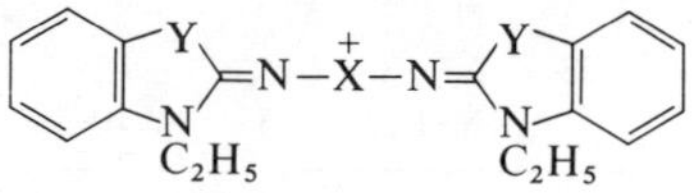

37

Y	X	$\lambda_{max}(m\mu)$
S	CH	370
S	N	485
—CH=CH—	CH	430
—CH=CH—	N	472

With increasing basicity (decreasing +M effect) of a resonance terminal in a carbocyanine ring, the red shift effect of negativizing a positive node (replacing CH by N) is decreased, as shown in Table 9.9 (62). The order of basicity is in fair agreement with that found by Brooker by his deviation method (63).

Many examples of negativizing a positive node are listed in Table 9.10 for diphenylmethane dyes. Thus in malachite green substitution of increasingly stronger electron-acceptor groups in the phenyl ring causes a red shift, λ_{max} increasing as follows: H, 620 mμ; 4-$(CH_3)_2HN^+$, 631 mμ; 4-$(CH_3)_3N^+$, 636 mμ; 3-O_2N, 638 mμ; and 4-O_2N, 645 mμ.

Other examples can be found among the heterocyclic dyes listed in

TABLE 9.9
Red Shift Effect of Negativizing a Positive Node (62)

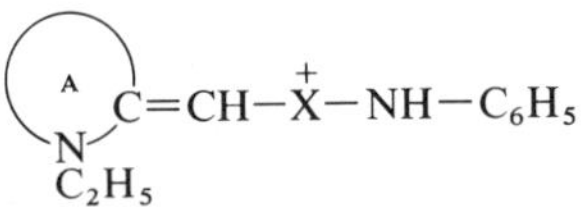

$\overset{+}{\text{A}}$: $C{=}CH{-}\overset{+}{X}{-}NH{-}C_6H_5$ (N–C_2H_5)

I

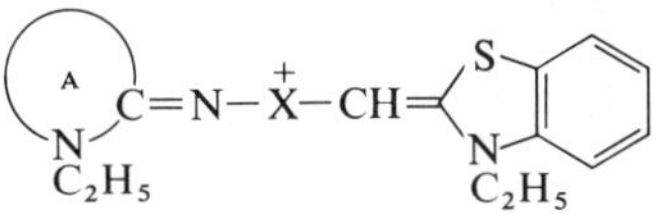

$C{=}N{-}\overset{+}{X}{-}CH{=}$ (benzothiazole, N–C_2H_5) (N–C_2H_5)

II

Nucleus A	λ_{max}(mμ)		Shift (mμ)
	X = CH	X = N	
	I		
1,3,3-Trimethylindolenine	409	438	29
1-Ethyl-2-quinoline	443	463	20
1-Ethyl-2-pyridine	412	423	11
1-Ethyl-4-quinoline	484	490	6
1-Ethyl-4-pyridine	427	432	5
	II		
3-Ethylbenzothiazole	465	495	30
3-Ethylbenzoselenazole	466	486	20
1-Ethyl-2-quinoline	488	497	9

TABLE 9.10
Spectral Effect of Positivizing and Negativizing the Central Positive Resonance Node of the 4,4′-Bisdimethylaminodiphenylcarbonium Ion

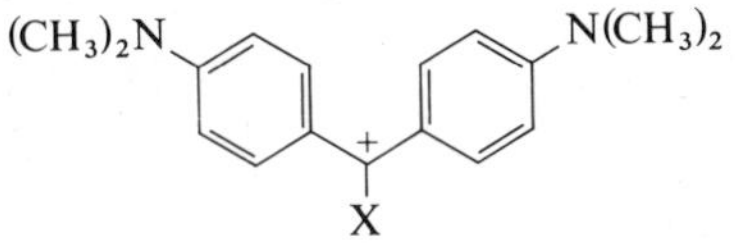

$(CH_3)_2N{-}C_6H_4{-}\overset{+}{C}(X){-}C_6H_4{-}N(CH_3)_2$

X	Solvent	λ_{max}(mμ)	mϵ	Ref.
H_2N	95% EtOH	434		64
HO		524		65
OC_2H_5		526		65
AcHN		Blue		
4-$(CH_3)_2NC_6H_4$	Water, pH 4	592	110.0	66
4-HOC_6H_4	AcOH	603	106.0	29
AcHN–(2-thienyl) 	Water	605		67
H	AcOH	608	148.0	29
4-$CH_3OC_6H_4$	AcOH	608	106.0	29

TABLE 9.10 (*cont'd*)

X	Solvent	λ_{max}(mμ)	mϵ	P$\epsilon\phi$.
H_3C–(S)	Water, pH 4	615	92.1	66
	Water	616		67
4-$CH_3CH{=}CHC_6H_4$	Water	617		68
4-$CH_3C_6H_4$	AcOH	617	108.0	29
	Water	620	93.9	66
4-$AcOC_6H_4$	AcOH	619		29
C_6H_5	AcOH	620	99.1	69
	Water, pH 4	623	83.5	66
CH_3 (S)		620		67
4-$C_6H_5COOC_6H_4$	AcOH	623	103.0	29
4-$CH_2{=}CHC_6H_4$	Water	623		68
(S)	Water	625		67
(S)	AcOH	627	88.7	69
4-$(CH_3)_2H\overset{+}{N}C_6H_4$		631	83.0	70
ϕ–(S)	Water	632		67
Br–(S)	Water	634		67
4–$(CH_3)_3\overset{+}{N}C_6H_4$		636		48
(O)	Water	637		67
3–$O_2NC_6H_4$	AcOH	638	87.0	29
	Water, pH 4	639	75.1	66
(S)	AcOH	645	75.6	69
4–$O_2NC_6H_4$	AcOH	645	83.0	29, 71
	Water, pH 4	648	55.9	66
N≡C–		>650		

TABLE 9.10 (*cont'd*)

X	Solvent	$\lambda_{max}(m\mu)$	$m\epsilon$	Ref.
Br, S	AcOH	661	81.0	69
O_2N, S	Water	662		67
Cl, S	AcOH	667	78.0	69
$C_6H_5C{\equiv}C$		675	63.0	72
$(CH_3)_2N$, $\overset{+}{N}$, $N(CH_3)_2$		~725		73

Table 9.11. For example, protonating the central positive resonance terminal of methylene blue causes a strong red shift since the NH^+ group is more strongly repellant to the resonating positive charge than the N group.

D. Positivizing a Positive Node

The violet shift effect of positivizing a positive node can be ascertained from the many examples listed in Table 9.10. Thus the increasing electron-donor strength of substituents at the central positive resonance node of the diphenylmethane cation causes a violet shift; for example, λ_{max} decreases as follows: H, 608 mμ; AcHN, blue; HO, 524 mμ; and H_2N, 434 mμ. The same phenomenon is seen for the 4-position of the unsubstituted ring of malachite green (Table 9.10); for example, H, λ_{max} 620 mμ; CH_3, 617 mμ; HO, 603 mμ; and $(CH_3)_2N$, 592 mμ.

This phenomenon applies to heterocyclic dyes (Table 9.11). For example, substituting the stronger electron-donor NH group for the oxygen of certain xanthene dyes causes a strong violet shift (e.g., λ_{max} 548 to 491 mμ).

Diarylcarbonyl compounds protonated at the oxygen atom could be expected to absorb at shorter wavelengths than analogous diarylcarbonium salts. Thus a solution of 3,3′-diguaiazulenyl ketone in 4 *N*

TABLE 9.11
Spectral Effect of Positivizing and Negativizing the Positive Resonance Nodes of Diphenylmethane and Diphenylamine Dyes

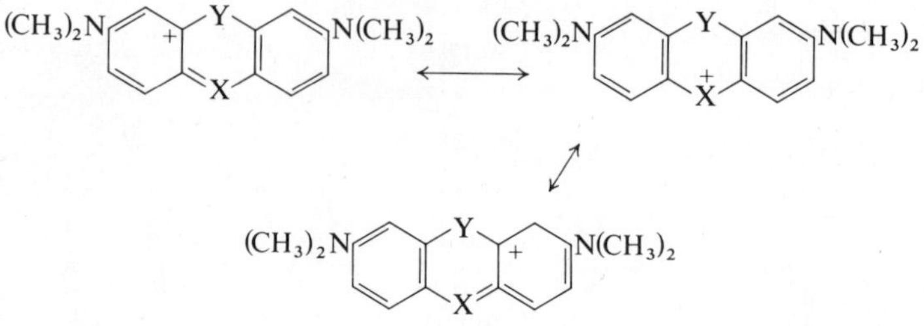

(CH$_3$)$_2$N ... Y ... N(CH$_3$)$_2$ (+) ⟷ (CH$_3$)$_2$N ... Y ... $\overset{+}{X}$ ... N(CH$_3$)$_2$ ⟷ (CH$_3$)$_2$N ... Y ... N(CH$_3$)$_2$ (+)

Substituents			
X	Y	λ_{max}(mμ)	Ref.
HC	NH	491	64
HOOCCH$_2$CH$_2$C	O	546	71
HC	O	548	64
C$_6$H$_5$C	O	550	71
HC	S	550	71
N≡C—C	NCH$_3$	Violet	74
N	NCH$_3$	Violet	74
N≡C—C	O	Blue	74
N	O	660	71
N	S	665	75
NH$^+$	S	742	75

ethanolic HCl absorbs at approximately 580 mμ, mϵ ~31.6(76). The analogous carbonium ion could be expected to absorb in the neighborhood of 640 mμ.

It is possible that, as the electron-acceptor (or electron-donor) strength of a negative (or positive) resonance node is increased, the resonance node can acquire the properties of a resonance terminal, and thus the chain of conjugation between the active resonance terminals is shortened. The result is an extreme violet shift and intensity decrease. This may partially explain the extremely large violet shifts of such compounds as **38**(77).

In **39**, $n = 0$ and 1, positivization of the positive resonance node and cross-conjugation effects probably play prominent roles in the violet shift caused by substitution of the oxygen by an NC$_6$H$_5$ group(77).

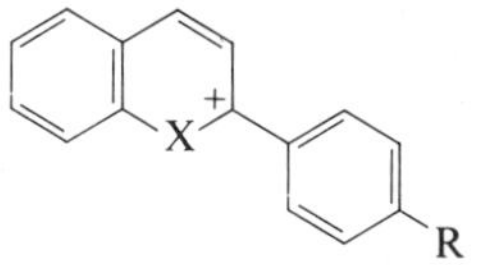

38

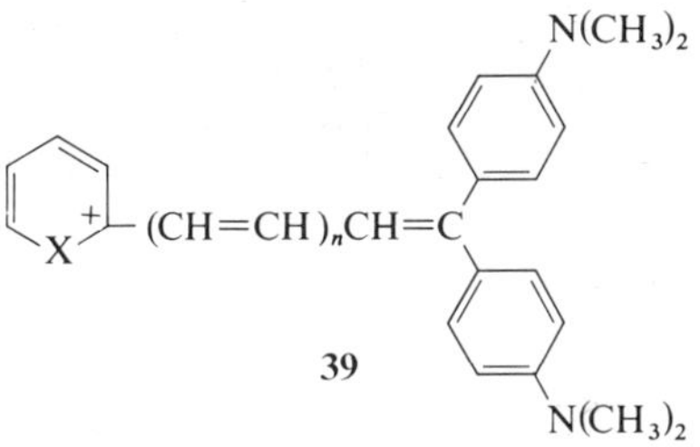

39

R	X	Color
H	O	Yellow
H	NCH_3	Colorless
$N(CH_3)_2$	O	Blue red
$N(CH_3)_2$	NCH_3	Orange

n	X	$\lambda_{max}(m\mu)$
0	O	736
0	NC_6H_5	494
1	O	852
1	NC_6H_5	590

E. Negativizing a Negative Node

The effect of substituting a more electronegative nitrogen atom for a methine negative resonance node is shown in compounds **40**(78). The resonance system of the trimethinecyanine dye is divided into two shorter resonance systems, with the ammonium nitrogen and the azeniate nitrogen as resonance terminals. When X = Y = N, the resonance chain is even shorter. As a result the shorter conjugated systems absorb at shorter wavelength and with lower intensity. The results have been confirmed with the analogous *N*-methyl trimethinecyanine dyes(79).

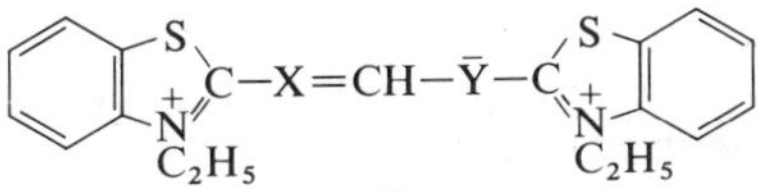

40

X	Y	Solvent	$\lambda_{max}(m\mu)$	$m\epsilon$
CH	CH	Methanol	558	148
N	CH	Methanol	466	63
N	N	Ethylene chloride	412	40

In a preceding section it was shown that negativizing a positive node in some trimethinecyanine dyes caused a red shift in the long wavelength maximum that decreased with increasing "basicity" of the resonance terminals in the heterocyclic nuclei(62). On the other hand, as shown in Table 9.12, negativization of a negative resonance terminal causes a

TABLE 9.12
Increase in the Violet Shift Effect of Negativizing a Negative Node on Increasing the "Basicity" of Resonance Terminals (62)

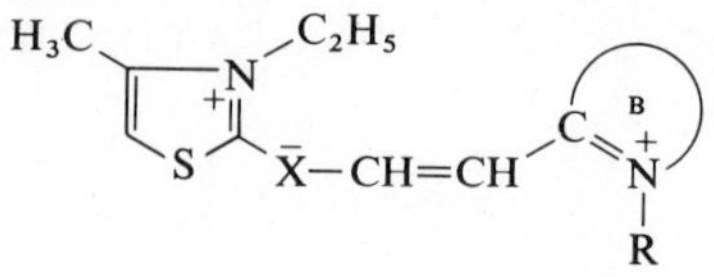

Nucleus B	$\lambda_{max}(m\mu)$ X=CH	$\lambda_{max}(m\mu)$ X=N	Shift $(m\mu)$
1,3,3-Trimethylindolenine	516	469	47
3-Ethylthiazoline	493	427	66
3-Ethylbenzoxazole	508	439	69
3-Ethylbenzoselenazole	554	473	81
3-Ethylbenzothiazole	550	472	78
3-Ethylnaphtho[1,2-*d*]thiazole	569	489	80
1-Ethyl-2-quinoline	583	500	83
3-Ethyl-4-methylthiazole	548	449	99
1-Ethyl-4-quinoline	645	534	111
1-Ethyl-2-pyridine	557	413	144

violet shift in the long wavelength maximum that increases with the increasing "basicity" of the resonance terminals (62). Many other examples of this phenomenon are known (61, 62).

In those types of molecules where an alkyl or phenyl substituent on a negative resonance node can cause greater steric hindrance in the ground state and thus a red shift, replacement of the alkyl group by an electronegative group can neutralize the red shift effect of the steric hindrance and the electron-donor group and cause a violet shift, as, for example, in **41** (80).

In other types of cationic resonance structures where even an electropositive substituent on a negative resonance node causes greater steric hindrance in the excited state and, thus, a violet shift, replacement of the electron-donor group by an electronegative substituent will reinforce the violet shift; an example is **42**.

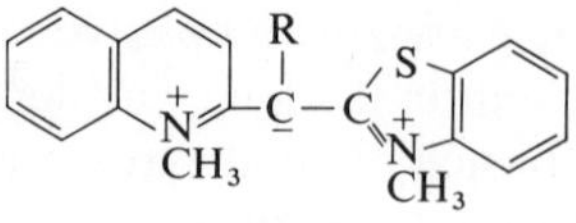

41

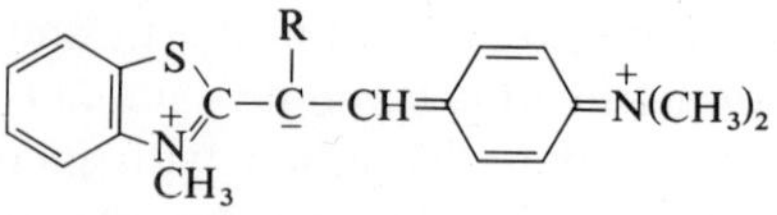

42

No.	R	λ_{max}(mμ)
41	H	486
	CH_3	545
	C_6H_5	552
	$COOC_2H_5$	474
42	H	525
	CH_3	460
	$COOC_2H_5$	418

An excellent example of negativizing a negative node has been reported for the compounds **43**(81). The long wavelength maximum shifts to shorter wavelengths with increasing electronegativity of the negative resonance node. Protonation of these molecules takes place on the negative resonance node, resulting in the formation of a dication that absorbs in the neighborhood of 320–340 mμ.

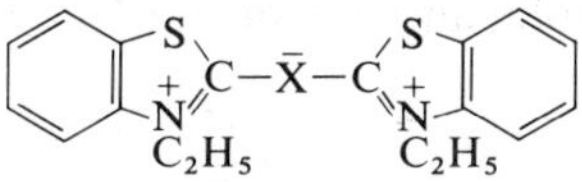

43

X	Electronegativity of X (82)	λ_{max}(mμ)	mϵ
P	2.15	472	48.1
C (H)	2.60	423	82.4
N	3.05	376	50.6

The spectral effect of negativizing and positivizing a negative resonance node can be demonstrated for some of the chromogens obtained in organic trace analysis. Thus in the analysis of aliphatic ketones (containing the structure RCH_2COCH_2R') with 2-hydroxy-1-naphthaldehyde as reagent, the electronegativity or electropositivity of substituents on the negative resonance node of the final chromogen (**44**) affects the long wavelength maximum(83). The negative resonance node is shown in the intermediate resonance structure **44**, with the negative charge on the node. A steric effect would also contribute to the violet shift.

Another intensely colored chromogen formed in organic trace analysis with 3-methyl-2-benzothiazolinone hydrazone as the reagent shows this interesting phenomenon. The chromogen **45a** is formed in the assay of formaldehyde(84), sugars(85, 86), ethanolamines(87), 21-hydroxy-20-

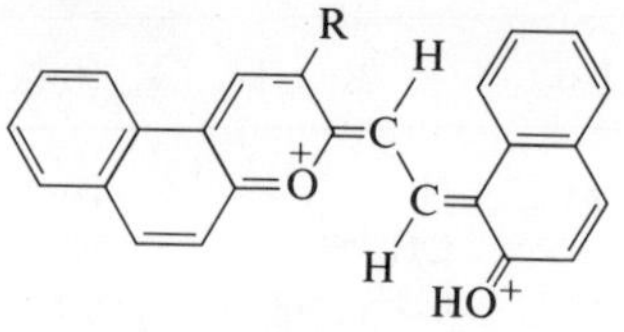

44

Test substance	R	$\lambda_{max}(m\mu)$
Acetone	H	584
2-Butanone	CH_3	598
Ethyl acetoacetate	$COOC_2H_5$	536

ketosteroids (85, 86), olefins (88), unsaturated acids (88), and other formaldehyde precursors. When the central negative resonance node of this chromogen is replaced with an aza nitrogen, as in **45b**, the wavelength maximum shifts to the violet and the intensity decreases (89, 90).

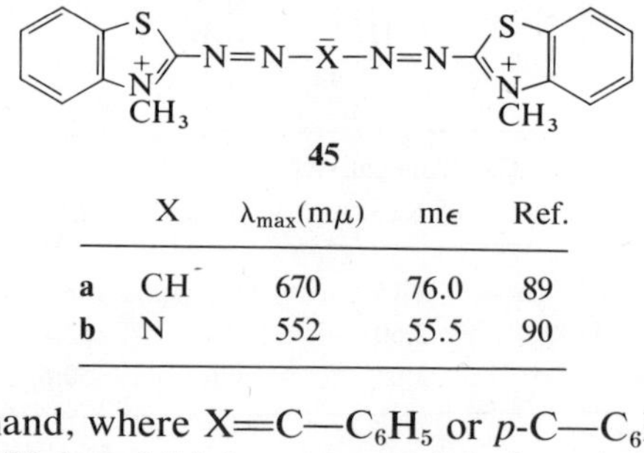

45

	X	$\lambda_{max}(m\mu)$	$m\epsilon$	Ref.
a	CH^-	670	76.0	89
b	N	552	55.5	90

On the other hand, where X=C—C_6H_5 or *p*-C—$C_6H_4NO_2$ (λ_{max} 670 and 660 mμ, mϵ 76.0 and 71.0, respectively) there is little effect on the wavelength position.

VI. *Ortho–Para* Effect

The *ortho–para* effect has been discussed in Section III of Chapter 6. Essentially the red shift effect in the present instance results when an electron-donor group is substituted in the *ortho* or *para* position of a benzene ring attached to a positive resonance terminal of a monomethine or polymethine cationic resonance structure. Some examples of this phenomenon have been discussed by Wizinger (91). With an electron-donor group in the *ortho* or *para* position to the resonance terminal the electron-donor strength of the terminal is increased considerably, as reflected by the red spectral shift.

Some examples of the *ortho* effect are shown in Table 9.13. Comparison of the diphenylamine, phenoxazine, and phenothiazine derivatives clearly shows the red shift effect with the increasing electron-donor

TABLE 9.13
Ortho Effect and the Long Wavelength Maximum of Some Cationic Resonance Structures in Acetic Acid (92)

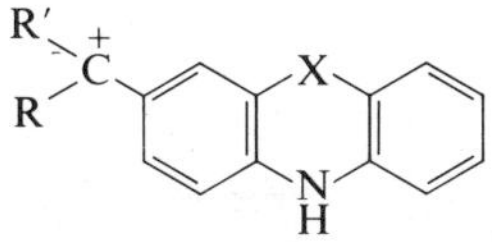

Substituents		λ_{max} (mμ) (mϵ)		
R	R′	X = H_2	X = O	X = S
C_6H_5	C_6H_5	493[a]	620	>620
4-$CH_3OC_6H_4$	C_6H_5	541	658	707
4-$CH_3OC_6H_4$	4-$CH_3OC_6H_4$	554	675	720[b]
4-$(CH_3)_2NC_6H_4$	C_6H_5	622 (83.2)	715 (28.8)	715 (20.9)
4-$(CH_3)_2NC_6H_4$	4-$(CH_3)_2NC_6H_4$	595 (77.6)[a]	596 (79.4)	605 (53.7)

[a]In ethanol.
[b]For X = Se, λ_{max} 720 mμ. For the *N*-methyl derivative where X = SO_2, λ_{max} 537 mμ (93).

strength of the *ortho* substituent to the NH resonance terminal. However, when the *ortho* substituent is an electron-acceptor group (e.g., X = SO_2 in Table 9.13), a violet shift results as compared with the phenothiazine analog (X = S). The base-weakening effect of the sulfone group is shown by the longer wavelength absorption of the diphenylamine analog, λ_{max} 554 mμ. In the bisdimethylamino derivatives the red shift effect of the *ortho* substituent is practically negligible.

Other compounds that show this effect are **46** and **47** (94).

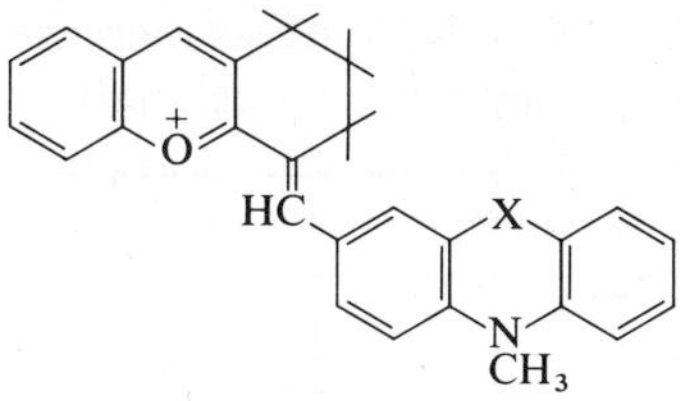

46

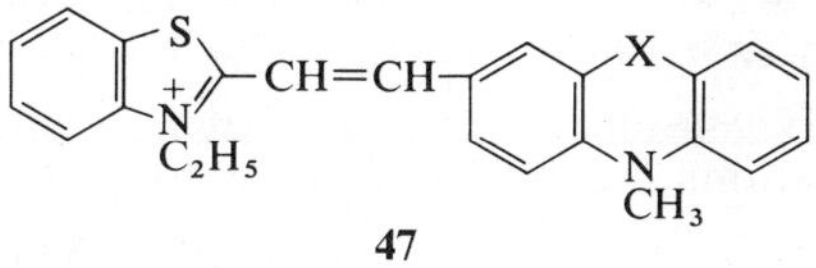

47

No.	X	λ_{max} (mμ)
46	H_2	660
	S	690
47	H_2	516
	S	561

The red shift is especially strong in the triarylmethane cations **48** and **49** (spectra in acetic acid) (91).

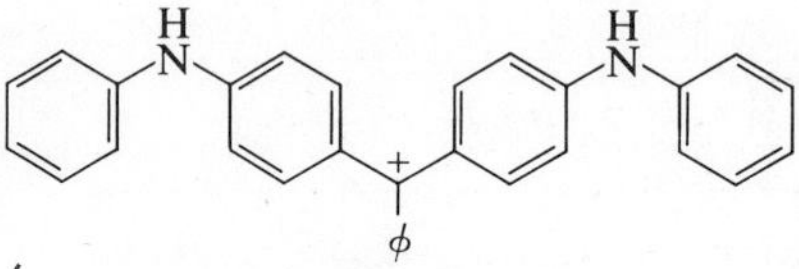

48

λ_{max} 444, 633 mμ
mϵ 14.5, 57.5

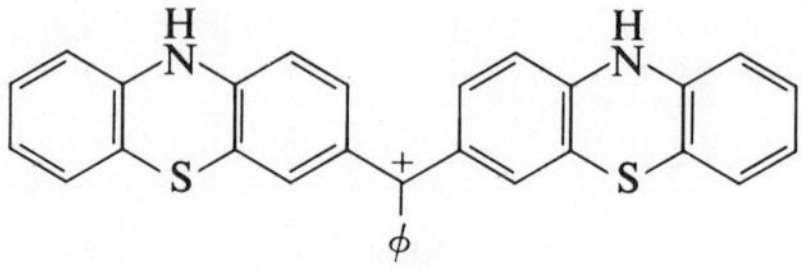

49

λ_{max} 440, 640br, 865 mμ
mϵ 11.4, 9.9, 39.8

The electron-donor group *ortho* to the resonance terminal does not have to be part of the ring system for its effect to be felt, as, for example, in **50** (spectra in acetic acid).

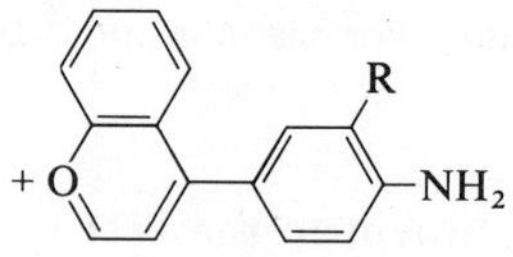

50

R	λ_{max} (mμ)
H	538
CH_3	550
CH_3O	573

A comparison of substitutions of electron-donor groups *meta*, *ortho*, and *para* to the resonance terminal shows that the red shift effect increases in that order, whereas the basicity of the compounds under study increases in the reverse order (Table 9.14). In the 2,5-polysubstituted cations the intensity of the long wavelength bands are low, as is to be expected of a compound in which the straight-line distance between resonance terminals is short, but the bands are very broad. The 2,2′,2″,4,4′,4″-hexamethoxytriphenylcarbinol is so strongly basic that it will dissolve in dilute acetic acid as the colored cation.

TABLE 9.14
Relative Basicity and the *Ortho–Para* Effect of Some Triphenylcarbonium Cations in Acetic–Perchloric Acid(91)

Triphenylcarbonium dye	Relative basicity	λ_{max} (mμ)	mϵ
2,4-Dimethoxy	57	502	17.0
3,4-Dimethoxy	43	517	20.9
2,5-Dimethoxy	6	680	0.96
2,2′,4,4′-Tetramethoxy	1200	565	25.1
2,2′,5,5′-Tetramethoxy	~ 9	725	5.0
2,2′,2″,4,4′,4″-Hexamethoxy	∞	559	31.6
2,2′,2″,5,5′,5″-Hexamethoxy	~ 11	669	8.1

Some of the factors that could influence the extent of the *ortho–para* effect are the relative electron-donor strengths of the resonance terminals and the length of the chain of conjugation between the terminals. The latter effect can be seen in the spectral data for the asymmetrical cationic resonance structures **51** in acetic acid(91).

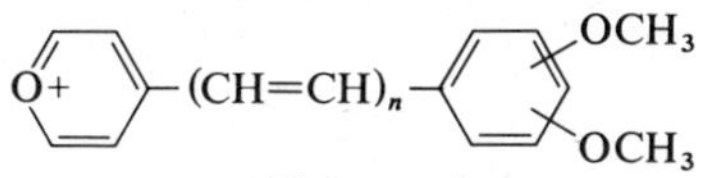

O+ —(CH=CH)$_n$— $(OCH_3)_2$

51

Position of CH_3O groups —λ_{max} (mμ)

n	*2,4*	*3,4*	*2,5*
0	446	452	482
1	522	512	510

The red shift effect of an electron-donor group in the *para* position to a resonance terminal is shown for **52** in acetonitrile(95,96).

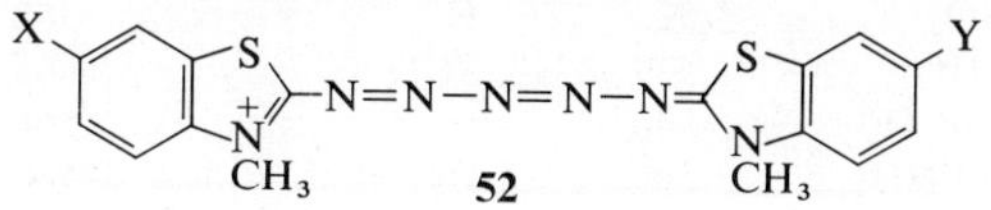

X, S, +, N, CH_3 —N=N—N=N—N= S, N, CH_3, Y

52

X	Y	λ_{max} (mμ)	mϵ
H	H	552	55.5
CH_3O	H	579	55.6
CH_3O	CH_3O	597	64.6

The magnitude of the violet shift effect of electron-acceptor groups and the red shift effect of electron-donor groups *para* to a resonance terminal depends on the relative electron-acceptor or electron-donor strength of the substituents (Table 9.15). A somewhat similar phenomenon is found in the azo-dye cations (Table 9.16). The 2′-amino isomer of the 4′-amino azo dye, (λ_{max} 610 mμ) absorbs at much longer wavelength – 643 mμ.

Benzopyrylium cations also show an increasing red shift with substitution of increasingly stronger electron-donor groups *para* to the resonance terminal; an example is **53** (100).

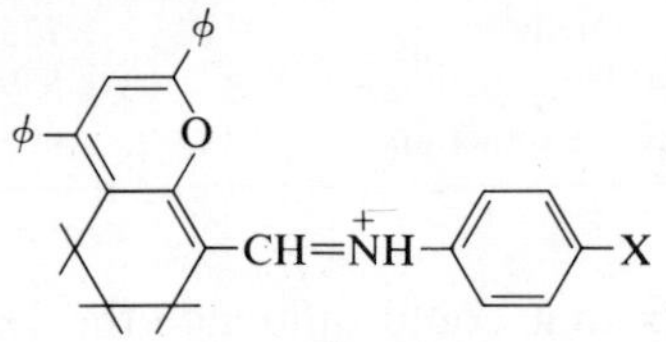

ϕ ϕ O CH=$\overset{+}{N}$H X

53

X	λ_{max} (mμ)
H	547
OCH_3	557
NHC_6H_5	580

TABLE 9.15
Spectral Effect of Substituents *Para* to the Resonance Terminal in 9-Acridinecarboxaldehyde Phenylhydrazone Cations (97,98)

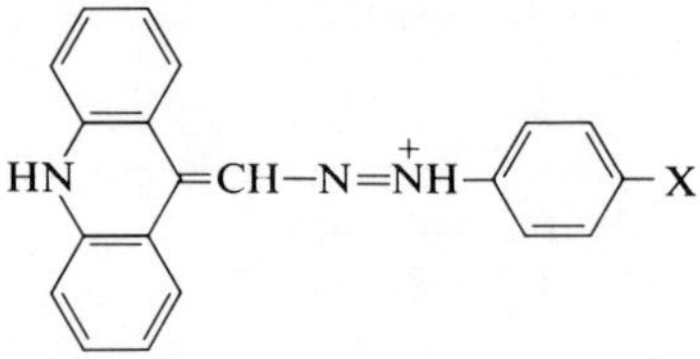

HN =CH–N=$\overset{+}{N}$H– X

X	λ_{max} (mμ)[a]	X	λ_{max} (mμ)[a]
NO_2	522	H	562
CHO	539	SCN	565
$COCH_3$	540	I	566
SO_2NH_2	540	CH_3	580
COOH	546	NHAc	599
$\overset{+}{N}H_3\,Cl^-$	550	OCH_3	610
Cl	554	OH	610
SO_3H	557	NHC_6H_5	635
Br	560	*p*-$NHC_6H_4OCH_3$	715

[a]In 0.02 *N* ethanolic HCl.

TABLE 9.16
Spectral Effect of Substituents *Para* to the Resonance Terminal in 4-Phenylazodimethylaniline Cations (99)

$$X-C_6H_4-\overset{H}{\underset{+}{N}}=N-C_6H_4-N(CH_3)_2$$

X	λ_{max} (mμ)[a]	X	λ_{max} (mμ)[a]
$\overset{+}{N}H(CH_3)_2$	505	i-C_3H_7	533
NO_2	508	C_6H_5	544
Ac	515	NHAc	545
COOH	515	OCH_3	548
H	516	SCH_3	559
SCN	516	NH_2	610[b]
CH_3	531	$N(CH_3)_2$	664[b]
C_2H_5	532		

[a]In 50% ethanol–1 *N* HCl, unless otherwise specified.
[b]In glacial acetic acid.

Table 9.17 shows other compounds in which either an electron-donor or an electron-acceptor group *para* to a resonance terminal will cause a red shift whose magnitude is dependent on the electron-donor or electron-acceptor strength of the substituent (101–105). The benzoxazole analogs of compound **I** in Table 9.17 show similar properties (103).

The neutralization of the *para* effect (stemming from dimethylamino groups *para* to the resonance terminals) by the increasing size of an alkyl substituent adjacent to the amino group is shown in Table 9.18. In this case steric hindrance has a similar effect (compound 6 is closely similar spectrally to compound 3) to protonation of the dimethylamino group (compound 7 is closely similar spectrally to No. 4).

A different interaction between the *para* effect and steric hindrance has been reported in the structures **54** (107). Because the resonance terminal on the left is more basic due to the *para* effects of the dimethylamino and nitro groups, structure **A** contributes mainly to the ground state and structure **B** to the excited state. When $R = CH_3$ in **54**, the steric hindrance is greater in the excited state. This results in a violet shift. When $R' = CH_3$, steric hindrance is greater in the ground state. This results in the red shift to 601 mμ. The electronic asymmetry of the dye is diminished, as is evident from the decrease in the deviation to 7 mμ (see Section VII.A on deviation).

In the spectrophotometric determination of malonaldehyde with aniline as reagent the chromogen **II** in Table 9.17, $n = 1$ and $X = H$, is obtained

TABLE 9.17
Effect of the Electron-Donor and Electron-Acceptor Strength of Substituents *Para* to a Resonance Terminal (101–104)

$$\text{(benzothiazoline, N-}C_2H_5\text{)}=CH-CH=\overset{+}{N}H-C_6H_4-X$$

I

$$X-C_6H_4-NH(CH=CH)_nCH=\overset{+}{N}H-C_6H_4-X$$

II

	λ_{max} (mμ)		
		II	
X	I	$n=1$	$n=2$
NO_2	442[a]	–	524
Ac	429[b]	409[c]	513
$COOC_2H_5$	425[d]	403[e]	506
COOH	423	400	498
H	412[f,g]	385[h]	485[i]
Cl	421	392	494
OCH_3	426[j]	405	509
$N(CH_3)_2$	488	–	548

[a] mϵ 36.3 (101).

[b] mϵ 63.1 (101).

[c] λ_{max} 395 mμ, mϵ 63.1 in ethanol (102).

[d] mϵ 83.2 (101).

[e] λ_{max} 386 mμ, mϵ 56.2 in ethanol (102).

[f] λ_{max} 414 mμ, mϵ 55.0 in methanol (105).

[g] The *N*-acetyl derivative absorbs at 364 mμ, mϵ 10.0, in methanol (105).

[h] λ_{max} 383 mμ, mϵ 50.0, in methanol (105).

[i] λ_{max} 485 mμ, mϵ 65.0, in methanol (105).

[j] For $X = SCH_3$, λ_{max} 436 mμ, mϵ 39.8, and for $X = OCH_3$, mϵ 47.9 (103).

TABLE 9.18
Effect of Steric Hindrance on the *Para* Effect(106)

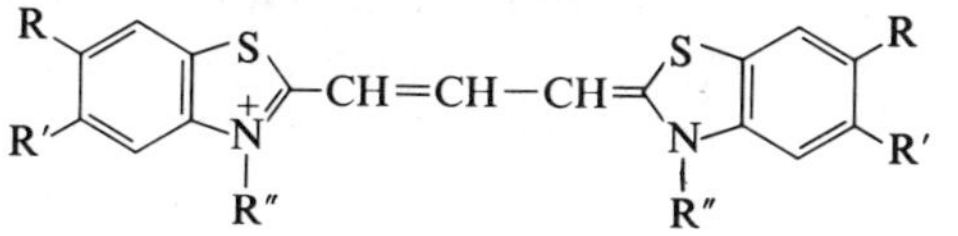

R, R′, S, $\overset{+}{N}$, R″ —CH=CH—CH= S, N, R″, R, R′

No.	Substituents R	R′	R″	λ_{max} (mμ)	mϵ
1	H	$N(CH_3)_2$	C_2H_5	608	8.2
2	CH_3	$N(CH_3)_2$	CH_3	575	10.1
3	$C(CH_3)_2$	$N(CH_3)_2$	CH_3	570	14.1
4	H	H	C_2H_5	558	16.5
5	CH_3	H	C_2H_5	563	14.5
6	$C(CH_3)_3$	H	CH_3	563	14.9
7	H	$\overset{+}{N}H(CH_3)_2$	C_2H_5	558	16.4

(108). This chromogen absorbs at λ_{max} 387 mμ, mϵ 48.6, under the reported analytical conditions, as compared with λ_{max} 383 mμ, mϵ 50.0, in methanol(105) as reported for the pure compound. The wavelength maximum of the chromogen could be shifted to longer wavelength by the substitution in the *para* position of the aniline reagent of electron-donor or electron-acceptor groups (see Table 9.17).

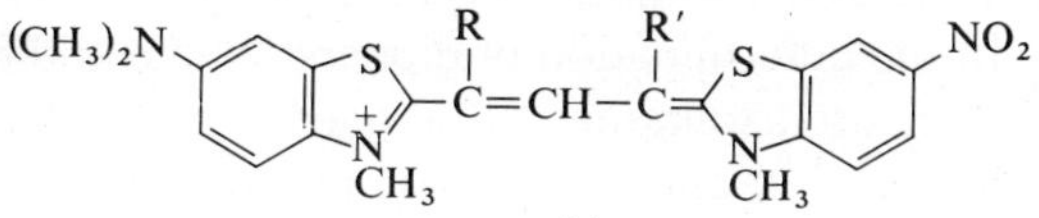

$(CH_3)_2N$, S, $\overset{+}{N}$, CH_3 —C(R)=CH—C(R′)= S, N, CH_3, NO_2

54a

↕

$(CH_3)_2N$, S, N, CH_3 =C(R)—CH=C(R′)— S, $\overset{+}{N}$, CH_3, NO_2

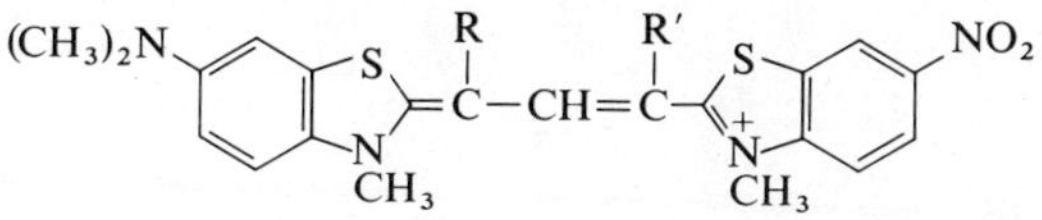

54b

R	R′	λ_{max} (mμ)	Deviation (mμ)
H	H	586	11
CH_3	H	556	44.5
H	CH_3	601	7

Other examples of the *ortho–para* effect in trace analysis are found in the determination of phenalkyl ethers with piperonal dichloride (109), phenols and aromatic hydrocarbons with various isatin reagents (Table 9.19) (52), and diphenylamines with 3-methyl-2-benzothiazolinone hydrazone (57). Thus in the piperonal determinations the red spectral shift

TABLE 9.19
Ortho–Para Effect and the Isatin Determinations of Aromatic Hydrocarbons and Phenols (52)

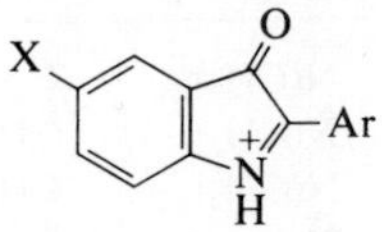

Ar	λ_{max} (mμ)		
	X = H	X = CH_3	X = OCH_3
Phenol	560		590
1-Naphthol	600		645
1-Anthrol	685		675, 730
Pyrene	685	697	725
Anthanthrene	860	865	895

caused by the *ortho–para* effect is seen when the chromogen from anisole (λ_{max} 527 mμ) is compared with the diphenylmethane chromogens from 1,2-diethoxybenzene and 1,4-dimethoxybenzene (λ_{max} 550 and 605 mμ, respectively). In the determination of diphenylamine and phenothiazine with 3-methyl-2-benzothiazolinone hydrazone the phenothiazine chromogen **55b** shows the red shift effect when compared with its diphenylamine analog **55a** (57).

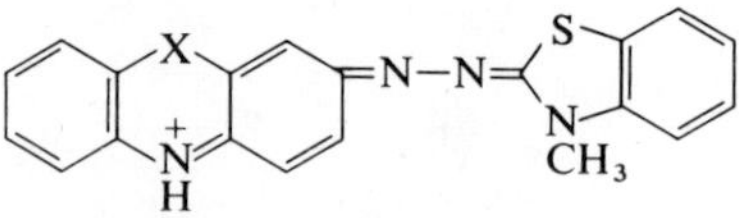

55

	X	λ_{max} (mμ)	mϵ'[a]
a	H_2	606	78
b	S	720	45
		765*s*	41

[a]mϵ' is mϵ obtained in analytical procedure.

The chromogens **II** in Table 9.17, $n = 2$, are obtained in the determination of pyridine and its derivatives through ring splitting to a glutaconaldehyde, which is then condensed with an aromatic amine. Thus pyridine can be determined through ring splitting with cyanogen bromide and condensation with aniline(110,111), benzidine(112), 2-naphthylamine(113), or *p*-aminoacetophenone(114, 115). Nicotinic acid and nicotinamide have been determined in a similar fashion with aniline(116, 117), *p*-aminoacetophenone(117, 118), sulfanilamide(119–120), sulfanilic acid(121–123), *m*- phenylenediamine(124), and procaine(125). Nicotine can also be determined by this reaction with aniline(126), 2-naphthylamine(127, 128), and benzidine(129) as the amine reagents.

The method has been applied to the estimation of *N*-(2-pyridyl) substituted antihistamines, such as tripelennamine, thenylpyramine, methapyrilene, pyrilamine, prophenpyridamine, and doxylamine, with aniline as the reagent(130), and to chlorpheniramine maleate, with sulfanilic acid as the reagent(131). β-Pyridylcarbinol can also be assayed by reaction with cyanogen bromide followed by *p*-aminoacetophenone(115). The chromogen in this case absorbs at 560 mμ—a longer wavelength than the analogous chromogen (λ_{max} 513 mμ) obtained from pyridine. This red shift is due to positivizing the negative node with the CH_2OH grouping.

VII. Resonance Terminals

A. Deviation

The spectrum of an asymmetrical cationic resonance structure is affected by the relative stabilities of the two limiting resonance structures (see **56**). If the energies of these two structures are of the same order of magnitude, the value of the wavelength maximum is close to the arithmetic mean of the values for the two corresponding symmetrical dyes; for example, compare **56** with the symmetrical dyes **57** and **58**(63, 115) all spectra in methanol).

Put in other terms, if the intramolecular or internal basicity of the two resonance terminals **A** and **B** of an asymmetrical cationic resonance structure (e.g., **56**) are nearly equivalent, the wavelength maximum of this compound would be almost equal to the arithmetic mean of the wavelength maxima of the two symmetrical cationic resonance structures, one containing **A** at each end of the conjugation chain (e.g., **57**) and the other containing **B** at each end of its conjugation chain (e.g., **58**). The difference in wavelength maxima between the asymmetrical structure and the

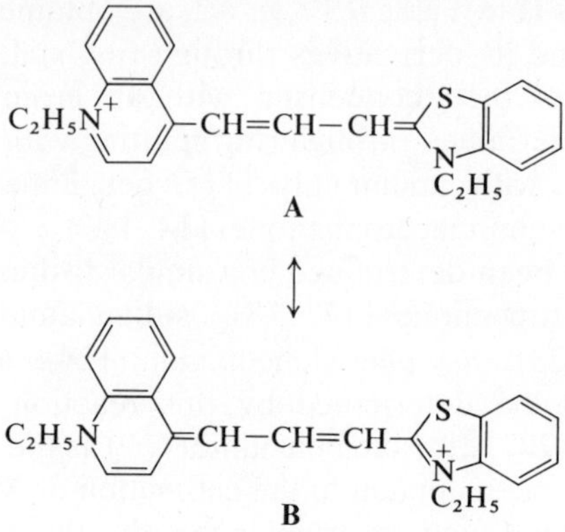

$C_2H_5\overset{+}{N}$ —CH=CH—CH= S N C_2H_5

A

↕

C_2H_5N =CH—CH=CH— S $\overset{+}{N}$ C_2H_5

B

56

λ_{max} 630 mμ

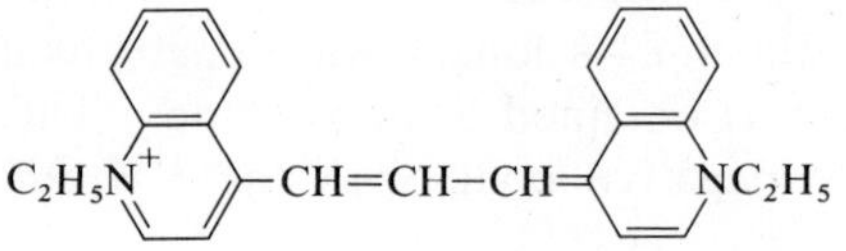

$C_2H_5\overset{+}{N}$ —CH=CH—CH= NC_2H_5

57

λ_{max} 705 mμ

S =CH—CH=CH— S N C_2H_5 $\overset{+}{N}$ C_2H_5

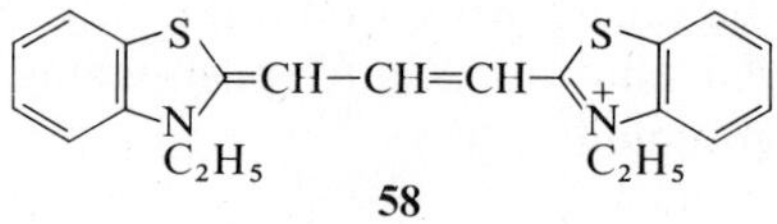

58

λ_{max} 557.5 mμ

arithmetic mean of its two symmetrical parents is called the deviation (132). Thus the deviation for **56** is

$$630-\left(\frac{705+557.5}{2}\right)=-1\ \text{m}\mu$$

This result signifies that the two resonance terminals in **56** are about equal in internal basicity.

The evidence indicates that, the greater the difference of basicity between the two resonance terminals of an asymmetrical cationic resonance structure—and therefore the greater the difference in energy between the two limiting resonance structures—the greater will be the deviation for a given conjugation chain length and the greater the degree of convergence with increasing chain length of the series. This has been shown to hold true for cyanine dyes (132).

These concepts have been most fully developed by Brooker and his co-workers (53, 63, 132), and have been described by Kiprianov (133, 134) and Schwarzenbach (135).

An example of a compound that exhibits a very large deviation is **59**, as shown by spectral comparison with its symmetrical parents **60** and **61** (136). For **59** the deviation is -203 mμ. In this compound there is a great disparity between the internal basicities, or "onium minus ide" stabilizations, of the two resonance terminals. The nitride terminal is much more strongly basic than the oxide one.

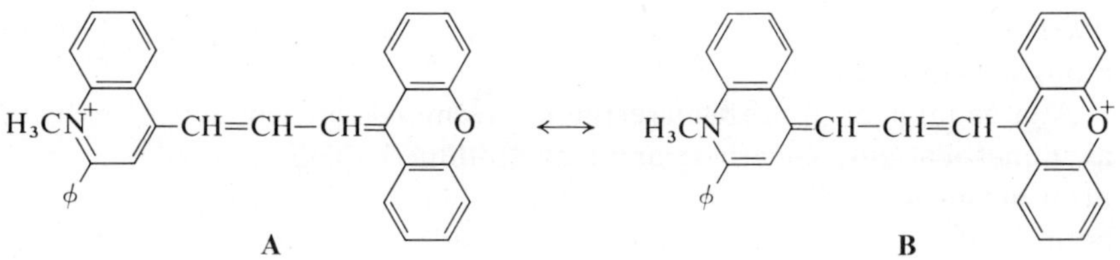

A ⟷ **B**

59

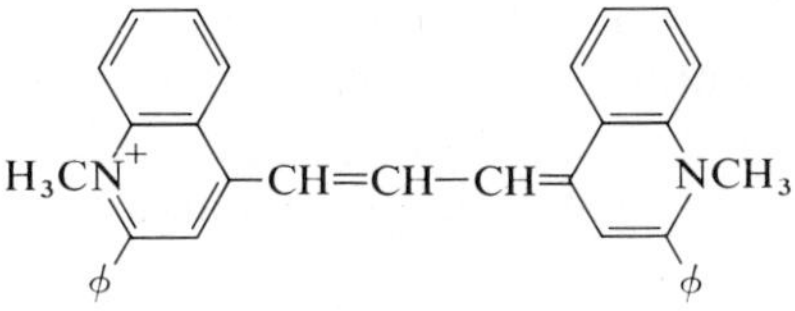

60

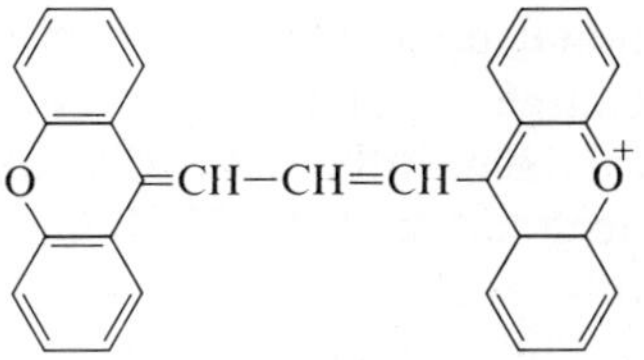

61

No.	Solvent	λ_{max} (mμ)
59	Nitromethane	510
60	Ethanol	723
61		703

Structure **59A** could be expected to contribute strongly to the ground state, whereas **59B** would contribute mainly to the excited state. If this is true, then the internal basicities of resonance terminals in the excited state should be widely different from those in the ground state.

The internal basicity or the "onium minus ide" stabilization of a positive resonance terminal in a cationic resonance structure is defined as the relative attraction of the uncharged form of a resonance terminal for a positive charge. Any gain in "onium minus ide" stabilization is affected by a variety of factors—such as steric hindrance of some part of the chromogen, the electron-donor strengths of the resonance terminals, the electron-acceptor strengths of the negative resonance nodes, the electron-donor strengths of the positive resonance nodes, the resonance stabilization of the ring systems present in the chromophore, and the length of conjugation between the terminals. Examples of "onium minus ide" resonance terminal systems include

$$-NH_2 \leftrightarrow =\overset{+}{N}H_2, \quad -NH- \leftrightarrow =\overset{+}{N}H- \quad -N= \leftrightarrow -\overset{+}{N}-,$$

$$-O- \leftrightarrow =\overset{+}{O}-, \quad -S- \leftrightarrow =\overset{+}{S}-, \quad =CH- \leftrightarrow -\overset{+}{C}H-$$

Intramolecular basicity will differ from intermolecular basicity, which is determined by the ease of proton (or cation) addition to the unsaturated atom of a compound.

A large number of nitrogen resonance terminals present in two series of asymmetrical pyrrolocarbocyanine and *p*-dimethylaminostyryl dyes have been arranged in the order of their internal basicities in Figure 9.2. In both series increasing deviation is considered to indicate increasing basicity of the variable resonance terminal. With a few exceptions the order of basicity in the two series is similar. Internal basicity tends to be augmented if a resonance terminal is part of a ring that acquires an additional double bond when the resonance terminal acquires the positive charge. For pyridine, N^{IV}–N^{III}, stabilization is very high; it is less for thiazole, still less for thiazoline, and practically zero for pyrrole, in which no additional double bond enters the ring; hence the deviations are in this order.

The resonance stabilization depends not only on the number of structures but also on the magnitude of their interaction(137). In both series the indole dyes have the lowest deviation, and hence the indole-nitrogen resonance terminal is the most weakly basic of the series; the 1,3-diethylbenzimidazole resonance terminal has the highest deviation and hence is the most basic of the series.

High energy quinonoid structures formed when a resonance terminal is in the onium form will decrease the internal basicity of the terminal. For this reason the dimethylamino nitrogen resonance terminal present in the compounds of Table 9.20 is less basic than the other nitrogen resonance terminals. Thus increasing the basicity of the benzothiazole nitrogen resonance terminal by appropriate substitutions on the terminal or in the ring shifts the wavelength maximum toward shorter wavelengths and increases the deviation(137). For both types of substitutions increasing the electron-donor strength of the benzothiazole resonance terminal increases the internal basicity of that terminal.

The much weaker electron-donor strength of the oxygen as compared with the nitrogen resonance terminal is reflected by the weaker internal basicity of the oxygen terminal.

The order of decreasing internal basicities for some nitrogen and oxygen resonance terminals is shown in Table 9.21 for a group of cationic

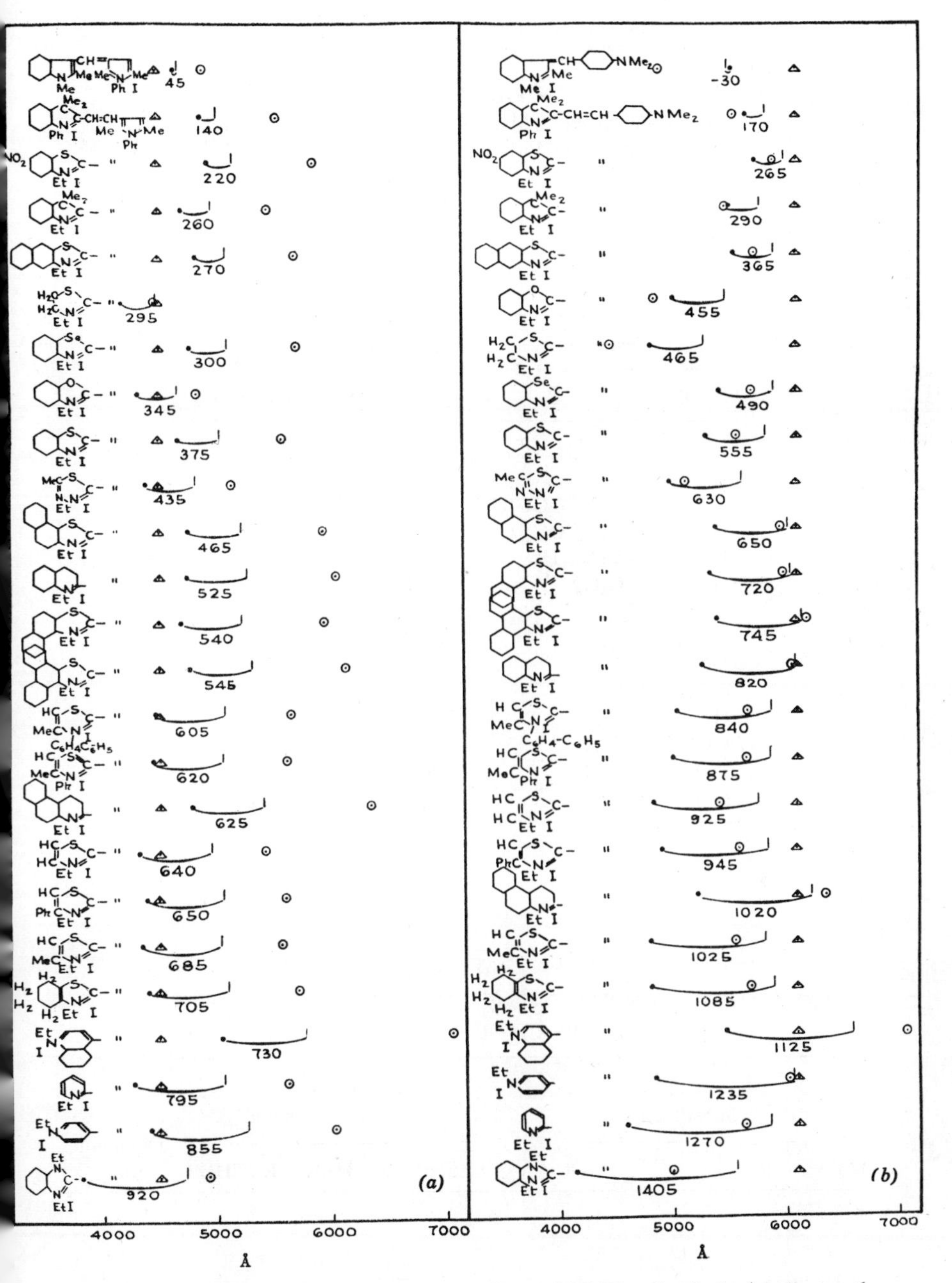

Fig. 9.2. Order of basicity of 25 heterocyclic nuclei(137). Symbols: | indicates the arithmetic mean between ⊙ and ◬; ⌣ indicates deviation, the magnitude of which is given in ångströms. (*a*) Values of λ_{max} in methanol are as follows: ● asymmetrical cyanine; ◬ symmetrical cyanine with two pyrrole nuclei; ⊙ second symmetrical cyanine. (*b*) Values of λ_{max} in nitromethane are as follows: ● styryl dye; ⊙ symmetrical cyanine; ◬ Michler's hydrol blue.

TABLE 9.20
Effect on Wavelength Maxima and Deviation of Increasing the Basicity of the More Basic Resonance Terminal (137)

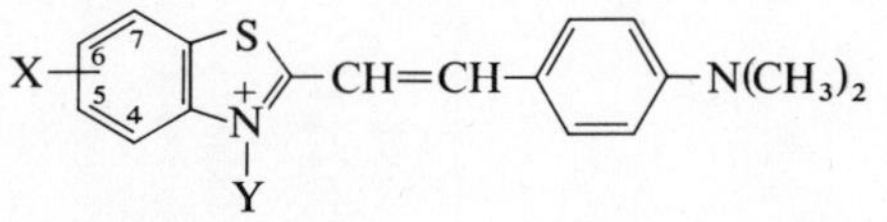

Substituents X	Substituents Y	λ_{max} (mμ)	Deviation (mμ)
H	$2'-C_6H_4NO_2$	561	−29
H	$4'-C_6H_4NO_2$	558	−33
H	$CH_2COOC_2H_5$	551	−36
H	$CH_2COC_6H_5$	545	−42
H	$CH_2C_6H_5$	542	−44
H	C_2H_5	528	−56
$4-(CH_2)_3$		525	−59
$6-NO_2$	C_2H_5	574	−27
$5-NO_2$	C_2H_5	556	−29
6-Cl	C_2H_5	540	−46
H	C_2H_5	528	−56
$6-CH_3$	C_2H_5	526	−62
$5-OC_2H_5$	C_2H_5	532	−62
$6-OC_2H_5$	C_2H_5	525	−67

TABLE 9.21
Deviation of the Long Wavelength Maxima of Some Trimethine Cations (136)

$$R{=}CH{-}\overset{+}{C}H{-}CH{=}R'$$

No.	Substituents R	Substituents R'	λ_{max} (mμ) Cation	Parent system R(CH)$_3$R	Parent system R'(CH)$_3$R'	Deviation (mμ)
1	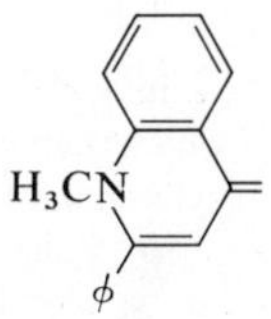 H_3CN (2-φ-quinolylidene)	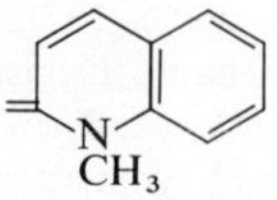 N-CH_3 (quinolylidene)	663	723	607	−2

TABLE 9.21 (*cont'd*)

No.	Substituents R	Substituents R′	λ_{max} (mμ) Cation	Parent system $R(CH)_3R$	Parent system $R'(CH)_3R'$	Deviation (mμ)
2	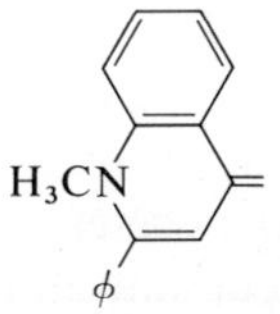 H_3CN, φ	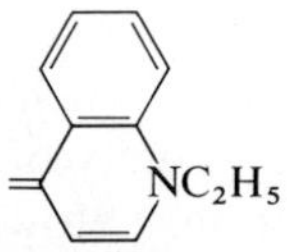 NC_2H_5	718	723	699	−2
3	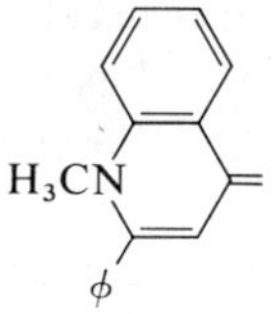 H_3CN, φ	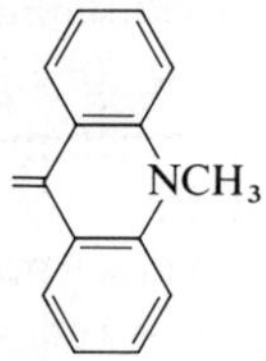 NCH_3	675	723	810	−91
4	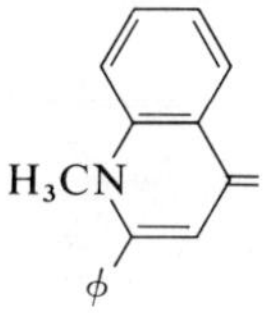 H_3CN, φ	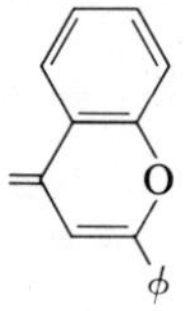 O, φ	598	723	705	−116
5	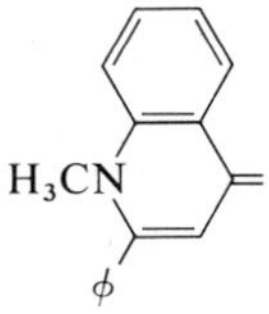 H_3CN, φ	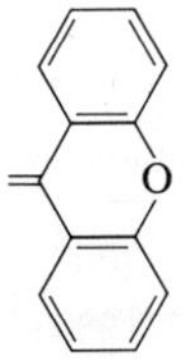 O	510	723	703	−203
6	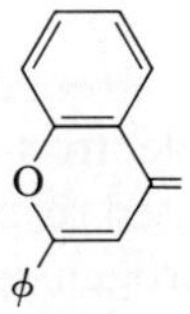 O, φ	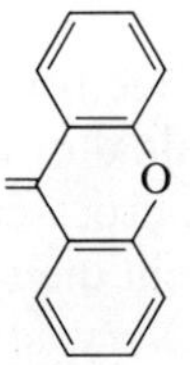 O	660	705	703	−44
7	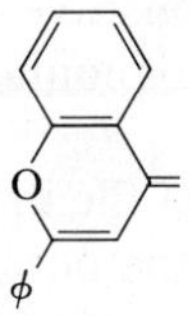 O, φ	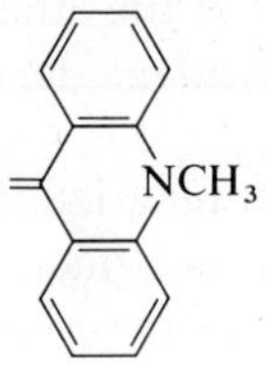 NCH_3	675	705	810	−83

resonance structures with one resonance terminal held constant. Thus in the series of dyes containing a 1-methyl-2-phenylquinoline nitrogen terminal connected to a varying resonance terminal by the same conjugation chain length, as the internal basicity of the varying resonance terminal is decreased the wavelength maximum shifts to the violet; for example, λ_{max} for compounds 2, 3, 4, and 5 is 718, 675, 598, and 510 mμ, respectively.

For compounds 6, 7, and 4 of Table 9.21, in which the flavylium oxygen terminal is held constant, the increasing basicity of the other resonance terminal is shown by the increasing deviation. However, the wavelength maxima of these asymmetric cations do not show a clear-cut shift to the violet. This is because in compound 6 the xanthene oxygen is less basic than the flavene oxygen, whereas the acridine nitrogen in No. 7 is more basic, and the quinoline nitrogen in No. 4 is much more basic. Thus the latter two compounds show the expected shifts.

An example of decreasing the basicity of one terminal toward energetic symmetry of the two terminals resulting in a red shift and a deviation decrease is shown by compounds **62** and **63** (spectra in acetic acid)(138).

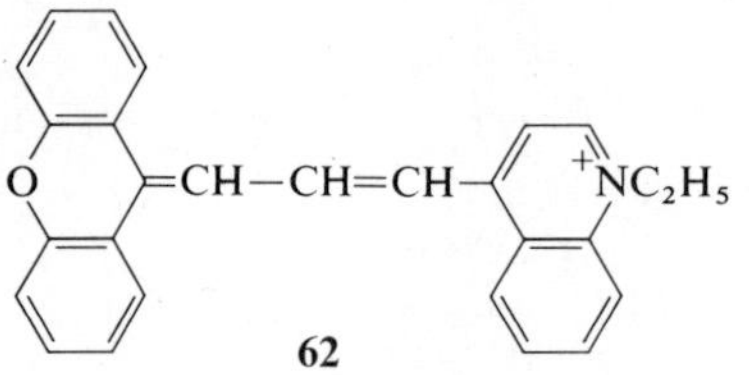

=CH−CH=CH− $\overset{+}{N}C_2H_5$ =CH−CH=CH− $\overset{+}{X}$

62 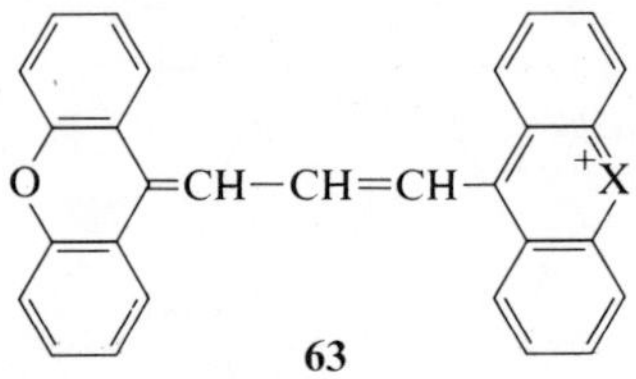 **63**

No.	X	λ_{max} (mμ)	Deviation (mμ)
62		502	−200
63	NCH_3	570	−188
	S	722	Probably ~ 0

In some compounds, depending on its electron-donor strength, the hetero atom could react as a resonance terminal or as an electron-donor substituent on a positive resonance node of a cross-conjugated resonance system. Examples are **38** and **39**(77), wherein the ring nitrogen appears to act as a resonance terminal, and as a very strong electron-donor substituent modifying the electron density of the adjacent positive resonance node; for example, in the oxygen derivatives the chain of conjugation between terminals appears longer.

Compounds **64** and **65** and **66** and **67** summarize much of the information on deviation. From previous data it is evident that the benzothiazole nitrogen terminal is about as basic as the quinoline nitrogen but more basic than the 4-dimethylaminophenyl nitrogen terminal.

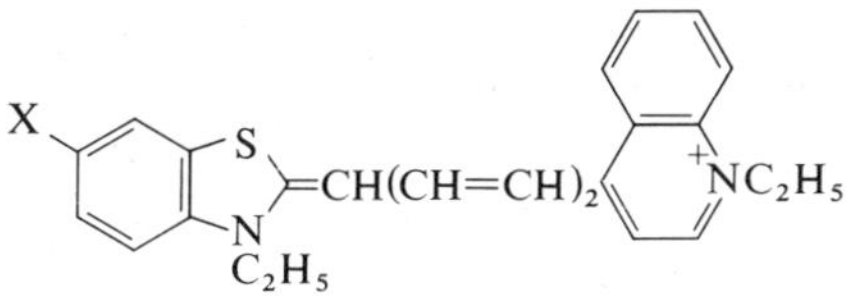

64, 65 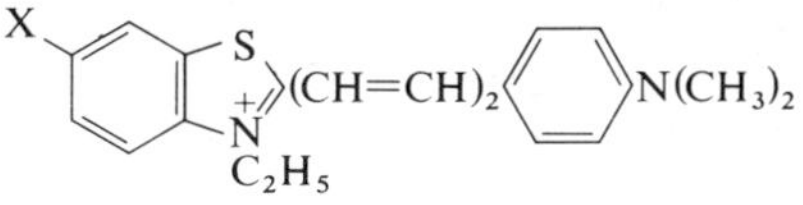 **66, 67**

No.	X	λ_{max} (mμ)	Deviation (mμ)
64	H	728	−4
65	NO_2	680	−68
66	H	570	−108
67	NO_2	651	−45

As can be seen from these compounds and the many others discussed in this section, any decrease or increase in the basicity of a resonance terminal resulting in a closer approach to energetic symmetry of the two terminals will result in a red shift and a deviation decrease. On the other hand, any decrease or increase in the basicity of a resonance terminal resulting in an increase in the energetic asymmetry of the two terminals will result in a violet shift and a deviation increase.

B. Some Simple Heterocyclic Systems

A somewhat simpler heterocyclic ring system is 2,1,3-benzoselenadiazole. The nitrogens in this molecule are strong electron attractors of the same order of strength as nitro oxygens (139). Salt formation can take place at one or both nitrogens. As shown in Table 9.22, derivatives of this ring system substituted in the 5-position show an increasing shift toward the visible of their long wavelength band with the following substituents: $H < CH_3 < OCH_3 < C_6H_5 < SCH_3 < NH_2 < N(CH_3)_2$ (139). Similar substituents in the 4-position were studied to a slight extent, and indications are that they would probably also absorb in the same order but at longer wavelength. This relative order of wavelength absorption has been found to hold for the bases, monocationic salts, and, except for the more strongly basic amino compounds, for the dicationic salts. In all cases, except for the more basic amino compounds, the dicationic salts absorbed at longest wavelength, and the bases absorbed at the shortest wavelength (see Figs. 8.10 and 8.11).

Structures involving a separation of charge are strong contributors to the excited state of the free bases. Salt formation at the ring nitrogen destroys these hypsochromic structures and thus lowers the relative energy of the excited state. The closing up of the energy levels results in the red shift in the spectrum. This is essentially an example of the

TABLE 9.22
Absorption Spectra of 5-Substituted Piaselenoles (139)

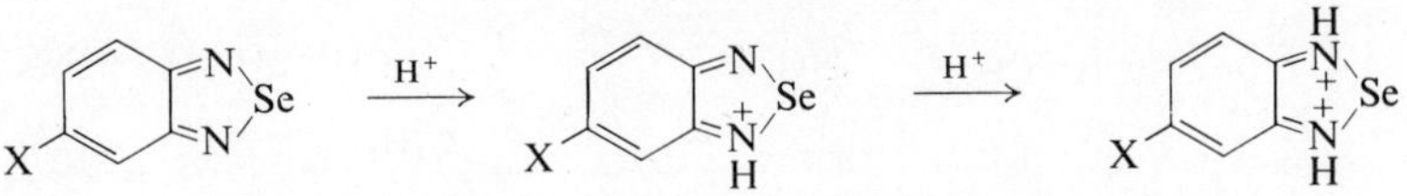

X	λ_{max} (mμ)[a] (log ϵ)		
	In 95% EtOH	In 50% H_2SO_4[b]	In 95% H_2SO_4[c]
H	231 (3.66)	222 (3.83)	222 (4.00)
	331 (4.22)	250*s* (3.25)	260*s* (3.05)
		341 (4.26)	350*s* (4.28)
		380*s* (3.29)	362 (4.36)
			435 (2.93)
CH_3		226 (3.94)	
		250*s* (3.28)	
		347 (4.32)	
		390*s* (3.35)	
OCH_3	240 (3.65)	234 (3.85)	234 (4.13)
	329 (4.04)	355 (4.15)	288 (3.16)
	358 (3.88)	386 (3.92)	365 (4.23)
			475 (3.69)
C_6H_5	238*s* (4.08)		257 (4.24)
	253 (4.34)		340 (3.95)
	346 (4.24)		402 (4.01)
	370*s* (3.94)		550 (3.92)
SCH_3	246 (4.19)	254 (4.09)	268 (4.24)
	340 (3.94)	292 (3.44)	290 (3.44)
	390 (3.99)	364 (3.93)	336 (3.75)
		448 (4.14)	394 (3.94)
			572 (4.08)
NH_2	236 (4.25)	236 (3.97)	222 (3.82)
	266*s* (3.64)	333 (4.10)	242*s* (3.48)
	324 (3.94)	459 (3.74)	337 (4.25)
	425 (3.80)		380*s* (3.35)
$N(CH_3)_2$	242 (4.26)	224 (4.10)	218 (3.75)
	247 (4.25)	246*s* (3.99)	244*s* (3.48)
	280*s* (3.62)	310*s* (3.8)	337 (4.21)
	326 (3.88)	334 (3.89)	380*s* (3.28)
	449 (3.83)	503 (3.77)	

[a] *s* = Shoulder.

[b] The spectral data for the amino compounds were determined in 50% EtOH–1.2 *N* hydrochloric acid.

[c] The spectral data for the amino compounds were the same in 50% sulfuric acid as in 95% sulfuric acid. In this case the second proton has added to the amino nitrogen.

Lewis rule: *the greater the fraction of a charge held by the resonance terminals, the further in the visible does the compound absorb* (48, 140). The Lewis rule appears to be a specialized version of the Forster–Knott rule (47), which has been discussed in Section V.A.

Quinoxaline derivatives substituted by electron-donor groups in an analogous fashion show the same red shift toward the visible with increasing electron-donor strength of the substituents (Table 9.23)(141).

TABLE 9.23
Electron-Donor Strength and Absorption Spectra of Some Quinoxaline Derivatives (141)

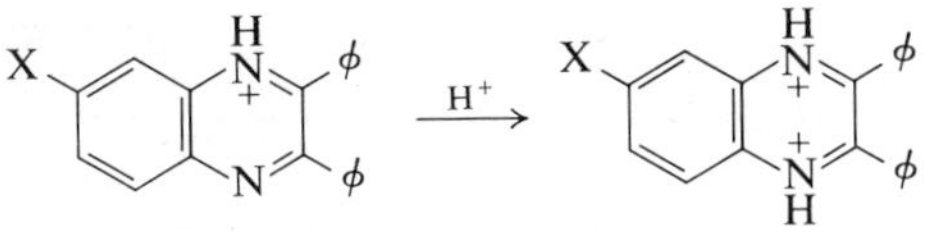

	In 6 *N* HCl–5% EtOH		In H_2SO_4	
X	$\lambda_{max}(m\mu)$	$m\epsilon$	$\lambda_{max}(m\mu)$	$m\epsilon$
H	390	13.5	473	14.5
CH_3O	420	16.2	518	19.5
C_6H_5	418	17.4	550	18.2
CH_3S	458	17.0	610	20.4
NH_2	492	10.0	–	–

Negativizing the positive resonance node, as shown in **68B**, shifts the long wavelength maximum to the red. The limiting resonance structures **68A** ↔ **C** are also shown.

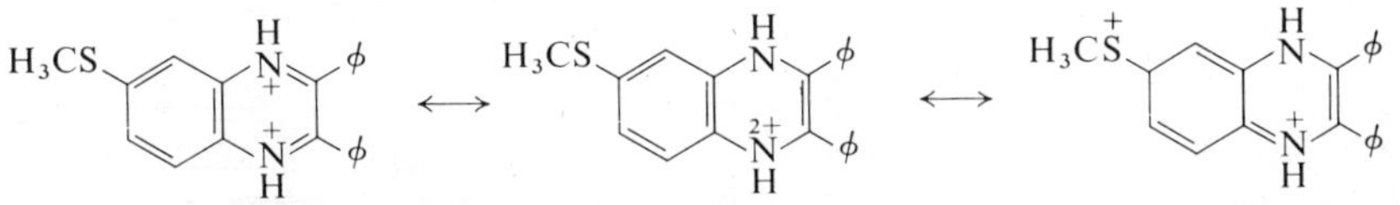

A **B** **C**

68

These heterocyclic series approach electron symmetry (equal basicity) of the resonance terminals as the electron-donor strength of the substituents increases and the wavelength maximum shifts to the red.

C. Diphenylmethane Cations

The effect of increasing electron-donor strengths of the resonance terminals in a diphenylcarbonium group on the long-wavelength maximum shows the high order of strength of the halogen atoms (Table 9.24). The other substituents show their usual order of strength. A somewhat similar order is found for the triphenylcarbonium salts listed in Table 9.25.

In the determination of benzene derivatives with piperonal chloride the wavelength maximum of the chromogen shifts to the red with increasing electron-donor strength of the substituents (Table 9.26). 2-Substituted

TABLE 9.24
Effect of Increasing the Electron-Donor Strength of a Resonance Terminal on a Cationic Resonance Band (142, 143)

$X-C_6H_4-\overset{+}{C}H-C_6H_4-X$

X	$\lambda_{max}(m\mu)$[a]	$m\epsilon$	X	$\lambda_{max}(m\mu)$[a]	$m\epsilon$
H	440	43.7	OCH_3	507	110.0
F	452	74.1	OC_6H_5	515	107.0
Cl	465	138.0	I	560	97.7
CH_3	472	74.1	$N(CH_3)_2$	610	89.1
Br	504	110.0			

[a] In aqueous sulfuric acid.

TABLE 9.25
Effect of Increasing the Electron-Donor Strengths of the Resonance Terminals in Triphenylcarbonium Salts

$(X-C_6H_4)_3C^+$

X	λ_{max} $(m\mu)$[a]	Ref.	X	λ_{max} $(m\mu)$[a]	Ref.
H[b]	404 431	142	NH_2	539	144
CH_3	452	142	C_6H_5[c]	542	145
Cl	465	142	$4-CH_3C_6H_4$[c]	575	145
OH	480	143	$N(CH_3)_2$	590	142, 143
OCH_3	483	142	$4-CH_3OC_6H_4$[c]	638	145
NHAc	525	143			

[a] In aqueous H_2SO_4 unless otherwise specified.
[b] In $AcOH-H_2SO_4$ λ_{max} 414 $m\mu$ (145).
[c] In $AcOH-H_2SO_4$.

TABLE 9.26
Wavelength Maxima Obtained with Piperonal Chloride in the Determination of Benzene Derivatives(109)

$$\text{(3,4-methylenedioxyphenyl)}-CHCl_2 + C_6H_5-X \xrightarrow{H^+} \text{(3,4-methylenedioxyphenyl)}-\overset{+}{C}H-C_6H_4-X$$

X	λmax (mμ)	X	λ_{max} (mμ)
CH_3	513	SCH_3	575
OCH_3	527	SC_6H_5	585
C_6H_5	560	SeC_6H_5	612

fluorene derivatives determined with piperonal chloride show the same phenomenon (Table 9.27).

Chromogens somewhat similar to the diphenylmethane cations are formed in the determination of azulene derivatives with 4-dimethylamino-benzaldehyde(148, 149). The pure cations (**69**) show the expected red shift with the increasing electron-donor strength of the substituent(150). The spectra were obtained in ethanol.

CH_3 $CH(CH_3)_2$ X– –CH CH_3

69

	X	λ_{max} (mμ)
a	H	452
b	CH_3O	518
c	$(CH_3)_2N$	640

In the determination of guaiazulene with 4-dimethylaminobenzaldehyde the chromogen **69c** was found to absorb at 640 mμ, mε 94.0(149). The alkyl groups in the azulene ring also have a red shift effect since the chromogen obtained from azulene absorbs at 620 mμ, mε 86.0.

D. Azo-Dye Cations

With the exception (due to solvent effect) of the methylthio derivative, the azobenzene dyes (Table 9.28) show the expected red shift with in-

TABLE 9.27
Wavelength Maxima Obtained with Piperonal Chloride in the Determination of Fluorene Derivatives (109, 146, 147)

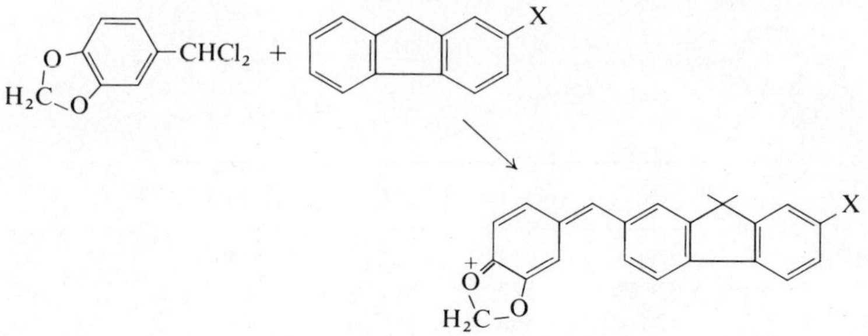

X	λ_{max} (mμ)[a]	X	λ_{max} (mμ)[a]
H	580	$NHSO_2CH_3$	605
OAc	580	4-$NHSO_2C_6H_4CH_3$	612
F	588	$NHCOOCH_2CH_2Cl$	620
CH_3	600	$NHCOOCH_2CH_2F$	620
$NHCOCH_3$	600	OH	623
$NHCOCH_2Cl$	600	$NHCOSCH_2CH_3$	625
$NHCOCH_2F$	600	$NHCOOCH_2CH_3$	627
$NHCOCCl_3$	600	OCH_3	630
$NHCOCHF_2$	600	OC_2H_5	630
$NHCOCF_3$	600	SCH_3	670
$NHCOC_6H_5$	600	$SCH_2C_6H_5$	680
C_2H_5	603		

[a]Probably substitution in 3-position also.

creasing electron-donor strength of the substituent. In some azo cations containing an amino nitrogen resonance terminal decreasing the electron density of this terminal shifts the wavelength maximum to the violet, as shown for **70** in methanol (153).

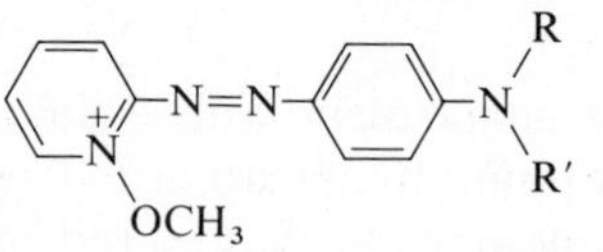

70

R	R′	λ_{max} (mμ)	mε
CH_2CH_3	CH_2CH_3	570	68.9
CH_3	CH_3	566	70.5
CH_3	CH_2CH_2CN	552	60.6

TABLE 9.28
Long Wavelength Maxima of Some Azo-Dye Cations

$X-C_6H_4-N=\overset{H}{\underset{+}{N}}-C_6H_5$

X	Solvent	λ_{max} (mμ)	mϵ[a]	Ref.
H	H_2SO_4	430	21.9	151
CH_3O	10% EtOH–65% H_2SO_4	466	41.7	152
C_2H_5O	10% EtOH–65% H_2SO_4	468	39.8	152
H_2N	6 *N* HCl in H_2O–EtOH (1 : 1)	502	36.7	4
CH_3HN	1.2 *N* HCl in H_2O–EtOH (1 : 1)	505	40.9	4
C_2H_5HN	1.2 *N* HCl in H_2O–EtOH (1 : 1)	514	36.3	4
$(CH_3)_2N$	6 *N* HCl in H_2O–EtOH (1 : 1)	518	47.3	4
$C_6H_5CH_2(CH_3)N$	1.2 *N* HCl in H_2O–EtOH (1 : 1)	526	41.2	4
CH_3S	75% Aqueous H_2SO_4	532	57.3	51

[a]The intensities of the amino derivatives are affected by the acid concentration.

In the *para*-substituted 1-phenylazoazulenes two azonium tautomers are possible. The one with the azulene part of the molecule as the site of the resonance terminal is favored (154). With the increase in electron-donor strength of the substituent the other azonium tautomer may also be present. However, as shown in Table 9.29, the long wavelength maxima of these compounds shift to the red with increased electron-donor strength of the substituent. This shift may be due mainly to the *para* effect of the substituent on the azonium resonance terminal of the predominant tautomer.

TABLE 9.29
Absorption Spectra of Some *Para*-Substituted Phenylazoazulene Cations (154)

$X-C_6H_4-\overset{H}{\underset{+}{N}}=N-C_{10}H_7 \rightleftarrows X-C_6H_4-N=\overset{H}{\underset{+}{N}}-C_{10}H_7$

X	λ_{max} (mμ)	mϵ	X	λ_{max} (mμ)	mϵ
$(CH_3)_2HN^+$	513	46.8	I	530	46.7
O_2N	519	58.0	CH_3	532	39.2
H	522	38.2	CH_3O	556	41.2
Cl	528	41.3	HO	561	40.1
Br	529	43.9	$(CH_3)_2N$	671	42.6

In one of the procedures for the determination of azulene with 4-azobenzenediazonium fluoborate the monocation **71** is obtained(155). The pure cation absorbs at λ_{max} 557 mμ, mϵ 52.1, in acid solution(156). Thus the phenylazo group is approximately equal to the methoxy group in its effect on the spectrum of the 1-phenylazoazulene cation (Table 9.29).

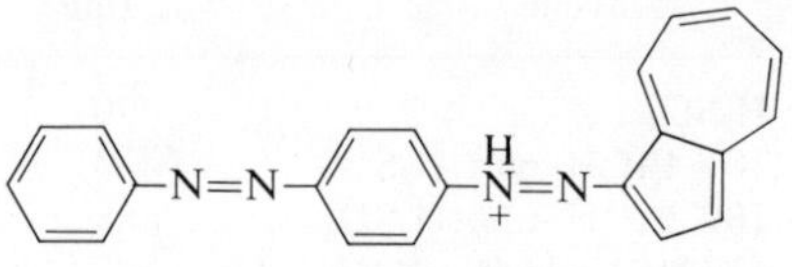

71

In the colorimetric determination of aniline derivatives with 4-azobenzenediazonium fluoborate the wavelength maximum of the chromogen **72** shifts to the red with increasing basicity of the amino nitrogen(157). The same phenomenon is found in the chromogens obtained in the determination of aniline and its derivatives with 3-methyl-2-benzothiazolinone hydrazone (Table 9.30).

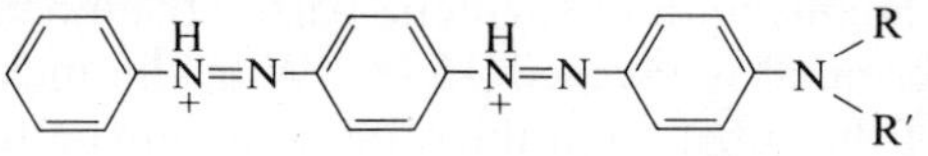

72

R	R′	λ_{max} (mμ)	mϵ'
H	H	610	8
H	CH_3	623	65
CH_3	CH_3	632	100

TABLE 9.30
Long Wavelength Maxima of the Chromogens Obtained in the Determination of Aniline Derivatives with MBTH(57)

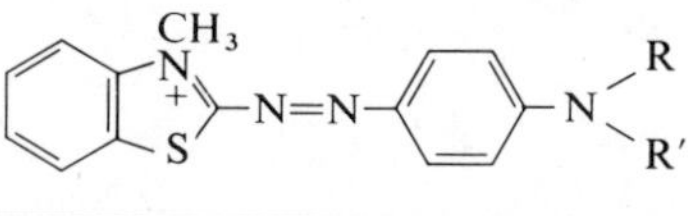

R	R′	λ_{max} (mμ)	mϵ'	R	R′	λ_{max} (mμ)	mϵ'
H	H	566	43	C_2H_5	C_2H_5	598	45
H	CH_3	580	76	CH_3	CH_2CHCH_2	600	87
H	C_2H_5	580	71	n-C_3H_7	n-C_3H_7	602	74
H	n-C_3H_7	582	73	H	C_6H_5	606	78
CH_3	CH_3	598	84	CH_3	$CH_2C_6H_5$	607	96
CH_3	C_2H_5		70	$CH_2C_6H_5$	$CH_2C_6H_5$	614	103

VIII. Conjugation-Chain Length

For many series of compounds lengthening of a chain of conjugation increases the sensitivity of a compound to oxidation and decomposition, and may make it more reactive. For all compounds under study solution in sulfuric acid could involve oxidation and/or sulfonation. Any study involving this solvent has to be pursued with care. Thus 1,6-diphenylhexatriene is believed to be immediately oxidized in 100% sulfuric acid to a dipositive ion (M^{2+})(158), whereas other workers cite evidence indicating that the compound is sulfonated twice in the time required for dissolution(159).

A. Convergence

In compounds that are not sterically hindered any increase in the chain of conjugation between resonance terminals results in a red shift of the long wavelength maximum. A series of such compounds is said to converge if their wavelength maxima appear to approach a limit with the increasing length of the conjugation chain—that is, the difference in wavelength maxima between each successive member decreases(160). However, in some of the series, especially where the resonance terminals have a weak basicity, the increase in the length of the conjugation chain (usually called a vinylene shift) results in a small steady shift to the red of the wavelength maxima. Examples of these two types are shown in Table 9.31.

Other series of cationic resonance structures that show small to moderate red shifts with increase in the conjugation-chain length are the ω-(4-methoxyphenyl)polyene aldehyde cations(167, 168), α,ω-diphenylpolyene cations(169), $C_6H_5(CH{=}CH)_nCHO^+H$(170), and $C_6H_5(CH{=}CH)_nCO^+HC_6H_5$(170).

With each increase in the conjugation-chain length the wavelength maxima of a series of cationic resonance structures shift to the red in larger increments than do the wavelength maxima of the comparable zwitterionic resonance structures(Table 9.32).

The determination of furylpolyenals with *N,N*-dimethyl-*p*-phenylenediamine results in the formation of chromogens that absorb at longer wavelengths with increasingly greater band widths and intensities as the conjugation-chain length increases (Table 9.31).

Examples are available of conjugation chains composed mainly of nitrogen atoms. In the polyaza cations **73**(171) increasing the chain length shifted the long wavelength maximum to the red. In all cases substitution of methoxy groups in a position *para* to the nitrogen resonance terminals caused a red shift that was larger with increasing chain length.

TABLE 9.31
Effect of the Vinylene Shift on the Wavelength Maxima of Chromogens Containing a Weakly Basic Resonance Terminal

$$Ar{-}(CH{=}CH)_n{-}{}^{+}CH{-}Ar'$$

Ar	Ar′	n	Solvent	λ_{max} (mμ)	Ref.
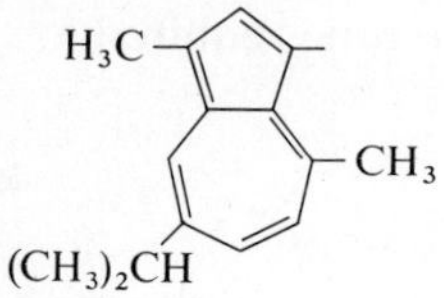 H_3C, CH_3, $(CH_3)_2CH$	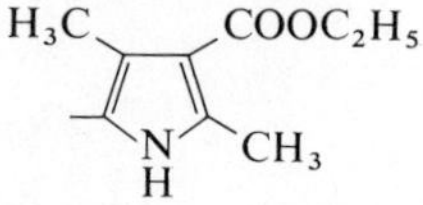 H_3C, $COOC_2H_5$, N H, CH_3	0	Chloroform	588[a]	161
		1		690[b]	
		2		771[c]	
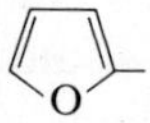 O	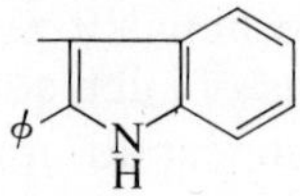 ϕ, N H	0	95% EtOH	447	162
		1		517	
		2		570	
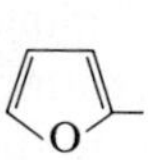 O	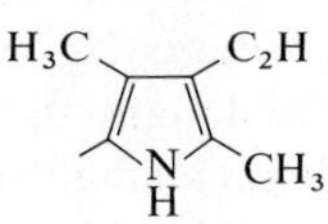 H_3C, C_2H, N H, CH_3	1		465	162
		2		500	
		3		530	
		4		565	
		5		600	
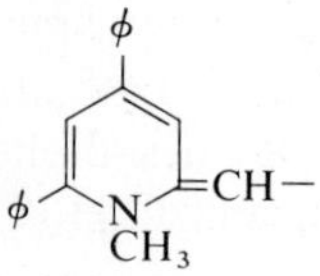 ϕ, ϕ, N CH_3 =CH—	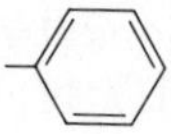	0	AcOH	360	163
		1		397	
		2		420	
		3		444	
		5		483	
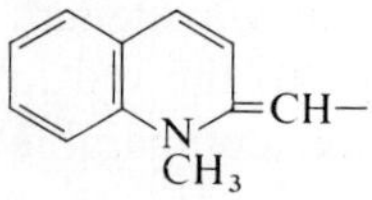 N CH_3 =CH—	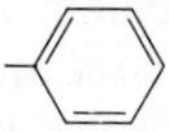	0	95% EtOH	379	163
		1		417	
		2	AcOH	450	
		3		483	
		5		515	
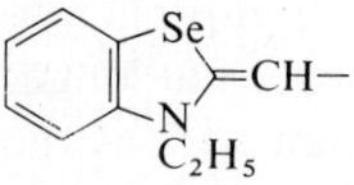 Se, N C_2H_5 =CH—	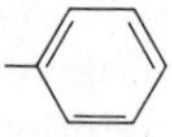	0	AcOH	384	163
		1		426	
		2		462	
		3		488	
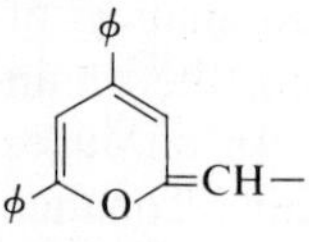 ϕ, ϕ, O =CH—	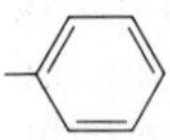	0	AcOH	458	163
		1		506	
		2		546	
		3		585	
		5		652	

TABLE 9.31 (*cont'd*)

Ar	Ar′	n	Solvent	λ_{max} (mμ)	Ref.
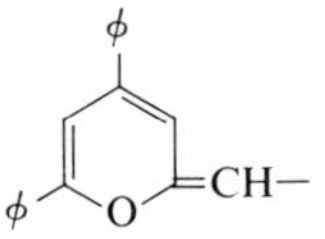 2,6-diphenyl-4H-pyran-4-ylidene=CH– (ϕ)	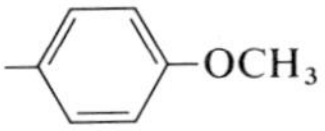 –C$_6$H$_4$–OCH$_3$	0	AcOH	502	164
		1		552	
		5		666	
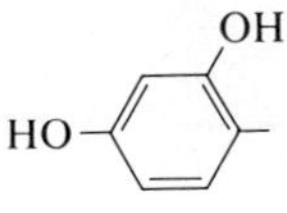 HO–C$_6$H$_3$(OH)–	 –C$_6$H$_4$–X; X = H	0	—[d]	402	165
		1		480	
	3,4-(OH)$_2$	0		525	
		1		579	
	4-OH-3,5-(OCH$_3$)$_2$	0		543	
		1		593	
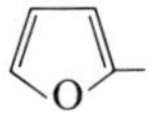 2-furyl	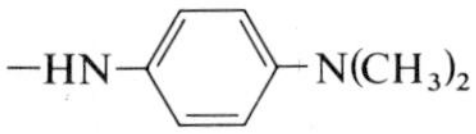 –HN–C$_6$H$_4$–N(CH$_3$)$_2$	0	Methanol	480	166
		1		510	
		2		530	
		3		542	
		4		553	
		5		570	

[a]mϵ 74.1 (161).
[b]mϵ 178 (161).
[c]mϵ 87.1 (161).
[d]Concentrated hydrochloric acid – 95% ethanol (1 : 2, V/V).

TABLE 9.32
Effect of the Vinylene Shift on the Wavelength Maxima of Zwitterionic Resonance Structures and Their Conjugated Acids (170)

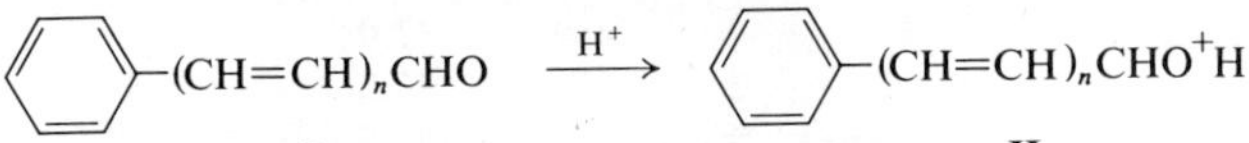

$$C_6H_5\text{-}(CH{=}CH)_nCHO \xrightarrow{H^+} C_6H_5\text{-}(CH{=}CH)_nCHO^+H$$

I → II

	I		II	
n	λ_{max} (mμ)[a]	mϵ	λ_{max} (mμ)[b]	mϵ
0	250	1.0		
1	287	2.3	~355	0.17
2	325	4.0	404	2.2
3	360	5.1	468	3.7
4	387	6.1	525	5.4
5	410	–	580	–
6	430	–	630	–

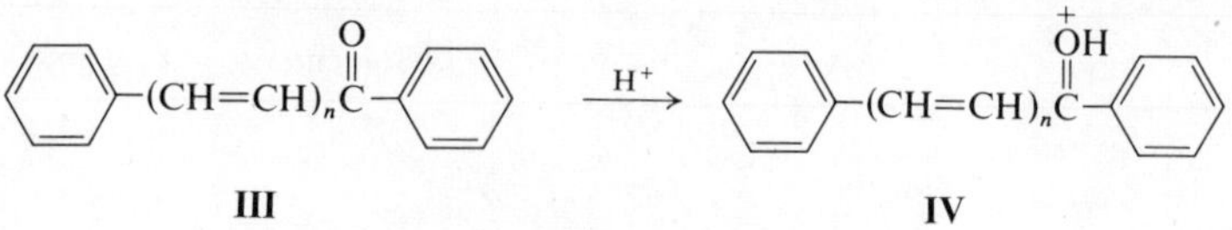

$$C_6H_5-(CH{=}CH)_n-C(=O)-C_6H_5 \xrightarrow{H^+} C_6H_5-(CH{=}CH)_n-C(=\overset{+}{O}H)-C_6H_5$$

III → IV

	III		IV	
n	λ_{max} (mμ)[a]	mϵ	λ_{max} (μ)[b]	mϵ
0	254	1.8	–	–
1	310	2.3	390	0.74
2	343	3.6	460	2.9
3	376	3.1	530	4.0
4	402	5.7	585	5.2
5	418	6.4	650	5.2
6	445	–	735	–

[a]In acetic acid.
[b]In 20% aqueous sulfuric acid.

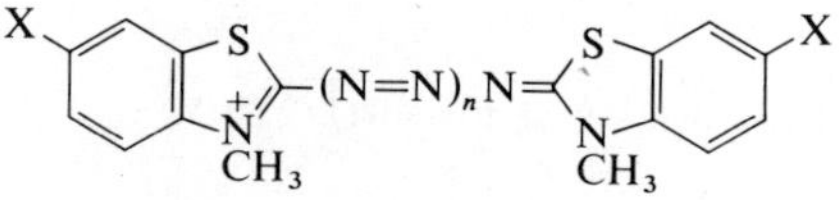

X-benzothiazolium(N-CH_3)$-(N{=}N)_n-N{=}$benzothiazoline(N-CH_3)-X

73

X	n	λ_{max} (mμ)	mϵ
H	0	375	51.0
H	1	484	42.0
H	2	553	57.5
CH_3O	0	384	47.5
CH_3O	1	512	44.8
CH_3O	2	597	64.6

B. Nonconvergence

Many of the vinylene-homologous series show nonconvergence; that is with each additional vinylene group the λ_{max} shifts to the red by a steady increment of approximately 100 mm (Table 9.33). A large quantity of data is available showing the nonconvergence of the wavelength maxima values of symmetrical(181) and some asymmetrical(182) vinylogous cyanines.

TABLE 9.33

Effect of the Vinylene Shift on the Wavelength Maxima of Chromogens Containing Nitrogen Resonance Terminals

$$Ar—(CH{=}CH)_n—\overset{+}{C}H—Ar'$$

Ar	Ar′	n	Solvent	λ_{max} (mμ)	mϵ	Ref.
$(CH_3)_2N$	$(CH_3)_2N$	1	95% EtOH[a]	315	52.5	172
		2		414	161.0	
		2	CH_2Cl_2	414	120.0	173
		3	95%EtOH[a]	514	263.0	172
		3	CH_2Cl_2	516	209.0	173
		4	CH_2Cl_2	619	288.4	
S N—	—N S	1	95% EtOH[a]	319	62.0	172
		2		418	168.0	
		3		520	263.0	
S N—	—N S	1	95% EtOH[a]	327	60.0	172
		2		427	167.0	
S N—	—N S	1	95% EtOH[a]	328	62.0	172
		2		421	162.0	
C_6H_5—NH—	—HN—C_6H_5	1	Methanol	383	50.0	105
		2		485	65.0[b]	
		2	95% EtOH–HCl	487	87.1	174

TABLE 9.33 (*cont'd*)

Ar	Ar′	n	Solvent	λ_{max} (mμ)	mϵ	Ref.
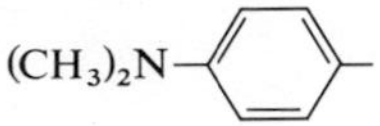 $(CH_3)_2N$–	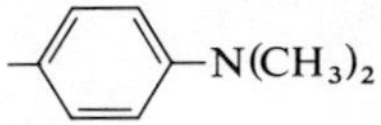 –$N(CH_3)_2$	0	Solvent	610		175[c]
		1		693		
		2		790		
		3		883		
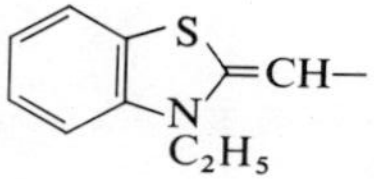 S, =CH–, N, C_2H_5	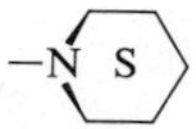 –N S	−1[d]	Methanol	295	8.0	105
		0		388	51.0	
		1		483	142.0	
		2		584	218.0	
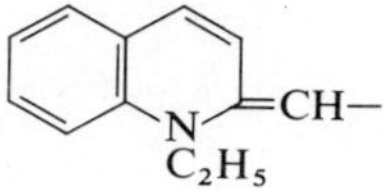 =CH–, N, C_2H_5	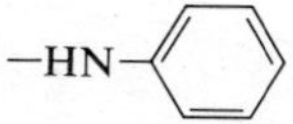 –HN–	0	Methanol	443	51.0	105
		1		528	95.0	
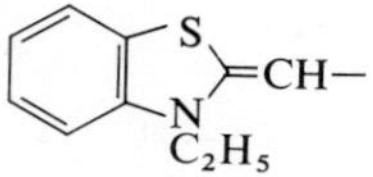 S, =CH–, N, C_2H_5	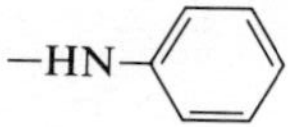 –HN–	−1[d]	Methanol	299	14.0	105
		0		414	55.0	
		1		516	107.0	
		2		613	76.0[b]	
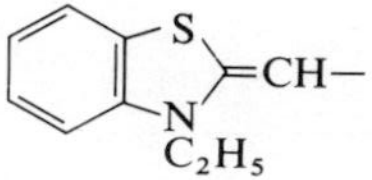 S, =CH–, N, C_2H_5	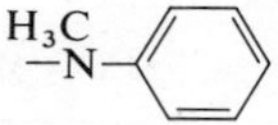 H_3C, –N–	−1[d]	Methanol	293	13.0	105
		0		400	46.0	
		1		497	107.0	
		2		598	133.0	
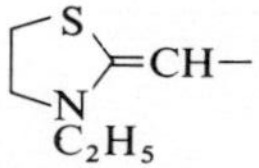 S, =CH–, N, C_2H_5	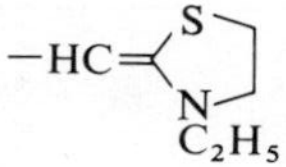 –HC=, S, N, C_2H_5	−1[e]	Methanol	335	37.3	177
		0		442	93.0	
		1		540	133.0	
		2		639	144.0	
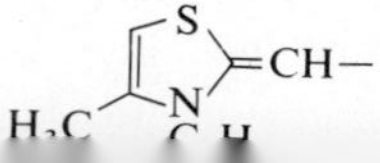 S, =CH–, N, H_3C	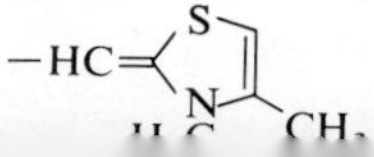 –HC=, S, N, CH_3	−1[e]	Methanol	409	42.5	177
		0		556	79.3	

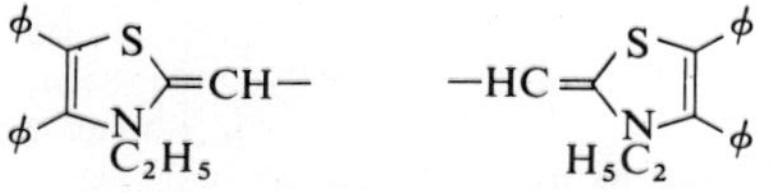

ϕ, S, =CH—, N, C_2H_5, ϕ	—HC=, S, ϕ, N, H_5C_2, ϕ	-1^{e}	Methanol	439	33.0	177
		0		589	90.3	
		1		663	139.0	
X, =CH—, N, C_2H_5	—HC=, X, N, C_2H_5	-1^{e}	Methanol	440	31.9	177
				459	30.5	
		0		541	90.0	
		1		638	144.0	
	X=CMe$_2$	2		727	78.5[b]	
	[X = S	-1^{e}	Methanol	423	84.5	105
		0		558	148.0	
		1		649		178
		2		747		
		4		1020		179
	[X = O	-1^{e}	Methanol	374		178
		0		482		
		1		576		
		2		681		
	[X = Se	-1^{e}	Methanol	430		178
		0		567		
		1		660		
		2		761		
S, =CH—, N, C_2H_5	—HC=, NC_2H_5	-1^{e}	Methanol	503		132
		0		630		
		1		728		

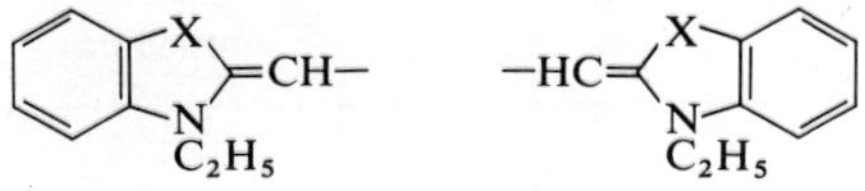

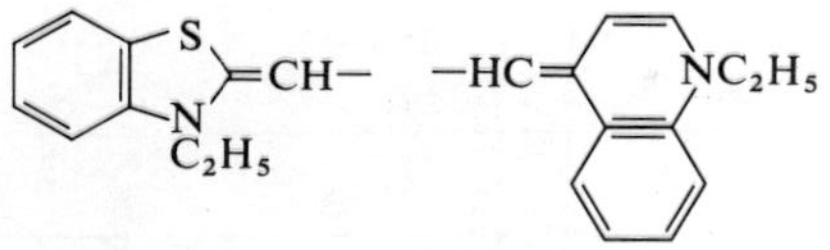

TABLE 9.33 (*cont'd*)

Ar	Ar′	n	Solvent	λ_{max} (mμ)	mϵ	Ref.
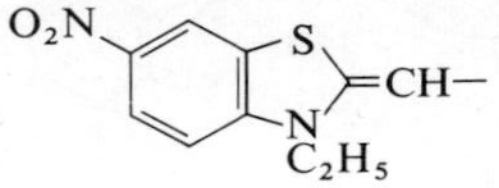 O_2N-benzothiazolyl(N-C_2H_5)=CH–	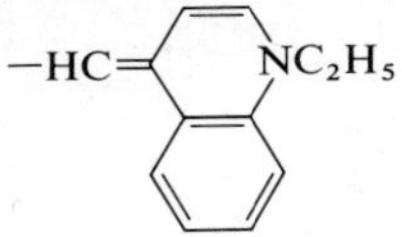 –HC=quinolyl-NC_2H_5	–1[e]	Methanol	505		180
		0		617		
		1		680		
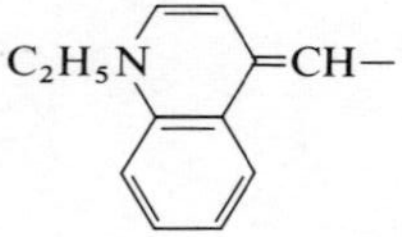 C_2H_5N-quinolyl=CH–	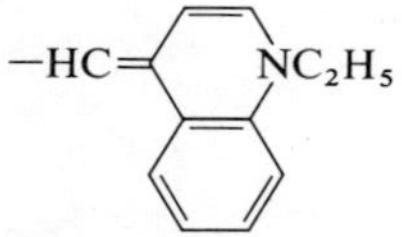 –HC=quinolyl-NC_2H_5	–1[e]		593		180
		0		704		
		1		810		
		2		932		
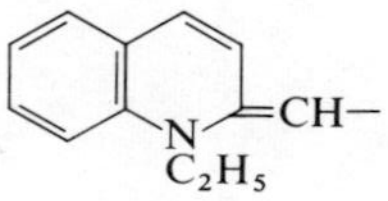 quinolyl(N-C_2H_5)=CH–	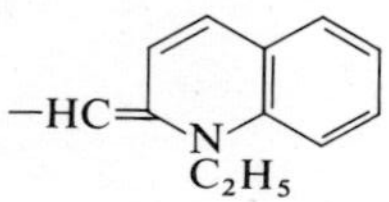 –HC=quinolyl(N-C_2H_5)	–1[e]		524	76.0	105
		0		604	194.0	
		1		708		180
		2		808		
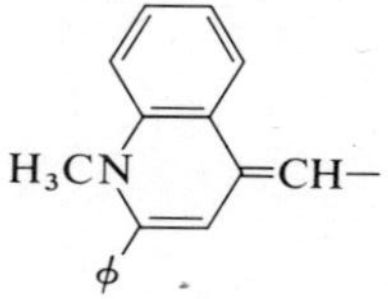 H_3CN-quinolyl(φ)=CH–	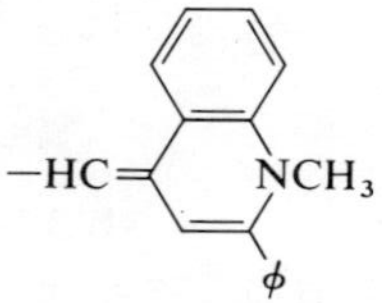 –HC=quinolyl(φ)-NCH_3	–1[e]	95% EtOH	608		136
		0		723		
		1		837		
		2	CH_3NO_2	960		136

[a]Temperature, 100°K.
[b]Low value due to hydrolysis of salt?
[c]Also analogous series with substituents on the chain showing similar red shifts (see also reference 176).
[d]Right hand nitrogen terminal attached directly to 2-position of heterocyclic ring on left.
[e]Monomethine cationic dye.

The cationic resonance structures in Table 9.33 are mostly nonconvergent since their resonance terminals have about the same internal basicity. Where the internal basicity of one of the resonance terminals in these compounds is decreased, the new series shows convergence. Thus in the series in Table 9.33, where the resonance terminals were benzothiazole and quinoline nitrogen atoms, introduction of a nitro group into the benzothiazole ring causes the degree of convergence to increase with increasing chain length. Substitution of an acetyl group for the hydrogen of an imino nitrogen resonance terminal in **74a** will decrease the basicity of the terminal and thus cause a greater degree of convergence in the new series, **74b**(105).

S, N (+), C_2H_5 —$(CH{=}CH)_n$N(X)—C_6H_5

74

	X	n	λ_{max} (mμ)	mϵ	$\Delta\lambda$ (mμ)
a	H	0	299	14	115
		1	414	55	102
		2	516	107	97
		3	613	76	
b	Ac	1	364	10	62
		2	426	35	35
		3	461	44	

When one of the nitrogen resonance terminals in a symmetrical cationic resonance structure is replaced by a much less basic grouping, a violet shift results and the vinylogous series acquires a greater degree of convergence, as in **75**(173). Decreasing the basicity of one of the terminals in these symmetrical compounds results also in decreased intensity.

$$(CH_3)_2\overset{+}{N}{=}CH{-}(CH{=}CH)_n{-}X$$

75

X	n	λ_{max} (mμ)	mϵ	$\Delta\lambda$ (mμ)
$N(CH_3)_2$	2	414	120.0	102
	3	516	209.0	103
	4	619	288.0	
Cl	3	368	45.7	50
	4	418	58.9	

Resonance terminals do not have to be strongly basic nitrogen atoms for a vinylogous series to be nonconvergent. As long as the cationic limiting resonance structures have energetic symmetry, compounds with oxygen or sulfur atoms as terminals show nonconvergence (Table 9.34).

TABLE 9.34
Long Wavelength Maxima of Symmetrical Cations Containing Oxygen or Sulfur Atoms as Terminals

$$Ar{-}(CH{=}CH)_n{-}\overset{+}{C}H{-}Ar$$

Ar	n	λ_{max} (mμ)	Ref.
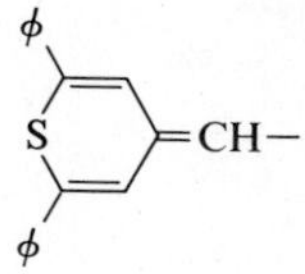 2,6-diphenylthiopyranylidene =CH−	−1[a]	628	183
	0	753	
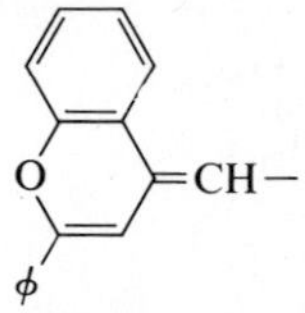 2-phenylbenzopyranylidene =CH−	−1[a]	600	136
	0	705	
	1	812	
	2	920	
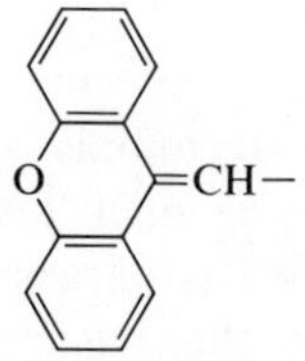 xanthylidene =CH−	−1[a]	648	138
	0	705	
	1	785	
	2	865	

[a]Monomethine derivative.

C. Deviation and Chain Length

In any vinylogous series increasing the disparity between the internal basicities of the two resonance terminals (i.e., increasing the energetic asymmetry of the two limiting resonance structures) results in an increasing deviation and a decreasing vinylene shift with lengthening of the chain (Table 9.35). The converse of this proposition is demonstrated in Table 9.36.

In the structure shown in Table 9.35, X = H and Y = C_2H_5, the resonance terminals are of about equal basicity. With decreasing basicity

TABLE 9.35
Effect of Increasing the Energetic Asymmetry of Resonance Terminals on Deviation and the Vinylene Shift (132)

Substituents					
X	Y	*n*	λ_{max} (mμ)[a]	$\Delta\lambda$ (mμ)	Deviation (mμ)
H	C_2H_5	0	503		5
		1	630	127	2
		2	728	98	4
H	C_6H_5	0	503		7
		1	627	124	9
		2	720	93	18
H	2-$C_6H_4NO_2$	0	499		
		1	615	116	23
		2	681	66	58
NO_2	C_2H_5	0	505		16
		1	617	112	28
		2	680	63	68

[a] All in methanol.

of one of the terminals a violet shift, a decreasing vinylene shift, and an increasing deviation ensue. These results are accentuated with the increasing length of the conjugation chain.

The opposite results are obtained in a compound such as that shown in Table 9.36, X = H and Y = C_2H_5. Decreasing the basicity of the more basic resonance terminal results in a red shift and decreasing convergence and deviation.

D. Polarization

The addition of a proton to the negative resonance terminal of a zwitterionic resonance structure (Z_n type) causes a red shift that increases with

TABLE 9.36
Effect of Decreasing and Energetic Asymmetry of Resonance Terminals on Deviation and the Vinylene Shift (132)

Substituents					
X	Y	n	λ_{max} ($m\mu$)[a]	$\Delta\lambda$ ($m\mu$)	Deviation ($m\mu$)
H	C_2H_5	1	528		56
		2	570	42	108
H	C_6H_5	1	547		42
		2	602	55	81
H	2-$C_6H_4NO_2$	1	561		29
		2	639	78	47
NO_2	C_2H_5	1	574		27
		2	651	77	45

[a] All in nitromethane.

the conjugation chain. The wavelength shift of phenylpolyenealdehydes dissolved in 20% aqueous sulfuric acid is attributed to this phenomenon. However, Krauss and Grund (5) have shown that increasing the polarization properties of the solvent causes an even greater red shift, which they attribute to the following type of excited state structure of decreased energy (results are shown in Table 9.37):

$$(SbCl_6)^- \quad (+\text{C}_6\text{H}_5)\!=\!(CH\!-\!CH)_n\!=\!CH\!-\!O^- \quad {}^+\overset{O}{\overset{\|}{C}}CH_3$$

76

The same type of phenomenon has been demonstrated for the diphenylpolyenes and vitamin A derivatives (184). A more thorough investigation of this phenomenon should prove useful.

TABLE 9.37
Effect of Strong Acid Media on the Long Wavelength Maxima of Some Phenylpolyene-aldehydes (5)

$C_6H_5-(CH=CH)_n-CHO$

	λ_{max} (mμ) (mε)				
n	In Chloroform	In 20% Aqueous H_2SO_4[a]	In $Ac^+(BF_3Cl)^-$	In $Ac^+FeCl_4^-$	In $Ac^+(SbCl_6)^-$
0	245 (10.5)				
1	289 (24.0)				318 (35.0)
2	324 (34.0)	404 (2.2)	396 (44.0)	448 (51.0)	455 (55.0)
3	353 (40.0)	468 (3.7)	442 (43.2)	540 (67.0)	545 (67.5)
4	380 (50.0)	525 (5.4)	490 (55.0)	593 (80.0)	612 (86.0)
5	407 (63.5)	580	550 (70.0)	625 (100.0)	670 (110.0
6	430 (67.5)	630	567 (74.5)	670 (110.0)	730 (124.0)

[a]Data from reference 170.

E. Polyaza Chains

The *p*-polyazobenzenes show the expected red shift and intensity increase with the lengthening of the conjugation chain. However, with the lengthening of the chain more *p*-quinonoid rings are present in the excited state. Consequently the series **77** is convergent (151).

$C_6H_5-\overset{H}{\underset{+}{N}}=N\left(-C_6H_4-\overset{H}{\underset{+}{N}}=N\right)_n-C_6H_5$

77

n	λ_{max} (mμ)	mε
0	430	21.9
1	502	55.0
2	560	79.4
3	604	83.2

A series of compounds that contain a chain consisting of one or three nitrogen atoms appeared to be nonconvergent (Table 9.38).

F. Resonance Terminals and the Vinylene Shift

An increase in the basicity of the resonance terminals of the diphenylpolyene cations results in a red shift and a shift toward nonconvergence (Table 9.39).

TABLE 9.38
Long Wavelength Maxima of Some Monoaza and Triaza Cations
(185)

$$Ar{=}N{-}(N{=}N)_n{-}\text{[3-methylbenzothiazolium-2-yl]}^{+}$$

Ar	$n=0$		$n=1$	
	λ_{max} (mμ)	mϵ	λ_{max} (mμ)	mϵ
CH_3, N, N, CH_3 (1,3-dimethylbenzimidazol-2-ylidene)	314	27.0	451	39.7
N, CH_3 (1-methylquinolin-2-ylidene)	395	35.0	468	50.0
S, N, CH_3 (3-methylbenzothiazol-2-ylidene)	375	51.0	484	42.0
H_3CO, S, N, CH_3 (methoxy-3-methylbenzothiazol-2-ylidene)	378	45.9	499	40.0

TABLE 9.39
Effect of the Basicity of Resonance Terminals on the Wavelength Maximum and the Vinylene Shift

$$X{-}C_6H_4{-}\overset{+}{C}H{-}(CH{=}CH)_n{-}C_6H_4{-}X$$

X	λ_{max}(mμ)					Ref.
	$n=0$	$n=1$	$n=2$	$n=3$	$n=4$	
H	441	498	561	622	688	186
4-OCH_3	512	581	655	725	792	186
2,4,6-$(OCH_3)_3$	518	562	648	–	–	187
4-$N(CH_3)_2$	606	693	790	883	–	186

In some symmetrical cations a decrease in basicity of the resonance terminals results in a red shift and an increased vinylene shift, as seen in **78**(188).

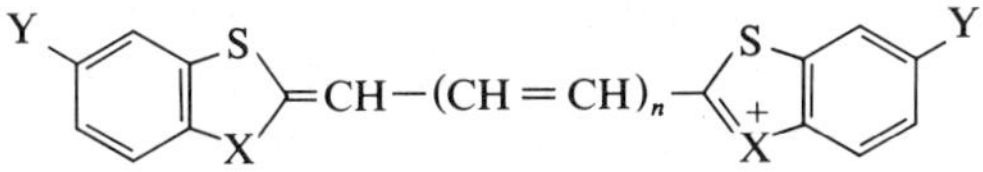

$$Y\text{-benzothiazoline}=CH-(CH=CH)_n-\text{benzothiazolium}^{+}\text{-}Y$$

78

Y	X	$\lambda_{max}(m\mu)$ $n=0$	$n=1$	$\Delta\lambda(m\mu)$
H	NC_2H_5	423	557	134
CH_3	S	514	666	152

The internal basicity of non-nitrogen resonance terminals in some compounds has the same order of strength as a dimethylaniline nitrogen terminal; an example is **79c**(189). In **79** (spectra in glacial acetic acid) as the basicity of the right hand terminal increases, the wavelength maximum shifts to the red, the vinylene shift increases to nonconvergence, and the deviation decreases from about −41 for **79a**, $n = 0$, to about +6 for **79c**, $n = 1$(189).

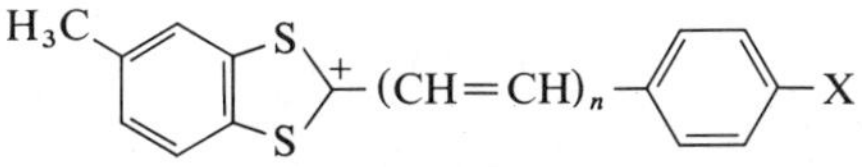

$$H_3C\text{-benzodithiolium}^{+}-(CH=CH)_n-C_6H_4-X$$

79

	X	n	$\lambda_{max}(m\mu)$	$\Delta\lambda(m\mu)$
a	H	0	388	
		1	446	58
b	OCH_3	0	440	
		1	512	72
c	$N(CH_3)_2$	0	536	
		1	644	108
		2	735	91

For compounds such as **79** the deviation concept has the difficulty that two sets of symmetrical parent cations can be picked for comparison. Thus two values can be obtained for the deviation.

G. Chain Length and Trace Analysis

Not much effort has been put into research on increasing chain length in a chromogen so as to shift the wavelength maximum to the red further away from background absorption and to increase the intensity.

The chain length between resonance terminals (and thus the wavelength position and the molar absorptivity) in a chromogen is dependent on the reagent or the test substance. Two examples of reagents by means of which the distance between resonance terminals in a chromogen can be modified are (*a*) *p*-dimethylaminobenzaldehyde and *p*-dimethylaminocinnamaldehyde, and (*b*) benzenediazonium and 4-azobenzenediazonium fluoborates.

p-Dimethylaminobenzaldehyde (**80**, $n = 0$) is a popular reagent in colorimetric analysis. It has been applied to the analysis of a large number of compounds, a few of which are discussed briefly in this section. The vinylog *p*-dimethylaminocinnamaldehyde (**80b**) has been substituted for the benzaldehyde in some analytical procedures with sometimes favorable and sometimes unfavorable results. Use of the vinylog usually results in a red shift and an intensity increase. For example, in the determination of enzymic demethylation of *N*-methylaniline to aniline the chromogen **81a** is formed(190).

$(CH_3)_2N-C_6H_4-(CH{=}CH)_nCHO$

80

a: $n = 0$; **b**: $n = 1$

$C_6H_5-\overset{+}{N}H{=}CH(CH{=}CH)_n-C_6H_4-N(CH_3)_2$

81

	n	$\lambda_{max}(m\mu)$	$m\epsilon$	Ref.
a	0	443	11.0	190
b	1	520	93.0	191

However, when aniline is determined with the vinylog, the wavelength maximum is shifted about 77 $m\mu$, and the apparent molar absorptivity is increased over eight times(191). This huge increase in intensity is due mainly to the greater yield of chromogen obtained in the second procedure. The two reagents have also been compared for the determination of aromatic amines in terms of wavelength position and intensity(192). Thus *p*-aminohippuric acid gave λ_{max} 448 mμ, $A = 0.347$, with **80a** and λ_{max} 545 mμ, $A = 1.091$, with **80b**. Other aromatic amines showed a similar phenomenon.

Succindialdehyde has been determined with both reagents (193). Comparison of the two chromogens in Table 9.40, R = H, shows that the one with the longer chain absorbs at longer wavelength, with slightly increased intensity, (193, 194). The latter effect results from a lower yield of the chromogen in which R = H and $n = 1$, because the procedure was developed for optimal results with *p*-dimethylaminobenzaldehyde.

TABLE 9.40

Spectral Properties of the Pyrrol Chromogens Obtained in the Determination of Various Compounds with *p*-Dimethylaminobenzaldehyde and *p*-Dimethylaminocinnamaldehyde(193, 194)

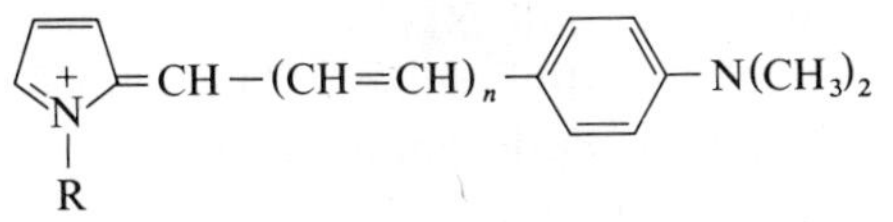

$$\text{(pyrrolium, N}^+\text{–R)}=CH-(CH=CH)_n-C_6H_4-N(CH_3)_2$$

Test substance	R	n	λ_{max}(mμ)	mϵ
Pyrrole	H	0	546	76
		1	628	91
Succindialdehyde	H	0	546	71
		1	628	75
Glycine	CH_2COOH	0	558	50
		1	636	58
Histamine	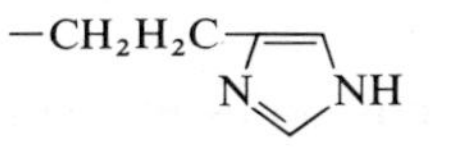 $-CH_2H_2C-$(imidazolyl, N, NH)	0	560	32
		1	640	59
Pyridoxamine	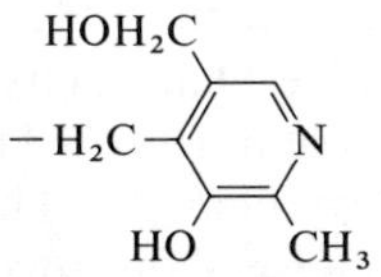 $-H_2C-$(pyridine ring with HOH_2C, HO, CH_3, N)	0	560	28
		1	640	49
Aniline	C_6H_5	0	557	44
		1	640	40
L-Tyrosine	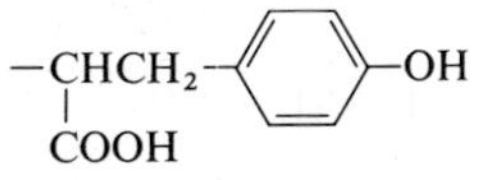 $-CH(COOH)CH_2-C_6H_4-OH$	0	560	29
		1	635	28
n-Butylamine	$(CH_2)_3CH_3$	0	558	21
		1	635	16

Spectral properties are listed in Table 9.40 for the chromogens that are formed from some of the compounds that have been determined with these reagents. In all cases the molar absorptivities obtained in Table 9.40 are derived from the natural molar absorptivity (which is greater for the vinylog) and the yield of chromogen. The compounds in the table are arranged to a large extent in decreasing yield. For all series the vinylog absorbs at approximately 80 mμ longer wavelength.

Indole derivatives show the same phenomenon. Thus tryptophan is determined by reaction with tryptophanase (195). The derived indole reacted with *p*-dimethylaminobenzaldehyde gives a chromogen that absorbs at 570 mμ or with the vinylog a chromogen that absorbs at 625 mμ. With the latter reagent the intensity was two to five times greater.

Another series of reagents that show the phenomenon of red shift and intensity increase with increasing chain length are benzenediazonium and 4-azobenzenediazonium fluoborates. The latter has been used in the analysis of various aromatic amines (157). The cations **82** are obtained in the determination of dimethylaniline. The data for **82a** are for the cationic tautomer in ethanolic hydrochloric acid.

$$C_6H_5-\overset{H}{\underset{+}{N}}=N-\left(C_6H_4-\overset{H}{\underset{+}{N}}=N\right)_n-C_6H_4-N(CH_3)_2$$

82

	n	λ_{max}(mμ)	mε
a	0	518	42·8 (196)
b	1	632	100·0

With some reagents vinylogous series of test substances can be determined through colorimetric measurement of the vinylogous chromogens. Azulene has been used to determine benzaldehyde and cinnamaldehyde as well as furfural and α-furylacrolein (197). The chromogens **83** show the expected red shift and intensity changes. The intensities were probably lower than the theoretical because of the lowered yield of chromogens and difficulties in removing impurities in the azulene.

$$\text{(C}_4\text{H}_3\text{X)}-(CH{=}CH)_n-CH{=}\text{(azulenium}^{+}\text{)}$$

83

X	n	λ_{max}(mμ)	mϵ
CH=CH	0	449	20
	1	507	32
O	0	490, 506	42, 42
	1	560, 563s	46, 45

Polyenals have been analyzed with N,N-dimethyl-p-phenylenediamine (Table 9.41). With increasing conjugation the long wavelength maxima shift to the red with increasing intensity. The latter effect has been verified for the α-furyl derivatives.

TABLE 9.41
Effect of Increasing Conjugation of Chromogens on Colors Obtained with N,N-Dimethyl-p-diaminobenzene(166)

$$(CH_3)_2N-C_6H_4-\overset{+}{N}H=CH(CH=CH)_n-R$$

	Color or λ_{max} (mμ) (mϵ)			
n	R = CH_3	R = C_6H_5	R = α-Furyl	
0	Colorless	Orange red	480 (2.14)	Orange red
1	Yellow	Red	510 (25)	Red
2	Red	Red	530 (35)	Violet red
3	Blue red	Blue red	542 (45)	Red violet
4	Violet	Violet	553 (52)	Violet
5	Blue	Blue violet	570 (66)	Blue
6	Blue	Blue		Blue

IX. Annulation

Fusion of a benzo ring to a cationic resonance structure usually causes a red shift and an intensity increase. However, a fusion (into a structure that contains two resonance terminals) either onto the conjugation chain or at one end of the chain can increase or decrease the energetic asymmetry of the limiting resonance structures and thus cause a violet or red shift.

A. Hydrocarbons

The position of resonance terminals in aromatic hydrocarbon cations usually cannot be assigned to particular atoms. Thus the positive charge appears to be diffused over the whole molecule.

An example of annulation in these types of molecules is the tropylium cation series **84**.

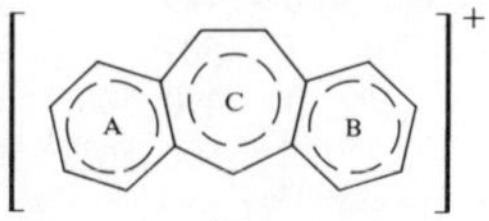

84

Ring	Solvent	λ_{max} (mμ)	mϵ	pK_{R^+}	Ref.
C	Aqueous acid	275	4·37	4·75	198
AC	60% H_2SO_4	426	1·8	1·6	199
ACB	98% H_2SO_4	540	3·24	−3·7	200

In this series annulation results in a substantial red shift and a large decrease in basicity.

A somewhat similar series are the thiapyrylium (**85**) and 1-thianaphthalenium (**86**) and cations (201).

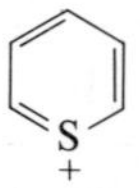

85

λ_{max} 245, 284 mμ in H_2O–1% $HClO_4$

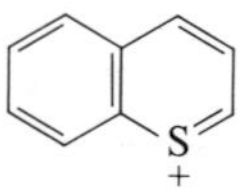

86

λ_{max} 258, 335, 385 mμ in AcOH–1% $HClO_4$

B. Aromatic Aldehydes

Annulation of the benzaldehyde cation results in a strong red shift, as shown in Table 9.42. A cation is apparently formed by proton addition to the carbonyl oxygen. Decomposition or sulfonation of the 9-anthral-

TABLE 9.42
Spectra of the Benzaldehyde Cation and Its Annulated Series in Sulfuric Acid

Aldehyde	λ_{max}(mμ) or color	mϵ
Benzaldehyde	298	35.2
1-Naphthaldehyde	425	10.3
9-Anthraldehyde	518	11.4
	553	13.3
5-Naphthacenealdehyde	Green	

dehyde salt does not take place in sulfuric acid since the aldehyde is recovered quantitatively on addition of water (202).

C. Cyanine Cations

In some series of compounds a violet shift results from annulation. Thus compound 2 in Table 9.21 has two resonance terminals of approximately equal basicity. The change from a quinoline nitrogen resonance terminal to an acridine nitrogen resonance terminal (e.g., compound 3 in Table 9.21) results in a decreased basicity of that terminal and an increase in energetic asymmetry. The result is an increased deviation and a violet shift.

Some fairly complex examples of the red shift effect of annulation are presented in Table 9.43. In the compounds in which $n = 0$: with annulation steric hindrance increases, as shown by the marked decrease in molar absorptivity. The red shift is believed to be due mainly to the Brunings–Corwin type of steric effect.

TABLE 9.43
Annulation and Steric Effects on the Long Wavelength Maxima of Some Monomethine and Trimethine Cations (33, 37)

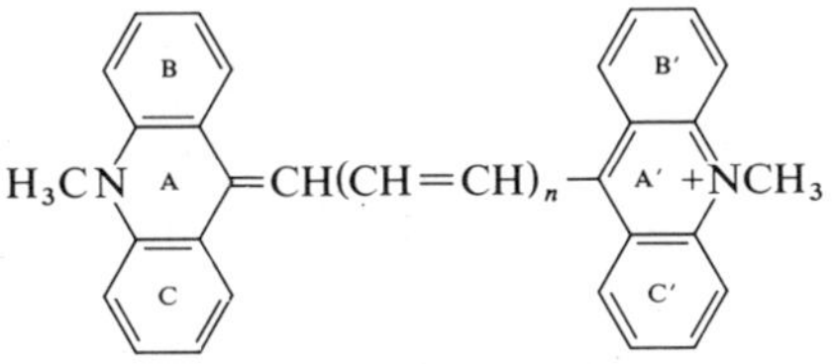

H_3CN (rings A, B, C) $=CH(CH=CH)_n-$ (rings A′, B′, C′) $+NCH_3$

Rings present	n	$\lambda_{max}(m\mu)$	$m\epsilon$
AA′	0	505	34
ABA′	0	530	87
ABA′B′	0	589	86
ABCA′B′	0	594	22
ABCA′B′C′	0	671	11
AA′	1	603	
ABA′B′	1	706	
ABCA′B′C′	1	810	

For the compounds in Table 9.43 in which $n = 1$ the steric effect plays a minor role. In these energetically symmetrical molecules the internal basicity of the resonance terminals decreases and the spectrum shifts strongly to the red with each annulation.

D. Diarylmethane Cations

The determination of aromatic hydrocarbons and phenols with benzal chloride reagents is based on the colorimetric measurement of the resulting diarylmethane chromogens (109). Annulation results in a red shift, as shown for the chromogens **87**, which are obtained in the determination of some polynuclear aromatic hydrocarbons with piperonal chloride (109).

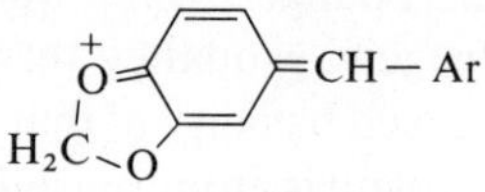

$\overset{+}{O}$=, =CH—Ar, H_2C, O

87

Hydrocarbon	$\lambda_{max}(m\mu)$
Benzene	Light yellow
Naphthalene	590
Anthracene	732

A red shift is shown in Table 9.44 for the chromogens obtained in the determination of pyrene, benzol[*a*]pyrene, and anthanthrene with various aromatic and heterocyclic aldehydes.

TABLE 9.44
Spectra of Chromogens Obtained in Determination of Pyrene, Benzo[*a*]pyrene and Anthanthrene with Various Aldehydes (203)

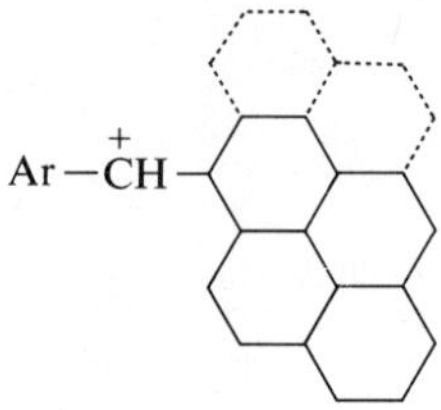

$Ar—\overset{+}{C}H—$

Aldehyde	$\lambda_{max}(m\mu)$		
	Pyrene	Benzo[*a*]pyrene	Anthanthrene
2-Furfural	670	743	787
2-Thenaldehyde	670	750	775
Piperonal	680	753	
Indole-3-aldehyde	690	750	790
3-Nitro-4-dimethyl-aminobenzaldehyde	702	770	845
9-Anthraldehyde	850	910	955

The chromogens obtained in the determination of phenol, 1-naphthol, and 1-anthrol with piperonal show the expected red shifts, as do their *N*-carbethoxyamino analogs (Table 9.45). *N*-Carbethoxy-2-naphthylamine and *N*-carbethoxy-2-anthrylamine reacted with piperonal showed the expected red shifts—namely, to 609 and 750 mμ, respectively (146).

TABLE 9.45
Spectra of Chromogens Obtained in the Determination of Phenols and *N*-Carbethoxyanilines with Piperonal (146, 204)

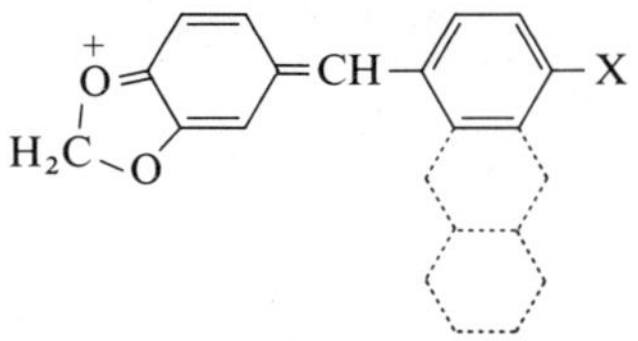

	X = OH	X = NHCOO C_2H_5	
Test substance	λ_{max} (mμ)	mϵ	λ_{max} (mμ)
Phenyl X	522	46	546
1-Naphthyl X	578	35	614
1-Anthryl X	651	90	700

E. Azo Dye Cations

With annulation 1-arylazoazulene and 1-arylazoguaiazulene cations show the red shift effect (Table 9.46). As shown in Table 9.47, annulation of the 4-phenylazophenol and 4-phenylazoaniline cations also causes a red shift toward longer wavelength (99). With increasing basicity of the resonance terminal X in the structure in Table 9.47 the extent of the red shift from annulation decreases, as shown by the $\Delta\lambda$ values of 102, 43, and 24 mμ for the OH, NH_2, and $N(CH_3)_2$ resonance terminals, respectively.

Many trace analysis methods for phenols and anilines depend on the formation of azophenolic or azoanilinic cations. Thus aromatic amines have been determined with 4-azobenzenediazonium tetrafluoborate (157) and with 3-methyl-2-benzothiazolinone hydrazone (57). A moderate red shift is shown in the annulation of the chromogens in Table 9.48 that are obtained with the former reagent (157). Unlike the other chromogens, the one derived from *N*,*N*-dimethyl-1-naphthylamine is probably sterically hindered at the amino nitrogen resonance terminal. Absorp-

TABLE 9.46
Annulation and the Long Wavelength Maxima of 1-Arylazoazulene and Guaiazulene Cations (205, 206)

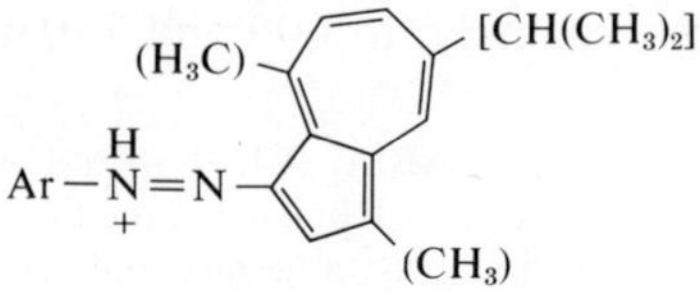

(H_3C) $[CH(CH_3)_2]$ $Ar-\overset{H}{\underset{+}{N}}=N-$ (CH_3)

Ar	Azulene $\lambda_{max}(m\mu)$[a]	Azulene $m\epsilon$	Guaiazulene $\lambda_{max}(m\mu)$[a]	Guaiazulene $m\epsilon$
Phenyl	522	34.8	538	38.6
1-Naphthyl	558	30.7	568	33.2
2-Naphthyl	554	41.8	566	43.0
1-Anthryl	619	20.9	623	22.1
2-Anthryl	610	32.7	616	37.3

[a]In ethanolic hydrochloric acid.

TABLE 9.47
Annulation and the Long Wavelength Maxima of 4-Phenylazophenols and 4-Phenylazoanilines (99)

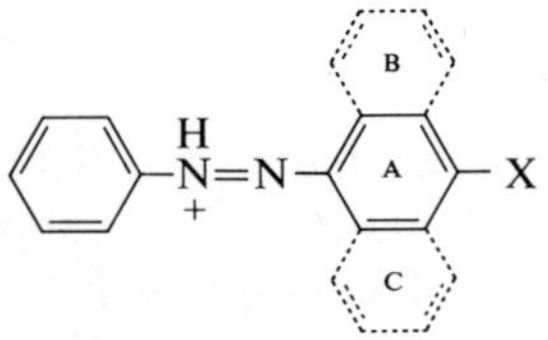

$-\overset{H}{\underset{+}{N}}=N-$ B A C X

Rings present	X	Solvent	$\lambda_{max}(m\mu)$ or color	$\Delta\lambda(m\mu)$
A	OH	H_2SO_4	476	
BA	OH	H_2SO_4	578	102
BAC	OH	H_2SO_4	Blue[a]	
A	NH_2	6 *N* HCl in H_2O–95% EtOH (1 : 1)	502	
BA	NH_2	0.1 *N* HCl in H_2O–95% EtOH (1 : 1)	545	43
BAC	NH_2	Acid	Blue[b]	
A	$N(CH_3)_2$	1 *N* HCl in H_2O–95% EtOH (1 : 1)	516	
BA	$N(CH_3)_2$	1.2 *N* HCl in H_2O–95% EtOH (1 : 1)	540	24

[a]In nitromethane–$AlCl_3$ $\lambda_{max} \simeq 775$ $m\mu$.
[b]Reference 207.

TABLE 9.48
Annulation and the Long Wavelength Maxima of Chromogens Obtained from the Determination of Aromatic Amines with 4-Azobenzenediazonium Tetrafluoborate(157)

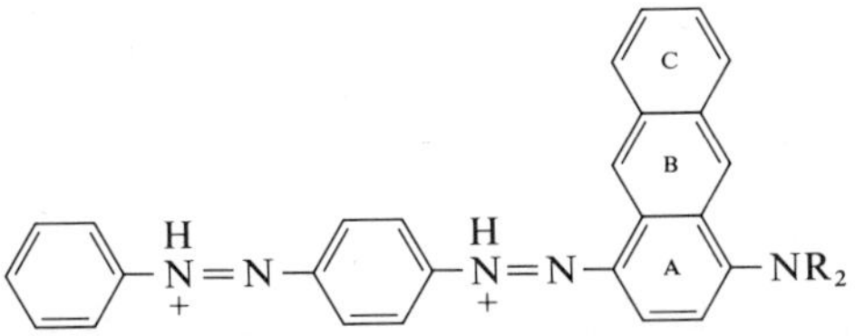

Amine	Rings present	λ_{max}(mμ)	mϵ'
Aniline	A	610	8
1-Naphthylamine	AB	642	95
1-Anthrylamine	ABC	675	42
N,N-Dimethylaniline	A	632	100
N,N-Dimethyl-1-naphthylamine	AB	642	22

tion at lower wavelengths with decreased intensity is probably the result of this steric effect. The decreased intensity also results probably from an increased amount of tautomer with protonation at the amino nitrogen.

3-Methyl-2-benzothiazolinone hydrazone is a fine reagent for the determination of aromatic amines(57). Brilliant colors and high intensities are obtained. The effect of annulation on the long wavelength maxima and molar absorptivities obtained in the analysis of amines with this reagent is shown in Table 9.49. With increasing basicity and steric twist of the amino nitrogen terminal the annulation shift decreases from 18 to 10 to −13. The intensities also show an adverse effect. This is probably due to the poorer yield of chromogen and the adverse effect of the steric twist of the amino nitrogen in the alkylated naphthylamines.

Somewhat similar to the azo dye cations are the chromogens obtained on reaction of hydrocarbons and phenols with various isatins(52). Annulation results in a red shift (Table 9.19). Isatin and nine of its derivatives have been used in the analysis of polynuclear aromatic hydrocarbons (52). Annulation in all the chromogens results in a red shift, as shown for **88**.

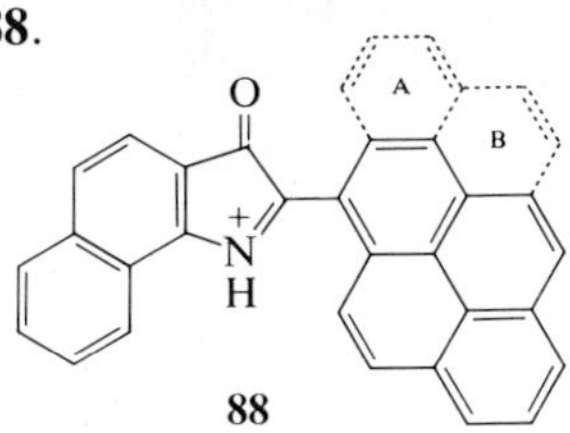

88

Compound	Rings present	λ_{max} (mμ)
Pyrene (P)	P	760
Benzo[*a*]pyrene	PA	870
Anthanthrene	PAB	940

TABLE 9.49
Annulation and the Spectra of Chromogens Obtained from the Determination of Amines with MBTH (57)

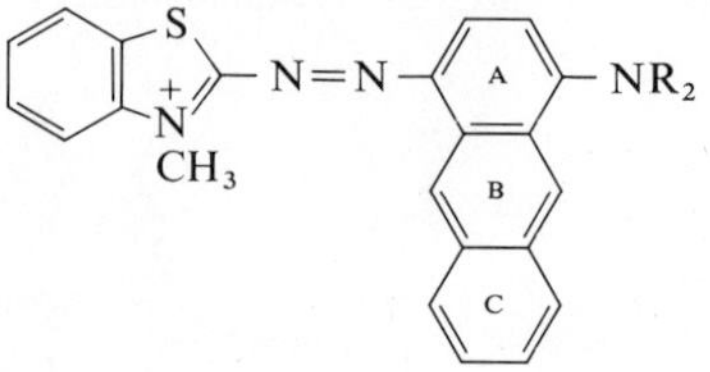

Amine	Rings present	$\lambda_{max}(m\mu)$	$m\epsilon'$	$\Delta\lambda(m\mu)$
Aniline	A	566	43	
1-Naphthylamine	AB	584	53	18
1-Anthrylamine	ABC	615	26	31
N-Methylaniline	A	580	76	
N-Methyl-1-naphthylamine	AB	590	53	10
N,N-Dimethylaniline	A	598	84	
N,N-Dimethyl-1-naphthylamine	AB	585	40	−13

X. Extraconjugation

By extraconjugation is meant an additional conjugation at one or both resonance terminals that is not between the terminals. Thus an unbroken chain of conjugation can be seen in one of the structures (**89**) contributing mainly to the excited state of this dye. In such a compound the methylamino nitrogens are the cationic resonance terminals.

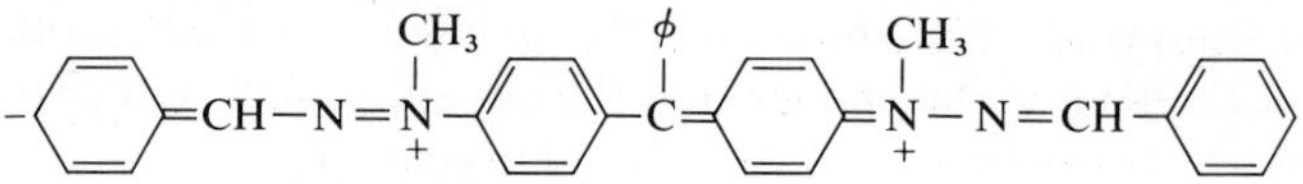

89

Extraconjugation results in a red shift. This phenomenon has been used to differentiate aliphatic and aromatic aldehydes with the hydrazine derivative $H_2NN(CH_3)C_6H_4C(OH)(C_6H_5)C_6H_4N(CH_3)NH_2$ as the reagent (208). The red shift effect is seen in compounds **90** (208).

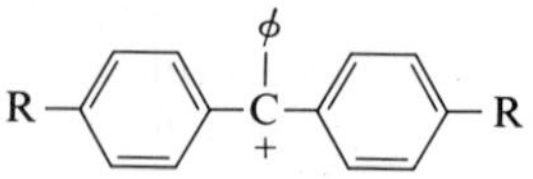

90

R	λ_{max} (mμ)
NH_2	562
NCH_3NH_2	590
$NCH_3N{=}CHCH_3$	645
$NCH_3N{=}CHC_6H_5$	700
$NCH_3N{=}CHCH{=}CHC_6H_5$	735

Another reagent that has been used for the determination of carbonyl compounds is 4-azobenzenehydrazinesulfonic acid, $C_6H_5N{=}NC_6H_4$-$NHNHSO_3H$(209). The chromogens obtained in the analysis show a red shift and an intensity increase with lengthening of the chain in the carbonyl compound (Table 9.50)(209). The series converge by $n = 2$ or 3. In the aldehyde hydrazones the millimolar absorptivity increases from about 83 for $n = 0$ to 93 for $n = 4$. The limiting resonance structures **A** and **B** of the chromogens are shown in Table 9.50. One of the coupled resonance forms that probably contributes to the extraconjugation is structure **C**.

TABLE 9.50
The Effect of Extraconjugation on the Spectra of Some Azobenzenehydrazones(209)

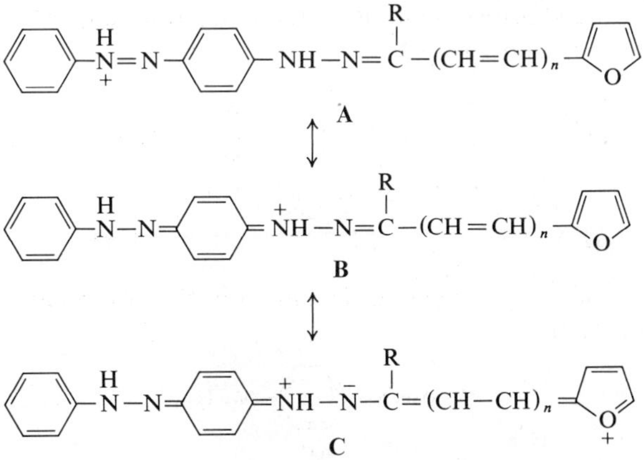

$$C_6H_5-\overset{+}{N}H{=}N-C_6H_4-NH-N{=}C(R)-(CH{=}CH)_n-C_4H_3O$$

A

↕

$$C_6H_5-NH-N{=}C_6H_4{=}\overset{+}{N}H-N{=}C(R)-(CH{=}CH)_n-C_4H_3O$$

B

↕

$$C_6H_5-NH-N{=}C_6H_4{=}\overset{+}{N}H-\overset{-}{N}-C(R){=}(CH-CH)_n{=}C_4H_3\overset{+}{O}$$

C

R	n	λ_{max} (mμ)	R	n	λ_{max} (mμ)
H	0	575	H	5	615
H	1	595	CH_3	0	515
H	2	610	CH_3	1	592
H	3	617	CH_3	2	605
H	4	612			

As shown below for **91** in methanol(78), polymethines exhibit similar extraconjugative effects.

$$R{=}CH{-}\overset{+}{C}H{-}CH{=}R$$

91

	R	λ_{max} (mμ)		R	λ_{max} (mμ)
a	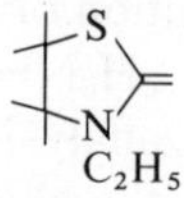 4,4,5,5-tetramethylthiazolinylidene, N-C_2H_5	445	**d**	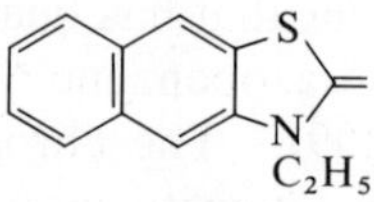 naphtho[2,3-d]thiazolinylidene, N-C_2H_5	570
b	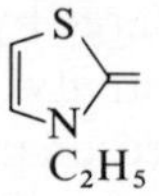 thiazolinylidene, N-C_2H_5	543	**e**	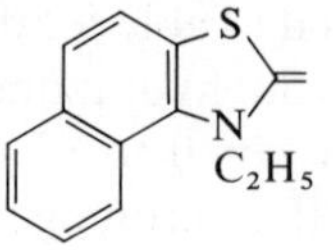 naphtho[1,2-d]thiazolinylidene, N-C_2H_5	595
c	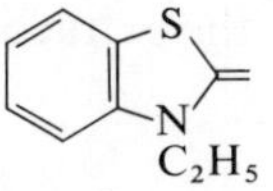 benzothiazolinylidene, N-C_2H_5	558			

If one considers the nitrogen atoms as the resonance terminals and a zwitterionic structure as an important contributor to extraconjugation, then **91b** has the shortest chain of extraconjugation (e.g., **92**), whereas **91e** has the longest, as shown in **93**.

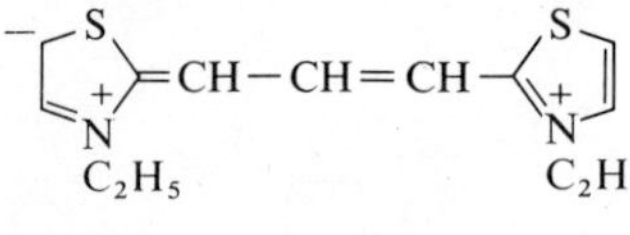

92 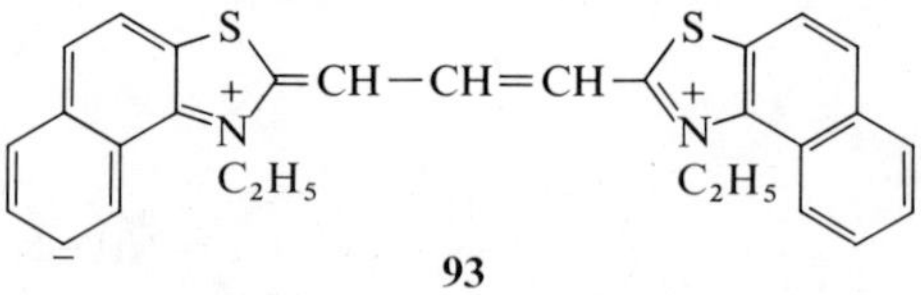   **93**

A similar extraconjugative effect is shown in the compounds **94** (spectra in methanol)(210).

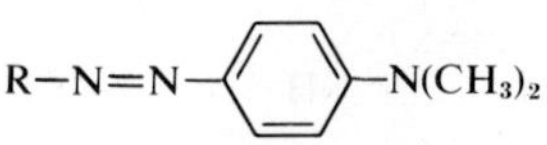

$$R{-}N{=}N{-}C_6H_4{-}N(CH_3)_2$$

94

R =	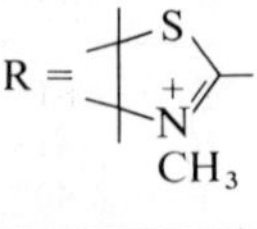 4,4,5,5-tetramethylthiazolinyl (N-CH_3, +)	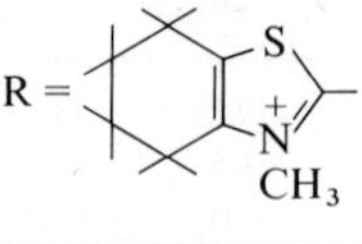 tetramethyl-tetrahydrobenzothiazolyl (N-CH_3, +)	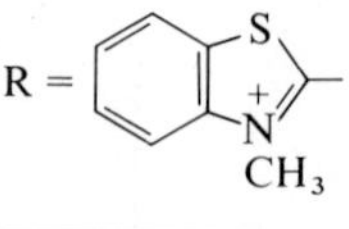 benzothiazolyl (N-CH_3, +)
λ_{max} (mμ)	533	595	595
mϵ	64.3	69.5	79.0

Other things being equal, the substitution of a phenyl group on a nitrogen resonance terminal will cause a red shift. Thus in the highly selective determination of formaldehyde with J-acid and phenyl J-acid the chromogen formed with the latter reagent absorbs at longer wavelength and with greater intensity than the J-acid chromogen, as seen in **95**(211).

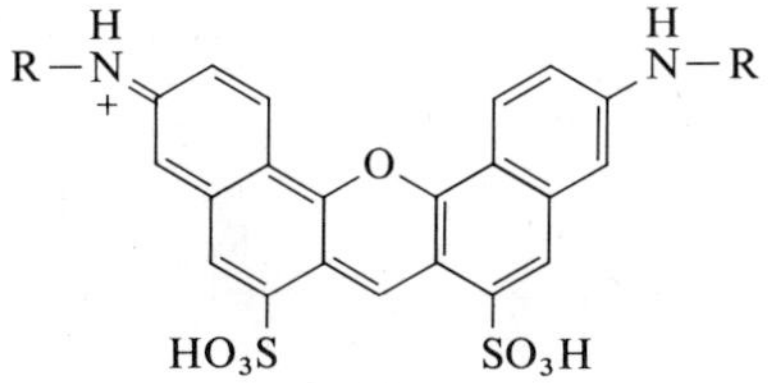

95

R	λ_{max} (mμ)	mε
H	612	34.0
C_6H_5	660	51.4

The red shift is usually greater the longer the chain of extraconjugation. This can be seen by comparing the spectra in methanol of **96** and **97**(212).

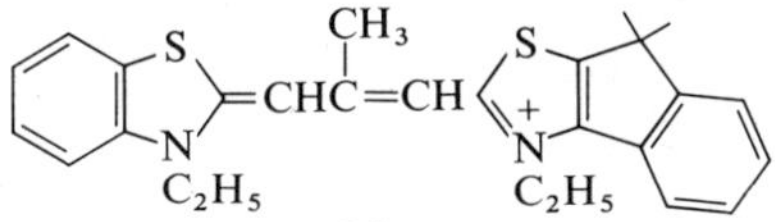

96

λ_{max} 540 mμ, mε 80

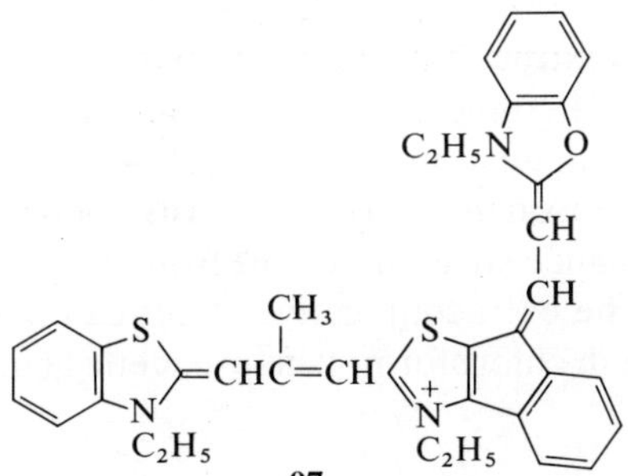

97

λ_{max} 474, 650 mμ, mε 105, 59

The band at 650 mμ in **97** is believed to be derived from the cationic resonance structure as modified by the extraconjugation of the benzoxazole ring attached to the nitrogen terminal through five carbon atoms.

The band at 474 mμ is believed to be derived from the conjugation system outside the cationic resonance structure proper. The dipolar structure **98** that contributes to the system absorbing at 474 mμ is probably stabilized by the plus charge on the adjacent cationic resonance terminal.

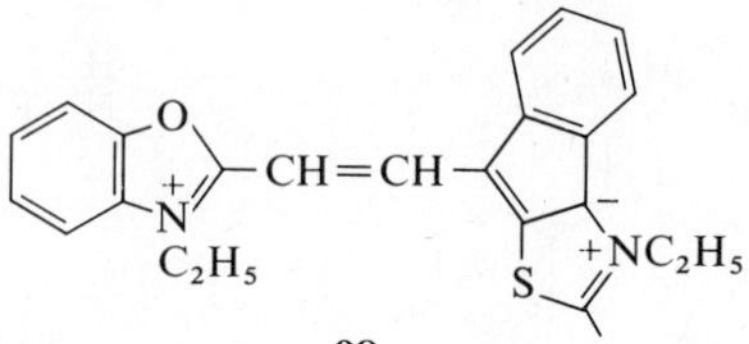

98

A form of extraconjugation is found in the homoconjugation of the nonclassical carbonium ion. Thus solution of the *cis*-carbinol **99** in 60% sulfuric acid gives the cation **100**, λ_{max} 386 and 427 mμ. (213). The *trans*-carbinol does not form this salt.

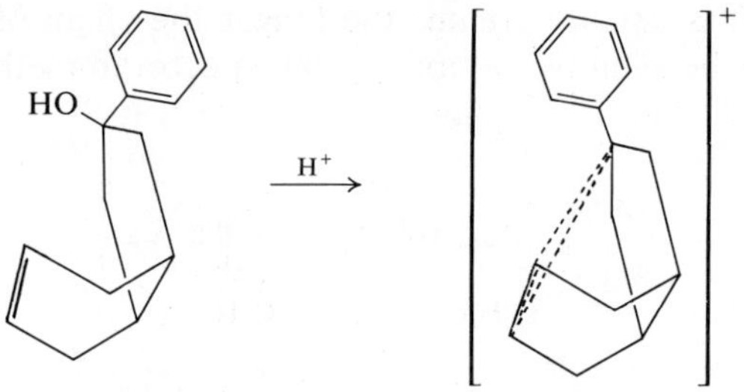

99 $\xrightarrow{H^+}$ **100**

One other type of extraconjugation is found in compounds that contain a protonated carbonyl connected to two conjugation systems. Thus in this type the cross-conjugation takes on properties of extraconjugation. The ketonic oxygen resonance terminal in **101** becomes a resonance node modifier in the resonance structure of **102** from which the long wavelength band is derived. The extraconjugative effect can be seen on comparing **101** and **102** [spectra in chloroform–trifluoroacetic acid (1 : 9)].

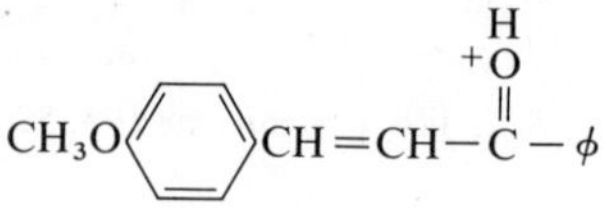

101

λ_{max} 481, mε 26.7

$$CH_3O-C_6H_4-CH=CH-\underset{}{\overset{OH}{C}}=CH-CH=C_6H_4=\overset{+}{O}CH_3$$

102

λ_{max} 396, 550 mμ, mε 14.0, 45.0

Another interesting extraconjugative effect is shown by the enamine salts of retinals with secondary amines (214). The marked red shifts on protonation and with increased extraconjugation indicate the possible relevance of this phenomenon to the chemistry of visual pigments. Naturally occurring visual pigments based on retinal show absorption maxima over the range 440–620 mμ. The absorption maxima of the Schiff base salts are shown in Table 9.51. The unprotonated retinal–secondary amine carbinols absorb at 330 mμ for the pyrroline derivative to 380 mμ for the diphenylamine, pyrrole, indole, and carbazole derivatives. In addition to the extraconjugation, the basicity of the resonance terminal probably also plays a role in this red shift.

This is in line with the views of Brooker et al. (137, 215), whose work indicates that the wavelength maximum of a symmetrical cyanine depends mainly on three factors: the conjugation-chain length, the basicity of the resonance terminal, and extraconjugation. A study of the absorption of symmetrical trimethine cyanines derived from thiazole nuclei has indicated that the susceptibilities of $\bar{\nu}$ (essentially $10^7/\lambda$, mμ) to the extraconjugation and to the basicity are in the ratio 1 : 10 (216).

TABLE 9.51
Long Wavelength Maxima of the Retinylidene Ammonium Salts (214)

(Structure: retinylidene chain ending in $CH=\overset{+}{N}<$)

Secondary amine	λ_{max} (mμ)
Piperidine	440
Piperazine	470
Indoline	510, 570*s*
Diphenylamine	530
Indole	610
Carbazole	620
Pyrrole	655

XI. Cross-Conjugation

A. Triphenylmethane cations

In polyenes, polyphenyls, and many of the cyanines the oscillation of electrons caused by the absorption of electromagnetic radiation is usually along one axis. In many other compounds there are two conjugation systems in two different dimensions. One portion of the conjugation chain and one resonance terminal are common to both systems in the triphenylmethane cations showing cross-conjugation. In agreement with the convention of Lewis and Calvin(140) the *X*-axis is taken to represent the length of the molecule, the *Y*-axis is along the width, and the *Z*-axis is perpendicular to the molecular plane.

Proof for the *Y*-band has been obtained for malachite green (no. 1 in Table 9.52) by two methods(217). The first depends on the production of oriented molecules in a rigid solvent; the second, on the use of polarized

TABLE 9.52
The *X*- and *Y*-Bands of Malachite Green Derivatives in Acetic Acid(29)

$(CH_3)_2N$ $\overset{+}{N}(CH_3)_2$
X
Y
R

R	*X*-Band		*Y*-Band	
	λ_{max} (mμ)	mϵ	λ_{max} (mμ)	mϵ
H	621	104	428	20
4-AcO	619		433	
4-CH_3	617	108	438	25
2-CH_3—4-CH_3O	614	124	460	22
2-CH_3—4-HO	610	121	465	24
4-CH_3O	608	106	465	34
4-HO	603	106	470	36

fluorescence. The more intense absorption of malachite green at 625 mμ, mϵ 67.6, is derived from electronic oscillations along the *X*-axis of the molecule, whereas the band at 423 mμ, mϵ 15.9, is derived from oscillations along the *Y*-axis.

As the electron-donor strength of the 4-R group in the structure shown in Table 9.52 is increased, the *X*-band shifts to the violet and the *Y*-band shifts to the red (29). The former shift is due to the positivization of the positive resonance node at the central carbon atom of the *X*-transition. The latter shift fits in with the *Y*-band's being derived from a cationic resonance structure containing a nitrogen and a phenyl or oxygen resonance terminal. In line with the assignments is the steric effect of a 2-methyl group on compounds in this series where R is OH or OCH_3 in causing the expected red shift of the *X*-band through decreased positivization of the resonance node and a violet shift of the *Y*-band through decreased interaction of the *Y*-transition terminals. Additional evidence is derived from the spectra of compounds in which R is 2-methyl or 2,6-dimethyl.

These substituents twist the phenyl group out of the plane of the *X*-transition involving two nitrogen terminals. Consequently with the increasing steric hindrance the *X*-band shifts to the red with increasing intensity, and the *Y*-band shifts to the violet with decreasing intensity (see Table 9.6).

Some work has been published on the effect of solvent polarity on the *X*- and *Y*-bands of the structure shown in Table 9.53. Both bands absorb at shortest wavelength in water and at longest wavelength in chloroform.

TABLE 9.53
Solvent Effect on the *X*- and *Y*-Bands (7) of

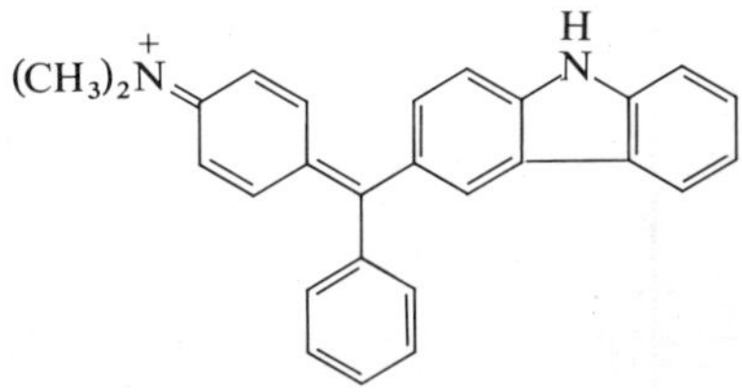

$(CH_3)_2\overset{+}{N}$ ... H N

Solvent	Molarity HCl	*X*-Band		*Y*-Band	
		λ_{max} (mμ)	mϵ	λ_{max} (mμ)	mϵ
Water	10^{-1}	562	35	408	11.6
Acetone	10^{-2}	579	49	412	15.4
Methanol	10^{-2}	585	50	416	15.2
Acetic acid	0	586	46	417	13.9
Chloroform	2×10^{-2}	605	51	428	13.6
Chloroform	10^{-4}	629	66	440	14.2

B. Polymethine Cations

Cross-conjugation of several series of tetraphenylpolymethine derivatives has been investigated(176). This effect is shown for the series in Figure 9.3. In the first member of the series a large amount of steric crowding is present so that, at most, only three of the benzene rings can be coplanar; the steric effect is shown by the decreased intensity of the X-band at 740 mμ, mϵ 36.0, as compared with the Y-band at ~643 mμ and the X-bands of the other members of the series. The X-band in these three compounds is associated with resonance between nitrogen terminals at opposite ends of the chain, whereas the Y-band is associated with resonance between two nitrogen terminals at one end of the chain. Thus in the structure in Figure 9.3, when $n = 1$, the Y-band is at 636 mμ, mϵ ~57, whereas the X-band is at 810 mμ, mϵ 183. When $n = 2$, the Y-band is at 663 mμ, mϵ ~41, and the X-band is at 910 mμ, mϵ 186.

This moderate red shift of the Y-band with increase in the conjugation-chain length is especially apparent in the three other series of tetraphenylpolymethines studied by Tuemmler and Wildi(176). In these other series only two phenyl groups at opposite ends of the chain have p-dimethylamino groups. Consequently the X-bands for the α, ω-diphenyl (Fig. 9.4), α,ω-di-4-chlorophenyl-, and α,ω-di-2,4-dichlorophenyl

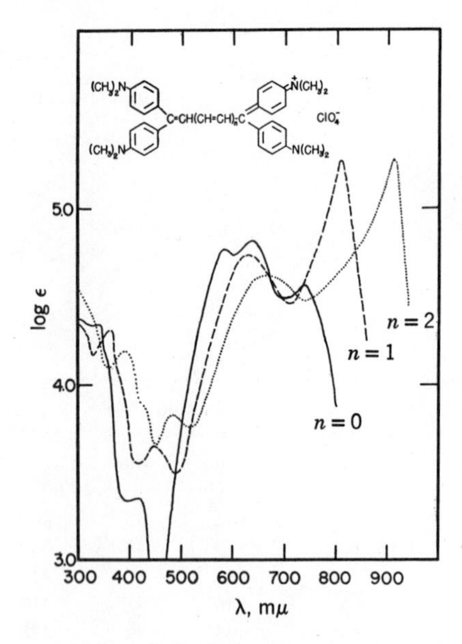

Fig. 9.3. Absorption spectra in chloroform containing 0.75% ethanol of some tetra-(p-dimethylaminophenyl)polymethine cations(176).

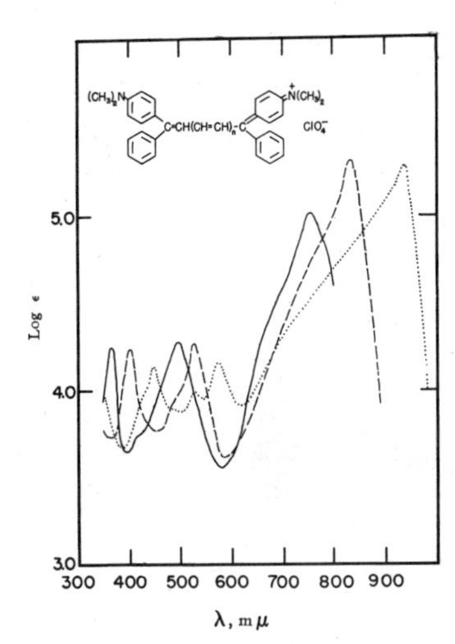

Fig. 9.4. Absorption spectra in chloroform containing 0.75% ethanol of some diphenyl-di-(*p*-dimethylaminophenyl)polymethine cations (176)

derivatives are in the same area as those of the comparable structure in Figure 9.3, $n = 0$, 1 and 2. On the other hand, the Y-bands range between 500 and 600 mμ, as would be expected of cationic resonance structures involving resonance between a phenyl and a dimethylaminophenyl group.

The third important absorption exhibited by these dyes lies in the 350 to 450 mμ region. These bands are believed to be second order X-bands (X'-bands) since they lie at a little more than twice the frequencies of the first order X-bands and have about one-tenth the intensities of the first order bands.

The trication **103** contains a fairly complicated system of cross-conjugation in which the pentamethine (**104**) and trimethine (**105**) chromophores are present several times (218). (The spectra shown below were measured in acetonitrile.)

$Me_2\overset{+}{N}$ ... Me_2N ... $\overset{+}{N}Me_2$ ... $Me_2\underset{+}{N}$ ... NMe_2 **103**

Me_2N ... Me ... $\overset{+}{N}Me_2$ **104**

Me_2N ... $\overset{+}{N}Me_2$ **105**

No.	λ_{max} (mμ)	mε
103	303, 379*s*, 418	45.7, ~45, 46.8
104	416	118
105	313	64.6

The dication **106** is a little more complicated in that it contains the three chromophores **105**, **107**, and **108**(218). Thus the bands of **106** at 308, 330 (shoulder), and 432 mμ are analogous to the bands of **105** at 313 mμ, **108** at 326 mμ, and **107** at 432 mμ. In most cases cross-conjugation has not affected the wavelength maximum of a chromophore but has drastically reduced the intensity.

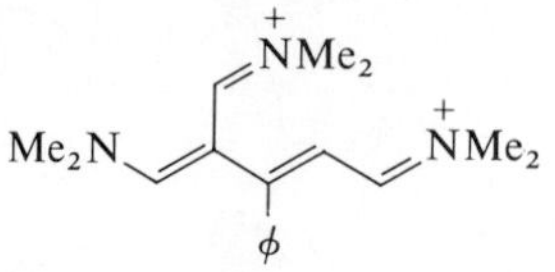

106 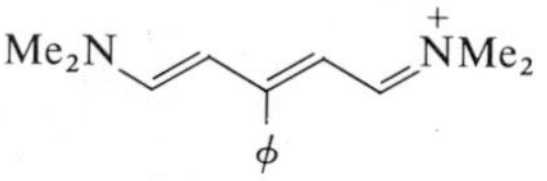 **107** 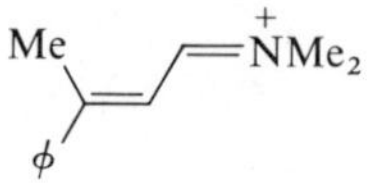 **108**

No.	λ_{max} (mμ)	mϵ
106	308	64.6
	330*s*	40.7
	432	7.08
107	432	97.7
108	326	33.9

Examples of cross-conjugation in the cyanine dyes are known. For example, **109a** has the astraphloxin chromophore **109b** and the dimethine chromophore **110**(105), similar to the one present in **109a**(219). The *N*,*N'*-diethyl analog of **109b** absorbs at 541 mμ, mϵ 90.0(178). Thus in the cross-conjugated compound **109a** the trimethine chromophore absorbs at shorter wavelength and decreased intensity than **109b**. This is due to the strong positivization of the central carbon positive resonance node. The other chromophore with the ring and piperidine nitrogens as resonance terminals absorbs in the same area as the comparison standard but with decreased intensity.

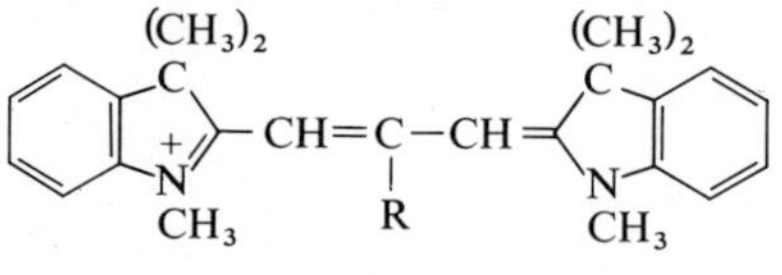

109 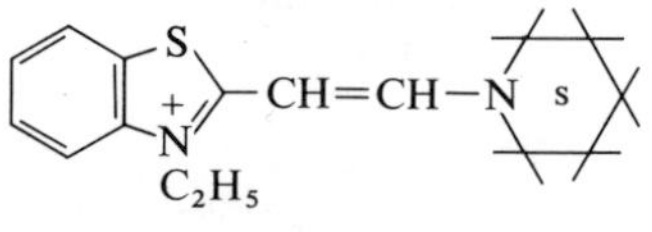 **110**

No.		λ_{max} (mμ)	mϵ
109a	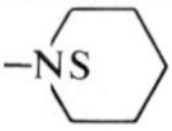 –N S	395, 457	24.0, 32.4
109b	H	545	
110		388	51.0

For some types of cross-conjugated cations the partial chromophores absorb at much shorter wavelengths than do the reference compounds. Such a state of affairs exists for the pentamethine derivative **111**(220).

The chromophore in **111** containing the *a* and *c* resonance terminals absorbs at approximately 614 mμ (at a much shorter wavelength than **112b**, λ_{max} 708 mμ), primarily due to the strong violet shift effect of the *b* nitrogen on the adjacent positive resonance node. The band at about 480 mμ is derived from the partial chromophores containing resonance terminals *a–b* and *b–c*. This absorption is at shorter wavelength than that of the reference standards **112a** and **113** (λ_{max} 524 and 560 mμ, respectively).

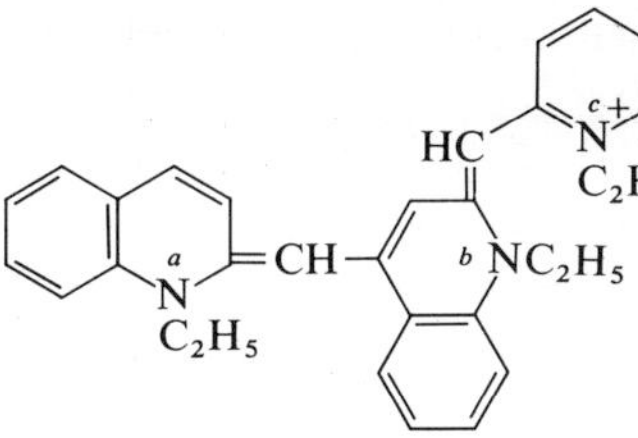

111

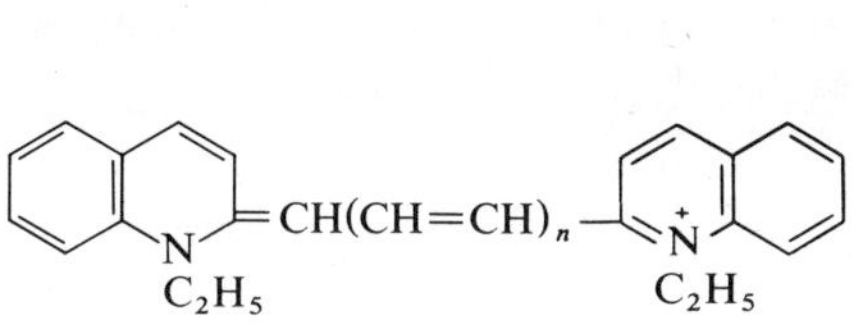

112

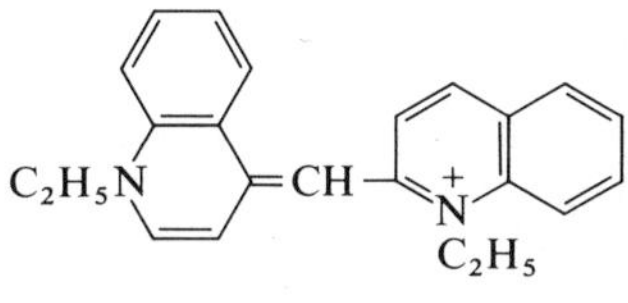

113

No.	*n*	λ_{max} (mμ)	mϵ
111		~480, ~614	~58, ~47
112a	0	524	76
112b	2	708	
113		~560	~81

The spectral properties of a cross-conjugated trinuclear dication **114** indicate that cross-conjugation in this molecule, as compared with the reference compound **115**, has caused a slight violet shift and a halving in intensity of the long wavelength band associated with the pentamethine chromophores (221).

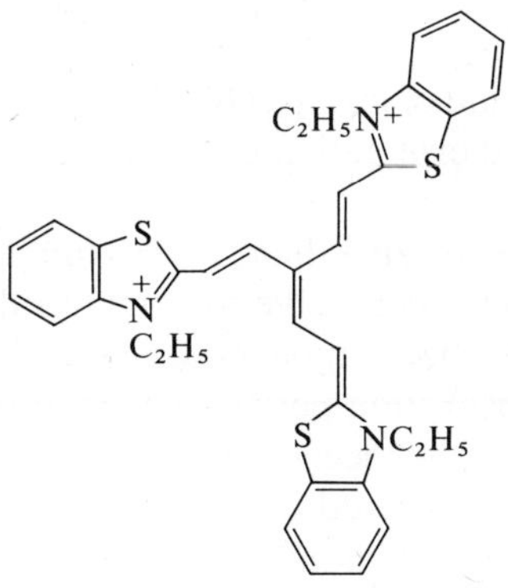

114

λ_{max} 639 mμ, mϵ 102

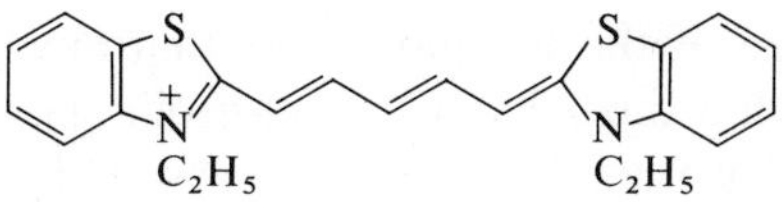

115

λ_{max} 653 mμ, mϵ ~200

XII. *Z*, or *J*, Band

The *Z*, or *J*, type of band is found in the spectra of polymers. The term "stacking" indicates the interaction between adjacent dye molecules when they are arranged in card pack fashion on the polymer(222, 223). The strength of the interaction varies with the nature of the cation and the polymer and is termed the stacking tendency. The electronic transition takes place between the parallel molecules, and the resultant band is called a *Z*-band. In this face to face interaction of identical coplanar molecules there is a very intimate proximity, with orbital overlap between molecules.

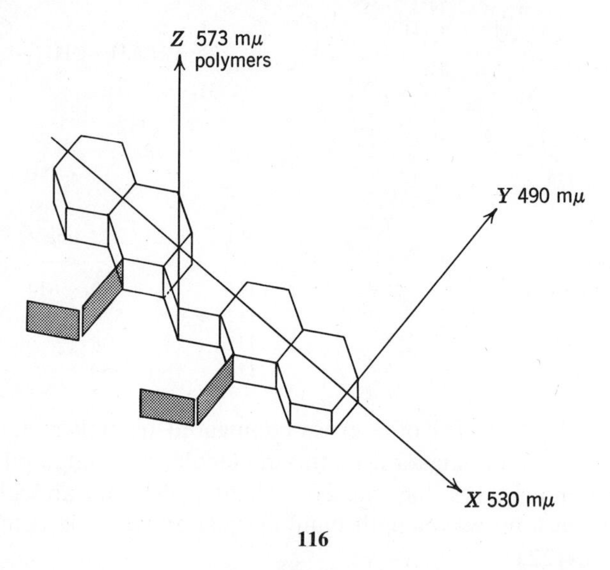

116

Jelley and Scheibe(224, 225) have studied *X*-, *Y*-, and *Z*-bands in pseudocyanines. In 1,1′-diethyl-2,2′-cyanine chloride (**116**) these bands have been correlated with the geometry and orientation of the molecules of the compound.

In very dilute aqueous solution the dye shows an *X*-band at 530 mμ and a *Y*-band at 490 mμ, but in concentrated solution the dye forms a fibrillar colloid and a new narrow and very intense band appears at about 573 mμ. This is a *Z*-band that arises from absorption perpendicular to the face of the flat molecules that are lying face to face. The band is also called the *J*-band in honor of Jelley, who discovered an intense, narrow absorption band at longer wavelength than the monomeric absorption band of a dye. The narrow band was derived from the polymeric dye attached to an anionic macromolecule.

Compound **116** undergoes reversible polymerization in solution(226). At concentration greater than 10^{-2} *M* in water a very sharp band appears at 573 mμ, with mϵ > 120. The short wavelength bands appear at different wavelengths than the monomer bands and are believed to arise from polymers also. At 5×10^{-4} *M* and lower concentrations of dye the 573 mμ band is absent and only the monomer bands at 524 and 490 mμ are present, with a shoulder at 462 mμ.

Large polymers that contain negative groups, such as SO_3^- and COO^-, react in dilute aqueous solutions with **116** in the aggregate form, as shown by the narrow long wavelength band at 567 or 572 mμ. Anionic macromolecules that show this phenomenon include chondroitinsulfuric acid, hyaluronic acid, polyethylenesulfonic acid, and heparin(227). The spectral changes are shown in Figure 9.5. Low concentrations of the pure dye in aqueous solutions are present as the monomer. With increasing concentrations of heparin the polymer band at 572 mμ appears and then increases in intensity.

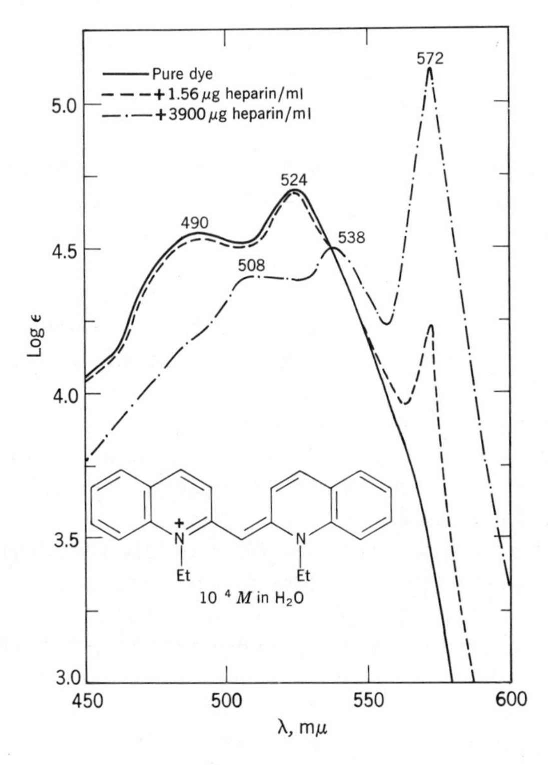

Fig. 9.5. Effect of heparin on the absorption spectrum of the 1,1′-diethyl-2,2′-cyanine cation.

Pseudoisocyanin interacts with densely positioned electronegative groups of mucopolysaccharides in tissues and solutions in the same way that it interacts with linearly positioned electronegative groups of synthetic products in solution (e.g., polyethylenesulfonic acids). The electrostatically bound polymerizate absorbs light at approximately 572 mμ in tissues or in solutions. This dye (**116**) is especially suited for the detection of mucopolysaccharides (228).

A method of detecting macromolecules by the spectral changes in the dye 1-ethyl-2-[3-(1-ethylnaphtho[1,2-*d*]thiazolin-2-ylidene)-2-methylpropenyl]naphtho[1,2-*d*]thiazolium bromide (**117**) on its aggregative complexation with very small amounts of various macromolecules has been reported (229).

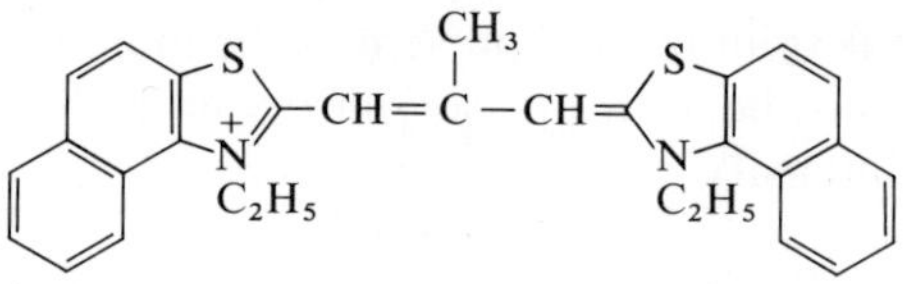

117

λ_{max} 575 mμ

Some of the macromolecules that form such complexes absorbing at 650 mμ include myoglobin, β-lactoglobulin, plasminogen, urease, fibrinogen, carbonic anhydrase, gelatin, cytochrome *c*, and oxidized ribonuclease (230). Thus macromolecules of biological origin could be detected on other planets through this type of absorption.

The formation of new absorption maxima by **117** in the presence of macromolecules is a function of the sign of electric charge on the macromolecule. Any change in the sign will cause extensive changes in the influence of the macromolecule on the absorption of the dye. Proteins showed a rapid decrease in the intensity of the induced dye absorption peak over different ranges of pH—for example, 8.6–7.6 for cytochrome *c* and 3.1–1.7 for urease (230).

On the other hand, interaction of **117** with DNA was virtually unaffected by pH changes, but alterations in wavelength occurred due to denaturation of DNA. Completely denatured DNA reacted with **117** to produce an absorption peak at 535 mμ derived from the dye aggregate. Native DNA binds the dye as a monomeric species (α_{max} 575 mμ). The ratio of the absorbance at 535 mμ to the absorbance at 570 mμ serves as an accurate and sensitive measure of the degree of denaturation.

When the molar ratio of dye to oxidized ribonuclease is 5.7 or less, only the 650 mμ maximum, derived from a very large polymer, is formed.

At greater molar ratios the 650 mμ maximum and several other maxima representing the monomeric and low polymeric forms are seen.

Results have been obtained with many proteins, peptides, and nucleotides. For a great many complexes that **117** forms with trace amounts of protein, the exact wavelength of the induced maximum is characteristic of the protein(231). The value in trace analysis stems from the data showing that a macromolecule concentration of less than 1 μg/ml results in the formation of a new absorption peak whose absorbance increases in an approximately linear manner with macromolecule concentration (230, 231).

The polymerization phenomenon has been studied with many dyes. The polarity of the solvent affects the degree of polymerization of the dye and thus its absorption spectra(232). This has been shown for methylene blue and thionine.

XIII. Intensity Determinants

A. Salt Formation

Other things being equal, the addition of a proton to the negative resonance terminal of a zwitterionic resonance structure (Z_n) causes a red shift and an intensity increase (see Tables 9.8 and 9.23). The carbonyl compounds in Table 9.32 seem to be an exception to this rule in the sense that, although a red shift occurs on protonation, the intensity decreases. For these compounds the acid solution may have been too weak to convert the carbonyl compounds to their salts quantitatively.

B. Straight-Line Distance between Terminals

The relationship between the straight-line distance between terminals and the intensity of the long wavelength band of zwitterionic resonance structures has been discussed in Chapter 7. Approximately the same relation holds here, as can be seen for the three monomethine dyes **118**, **119** and **120**(233) as compared with the pseudoisocyanine **116** [λ_{max} 524 mμ, mε 76.0, in methanol(105).]. As the straight-line distance between resonance terminals is decreased, the molar absorptivity decreases. The same results are seen for the malonaldehyde derivatives **121** and **122**(234) and **123**, **124** and **125**(235).

The intensity is about twice as large in the *trans, trans* form (**121**), as it is in the *cis, cis* form (**122**) (both measured at pH 2.2). In the same way the intensity of the long wavelength bands derived from the transition involving the nitrogen resonance terminals of the all-*trans* form **123**, the *trans cis* form **124**, and the all-*cis* form **125** decreases in that

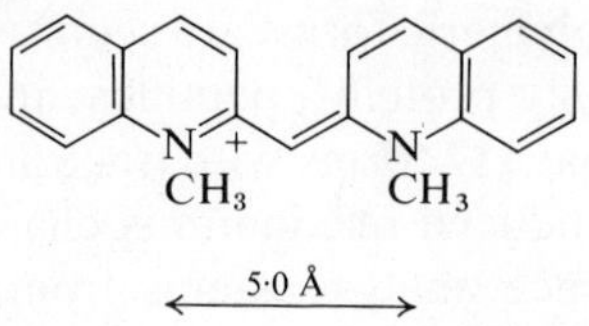

5·0 Å

118

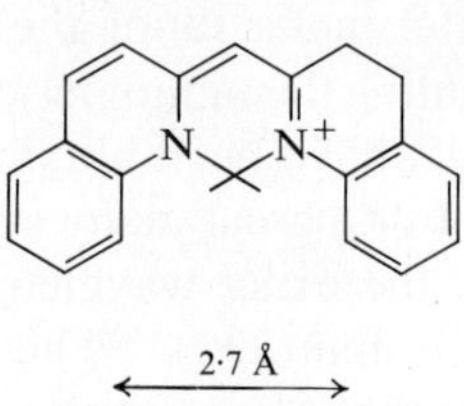

2·7 Å

119

4·4 Å

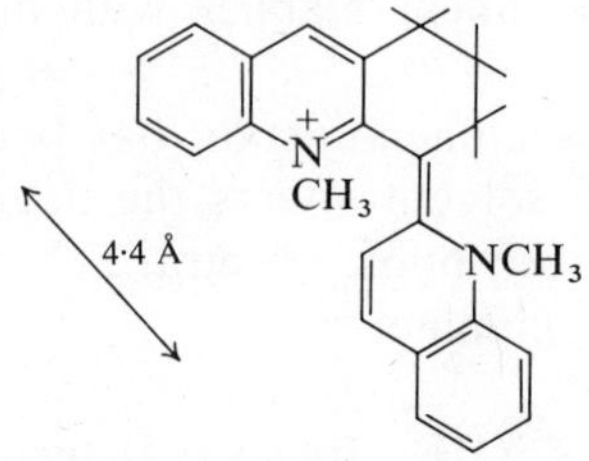

120

No.	λ_{max} (mμ)	mϵ
118	570	50.5
119	523, 562	23.8, 16.4
120	573	34.0

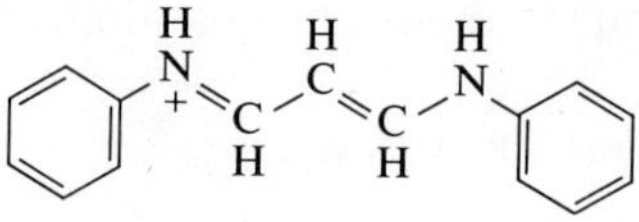

121

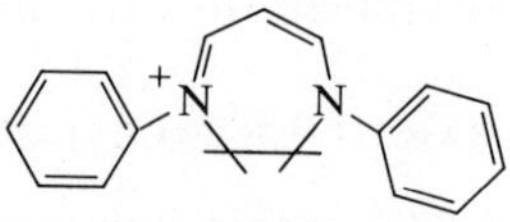

122

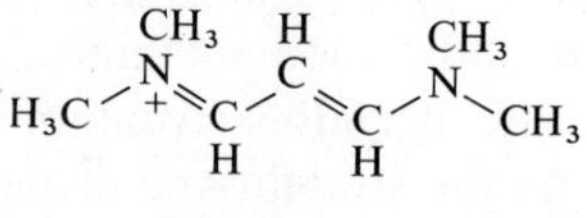

123

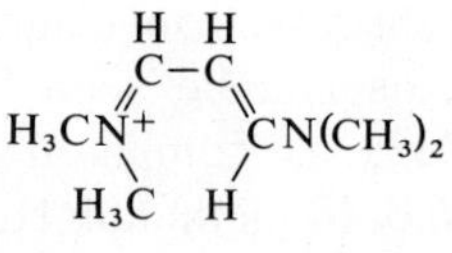

124

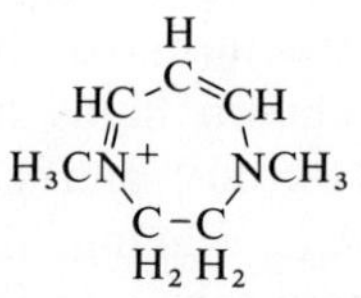

125

No.	λ_{max} (mμ)	mϵ
121	242, 384	13.8, 56.8
122	230, 380	6.9, 25.2
123	314	52.5
124	340	31.0
125	352	17.0

order as does the straight-line distance between the terminals. Feldman et al.(235) have reported that at 100°K the oscillator strengths for **123**, **124**, and **125** are 0.76, 0.58, and 0.28, respectively. These authors have discussed other examples of the decrease of long wavelength intensity and oscillator strength with a decrease in the straight-line distance between resonance terminals.

C. Steric Hindrance

Strong intensity decreases are found in the crystal violet dyes when methyl groups are substituted in the 2-position (Table 9.6). The intensity decrease is especially strong when methyl groups are substituted *ortho* to the dimethylamino group(s) of crystal violet, Michler's hydrol blue, and malachite green. Substitution of an alkyl group on the central carbon atom of a sterically crowded monomethine cation results in an intensity decrease (Table 9.8).

Steric hindrance can also cause an increase in intensity. Substitution of an alkyl group in the 2-position of the nonaminated phenyl ring of malachite green twists this ring out of the plane of the molecule. The intensity of the long wavelength band then increases toward that of Michler's hydrol blue (Table 9.6). The increase is substantial, $m\epsilon$ 104 to 123.

A similar phenomenon is seen in some trimethine dyes (Table 9.18). When an alkyl group is substituted *ortho* to the dimethylamino group of compound 1 in Table 9.18, the intensity increases to that of the deaminated (No. 4) or protonated (No. 7) molecule.

D. Chain Length and Intensity

Other things being equal, the intensity of the long wavelength band increases with increasing length of chain between resonance terminals (Tables 9.32, 9.33, and 9.40). In the latter table there appear to be some exceptions to the rule, but these are probably spurious since the millimolar absorptivities are dependent on the yield of chromogen obtained in the determination of various test substances with *p*-dimethylaminobenzaldehyde and *p*-dimethylaminocinnamaldehyde as well as the true molar absorptivities of the pure chromogens. The yields of chromogen obtained with the cinnamaldehyde derivative are believed to be lower than those obtained with the benzaldehyde derivative.

Another exception, the [2-bis(1-ethylquinolyl)] cyanine dyes (**126**), has been reported (178).

126

n	λ_{max} (mμ)	mε
0	519	70.3
1	596	205.0
2	704	160.0
3	808	127.0

Such exceptions need further study to determine whether increasing hydrolysis or some interesting electronic factors are at the root of this intensity decrease.

E. Resonance Nodes and Intensity

Other things being equal, it should be possible by positivizing a negative node or negativizing a positive node to increase the intensity. On the other hand, negativizing a negative node or positivizing a positive node should decrease the intensity. An example of the latter phenomenon is shown in compounds **40**. In this respect Brooker *et al.*(78) have stated that the effect of dividing a long resonating chain into shorter fragments is to shift the absorption toward the violet and to reduce the intensity.

F. Solvent Effect on Tautomeric Equilibrium

In zwitterionic resonance structures in which the negative and positive resonance terminals have about the same order of basic strength, protonation takes place at both terminals with the formation of tautomeric monocations. An example of solvent-sensitive intensities is found in the spectra of the *N*, *N*-dimethyl-4-phenylazoaniline cations in various acid solutions (Table 9.54). For some of the azo cations the intensity

TABLE 9.54
Tautomeric Equilibrium and Spectral Intensity as Affected by the Acidity of 50% Aqueous Ethanol (4)

$C_6H_5-\overset{+}{N}H=N-C_6H_4-N(CH_3)_2 \rightleftharpoons C_6H_5-N=N-C_6H_4-\overset{+}{N}H(CH_3)_2$

I II

I		II			
λ_{max} (mμ)	mε	λ_{max} (mμ)	mε	Normality HCl	E_I/E_{II}
517	34.0	320	10.1	0.6	3.4
516	35.5	320	9.8	1.2	3.6
516	40.3	320	8.1	3.0	5.0
518	47.3	323	5.26	6.0	9.0

changes even more dramatically with change in acid concentration. Thus 4-phenylazoaniline in *N* HCl absorbs at 500 mμ, mϵ 12.2 (proton to β-azo nitrogen) and 319 mμ, mϵ 16.8 (proton to the amino nitrogen); in 6 *N* HCl it absorbs at 502 and 319 mμ, mϵ 36.7 and 9.16, respectively. Thus the tautomeric equilibrium is shifted by changes in solvent acidity.

References

1. Y. Hirshberg, E. B. Knott, and E. Fischer, *J. Chem. Soc.*, 3313 (1955).
2. R. Wizinger and H. Wenning, *Helv. Chim. Acta*, **23**, 247 (1940).
3. D. S. Noyce and M. J. Jorgenson, *J. Am. Chem. Soc.*, **84**, 4312 (1962).
4. E. Sawicki, *J. Org. Chem.*, **22**, 621 (1957).
5. W. Krauss and H. Grund, *Z. Elektrochem.*, **58**, 767 (1954).
6. A. Wasserman, *J. Chem. Soc.*, 3228 (1958).
7. G. Branch, B. Tolbert, and W. Lowe, *J. Am. Chem. Soc.*, **67**, 1693 (1945).
8. G. Zinner and W. Deucker, *Naturwissenschaften*, **49**, 300 (1962).
9. E. Sawicki, T. W. Stanley, and J. D. Pfaff, *Anal. Chim. Acta*, **28**, 156 (1963).
10. E. Sawicki, T. W. Stanley, J. D. Pfaff, and J. Ferguson in *Analytical Chemistry 1962*, Elsevier, Amsterdam, 1963.
11. A. W. Archer and E. A. Haugas, *J. Pharm. Pharmacol.*, **12**, 754 (1960).
12. J. R. Platt, *J. Chem. Phys.*, **25**, 80 (1956).
13. P. E. Blatz and D. L. Pippert, *J. Am. Chem. Soc.*, **90**, 1296 (1968).
14. W. Dilthey and H. Wübken, *Ber.*, **61**, 963 (1928).
15. E. Sawicki, *Chemist-Analyst*, **48**, 4 (1959).
16. W. Theilacker, W. Berger, and P. Popper, *Chem. Ber.*, **89**, 970 (1956).
17. A. Baeyer and V. Villiger, *Ber.*, **37**, 2861 (1904).
18. A. Baeyer and V. Villiger, *Ber.*, **35**, 3013 (1902).
19. C. Marvel, J. Whitson, and H. Johnston, *J. Am. Chem. Soc.*, **66**, 415 (1944).
20. A. Morton and W. Emerson, *J. Am. Chem. Soc.*, **59**, 1947 (1937).
21. L. Anderson and D. Thomas, *J. Am. Chem. Soc.*, **65**, 234 (1943).
22. W. Theilacker and W. Schmid, *Ber.*, **84**, 204 (1951).
23. A. Morton and L. McKenney, *J. Am. Chem. Soc.*, **61**, 2905 (1939).
24. W. Dilthey, *J. Prakt. Chem.*, **109**, 317 (1925).
25. R. Wizinger and G. Renckhoff, *Helv. Chim. Acta*, **24**, 369E (1941).
26. E. Sawicki and V. T. Oliverio, *J. Org. Chem.*, **21**, 183 (1956).
27. A. I. Kiprianov, G. G. Dyadyusha, and F. A. Mikhailenko, *Russ. Chem. Rev.*, **35**, 361 (1966).
28. G. L. Eian and C. A. Kingsbury, *J. Org. Chem.*, **32**, 1864 (1967).
29. C. C. Barker, in G. W. Gray, Ed., *Steric Effects in Conjugated Systems*, Butterworths, London, 1958.
30. C. C. Barker, G. Hallas, and A. Stamp, *J. Chem. Soc.*, 3790 (1960).
31. C. C. Barker and G. Hallas, *J. Chem. Soc.*, 1395 (1961).
32. C. C. Barker, M. H. Bride, G. Hallas, and A. Stamp, *J. Chem. Soc.*, 1285 (1961).
33. L. G. S. Brooker, F. L. White, R. H. Sprague, S. G. Dent, Jr., and G. Van Zandt, *Chem. Rev.*, **41**, 325 (1947).
34. A. Evans, J. Jones, and J. Thomas, *J. Chem. Soc.*, 1824 (1955).
35. R. Dickinson and I. Heilbron, *J. Chem. Soc.*, 1704 (1927).
36. I. Heilbron, R. Heslop, and F. Irving, *J. Chem. Soc.*, 450 (1933).
37. A. I. Kiprianov and F. A. Mikhailenko, *Zh. Obshch. Khim.*, **31**, 781 (1961).

38. K. Brunings and A. Corwin, *J. Am. Chem. Soc.*, **64**, 593 (1942).
39. R. Jeffreys and E. Knott, *J. Chem. Soc.*, 1028 (1951).
40. G. G. Dyadyusha, T. M. Verbovskaya, and A. I. Kiprianov, *Ukrain. Khim. Zh.*, **32**, 357 (1966).
41. H. Gilman and G. Dunn, *J. Am. Chem. Soc.*, **72**, 2178 (1950).
42. U. Wannagat and F. Brandmair, *Z. Anorg. Allgem. Chem.*, **280**, 223 (1955).
43. G. Wittig, *Ber.*, **88**, 962 (1955).
44. S. Hunig, H. Schweeberg, and H. Schwarz, *Ann.*, **587**, 132 (1955).
45. F. Mason, *J. Chem. Soc.*, **123**, 1546 (1923).
46. T. Forster, *Z. Elektrochem.*, **45**, 548 (1939).
47. E. B. Knott, *J. Chem. Soc.*, 1024 (1951).
48. G. N. Lewis, *J. Am. Chem. Soc.*, **67**, 770 (1945).
49. B. Das, K. Patnaik, and M. K. Rout, *J. Indian Chem. Soc.*, **37**, 603 (1960); through *Chem. Abstr.*, **56**, 11752 (1962).
50. I. Ushenko, *Ukrain. Khim. Zh.*, **21** 738 (1955); through *Chem. Abstr.*, **50**, 16753 (1956).
51. E. Sawicki, T. W. Stanley, and W. C. Elbert, *Mikrochim. Acta*, 505 (1961).
52. E. Sawicki, T. W. Stanley, T. R. Hauser, and R. Barry, *Anal. Chem.*, **31**, 1664 (1959).
53. L. G. S. Brooker and R. H. Sprague, *J. Am. Chem. Soc.*, **63**, 3203 (1941).
54. L. Horwitz, *J. Am. Chem. Soc.*, **77**, 1687 (1955).
55. E. Sawicki, T. R. Hauser, T. W. Stanley, W. C. Elbert, and F. T. Fox, *Anal. Chem.*, **33**, 1574 (1961).
56. D. Jerchel and W. Edler, *Ber.*, **88**, 1284 (1955).
57. E. Sawicki, T. W. Stanley, T. R. Hauser, W. C. Elbert, and J. L. Noe, *Anal. Chem.*, **33**, 722 (1961).
58. E. Sawicki, T. W. Stanley, and W. C. Elbert, *Microchem. J.*, **5**, 225 (1961).
59. S. Hunig and K. Fritsch, *Ann.*, **609**, 143 (1957).
60. E. C. Kirby and D. H. Reid, *J. Chem. Soc.*, 163 (1961).
61. H. Balli and F. Kersting, *Ann.*, **663**, 96 (1963).
62. E. B. Knott and L. A. Williams, *J. Chem. Soc.*, 1586 (1951).
63. L. G. S. Brooker, *Rev. Mod. Physics*, **14**, 275 (1942).
64. T. Forster, *Z. Physik. Chem.*, **48**, 12 (1940).
65. H. Meerwein, W. Florian, N. Schon, and G. Stopp, *Ann.*, **641**, 1 (1961).
66. C. D. Ritchie, W. F. Sager, and E. S. Lewis, *J. Am. Chem. Soc.*, **84**, 2349 (1962).
67. G. A. Grant, R. Blanchfield, and D. M. Smith, *Can. J. Chem.*, **36**, 242 (1958).
68. G. Manecke and G. Kossmehl, *Ber.*, **93**, 1899 (1960).
69. V. V. Ghaisas, B. J. Kane, and F. F. Nord, *J. Org. Chem.*, **23**, 560 (1958).
70. R. Cigen, *Acta Chem. Scand.*, **12**, 1456 (1958).
71. G. N. Lewis and J. Bigeleisen, *J. Am. Chem. Soc.*, **65**, 1207 (1943).
72. C. Dufraisse, J. Le Francq, and P. Barbier, *Rec. Trav. Chim.*, **69**, 380 (1950).
73. R. Wizinger, *Mededel. Vlaam. Chem. Ver.*, **19**, 65 (1957).
74. P. Ehrlich and L. Benda, *Ber.*, **46**, 1931 (1913).
75. G. N. Lewis and J. Bigeleisen, *J. Am. Chem. Soc.*, **65**, 1144 (1943).
76. D. H. Reid, W. H. Stafford, and W. L. Stafford, *J. Chem. Soc.*, 1118 (1958).
77. R. Wizinger, *J. Prakt. Chem.*, **157**, 154 (1941).
78. L. G. S. Brooker, F. L. White, and R. H. Sprague, *J. Am. Chem. Soc.*, **73**, 1087 (1951).
79. J. Voltz, *Angew. Chem., Intern. Ed.*, **1**, 532 (1962).
80. F. M. Hamer, *J. Chem. Soc.*, 1480 (1956).
81. K. Dimroth and P. Hoffmann, *Ber.*, **99**, 1325 (1966).
82. M. L. Huggins, *J. Am. Chem. Soc.*, **75**, 4123 (1953).

83. E. Sawicki and T. W. Stanley, *Anal. Chem.*, **31**, 122 (1959).
84. E. Sawicki, T. R. Hauser, T. W. Stanley, and W. C. Elbert, *Anal. Chem.*, **33**, 93 (1961).
85. J. Bartos, *Ann. Pharm. Franc.*, **20**, 650 (1962).
86. E. Sawicki, R. Schumacher, and C. R. Engel, *Microchem. J.*, **12**, 377 (1967).
87. E. Sawicki and C. R. Engel, *Chemist-Analyst*, **56**, 7 (1967).
88. E. Sawicki, C. R. Engel, and M. Guyer, *Anal. Chim. Acta*, **39**, 505 (1967).
89. S. Hunig and K. Fritsch, *Ann.*, **609**, 172 (1957).
90. S. Hunig, H. Balli and H. Quast, *Angew. Chem.*, **74**, 28 (1962).
91. R. Wizinger, *Chimia*, **19**, 339 (1965).
92. R. Wizinger and S. Chatterjee, *Helv. Chim. Acta*, **35**, 316 (1952).
93. L. Roosens, *Ind. Chim. Belg.*, **20**, 645 (1955).
94. L. Roosens, *Communication au 28e congrès international de chimie industrielle, Bruxelles, 1955.*
95. S. Hunig, H. Balli, and H. Quast, *Angew. Chem.*, **1**, 47 (1962).
96. S. Hunig and H. Quast, *Ann.*, **711**, 139 (1968).
97. A. A. Kharkharov, *Izvest. Akad. Nauk SSSR, Otdel. Khim. Nauk*, 326 (1955); through *Chem. Abstr.*, **49**, 12126 (1955).
98. A. A. Kharkharov, *Izvest. Akad. Nauk SSSR, Otdel. Khim. Nauk*, 880 (1955); *Chem. Abstr.*, **50**, 9422 (1956).
99. E. Sawicki, *J. Org. Chem.*, **22**, 1084 (1957).
100. H. Kirner and R. Wizinger, *Helv. Chim. Acta*, **44**, 1766 (1961).
101. A. Van Dormael and J. Nys, *Bull. Soc. Chim. Belges*, **61**, 614 (1952).
102. J. Nys and A. Van Dormael, *Bull. Soc. Chim. Belges*, **65**, 809 (1956).
103. A. Van Dormael, J. Jaeken, and J. Nys, *Bull. Soc. Chim. Belges*, **58**, 477 (1949).
104. N. E. Grigoreva, I. K. Gintse, and A. Severina, *Zh. Obshch. Khim.*, **26**, 3096 (1956); through *Chem. Abstr.*, **51**, 7145 (1957).
105. L. G. S. Brooker, F. L. White, G. H. Keyes, C. P. Smyth, and P. F. Oesper, *J. Am. Chem. Soc.*, **63**, 3192 (1941).
106. A. I. Kiprianov and I. N. Zhmurova, *Zh. Obshch. Khim.*, **23**, 493 (1953); *Chem. Abstr.*, **48**, 3963 (1954).
107. A. I. Kiprianov and F. A. Mikhailenko, *Zh. Obshch. Khim.*, **31**, 786 (1961); *Chem. Abstr.*, **55**, 24020 (1961).
108. E. Sawicki, T. W. Stanley, and H. Johnson, *Anal. Chem.*, **35**, 199 (1963).
109. E. Sawicki, R. Miller, and T. R. Hauser, *Anal. Chem.*, **30**, 1130 (1958).
110. G. V. Semukhina, V. D. Barskii, and V. V. Noskov, *Zh. Anal. Khim.*, **19**, 1155 (1964); through *Anal. Abstr.*, **13**, 227 (1966).
111. S. B. Tallantyre, *J. Soc. Chem. Ind.*, **49**, 466T (1930).
112. R. I. Alekseev, *Zavodskaya Lab.*, **8**, 807 (1939).
113. H. von Euler, F. Schlenk, H. Heiwinkel, and B. Hogberg, *Z. Physiol. Chem.*, **256**, 208 (1938).
114. L. J. Harris and W. D. Raymond, *Biochem. J.*, **33**, 2037 (1939).
115. E. G. Wollish, G. P. Kuhnis, and R. T. Price, *Anal. Chem.*, **21**, 1412 (1949).
116. S. Banerjee, N. C. Ghosh, and G. Bhattacharya, *J. Biol. Chem.*, **172**, 495 (1948).
117. F. D. Snell and C. T. Snell, *Colorimetric Methods of Analysis*, Vol. IV, Van Nostrand, New York, 1954, p. 247. Listing of many references to this reaction with aniline or *p*-aminoacetophenone.
118. Y. L. Wang and E. Kodicek, *Biochem. J.*, **37**, 530 (1943).
119. Y. Raoul and O. Crépy, *Bull. Soc. Chim. Biol.*, **23**, 362 (1941).
120. Y. Raoul, *Ann. Pharm. Franc.*, **1**, 17 (1943).
121. J. P. Sweeney and W. L. Hall, *Anal. Chem.*, **23**, 983 (1951).

122. J. P. Sweeney, *J. Assoc. Offic. Agr. Chemists*, **34**, 380 (1951).
123. *Official Methods of Analysis*, Association of Official Agricultural Chemists, Washington, D.C., 1960, p. 660.
124. A. E. Teeri and S. R. Shimer, *J. Biol. Chem.*, **153**, 307 (1944).
125. R. Strohecker and H. M. Henning, *Vitamin Assay*, Verlag Chemie, GMBH, Weinheim, Germany, 1965, p. 193.
126. E. Werle and H. N. Becker, *Biochem. Z.*, **313**, 182 (1942).
127. F. D. Snell and C. T. Snell, *Colorimetric Methods of Analysis*, Vol. IV, Van Nostrand, New York, 1954, p. 462. Listing of many references.
128. A. R. Trim, *Biochem. J.*, **43**, 57 (1948).
129. A. C. Corcoran, O. M. Helmer, and I. H. Page, *J. Biol. Chem.*, **129**, 89 (1939).
130. H. M. Jones and E. S. Brody, *J. Am. Pharm. Assoc., Sci. Ed.*, **38**, 579 (1949).
131. J. Hudanick, *J. Pharm. Sci.*, **53**, 332 (1964).
132. L. G. S. Brooker, G. H. Keyes, and W. W. Williams, *J. Am. Chem. Soc.*, **64**, 199 (1942).
133. A. I. Kiprianov, *Dopovodi Akad. Nauk V.R.S.R.*, No. 12, 3 (1940); through *Chem. Abstr.*, **38**, 1427 (1944).
134. A. I. Kiprianov and G. T. Pilyugin, *Byul. Vses. Khim. Obshchestva*, in *D. I. Mendeleeva*, No. 3–4, 60 (1939); through *Chem. Abstr.*, **34**, 4663 (1940).
135. G. Schwarzenbach, *Z. Elektrochem.*, **47**, 40 (1941).
136. R. Wizinger and W. Haldemann, *Ber.*, **93**, 1533 (1960).
137. L. G. S. Brooker et al., *J. Am. Chem. Soc.*, **67**, 1875 (1945).
138. R. Wizinger and U. Arni, *Ber.*, **92**, 2309 (1959).
139. E. Sawicki and A. Carr, *J. Org. Chem.*, **22**, 507 (1957).
140. G. N. Lewis and M. Calvin, *Chem. Rev.*, **25**, 273 (1939).
141. E. Sawicki, B. Chastain, H. Bryant, and A. Carr, *J. Org. Chem.*, **22**, 625 (1957).
142. N. C. Deno, J. J. Jaruzelski, and A. Schriesheim, *J. Am. Chem. Soc.*, **77**, 3044 (1955).
143. N. C. Deno and W. Evans, *J. Am. Chem. Soc.*, **79**, 5804 (1957).
144. N. C. Deno, J. J. Jaruzelski, and A. Schriesheim, *J. Org. Chem.*, **19**, 155 (1954).
145. A. Morton and W. Emerson, *J. Am. Chem. Soc.*, **60**, 284 (1938).
146. E. Sawicki, *Chemist-Analyst*, **47**, 9 (1958).
147. E. Sawicki, T. W. Stanley, and T. R. Hauser, *Chemist-Analyst*, **47**, 69 (1958).
148. K. H. Muller and H. Hoverlagen, *Deut. Apotheker-Ztg.*, **100**, 309 (1960).
149. E. Sawicki, T. W. Stanley, and W. C. Elbert, *Anal. Chem.*, **33**, 1183 (1961).
150. D. H. Reid, W. H. Stafford, W. L. Stafford, G. McLennan, and A. Voigt, *J. Chem. Soc.*, 1110 (1958).
151. H. Dahn and H. Castelmur, *Helv. Chim. Acta*, **36**, 638 (1953).
152. H. H. Jaffé, S. Yeh, and R. W. Gardner, *J. Mol. Spectroscopy*, **2**, 120 (1958).
153. T. A. Liss, *J. Org. Chem.*, **32**, 1146 (1967).
154. F. Gerson, J. Schulze, and E. Heilbronner, *Helv. Chim. Acta*, **41**, 1463 (1958).
155. E. Sawicki, T. W. Stanley, and W. C. Elbert, in N. Cheronis, Ed., *Microchemical Techniques*, Interscience, New York, 1962, pp. 633–642.
156. F. Gerson and E. Heilbronner, *Helv. Chim. Acta*, **42**, 1877 (1959).
157. E. Sawicki, J. L. Noe, and F. T. Fox, *Talanta*, **8**, 257 (1961).
158. E. De Boer and van der Meij, *Proc. Chem. Soc.* (*London*), 139 (1961).
159. J. A. Leisten and P. R. Walton, *Proc. Chem. Soc.* (*London*), 60 (1963).
160. L. G. S. Brooker, in *Nuclear Chemistry and Theoretical Organic Chemistry*, Interscience, New York, 1945.
161. A. Treibs, R. Zimmer-Galler, and C. Jutz, *Ber.*, **93**, 2542 (1960).
162. M. Strell, A. Kalojanoff, and L. Brem-Rupp, *Ber.*, **87**, 1019 (1954).

163. R. Wizinger and H. Sontag, *Helv. Chim. Acta*, **38**, 363 (1955).
164. R. Wizinger and P. Kolliker, *Helv. Chim. Acta*, **38**, 372 (1955).
165. J. Pew, *J. Am. Chem. Soc.*, **73**, 1678 (1951).
166. S. Hunig, J. Utermann, and G. Erlemann, *Ber.*, **88**, 708 (1955).
167. A. Wassermann, *J. Chem. Soc.*, 983 (1959).
168. A. Wassermann, *J. Chem. Soc.*, 1014 (1958).
169. R. Kuhn and A. Winterstein, *Helv. Chim. Acta*, **11**, 144 (1928).
170. J. F. Thomas and G. Branch, *J. Am. Chem. Soc.*, **75**, 4793 (1953).
171. H. Quast and S. Hunig, *Ann.*, **711**, 157 (1968).
172. G. Scheibe, J. Heiss, and K. Feldmann, *Ber. Bunsenges. Physik. Chem.*, **70**, 52 (1966).
173. H. E. Nikolajewski, S. Dahne, and B. Hirsch, *Ber.*, **100**, 2616 (1967).
174. N. E. Grigoreva, I. K. Gintse, and Z. M. Afanaseva, *Zh. Obshch. Khim.*, **29**, 849 (1959).
175. H. Schmidt and R. Wizinger, *Ann.*, **623**, 204 (1959).
176. W. B. Tuemmler and B. S. Wildi, *J. Am. Chem. Soc.*, **80**, 3772 (1958).
177. A. Leifer, M. Boedner, P. Dougherty, A. J. Fusco, M. Koral, and J. E. Lu Valle, *Appl. Spectry.*, **20**, 150 (1966).
178. A. Leifer, D. Bonis, M. Boedner, F. Dougherty, A. J. Fusco, M. Koral, and J. E. Lu Valle, *Appl. Spectry.*, **21**, 71 (1967).
179. W. Dieterle and O. Riester, *Z. Wiss. Phot.*, **36**, 18, 141 (1937).
180. A. Van Dormael, *Ind. Chim. Belg.*, **24**, 463 (1959).
181. N. I. Fisher and F. M. Hamer, *Proc. Roy. Soc.* (*London*), **A154**, 703 (1936).
182. B. Beilenson, N. I. Fisher, and F. M. Hamer, *Proc. Roy. Soc.* (*London*), **A163**, 138 (1937).
183. R. Wizinger and P. Ulrich, *Chimia*, **7**, 92 (1953).
184. W. Krauss and H. Grund, *Z. Elektrochem.*, **58**, 142 (1954).
185. H. Quast and S. Hunig, *Ber.*, **99**, 2017 (1966).
186. J. Sondermann and H. Kuhn, *Ber.*, **99**, 2491 (1966).
187. G. Zinner and R. Uhlig, *Angew. Chem.*, **73**, 467 (1961).
188. L. Soder and R. Wizinger, *Helv. Chim. Acta*, **42**, 1733 (1959).
189. L. Soder and R. Wizinger, *Helv. Chim. Acta*, **42**, 1779 (1959).
190. D. Kupfer and L. L. Bruggeman, *Anal. Biochem.*, **17**, 502 (1966).
191. M. Pesez and J. Bartos, *Bull. Soc. Chim. France*, 3802 (1966).
192. S. Sakai, K. Suzuki, H. Mori, and M. Fujino, *Japan Analyst*, **9**, 862 (1960).
193. E. Sawicki and J. D. Pfaff, *Chemist-Analyst*, **55**, 6 (1966).
194. E. Sawicki and H. Johnson, *Chemist-Analyst*, **55**, 101 (1966).
195. T. A. Scott, *Biochem. J.*, **80**, 462 (1961).
196. G. Cilento, E. C. Miller, and J. A. Miller, *J. Am. Chem. Soc.*, **78**, 1718 (1956).
197. E. Sawicki and C. R. Engel, *Anal. Chim. Acta*, **32**, 315 (1967).
198. W. von E. Doering and L. H. Knox, *J. Am. Chem. Soc.*, **76**, 3203 (1954).
199. H. H. Rennhard, E. Heilbronner, and A. Eschenmoser, *Chem. Ind.* (*London*), 415 (1955).
200. G. Berti, *J. Org. Chem.*, **22**, 230 (1957).
201. I. Degani, R. Fochi, and C. Vincenzi, *Tetrahedron Letters*, 1167 (1963).
202. F. Greene, S. Ocampo, and L. Kaminski, *J. Am. Chem. Soc.*, **79**, 5957 (1957).
203. E. Sawicki and R. Barry, *Talanta*, **2**, 128 (1959).
204. T. W. Stanley, E. Sawicki, H. Johnson, and J. D. Pfaff, *Mikrochim. Acta*, 48 (1965).
205. F. Gerson and E. Heilbronner, *Helv. Chim. Acta*, **41**, 1444 (1958).
206. A. Morikofer and E. Heilbronner, *Helv. Chim. Acta*, **42**, 1909 (1959).
207. I. Pisovschi, *Ber.*, **41**, 1434 (1908).

208. L. Kuhn and L. De Angelis, *J. Am. Chem. Soc.*, **71**, 3084 (1949).
209. S. Hunig and J. Utermann, *Ber.*, **88**, 423 (1955).
210. S. Hunig and F. Muller, *Ann.*, **651**, 73 (1962).
211. E. Sawicki, T. R. Hauser, and S. McPherson, *Anal. Chem.*, **34**, 1460 (1962).
212. R. H. Sprague and G. De Stevens, *J. Am. Chem. Soc.*, **81**, 3095 (1959).
213. G. Leal and R. Pettit, *J. Am. Chem. Soc.*, **81**, 3160 (1959).
214. J. Toth and B. Rosenberg, *Vision Res.*, **8**, 1471 (1968).
215. L. G. S. Brooker and W. T. Simpson, *Ann. Rev. Phys. Chem.*, **2**, 121 (1951).
216. A. Gasco, E. Barni, and G. Di Modica, *Tetrahedron Letters*, 5131 (1968).
217. G. N. Lewis and J. Bigeleisen, *J. Am. Chem. Soc.*, **65**, 2102 (1943).
218. C. Jutz, W. Muller, and E. Muller, *Ber.*, **99**, 2479 (1966).
219. M. Coenen, *Ann.*, **633**, 102 (1960).
220. L. G. S. Brooker and L. A. Smith, *J. Am. Chem. Soc.*, **59**, 67 (1937).
221. C. Reichardt, *Tetrahedron Letters*, 4327 (1967).
222. A. L. Stone and D. F. Bradley, *J. Am. Chem. Soc.*, **83**, 3627 (1961).
223. D. F. Bradley and M. K. Wolf, *Proc. Natl. Acad. Sci., U.S.*, **45**, 944 (1959).
224. E. Jelley, *Nature*, **139**, 631 (1937).
225. G. Scheibe, *Angew. Chem.*, **52**, 631 (1939).
226. H. Zimmermann and G. Scheibe, *Z. Elektrochem.*, **60**, 566 (1956).
227. W. Appel and G. Scheibe, *Z. Naturforschung*, **13b**, 359 (1958).
228. A. Schauer and G. Scheibe, *Histochemie*, **1**, 190 (1959).
229. R. E. Kay, E. R. Walwick, and C. K. Gifford, *J. Phys. Chem.*, **68**, 1896 (1964).
230. R. E. Kay, E. R. Walwick, and C. K. Gifford, *J. Phys. Chem.*, **68**, 1907 (1964).
231. *Experimental Studies for the Detection of Protein in Trace Amounts*, NASA Publ. CR–466, Clearinghouse for Federal Scientific and Technical Information, Springfield, Va.
232. E. Rabinowitch and L. Epstein, *J. Am. Chem. Soc.*, **63**, 69 (1941).
233. G. Scheibe, *Z. Elektrochem. Angew. Physik. Chem.*, **47**, 77 (1941).
234. B. Eistert and F. Haupter, *Ber.*, **93**, 264 (1960).
235. K. Feldmann, E. Daltrozzo, and G. Scheibe, *Z. Naturforsch.*, **22b**, 722 (1967).

CHAPTER 10

ABSORPTION SPECTRA OF ANIONIC RESONANCE STRUCTURES
I. SOLVENT EFFECTS

The absorption spectra of conjugated anions are usually very sensitive to changes in solvent composition. Consequently solvent properties are of importance to the analyst investigating an analytical method in which a conjugated anion is the chromogen.

I. Basic Solvent Systems

A. Reactions in Basic Solution

When an unsaturated organic compound is dissolved in alkaline solution, at least six main structural changes could take place that would affect the electronic spectrum of the molecule:

1. A compound containing acidic hydrogen could be deprotonated to an anion; for example,

$$O_2NC_6H_4CH_2CN \rightarrow O_2NC_6H_4\bar{C}HCN \tag{10.1}$$

2. An aromatic compound lacking acidic hydrogen and containing one or more powerful electron-acceptor groups could be acidic enough to add an activated anion and thus form a resonating anionic structure, for example,

$$C_6H_5\text{–}NO_2 + (OR)^- \longrightarrow C_6H_5(H)(OR)\text{=}NO_2^- \tag{10.2}$$

3. An organic onium salt containing acidic hydrogen could be deprotonated to a betaine, for example,

$$C_6H_5COCH_2N^+C_5H_5 \xrightarrow{OH^-} C_6H_5CO\bar{C}HN^+C_5H_5 \tag{10.3}$$

4. A heterocyclic onium salt could add an anion to form a neutral, less unsaturated compound, as shown, for example, below.

$$\text{thioxanthylium}^{+} \xrightarrow{OH^-} \text{9-hydroxythioxanthene} \qquad (10.4)$$

5. The compound could be oxidized via the anion, for example, fluorene anion to fluorenone.

6. The compound could be hydrolyzed, for example, an aromatic acylamine to the aromatic amine.

7. Compounds with a 1-alkynylcyclohexene system could be aromatized and conjugated enynes could be isomerized into conjugated trienes (1).

Many other reactions could take place, but the ones of prime interest in this treatise are those that involve the formation of an anion.

B. Solvents

1. *Introduction*

Protic solvents—such as water, methanol, trifluoroethanol, and formamide—are hydrogen donors. Anions are hydrogen bonded by these solvents. Most anions are more solvated in protic solvents than in dipolar aprotic solvents (2). However, water usually present in the hygroscopic dipolar aprotic solvents, is a very weak general hydrogen-bond donor, but phenol is a strong donor in dipolar aprotic solvents (3).

Dipolar aprotic solvents are defined as solvents, with dielectric constants greater than 15, that cannot donate suitably labile hydrogen atoms to form strong hydrogen bonds with an appropriate species (4). They have dipole moments in the range of 2.7–4.7 and dielectric constants in the approximate range of 21–47.

Dimethylformamide (DMF), dimethyl sulfoxide (DMSO), dimethylacetamide (DMAC), *N*-methyl-2-pyrrolidone, tetrahydrothiophene 1-dioxide (sulfolane), acetone, dimethylsulfone, *N,N*-dimethylethylene urea, nitromethane, acetonitrile, monoglyme, nitrobenzene, sulfur dioxide, and propylene carbonate are some common dipolar aprotic solvents. Anions are much stronger bases, relative to neutral bases, in these solvents than in the protic solvents. Anions are not hydrogen bonded to these solvents and thus are much less solvated than they would be in protic solvents and are more reactive (4). The properties of some of these solvents are shown in Table 10.1. Dipolar aprotic solvents are conveniently purified by passage through type 5A molecular sieves (5–7).

A very large number of organic compounds are soluble in such solvents as dimethylformamide, dimethyl sulfoxide, and acetone. A pos-

TABLE 10.1
The Properties of Some Dipolar Aprotic Solvents (2)

Property	DMSO	DMF	DMAC	Acetonitrile
Melting point, °C	18.55	−61	−20	−45
Boiling point, 760 mm Hg, °C	189	152.5	165.5	81.6
Refractive index, n_D^{20}	1.4783	1.4272	1.4356	1.3416
Dielectric constant, 25°C	48.9	37.6	37.8	37.5
Dipole moment, Debye units	4.3	3.82	3.79	3.37
Polarity (*Z* value)	71.1	68.5	–	71.3

sible reason for the strong solvent action of dipolar solvents is their large dipole moments and their polarizability (2), for example,

$$(CH_3)_2N{-}CHO \rightleftarrows (CH_3)_2\overset{+}{N}{=}CH{-}O^- \qquad (10.5)$$

Thus a strong dipole interaction can take place between polar solvents and solutes. Since the dielectric constants of these solvents are high, the solutes readily dissociate into ions. The dipolar aprotic solvents are strong hydrogen bond acceptors that interact strongly with dipolar hydrogen bond donors, such as alcohols, phenols, amines, thiols, proteins, carbohydrates, and steroids.

The polarizability of these and other polar solutes is believed to be an important factor in determining the solubility of these nonionic species in such solvents as dimethylformamide and dimethyl sulfoxide (2).

2. *Oxidation*

The majority of organic compounds are recovered unchanged from their solutions in the dipolar aprotic solvents if the temperature has not exceeded 100°C. However, in the presence of alkali, air oxidation takes place with ease. Thus potassium *tert*-butoxide in dimethyl sulfoxide or dimethylformamide strongly catalyzes autoxidation at room temperature (8, 9), via the carbanions, of substituted toluenes, picolines, pyrrole, diphenylmethane, triphenylmethane, fluorene, and xanthene to alcohols, ketones, or carboxylic acids in high yield. Aromatic amines give azobenzenes with this reagent (8). The chemiluminescent oxidation of the luminols (10), lucigenin (11), and the bisquaternary salts of 1,1′-bisisoquinolines (12) takes place in alkaline solution. Aliphatic aldehyde 2,4-dinitrophenylhydrazone anions in alkaline solution are gradually destroyed (13).

Thus any analysis that involves anions must be based on prior studies of the stability of the anion to light and air. Anions in protic solvents are hydrogen bonded by these solvents and hence are highly solvated and protected to some extent from oxygen. However, anions in dipolar aprotic solvents are, at most, weakly hydrogen bonded to the solvent and hence are much less solvated and much more open to reaction. The high electron density on the bare oxygen atom of the aprotic solvent results in ease of reaction with oxygen or with compounds with active hydrogen. Thus potassium *tert*-butoxide in dimethyl sulfoxide displaces some carbanions from hydrogen 10^{12} times faster than does potassium methoxide in methanol(5, 14).

3. *Alkaline Solvents*

Many strong organic bases are available, such as tetramethylammonium hydroxide (as solid or in water or methanol), tetraethylammonium hydroxide (in water or methanol), benzyltrimethylammonium hydroxide or methoxide (in methanol), and tetra-*n*-butylammonium hydroxide in methanol. These bases are soluble in water, ethanol, dimethyl sulfoxide, or dimethylformamide. The tetrabutyl base is even soluble to some extent in the less polar solvents, such as *o*-dichlorobenzene.

Other very strongly basic solvent systems that have been used to form anions from very weak CH, NH, or OH acids include potassium *tert*-butoxide in dimethyl sulfoxide or dimethylformamide(8, 9) and sodium hydride in dimethyl sulfoxide ($Na^+ \, ^-CH_2SOCH_3$, or dimsyl sodium) (15, 16). Dimsyl sodium is equivalent to an alkaline solution of $H_- = 30$ or more. It is made by dissolving powdered sodium hydride in dimethyl sulfoxide at 65–70°C under nitrogen, with stirring, over 45 min. The reagent decomposes above 80°C, but 2 *M* solutions can be stored at room temperature if protected from air(15). Other strong base systems that can generate carbanions from active methylene compounds are formed by dissolving either sodium hydride in dipolar aprotic or dipolar aprotic–inert solvent mixtures(17, 18) or potassium *tert*-butoxide in a dipolar aprotic solvent(5, 14). These types of basic solvents should see much more use in organic trace analysis.

C. Solvent Basicity

Preliminary research indicates that the long wavelength maximum of a conjugated solute anion shifts to the red with increasing basicity of the solvent system(19). Consequently knowledge of the basicity of the solvent will be of prime importance to the spectroscopist and the research analyst.

The strongest base that can exist in water is the solvated hydroxide ion, but much stronger base systems are available in such dipolar aprotic solvents as dimethylformamide and dimethyl sulfoxide.

Solutions with H_- values up to 19 have been obtained by using a small fixed concentration of hydroxide ion in aqueous solutions of aprotic polar solvents, such as sulfolane, pyridine, and dimethyl sulfoxide (20–22). The addition of dimethyl sulfoxide to aqueous solutions of 0.011 *M* tetramethylammonium hydroxide causes H_- to rise from 12 in water to 26 in 99.5 mole % dimethyl sulfoxide, an increase of 14 powers of 10 in basicity (as measured by its effect on arylamine ionization) (Table 10.2) (23).

The term H_- describes the ability of a solvent to remove a proton from a neutral acid and is defined as follows (20, 24):

$$H_- = -\log \frac{a_{H^+} f_{A^-}}{f_{HA}} = pK_a + \log \frac{[A^-]}{[HA]} \tag{10.6}$$

where a_{H^+} is the hydrogen ion activity in the solution, f is the activity coefficient of the weak acid, pK_a is the negative logarithm of the thermodynamic ionization constant of the acid in water, and $[A^-]/[HA]$ is the measured ionization ratio of the acid. In dilute aqueous solutions H_- becomes identical with pH.

One approach to understanding the enhanced basicity of the hydroxide ion in dipolar aprotic solvents is to consider the equilibrium between the hydroxide ion and the weak acid HA in which the hydroxide ion is considered to be intimately solvated by a number of water molecules (23):

$$OH^- (aq) + HA \leftrightarrows A^- + (n+1) H_2O \tag{10.7}$$

The number n can be considered either as the hydration number of the hydroxyl ion or, if the indicator or its anion is also solvated, as the difference in solvation numbers between the left and right sides of equation 10.7. The effect of adding a dipolar aprotic solvent to the system is to drive the equilibrium to the right and thus to increase the basicity of the hydroxyl ion.

Dolman and Stewart (23) have pointed out that the large increases in H_- on the addition of a dipolar aprotic solvent to a system must be due to increases in the activity of the hydroxide ion and to decreases in the activity of water. The decrease in the activity of water is due not only to a dilution effect but also to the strong hydrogen bond formation between the dipolar aprotic solvent and water. Thus the increase in the basicity when a dipolar aprotic solvent is added to aqueous solutions of hydroxide ion is due partly to the reduced activity of water

TABLE 10.2
H_- Functions for Various Solvent Systems

Solvent	Alkali	Range		Ref.
		Concentration	H_-	
Methanol	$NaOCH_3$	5×10^{-4}–3.75 *M*	10.36–16.64	27–30
Water	N_2H_4	12.5–45 mole %	10.87–14.60	31, 32
	$HOCH_2CH_2NH_2$	5–100 mole %	11.17–14.27	33
	$H_2NCH_2CH_2NH_2$	5–30 mole %	11.74–15.50	34
Water–ethanol (0–100 mole % of EtOH)	$NaOC_2H_5$	10^{-3} or 5×10^{-3} *M*	11.74–13.35	35
Water–isopropanol (0–100 mole % of *i*-PrOH)	$NaOCH(CH_3)_2$	10^{-3} or 5×10^{-3} *M*	11.74–15.71	35
Water–*tert*-pentanol (0–100 mole % of *tert*-PeOH)	$NaOC(C_2H_5)(CH_3)_2$	10^{-3} or 5×10^{-3} *M*	11.74–16.70	35
Water–*tert*-butanol (0–100 mole % of *tert*-BuOH)	$NaOC(CH_3)_3$	10^{-3} or 5×10^{-3} *M*	11.74–17.85	35
Water–dioxane (0–40 mole % of dioxane)	$NaOC_2H_5$	10^{-3} or 5×10^{-3} *M*	11.74–13.01	35
Water	$C_6H_5CH_2N^+(CH_3)_3(OH)^-$	0.01–2.38 *M*	11.98–16.20	20
	$NaOH$	10^{-2}–12 *M*	12.01–16.70	36
Water–dimethyl sulfoxide (5–95 mole % of DMSO)	$(CH_3)_4NOH$	0·011 *M*	12.18–22.5	20, 21
Methanol–dimethyl sulfoxide (0–95 mole % of DMSO)	$NaOCH_3$	0.025 *M*	12.23–19.37	37
Water–pyridine (1–65 mole % of pyridine)	$(CH_3)_4NOH$	0.11 *M*	12.27–15.40	20, 21
Water–sulfolane (1–95 mole % of sulfolane)	$(CH_3)_4NOH$	0.011 *M*	12.36–19.28	20, 21
Ethanol	$NaOC_2H_5$	10^{-3}–10^{-1} *M*	12.57–14.57	38, 39

Water	$H_2NCH_2CH_2NH_2$	9.2–31 mole %	12.76–15.48	20
Methanol	$LiOCH_3$	2×10^{-1}–14.73 *M*	12.78–14.73	24
	$KOCH_3$	2×10^{-1}–3.5 *M*	12.91–16.93	38, 39
Water	N_2H_4	23.2–95.0 mole %	13.06–16.00	20
Water–dimethyl sulfoxide				
(10.32–99.59 mole % of DMSO)	$(CH_3)_4NOH$	0.011 *M*	13.17–26.19	22
Water	LiOH	1–5 *M*	13.43–14.31	20
Water–sulfolane		Varying		
(9.0–95.0 mole % of sulfolane)	$C_6H_5N(CH_3)_3OH$	0.0094–0.0707 *M*	13.44–19.70	24, 40
Water	LiOH	1–5 *M*	13.48–14.33	20, 21
Aqueous pyridine (30 mole %)	$C_6H_5CH_2N^+(CH_3)_3OH^-$	2×10^{-2}–2 *M*	13.89–17.64	20, 21
Water	NaOH	1–10 *M*	14.16–16.40	24, 41
	KOH	1–10 *M*	14.17–17.45	24, 41
2-Diethylaminoethanol	$NaOCH_2CH_2N(C_2H_5)_2$	5×10^{-3}–1 *M*	14.24–16.69	42
2-Aminoethanol	$HOCH_2CH_2N^+H_3Cl$	0–0.5 *M*	14.37–12.29	43
	$NaOCH_2CH_2NH_2$	0–0.5 *M*	14.37–16.60	33
Isopropanol	$NaOCH(CH_3)_2$	5×10^{-4}–10^{-1} *M*	14.65–16.95	44, 45
Aqueous pyridine (50 mole %)	$C_6H_5CH_2N^+(CH_3)_3OH^-$	0.01–2.12 *M*	15.67–18.91	20
tert-Pentanol	$NaOC(C_2H_5)(CH_3)_2$	10^{-3}–10^{-1} *M*	16.09–18.09	46
tert-Butanol	$NaOC(CH_3)_3$	10^{-3}–10^{-1} *M*	17.14–19.14	35, 47
Ethylenediamine			~ 19.0	34
Dimethyl sulfoxide	$NaOC(CH_3)_3$		Near 30	25
	$NaCH_2SOCH_3$		30 or more	16

and its subsequent mass-action effect on the equilibrium between hydroxide ion and acid, and partly to the increased activity of the hydroxide ion. The decrease in the number of water molecules solvating the hydroxide ion brings about a great increase in its activity and therefore in the basicity of the solution(23).

In comparison dimsyl sodium in dimethyl sulfoxide has an H_- of 30 or more(16), whereas a solution of potassium *tert*-butoxide is almost as strong a base as dimsyl sodium(25). A wide range of basicities is also available with *tert*-butanol–dimethyl sulfoxide(8) and toluene–dimethyl sulfoxide(26) mixtures. In the former mixture solvation of *tert*-BuO^- by *tert*-butanol reduces reactivity; in the latter decrease in the dielectric constant encourages ion pairs that are less reactive.

Table 10.2 lists a large number of H_- functions for various solvent systems, many of which are worthy of study and incorporation into future analytical methods by organic analysts.

II. pK of Conjugated Acids

The analysis of conjugated, noncarboxy, nonphenolic acids has been little studied from the viewpoint of their anionic spectra. The ultraviolet–visible absorption spectra of these anions is an additional criteria in their characterization and determination, especially since the anion usually absorbs at longer wavelength than the neutral compound. Knowledge of solvent basicity (Table 10.2), and the acidic strength (Table 10.3) of these compounds should be of value to the research analyst.

TABLE 10.3
The Ionization Constants of Some Noncarboxy, Nonphenolic Acids

Compound	Solvent	pK_a	Ref.
Hexacyanoisobutylene	$HClO_4$	<-8.5	48
Pentacyanopropene	$HClO_4$	<-8.5	48
Tetracyanopropene	$HClO_4$	<-8.0	48
Bis(tricyanovinyl)amine	$HClO_4$	-6.07	48
Trycyanovinyl alcohol	$HClO_4$	-5.3	48
Cyanoform	$HClO_4$	-5.13	48
Hexacyanoheptatriene	$HClO_4$	-3.55	48
Methyl dicyanoacetate	$HClO_4$	-2.78	48
p-(Tricyanovinyl)phenyldicyano-methane	$HClO_4$	0.60	48
2,4,6,2′,4′,6′-Hexanitrodiphenyl-amine	Aqueous buffer	2.63	20
Bromomalononitrile		~5	48

TABLE 10.3 (*cont'd*)

Compound	Solvent	pK$_a$	Ref.
Tris(7*H*-dibenzo[*c*,*g*]fluorenylidenemethyl)methane[a]	DMSO	5.9	49
Ethyl bromocyanoacetate		~6	48
2,4,6,2′,4′-Pentanitrodiphenylamine	Aqueous buffer	6.72	20
Hydrazoic acid		7.9	50
Nitroethane		8.5	51
2,4,6,4′-Tetranitrodiphenylamine	Aqueous buffer	8.88	20
2,4-Pentanedione		9.0	51
Ethyl cyanoacetate		>9	48
2,4,6,3′-Tetranitrodiphenylamine	Aqueous buffer	9.15	20
2,4,6-Trinitrodiphenylamine	Aqueous buffer	10.38	20
Nitromethane	Water	~10	52
(Biph C=CH)$_3$CH[b]	DMSO	10.4	49
2,4,2′,4′-Tetranitrodiphenylamine	Aqueous buffer	10.82	24
2,4,6-Trinitro-4′-aminodiphenylamine		10.82	24
3,6-Dinitrocarbazole		10.83	21, 32, 33
Malononitrile		11.19	51
9-Cyanofluorene		11.4	53
Benzoylacetone	DMSO	12.1	50
Tris(*p*-nitrophenyl)methane	DMSO	12.2	50
2,4,6-Trinitroaniline		12.2	24
(Biph C=CH)$_2$CH$_2$[b]	DMSO	12.3	49
2,4,4′-Trinitrodiphenylamine	Aqueous DMSO	12.35	20, 21
3-Formylindole	Aqueous KOH	12.36	54
Benzaldehyde 2,4-dinitrophenylhydrazone		12.47	30, 38
2,4,3′-Trinitrodiphenylamine	Aqueous DMSO	12.65	20, 21
Propanal 2,4-dinitrophenylhydrazone		12.80	30, 38
Acetone 2,4-dinitrophenylhydrazone		12.80	30, 38
Acetophenone 2,4-dinitrophenylhydrazone		12.81	30, 38
Cyclohexanone 2,4-dinitrophenylhydrazone		12.89	30, 38
Methylfluorene-9-carboxylate		12.88	53
Hydrocyanic acid		12.9	50
3-Acetylindole	Aqueous KOH	12.99	54
Cyclopentanone 2,4-dinitrophenylhydrazone		13.03	30, 38
Benzophenone 2,4-dinitrophenylhydrazone		13.57	30, 38
6-Bromo-2,4-dinitroaniline	Aqueous DMSO	13.63	20, 21
Cyclohexanone 4-nitrophenylhydrazone		13.68	38

TABLE 10.3 (*cont'd*)

Compound	Solvent	pK_a	Ref.
Benzaldehyde 4-nitrophenylhydrazone		13.78	38
2,4-Dinitrodiphenylamine	Aqueous DMSO	13.84	20, 21
Propanal 4-nitrophenylhydrazone		13.87	38
Fluoraden	DMSO–NaOAc–AcOH	14.0	50
Acetone 4-nitrophenylhydrazone		14.03	38
4,4′-Dinitrodiphenylamine	Aqueous DMSO	14.08	20, 21
Cyclopentanone 4-nitrophenylhydrazone		14.08	38
3-Nitrocarbazole		14.10	20
Biph C═C(C_6H_5)CH Biph[b]	DMSO–NaOAc–AcOH	14.1	50
Imidazole	Aqueous KOH	14.17	54
Biph CH—CH═C Phen[b,c]	DMSO–NaOAc–AcOH	14.2	50
Pyrazole	Aqueous KOH	14.21	54
4,4′,4″-Trinitrotriphenylmethane		14.32	53
Benzophenone 4-nitrophenylhydrazone		14.34	38
Histidine		14.37	54
Acetophenone 4-nitrophenylhydrazone		14.39	38
Biph C═CH—CH Biph[b]	DMSO–NaOAc–AcOH	14.4	50
2,4-Dinitro-4′-aminodiphenylamine	Aqueous sulfolane	14.64	20
3,4′-Dinitrodiphenylamine		14.66	22
5-Nitroindole	Aqueous KOH	14.75	54
2,4-Dinitroaniline	DMSO	14.8	50
7-Phenyl-7*H*-dibenzo[*c*,*g*]fluorene	DMSO–EtOH	14.8[d]	55
Biph C═CHC(C_6H_5)═CHCH Biph[b]	DMSO–NaOAc–AcOH	14.9	50
4-Nitro-3′-trifluoromethyldiphenylamine	Aqueous DMSO–Me_4NOH	14.96	22
Cyclopentadiene		15.0	56
Indoxyl sulfate	Aqueous KOH	15.23	54
5-Cyanoindole	Aqueous KOH	15.24	54
2,6-Dichloro-4-nitroaniline		15.55	24
Formanilide		15.56	44
3-Carboxyindole	Aqueous KOH	15.59	54
p-Bromobenzanilide		15.73	44
p-Nitrobenzamide		15.85	44
4,4′-Dinitrodiphenylmethane		15.85	53
Nitromethane	Aqueous DMSO	15.9	50
4-Nitrodiphenylamine	Aqueous sulfolane	15.9	20, 21
Methyl lactate		15.99	44
Gramine	Aqueous KOH	16.0	54
4-Nitro-2,5-dichloroaniline	Aqueous sulfolane	16.05	20
5-Bromoindole	Aqueous KOH	16.13	54
5-Fluoroindole	Aqueous KOH	16.30	54
4-Fluoroindole	Aqueous KOH	16.30	54

TABLE 10.3 (*cont'd*)

Compound	Solvent	pK_a	Ref.
Indole-3-carbinol	Aqueous KOH	16.50	54
Benzanilide		16.53	44
Skatole	Aqueous KOH	16.60	54
Tryptamine	Aqueous KOH	16.60	54
Ethylene glycol		16.68	44
6-Methoxytryptophan	Aqueous KOH	16.70	54
7*H*-Dibenzo[*c,g*]fluorene	DMSO–EtOH–$NaOCH_3$	16.8	55
L-Tryptophan	Aqueous KOH	16.82	54
4-Methyltryptophan	Aqueous KOH	16.90	54
5-Benzyloxygramine	Aqueous KOH	16.90	54
Indole-3-acetic acid	Aqueous KOH	16.90	54
6-Tryptophanol	Aqueous KOH	16.91	54
Indole	Aqueous KOH	16.97	54
5-Methoxy-2-carboxyindole	Aqueous KOH	17.03	54
2-Carboxyindole	Aqueous KOH	17.13	54
p-Bromobenzamide	Aqueous KOH	17.13	54
Phenoxyacetamide		17.20	44
Formamide		17.20	44
4-Chloro-2-nitroaniline	Aqueous sulfolane	17.22	20
2-Ethoxyethanol		17.23	44
2-Phenoxyethanol		17.27	44
13-Phenyl-13*H*-dibenzo[*a,i*]fluorene		17.3	55
Phenylacetanilide		17.31	44
2,4′-Dinitrodiphenylmethane		17.38	53
2-Methoxyethanol		17.41	44
4-Nitro-2,6-di-*tert*-butylaniline		17.4	22
2-Aminoethanol		17.42	44
13*H*-Dibenzo[*a,i*]fluorene	DMSO–EtOH–$NaOCH_3$	17.5	55
Pyrrole	Aqueous KOH	17.51	54
2-Nitrodiphenylamine		17.57	20
Acetanilide		17.59	44
3,4′-Dinitrodiphenylmethane		17.62	53
Methanol		17.71	44
Benzyl alcohol		17.73	44
Allyl alcohol		17.88	44
2-Nitroaniline	Aqueous sulfolane	17.88	20
Benzhydrol		17.99	44
1-Methoxy-2-propanol		18.05	44
Indene		18.2	57
Water		18.23	44
Ethanol		18.33	44
9-Phenylfluorene		18.5	55
Phenylacetylene		18.5	58
4-Nitro-*N*-ethylaniline	Aqueous DMSO–Me_4NOH	18.58	22
4-Methylsulfonyldiphenylamine		18.80	22

TABLE 10.3 (*cont'd*)

Compound	Solvent	pK_a	Ref.
2,5-Hexanedione		18.70	59
4-Nitroaniline		18.91	22[e]
12-Phenyl-12*H*-dibenzo[*b,h*]fluorene		19.1	55
5-Hydroxytryptophan	Aqueous KOH	19.2	54
2,3,5,6-Tetrachloroaniline		19.22	22
7*H*-Benzo[*c*]fluorene	Cyclohexylamine	19.4	60
Isopropanol		19.43	44
3-Nitrodiphenylamine		19.53	22
3,4′-Dichlorodiphenylamine		19.73	22
2,2′-Dipyridylamine	Aqueous DMSO–Me_4NOH	19.91	22
Acetone	DMSO	20.0	52
11*H*-Benzo[*a*]fluorene	Cyclohexylamine	20.0	60
4,5-Methylenephenanthrene	DMSO	20.0	50
3-Trifluoromethyldiphenylamine		20.48	22
Fluorene	DMSO	20.5[f]	50
3-Chlorodiphenylamine	Aqueous DMSO–Me_4NOH	20.73	22
4-Chlorodiphenylamine	Aqueous DMSO–Me_4NOH	21.33	22
12*H*-Dibenzo[*b,h*]fluorene	DMSO–EtOH–$NaOCH_3$	21.4	55
Benzylsulfone		22.0	61
3-Methoxydiphenylamine		22.2	62
Diphenylamine		22.4	22, 62
4-Cyanoaniline		22.7	22, 62
Trisdiphenylmethane	DMSO + dimsyl K	22.8	57
5*H*-Benzo[*b*]fluorene	Cyclohexylamine	23.2	60
4-Methoxydiphenylamine	Aqueous DMSO–Me_4NOH	23.22	22, 62
α-Methylbenzylsulfone		23.5	61
9-Phenylxanthene	Aqueous DMSO	24.3	57
3-Cyanoaniline		24.64	22, 62
Diphenylyldiphenylmethane	DMSO + dimsyl K	25.3	57
3-Chloroaniline		25.63	57
Acetylene		26.0	63
1,1,3-Triphenylpropene	Cyclohexylamine	26.5	60
Aniline		~27	22
Methylphenylsulfone	DMSO	27.0	61
Xanthene		27.1	57
Triphenylmethane	Aqueous DMSO	27.2	57
4-Benzyldiphenyl		27.2	57
9-Methylanthracene		27.7	57
Methylsulfone	DMSO	28.5	61
Diphenylmethane	DMSO	28.6	57
9,9-Dimethyl-10-phenyldihydro-anthracene	Cyclohexylamine	29.0	60
Trimethylenesulfone	DMSO	>30	61
Tetramethylenesulfone	DMSO	>31	61
Pentamethylenesulfone	DMSO	>31	61
p-Biphenylyldiphenylmethane	Cyclohexylamine	31.2	60

TABLE 10.3 (*cont'd*)

Compound	Solvent	pK_a	Ref.
Dimethyl sulfoxide		31.3[g]	57
Triphenylmethane	Cyclohexylamine	32.5[h]	60
Ammonia		~34[i]	62
Diphenylmethane	Cyclohexylamine	34.1	60
Methane		~58	64

[a]Tautomeric mixture.
[b]Biph is

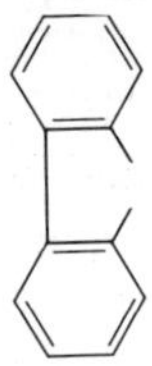

[c]Phen is

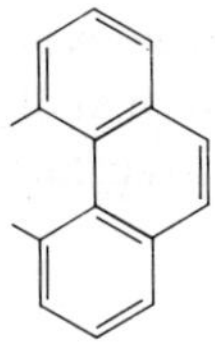

[d]pK_a 15.9(50).
[e]Acidities of many 4-nitroanilines reported.
[f]pK_a 22.9(60).
[g]pK_a 41(52).
[h]pK_a 40(52), 28(50).
[i]pK_a 35(64).

A. Solvent and *p*K of Acid

The data in the literature indicate that acidities in different solvents are not strictly comparable(50). For example, tris(*p*-nitrophenyl) methane is a stronger acid in dimethyl sulfoxide than is nitromethane, whereas in aqueous solution the acidities are completely reversed. The same is true for acetic acid and *p*-nitrophenol. Another example is the ΔpK_{HA}^{d} of salicyclic and benzoic acids, which is 1.2 in water and 4.3 in dimethyl sulfoxide(65). Acidic anions that have a delocalized charge have a smaller difference in acidity between water and dimethyl sulfoxide than do acidic anions with a localized charge. For example, the ΔpK_{HA}^{d} for 4-chloro-2,6-dinitrophenol between water and dimethyl sulfoxide is 0.5, whereas for 3,5-dinitrobenzoic acid, which has almost the same pK_{HA}^{d} of 2.8 in water as the former phenol ($pK_w = 3.0$), ΔpK_{HA}^{d} is 3.85. The conclusion of Kolthoff and co-workers (65) is that the 3,5-dinitrobenzoate ion is strongly hydrogen bonded in water.

Another striking example is the report that the dissociation constant of picric acid is 500 times as great in dimethyl sulfoxide as it is in water, whereas the pK of benzoic acid is 10^6 times as great in water as it is in dimethyl sulfoxide. Kolthoff and Reddy (66) attribute this to the unusual stability of the picrate ion in dimethyl sulfoxide, whereas Davis and Paabo (67) attribute the greater acidity of carboxylic acids in water as due to the hydrogen bonding of both carboxylate oxygens with water. Thus it is obvious that no close correlation exists between acidities in dipolar aprotic and protic solvents (65, 68).

It is difficult to compare the acidities listed in Table 10.3. In some cases there is a widespread variation in results obtained by different investigators. The results reported by an individual investigator are usually reliable, but in comparing the results of different investigators one should bear in mind that the use of a different standard acid can give somewhat different values for the same compound. However, the values reported in Table 10.3 are of definite value to the research analyst. From the tables the appropriate solvent system can be ascertained for the analysis of the appropriate acid.

B. Substituent Effects

The pK_{HA} values of CH_4, CH_3NO_2, $CH_2(NO_2)_2$, and $CH(NO_2)_3$ have been estimated to be approximately 40, 11, 4, and 0, respectively (56, 64, 69, 70). The first nitro group has the largest effect; subsequent groups aid in delocalizing the charge. The cyano group, which is a weaker electron acceptor, has a more evenly proportioned effect, the pK_{HA} values of CH_4, CH_3CN, $CH_2(CN)_2$, and $CH(CN)_3$ being $\geqslant 40$, 25, 12, and ~ 0 (51, 56, 71).

The acidities and basicities of some anilines and diphenylamines have been compared (62). Except for the 2- or 4-nitro derivatives, there is a linear relation between the acidities and basicities of diphenylamines, as given by the following equation:

$$pK_{HA} = 21.4 + 1.30\,(pK_{BH^+}) \qquad (10.8)$$

Similarly anilines without an *ortho* or *para* nitro group show a similar linear relation, with a slope of 1.30. Those amines with at least one nitro group in a conjugated position fall on a different, curved line, with a slope of less than unity. A slope of 0.6 has been reported (21). As shown in Table 10.4, for most of the anilines and diphenylamines as the basicity decreases the acidity increases. The first conjugating nitro

TABLE 10.4
Values of pK_{BH^+} and pK_{HA} for Anilines and Diphenylamines (62)

Substituent	pK_{BH^+}	pK_{HA}
Anilines		
3-Chloro	3.5	25.6
3-Trifluoromethyl	3.2	25.4
3-Cyano	2.8	24.6
4-Cyano	1.7	22.7
3,4-Dichloro	3.0	24.6
3,5-Dichloro	2.4	23.6
2,3-Dichloro	1.8	23.1
2,4-Dichloro	2.0	23.5
2,5-Dichloro	1.5	22.7
2,6-Dichloro	0.4	22.4
4-Nitro	1.0	~18·9
2-Nitro	−0.3	17.9
2-Nitro-4-chloro	−1.1	17.1
4-Nitro-2,5-dichloro	−1.8	16.1
4-Nitro-2,6-dichloro	−3.3	15.6
2,4-Dinitro	−4.3	15.0
2,4-Dinitro-6-bromo	−6.7	13.6
2,4,6-Trinitro	−10.1	12.2
Diphenylamines		
4-Methoxy	1.4	23.2
4-Methyl	1.2	23.0
None	0.8	22.4
3-Methoxy	0.4	22.2
4-Chloro	0.0	21.3
3-Chloro	−0.5	20.7
3-Trifluoromethyl	−0.8	20.5
3,4′-Dichloro	−1.2	19.7
3-Nitro	−1.6	19.5
4-Methylsulfonyl	−2.5	18.8
2-Nitro	−4.1	17.9
4-Nitro	−3.1	15.7
4-Nitro-3′-trifluoromethyl	−4.5	15.0
3,4′-Dinitro	−5.2	14.7
4,4′-Dinitro	−6.2	14.1

group has an anomalously large effect on the acidity, but not on the basicity, of aromatic amines.

The effects of ring substituents on the acidity of anilines and diphenylamines are similar to those found in the analogous ionization of phenols (Table 10.5).

TABLE 10.5
Comparison of the Acidities of Anilines and Phenols (21)

Substituent	pK_a Aniline	pK_a Phenol	ΔpK_a
4-Nitro	18.37	7.15	11.22
2-Nitro	17.88	7.08	10.80
2-Nitro-4-chloro	17.22	6.36	10.86
2,4-Dinitro	15.00	4.07	10.93
2,6-Dichloro-4-nitro	15.55	3.48	12.17
2,4,6-Trinitro	12.20	0.3	11.9

C. Anion Addition

Addition of anions to some nitro compounds may be a competing process with ionization by proton loss. This is particularly true of amines with more than two nitro groups in one ring. Thus 2,4,6-trinitroaniline in dimethyl sulfoxide reacts with sodium methoxide predominantly by methoxide ion addition rather than by proton dissociation, whereas 2,4-dinitroaniline and 2,4-dinitrodiphenylamine ionize by proton loss (71). In the conjugated mononitroanilines reaction with hydroxide ion is by proton loss rather than by hydroxide addition. In 3-nitroaniline the nitro group displays its weakest acid-strengthening effect so that deprotonation does not take place but hydroxyl ion addition at a position *ortho* or *para* to the nitro group is postulated as taking place. Both 3-nitroaniline and nitrobenzene in 97 mole % dimethyl sulfoxide containing 0.011 *M* hydroxide ion produce broad absorption in the 300 to 550 mμ region, which is assumed to be indicative of hydroxyl ion addition (22). Substitution of an *N*-phenyl group into 3-nitroaniline increases the acidity of the NH by nearly 5 pK units; hence 3-nitrodiphenylamine reacts by dissociation rather than by addition of a hydroxy group.

The polynitroarenes ionize by anion addition (30, 34, 35, 72). A J_--function has been used to describe the basicity of some of the solvents used in these studies. The assumption is that J_- and H_- are identical in dilute aqueous hydroxide (up to 0.1 *M*) (24). Some of the values are listed in Table 10.6. Many of these solvent systems are used in trace analysis work. The pK_a values of some polynitro compounds are presented in Table 10.7.

Evidence has been presented indicating that one of the important reactions is simple anion addition to the polynitro compound (73–77), as, for example, in the following:

$$1,3,5\text{-}(NO_2)_3C_6H_3 \xrightarrow{X^-} [\text{1-H, 1-X, 2,4-}(NO_2)_2\text{-6-}NO_2^- \text{ cyclohexadienyl anion}] \quad (10.9)$$

As will be shown, the anion formed absorbs at much longer wavelength than the original anion or the polynitro derivative.

III. Spectra of Conjugated Anions

The addition of alkali to solutions of many types of organic compounds results in a red spectral shift due to the formation of an anion to whose

TABLE 10.6
The J_--Function for Aqueous and Alcoholic Bases

	J_--Function		
Molarity of base	Aqueous NaOH[a]	$NaOCH_3$ in methanol[b]	$NaOC_2H_5$ in ethanol[c]
1×10^{-3}	11.00	11.66	13.75
5×10^{-3}	11.70	12.36	14.45
1×10^{-2}	12.00	12.66	14.75
5×10^{-2}	12.70	13.36	15.45
1×10^{-1}	13.00	13.66	15.75

[a]Data from references 35 and 72.
[b]Data from references 30 and 72.
[c]Data from reference 30.

TABLE 10.7
Apparent "pK_a" Values of Polynitroarenes in Various Solvents

Compound	Aqueous $H_2NCH_2CH_2NH_2$[a]	Aqueous NaOH[b]	$NaOCH_3$ in methanol[c]	$NaOC_2H_5$ in ethanol[d]
1,3,5-Trinitrobenzene	13.53	13.57	12.45	12.37
2,4,6-Trinitrotoluene[e]	13.60	13.69	12.81	12.43
2,4,6-Trinitro-*m*-xylene	14.98	—	13.94	13.87
1,3-Dinitrobenzene	15.70	—		
2,4-Dinitrotoluene	15.99	—		

[a]Data from reference 34.
[b]Data from references 35 and 72.
[c]Data from references 30 and 72.
[d]Data from reference 30.
[e]pK_a 13.73 in aqueous hydrazine (32).

hybrid structure anionic resonance forms contribute strongly. Such an anion can be formed by the following methods:

1. Deprotonation of a compound that contains acidic hydrogen.
2. Addition of an anion to a neutral compound.
3. One electron addition to a neutral compound to form a free-radical anion.
4. Addition of two electrons to a neutral compound or one electron to a free-radical anion to form a dianion.
5. Proton addition to a dianion.

Examples of these various methods will be given, although it should be emphasized that innumerable examples are known of the first method and a fair number of the second, but a limited number of the last three.

A. Arenes

Formation of anions from aromatic hydrocarbons lacking acidic hydrogen is accomplished under anaerobic conditions in the presence of an extremely powerful base—for example, an alkali metal or a hydrocarbon derivative of such a metal. The variety of anions and spectra that can be obtained are shown below for tetracene.

1 $\xrightarrow{+e}$ 2 $\xrightarrow{+e}$ 4 $\xrightarrow{+H^+}$ 3 (10.10)

No.	Solvent	λ_{max} (mμ) and relative intensity	Ref.
1	Benzene	396 (2,2), 418 (4.4), 445 (8.1), 476 (9.6)	78
2	Benzene	360 (20), 403 (20), 704 (7), 794 (22)	79
3	Diglyme	253 (4.5), 314 (2), 485 (17.5), 610 (4)	80
4	Diglyme	353 (34), 398 (5), 498 (7), 617 (17)	79

The complexity of the spectra of the negative ions of naphthacene and 5,12-dihydronaphthacene in 1,2-dimethoxyethane has been emphasized, demonstrated, and discussed(81). The spectra of the mononegative ions of naphthalene, anthracene, tetracene, phenanthrene, diphenyl, *p*-terphenyl, *p*-quaterphenyl, pyrene, perylene, triphenylene, and coronene and the dinegative ions of anthracene, tetracene, *p*-terphenyl, *p*-quaterphenyl, and perylene have been obtained (79).

The electronic absorption spectra of the primary proton adducts (the carbanions MH^-) of the dinegative ions of naphthalene, anthracene, tetracene, and pyrene have been measured and shown to be closely similar to the spectra of the corresponding carbonium ions MH^+ (80). The close spectral similarity between the carbonium ions and carbanions is shown for tetracene in Figure 10.1. In these comparisons the carbanion absorbs at longer wavelength than the carbonium ion. The red solvent shift of the carbanion is probably due to the longer wavelength absorbance in the highly basic solvent. The wavelength position of an anion is very sensitive to solvent basicity.

B. Diphenyl Methanes

Since benzoic acid is about 10^{30} times and phenol is about 10^{24} times as strong an acid as diphenylmethane, it is obvious that a very basic solvent is necessary to form the anion of the diphenylmethanes. Some of the solvent systems that have been used to ionize arenes with acidic hydrogen include dimsyl sodium in dimethyl sulfoxide (16), alkali metal cyclohexylamide in cyclohexylamine (60), butyl lithium in an ether (82, 83), a sodium

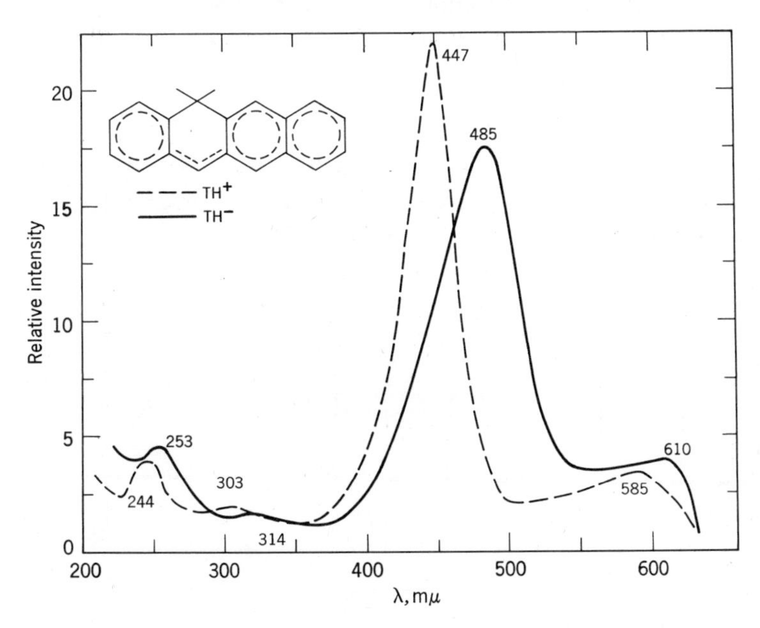

Fig. 10.1. Absorption spectra of the tetracene carbanion and carbonium ion.

alkoxide in ethanol plus dimethyl sulfoxide (55), or a strong base in a dipolar aprotic solvent (84).

The absorption spectra for some of these anions are shown in Table 10.8. Such spectra can be useful in characterizing these types of compounds, providing that air is carefully excluded.

The importance of the five-membered ring in these compounds is apparent from the difference of 14 pK units between triphenylmethane and 9-phenylfluorene. Of this difference in acidity about one-third comes from the increased coplanarity of the rings in the latter carbanion and two-thirds from the distinctive anion-stabilizing ability of the five-membered ring (60). Of course, the more acidic the compound, the more easily it can be handled in analysis.

The strong red shift that occurs on deprotonation of the diphenylmethanes is shown for fluorene in Figure 10.2.

The mononegative and the positive radical ions of an aromatic hydrocarbon, an even alternant system, have very similar absorption spectra. Other things being equal, a parallel similarity should be obtained between the spectra of the anion and the cation of an odd alternant system, such as that in the diarylmethanes.

A change in strong acid does not alter the spectra of arylcarbonium ions, but the absorption wavelengths of conjugated anions vary with the change in cation or solvent, the variation increasing with increasing localization of the negative charge in the anion. The charge is delocalized over all the

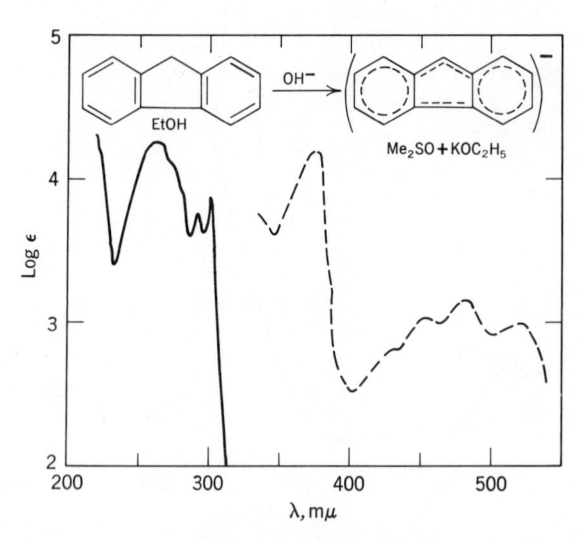

Fig. 10.2. Absorption spectra of fluorene and its anion (55).

TABLE 10.8
Absorption Spectra of Diarylmethane Anions

Compound	Cation	Solvent	λ_{max} (mμ)[a]	mϵ	Ref.
Diphenylmethane	Na	DMSO	454	29.0	16
	Cs	Cyclohexyl-amine	443	37.2	60
Triphenylmethane	Na	DMSO	503	30.0	16
	Cs	Cyclohexyl-amine	488	28.5	60
p-Biphenyldiphenylmethane	Cs	Cyclohexyl-amine	573	44.3	60
1,3,3-Triphenylpropene	Li	Cyclohexyl-amine	470 556	21.4 46.2	60
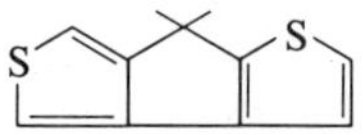 S S	Li	Cyclohexyl-amine	363 453*s*	10.0 2.55	82
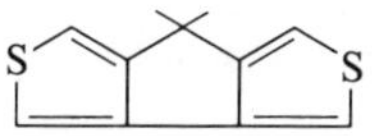 S S	Li	Cyclohexyl-amine	540	0.34	82
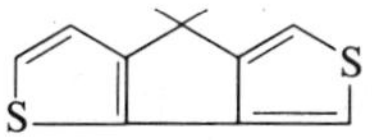 S S	Li	Cyclohexyl-amine	520	3.4	82
4,4′-Dinitrodiphenylmethane	K	Acetone	750		84
Fluorene[b]	K	DMSO	371 452 483 533	15.5 1.0 1.4 1.0	55
13*H*-Dibenzo[*a*,*i*] fluorene[b]	K	DMSO	342 355*s* 444 468	52.0 47.0 13.0 16.0	55
12*H*-Dibenzo[*b*,*h*] fluorene[b]	K	DMSO	332 362*s* 395*s* 462*s* 488	69.0 16.0 2.5 18.0 50.0	55

TABLE 10.8 (*cont'd*)

Compound	Cation	Solvent	λ_{max} (mμ)[a]	mϵ	Ref.
7*H*-Dibenzo[*c*,*g*]-fluorene[b]	K	DMSO	351	15.0	55
			383*s*	20.0	
			411*s*	6.5	
			445	5.6	
			500*s*	1.7	
9-Phenylfluorene[b]	K	DMSO	372	22.0	55
			408	25.0	
			488	25.0	
			536	18.0	
13-Phenyl-13*H*-dibenzo-[*a*,*i*]fluorene	K	DMSO	345	59.0	55
			450*s*	14.0	
			472	18.0	
12-Phenyl-12*H*-dibenzo-[*b*,*h*]fluorene	K	DMSO	333	69.0	55
			362*s*	18.0	
			437	32.0	
			463	24.0	
			490	46.0	
7-Phenyl-7*H*-dibenzo-[*c*,*g*]fluorene	K	DMSO	356	60.0	55
			377*s*	42.0	
			453	7.5	
			480*s*	63.0	
			513	5.0	
4*H*-Cyclopenta[*d*,*e*,*f*] phenanthrene	Li	Cyclohexyl-amine	505	7.36	60
9,10-Dihydro-9,9-dimethyl-10-phenylanthracene	Li	Cyclohexyl-amine	454	31.5	60
Tris(biphenylenevinyl)-methane			645	100.0	85
Bisbiphenylenepenta-diene			633		85
1,1,5,5-Bisbiphenylene-3-(diphenylvinyl)pentadiene			676		85
1,1,5,5-Bisbiphenylene-3-phenylpentadiene			686	82.0	85
Tris(7*H*-dibenzo[*c*,*g*]-fluorenylidenemethyl)-methane	Na	DMSO	697	141.3	49

[a]*s* = Shoulder.
[b]See also reference 60.

conjugated atoms of an even alternant anion, but in the ground state of odd alternant ions the charge is confined to the starred atoms (e.g., the negative resonance nodes in an anion), the exocyclic carbon atom in the arylmethyl ions always carrying the largest fraction of the charge. Thus the influence of ion-pairing and solvation forces on the transition energies is more marked in the arylmethyl anions than in the aromatic hydrocarbon negative radical ions, accounting for the poorer resemblance between the spectra of corresponding anions and cations in the odd alternant series (Table 10.9) than for the even alternant systems (86).

TABLE 10.9
The Long Wavelength Absorption Band Maxima in the Spectra of Diarylmethyl and Triarylmethyl Ions (86)

	Cation		Anion	
Methyl ion	λ_{max} (mμ)	mϵ	λ_{max} (mμ)	mϵ
Tris-*p*-biphenylyl	510	105.0	580	72.5
Tris-*m*-biphenylyl	500	10.1	485	45.5
Bis-*p*-biphenylyl	555	71.4	545	156.0
Bis-*p*-biphenylylmethyl	515	64.0	590	145.0

C. Cyanocarbons

The remarkable strength of these acids is due to resonance stabilization in the anion that is not possible in the protonated form, as, for example, in **5**.

$$(NC)_2C{=}CH{-}\overset{-}{C}(CN)_2 \leftrightarrow (NC)_2\overset{-}{C}{-}CH{=}C(CN)_2$$

5

This strong acidity is emphasized by the report that the sodium salt of pentacyanopropene is probably present in anhydrous sulfuric acid as the anion (48).

The absorption of these anionic resonance structures at very long wavelengths is due to the very high order of the electron-acceptor strength of the resonance terminal carbon to which two cyano groups are attached. For some of the most powerful electron-acceptor groups the electron-acceptor strength decreases in the following series (87):

$$—SO_2CF_3 > —C(CN){=}C(CN)_2 > —CH{=}C(CN)_2 \simeq —NO_2$$

The seemingly greater electron-acceptor strength of the tricyano group as compared with the dicyano group is probably due to the negativization of the positive resonance node by the lone cyano group. The absorption spectra of some of these anions are listed in Table 10.10.

TABLE 10.10
Ultraviolet Absorption Spectra of Cyanocarbon Acid Anions (88)

Anion	Solvent	λ_{max} (mμ)	mϵ
$[C(CN)_3]^-$	Water	211	37.4
$[C(CN)_2C(NHCH_3)C(CN)_2]^-$	Water	306	22.8
$[C(CN)_2C(NH_2)C(CN)_2]^-$	95% ETOH	310	30.5
$\{(CN)_2C[N(CH_3)_2]C(CN)_2\}^-$	Water	310	22.0
$[C(CN)_2C(OC_2H_5)C(CN)_2]^-$	Water	327	29.7
$[C(CN)_2C(OCH_2CH_2OH)C(CN)_2]^-$	Water	327	30.7
$\{[C(CN)_2C[{=}C(CN)_2]C(CN)_2\}^{2-}$	Water	335	32.7
$[C(CN)_2CHC(CN)_2]^-$		344	
$[C(CN)_2C(Cl)C(CN)_2]^-$	Water	353	32.0
$[C(CN)_2C(Br)C(CN)_2]^-$	Water	356	28.3
$[C(CN)_2C(C_6H_5)C(CN)_2]^-$	95% ETOH	357	22.8
$\{C(CN)_2C[C_6H_4N(CH_3)_2]C(CN)_2\}^-$	95% ETOH	360	32.3
$[C(CN)_2C(SCH_3)C(CN)_2]^-$	Water	361	26.1
$[C(CN)_2C(NHNO)C(CN)_2]^-$	Acetone	362	19.1
$[C(CN)_2C(CN)C(CN)_2]^-$	Water	393	22.6
		412	22.1
$[C(CN)_2C(CN)NC(CN)C(CN)_2]^-$	Acetone	440	38.0
		464	45.3
$[C(CN)_2C(CN)NNC(CN)C(CN)_2]^{2-}$	Acetone	456	15.3
		478	14.5
$[p\text{-}C(CN)_2C(CN)C_6H_4C(CN)_2]^-$ [a]	95% EtOH	292	7.7
		350	5.6
		625	52.0

[a] Data from reference 89.

The large red shift that occurs on ionization of the cyanocarbon acids is shown in Table 10.11.

D. Phenols

Many analytical methods based on the absorption spectrum of the anion have been developed for phenols—for example, monocyclic phenols in gasolines (90), eugenol as based on difference spectra in alkaline and acid solutions (91), and naphthols after nitration (92).

The absorption spectra of some phenols and their anions are presented

TABLE 10.11
Long Wavelength Maxima of Some Cyanocarbon Acids (48)

Cyanocarbon	λ_{max} (mμ)	
	Neutral	Anion
Cyanoform	< 200	210
Methyl dicyanoacetate	< 200	235
Tricyanovinyl alcohol	275	295
Bis(tricyanovinyl)amine	366	467
p-(Tricyanovinyl)phenyl-dicyanomethane	332	607
Hexacyanoheptatriene	347	645

in Table 10.12. In comparison with the neutral compound the phenolic anion absorbs at longer wavelength, with approximately the same intensity. The longer the chain of conjugation of a phenolic anion, the further into the visible does it absorb. The substitution of an electron-acceptor group as a resonance terminal in a phenolic anion causes a red shift.

The resonance terminals for $HOC_6H_4—\overset{O^-}{\underset{+}{N}}=C_6H_4=O$ in Table 10.12 are the O^- and the quinonoid oxygen. Replacement of the hydroxy group by a hydrogen atom has little effect on the spectrum, since the latter

TABLE 10.12
Absorption Maxima of Phenols

Phenol	In 0.1 or 0.01 *N* HCl		In 0.1 *N* NaOH		
	λ_{max} (mμ)[a]	mε	λ_{max} (mμ)	mε	Ref.
Phenol	270	1.5	234	9.7	93
			287	2.6	
o-Cresol	270	1.6	237	9.2	93
			289	3.2	
m-Cresol	271	1.4	238	8.4	93
			288	2.6	
p-Cresol	277	1.7	237	8.8	93
			295	2.6	
2,3-Xylenol	271	1.2	239	8.2	93
			289	2.9	
2,4-Xylenol	277	1.9	238	8.4	93
			296	3.2	

TABLE 10.12 (*cont'd*)

	In 0.1 or 0.01 *N* HCl		In 0.1 *N* NaOH		
Phenol	λ_{max} (mμ)[a]	mϵ	λ_{max} (mμ)	mϵ	Ref.
2,5-Xylenol	274	1.8	239	8.0	93
			291	3.5	
2,6-Xylenol	270	1.2	239	8.5	93
			288	3.6	
3,4-Xylenol	277	1.8	237	8.7	93
			295	2.8	
3,5-Xylenol	272	1.2	240	8.0	93
			290	2.6	
o-Phenylthiophenol	245	10.5	243	14·5	94
	282[b]	5.4	309	6.5	
p-Phenylthiophenol	232	12.3	260	19.5	94
	250	14.5	275*s*	12.6	
	270[c]	7.2			
3-Hydroxybiphenyl	250	13.7	305	4.13	95
4-Hydroxybiphenyl	260	17.8	288	20.1	95
1-Naphthol	298[d]	5.4	356	5.6[e]	19
1-Anthrol	395*fs*[d]	4.0	482	4.0[e]	19
2-Hydroxyfluorene	268	19.9	294	20.8	95
	306	6.2			
2-Nitrosophenol	310	6.3	330	7.9	96
	400	2.5	470	5.0	
	690[f]	0.016	660[f]	0.04	
4-Nitrophenol	226	7.1	228	5.9	97
	316	10.0	400	18·2	
4′-Nitro-4-hydroxy-biphenyl	232	12.5	265	12.2	95
	340	14.2	400	15.1	
3′-Nitro-4-hydroxy-biphenyl	263	24.8	294	23.4	95
2-Nitro-7-hydroxy-fluorene	250	9.1	270	12.5	95
	370	16.7	430	17.6	
1-Nitroso-2-naphthol	311	12.6	330	8.9	96
	414	5.0	431	7.9	
			697[f]	0.071	
1-Hydroxyanthraquinone	402[d]	5.5	561[e]	5.9	19
2-Hydroxyanthraquinone	375[d]	2.9	548[e]	5.3	19
$HOC_6H_4{-}\overset{O^-}{\underset{+}{N}}{=}C_6H_4{=}O$	382[c]	16.0	672[g]	35.0	98
1,4-Dihydroxyanthra-quinone	470[h]	19.1	560[i]	22.4	99
2-Hydroxy-7-acetyl-fluorene	234	13.2	252	13.2	
	331[c]	30.2	384[j]	28.2	

TABLE 10.12 (*cont'd*)

	In 0.1 or 0.01 *N* HCl		In 0.1 *N* NaOH		
Phenol	λ_{max} ($m\mu$)[a]	$m\epsilon$	λ_{max} ($m\mu$)	$m\epsilon$	Ref.
2-Hydroxypteridine	230	7.6	260	7.1	100
	307[k]	6.8	375[l]	6.0	
4-Hydroxypteridine	230	9.6	242	17.0	100
	265	3.5	333[n]	6.2	
	310[m]	6.6			
4′-Hydroxyaurone	260	22.9	358	5.0	101[o]
	346	12.0	487[g]	44.7	
	405[c]	29.5			

[a]*s* = Shoulder; *fs* = fine structure.
[b]In HCl.
[c]In ethanol.
[d]In dimethylformamide.
[e]In dimethylformamide with 2% of 10% aqueous tetraethylammonium hydroxide.
[f]$n \rightarrow \pi^*$ band.
[g]In alkaline ethanol.
[h]In methanol.
[i]In methanolic 0.1 *N* alkali.
[j]In ethanolic 0.1 *N* KOH.
[k]At pH 7.1.
[l]At pH 13.
[m]At pH 5.6.
[n]At pH 10.0
[o]Also spectra of many other hydroxyaurones.

compound absorbs at 371 $m\mu$, $m\epsilon$ 25.0. However, deprotonation causes a 290 $m\mu$ red shift, indicating a longer chain of conjugation with the end oxygen atoms as resonance terminals.

An example of a phenolic anion that absorbs at very long wavelength in alkaline solution mainly because of the long chain of conjugation involved in the resonance structures is **6**(102).

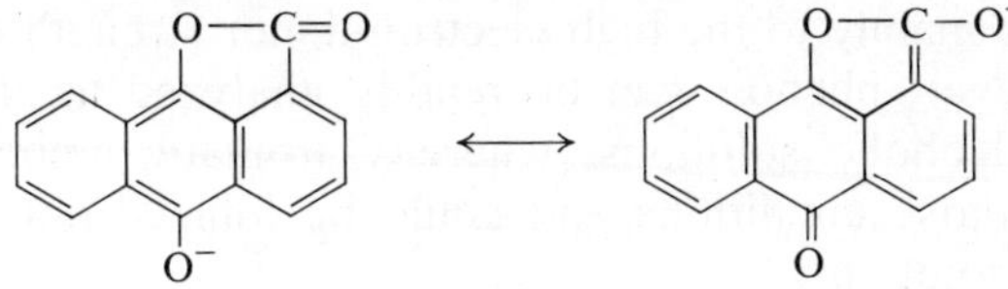

6

λ_{max} 790 $m\mu$, $m\epsilon$ 7.6

E. Aromatic Amines

Aromatic amine anions absorb at much longer wavelengths than do phenolic anions due mainly to the fact that the —NH⁻ group is a much more powerful electron donor than is the —O⁻ group. However, the phenols are approximately 10^{16} to 10^{17} times more acidic than the anilines. This relationship can be seen in the comparison of 3-chlorophenol (**7**) and 3-chloroaniline (**9**)

$$\text{3-ClC}_6\text{H}_4\text{–OH}\ (\mathbf{7}) \longrightarrow \text{3-ClC}_6\text{H}_4\text{–O}^-\ (\mathbf{8}) \qquad (10.11)$$

$$\text{3-ClC}_6\text{H}_4\text{–NH}_2\ (\mathbf{9}) \longrightarrow \text{3-ClC}_6\text{H}_4\text{–NH}^-\ (\mathbf{10}) \qquad (10.12)$$

No.	Solvent	λ_{max} (mμ)	mε	pK[a]	Ref.
7	0.1 *N* HCl	212, 275	6.3, 1.74	pK_a 9.02	103
8	0.1 *N* NaOH	239, 292	9.78, 3.5		104
9	EtOH	240, 292	6.3, 2.0	pK_{HA} 25.63	105
10	DMSO	386	2.7		22

[a] Data for pK_a and pK_{HA} are from references 104 and 22, respectively.

Only a small portion of the wavelength difference between the phenol and aniline anions is due to solvent effect since phenol absorbs at 310 mμ in a very basic dipolar aprotic solvent, such as dimethylformamide containing 2% of 10% aqueous tetraethylammonium hydroxide(19) as compared with 292 mμ in aqueous alkali. Thus the strong red shift in going from a neutral aniline to its anion as compared with a similar shift for the phenols is due mainly to the high electron-donor strength of the —NH⁻ group. However, phenols can be readily analyzed as their anions in aqueous or alcoholic solutions, whereas aromatic amines would not ionize under those conditions and could be ionized only as anions in strongly basic solutions.

In the polynuclear aromatic amines deprotonation causes a strong red shift, as shown below for 1-naphthylamine (**11**). Other polynuclear aromatic amines show a similar phenomenon(106).

NH_2 → NH^- (10.13)

11 **12**

No.	Solvent	λ_{max} (mμ)	mϵ	Ref.
11	EtOH	240	22.9	107
		322	5.1	
12	DMSO	406	8.3	16

The *N*-acyl aromatic amines would be expected to be more acidic than the parent amine, but their anions absorb at shorter wavelengths than the parent compounds due to the decreased electron-donor strength of the nitrogen resonance terminal. Examples of such anions are *N*-tosyl-l-aminoanthracene (see Fig. 10.3)(106), and the acylamino derivatives listed in Table 10.13. The nitroanilines are also more acidic than the parent anilines, and their anions absorb at longer wavelengths than the parent aniline anions due to the lengthening of the chain of conjugation and the introduction of a powerful negative resonance terminal (e.g., compare **13** and **14**).

O_2N—C₆H₄—$\bar{N}H$ ⟷ $\bar{O}_2N$=C₆H₄=NH   C₆H₅—$\bar{N}H$ ⟷ C₆H₅=NH

13 **14**

λ_{max} 467 mμ, mϵ 32.3 λ_{max} ~ 380 mμ

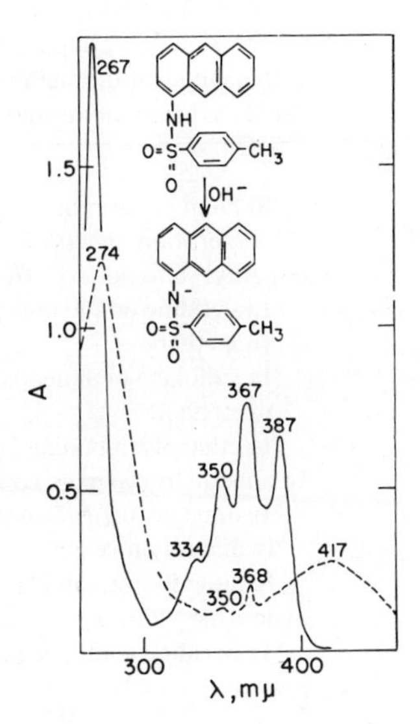

Fig. 10.3. Absorption spectra of *N*-tosyl-1-anthramine: 10^{-4} *M* in neutral dimethylformamide (——) and 5×10^{-5} *M* in dimethylformamide–40% aqueous tetraethylammonium hydroxide (1 : 1) (---) (106).

TABLE 10.13
Spectra of Electronegatively Substituted Aromatic Amines and Their Anions[a]

Compound	Neutral molecule λ_{max} (mμ)	Neutral molecule mϵ	Anion[b] λ_{max} (mμ)	Anion[b] mϵ	Ref.
2-Nitroaniline[c]	410	4.5	515	8.4	20
4-Chloro-2-nitroaniline	417[d]	3.5	516[e]	6.9	20
4-Nitroaniline	378[c]	16.9	467[e]	32.3	20
2,6-Dichloro-4-nitroaniline[c]	368	13.7	467	38.8	20
2,5-Dichloro-4-nitroaniline[c]	368	11.8	458	37.2	20
N-Perfluorobutyroyl-4-	225	9.1	235	7.5	
nitroaniline	299[f]	14.0	358[g]	12.7	
2,4-Dinitroaniline	336[d]	14.9	388	21.1	20
			535[c]	15.3	
6-Bromo-2,4-dinitroaniline	350	12.0	395	23.2	20
			527[e]	12.7	
2,4-Dinitroacetanilide			347[h]	9.7	108
2,4,6-Trinitroaniline	328	11.0	412	23.4	20
	417[e]	6.0			
1-Aminoanthraquinone	481[i]	6.5	406	3.2	
			682[j]	6.1	
2-Aminoanthraquinone	450[i]	3.7	430	7.7	
			675[j]	3.8	
2-Amino-1-chloroanthraquinone[k]			650	1.5	109
2-Amino-1-nitroanthraquinone[k]			650	4.7	109
2-Acetamidoanthraquinone[k]			515	6.9	109
2-Benzamidoanthraquinone[k]			527	10.2	109
2-*p*-Tosylaminoanthraquinone[k]			500	7.0	109

[a] Solvent is water unless otherwise noted.

[b] Absorption spectra of many chloronitroaniline and nitroaniline anions in dimethyl sulfoxide reported (22).

[c] In pyridine or aqueous pyridine.

[d] In methanol.

[e] In sulfolane or aqueous sulfolane.

[f] In ethanol.

[g] In ethanol containing 2% of 10% aqueous tetraethylammonium hydroxide. In dimethylformamide containing the same alkali λ_{max} 401 mμ, mϵ 20.7.

[h] In aqueous 0.1 *N* NaOH.

[i] In dimethylformamide.

[j] In dimethylformamide containing 1% of 25% aqueous tetraethylammonium hydroxide.

[k] In pyridine containing methanolic KOH.

F. Diarylamines

Diphenylamine absorbs at 376 mμ, mϵ 28.5, in 99.5 mole % dimethyl sulfoxide containing 0.011 M tetramethylammonium hydroxide(22). The aniline anion absorbs in the same region, with a much lower intensity. Hence it appears that the nitrogen atom is a resonance terminal in both cases.

The carbazole anion absorbs at longer wavelength than neutral carbazole (see Fig. 10.4)(110) but at shorter wavelength than the fluorene anion (Fig. 10.2).

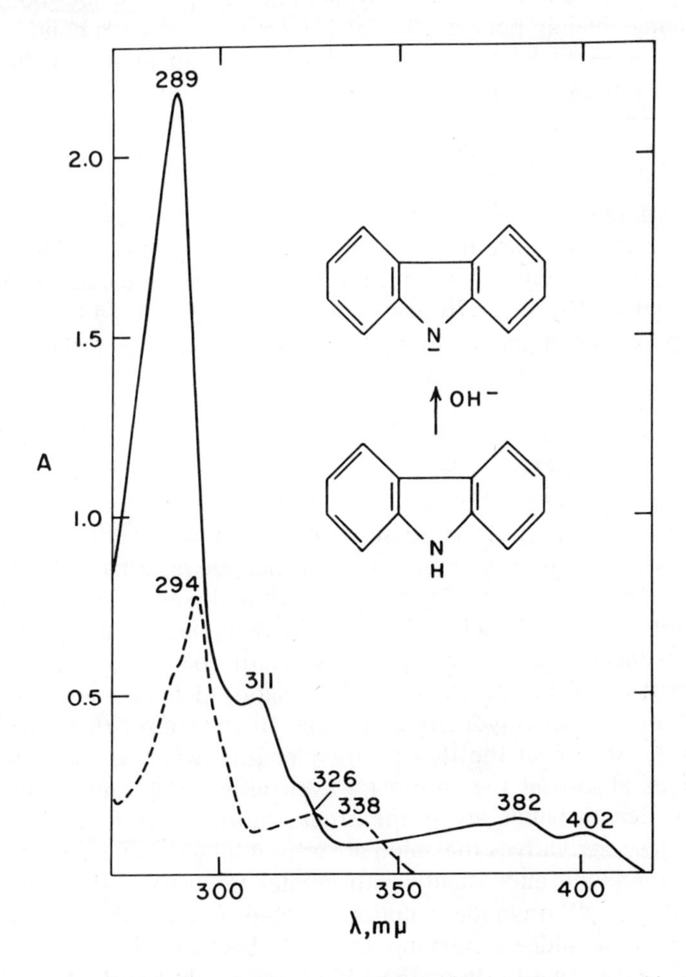

Fig. 10.4. Absorption spectra of carbazole in dimethylformamide (---) and in dimethylformamide containing 1% of 25% aqueous tetraethylammonium hydroxide (——)(110).

The absorption spectra of some nitrodiphenylamines and anilinoquinones are listed in Table 10.14. The strong red shift on deprotonation is noteworthy. Examination of the spectra of the diphenylamine anion and its derivatives discloses that electron-donor groups and chloro, trifluoromethyl, and mesyl groups have little effect on the spectra, but the nitro group causes a strong red shift (22).

An orange solution of 2-phenylaminoanthraquinone in pyridine shows a λ_{max} of 475 mμ, mϵ 6.4; it becomes green with a λ_{max} of 710 mμ, mϵ 7.2, on the addition of a methanolic solution of potassium hydroxide (109). Under the same conditions 1-phenylaminoanthraquinone does not show a color change mainly because the intramolecular hydrogen bond has considerably decreased the acidity of the NH group in this molecule as compared with the 2-isomer.

G. Hydrazones

The hydrazones that have been studied most thoroughly in terms of their anionic spectra are the 2,4-dinitrophenylhydrazones. The absorption spectra of aldehyde 2,4-dinitrophenylhydrazones (Table 10.15) show that the anion absorbs at 70 to 140 mμ longer wavelength than does the neutral parent compound. Many other spectra of the neutral and anionic forms of 2,4-dinitrophenylhydrazones have been reported—for example, for 40 carbonyl derivatives (13), 96 carbonyl derivatives (112), 11 *n*-alka-2-enal and 12 *n*-alka-2,4-dienal derivatives (114), 18 dicarbonyl compounds (115), and 8 benzaldehyde and 9 acetophenone derivatives (116).

Of the various carbonyl 2,4-dinitrophenylhydrazone anions, the aliphatic aldehyde ones are least stable, as shown by their decreasing absorbance with time at their long-wavelength maxima (13). Caution must also be exercised in determining carbonyl compounds through the absorption spectra of their 2,4-dinitrophenylhydrazone anions, since in many solvents they are not completely ionized (19).

Comparison of the spectra of some aromatic aldehyde 2-nitro-, 4-nitro-, and 2,4-dinitrophenylhydrazones (Table 10.16) shows that the 2-nitro derivatives absorb at the longest wavelengths, whereas the 2,4-dinitro derivatives absorb at the shortest wavelengths. The intensities of the long wavelength bands are in the order 4-nitro $>$ 2,4-dinitro $>$ 2-nitro. For all these derivatives the anion absorbs at longer wavelength than the neutral compound and usually with greater intensity; for example, the yellow dimethylformamide solution of 2-nitrobenzalphenylhydrazone, which has a shoulder at 405 mμ (mϵ 8.4), becomes dark green, with a λ_{max} of 717 mμ, (mϵ 17.0), on the addition of alkali (Fig. 10.5). Such an extreme red shift is of obvious value in trace analysis.

TABLE 10.14
Spectral Data for Some Imino Derivatives[a]

Compound or Derivative	BH		B⁻		Ref.
	λ_{max} (mμ)	mϵ	λ_{max} (mμ)	mϵ	
Indole	270	5.0	289	4.3	54
Carbazole	326	3.4	382	2.6	110
	338[b]	3.1	402[c]	2.2	
Diphenylamine	285[d]	20.4	376[e]	28.5	22, 111
2-Nitro	435[f]	6.4	545[f]	9.2	20
4-Nitro	400[g]	19.6	508[g]	34.6	20
2,4-Dinitro-4′-amino	375[f]	18.3	495[f]	18.3	20
4,4′-Dinitro	416[g]	35.4	580	37.0	20
2,4-Dinitro	362[g]	16.8	495	16.0	20
2,4,3′-Trinitro	360[f]	17.9	450[f]	18.7	20
2,4,4′-Trinitro	385[g]	21.9	520[g]	27.4	20
2,4,6-Trinitro-4′-amino	420[f]	11.8	470[f]	19.3	20
2,4,2′,4′-Tetranitro	413	21.2	510	29.0	20
2,4,6-Trinitro	372	13.1	450	19.0	20
2,4,6,3′-Tetranitro	337[h]	20.7	441[g]	26.0	20
2,4,6,4′-Tetranitro	377[h]	17.5	465	23.8	20
2,4,6,2′,4′-Pentanitro	378[i]	18.9	480	25.6	20
2,4,6,2′,4′,6′-Hexanitro	390	16.0	425	23.0	20
2-Anilino-3-chloro-1,4-naphthoquinone	482[b]	4.5	600[c]	5.4	19
2-Anilinoanthraquinone	475[g]	6.4	710[j]	7.2	109
3-Nitrocarbazole	279	24.6	488	19.0	20
	305	41.3			
	364	10.5			

[a]Solvent is water unless otherwise noted and alkaline for B⁻.

[b]In dimethylformamide.

[c]In dimethylformamide containing 2% of 10% aqueous tetraethylammonium hydroxide.

[d]In methanol.

[e]In dimethyl sulfoxide containing 0.011 *M* tetramethylammonium hydroxide.

[f]In sulfolane or aqueous sulfolane.

[g]In pyridine or aqueous pyridine.

[h]In benzene.

[i]In ethanol.

[j]In pyridine containing a methanolic solution of potassium hydroxide.

TABLE 10.15
Main Maxima of 2,4-Dinitrophenylhydrazones in Neutral and Alkaline Solution

Dinitrophenylhydrazone of	Neutral solution			Alkaline solution			Ref.
	λ_{max} (mμ)	mϵ	Solvent[a]	λ_{max} (mμ)	mϵ	Solvent[a,b]	
Formaldehyde	348	19.5	CA	428[c]	18.0	CAA	112
$CH_3(CH_2)_nCHO$, $n = 0$–10	357 ± 1	22 ± 1	CA or C	427 ± 3[c]	21.5 ± 3.5[d]	CAA or A	13, 112
3,4,4-Trimethylpentanal	358	21.0	CA	434[c]	21.0	CAA	112
2-Hydroxy-2-methylpropionaldehyde	354	24.0	CA	429[c]	19.0	CAA	112
5-Ethoxycarbonylpentanal	357	22.0	CA	431[c]	20.0	CAA	112
Tetrahydropyran-2-aldehyde	356	23.0	CA	436[c]	18.0	CAA	112
2,6,6-Trimethylcyclohex-2-enal	360	24.5	CA	435[c]	24.5	CAA	112
Phenylacetaldehyde	355	22.5	C	435[c]	22.5	A	13
Hydrocinnamaldehyde	358	22.8	C	435[c]	25.0	A	13
Acrolein	366	27.6	C	459[c]	29.5	CAA	114
$CH_3(CH_2)_nCH{=}CHCHO$, $n = 0$–7	374.5 ± 1.5	29.3 ± 0.6	C	458.5 ± 0.5	30.0 ± 0.8	CAA	114
Tiglaldehyde	376	30.0	C	458	28.9	A	13
Cyclohex-1-enealdehyde	376	28.5	CA	458	28.5	CAA	112
2,6,6-Trimethylcyclohex-1-enealdehyde	382	28.5	CA	458	28.5	CAA	112
n-Octa-2-ynal	357	22.5	CA	451	6.0	CAA	112
n-Penta-2,4-dienal	384	35.6	C	481	40.0	CAA	114
$CH_3(CH_2)_n(CH{=}CH)_2CHO$, $n = 0$–6	391	37.1 ± 0.5	C	480.5 ± 0.5	40 ± 1	CAA	114
Benzaldehyde	378	28.3	C	462	33.0	A	13
2-Nitrobenzaldehyde	381	29.0	C	—	—		113
4-Nitrobenzaldehyde	382	36.0	C	—	—		113
2-Hydroxybenzaldehyde	381	29.6	C	475	35.0	A	13
4-Methoxybenzaldehyde	390	30.9	C	460	33.0	A	13
3-Methoxy-4-hydroxybenzaldehyde	393	27.0	CA	478	27.5	CAA	112
4-Dimethylaminobenzaldehyde	434	30.0	C	478	38.8	A	13

4-Diethylaminobenzaldehyde	442	32.0		—	—		113
Cinnamaldehyde	390	39.0	C	486	43.0	A	13
2-Furaldehyde:							
Yellow form	379	38.0	C	475	27.4	A	13
Red form	386	26.5	C	468	26.2	A	13
5-Methyl-2-furaldehyde	389	26.2	C	478	27.2	A	13
2-Furanacrylaldehyde	400	37.9	C	490	41.0	A	13
Bis-2,4-*dinitrophenylhydrazone of*							
Glutaraldehyde	355	45.0	C	435	43.0	A	115
β-Methylglutaraldehyde	355	46.0	C	435	45.0	A	115
α-Hydroxyadipaldehyde	350	43.0	C	430	36.0	A	115
Glyoxal	438	44.0	C	575	66.0	A	115
Pyruvaldehyde	435	41.0	C	565	60.0	A	115

[a]CA = 10% $CHCl_3$–EtOH (v/v); CAA = 0.01 *N* NaOH in 10% $CHCl_3$–EtOH; C = $CHCl_3$; A = 0.025 *N* NaOH in EtOH.
[b]Spectra determined in fresh solutions.
[c]Also a band of about one-half the intensity at approximately 520 mμ.
[d]Reference 113.

TABLE 10.16
Long Wavelength Maxima of Some Aromatic Aldehyde Nitrophenylhydrazones and Dinitrophenylhydrazones in Weakly Acidic and Alkaline Dimethylformamide

	λ_{max} (mμ) (mϵ)					
	2-Nitrophenylhydrazine		4-Nitrophenylhydrazine		2,4-Dinitrophenylhydrazine	
Aldehyde	Acidic[a]	Alkaline[b]	Acidic	Alkaline	Acidic	Alkaline
Benzaldehyde	448 (9.4)	585 (14)	413 (30)[c]	553 (48)[c]	391 (30)	493 (36)
1-Naphthaldehyde	450 (11)	600 (17)	425 (35)	580 (53)	360 (27)	460 (30)
						510 (30)
9-Anthraldehyde	455 (14)	630 (23)	—	—	362 (30)	457 (28)
						500 (25)
4-Dimethylaminocinnamaldehyde	367 (28)	435 (24)	437 (44)	375 (16)	358 (25)	515 (35)
	465 (14)	606 (17)		575 (62)	410*s* (16)	
2-Nitrobenzaldehyde	432 (13)	650 (18)	420 (33)	600 (43)	292 (26)	545 (34)
4-Nitrobenzaldehyde	382 (13)	665 (45)	428 (42)	650 (60)	406	575
	443 (20)					

[a]0.02 ml of concentrated hydrochloric acid to every 10 ml of dimethylformamide.
[b]Dimethylformamide containing 2% of 10% aqueous tetraethylammonium hydroxide (V/V).
[c]Data for acetophenone 2,4-dinitrophenylhydrazone.

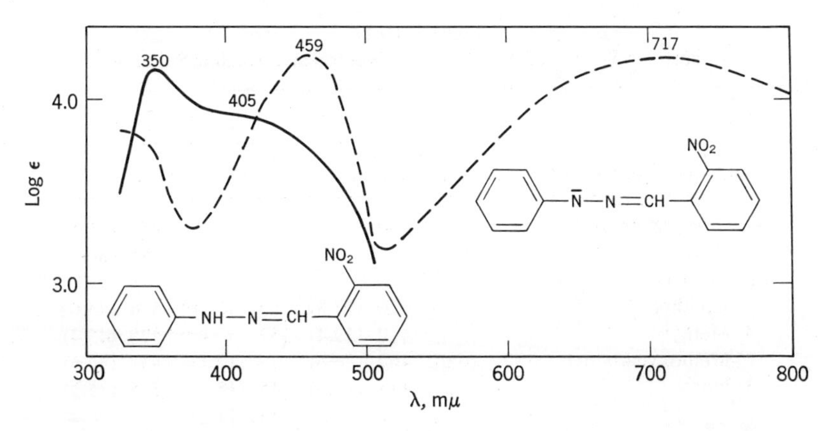

Fig. 10.5. Absorption spectra of 2-nitrobenzalphenylhydrazone in dimethylformamide (——) and dimethylformamide containing 2% of a 10% aqueous solution of tetraethylammonium hydroxide (- - -).

H. Azophenols and Azoanilines

1. Azophenols

The red-shift effect of 4′-substituents increases in the order H < $COOC_2H_5$ < Ac < NO_2 for the phenylazophenol anions (Table 10.17). Similarly in these dyes the red shift effect of the position of substitution of a nitro group decreases in the order 4′-NO_2 ≫ 2′-NO_2 > 3′-NO_2. The anion absorbs approximately 110–200 mμ further into the visible than does its neutral compound.

2. Azoanilines

Comparison of the absorption spectra of 2-, 3-, and 4-aminoazobenzenes in alkaline dimethylformamide shows that the 4-isomer absorbs at a much longer wavelength, probably because the 2- and 3-isomers are not deprotonated. Although 4-aminoazobenzenes and 4-alkylaminoazobenzenes do not seem to ionize completely in alkaline dimethylformamide, their spectra in this same solvent show the presence of two overlapping bands near 550 mμ. Some examples of this phenomenon are *N*-methyl-, *N*-ethyl, 3′,*N*-dimethyl-, 3,4′-dimethyl-, 2′,3-dimethyl-, and 2,3′-dimethyl-4-phenylazoanilines. The spectra of some neutral azoanilines and their anions are listed in Table 10.18. For all compounds there is a strong red shift and an intensity increase on ionization, as shown in Figure 10.6.

An *N*-acyl group increases the acidity of a 4-phenylazoaniline. Consequently the *N*-acyl derivatives ionize completely in alkaline dimethyl-

TABLE 10.17
Absorption Spectra of Azophenols in Neutral and Alkaline Solution

	λ_{max} (mμ)[a] (mϵ)			
Compound or derivative	Neutral solution[b]		Alkaline solution[c]	
4-Hydroxyazobenzene	353 (24.4)	400*s*[d] (1.4)		480 (38.5)
4′-Nitro	390 (26.6)			595 (60.2)
4-Phenylazo-1-naphthol		410 (17.8)	352 (6.40)	531 (44.0)
4′-Methyl		410 (19.4)	352 (7.40)	528 (45.4)
2′-Methoxycarbonyl	332 (6.0)	460 (26.6)	346 (12.5)	537 (44.6)
3′-Nitro		443 (16.0)	345*s* (15.2)	555 (35.2)
			375*s* (11.2)	
2′-Nitro		450 (17.8)	350*s* (16.0)	566 (43.6)
			375*s* (13.2)	
4′-Ethoxycarbonyl		461 (29.5)	338*s* (8.6)	
			375*s* (5.5)	575[e] (57.0)
4′-Acetyl		462 (31.2)		586 (60.0)
4′-Nitro	350 (8.2)	473 (49.0)	450 (2.0)	649 (78.2)
10-Phenylazo-9-anthrol	356 (5.2)	452 (17.5)	372 (16.7)	573 (28.8)
			449 (17.0)	603 (29.4)
2-Phenylazofluorene				
4′-Hydroxy	376 (35.6)	390*s* (32.8)		511 (36.1)
3-Hydroxy	375*s* (22.8)	404 (24.8)	363 (20.0)	550 (22.2)
3-Hydroxy-3′-nitro	382*s* (21.6)	416 (24.0)		563 (25.8)
3-Hydroxy-2′-nitro		420 (21.6)		570 (24.8)
3-Hydroxy-4′-nitro		436 (27.0)	340 (16.6)	613[f] (32.0)

[a]*s* = Shoulder.

[b]Contains 1 drop of concentrated hydrochloric acid in 100 ml of dimethylformamide.

[c]Contains 2% of a 10% aqueous solution of tetraethylammonium hydroxide in dimethylformamide.

[d]Shoulder derived from $n \rightarrow \pi^*$ transition.

[e]Fades quickly, the greatest changes taking place in the ultraviolet region.

[f]Solvent contains 2% of a 10% methanolic solution of benzyl tetramethylammonium methoxide.

formamide. The data in Table 10.19 show that for the *N*-acyl derivatives of 2′,3-dimethyl-4-phenylazoaniline the long wavelength maxima of the anions shift to the shorter wavelength in the following order of acyl and carbethoxy groups:

$$CH_3CH_2O_2C— > FCH_2CH_2O_2C— > CH_3CO— > CF_3CH_2O_2C— > FCH_2CO— > F_3CCO—$$

TABLE 10.18
Absorption Spectra of Azoanilines in Neutral and Alkaline Solution

	λ_{max} (mμ) (mε)			
Compound or derivative	Neutral solution[a]		Alkaline solution[b]	
4-Aminoazobenzene		397 (21.7)	401 (15.5)	545 (29.5)
				530*s* (27.5)
N-Phenyl		416 (21.6)		568 (63.8)
4′-Methylthio-*N*-phenyl[c]	333 (7.5)	433 (24.0)	366 (11.2)	588 (72.0)
4′-Nitro-*N*-phenyl		482 (31.4)	345 (6.0)	673 (83.0)
			455 (8.0)	
4-Phenylazo-1-naphthylamine	340 (3.0)	455 (24.4)	371 (8.0)	565 (62.8)
	353 (32.0)		434 (5.6)	
1-4′-Phenylazophenylazo-2-phenylaminonaphthalene	357 (20.8)	542 (30.4)	375*s* (20.8)	706 (51.6)
	377 (21.2)			

[a]Contains 1 drop of concentrated hydrochloric acid in 100 cc of dimethylformamide.

[b]Contains 2% of a 10% aqueous solution of tetraethylammonium hydroxide in dimethylformamide.

[c]Unstable in alkaline solution.

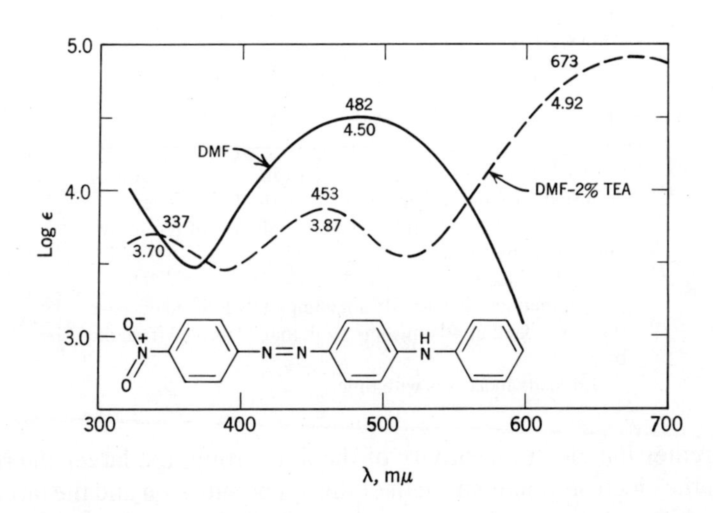

Fig. 10.6. Absorption spectra of 4-4′-nitrophenylazodiphenylamine in dimethylformamide (——) and dimethylformamide containing 2% of a 10% aqueous solution of tetraethylammonium hydroxide (– – –).

TABLE 10.19
The Long Wavelength Bands of *N*-Acyl-2′,3-Dimethyl-4-phenylazoanilines in Neutral and Alkaline Solution

—N=N— —NH—X
CH_3 CH_3

	DMF[a]		Alkaline DMF[b]	
X	λ$_{max}$ (mμ)	mε	λ$_{max}$ (mμ)	mε
H	375*s*[c]	25.3	535*s*[c]	12.4[d]
	404	29.0	555	13.0
4-$COC_6H_4OCH_3$	350	18.8	500	34.6
4-$COC_6H_4NO_2$	350	23.9	495	27.4
$COOC_2H_5$	354	19.6	493	34.8
COC_6H_5	350	21.6	492	30.0[e]
2-$COC_6H_4OC_2H_5$	360	22.0	489	28.6
2-CO—Furyl	350	19.0	482	25.0[e]
$COOCH_2CH_2F$	355	18.9	481	28.6
$COCH_3$	355	20.2	467	24.0
SO_2CH_3	346	22.8	459	28.9
4-$SO_2C_6H_5CH_3$	350	23.8	455	26.8
$COOCH_2CF_3$	352	18.6	455	22.3
$COCH_2F$	346	16.8	453	17.6
4-$SO_2C_6H_4NO_2$	343	22.9	434	25.8
$COCF_3$	337	16.8	395	12.9
			420	12.9
COC_3F_7	336	19.8	417	18.6

[a]All acylamines also contained in $n \rightarrow \pi^*$ band at approximately 445 mμ in neutral dimethylformamide.

[b]Contains 2% of a 10% aqueous solution of tetraethylammonium hydroxide.

[c]*s* = Shoulder.

[d]Compound is apparently not completely ionized because there is a band at 404 mμ, mε 19.4, that is derived from the base.

[e]Intensity increases with time.

The greater the electronegativity of the acyl group, the larger the reduction in the electron-donor strength of the amino nitrogen and the larger the violet shift. An example of the effect of alkali on the spectrum of an acylaminoazo dye is shown in Figure 10.7. Here again we see the presence of two overlapping long wavelength bands.

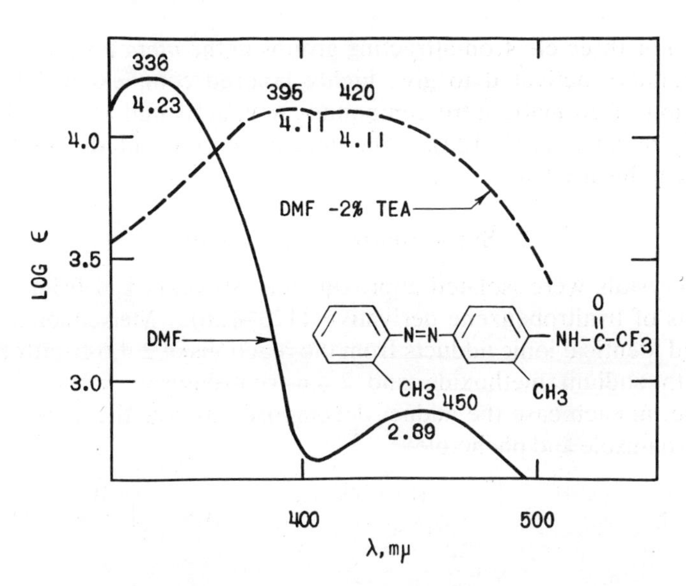

Fig. 10.7. Absorption spectra of *N*-trifluoroacetyl-4-*o*-tolylazo-*o*-toluidine in dimethylformamide (——) and dimethylformamide containing 2% of a 10% aqueous solution of tetraethylammonium hydroxide (– – –).

I. Anions Formed through Anion Addition

1. Introduction

There are many compounds that give brilliant colors in alkaline solution even though they do not contain acidic hydrogen. At least two electron-attracting substituents are usually found in the *meta* positions in these molecules. The long wavelength bands are then classified as anionic resonance bands. These color reactions are related to the reaction of *m*-dinitrobenzene with activated anions in alkaline solution to give brilliantly colored anions. When a small amount of a methanolic potassium hydroxide solution is added to an acetone solution of *m*-dinitrobenzene (λ_{max} 227 mμ), a violet solution (λ_{max} 559 mμ, mϵ 23.1) is obtained(108). The *ortho* and *para* isomers do not give this reaction.

Canback(108, 117) has postulated that various types of *m*-dinitrobenzenes react with a series of anions, such as CH_2COR^-, OR^-, OH^-, NR_2^-, CH_2CN^-, and $CH_2NO_2^-$, to give colored addition products. This type of addition has also been observed with various nitroalkanes that contain the $CHNO_2$ group(118) and the cyanide ion(119) in their interaction with *m*-dinitrobenzenes. This type of addition has been shown to belong to a general reaction involving a compound that contains an aromatic ring with two or three electron-attracting groups in the *meta* positions and an anion suitably activated to give highly colored complexes(120). The interaction of aromatic nitro compounds with activated anions has been critically reviewed(121, 122). The information presented complements the data in this treatise.

2. Sym-*Trinitrobenzene Anions*

Red crystals were isolated approximately 80 years ago from alkaline solutions of trinitrobenzene derivatives(123–126). Meisenheimer(127) prepared identical ionic adducts from the reactions of 2,4,6-trinitrophenetole with sodium methoxide and 2,4,6-trinitroanisole with potassium ethoxide; in each case the adduct decomposed to give the same mixture of trinitroanisole and phenetole:

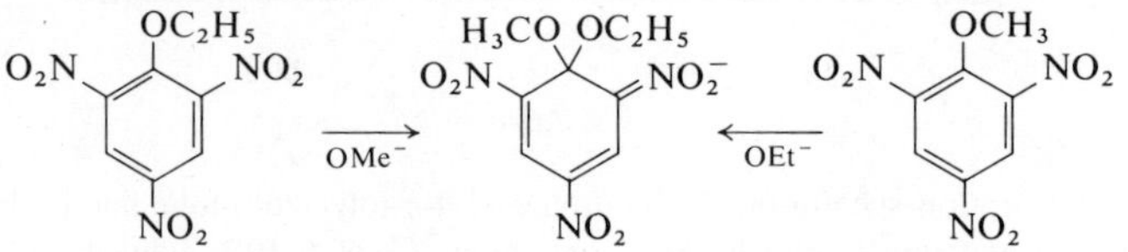

OC_2H_5, O_2N, NO_2, NO_2 — **15** $\xrightarrow{OMe^-}$ H_3CO OC_2H_5, O_2N, NO_2^-, NO_2 — **16** $\xleftarrow{OEt^-}$ OCH_3, O_2N, NO_2, NO_2 — **17** (10.14)

The visible absorption spectra of both complexes have been shown to be identical, with maxima at 420, 478, and 494 mμ(128). Further confirmation has been obtained with proton(75) and nuclear magnetic resonance spectral(129) and crystal structure(73) studies.

In the presence of alkali a ketone with an adjacent CH grouping can form an anion that will readily add to 1,3,5-trinitrobenzene to give the highly colored anion **18**, whose spectrum was obtained in methanol(77).

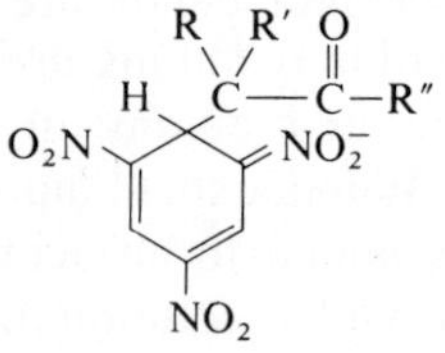

R R′ O
H C—C—R″
O_2N NO_2^-
NO_2

18

$R = R' = H; R'' = CH_3$

λ_{max} 465, 552 mμ, mε 23.9, 11.2

The spectrum of this compound resembles the spectrum of potassium ethyl 1-methoxypicrate in ethanol ($\lambda_{max} \simeq$ 410 and 484 mμ, mε 19.0 and 15.0, respectively), except that the latter compound absorbs at shorter wavelength.

3. *m-Dinitrobenzene and Analogous Anions*

The ease of complex formation with dinitro-substituted aromatic ethers depends on the polarity of the solvent and on the electropositivity of the metal in the methoxide used(74). For 2- or 4-alkoxy-substituted 1,3-dinitrobenzenes that have been studied the attack by alkoxide always occurs at a carbon atom substituted by an alkoxy group(76). The spectral evidence can be summarized in the data for **19**(74) and **22**(76) and in the formation of the ethers by reactions 10.15 and 10.16.

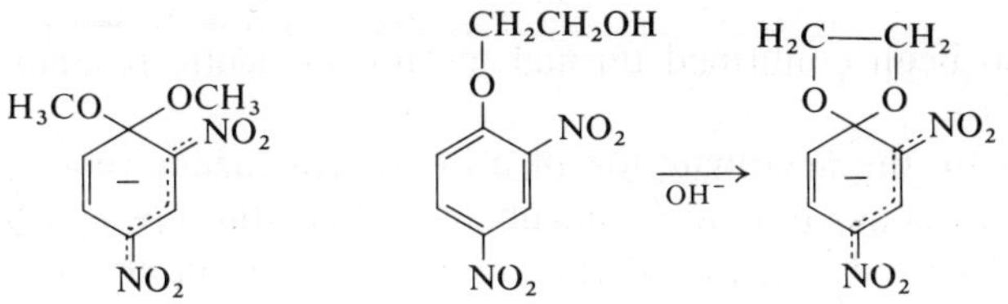

H_3CO, OCH_3, NO_2, NO_2 — **19**; CH_2CH_2OH, O, NO_2, NO_2 — **20** $\xrightarrow{OH^-}$ H_2C—CH_2, O, O, NO_2, NO_2 — **21** (10.15)

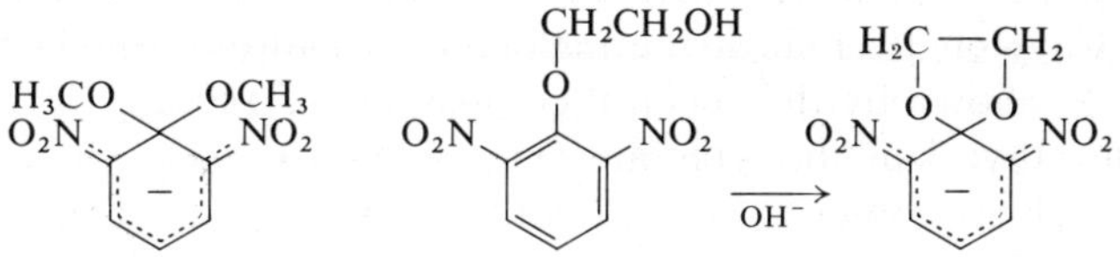

H_3CO, OCH_3, O_2N, NO_2 — **22**; CH_2CH_2OH, O, O_2N, NO_2 — **23** $\xrightarrow{OH^-}$ H_2C—CH_2, O, O, O_2N, NO_2 — **24** (10.16)

No.	Solvent	λ_{max} (mμ)	Ref.
19	DMF	495 (mϵ 20)	74
21	DMSO	~ 500	76, 130
22	DMSO	585	76
24	DMSO	585	76

A novel intramolecular cyclization that occurs when two potential nucleophilic sites are present in the attacking carbanion has been described and is shown in equation 10.17. The addition of triethylamine to a saturated solution of 1,3,5-trinitrobenzene in dibenzylketone or 2,5-pentanedione forms the usual anion adduct, which on standing cyclizes to the dicyclic anion(131). The visible spectra of the latter anions show one absorption band at approximately 500 mμ that is characteristic of a dinitropropenide structure(132).

The longer chain of conjugation between the resonance terminals explains the approximately 90 mμ longer wavelength absorption of **22** and its analog **24** as compared with **19** and its analog **21**. These structures

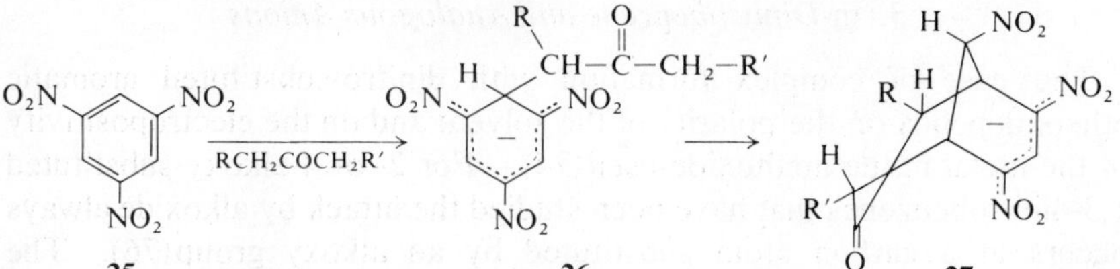

$$\underset{\mathbf{25}}{1,3,5\text{-}(O_2N)_3C_6H_3} \xrightarrow{RCH_2COCH_2R'} \underset{\mathbf{26}\;\lambda_{max}\sim 410,\,550\,m\mu}{[\text{H, CH(R)–C(=O)–CH}_2\text{–R}'\text{ adduct}]^-} \longrightarrow \underset{\mathbf{27}\;\lambda_{max}\,500\,m\mu}{[\text{bicyclic adduct}]^-} \quad (10.17)$$

have also been confirmed through proton magnetic resonance spectral studies (76).

Attack by the acetonate ion of a *m*-dinitrobenzene-type compound is believed to occur only at an unsubstituted position (76). Thus no reaction is observed when triethylamine is added to an acetone solution of 1,3,5-trichloro-2,4,6-trinitrobenzene. Similarly dinitromesitylene in alkaline acetone solution shows no absorption in the visible region (108).

The anionic resonance structures formed by acetonate addition absorb at longer wavelengths than do the analogous structures formed by alkoxide addition, as shown by the spectral comparison of **18** (λ_{max} 552 mμ) with potassium ethyl 1-methoxypicrate (λ_{max} 485 mμ) (77). The EtS^- complexes, which are somewhat more readily formed than the OH^- or CH_3O^- complexes, absorb at somewhat longer wavelengths; this is shown in Table 10.20, which lists the absorption spectra of various 4-substituted complexes of 1,3-dinitrobenzene.

Nuclear magnetic resonance spectral evidence has been presented indicating that the acetonate anion adds to the 4(6) position of *m*-dinitrobenzene (134). However, Pollitt and Saunders (133) have presented absorption spectral evidence that indicates that the acetonate anion adds to *m*-dinitrobenzene in the 2- and 4-positions (equation 10.18), to 2-substituted *m*-dinitrobenzenes in the 4-position (e.g., equation 10.19), and to the 4,6-disubstituted *m*-dinitrobenzenes in the 2-position (e.g., equation 10.20) (133).

$$\underset{\mathbf{28}}{1,3\text{-}(O_2N)_2C_6H_4} \xrightarrow[\text{acetone}]{OH^-} \underset{\mathbf{29}\;\lambda_{max}\,700\,m\mu\text{ in DMF}}{[\text{2-H, 2-CH}_2\text{–C(=O)–CH}_3\text{ adduct}]^-} + \underset{\mathbf{30}\;\lambda_{max}\,580\,m\mu\text{ in DMF}}{[\text{4-H, 4-CH}_2\text{–C(=O)–CH}_3\text{ adduct}]^-} \quad (10.18)$$

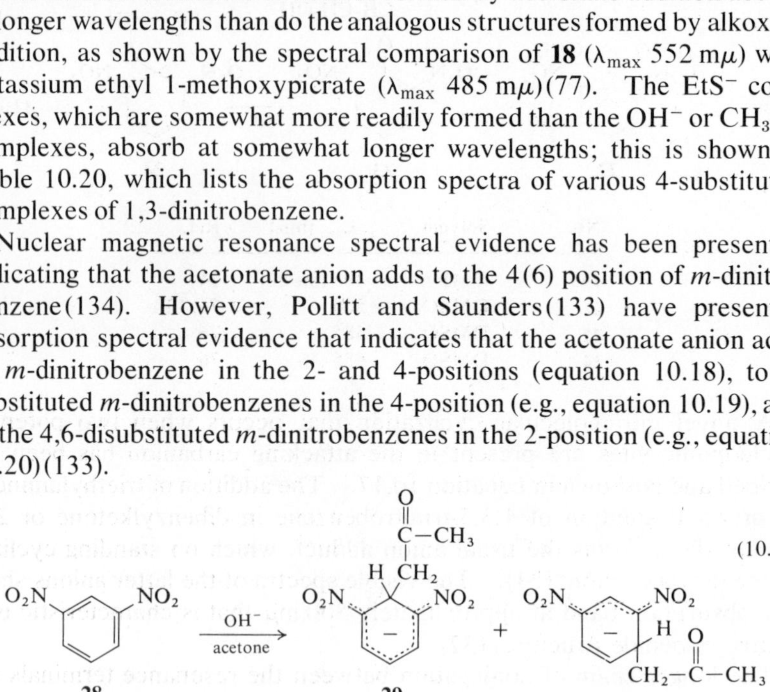

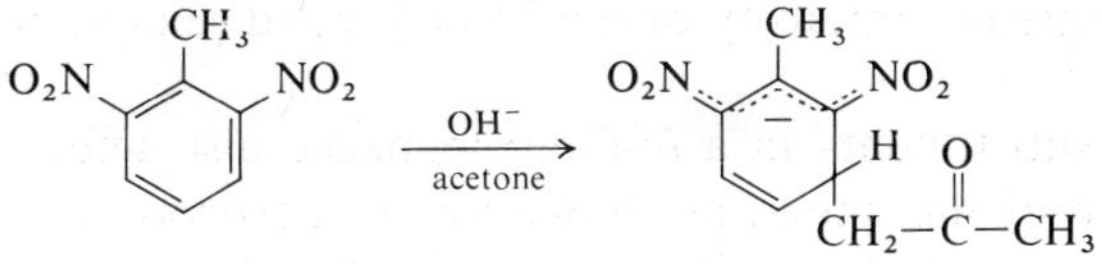

$$\text{31} \xrightarrow[\text{acetone}]{OH^-} \text{32} \qquad (10.19)$$

λ_{max} 555 mμ

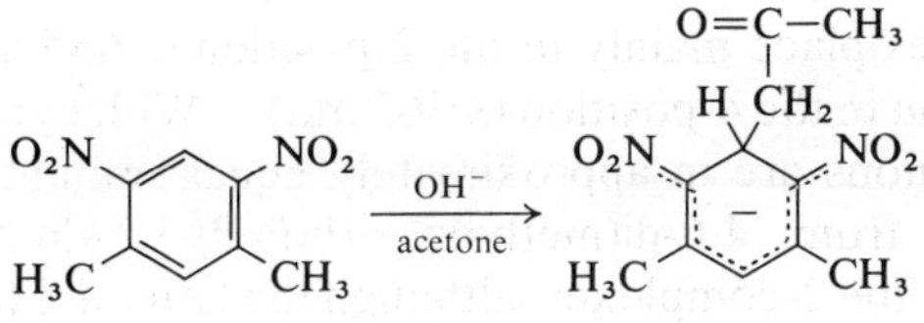

$$\text{33} \xrightarrow[\text{acetone}]{OH^-} \text{34} \qquad (10.20)$$

λ_{max} 651 mμ

The type of the anion also plays a role in the position of substitution on a *m*-dinitrobenzene molecule. Thus in the presence of dimethylformamide and sodium hydroxide the acetonate anion adds predominantly to the 4-position, whereas the dibutyl phosphite anion adds mainly to the 2-position(133). The predominance of the 2-complex is believed to be due to the acidity of the dibutyl phosphite combined with the high reactivity of the phosphonate anion. The predominance of the 2-complex is also

TABLE 10.20
Long Wavelength Maxima of Some 1,3-Dinitrobenzene Anions in Dimethylformamide(133)

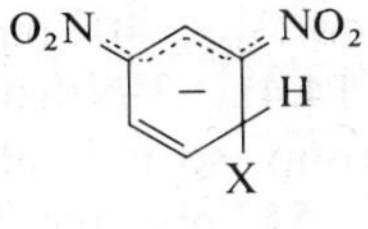

X	λ_{max} (mμ)	X	λ_{max} (mμ)
$-OCH_3$	531	2-oxo-tetrahydrothiophen-3-yl	585
$-OH$	536	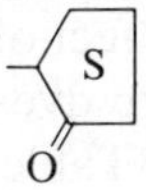	
$-C(CH_3)(CO_2C_2H_5)_2$	563		
$-CH(CO_2C_2H_5)_2$	570		
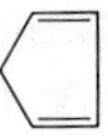 cyclopentenyl	579	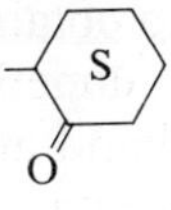 2-oxo-tetrahydrothiopyran-3-yl	585
$-CH_2COCH_3$	580	$-P(O)(O-n\text{-}Bu)_2$	586

shown by the greater intensity of the 715 mμ band relative to the 586 mμ band.

The other substituents in a *m*-dinitrobenzene also affect the position of substitution of the anion, as shown for the absorption spectra of 4,6-di-(methoxycarbonyl)-1,3-dinitrobenzene in dimethylformamide containing sodium hydroxide and an appropriate anion. Thus with dibutyl phosphite or cyclopentadiene substitution takes place in the 4-position mainly, as shown by the prominent band near 550 mμ. With acetone substitution takes place mainly in the 2-position (λ 685 mμ), but there is some substitution in the 4-position (λ 562 mμ). With cyclopentanone the two isomeric anions are in approximately equal amounts. The 4-complexes formed from 4,6-di(methoxycarbonyl)-1,3-dinitrobenzene are more stable than the 2-complexes, although the latter are formed faster.

4. Azo-Dye Anions

In contrast to the 2- and 4-isomers, 3-nitroazobenzenes form brilliantly colored solutions in alkaline dimethylformamide. The acid-weakening effect of a 4-dimethylamino group and the partial canceling of this effect by the twisting of this group out of the plane of the molecule with the help of an adjacent 3-methyl group can be seen in the visible spectra of the 3′-nitroazobenzene derivatives (Fig. 10.8).

3′-Nitro-4-dimethylaminoazobenzene is a very weak acid and is present in alkaline dimethylformamide mainly as the uncombined compound (λ_{max} 450 mμ, mϵ 27.2), with a small amount of the complex, as shown by the shoulder at 560 mμ. A similar solution of 3′-nitro-3-methyl-4-dimethylaminoazobenzene contains approximately equivalent amounts of the uncombined compound (λ_{max} 402 mμ, mϵ 13.4) and the anionic complex (λ_{max} 561 mμ, mϵ 14.4). 3-Nitroazobenzene lacks the acid-weakening dimethylamino group, so it is almost completely present in solution as the complex (λ_{max} 553 mμ, mϵ 26.8). From the position of the wavelength maxima the hydroxyl ion must add to either the 4′- or 6′-positions. A mixture of these two isomers could be present.

The acid-weakening effect of a 4-dialkylamino group is not apparent in many of these 3′-nitro azo dyes when acetone is present, as shown by the high intensities obtained (Table 10.21).

The absorption spectra obtained from the addition of alkali to a solution of *m*-bisazobenzene in dimethylformamide containing 1% acetone is especially striking. A band at 666 mμ increases in intensity, reaches a maximum intensity of about mϵ 70 in 4 min, and then gradually decreases. At the same time a band of about equal intensity at 375 mμ increases in intensity but keeps on increasing even after 18 min.

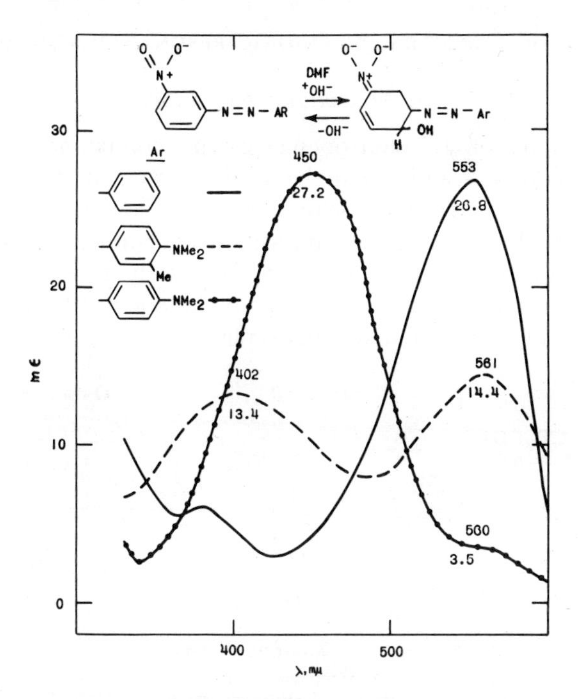

Fig. 10.8. Absorption spectra in dimethylformamide containing 2% of a 10% aqueous solution of tetraethylammonium hydroxide: 3-nitroazobenzene (——); *N,N*-dimethyl-4-3′-nitrophenylazoaniline (· · · · ·); and *N,N*-dimethyl-4-3′-nitrophenylazo-*o*-toluidine (---).

TABLE 10.21
Absorption Spectra of 3-Nitroazobenzene Derivatives and Analogous Compounds in Alkaline Solution[a]

(Structure: 3-NO_2-C_6H_4—N=N—C_6H_4—X, ring positions 1–6)

X	λ_{max} (mμ)[b]	mϵ	X	λ_{max} (mμ)	mϵ
4-$(C_2H_5)_2N$	575	37	3-Acetylazobenzene	~570[c]	
4-$CH_3C_2H_5N$	575	36		~545*s*	
4-$(CH_3)_2N$	575	37	3-Nitrobenzalaniline	Red[c]	
2-CH_3-4-$(CH_3)_2N$	575	37	*m*-Bisazobenzene	666	70
3-CH_3-4-$(CH_3)_2N$	581	33	2-Nitrofluorenone	548	27
H	588	29		577	29

[a]The solvent consists of dimethylformamide containing 2% of a 10% aqueous solution of tetraethylammonium hydroxide and 1% acetone.

[b]The long wavelength bands of these compounds increase in intensity with time, reaching a maximum in 2–5 min and then slowly decreasing in intensity. Since the shorter wavelength bands were much less stable, they are not reported; their intensity increased with time.

[c]Color too unstable for the determination of the wavelength maximum.

5. *Miscellaneous Anions*

The reactions of 1,3-dinitrobenzene-type compounds with various anions in alkaline solution can be fairly complicated. Thus various isomers can be formed in addition to dianion addition, oxidation–reduction reactions, and other types of decomposition and condensation.

Under Zimmerman conditions (approximately equimolar or excess 1,3-dinitrobenzene derivative and a methyl ketone in ethanolic potassium hydroxide) the following reaction takes place(135):

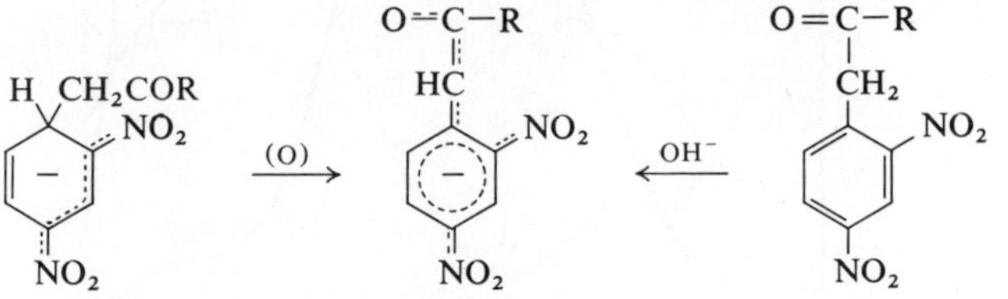

H CH₂COR, NO₂, NO₂ **35** —(O)→ O=C—R, HC, NO₂, NO₂ **36** ←OH⁻— O=C—R, CH₂, NO₂, NO₂ **37** (10.21)

R	λ_{max} (mμ)	mε
CH_3	490	28.0
C_6H_5	515	32.0

The millimolar absorptivities in reaction 10.21 were obtained from the 2,4-dinitrobenzylketones.

Formation of charge-transfer complexes (**38**) has been suggested to account for the frequent observation of rapidly formed transient intermediates,(136–140). The radical anion **39** has been obtained on treatment of 1,3-dinitrobenzene with alkali(141, 142); similar intermediates have been proposed in the reactions of trinitroaromatics(137, 143).

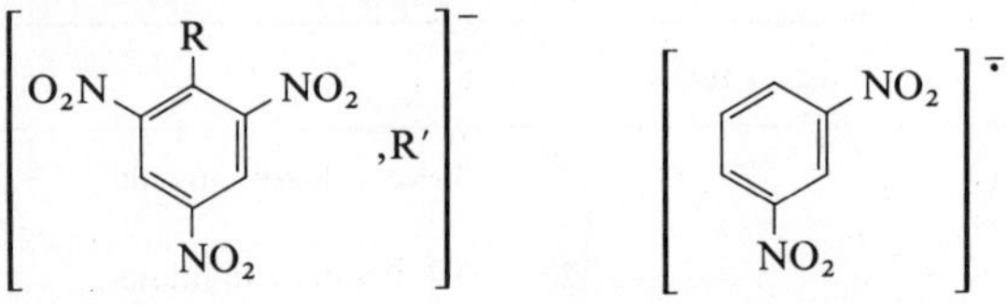

[R, O_2N, NO_2, NO_2 ,R′]⁻ **38**   [NO_2, NO_2]$^{\overline{\cdot}}$ **39**

Zwitterionic intermediates, such as **40**(136, 144, 145) and **41**(75), have also been proposed to account for the reaction of polynitroaromatic compounds with amines. Polyanions, such as **42**(75), **43**(75), **44a**(75, 76) and **44b**(76), have also been reported. The many complex aspects of the interaction of aromatic nitro compounds with bases have been critically discussed(121).

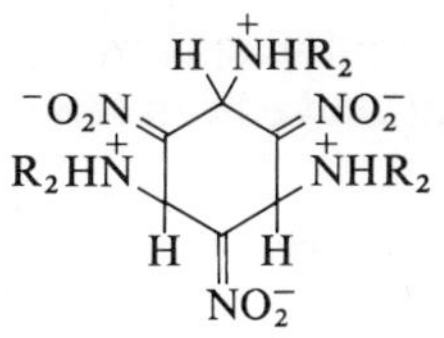

40

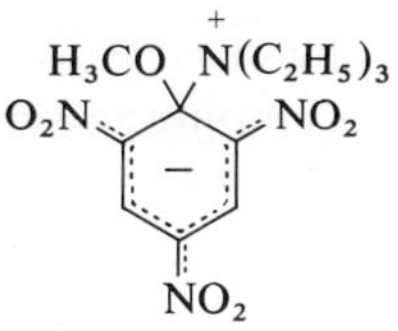

41

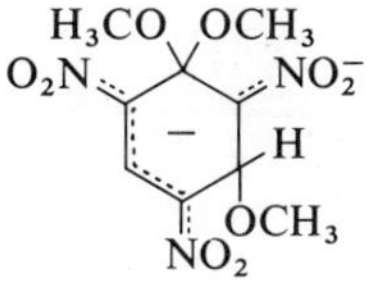

42

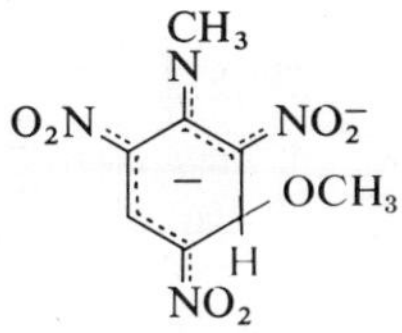

43

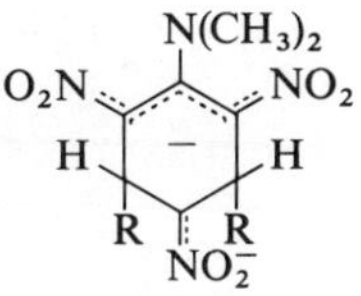

44

a: R = OCH_3
b: R = CH_2COCH_3

Compounds that contain an unsaturated carbon atom (which is not part of a ring and has a low electron density) can form conjugated anions through the addition of a small activated anion, as shown, for example, in the following reaction (146):

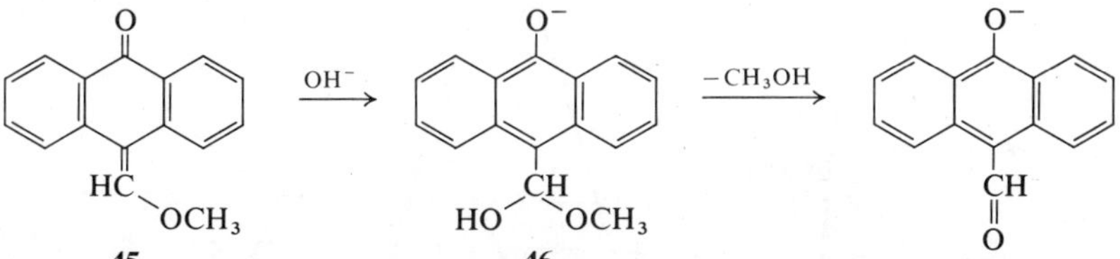

45 —OH^-→ 46 —$-CH_3OH$→ 47

Yellow fluorescence

(10.22)

Compound **46** would be closely similar to the anion of 9-hydroxyanthracene in absorption and fluorescence spectra. However, **46** is unstable and forms the aldehyde derivative after a few minutes.

J. Spectral Comparison of Zwitterionic, Cationic, and Anionic Resonance Structures

Where an NH group is attached to a conjugated system, zwitterionic, anionic, and cationic resonance structures can be formed at the appropriate conditions of alkalinity and acidity. An example is the chromogen **48** (λ_{max} 387 mμ, mϵ 48.6), obtained in the determination of malonaldehyde and its precursors (147). The absorption spectra of this trimethine derivative and its anionic and cationic salts have been reported (148):

$$C_6H_5\overset{+}{N}H{=}CH{-}CH{=}CH{-}NHC_6H_5 \underset{H^+}{\overset{-H^+}{\rightleftarrows}} C_6H_5NH{-}CH{=}CH{-}CH{=}NC_6H_5 \quad (10.23)$$

λ_{max} 389 mμ, mϵ 56.0 λ_{max} 360 mμ, mϵ 42.5

48 $H^+ \upharpoonleft\downharpoonright OH^-$ **49**

$$C_6H_5\bar{N}{-}CH{=}CH{-}CH{=}NC_6H_5$$

λ_{max} 437 mμ, mϵ 61.7

50

In the determination of 1-naphthol with isatin the cation (λ_{max} 600 mμ, m$\epsilon \sim 15$) of 2-(4-hydroxy-1-naphthyl)-3-pseudoindolone is obtained (148). The long wavelength maxima of the neutral, cationic, and anionic structures of this compound in *o*-dichlorobenzene solution shift toward the red in that order (Fig. 10.9)(148). Another compound that shows a similar spectral shift is 1-(4-phenylazophenylazo)-2-anilinonaphthalene (Fig. 10.10). Even in strong acid solution, where a dication is formed,

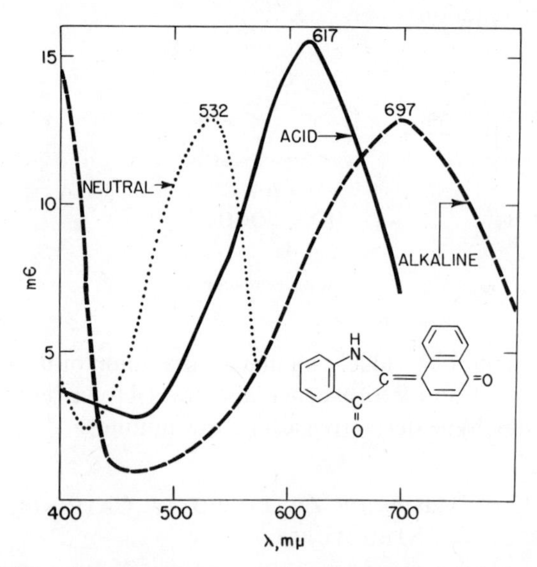

Fig. 10.9. Visible absorption spectra of the chromogen 2-(4-hydroxy-1-naphthyl)-3-pseudoindolone, obtained in the determination of 1-naphthol with isatin (148). In *o*-dichlorobenzene (· · · ·); in *o*-dichlorobenzene saturated with hydrogen chloride (——); and in *o*-dichlorobenzene containing 0.5% of a 40% methanolic solution of benzyltrimethylammonium methoxide (– – –).

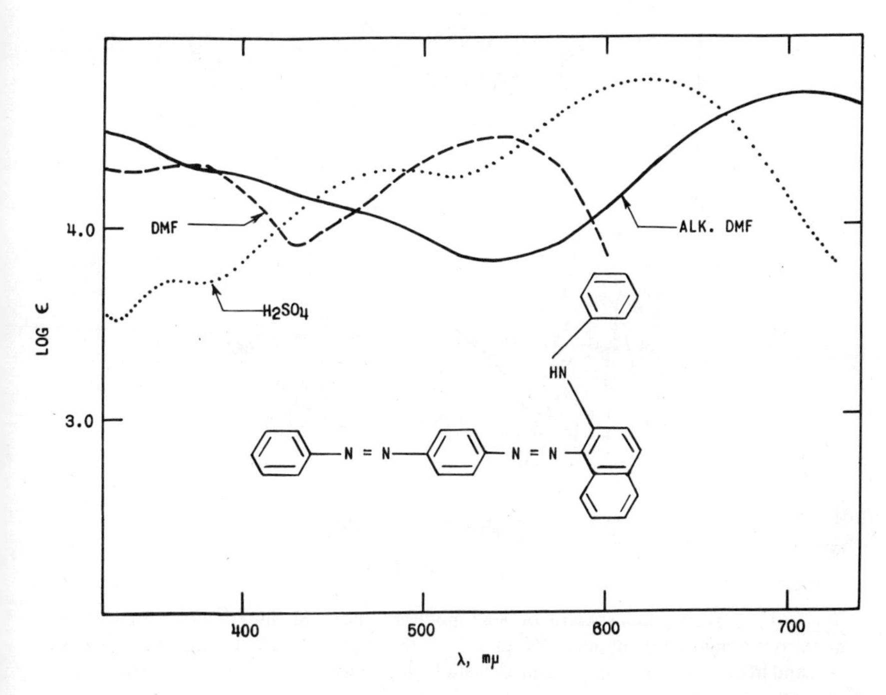

Fig. 10.10. Visible absorption spectra of 1-(4-phenylazophenylazo)-2-anilinonaphthalene in dimethylformamide (– – –), sulfuric acid (· · · ·), and in dimethylformamide containing 2% of a 10% aqueous solution of tetraethylammonium hydroxide (——).

the compound absorbs at shorter wavelengths than the anion. For the more basic amines the cation absorbs at much shorter wavelength than does the neutral compound, whereas the anion absorbs at very much longer wavelength. This phenomenon is shown for 8-aminofluoranthene in Figure 10.11.

For some compounds the red shift is in the following order: neutral < anion < cation. An example is 1,4-naphthoquinone phenylhydrazone (148):

$$\mathbf{51} \xleftarrow{H^+} \mathbf{52} \xrightarrow{OH^-} \mathbf{53} \qquad (10.24)$$

51	52	53
λ_{max} 578 mμ	λ_{max} 464 mμ	λ_{max} 531 mμ

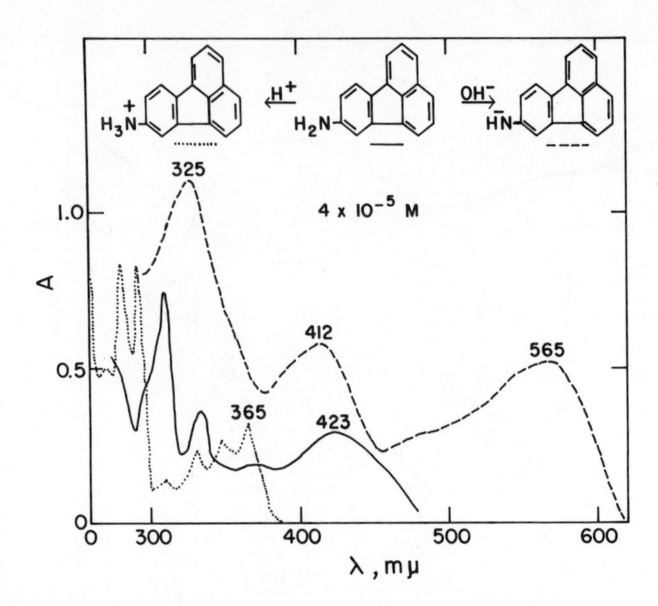

Fig. 10.11. Absorption spectra of 8-aminofluoranthene in dimethylformamide (——), in dimethylformamide containing 1% of 29% methanolic tetraethylammonium hydroxide (– – –), and in dimethylformamide–concentrated hydrochloric acid (1 : 1) (· · · ·)(106).

The spectra were obtained in sulfuric acid, dimethylformamide, and dimethylformamide containing 2% of a 10% aqueous solution of tetraethylammonium hydroxide, respectively.

K. Effect of Cation

The position of absorption of conjugated anions in benzene solution is affected by the structure of the ammonium cation (Table 10.22). The formation of hydrogen bonded ion pairs, where the hydrogen of the ammonium cation is attached to the oxygen of the anion, shifts the long wavelength band toward the violet, as shown for bromophthalein magenta E and tribenzylamine oxide (**54**).

54

TABLE 10.22
Effect of Cations on the Spectra in Benzene of Some Anionic Resonance Structures (67, 149)

Picrate anion			Bromophthalein magenta E anion		
Cation	λ_{max} (mμ)	mϵ	Cation	λ_{max} (mμ)	mϵ
H[a]	340	3.5	H[a]	400	23
$(\phi CH_2)_3\overset{+}{N}OH$	340	15.0	$(\phi CH_2)_3\overset{+}{N}OH$	520	26
	393	7.3			
$(C_2H_5)_4N^+$	364	16.8	$(C_2H_5)_3\overset{+}{N}H$	542	40
	423	8.9	Bu_4N^+	603	81

[a]Neutral phenol.

A change of strong mineral acid does not alter the spectra of arylcarbonium ions, but the absorption wavelengths of conjugated anions vary with the alkali metal cation and with the particular ether used as solvent, the variation increasing with increasing localization of the negative charge in the anion (86, 150). The long wavelength maxima of many anions shift to the red with increasing size of the alkali metal ion (150, 151). An example of this type is di-2-quinolylmethane, which occurs in the colorless (**55**) and red, chelated (**56a**) forms (151).

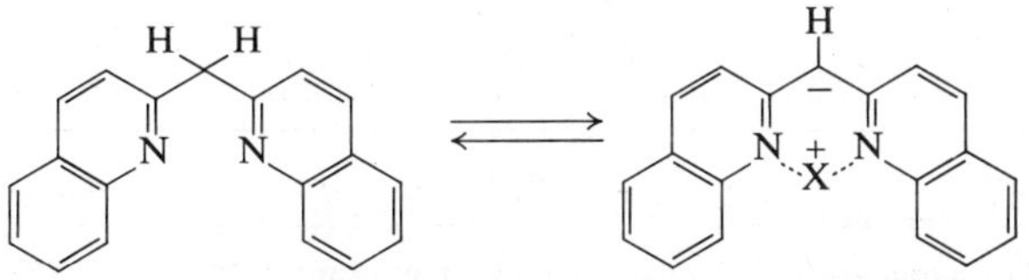

H H N N ⇄ H N + X N (10.25)

55 **56**

a: X = H; **b**: X = alkali metal

Alkali metal chelate complexes can be formed, as in **56b**. As shown in Table 10.23, the long wavelength absorption bands are displaced to the red with increasing size of the alkali metal ion. The metal ion is believed to be localized between the nitrogen atoms. A straight line is obtained when the positions of the absorption maxima are plotted against the ionic radii of the alkali metal ions.

IV. Solvent Effect on Spectra

The position of the long wavelength maxima of conjugated anions is sensitive to the composition of the solvent, as shown by the absorption

TABLE 10.23
Relation between the Ionic Radii of Metal Ions and the Long-Wavelength Maxima of Di-2-quinolylmethane in Dioxane(151)

Ion	Ionic radius (Å)	λ_{max} (mμ)
H^+	–	510
Li^+	0.60	530
Na^+	0.95	543
K^+	1.33	558
Cs^+	1.69	572

spectra of 4′-nitro-4-hydroxyazobenzene in various alkaline solutions (Table 10.24). Since it has been shown that the order of decreasing basicity of some alkaline solvents is dimethylformamide > acetone > *n*-butanol > 2-methoxyethanol > methanol, and since the wavelength maximum of the azo dye in Table 10.24 shifts to the violet in this series, it would appear that an increase in solvent basicity results in a red shift of the long wavelength maximum of the anionic resonance structure. The evidence for this appears to be unequivocal. Thus the anion is at longest wavelength in dipolar aprotic solvents, which have been shown to be

TABLE 10.24
Absorption Spectra of 4′-Nitro-4-hydroxyazobenzene in Alkaline Solution(19)

Solvent[a]	λ_{max} (mμ)	mϵ	Solvent[a]	λ_{max} (mμ)	mϵ
Dimethylformamide	605	60.0	Chloroform	538	35.0
Pyridine	605	49.0	Isoamyl alcohol	520	37.8
N-Methyl-2-pyrrolidone	600	58.8	2-Methoxyethanol	520	35.4
Dimethyl sulfoxide	590	55.6	Formamide	520	33.6
Dimethylacetamide	582	48.0	*n*-Propanol	515	35.0
2-Pyrrolidone	568	40.8	Benzyl alcohol	517	32.0
Acetone	565	51.0	Ethanol	510	35.0
Dimethylaniline	560	40.4	2,2,2-Trichloro-		
Aniline	560	40.0	ethanol	495	26.0
Nitropropane	560	41.4	Ethylene glycol	490	42.0
Dimethylmalonate	560	38.8	Methanol	490	31.0
Nitromethane	555	44.2	Water	485	30.3
Xylene	543	39.0	2,2,2-Trifluoro-		
n-Butanol	540	37.0	ethanol	450	24.0

[a]All solutions contain 2% of a 40% methanolic solution of *N*-benzyltrimethylammonium methoxide.

strongly basic, and at shortest wavelength in hydroxylic solvents, which have been shown to be weakly basic (24).

With increasing basicity of a hydroxylic solvent the long wavelength maximum of the 4-nitrophenylazophenol anion shifts to the red (Table 10.25).

TABLE 10.25
Correlation between Solvent Basicity and the Position and Intensity of the Long Wavelength Maximum of the 4-Nitrophenylazophenol Anion

$O_2N-C_6H_4-N{=}N-C_6H_4-O^-$

Solvent	H_- [a]	λ_{max} ($m\mu$) [b]	$m\epsilon$
Water	12.01	440	30.0
Methanol	12.66	485	31.0
Ethanol	14.57	505	35.6
Isopropanol	16.95	523	39.8
tert-Pentanol	18.09	540	35.6
tert-Butanol	19.14	538	38.6
Dimethylformamide	>20	564	47.3

[a] Solutions containing 0.1 *M* alkali metal alkoxide (24).

[b] Solutions containing 0.1 *M* tetrabutylammonium hydroxide.

There is an abundance of data showing that an increase in the percentage of water in alkaline mixtures of water and dipolar aprotic solvents decreases the solvent basicity (20, 21, 37, 40) and shifts the long wavelength maxima of a large variety of phenolic and other anions to the violet (19). Examples of compounds that show a red shift and intensity increase with an increase in solvent basicity are 4-hydroxychalcone shown in Figure 10.12, and the fluorene derivative in Figure 10.13.

The absorption spectra of acylamino and anilino derivatives of anthraquinone and fluorenone in pyridine containing methanolic potassium hydroxide have been reported (109). The red shift with increasing solvent basicity is shown by these compounds, an example of which is *N*-benzoyl-1-nitro-2-aminoanthraquinone (**57**) (109).

The anions formed from *m*-dinitrobenzene compounds show the same phenomena of red shift with increasing solvent basicity (Table 10.26). The anion formed from addition of acetonate ion to *m*-dinitrobenzene shows a strong red shift, whereas the anion from 2,4-dinitrobenzyl methyl

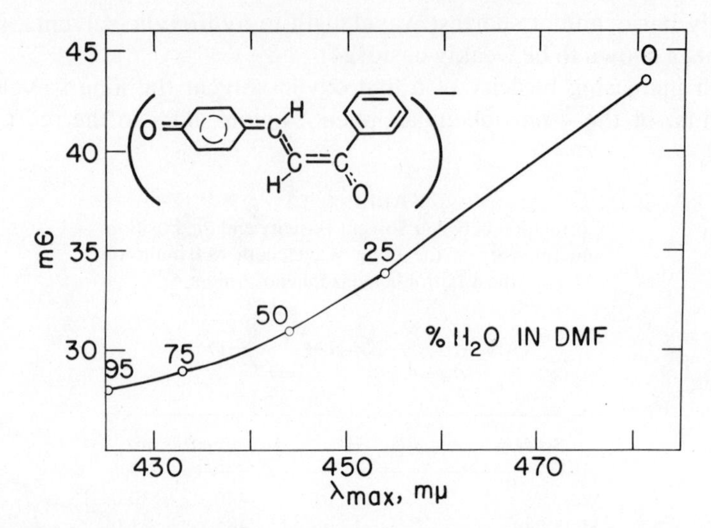

Fig. 10.12. The effect of water on the long-wavelength maximum and millimolar absorptivity of 4-hydroxychalcone in dimethylformamide containing 2% of a 10% aqueous solution of tetraethylammonium hydroxide.

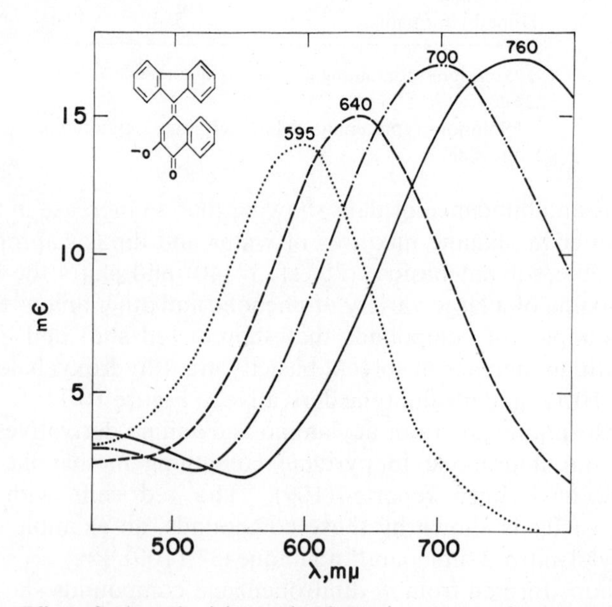

Fig. 10.13. Effect of solvent basicity on the absorption spectra of 4-(9-fluorenylidene)-2-hydroxy-1,4-naphthoquinone in the following solvents containing 2% of 10% aqueous tetraethylammonium hydroxide: 10% aqueous methanol (· · · ·), 2-methoxyethanol (– – –), acetone (— · · —), and dimethylformamide (——).

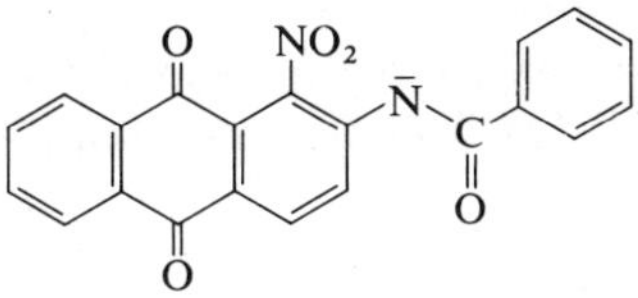

57

Solvent[a]	λ_{max} (mμ)	mϵ
Pyridine	520	9.2
Acetone	505	9.1
Morpholine	485	8.15
Ethanol	470	5.7

[a]Containing methanolic potassium hydroxide.

TABLE 10.26
Solvent Effect on the Spectra of Some *m*-Dinitrobenzene Anions (152,153)

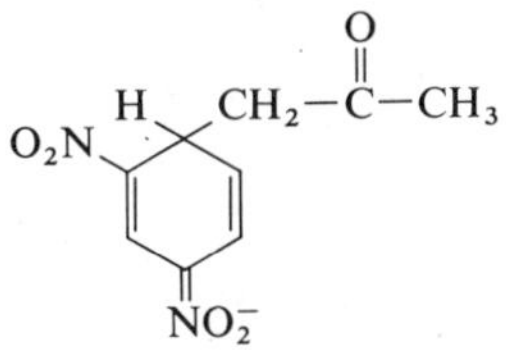

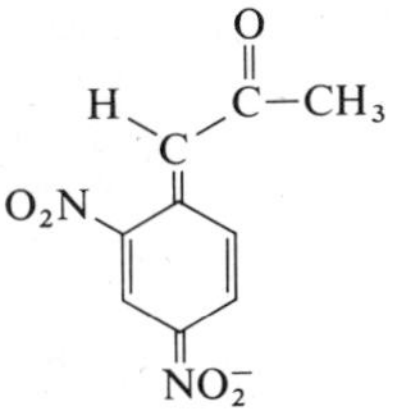

	I		II	
Solvent[a]	λ_{max} (mμ)	mϵ[b]	λ_{max} (mμ)	mϵ[b]
Water	480	6.7	487	0.58
Methanol	536	7.7	485	0.42
Ethanol	555	10.8	490	1.6
Tetrahydrofuran	555	11.9	493	1.7
2-Methoxyethanol	563	9.8	494	15.0
Isopropanol	565	1.9	495	4.0
Acetone	579	19.2	493	5.6
Dimethylformamide	582	18.1	495	28.0
Pyridine	580	5.6	500	4.0

[a]Potassium salt dissolved in solvent.
[b]Millimolar absorptivities are low for many solvents probably due to hydrolysis and decomposition reactions.

ketone shows this effect to a much smaller extent. On the other hand, the trinitrobenzene anions **58** are reported as not showing a red shift with similar solvent changes(152). These interesting exceptions need much further study.

H R

O_2N NO_2

NO_2

58

$R = CH_2COCH_3$ or $CH_2COC_6H_5$

Since many of the compounds that have been discussed in this chapter are weak acids, analysis of these types of chromogens can be affected in terms of sensitivity by the type of solvent system used in the analysis. If the solvent is too weakly basic for a chromogen to ionize completely, the intensity of the anionic band is decreased. This can be seen for 4-phenylazodiphenylamine (Fig. 10.14). Analysis of this chromogen in

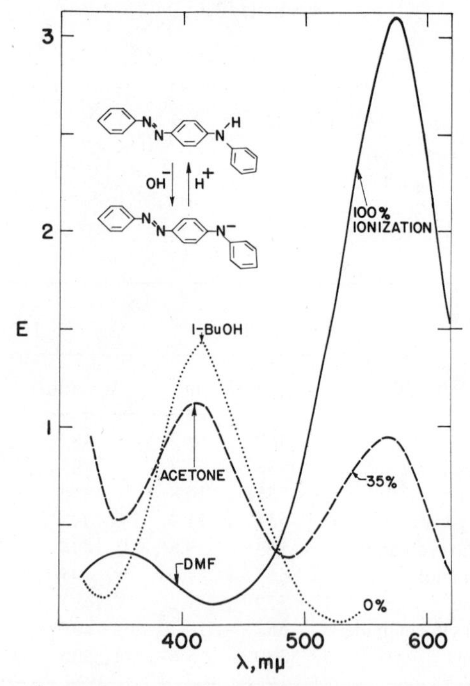

Fig. 10.14. Absorption spectra showing the percentage of ionization of 4-phenylazodiphenylamine (5×10^{-5} M) in solvents containing 2% of 10% aqueous tetraethylammonium hydroxide: dimethylformamide (——), acetone (– – –), and 1-butanol (· · · ·).

solvents less basic than the alkaline acetone of Figure 10.14 would result in a considerable loss of sensitivity due largely to the fact that in weakly basic solvents much of the compound would be present in the neutral form, which absorbs at lower wavelengths.

References

1. J. P. C. M. Van Dongen, A. J. De Jong, H. A. Selling, P. P. Montijn, J. H. Van Boom, and L. Brandsma, *Rec. Trav. Chim.*, **86**, 1077 (1967).
2. A. J. Parker, in R. A. Raphael, E. C. Taylor, and H. Wynberg, Eds., *Advances in Organic Chemistry*, Interscience, New York, 1965, pp. 1–46.
3. A. J. Parker, *Australian J. Chem.*, **16**, 585 (1965).
4. A. J. Parker, *Quart. Rev.* (*London*), 163 (1962).
5. D. J. Cram, B. Rickborn, C. A. Kingsbury, and P. Haberfield, *J. Am. Chem. Soc.*, **83**, 3678 (1961).
6. H. Fainberg, S. G. Smith, and S. Winstein, *J. Am. Chem. Soc.*, **83**, 618 (1961).
7. J. E. Prue and P. J. Sherrington, *Trans. Faraday Soc.*, **57**, 1796 (1961).
8. G. A. Russell, E. G. Janzen, H. D. Becker, and F. J. Smentowski, *J. Am. Chem. Soc.*, **84**, 2652 (1962).
9. W. Bartok, D. D. Rosenfeld, and A. Schriesheim, *J. Org. Chem.*, **28**, 410 (1963).
10. Y. Omote, T. Mikaye, S. Ohmori, and N. Sugiyama, *Bull. Chem. Soc. Japan*, **39**, 932 (1966).
11. K. Maeda and T. Hayashi, *Bull. Chem. Soc. Japan*, **40**, 169 (1967).
12. S. F. Mason and D. R. Roberts, *Chem. Commun.*, (*London*), 476 (1967).
13. L. A. Jones, J. C. Holmes, and R. B. Seligman, *Anal. Chem.*, **28**, 191 (1956).
14. D. J. Cram, C. A. Kingsbury, and B. Rickborn, *J. Am. Chem. Soc.*, **83**, 3688 (1961).
15. E. J. Corey and M. Chaykowski, *J. Am. Chem. Soc.*, **84**, 866 (1962).
16. G. G. Price and M. C. Whiting, *Chem. Ind.* (*London*), 775 (1963).
17. H. E. Zaugg, D. Dunnigan, R. Michaels, L. Swett, T. Wang, A. Sommers, and R. De Net, *J. Org. Chem.*, **26**, 644 (1961).
18. H. E. Zaugg, B. W. Horrom, and S. Borgwardt, *J. Am. Chem. Soc.*, **82**, 2895 (1960).
19. E. Sawicki, T. R. Hauser, and T. W. Stanley, *Anal. Chem.*, **31**, 2063 (1959).
20. R. Stewart and J. P. O'Donnell, *Can. J. Chem.*, **42**, 1681 (1964).
21. R. Stewart and J. P. O'Donnell, *Can. J. Chem.*, **42**, 1694 (1964).
22. R. Stewart and J. P. O'Donnell, *J. Am. Chem. Soc.*, **84**, 493 (1962).
23. D. Dolman and R. Stewart, *Can. J. Chem.*, **45**, 911 (1967).
24. K. Bowden, *Chem. Rev.*, **66**, 119 (1966).
25. C. H. Snyder, *Chem. Ind.* (*London*), 121 (1963).
26. M. F. Grunden, H. B. Henbest, and M. D. Scott, *J. Am. Chem. Soc.*, **85**, 1855 (1963).
27. R. A. M. O'Ferrall and J. H. Ridd, *J. Chem. Soc.*, 5030 (1963).
28. F. Peure and R. Schaal, *Bull. Soc. Chim. France*, 2636 (1963).
29. J. H. Ridd, *Chem. Ind.* (*London*), 1268 (1957).
30. R. Schaal and G. Lambert, *J. Chim. Phys.*, **59**, 1151, 1164, 1170 (1962).
31. N. C. Deno, *J. Am. Chem. Soc.*, **74**, 2039 (1952).
32. R. Schaal and P. Favrier, *Bull. Soc. Chim. France*, 2011 (1959).
33. F. Masure and R. Schaal, *Bull. Soc. Chim. France*, 1138, 1141 (1956).
34. R. Schaal, *J. Chim. Phys.*, **52**, 784, 796 (1955).
35. K. Bowden, *Can. J. Chem.*, **43**, 2624 (1965).
36. J. T. Edward and I. C. Wang, *Can. J. Chem.*, **40**, 399 (1962).

37. R. Stewart, J. P. O'Donnell, D. J. Cram, and B. Rickborn, *Tetrahedron*, **18**, 917 (1962).
38. R. Schaal and C. Gadet, *Bull. Soc. Chim. France*, 2154 (1961).
39. R. S. Stearns and G. W. Wheland, *J. Am. Chem. Soc.*, **69**, 2025 (1947).
40. R. L. Burwell and C. H. Langford, *J. Am. Chem. Soc.*, **82**, 1503 (1960).
41. G. Schwarzenbach and R. Sulzberger, *Helv. Chim. Acta*, **27**, 348 (1944).
42. R. Gaboriaud, B. Monnaye, and R. Schaal, *Compt. Rend.*, **254**, 4027 (1962).
43. C. Jacquinot-Vermesse and R. Schaal, *Compt. Rend.*, **254**, 2679 (1961).
44. J. Hine and M. Hine, *J. Am. Chem. Soc.*, **74**, 5266 (1952).
45. C. Jacquinot-Vermesse, R. Schaal, and P. Rumpf, *Bull. Soc. Chim. France*, 2030 (1960).
46. C. Jacquinot-Vermesse, R. Schaal, and P. Souchay, *Bull. Soc. Chim. France*, 141 (1960).
47. D. Bethel, *J. Chem. Soc.*, 666 (1963).
48. R. G. Boyd, *J. Phys. Chem.*, **67**, 737 (1963).
49. R. Kuhn and D. Rewicki, *Angew. Chem.*, Intern. Ed. **6**, 635 (1967).
50. C. D. Ritchie and R. E. Uschold, *J. Am. Chem. Soc.*, **89**, 1721 (1967).
51. R. G. Pearson and R. L. Dillon, *J. Am. Chem. Soc.*, **75**, 2439 (1953).
52. D. J. Cram, *Chem. Eng. News*, **41**, 92 (1963).
53. K. Bowden and R. Stewart, *Tetrahedron*, **21**, 261 (1965).
54. G. Yagil, *Tetrahedron*, **23**, 2855 (1967).
55. R. Kuhn and D. Rewicki, *Ann.*, **704**, 9 (1967).
56. D. J. Cram, *Fundamentals in Carbanion Chemistry*, Academic Press, New York, 1965, p. 19.
57. E. C. Steiner and J. M. Gilbert, *J. Am. Chem. Soc.*, **87**, 382 (1965).
58. W. H. Okamura and T. J. Katz, *Tetrahedron*, **23**, 2941 (1967).
59. R. P. Bell, *Trans. Faraday Soc.*, **39**, 253 (1943).
60. A. Streitwieser, Jr., J. I. Brauman, J. H. Hammons, and A. H. Pudjaatmaka, *J. Am. Chem. Soc.*, **87**, 384 (1965).
61. F. G. Bordwell, R. H. Imes, and E. C. Steiner, *J. Am. Chem. Soc.*, **89**, 3905 (1967).
62. R. Stewart and D. Dolman, *Can. J. Chem.*, **45**, 925 (1967).
63. W. S. Wooding and W. C. E. Higginson, *J. Chem. Soc.*, 774 (1952).
64. R. P. Bell, *The Proton in Chemistry*, Cornell University Press, Ithaca, N.Y., 1959, Chapters VI and VII.
65. I. M. Kolthoff, M. K. Chantooni, Jr., and S. Bhowmik, *J. Am. Chem. Soc.*, **90**, 23 (1968).
66. I. M. Kolthoff and T. B. Reddy, *Inorg. Chem.*, **1**, 189 (1962).
67. M. M. Davis and M. Paabo, *J. Am. Chem. Soc.*, **82**, 5081 (1960).
68. B. W. Clare, D. Cook, E. C. F. Ko, Y. C. Mac, and A. J. Parker, *J. Am. Chem. Soc.*, **88**, 1911 (1966).
69. T. N. Hall, *J. Org. Chem.*, **29**, 3587 (1964).
70. S. S. Novikov, V. I. Slovetskii, V. A. Tartakovskii, S. A. Shevelev, and A. A. Fainzil'berg, *Doklad. Akad. Nauk SSSR*, **146**, 104 (1962).
71. M. R. Crampton and V. Gold, *Proc. Chem. Soc.* (*London*), 298 (1964).
72. V. Gold and C. H. Rochester, *J. Chem. Soc.*, 1692, 1697, 1722, 1727 (1964).
73. H. Ueda, N. Sakabe, J. Tanaka, and A. Furusaki, *Nature*, **215**, 956 (1967).
74. W. E. Byrne, E. J. Fendler, J. H. Fendler, and C. E. Griffin, *J. Org. Chem.*, **32**, 2506 (1967).
75. K. L. Servis, *J. Am. Chem. Soc.*, **89**, 1508 (1967).
76. C. Foster, C. A. Fyfe, P. H. Emslie, and M. I. Foreman, *Tetrahedron*, **23**, 227 (1967).
77. R. Foster and C. A. Fyfe, *J. Chem. Soc., Sect. B*, 53 (1966).

78. R. G. Douris, *Ann. Chim. (Paris)*, **4**, 479 (1959).
79. P. Balk, G. J. Hoijtink, and J. W. H. Schreurs, *Rec. Trav. Chim.*, **76**, 813 (1957).
80. P. Balk, and G. J. Hoijtink, *J. Am. Chem. Soc.*, **87**, 4529 (1965).
81. J. Eloranta and H. Joela, *Acta Chem. Scand.*, **20**, 1626 (1966).
82. M. J. Janssen and J. De Jong, *Rec. Trav. Chim.*, **86**, 1246 (1967).
83. A. Streitwieser and J. I. Brauman, *J. Am. Chem. Soc.*, **85**, 2633 (1963).
84. C. Porter, *Anal. Chem.*, **27**, 805 (1955).
85. R. Kuhn and H. Fischer, *Angew. Chem.*, **76**, 146 (1964).
86. R. Grinter and S. F. Mason, *Proc. Chem. Soc. (London)*, 386 (1961).
87. W. A. Sheppard and R. M. Henderson, *J. Am. Chem. Soc.*, **89**, 4446 (1967).
88. W. J. Middleton, E. L. Little, D. D. Coffman and V. A. Engelhardt, *J. Am. Chem. Soc.*, **80**, 2795 (1958).
89. J. K. Williams, *J. Am. Chem. Soc.*, **84**, 3478 (1962).
90. M. J. Murray, *Anal. Chem.*, **21**, 941 (1949).
91. J. C. Demetrius and J. E. Sinsheimer, *J. Am. Pharm. Assoc., Sci. Ed.*, **49**, 522 (1960).
92. M. P. Babkin, *Zh. Anal. Khim.*, **19**, 1271 (1964).
93. E. F. G. Herington and W. Kynaston, *Trans. Faraday Soc.*, **53**, 138 (1957).
94. L. Ruzzier and A. Tundo, *Boll. Sci. Fac. Chim. Ind. Bologna*, **14**, 6 (1956).
95. V. P. Kreiter, W. A. Bonner, and R. H. Eastman, *J. Am. Chem. Soc.*, **76**, 5770 (1954).
96. H. Shimura, *J. Chem. Soc. Japan*, **76**, 867 (1955).
97. K. S. Dodgson and B. Spencer, *Biochem. J.*, **53**, 444 (1953).
98. C. Pedersen, *J. Am. Chem. Soc.*, **79**, 5014 (1957).
99. C. J. P. Spruit, *Rec. Trav. Chim.*, **68**, 309 (1949).
100. A. Albert, D. Brown, and G. Cheeseman, *J. Chem. Soc.*, 474 (1951).
101. T. Geissman and J. Harborne, *J. Am. Chem. Soc.*, **78**, 832 (1956).
102. J. Barltrop and K. Morgan, *J. Chem. Soc.*, 4245 (1956).
103. J. N. Smith, B. Spencer, and R. T. Williams, *Biochem. J.*, **47**, 284 (1950).
104. C. M. Judson and M. Kilpatrick, *J. Am. Chem. Soc.*, **71**, 3110 (1949).
105. P. Grammaticakis, *Bull. Soc. Chim. France*, **18**, 534 (1951).
106. E. Sawicki, H. Johnson, and K. Kosinski, *Microchem. J.*, **10**, 72 (1966).
107. Y. Hirshberg and R. N. Jones, *Can. J. Res.*, **27B**, 437 (1949).
108. T. Canback, *Farm. Revy*, **48**, 217 (1949).
109. W. Bradley and E. Leete, *J. Chem. Soc.*, 2129 (1951).
110. E. Sawicki, T. R. Hauser, T. W. Stanley, W. C. Elbert, and F. T. Fox, *Anal. Chem.*, **33**, 1574 (1961).
111. H. Walba and G. E. K. Branch, *J. Am. Chem. Soc.*, **73**, 3341 (1951).
112. C. J. Timmons, *J. Chem. Soc.*, 2613 (1957).
113. G. D. Johnson, *J. Am. Chem. Soc.*, **75**, 2720 (1953).
114. F. Stitt, R. B. Seligman, F. E. Resnick, E. Gong, E. L. Peppen, and D. A. Forss, *Spectrochim. Acta*, **17**, 51 (1961).
115. L. A. Jones and C. K. Hancock, *J. Am. Chem. Soc.*, **82**, 105 (1960).
116. L. A. Jones and N. L. Mueller, *J. Org. Chem.*, **27**, 2356 (1962).
117. T. Canback, *Svensk. Farm. Tid.*, **54**, 1 (1950) and previous publications.
118. C. A. Fyfe, *Can. J. Chem.*, **46**, 3047 (1968).
119. E. Buncel, A. R. Norris, and W. Proudlock, *Can. J. Chem.*, **46**, 2759 (1968).
120. E. Sawicki, *Anal. Chem.*, **24**, 1204 (1952).
121. E. Buncel, A. R. Norris, and K. E. Russell, *Quarterly Rev., (London)*, **22**, 123 (1968).
122. R. Foster and C. A. Fyfe, *Rev. Pure Appl. Chem.*, **16**, 61 (1966).
123. P. Hepp. *Ann.*, **215**, 344 (1882).
124. V. Meyer, *Ber.*, **27**, 3153 (1894).

125. C. A. Lobry de Bruyn, *Rec. Trav. Chim.*, **14**, 39 (1885).
126. A. Hantzch and H. Kissel, *Ber.*, **32**, 3137 (1899).
127. J. Meisenheimer, *Ann.*, **323**, 205 (1902).
128. R. Foster, *Nature*, **176**, 746 (1955).
129. M. R. Crampton and V. Gold, *J. Chem. Soc.*, 4293 (1964).
130. R. J. Pollitt and B. C. Saunders, *J. Chem. Soc.*, 1132 (1964).
131. R. Foster, M. I. Foreman, and M. J. Strauss, *Tetrahedron Letters*, 4949 (1968).
132. R. Foster and C. A. Fyfe, *Tetrahedron*, **21**, 3363 (1965).
133. R. J. Pollitt and B. C. Saunders, *J. Chem. Soc.*, 4615 (1965).
134. C. A. Fyfe and R. Foster, *Chem. Commun.*, 1219 (1967).
135. R. Foster and R. K. Mackie, *Tetrahedron*, **18**, 1131 (1962).
136. R. Foster, *J. Chem. Soc.*, 3508 (1959).
137. R. Foster and R. K. Mackie, *J. Chem. Soc.*, 3843 (1962).
138. R. E. Miller and W. F. K. Wynne-Jones, *J. Chem. Soc.*, 2375 (1959).
139. J. B. Ainscough and E. F. Caldin, *J. Chem. Soc.*, 2528, 2540, 2546 (1956).
140. C. R. Allen, A. J. Brooks, and E. F. Caldin, *J. Chem. Soc.*, 2171 (1961).
141. G. A. Russell, E. G. Janzen, and E. T. Strom, *J. Am. Chem. Soc.*, **86**, 1807 (1964).
142. P. H. Rieger and G. K. Fraenkel, *J. Chem. Phys.*, **39**, 609 (1963).
143. R. Foster and R. K. Mackie, *Tetrahedron*, **16**, 119 (1961).
144. G. N. Lewis and G. T. Seaborg, *J. Am. Chem. Soc.*, **62**, 2122 (1940).
145. M. M. Labes and S. D. Ross, *J. Org. Chem.*, **21**, 1049 (1956).
146. L. Nedelec and J. Rigaudy, *Bull. Soc. Chim. France*, 1204 (1960).
147. E. Sawicki, T. W. Stanley, and H. Johnson, *Anal. Chem.*, **35**, 199 (1963).
148. E. Sawicki, T. W. Stanley, T. R. Hauser, and R. Barry, *Anal. Chem.*, **31**, 1664 (1959).
149. M. M. Davis and H. B. Hetzer, *J. Am. Chem. Soc.*, **76**, 4247 (1954).
150. H. V. Carter, B. J. McClelland, and E. Warhurst, *Trans. Faraday Soc.*, **56**, 455 (1960).
151. H. J. Friedrich and G. Hohlneicher, *Angew. Chem., Intern. Ed.*, **1**, 330 (1962).
152. M. Kimura, M. Kawata, and M. Nakadate, *Chem. Ind.* (*London*), 2065 (1965).
153. M. Kimura, M. Kawata, M. Nakadate, N. Obi, and M. Kawazoe, *Chem. Pharm. Bull.*, **16**, 634 (1968).

CHAPTER 11

ABSORPTION SPECTRA OF ANIONIC RESONANCE STRUCTURES II. INTRAMOLECULAR WAVELENGTH AND INTENSITY DETERMINANTS

I. Steric Hindrance

A decrease in the uniplanarity of a conjugated system containing alternating single and multiple bonds by the substitution of bulky groups in the molecule results in changes in the electronic absorption spectrum. One type of spectral steric effect involves transitions between nonplanar ground states and excited states twisted even further out of planarity. For example, 4-nitrobenzylcyanide has a λ_{max} of 262 mμ, mϵ 66.0, in ethanol and a λ_{max} of 560 mμ, mϵ 25.7, in 2 *N* ethanolic sodium ethoxide solution, whereas the sterically hindered 4-nitro-3,5-dimethylbenzylcyanide has a λ_{max} of 250 mμ, mϵ 14.8, in ethanol and a λ_{max} of 520 mμ, mϵ 13.5, in the alkaline ethanolic solution(1). Thus in both the neutral compounds and their anions the sterically hindered compound absorbs at shorter wavelength and lower intensity.

In some compounds the neutral compound has more steric hindrance in the excited state than does its anion (Table 11.1). Thus substitution of large alkyl groups *ortho* to a phenolic hydroxy group causes a red shift and an intensity increase in comparison with the analogous nonalkylated anion. The decrease in steric crowding can be seen in the ionization of 3,5-di-*tert*-butyl-4-hydroxybenzaldehyde (**1**). In the excited state the bond between the benzene ring and the phenolic oxygen would be shortened, and the steric crowding would be decreased further.

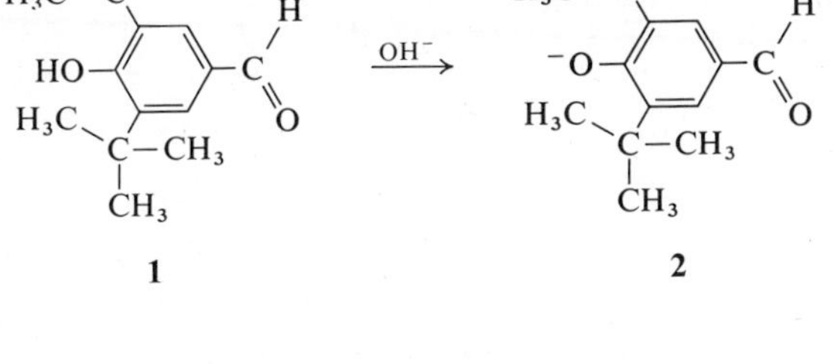

$$\mathbf{1} \xrightarrow{OH^-} \mathbf{2} \qquad (11.1)$$

TABLE 11.1

Effect of Steric Hindrance on the Spectra of Some Phenols and Their Anions

	Neutral compound			Anion			
Compound[a]	Solvent	λ_{max} (mμ)[b]	mϵ	Solvent	λ_{max} (mμ)[c]	mϵ	Ref.
4-Hydroxybenzaldehyde	0.1 *N* HCl	221	12.0	0.1 *N* NaOH	238	7.5	2
		284	16.0		330	27.9	
3,5-Di-*tert*-butyl-4-hydroxy-	Aqueous acid	233	14.0	Aqueous alkali	258	8.0	3
benzaldehyde		290	12.0		367	32.0	
Ethyl 4-hydroxybenzoate	95% EtOH	255	20.0	0.01 *M* NaOEt	300	25.0	4
3,5-Di-*tert*-butyl-4-hydroxy-	95% EtOH	262	14.1	0.01 *M* NaOEt	240	9.1	4
benzoate					330	29.5	
4-Hydroxybenzonitrile	95% EtOH	247	20.0	0.01 *M* NaOEt	280	20.0	4
3,5-Di-*tert*-butyl-4-hydroxy-	95% EtOH	252	15.1	0.01 *M* NaOEt	232	6.6	4
benzonitrile					287	28.8	
4-Nitrophenol	0.002 *N* HCl	226	7.2	1 *N* NaOH	226*fs*	6.3	5
		317	10.0		402	18.2	
4-Nitro-3,5-xylenol	0.002 *N* HCl	251	2.5	1 *N* NaOH	271	3.6	5
		281	2.2		325	2.2	
		359	1.7		404	7.2	5
4-Methylsulfonylphenol	0.002 *N* HCl	238	15.1	1 *N* NaOH	266	20.0	
		276	0.55			15.5	5
4-Methylsulfonyl-3,5-xylenol	0.002 *N* HCl	243	12.3	1 *N* NaOH	270		
		270*s*	1.3				
		280*s*	0.5				

[a] Interrelations of the benzene bands of benzene compounds discussed (6).

[b] *s* = Shoulder.

[c] *fs* = Fine structure.

Where alkyl groups are substituted *ortho* to a nitro group in a 4-nitrophenol anion, the result is a decrease in intensity (5). On the other hand, in a 4-methanesulfonylphenol the mesomeric effect of the CH_3SO_2— group is not hindered by *o*-methyl groups.

A study has been made of the absorption spectra of phenols in alcohol and in alkaline alcohol (7). Unhindered phenols were completely ionized in 0.01 *N* sodium hydroxide solution. Partially hindered phenols (containing one *ortho tert*-butyl group) were completely ionized in 0.05 *N* NaOH, whereas hindered phenols (containing two *ortho tert*-butyl groups) were not completely ionized even in 0.5 *N* NaOH. In going from alcohol to basic solution six unhindered phenols had a change in wavelength of 17.5 mμ, four partially hindered phenols had a change in wavelength of 20 mμ, and four hindered phenols had a change in wavelength of 31 mμ. Here again the sterically hindered phenols in the excited state are less hindered than the compounds in the ground state. Consequently the red shift is greater for the hindered phenols.

The anion of 4,α,α-trinitrotoluene absorbs at 368 mμ, mϵ 15.9 (8), whereas the anions of other dinitromethanes, such as α,α-dinitrotoluene (9) and 1,1-dinitroethane (10) (λ_{max} 373 and 380 mμ, mϵ 15.1 and 15.9, respectively), absorb at slightly longer wavelength. This indicates that the 4-nitrophenyl group in the trinitro derivative is out of plane with the $(NO_2)_2C$— grouping and that the two latter nitro groups are the actual resonance terminals in all these compounds. A rotated conformer is anticipated from steric considerations because of the proximity of the *ortho* hydrogen atoms to the oxygen atoms of the nitro groups (8,9).

II. Tautomerism

Tautomerism has been studied to only a slight extent in anionic resonance structures. Phloroglucinol anions are believed to be capable of tautomerism as shown in equation 11.2 (11).

HO, OH, OH **3** —pK 8·0→ HO, O⁻, OH **4** ⇌ HO, O⁻, O **5** —pK 9·7→ ⁻O, O⁻, O **6**

(11.2)

No.	pH	λ_{max} (mμ)
3	6.3	214, 223*s*, 267
4 → **5**	8.68	224*s*, 254, 277, 350
6	11.0	227, 254, 350

Comparison with the spectrum of the dianion (7) of filicinic acid (12) in 0.01 *N* sodium hydroxide solution indicates that the equilibrium is primarily on the aliphatic side.

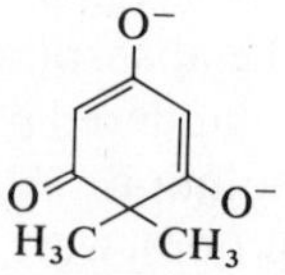

O⁻

O O⁻

H_3C CH_3

7

λ_{max} 250, 350 mμ, mε 25.0, 10.0

Since the anions of the methylated derivatives of benzohydroxamic acid all exhibit only one absorption band, whereas the benzohydroxamic acid anion (λ_{max} 215, 268 mμ; mε 9.5, 5.4, respectively in 0.005 *N* NaOH) and its *para*-substituted derivatives exhibit two bands, it has been postulated that these latter bands are derived from tautomers (13).

III. Resonance Nodes

In anionic resonance structures, as in cationic and zwitterionic resonance structures, negativizing a negative node or positivizing a positive node causes a violet shift, whereas positivizing a negative node or negativizing a positive node causes a red shift in the long wavelength band. Some examples of these effects are shown in Table 11.2 for pairs of compounds in which the CH group is replaced by a nitrogen atom. When a CH group at a negative resonance node is replaced by a more negative grouping (e.g., N), the long wavelength maximum of an anionic resonance structure shifts to the violet (Table 11.2). Such a substitution at a positive resonance node causes a red shift.

Similar effects are shown for the derivatives of 4′-nitro-4-hydroxystilbene in Table 11.3. Thus negativizing the negative node by replacing a CH group by a nitrogen atom causes a violet shift, and negativizing the positive node causes a red shift to 448 mμ. The azo derivative is an exception to the rule. This may be due to the fact that the azo group is highly electronegative and consequently the β-nitrogen atom may have more of the properties of a negative resonance terminal than a negative resonance node.

Replacement of the hydrogen in a CH resonance node by electronegative and electropositive groups can cause red or violet shifts in the long wavelength maximum, depending on the factors already discussed. Thus it can be seen from Table 11.4 that electronegative groups substituted at the positive resonance node shift the wavelength maximum to the red,

TABLE 11.2
Spectral Effect of Replacing a Chain CH Group by a Nitrogen Atom

Anion[a]	X = CH				X = N			
	Solvent	λ_{max} (mμ) or color	mϵ	Ref.	Solvent	λ_{max} (mμ) or color	mϵ	Ref.
Negativizing a Negative Node								
$C_6H_5-\bar{X}-C_6H_5$	DMSO	454	29.0	14	DMSO	376	28.9	14
(fluorenyl-type, $\bar{X}$ bridging two benzene rings)	DMSO	533*fs*[b]	1.0	15	DMSO	402*fs*[b]	2.2	16
$O_2N-C_6H_4-\bar{X}-C_6H_4-NO_2$	DMF	765	61.0	[c]	Isopropanol	585	51.3	17
$O_2N-C_6H_3(NO_2)-\bar{X}-C_6H_3(NO_2)-NO_2$		Blue		18, 19		Red orange		18, 19
$O{=}CH(CH{=}CH)_2-\bar{X}-(CH{=}CH)_2CH{=}O$		600		12		550	126.0	12
Negativizing a Positive Node								
$^{-}O-C_6H_4-\overset{+}{X}-C_6H_4-O^{-}$		Red		19		Blue		19
$(C_6H_5)_2\bar{C}-\overset{+}{X}-\bar{N}-CH{=}C(C_6H_5)_2$	DMF	502		20	DMF	528		20

TABLE 11.2 (*cont'd*)

Anion[a]	X = CH				X = N			
	Solvent	λ_{max} (mμ) or color	mϵ	Ref.	Solvent	λ_{max} (mμ) or color	mϵ	Ref.
$\bar{C}-\overset{+}{X}-\bar{N}-CH=C$ (perinaphthylidene groups)	DMF	561	164.0	20	DMF	603	76.0	20
$^{-}C-\overset{+}{X}-\bar{N}-CH=C$ (fluorenylidene groups)	DMF	556	158.0	20	DMF	598	78.9	20

[a]Structures are shown with charge on the resonance node.
[b]*fs* = fine structure.
[c]Unpublished results by E. Sawicki.

TABLE 11.3
Spectral Effect of Substitution at Positive and Negative Resonance Nodes (20)

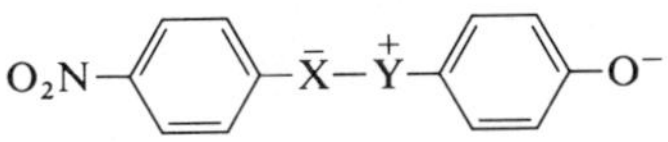

$O_2N-C_6H_4-\bar{X}-\overset{+}{Y}-C_6H_4-O^-$

X	Y	λ_{max} (mμ)[a]
CH	CH	437
N	CH	406
CH	N	448
N	N	505[b]

[a] In alkaline ethanol.
[b] In alkaline dimethylformamide λ_{max} 600 mμ, mϵ 60.2 (21).

TABLE 11.4
Spectral Effect of Substitution on a Carbon Atom
Substitution at a Positive Resonance Node

$(NC)_2\bar{C}-\overset{+}{C}X-\bar{C}(CN)_2$

I

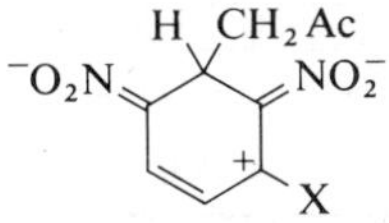

II

I in water (22)			II in acetone–NaOH (23)	
X	λ_{max} (mμ)	mϵ	X	λ_{max} (mμ)
CN	393	22.6	H	692
	412	22.1	CN	705
Br	356	28.3	CO_2CH_3	685
Cl	353	32.0	Br	675
H	344		Cl	666
OC_2H_5	327	29.7	CH_3	665
SCH_3	361	26.1	F	641
NH_2[a]	310	30.5		
$N(CH_3)_2$	310	22.0		
$NHCH_3$	306	22.8		

TABLE 11.4 (*cont'd*)

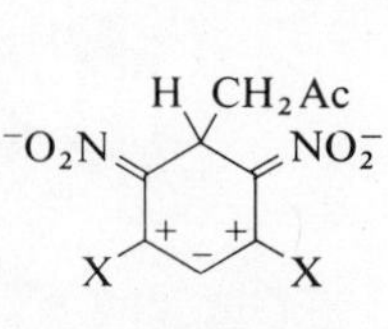

III 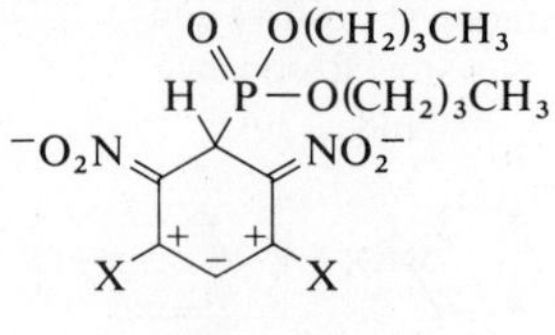 IV 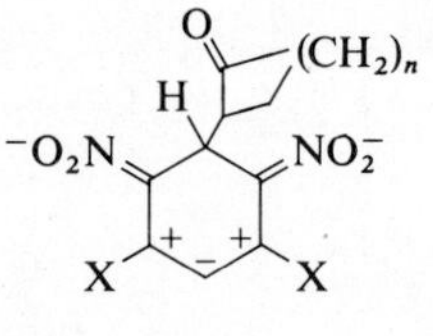 V

X	III in acetone–NaOH	IV in DMF–NaOH	V in DMF–NaOH $n = 2$	V in DMF–NaOH $n = 3$
H	692	715	585	585
CO_2CH_3	685	700	694	660
CH_3	651			
Cl	634	658	618	613
OCH_3	556	561	552	542

λ_{max} (mμ) (23)

Substitution at a Negative Resonance Node

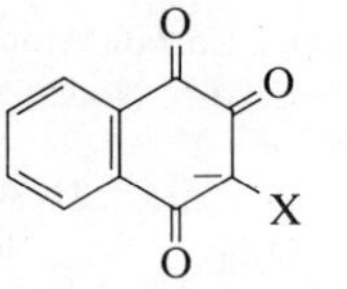

VI 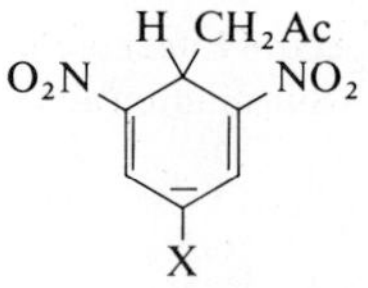 VII

VI (24) X	λ_{max} (mμ)	mϵ	VII in acetone–NaOH (23) X	λ_{max} (mμ)
COC_6H_5[b]	440	2.4	H	692
$CH_2\overset{+}{N}H(CH_3)_2$	445	2.3	CN	631
$CH_2N(CH_3)_2$	468	2.4	CO_2CH_3	625
$(CH_2)_2CHCH_3$	490	2.6	$CONH_2$	660
$CH{=}CHCH_3$[b]	528	2.3		
$CH{=}CHC_6H_5$[b]	548	2.6		

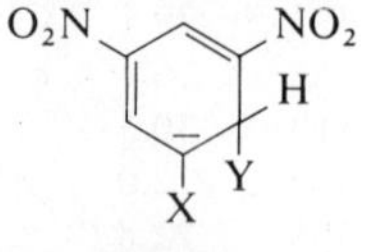

VIII

TABLE 11.4 (*cont'd*)

	λ_{max} (mμ) (23)				
	Y = CH_2Ac[c]		Y = OH	Y = OCH_3	Y = SC_2H_5
X	Acetone–NaOH	DMF–NaOH	DMF–NaOH	DMF–$NaOCH_3$	DMF–$NaSC_2H_5$
H	576	580	536	531	566
NO_2[d]	570		518	516	570
CN	549		502	498	540
CO_2CH_3	561	572	503	508	542
$CONH_2$	580	580	531	525	555
Cl		570	524	514	560
CH_3	580	588			
OCH_3	590	603	560	548	570
$N(CH_3)_2$	614	625	601	594	635
NH_2	624				

[a]In 95% ethanol.
[b]Cross-conjugation is also involved here in addition to the red shift found in all these quinones derived from the coupling effect.
[c]Resonance node is found in the chain of extraconjugation.
[d]Absorption would ordinarily be expected at shorter wavelengths.

whereas electron-donor groups substituted in the same position shift the wavelength maximum to the violet.

On the other hand, substitution of an electronegative group at a negative resonance node causes a violet shift, and, except for the nitro group, the shift increases with increasing electron-acceptor properties of the group. Substitution by an electron-donor group causes a red shift that increases with increasing electron-donor strength of the group (Table 11.4).

Exceptions are the $N(CH_3)_2$ group as compared with the NH_2 and the NO_2 groups. The weaker red shift effect of the stronger electron-donor $N(CH_3)_2$ group as compared with the NH_2 group may be due to the larger spatial requirements of the $N(CH_3)_2$ group. The consequent steric strain could push this group out of the plane of the molecule. The weaker violet shift effect of the NO_2 group may be derived from the symmetry properties of the 1,3,5-trinitrobenzene derivative.

In addition to the compounds shown in Table 11.4 many other 5-substituted 1,3-dinitrobenzenes that are complexed with an anion in the 4-position showed a violet shift in wavelength maxima as the electron-donor strength of the 5-substituent decreased and (except for the nitro group) showed an increased violet shift with an increase of the electron-acceptor strength of the 5-substituent. Thus in the compounds **8** the wavelength maximum shifts to the violet in the order

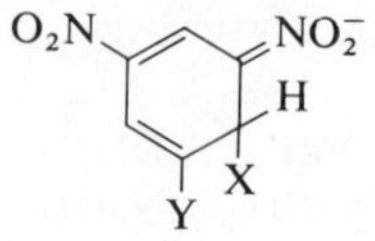

8

$Y = N(CH_3)_2 > OCH_3 > H > Cl > CONH_2 > CO_2CH_3 > CN$

when X^- was any of the following anions (23):

$CH_3COCH_2^-$, HO^-, CH_3O^-, $C_2H_5S^-$, $(C_2H_5O_2C)_2CH^-$, $C_2H_5O_2C(CH_3)C^-$, $(CH_2)_3{-}CO{-}CH^-$,

$(CH_2)_4{-}CO{-}CH^-$, —H, $(n\text{-}BuO)_2(O)P^-$

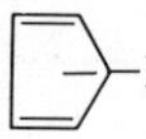

It must be emphasized that the formation of the polynitro anions involves some complexities that are as yet not too well understood. This is because several species are sometimes present in solution at the same time, and some of these species are decomposing or changing with time. However, in spite of these uncertainties, the results obtained are, with few exceptions, in line with what would be expected in negativizing or positivizing a negative node.

The red shift effect of increasing negativization of a positive node in indophenol and its analogs is shown for **9**.

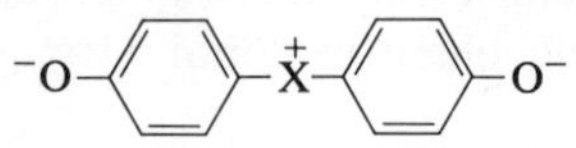

9

X	Solvent	Color or λ_{max} (mμ)	mϵ	Ref.
—CH—		Red		
—N—	Dioxane–NaOH	625	31.6	25
—N(O)—	Alkaline EtOH	672	35.0	26

IV. *Ortho–Para* Effect

In anionic resonance structures that contain a benzene ring the presence of an electron-donor substituent *ortho* or *para* to the resonance terminal directly attached to the ring causes a red shift, the shift increasing with increasing electron-donor strength of the substituent.

Examples of the *ortho* effect can be seen in the diphenylamine anions (**10**)(27):

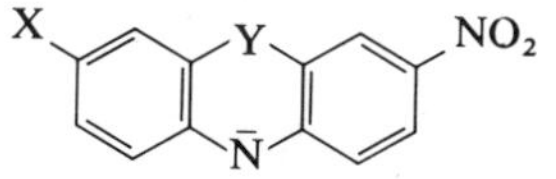

10

X	Y	Color or λ_{max} (mμ)	X	Y	λ_{max} (mμ)
H	H_2	Red	NO_2	H_2	516
H	O	615	NO_2	O	680
H	S	620	NO_2	S	690

p-Nitrophenol anions (**11**) show the *ortho* effect, and *o*-nitrophenol anions (**12**) show the *para* effect.

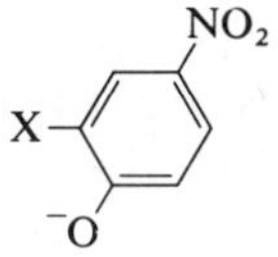

11 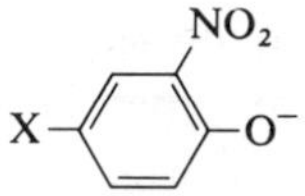 **12**

No.	X	λ_{max} (mμ)	mϵ	Solvent	Ref.
11	H	402	17.0	0.1 *N* NaOH	28
	O^-	510	10.5	0.05 *N* NaOH	29
12	H	412	4.5	0.1 N NaOH	28
	O^-	540	3.9	0.05 N NaOH	29

A weaker electron-donor causes a smaller red shift, as shown by the absorption of 2-nitrophenol at 427 mμ and 2-nitro-4-methylphenol at 455 mμ in alkaline ethanol(30).

The spectral effect of increasing the electron-donor strength of the substituent *para* to the O^- resonance terminal is shown for **13** in alkaline ethanol(30).

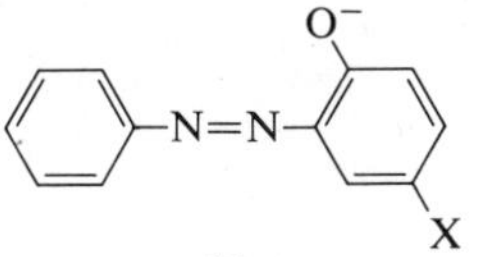

13

X	λ_{max} (mμ)	X	λ_{max} (mμ)
H	460	OCH_3	502
CH_3	481	O^-	572

The *ortho–para* effect is shown by the chromogens obtained by reaction of aromatic diazonium salts with dihydroxybenzenes (30).

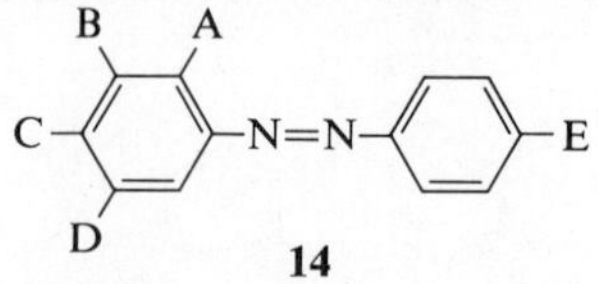

14

A	B	C	D	E	λ_{max} (mμ)
O^-	H	O^-	H	H	473
H	O^-	O^-	H	H	501
O^-	H	H	O^-	H	572
O^-	H	O^-	H	NO_2	574
H	O^-	O^-	H	NO_2	613
O^-	H	H	O^-	NO_2	655

V. Resonance Chain Length

Like the cationic and zwitterionic resonance structures, the anionic resonance structures usually show a red spectral shift with a lengthening of the conjugation chain that connects the two resonance terminals. Examples of such series are shown in Table 11.5. Both convergent and nonconvergent series are shown.

TABLE 11.5
Spectral Effect of Resonance-Chain Length in Anionic Resonance Structures

Compound	Solvent	n	λ_{max} (mμ)	Ref.
$(NC)_2C{=}CH(CH{=}CH)_n\bar{C}(CN)_2$	Methanol	0	343	31
		1	444	
		2	535	
$C_6H_5(CH{=}CH)_n\bar{C}HC_6H_5$[a]		0	501	32
		1	535	
		2	568	
		3	600	
		4	635	
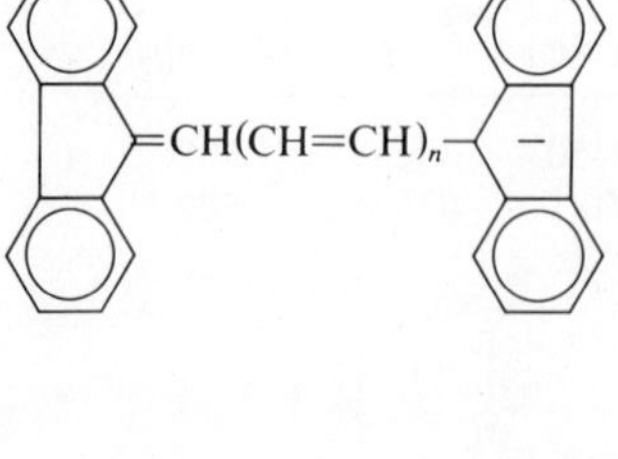 (fluorenylidene)$=CH(CH{=}CH)_n-$(fluorenyl anion)	DMF–NaOH	0	557[b]	33
		1	633[c]	
		2	740	

TABLE 11.5 (*cont'd*)

Compound	Solvent	n	$\lambda_{max}(m\mu)$	Ref.
3-ethylrhodanine–C=CH(CH=CH)$_n$–C (oxazolidinethione, $^-$O) dimethine: S—C=CH(CH=CH)$_n$—C—S; S=C, C=O, N, C_2H_5; $^-$O, C, N, C=S, C_2H_5		0	542	34
		1	613	
		2	714	
O_2N–C_6H_4–$\bar{N}$N=CH(CH=CH)$_n$$C_6H_5$	DMF-TEA[d]	0	553[c]	
		1	575[f]	
$^-$O, NC, C_6H_5 cyclopentadiene –(CH=CH)$_n$CH= furanone (O, O, CN)	DMF	0	605[g]	35
		1	716[h]	
thiobarbiturate (H, N, O, S=, N, H, O) =CH(CH=CH)$_n$– thiobarbiturate anion ($^-$O, H, N, =S, N, H, O)		0	452	36
		1	532	
		2	622	

[a]At −60°C. Stable only at low temperatures.
[b]$m\epsilon$ 63.
[c]$m\epsilon$ 200.
[d]Dimethylformamide containing 2% of 10% aqueous tetraethylammonium hydroxide.
[e]The acetophenone derivative; $m\epsilon$ 48.
[f]$m\epsilon$ 66.
[g]$m\epsilon$ 115.
[h]$m\epsilon$ 120.

Cyanide groups have, at the most, weak resonance properties but a strong inductive effect. Consequently in the symmetrical tetracyano-polymethine anions the end carbon atoms to which the cyanide groups are attached have the highest electron density in the molecule and are the resonance terminals, as, for example, in **15** (31).

$$(NC)_2C^-\text{—}(CH{=}CH)_nCH{=}C(CN)_2 \leftrightarrow (NC)_2C{=}(CH\text{—}CH{=})_nCH\text{—}\bar{C}(CN)_2$$

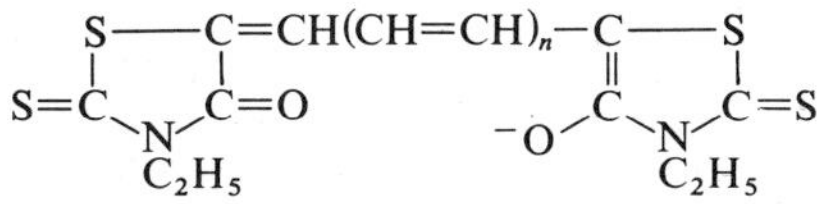

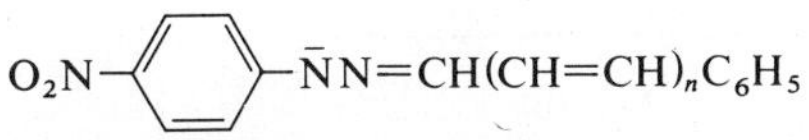

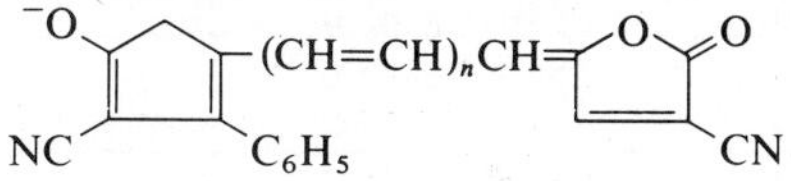

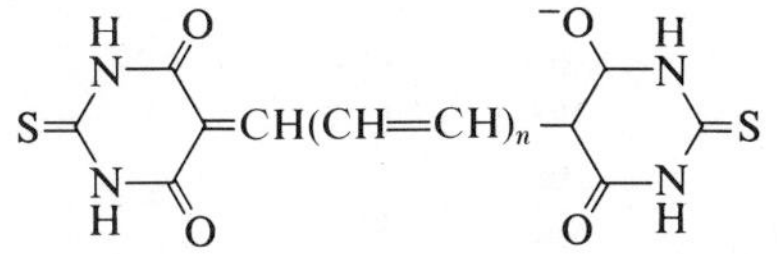

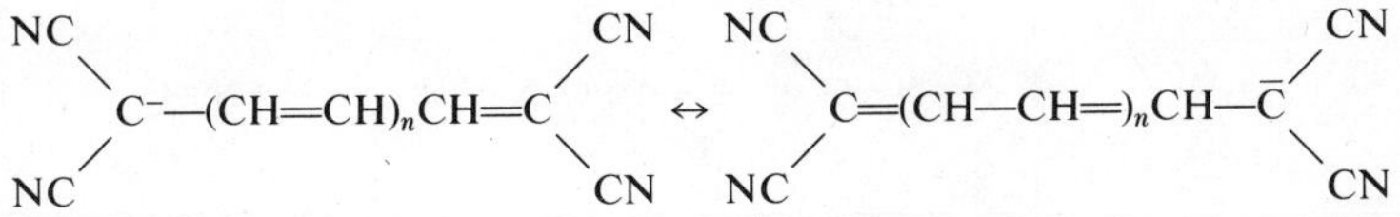

15

As shown by the data in Table 11.6, in the anions of the nitrated phenyl-hydrazones of benzaldehydes the nitro compounds with the longer chain of conjugation absorb at longer wavelength and lower intensity. Thus

TABLE 11.6
Spectra of Nitro Derivatives of Benzaldehyde Phenylhydrazones in Alkaline Dimethylformamide[a]

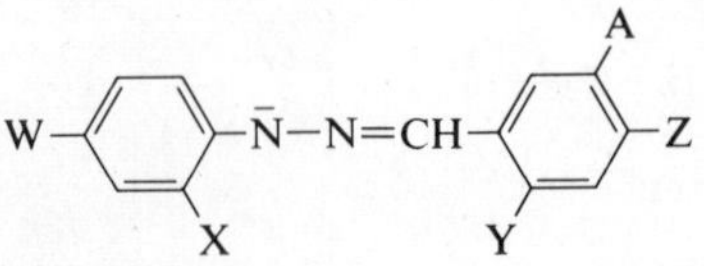

W–C₆H₃(X)–N̄–N=CH–C₆H₃(A)(Y)–Z

W	X	Y	Z	A	λ_{max} (mμ)[b]	mϵ
NO_2	H	H	H	H	553[c]	48.0
H	NO_2	H	H	H	585	13.6
H	H	NO_2	H	H	717	17.0
H	H	H	NO_2	H	689	49.2
H	NO_2	NO_2	H	H	650	18.0
NO_2	H	NO_2	H	H	600	42.6
H	NO_2	H	H	NO_2	585	15.4
NO_2	H	H	H	NO_2	566	57.0
H	NO_2	H	NO_2	H	665	44.8
NO_2	H	H	NO_2	H	651	60.4
NO_2	NO_2	H	NO_2	H	575	34.0
NO_2	NO_2	H	H	NO_2	520	21.0

[a]Containing 2% of a 10% aqueous solution of tetraethylammonium hydroxide.

[b]Spectra of fresh solutions since many of these colors fade with time.

[c]Acetophenone derivative.

2-nitro derivatives absorb at longer wavelength than do the 4-nitro derivatives.

Sometimes a longer chain of conjugation shifts the long wavelength maximum to the violet, as shown for **16** compared with **17** (37). In **16** one of the aza nitrogens becomes more like a resonance terminal than a resonance node.

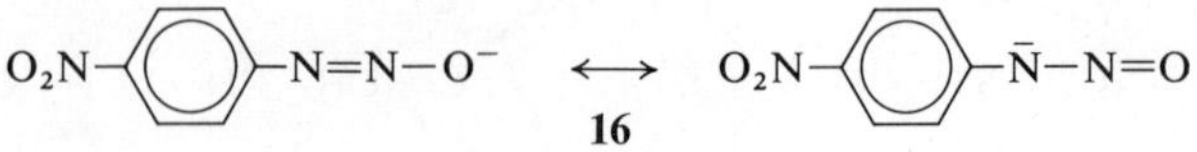

O_2N–C₆H₄–N=N–O⁻ ⟷ O_2N–C₆H₄–N̄–N=O

16

λ_{max} 330 mμ, mϵ 15.0

O_2N–C₆H₄–O⁻ ⟷ ⁻O_2N=C₆H₄=O

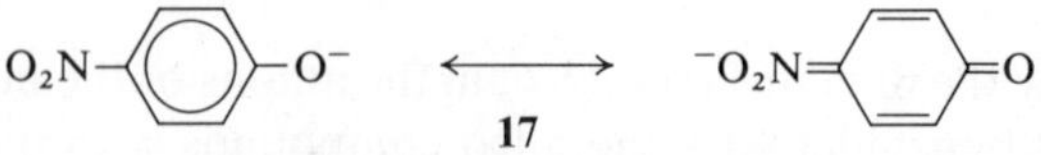

17

λ_{max} 402 mμ, mϵ 15.0

The formation of isomeric conjugated anions from the addition of activated anions to the 2- and 4-positions of 1,3-dinitrobenzene and analogous compounds has been discussed in Chapter 10. The compound formed by addition at the 2-position absorbs at much longer wavelength than the isomer formed by addition at the 4-position. This is because the latter isomer has a shorter chain of conjugation between the resonance terminals. This phenomenon can be seen by comparing structures **19** and **22** in Chapter 10, equations 10.15 and 10.16, equations 10.19 and 10.20, and the spectra of the isomers formed in equation 10.18.

With the same reagent vinylogous series of test substances can be determined through colorimetric measurement of the resulting vinylogous chromogens. An example is 2-thiobarbituric acid, which has been used to determine the following series: formic acid, malonaldehyde, glutaconic dialdehyde, and their precursors (36).

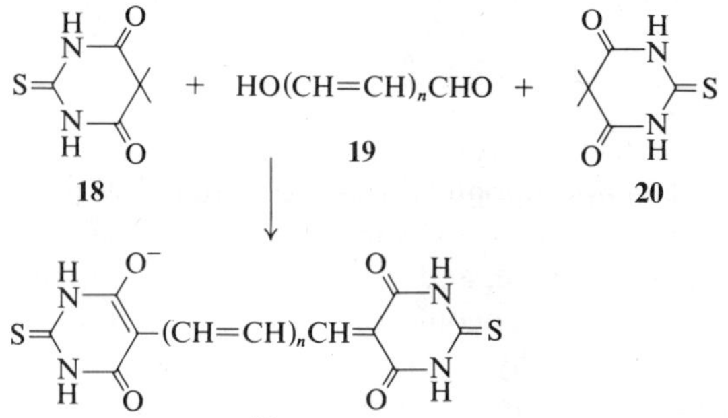

18 + $HO(CH{=}CH)_nCHO$ (**19**) + **20** → **21** [$(CH{=}CH)_nCH{=}$ linking two 2-thiobarbituric acid rings, one as O^- anion]

VI. Resonance Terminals

Increasing the electron-acceptor strength of a resonance terminal in an anionic resonance structure causes a red shift (Table 11.7). An example of electronegative terminal groups with increasing electron-acceptor strength is the series:

$$H < -SO_2CH_3 < -CONH_2 < COOCH_3 < -COCH_3 < -COCF_3 < -NO_2 \simeq -CH{=}C(CN)_2 < -C(CN){=}C(CN)_2 < -SO_2CF_3(?)$$

The electron-acceptor strengths of the groups more strongly electronegative than the nitro groups have been discussed by Sheppard and Henderson (42, 43). The supposedly greater electronegativity of the $-SO_2CF_3$ group than that of the NO_2 group needs more thorough investigation.

TABLE 11.7
Spectral Effect of Increasing Electron-Acceptor Strength of the Resonance Terminals in Anionic Resonance Structures
$X—\bar{C}H—Y$

X	Y	λ_{max} (mμ)	mε	Solvent	Ref.
CH_3SO_2	SO_2CH_3	256*s*	0.0124	$NaOC_2H_5$	38
		< 210	> 1.59		
CH_3CO	CN	252	12.0	0.01 *N* NaOH	39
$C_6H_5SO_2$	$SO_2C_6H_5$	267	7.08	$NaOC_2H_5$	38
$C_6H_5SO_2$	$COOC_2H_5$	275	3.55	$NaOC_2H_5$	38
$C_6H_5SO_2$	$COCH_3$	278	7.59		38
CH_3CO	$COCH_3$	293	21.9	5 *N* NaOH	40
C_6H_5CO	CN	308	9.33	0.01 *N* NaOH	39
C_6H_5CO	$COCH_3$	321	20.0	5 *N* NaOH	40
C_6H_5CO	COC_6H_5	348	21.4	5 *N* NaOH	40
O_2N	NO_2	360	21.9	5×10^{-3} *N* NaOH	41

Unsaturated organic compounds containing OH, SH, NHR, and CHR_2 groups can ionize in appropriately alkaline media. With ionization there will usually be a red shift of the long wavelength band. For example, *p*-nitrophenol, *p*-nitroaniline, and *p*-nitrotoluene absorb at 318, 381, and 284 mμ, respectively, in aqueous solution (2), and at 430, 465, and 520 mμ, respectively, in dimethylformamide containing 1% of a 10% aqueous solution of tetraethylammonium hydroxide (44). In alkaline solution the following structures contribute to the resonance hybrid:

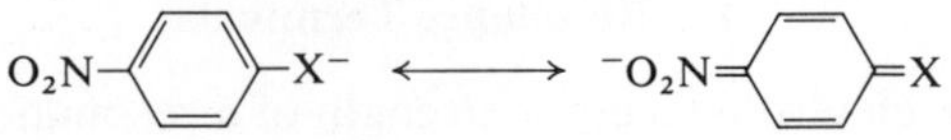

$$O_2N—C_6H_4—X^- \longleftrightarrow {}^-O_2N{=}C_6H_4{=}X$$

A **22** **B**

Such structures as **22A** and **B** are of decreasingly lower and increasingly higher energy, respectively, in the series X = CH_2, NH, O because of the increasing electron-acceptor strength of the groups X = $C{=}CH_2$ < C=NH < C=O and the decreasing electron-donor strength of the series X = CH_2^- > NH^- > O^-.

Nitrophenyl anions have been studied fairly extensively. In the determination of malonaldehyde and its precursors with *p*-nitroaniline and with 4-aminoacetophenone the nitro chromogen **23b** absorbs at much longer wavelength and with greater intensity than the acetyl chromogen **23a** (spectra in dimethylformamide) (44).

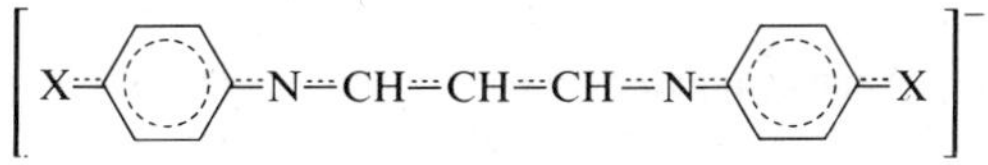

$$\left[X\!=\!\langle\bigcirc\rangle\!=\!N\!=\!CH\!=\!CH\!=\!CH\!=\!N\!=\!\langle\bigcirc\rangle\!=\!X\right]^-$$

23

	X	λ_{max} (mμ)	mϵ
a	Ac	504	66.7
b	NO_2	580	74.6

In comparison with these anions the *p*-nitroaniline anion absorbs at 465 mμ in alkaline dimethylformamide (containing 2% of 10% aqueous tetraethylammonium hydroxide). Acylation of the amino group decreases the electron-donor strength of the negatively charged nitrogen atom, as shown for compounds **24**. This results in a violet shift that is approximately greater the greater the acidity of the molecule and the greater the decrease in the electron-donor strength of the negatively charged nitrogen atom. The long wavelength maxima of the *ortho* and *meta* isomers are at equivalent or shorter wavelengths, with much lower intensity.

$$O_2N-C_6H_4-\bar{N}-X$$

24

X	λ_{max} (mμ)	mϵ	X	λ_{max} (mμ)	mϵ
H	467	32.3	$COOCH_2CF_3$	431	24.6
4-$COC_6H_4NO_2$	460	30.4	4-$SO_2C_6H_4CH_3$	423	26.4
COC_6H_5	457	32.6	4-$SO_2C_6H_4NO_2$	416	25.0
$COOC_2H_5$	448	31.0	$COCF_3$	401	19.9
$COCH_3$	445	25.5	COC_3F_7	401	20.7

The effect of an *N*-acyl group on the absorption spectrum of a 4-aminoazobenzene dye in alkaline solution is striking. As shown in Table 11.8, for *N*-substituted 4-phenylazoaniline anions the spectra shift to the violet in the order:

$C_6H_5 > CH_3 > H > CONHC_2H_5 > COC_6H_5 > COOC_2H_5 > COOCH_2CH_2F > COCH_3 > p\text{-}SO_2C_6H_4CH_3 > COOCH_2CF_3 > COCH_2F > COCF_3$

The *N*-acyl derivatives of 4-(2-tolylazo)-2-toluidine are grouped in Table 11.9 to show the effect of the electronegativity of the acyl group. The wavelength maxima of the acetyl, monofluoroacetyl, and trifluoroacetyl derivatives are at 477, 453, and 420 mμ, respectively. Apparently

TABLE 11.8
4-Aminoazobenzene Dyes in Neutral and Alkaline Solution

$$C_6H_5\text{-}N{=}N\text{-}C_6H_4\text{-}NHX \xrightarrow{OH^-} C_6H_5\text{-}N{=}N\text{-}C_6H_4\text{-}\bar{N}\text{-}X$$

X	Neutral solution[a]		Alkaline solution[b]	
	λ_{max} (mμ)	mϵ	λ_{max} (mμ)	mϵ
C_6H_5	419	21.6	568	63.4
CH_3	410	17.8	554[c]	
H	397	21.7	545[c]	
$CONHC_2H_5$[d]	371	30.4	527	39.6
p-$COC_6H_4NO_2$	360	24.6	489	25.7
COC_6H_5	360	29.6	485	33.5
$COOC_2H_5$	355	19.7	478	35.4
$COCH_3$	356	19.2	467	26.0
SO_2CH_3	348	25.8	455	29.6
p-$SO_2C_6H_4CH_3$	348	21.5	450	24.8
p-$SO_2C_6H_4NO_2$	345	25.8	426	26.8
$COCF_3$	—	—	420	15.6
			395*s*	14.0

[a]Dimethylformamide neutralized with a drop or two of concentrated HCl.

[b]Dimethylformamide containing 2% of a 10% aqueous solution of tetraethylammonium hydroxide.

[c]Not all of the compound was ionized.

[d]Except for the first three compounds, the spectra of all neutral compounds showed $n \rightarrow \pi^*$ bands at approximately 440 mμ with mϵ 0.7–2.0.

the greater the electronegativity of the acyl group, the larger the violet shift of the anion and the larger its decrease in intensity and the smaller the red shift on ionization of the neutral compound. This means that, the more electronegative the acyl group, the more acidic the molecule and the greater the interference with the anionic resonance system. An example of the effect of ionization on the spectrum of an *N*-acyl dye is shown in Figure 11.1.

The spectra of dimethylformamide solutions of 4-phenylazo-1-naphthol dyes (**25**) show a red shift of about 120 mμ on addition of alkali. Increasing electron-acceptor strength of a 4′-substituent in these dyes causes a red shift in the long wavelength band.

All the compounds in Table 11.10, except 2-nitrobenzaldehyde phenylhydrazone, show a red shift of 90–157 mμ on ionization. The exception shows a red shift of 317 mμ. The anion of this compound also absorbs at

TABLE 11.9
Spectral Effect of *N*-Acyl Groups

(2-methylphenyl)–N=N–(3-methylphenyl)–NH–X $\xrightarrow{OH^-}$ (2-methylphenyl)–N=N–(3-methylphenyl)–$\bar{N}$–X

X	Neutral solution[a]		Alkaline solution[b]	
	λ_{max} (mμ)	mϵ	λ_{max} (mμ)	mϵ
H	404	29.0	555	13.0+[c]
$COOC_2H_5$	354	19.6	493	34.8
$COOCH_2CH_2F$	355	18.9	481	28.6
$COCH_3$	355	20.2	467	24.0
$COOCH_2CF_3$	352	18.6	455	22.3
$COCH_2F$	346	16.8	453	17.6
$COCF_3$	337	16.8	420	12.9

[a] Dimethylformamide neutralized with a drop or two of HCl.
[b] Dimethylformamide containing 2% of a 10% aqueous solution of tetraethylammonium hydroxide.
[c] Partially ionized.

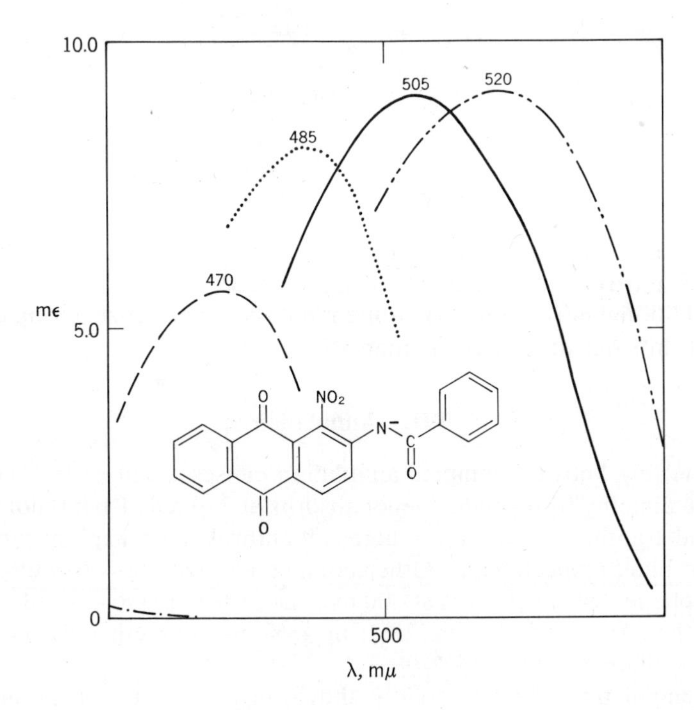

Fig. 11.1. Visible absorption spectra of potassium 1-nitro-2-benzoylaminoanthraquinone in pyridine (—· ·—), acetone(——), morpholine (· · · ·), ethanol (- - - -), and water (— · —).

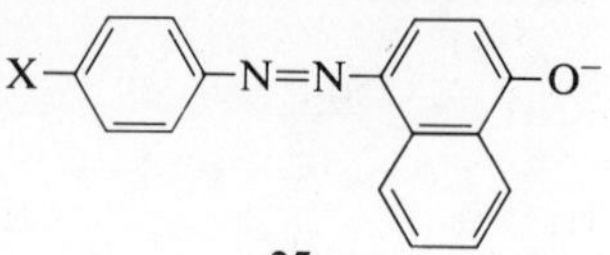

X–C6H4–N=N–(1-naphthyl)–O⁻

25

X	λ_{max} (mμ)	mϵ
CH_3	528	45.4
H	531	44.0
$COOC_2H_5$	575	56.6
$COCH_3$	586	60.0
NO_2	650	80.0

longer wavelength than any other compound in Table 11.10. The addition of nitro groups to the phenylhydrazone portion of the 2-nitrobenzaldehyde phenylhydrazone anion causes a strong violet shift. This may be a general phenomenon, as seen for **26**.

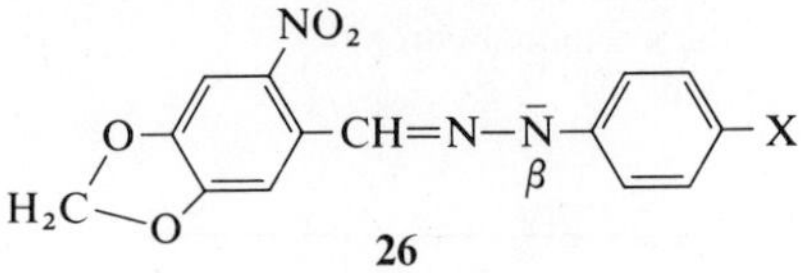

26

X	λ_{max} (mμ)	mϵ
H	703	22.7
2-NO_2	688	24.2
4-NO_2	627	36.0

In the 2-nitro- and 4-nitrobenzaldehyde phenylhydrazones the benzaldehyde nitro oxygen and the β-nitrogen atom are the resonance terminals. Any additional nitro group substituted into the other benzene ring causes a violet shift and an increase in intensity.

VII. Annulation

In the few known examples annulation causes a red shift. Thus in alkaline dimethylformamide phenol absorbs at 310 mμ, 1-naphthol at 351 mμ, and 1-anthrol at 510 mμ, whereas 9-anthrol and 5-naphthacenol are red and blue, respectively. Other compounds that show this effect are *p*-nitrophenol, which absorbs at 400 mμ, mϵ 18.6, at pH 8(45) and 4-nitro-1-naphthol, λ_{max} 455 mμ, mϵ 28.2, in 48% aqueous ethanol containing 0.09 *N* sodium hydroxide(46).

The annulation effect is vividly shown in the spectra of the alkaline solutions of 4-phenylazophenol, 4-phenylazo-1-naphthol, and 10-phenyl-

TABLE 11.10
Absorption Spectra of Some Benzaldehyde Nitro and Dinitrophenylhydrazone Anions

$Ar{-}CH{=}N{-}\bar{N}{-}Ar'$

Substituents		In DMF		In alkaline DMF[a]			
Ar	Ar′	λ_{max} (mμ)	mϵ	λ_{max} (mμ)	mϵ	$\Delta\lambda$ (mμ)	Δmϵ
Phenyl	2-Nitrophenyl	448	9.4	585	13.6	137	4.2
Phenyl[b]	4-Nitrophenyl	413	30.2	553	48.0	140	17.8
Phenyl	2,4-Dinitrophenyl	391	30.2	493	35.8	102	5.6
2-Nitrophenyl	Phenyl	350[c]	14.3	717	17.0	317	2.7
2-Nitrophenyl	2-Nitrophenyl	434	12.6	650	18.0	116	5.4
2-Nitrophenyl	4-Nitrophenyl	418	33.3	600	42.6	182	9.3
2-Nitrophenyl	2,4-Dinitrophenyl	392	26.4	542	34.0	150	7.6
3,4-Dimethoxyphenyl	2-Nitrophenyl	457	10.8	593	13.6	136	2.8
3,4-Dimethoxyphenyl	4-Nitrophenyl	425	36.4	565	55.2	140	18.8
3,4-Dimethoxyphenyl	2,4-Dinitrophenyl	404	30.2	494	36.6	90	6.4
4-$(CH_3)_2NC_6H_4$-CH=CH—	2-Nitrophenyl	465	14.4	606	17.0	141	2.6
4-$(CH_3)_2NC_6H_4$-CH=CH—	4-Nitrophenyl	437	43.6	575	62.0	138	18.4
4-$(CH_3)_2NC_6H_4$-CH=CH—	2,4-Dinitrophenyl	358[c]	25.0	515	34.8	157	9.8
1-Naphthyl	2-Nitrophenyl	450	11.2	600	16.8	150	5.6
1-Naphthyl	4-Nitrophenyl	425	35.0	580	52.6	155	17.6
1-Naphthyl	2,4-Dinitrophenyl	362	27.4	510[d]	30.0	148	2.6

[a]Dimethylformamide containing 2% of 10% aqueous tetraethylammonium hydroxide.
[b]Acetophenone derivative.
[c]Shoulder at 400 mμ.
[d]Also a band at 450 mμ, mϵ 30.0.

azo-9-anthrol, which are orange, violet and blue, respectively (Fig. 11.2). This annulation red shift is related to the increasing stability in the excited state of the quinonoid structures (47). Thus *p*-benzoquinone, 1,4-naphthoquinone, and 9,10-anthraquinone have reduction potentials of 0.71, 0.49, and 0.15 volt, respectively. The anions of 4-aminoazobenzene, 4-phenylazo-1-naphthylamine, and 10-phenylazo-9-anthramine are red, violet, and blue, respectively. Thus they also show the same phenomenon.

In the determination of phenol with diazotized *p*-nitroaniline the obtained azophenol anion absorbs at 485 mμ, mε 27.0, whereas the azo-dye anion from 1-naphthol absorbs at 577 mμ, mε 18.0 (48). The analogous chromogen anions obtained by reaction of *p*-aminoacetophenone with phenol and 1-naphthol absorb at 476 and 526 mμ, respectively, in alkaline ethanol (49).

VIII. Cross-Conjugation

In the 2,4-dinitrophenylhydrazone anions of aldehydes and ketones the two nitro groups are, in some respects, competitive resonance terminals.

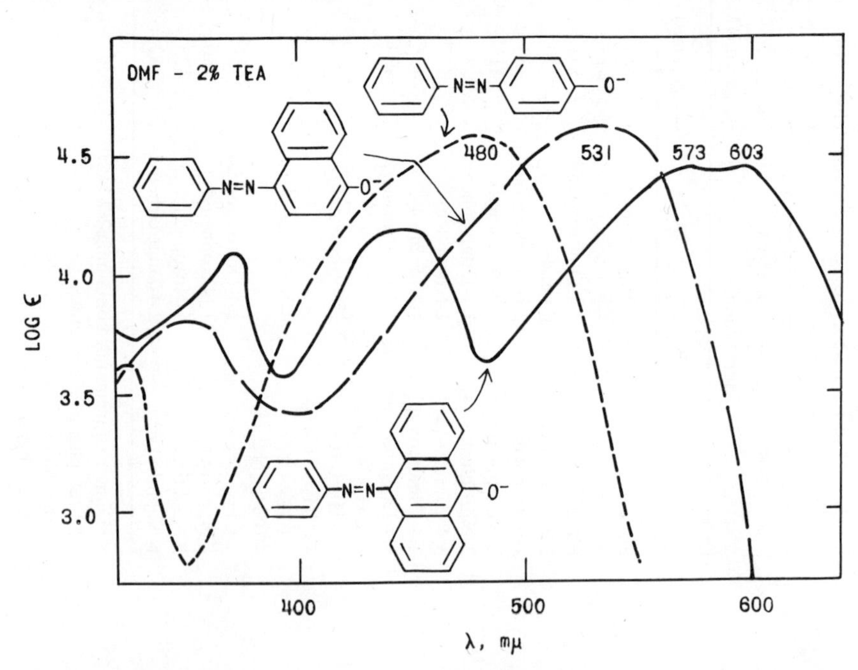

Fig. 11.2. Visible absorption spectra of 4-phenylazophenol (-----), 4-phenylazo-1-naphthol (— —), and 9-phenylazo-10-anthrol (——) in dimethylformamide containing 2% of a 10% aqueous tetraethylammonium hydroxide.

Thus this cross-conjugation causes a violet shift of the long wavelength maxima of the dinitro anions as compared with the analogous bands of 2- and 4-nitrophenylhydrazone anions. The 2-nitro derivatives, which have the longest chain of conjugation, absorb at the longest wavelength with the lowest intensity; the 4-nitro analogs absorb with the greatest intensity, and the 2,4-dinitro analogs absorb at the shortest wavelength (Table 11.10).

Except for the 1-naphthaldehyde derivative, all 2-nitrophenylhydrazones show the smallest increase in intensity on formation of the anion in comparison with the 4-nitro and 2,4-dinitro analogs.

Another effect of cross-conjugation is seen in dinitrophenol anions. In the 2,4-dinitrophenol anions two long wavelength bands are obtained, the less intense longer wavelength one being associated with the 2-nitro oxygen resonance terminal, and the more intense shorter wavelength one being associated with the 4-nitro oxygen resonance terminal. However, both bands show a violet shift effect in comparison with the comparable bands of the anions of 2-nitro- and 4-nitrophenol. Picric acid shows the same phenomenon. The band at longest wavelength is called the *X*-band; the other band is called a *Y*-band.

As is to be expected, the 2,6-dinitrophenolate anion shows the presence only of the long wavelength *X*-band. Lewis and Calvin's *Y*- and *X*-bands are believed to correspond to Doub and Vandenbelt's primary and secondary bands, respectively (2, 6).

IX. Extraconjugation

Examples of extraconjugation are also available in the 2,4-dinitrophenylhydrazones. As shown in Table 11.11, increase in the chain of extraconjugation shifts the long wavelength maximum to the red, with

TABLE 11.11
Solvent Effect on Some Dinitrophenylhydrazones (50)

	Solvent (0.01 *N* in NaOH)			
	10% $CHCl_3$ in EtOH		Dimethylformamide	
Aldehyde	λ_{max}(mμ)	mε	λ_{max} (mμ)	mε
Acetaldehyde	424	18.0	439	20.5
Crotonaldehyde	456	29.5	464	30.5
Penta-2,4-dienal	480	34.5	494	39.0
Benzaldehyde	464	30.0	488	34.5
Cinnamaldehyde	488	41.5	508	44.0

an increase in intensity. Another example that shows the same phenomenon is demonstrated in Figure 11.3(51). The absorption maxima of a large number of *n*-alka-2-enal (**27a**) and *n*-alka-2,4-dienal (**27b**) 2,4-dinitrophenylhydrazone anions in 95% ethanol containing 10% by volume of chloroform and 0.01 *N* in sodium hydroxide show this phenomenon.

The two long wavelength bands found in the anions of many of these compounds (see Fig. 11.3) are probably derived from the two cross-conjugated systems—the longest wavelength maximum derived from the 2-nitro terminal and the more intense shorter wavelength maximum derived from transitions involving the 4-nitro terminal.

X. Coupling

The coupling of two trimethinoxonols across two principal valences, as in **28**, results in a strong red shift, as compared with such a trimethinoxonol as the malonaldehyde anion(**29**).

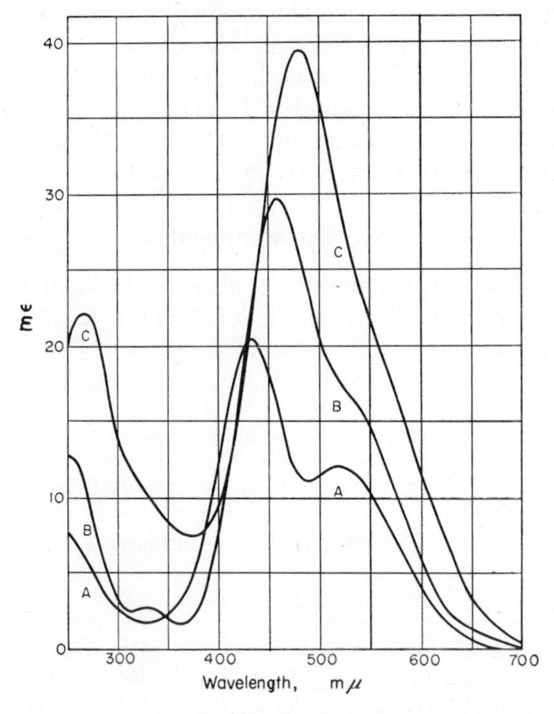

Fig. 11.3. Absorption spectra of the 2,4-dinitrophenylhydrazones of (A) *n*-heptanal, (B) *n*-hepta-2-enal, and (C) *n*-hepta-2,4-dienal. Solvent is 95% ethanol, 0.001 *N* NaOH, and containing 10% (by volume) of chloroform(51).

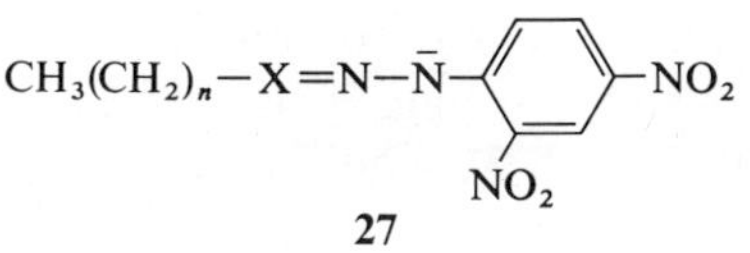

$CH_3(CH_2)_n-X=N-\bar{N}-C_6H_3(NO_2)_2$

27

	X	n	$\lambda_{max}(m\mu)$	$m\epsilon$
a	CH	1–9	458–459	29.2–30.9
b	CH=CHCH	1–8	480–481	39.1–41.1

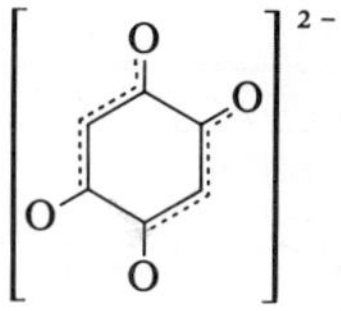

28

$[O \doteq CH \doteq CH \doteq CH \doteq O]^-$

29

No.	Solvent or PH	$\lambda_{max}(m\mu)$	$m\epsilon$	Ref.
28	0.1 N NaOH	315,496		52
29	pH 11.3	267	30.0	53
		350	0.22	

According to calculations, the long wavelength band of **28** corresponds to a $\pi \rightarrow \pi^*$ transition (52). Because of its low intensity, the long wavelength band of **29** may be derived from an $n \rightarrow \pi^*$ transition.

Various other types of coupled oxonols have been described by Dahne and Leupold (52).

The coupling of two pentamethineoxonols, as in **30** (52), across two principal valences show the red shift effect of coupling (see **32a** and **b**) and of the lengthening chain of conjugation between the resonance terminals. Compound **30** is the dianion of 1,5-dihydroxy-2,6-naphthoquinone, whereas **31** (spectrum in alkaline methanol) is the dianion of 2,6-dihydroxy-1,5-anthraquinone.

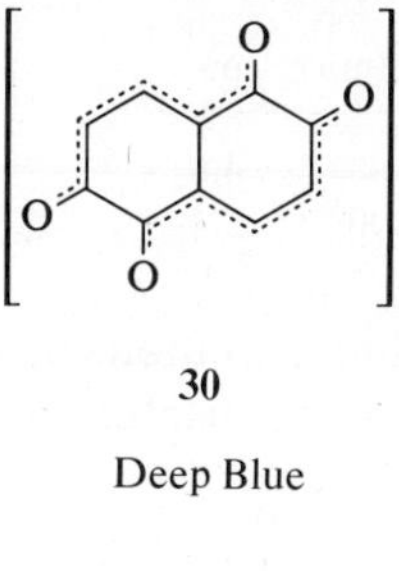

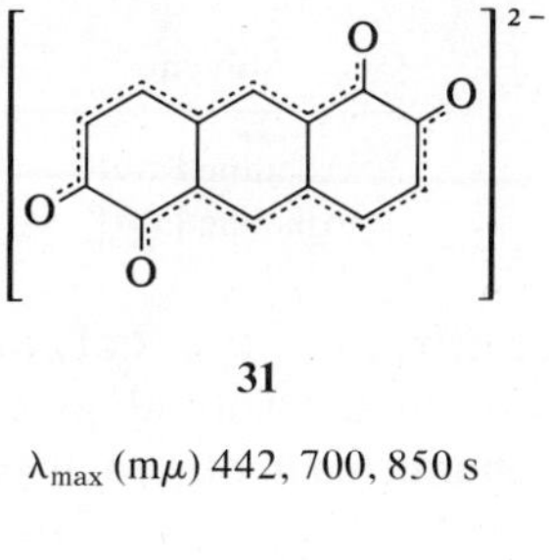

30

Deep Blue

31

λ_{max} $(m\mu)$ 442, 700, 850 s

The pentamethineoxonol corresponding to **30** is the glutacondialdehyde anion (**32a**), which is pale yellow. The heptamethineoxonol anion (**32b**) corresponding to **31** would be expected to absorb at around 460mμ.

$$O{=\!=}CH{=\!=}CH{=\!=}(CH{=\!=}CH)_n{=\!=}CH{=\!=}O)^-$$

32

	n	λ_{max}(mμ)	Ref.
a	1	365	54, 55
b	2	~460	

In many polyhydroxyquinone anions it is possible to formulate two or more polymethine structural units of different chain lengths. An example is alizarin, which has a λ_{max} of 520 mμ (fine structure) in 2 *N* aqueous sodium hydroxide solution (52). Considering their short chain of conjugation, especially long wavelength absorption is shown by the dianions of 1,4-dihydroxynaphthoquinone, λ_{max} 585 and 614 mμ in aqueous 0.1 *N* sodium hydroxide (56), and 1,4-dihydroxyanthraquinone, λ_{max} 560 mμ, mε 22.4, in methanolic 0.1 *N* alkali (57), as compared with their isomers.

Polyene–polymethine coupling can also cause a red shift effect (52). An example is the 2-anilino-3-chloro-1,4-naphthoquinone anion, (**33**) in solvents containing 2% of a 10% aqueous solution of tetraethylammonium hydroxide (21).

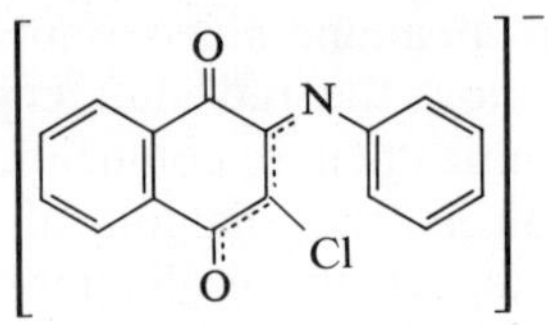

O N Cl O

33

Solvent	λ_{max}(mμ)	mε
Alkaline EtOH	563	4.6
Alkaline DMF	600	5.4

The presence of various types of coupling variations in the chromogens obtained in colorimetry should prove to be useful property in trace analysis of the chemicals in our environment.

XI. Intensity Determinants

A. Effect of Ionization

In most cases formation of the anion from the neutral compound results in an intensity increase, as shown for phenols in Table 10.12, nitroanilines in Table 10.13, 8-aminofluoranthene in Figure 10.11, diphenylamines in Table 10.14, azophenols in Table 10.17, azoanilines in Tables 10.18 and 10.19, and nitrophenylhydrazones in Table 10.16. 2,4-Dinitrophenylhydrazones show mainly small changes in intensity, Table 10.15. However, the dinitrophenylhydrazones of glyoxal and pyruvaldehyde show a strong increase in intensity on formation of the dianion.

B. Straight-Line Distance between Terminals

Other factors being equal, lengthening the chain of conjugation in anionic resonance structures increases the intensity, as shown in Tables 11.5 and 11.11, and Figure 11.3.

Among isomers the one that has the shortest straight-line distance between resonance terminals also has the lowest intensity. For example, the 2-nitrophenol anion (**34**) (28) absorbs at longer wavelength with lower intensity than the 4-isomer (**35**) (28). The same phenomenon is seen for 2-nitroaniline (**36a**) as compared with 4-nitroaniline (**37a**), and for 2-nitrodiphenylamine (**36b**) as compared with the 4-isomer **37b** (58).

O^-, NO_2 — **34** O_2N—C₆H₄—O^- — **35** NO_2, N^-—R — **36** O_2N—C₆H₄—$\bar{N}$—R — **37**

No.	Solvent	R	$\lambda_{max}(m\mu)$	$m\epsilon$
34	0.1 *N* NaOH		412	4.5
35	0.1 *N* NaOH		402	17.0
36a	Alkaline pyridine	H	515	8.4
36b	Alkaline sulfolane	C_6H_5	545	9.2
37a	Alkaline pyridine	H	467	32.3
37b	Alkaline pyridine	C_6H_5	508	34.6

The low absorption intensity of the quinizarin dianion (**38**) in dimethylformamide and in water containing 2% of methanolic 1 *N* tetrabutylammonium hydroxide is due to the short straight-line distance between resonance terminals.

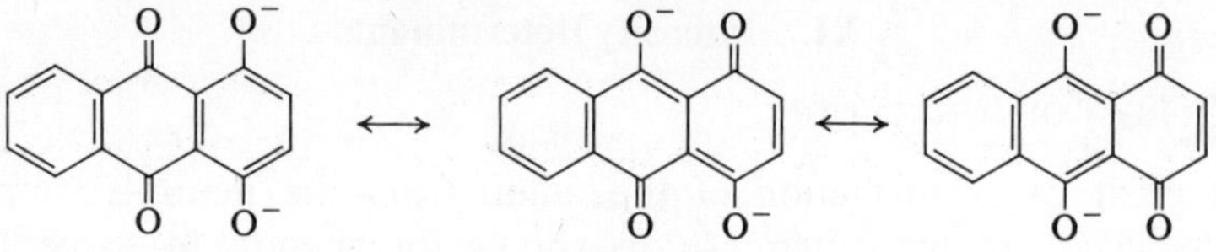

38

Solvent	λ_{max} (mμ)	mϵ
Aqueous alkali	557, 592	11.0, 10.4
Alkaline DMF	600, 643	9.0, 12.5

The isomeric nitro derivatives of the benzaldehyde phenylhydrazones show the same phenomenon—the 2-and 2′-nitro derivatives absorb at longer wavelength and with lower intensity than their 4- and 4′-isomers (see Tables 10.16, 11.6 and 11.10, and Figure 10.5).

C. Resonance Terminal Effects

Other things being equal, increasing the electron-acceptor strength of the other resonance terminal in a phenolic anion results in a red shift (see data for compounds **25**).

In some chromogens there is a competing interaction, as shown in **39B** ↔ **C**, between a terminal and a subsidiary terminal outside the main conjugation system, as shown for *N*-trifluoroacetyl-4-nitroaniline (**39**).

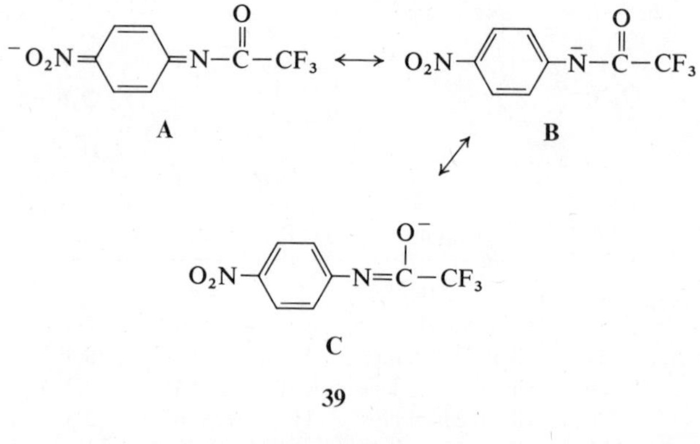

$^{-}O_2N{=}C_6H_4{=}N{-}C({=}O){-}CF_3 \longleftrightarrow O_2N{-}C_6H_4{-}\bar{N}{-}C({=}O){-}CF_3$

A B

$O_2N{-}C_6H_4{-}N{=}C(O^{-}){-}CF_3$

C

39

This type of interaction results in a decreased availability of electrons on the amino nitrogen for the **39A** ↔ **B** type of resonance. Thus the intensity decreases with decreasing availability of electrons on the resonance terminal common to two conjugation systems (see data for compounds **24** and Tables 11.8 and 11.9).

In some azophenolic anions the presence of a nitro group at one end of the chain can cause a strong red shift, depending on whether or not the nitro group is in the 4′-position (see Table 11.12). The data appear to indicate that the 4′-nitro group in this compound is a resonance terminal, whereas in the non-nitrated and in the 2′- and 3′-nitro compounds the β-aza nitrogen is the terminal.

TABLE 11.12
Spectra of Nitro Derivatives of 4-Phenylazo-1-naphthol

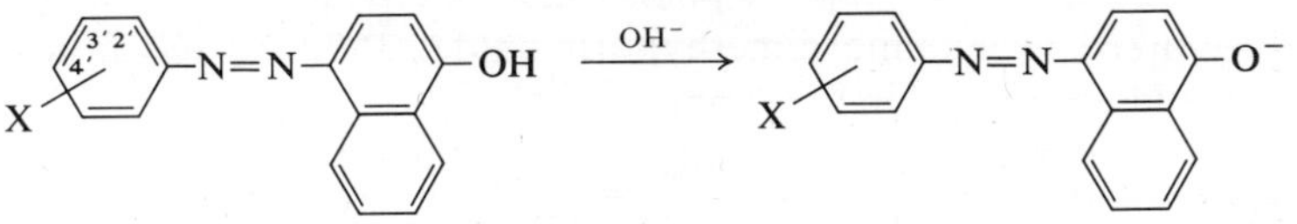

X	In DMF		In alkaline DMF[a]	
	λ_{max} (mμ)	mϵ	λ_{max} (mμ)	mϵ
H	412	–	531	44
2′-NO_2	450	17.8	566	44
3′-NO_2	443	16.0	555	35
4′-NO_2	473	49.0	650	80

[a]Containing 2% of a 10% aqueous solution of tetraethylammonium hydroxide.

D. Resonance Nodes

Other things being equal, any change in a resonance node that causes a violet shift will result in an intensity decrease also, whereas a red shift would result in an intensity increase. This is what one would expect, but the exceptions shown in the negativizing of the positive nodes of the compounds in Table 11.2 indicate that there are other factors involved here.

E. Ortho–Para Effect

We shall show one example of this effect, and this is in the nitrophenols. As expected, the nitrohydroquinone dianion [λ_{max} 540 mμ, mϵ 3.9(28)] absorbs at longer wavelength than the 2-nitrophenol anion (**34**). But the *para* effect in the dianion also decreases the intensity.

The same phenomenon is seen in the 4-nitropyrocatechol dianion [λ_{max} 510 mμ, mϵ 10.2 (28)], which absorbs at longer wavelength and with lower intensity than the 4-nitrophenol anion (**35**). This is an example of the *ortho* effect.

Thus the *ortho–para* effect has an intensity decreasing effect in addition to its red shift effect.

F. Solvent Basicity Effect

In analytical work involving weakly acidic compounds the use of a solvent system that is too weakly basic can result in a lowered proportion of chromogen in the anionic form. An example of this phenomenon is seen in Figure 10.14. The 4-phenylazodiphenylamine anion is completely ionized in alkaline dimethylformamide, 35% ionized in alkaline acetone, and un-ionized in l-butanol.

Where a compound is completely ionized, the intensity increases with an increase in the basicity of the solvent (see compounds **57** in Chapter 10, Tables 10.24 and 10.25, and Figures 10.12 and 10.13).

An interesting exception to the rule is the quinizarin dianion (**38**). The long wavelength $0 \rightarrow 0$ band increases somewhat in intensity with solvent basicity increase, but the shorter wavelength $0 \rightarrow 1$ band shows an intensity decrease under the same conditions.

XII. Analytical Applications

Many organic compounds present in the human environment that are determined by means of the anion have to be separated before spectral measurement. With the help of organic synthetic reactions many test substances can be changed into brilliantly colored anions that are thus more readily determined before separation as a family of compounds or after separation as individual components. A few examples of these types are shown in Table 11.13.

The 2, 4-dinitrophenylhydrazone anions obtained in the determination of aliphatic carbonyl compounds with 2, 4-dinitrophenylhydrazine are a good example of cross-conjugation. These anions show the presence of two bands. The more intense shorter wavelength band is derived from the resonance structure involving the 4-nitro group, whereas the less intense longer wavelength band is derived from the resonance structure involving the 2-nitro group as a resonance terminal (76).

For some of the compounds in Table 11.13 enzymic hydrolysis is used to form the chromogens; for others, such as dinobuton, hydrolysis takes place on the alumina column. Other reactions involve oxidation and formation and measurement of the derived phenolic anions. Many of the other analyses involve the formation of azo dye, hydrazone, formazan, and *m*-dinitrobenzene anions.

TABLE 11.13
Trace Analysis Involving Chromogen Anions

Test substance	Reagent	Chromogen[a]	λ_{max} (mμ)	mϵ[b]	Ref.
β-Glucuronidase	Phenolphthalein β-glucuronide	Phenolphthalein	545		59, 60
Acid phosphatase	Na 4-nitrophenyl phosphate	4-Nitrophenol	400	(17.0)	61
Isomaltase	4-Nitrophenol-α-D-glucopyranoside	4-Nitrophenol	400	(17.0)	62
Aryl sulfatase	Nitrocatechol sulfate	4-Nitrocatechol	510	(11.2)	60, 63
Dinobuton	Alumina	2-Isobutyl-2,4-DNP	378	10.6	64
			425	8.7	
Cholinesterase	2-Nitrophenylbutyrate	2-Nitrophenol	415	3.6	65
4-Hydroxy-3-methoxy-mandelic acid	Air	Vanillin	348[c]	(28.2)	66
	Periodate + 2,4-DNP hydrazine	Vanillin 2,4-DNP hydrazone	480		67
Phenylephrine	Periodate	3-Hydroxybenzaldehyde	240[d]	(24.0)	68
Homogentisic acid	Air + 2,4-DNP hydrazine	1,4-Benzoquinone-2-acetic acid 2,4-DNP hydrazone	575	24.9	69
Urea	Phenol + NaOCl	Indophenol	630[e]		70
Glutamine	Glutaminase, phenol, NaOCl	Indophenol	630		71
Ornithine carbamoyltransferase	Citrulline, phenol, NaOCl	Indophenol	640		72
Aniline	4-Phenylazobenzenediazonium fluoborate	4-Amino-*p*-bisazobenzene	600[f]	21.0	73
	4-Nitrobenzenediazonium fluoborate	4-Phenylazoaniline	570[f]	60.0	21, 73
			490[g]	14.0	21
2,6-Dichloro-4-nitroaniline	Alkaline DMF	Anion	515		74
Formaldehyde	2-Hydrazinobenzothiazole	1,5-Di-(2-benzothiazolyl) formazan	582	48.0	75
Aliphatic carbonyl compounds	2,4-DNP hydrazine	2,4-DNP hydrazone	428[h]	17.1	76
			425[i]	20.6	77
Glyoxal	4-Nitrophenylhydrazine	Glyoxal bis-4-nitrophenyl-hydrazone	702[j]	97.0	78
	2,4-DNP hydrazine	Glyoxal bis-2,4-DNP hydrazone	608[j]	65.0	78

TABLE 11.13 (*cont'd*)

Test substance	Reagent	Chromogen[a]	λ_{max} (mμ)	mϵ[b]	Ref.
Delnav	2,4-DNP hydrazine	Glyoxal bis-2,4-DNP hydrazone	614[j]		79
Glycoaldehyde	2,4-DNP hydrazine	Glyoxal bis-2,4-DNP hydrazone	560[k]		80
Isocitrate lyase	2,4-DNP hydrazine	Glyoxylate-2,4-DNP hydrazone	540		81
Leucine aminotransferase	2,4-DNP hydrazine	4-Methyl-2-oxo-pentanoate 2,4-DNP hydrazone	440		82
Alcohols, phenols, etc.	3,5-Dinitrobenzoyl chloride	3,5-Dinitrobenzoate acetonate anion	555		83
Alcohols	3,5-Dinitrobenzoyl chloride	Alkyl 3,5-dinitrobenzoate anion	510	10.0[l]	84
Xylene	Nitric acid	Dinitroxylene anion	470 570		85
17-Ketosteroids	*m*-Dinitrobenzene	Dinitrobenzene-ketosteroid anion	530		86
Digitoxin	2,2′,4,4′-TNDP	TNDP complex	620	26.2	87
Creatinine	2,2′,4,4′-TNDP	TNDP complex	555	15.5	88
Furazolidone	Alkaline DMF	Furazolidone complex	600		89
Quarternary amines	Picric acid	Picrate	375	18.3	90
Sulfonyl halides	Alkaline pyridine	$ArSO_2N{=}CH(CH{=}CH)_2O^-$	395	41.0	91
Dihydroxyurea	Alkaline alcohol		400	6.2	92
Oximes	4-Nitrobenzaldehyde	4-Nitrobenzaldehyde oxime	368	14.0	93

[a] Abbreviations: DNP = dinitrophenyl (or dinitrophenol); TNDP = tetranitrodiphenyl.
[b] Values in parentheses refer to the millimolar absorptivity of the pure chromogen in aqueous solution.
[c] Determined at 365 mμ.
[d] In alcohol. Also band at 364 mμ, mϵ 2.4.
[e] In alkaline dioxane λ_{max} 625 mμ, mϵ 31.6(25).
[f] In dimethylformamide.
[g] In water.
[h] Acetone derivative in 78% alkaline methanol. Subsidiary maximum at 530 mμ.
[i] Acetaldehyde derivative in benzene–methanol (96:4). Subsidiary band at 530 mμ, mϵ 8.6.
[j] In aqueous dimethylformamide.
[k] In alcoholic benzene.

It must be emphasized that a large number of organic compounds have been determined as highly colored chromogen anions after reaction with an appropriate reagent. For example, phenylalanine and tyrosine have been determined as chromogen anions by reaction with phenylalanine ammonia lyase (94), as have D-malate with D-2-hydroxy acid oxidoreductase (95), chondroitin sulfate with *p*-nitrobenzenediazonium fluoborate (96), ketosteroids with 4-nitrophenylhydrazine (97), nitroalkanes with *m*-dinitrobenzene (98), diphenylmethane after nitration (99), carbazole and benzocarbozoles with alkali (100), polynuclear aromatic amines with alkali (101), glyoxal with 4-nitrophenylhydrazine or 2,4-dinitrophenylhydrazine (78), malonaldehyde and its precursors with aromatic amines (44), aliphatic aldehydes with 2-hydrazinobenzothiazole (75), and nitrite precursors with *p*-phenylazoaniline (102) or benzaldehyde 2-benzothiazolyl hydrazone and *p*-phenylazoaniline (103).

An example of the analytical formation of a long chain conjugated anion is shown in Figure 11.4, where benzaldehyde phenylhydrazone is determined by reaction with 4-nitrobenzenediazonium fluoborate (104).

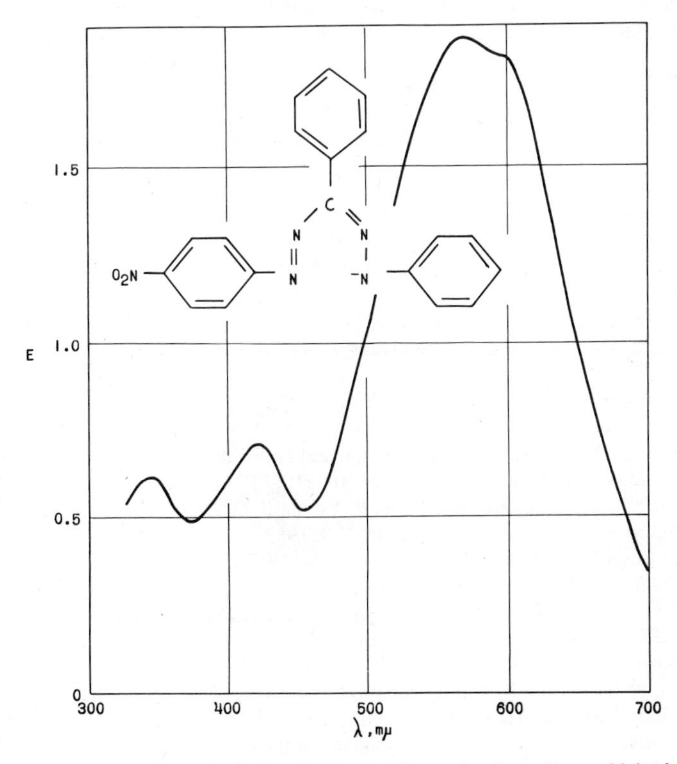

Fig. 11.4. Absorption spectrum obtained following determination of benzaldehyde phenylhydrazone with *p*-nitrobenzenediazonium fluoborate (104).

References

1. A. Bruylants, E. Brayl, and A. Schonne, *Helv. Chim. Acta,* **35**, 1127 (1952).
2. L. Doub and J. M. Vandebelt, *J. Am. Chem. Soc.*, **69**, 2714 (1947).
3. T. Campbell and G. Coppinger, *J. Am. Chem. Soc.,* **74**, 1469 (1952).
4. L. A. Cohen, *J. Org. Chem.,* **22**, 1333 (1957).
5. H. Kloosterziel and H. Backer, *Rec. Trav. Chim.,* **72**, 185 (1953).
6. L. Doub and J. M. Vandenbelt, *J. Am. Chem. Soc.,* **71**, 2414 (1949); ibid., **77**, 4535 (1955).
7. N. Coggeshall and A. S. Glessner, Jr., *J. Am. Chem. Soc.,* **71**, 3150 (1949).
8. M. J. Kamlet and D. J. Glover, *J. Org. Chem.,* **27**, 537 (1962).
9. K. Torssell, C. Lagercrantz, and S. Wold, *Arkiv Kemi,* **29**, 219 (1968).
10. M. F. Hawthorne, *J. Am. Chem. Soc.,* **78**, 4980 (1956).
11. H. Kohler and G. Scheibe, *Z. Anorg. Chem.,* **285**, 221 (1956).
12. G. Schwarzenbach, K. Lutz, and E. Felder, *Helv. Chim. Acta,* **27**, 576 (1944).
13. R. E. Plapinger, *J. Org. Chem.,* **24**, 802 (1959).
14. G. G. Price and M. C. Whiting, *Chem. Ind.* (*London*) 775 (1963).
15. R. Kuhn and D. Rewicki, *Ann.,* **704**, 9 (1967).
16. E. Sawicki, T. R. Hauser, T. W. Stanley, W. C. Elbert, and F. T. Fox, *Anal. Chem.,* **33**, 1574 (1961).
17. R. Schaal and C. Jacquinot-Vermesse, *Compt. Rend.*, Ser. B, 2201 (1959).
18. R. Wizinger, *J. Prakt. Chem.,* **157**, 129 (1941).
19. G. Schwarzenbach and R. Weber, *Helv. Chim. Acta,* **25**, 1628 (1942).
20. E. B. Knott, *J. Chem. Soc.,* 1024 (1951).
21. E. Sawicki, T. R. Hauser, and T. W. Stanley, *Anal. Chem.,* **31**, 2063 (1959).
22. W. J. Middleton, E. L. Little, D. D. Coffman, and V. A. Engelhardt, *J. Am. Chem. Soc.,* **80**, 2795 (1958).
23. R. J. Pollitt and B. C. Saunders, *J. Chem. Soc.,* 4615 (1965).
24. M. Ettlinger, *J. Am. Chem. Soc.,* **72**, 3085 (1950).
25. D. N. Kramer, R. M. Gamson, and F. M. Miller, *J. Org. Chem.,* **24**, 1742 (1959).
26. C. Pedersen, *J. Am. Chem. Soc.,* **79**, 5014 (1957).
27. R. Wizinger and S. Chatterjea, *Helv. Chim. Acta,* **35**, 316 (1952).
28. J. N. Smith, *J. Chem. Soc.,* 2861 (1951).
29. D. Robinson, J. Smith, and R. Williams, *Biochem. J.,* **50**, 221 (1951).
30. R. Wizinger, *Chimia,* **19**, 339 (1965).
31. M. Strell, W. Braunbruck, W. Fuhler, and O. Huber, *Ann.,* **587**, 177 (1954).
32. K. Hafner and K. Goliasch, *Angew. Chem.,* **74**, 118 (1962).
33. R. Kuhn, H. Fischer, F. A. Neugebauer, and H. Fischer, *Ann.,* **654**, 64 (1962).
34. A. Van Dormael, *Ind. Chim. Belg.,* **24**, 463 (1958).
35. J. A. Ford, Jr., C. W. Wilson, and W. R. Young, *J. Org. Chem.,* **32**, 173 (1967).
36. H. Schmidt, *Naturwissenschaften,* **46**, 379 (1959).
37. R. L. Fevre and J. Sousa, *J. Chem. Soc.*, 744 (1957).
38. E. Fehnel and M. Carmack, *J. Am. Chem. Soc.,* **71**, 231 (1949).
39. P. Russell and J. Mentha, *J. Am. Chem. Soc.,* **77**, 4245 (1955).
40. B. Eistert, E. Merkel, and W. Reiss, *Ber.,* **87**, 1513 (1954).
41. G. Kortüm, *Z. Physik. Chem.*, **43**, 271 (1939).
42. W. A. Sheppard and R. M. Henderson, *J. Am. Chem. Soc.,* **89**, 4446 (1967).
43. W. A. Sheppard, *J. Am. Chem. Soc.,* **85**, 1314 (1963).
44. E. Sawicki, T. W. Stanley, and H. Johnson, *Anal. Chem.,* **35**, 199 (1963).
45. C. J. Martin, J. Golubow, and A. E. Axelrod, *J. Biol. Chem.,* **234**, 294 (1959).
46. K. C. Schreiber and M. C. Kennedy, *J. Am. Chem. Soc.,* **78**, 153 (1956).

47. J. B. Conant and L. Fieser, *J. Am. Chem. Soc.*, **45**, 2194 (1923).
48. T. W. Stanley, E. Sawicki, H. Johnson, and J. D. Pfaff, *Mikrochim. Acta*, 48 (1965).
49. J. Hewitt, G. Mann, and F. Pope, *J. Chem. Soc.*, 2193 (1914).
50. C. J. Timmons, *J. Chem. Soc.*, 2613 (1957).
51. F. Stitt, R. B. Seligman, F. E. Resnik, E. Gong, E. L. Pippen, and D. A. Forss, *Spectrochim. Acta*, **17**, 51 (1961).
52. S. Dahne and D. Leupold, *Angew. Chem., Intern. Ed.*, **5**, 984 (1966).
53. F. Mashio and Y. Kimura, *Nippon Kagaku Zasshi*, **81**, 434 (1960).
54. G. Scheibe, D. Bruck, and F. Dorr, *Ber.*, **85**, 867 (1952).
55. S. S. Malhotra and M. C. Whiting, *J. Chem. Soc.*, 3812 (1960).
56. R. A. Morton and W. T. Earlam, *J. Chem. Soc.*, 159 (1941).
57. C. J. P. Spruit, *Rec. Trav. Chim.*, **68**, 309 (1949).
58. R. Stewart and J. P. O'Donnell, *Can. J. Chem.*, **42**, 1681 (1964).
59. F. Kiermeier and J. Gull, *Naturwissenschaften*, **53**, 613 (1966).
60. D. Utermann, F. Lorenzen, and H. Hilz, *Klin. Wochschr.*, **42**, 352 (1964).
61. F. Gunther and O. Burckhart, *Deut. Lebensm. Rundschau*, **63**, 41 (1967).
62. G. Terui, H. Okada, and Y. Oshima, *Technol. Rept. Osaka Univ.*, **9**, 237 (1959).
63. T. R. Ittyerah, M. E. Dumm, and B. K. Bachhawat, *Clin. Chim. Acta*, **17**, 405 (1967).
64. H. Crossley and V. P. Lynch, *J. Sci. Food Agr.*, **19**, 57 (1968).
65. G. Szasz, *Clin. Chim. Acta.*, **14**, 646 (1968).
66. M. Sandler and C. R. J. Ruthven, *Biochem. J.*, **80**, 78 (1961).
67. R. J. Georges, *Clin. Chim. Acta*, **10**, 583 (1964).
68. R. B. Bruce and J. E. Pitts, *Biochem. Pharmacol.*, **17**, 335 (1968).
69. R. E. Stoner and B. B. Blivaiss, *Clin. Chem.*, **11**, 833 (1965).
70. R. L. Searcy and F. M. Cox, *Clin. Chim. Acta*, **8**, 810 (1963).
71. H. Kluge, H. Hermann, and V. Wieczorek, *Z. Klin. Chem.*, **5**, 86 (1967).
72. F. M. L. Moore, *Clin. Chim. Acta*, **15**, 103 (1967).
73. E. Sawicki, T. W. Stanley, and T. R. Hauser, *Chemist-Analyst*, **48**, 30 (1959).
74. K. Groves and K. S. Chough, *J. Agr. Food Chem.*, **14**, 668 (1966).
75. E. Sawicki and T. R. Hauser, *Anal. Chem.*, **32**, 1434 (1960).
76. C. F. Wells, *Tetrahedron*, **22**, 2685 (1966).
77. K. J. Monty, *Anal. Chem.*, **30**, 1350 (1958).
78. E. Sawicki, T. R. Hauser, and R. Wilson, *Anal. Chem.*, **34**, 505 (1962).
79. C. L. Dunn, *J. Agr. Food Chem.*, **6**, 203 (1958).
80. R. A. Basson and T. A. du Plessis, *Analyst*, **92**, 463 (1967).
81. T. J. Jacks and N. A. Alldridge, *Anal. Biochem.*, **18**, 378 (1967).
82. R. T. Taylor and W. T. Jenkins, *J. Biol. Chem.*, **241**, 4391 (1966).
83. G. R. Umbreit and R. L. Houtman, *J. Pharm. Sci.*, **56**, 349 (1967).
84. M. Pesez and J. Bartos, *Talanta*, **5**, 216 (1960).
85. J. Trzeszczynski and Z. Luczak, *Chemia Anal.*, **11**, 1165 (1966).
86. K. Larsen, *Acta Endocrin.*, **57**, 228 (1968).
87. G. Rabitzsch and U. Tambor, *Pharmazie*, **11**, 668 (1967).
88. M. Nakadate, C. Matsuyama, and M. Kimura, *Chem. Pharm. Bull.*, **12**, 1138 (1964).
89. R. D. Hollifield and J. D. Conklin, *J. Pharm. Sci.*, **57**, 325 (1968).
90. K. Gustavii and G. Schill, *Acta Pharm. Suecica*, **3**, 241 (1966).
91. M. R. F. Ashworth and G. Bohnstedt, *Anal. Chim. Acta*, **36**, 196 (1966).
92. W. N. Fishbein, *Anal. Chim. Acta*, **40**, 269 (1968).
93. D. P. Johnson, *Anal. Chem.*, **40**, 646 (1968).
94. K. Uchiyama, H. Yamada, T. Tochikura, and K. Ogata, *Agr. Biol. Chem.*, **32**, 764 (1968).
95. J. S. Britten, *Anal. Biochem.*, **24**, 330 (1968).

96. S. Ogawa, M. Morita, A. Nakajima, and A. Yoshida, *Yakugaku Zasshi,* **88**, 866, 876 (1968).
97. M. Kimura, T. Nishina, and T. Sakamoto, *Chem. Pharm. Bull.,* **15**, 454 (1967).
98. M. R. F. Ashworth and E. Gramsch, *Mikrochim, Acta,* 358 (1967).
99. M. R. F. Ashworth and R. Schupp, *Mikrochim. Acta,* 366 (1967).
100. D. F. Bender, E. Sawicki, and R. M. Wilson, Jr., *Intern. J. Air Water Poll.,* **8**, 633 (1964).
101. E. Sawicki, H. Johnson, and K. Kosinski, *Microchem. J.,* **10**, 72 (1966).
102. E. Sawicki, T. W. Stanley, and W. C. Elbert, *Anal. Chem.,* **34**, 297 (1962).
103. E. Sawicki, T. W. Stanley, and W. C. Elbert, *Mikrochim. Acta,* 891 (1961).
104. E. Sawicki, T. W. Stanley, and T. R. Hauser, *Chemist-Analyst,* **47**, 87 (1958).

INDEX